2012 IMC®
CODE AND COMMENTARY

INTERNATIONAL
CODE COUNCIL®

2012 International Mechanical Code® Commentary

First Printing: April 2012

ISBN: 978-1-60983-071-7 (soft-cover edition)

COPYRIGHT © 2012
by
INTERNATIONAL CODE COUNCIL, INC.

PRINTED IN THE U.S.A.

PREFACE

Internationally, code officials recognize the need for a modern, up-to-date mechanical code addressing the design and installation of mechanical systems through requirements emphasizing performance. The *International Mechanical Code* is designed to meet these needs through model code regulations that safeguard the public health and safety in all communities, large and small.

The principal purpose of this Commentary is to provide a basic volume of knowledge and facts relating to the code. The Commentary provides it in a small package and at reasonable cost thorough coverage of many issues likely to be dealt with when using the *International Mechanical Code*—and then supplements that coverage with historical and technical background. Reference lists, information sources and bibliographies are also included.

Strenuous effort has been put into keeping the vast quantity of material accessible and its method of presentation useful. With a comprehensive yet concise summary of each section, the Commentary is a convenient reference for mechanical regulations. In the chapters that follow, discussions focus on the full meaning and implications of the code text. Guidelines suggest the most effective method of application, and the consequences of not adhering to the code text. Illustrations are provided to aid understanding; they do not necessarily illustrate the only methods of achieving code compliance.

The format of the Commentary includes the full text of each section, table and figure in the code, followed immediately by the commentary applicable to that text. At the time of printing, the Commentary reflects the most up-to-date text of the 2012 *International Mechanical Code*. Each section's narrative includes a statement of its objective and intent, and usually includes a discussion about why the requirement commands the conditions set forth. Code text and commentary text are easily distinguished from each other. All code text is shown as it appears in the *International Mechanical Code*, and all commentary is indented below the code text and begins with the symbol ❖.

Readers should note that the Commentary is to be used in conjunction with the *International Mechanical Code* and not as a substitute for the code. **The Commentary is advisory only;** the code official alone possesses the authority and responsibility for interpreting the code.

Comments and recommendations are encouraged, for through your input, we can improve future editions. Please direct your comments to the Codes and Standards Development Department at the Chicago District Office.

TABLE OF CONTENTS

Chapter 1:
Scope and Administration

General Comments

This chapter contains provisions for the application, enforcement and administration of subsequent requirements of the code. In addition to establishing the scope of the code, Chapter 1 identifies which mechanical equipment, appliances and systems it covers. Sections 101 and 102 establish the scope and applicability of the code and address existing equipment and systems. Section 103 establishes the department of mechanical inspection and the appointment of department personnel. Section 104 contains the duties and authority of the code official for rulemaking, permits, inspections and right of entry. Section 105 deals with approval of modifications, alternative materials, methods and equipment. Section 106 states when permits, construction document submittals, permit issuance and fees are required. Section 107 includes inspection duties of the code official or an inspection agency that has been approved by the code official, mechanical system testing, contractor responsibilities, notice of approval and temporary connections for testing mechanical systems. Administrative provisions for violations are addressed in Section 108, including provisions covering unlawful acts, violation notices, prosecution, penalties, stop work orders and unsafe mechanical systems. Section 109 establishes the board of appeals and includes provisions for application for appeal, membership of the board, board meeting notices, open hearing requirements, when postponements are in order, appeals board decisions and court review.

The law of building regulation is based on the police power of the state. This police power is the source of all authority to enact building regulations. Police power is the power of the state to legislate for the general welfare of its citizens. This power enables passage of such laws as a mechanical code. It is from the police power delegated by the state legislature that local governments are able to enact building regulations. If the state legislature has limited this power in any way, the municipality may not exceed these limitations. Although the municipality may not further delegate its police power by delegating the burden of determining code compliance to the building owner, contractor or architect, it may turn over the administration of building regulations to a municipal official, such as a code official, if he or she is given sufficient criteria to clearly establish the basis for deciding whether or not a proposed building, including its mechanical systems, conforms to the code.

Chapter 1 of the code is largely concerned with maintaining "due process of law" in enforcing the performance criteria contained in the body of the code. Only through careful observation of the administrative provisions can the code official demonstrate "equal protection under the law." Although the administrative and enforcement section of a code is geared toward the code official, the provisions also establish the rights and privileges of the design professional, the contractor and the building owner. The position of the code official is to review the proposed and completed work and to determine whether a mechanical system installation conforms to the code requirements. The design professional is responsible for the design of a safe mechanical system. The contractor is responsible for installing the system in strict accordance with the plans. During the construction of a mechanical system, the code official reviews the activity to see that the spirit and intent of the law are being met and that the mechanical system provides adequate protection of public health. As a public servant, the code official enforces the code in an unbiased, proper manner. Every individual is guaranteed equal enforcement of the code. Furthermore, design professionals, contractors and building owners have the right of due process for any requirement in the code.

Purpose

A mechanical code, as with any other code, is intended for adoption as a legally enforceable document to safeguard health, safety, property and public welfare. A mechanical code cannot be effective without adequate provisions for its administration and enforcement. The official charged with the administration and enforcement of mechanical regulations has a great responsibility, and with this responsibility goes authority. No matter how detailed the mechanical code may be, the code official must, to some extent, exercise judgment in determining code compliance. The code official has the responsibility for establishing that the homes in which the citizens of the community reside and the buildings in which they work are designed and constructed to be reasonably free from hazards associated with the presence and use of mechanical equipment, appliances and systems. The code intends to establish a minimum acceptable level of safety.

PART 1–SCOPE AND APPLICATION

SECTION 101
GENERAL

[A] 101.1 Title. These regulations shall be known as the *Mechanical Code* of **[NAME OF JURISDICTION]**, hereinafter referred to as "this code."

❖ This section identifies the adopted regulations by the insertion of the name of the adopting jurisdiction into the code. This is one of several places in the code that the adopting agency must "fill in the brackets" to insert information that is specific to the local jurisdiction (see the sample ordinance in the front of the code book).

[A] 101.2 Scope. This code shall regulate the design, installation, maintenance, *alteration* and inspection of mechanical systems that are permanently installed and utilized to provide control of environmental conditions and related processes within buildings. This code shall also regulate those mechanical systems, system components, *equipment* and appliances specifically addressed herein. The installation of fuel gas distribution piping and *equipment*, fuel gas-fired appliances and fuel gas-fired *appliance* venting systems shall be regulated by the *International Fuel Gas Code*.

> **Exception:** Detached one- and two-family dwellings and multiple single-family dwellings (townhouses) not more than three stories high with separate means of egress and their accessory structures shall comply with the *International Residential Code*.

❖ This section describes the types of mechanical systems covered by the code. The code is applicable from the initial design of mechanical systems, through installation and construction and into the maintenance of operating systems. The scope of the code is primarily focused on heating, ventilating and air-conditioning (HVAC) systems, those items specifically addressed in the code and those building "service" items that make a building comfortable, functional and safe. The code is intended to cover all mechanical appliances, equipment and systems that are specifically intended, designed and necessary for the general safety and well-being of the occupants of a building. The code intends to regulate the installation, operation and maintenance of any and all equipment and appliances that can affect the health, safety and welfare of building occupants.

Note that this section references the *International Fuel Gas Code*® (IFGC®) for fuel-gas-related regulations. This is a result of a written cooperative agreement between the International Code Council® (ICC®) and the American Gas Association (AGA) to promulgate the IFGC. All regulations for the installation of fuel gas distribution piping and equipment, fuel gas-fired appliances and fuel gas-fired appliance venting systems have been removed from the code. See the IFGC for fuel-gas-related requirements.

The exception sends the user to the *International Residential Code*® (IRC®) for one- and two-family dwellings and townhouses not more than three stories in height. It is the intent of the ICC family of codes that the *International Mechanical Code*® (IMC®) be applied to structures not within the scope of the IRC. For example, a four-story single-family dwelling would be subject to the provisions of the IMC.

[A] 101.2.1 Appendices. Provisions in the appendices shall not apply unless specifically adopted.

❖ This section clarifies that the appendices are not part of the code unless specifically included in the adopting ordinance of the jurisdiction. Otherwise, the appendices are not intended to be enforceable.

[A] 101.3 Intent. The purpose of this code is to provide minimum standards to safeguard life or limb, health, property and public welfare by regulating and controlling the design, construction, installation, quality of materials, location, operation and maintenance or use of mechanical systems.

❖ The intent of the code is to set forth requirements that establish the minimum acceptable level to safeguard life or limb, health, property and public welfare. The intent becomes important in the application of such sections as Sections 102, 104.2, 105.2 and 108, as well as any enforcement-oriented interpretive action or judgement. Like any code, the written text is subject to interpretation. Interpretations should not be affected by economics or the potential impact on any party. The only consideration should be protection of the public health, safety and welfare.

[A] 101.4 Severability. If a section, subsection, sentence, clause or phrase of this code is, for any reason, held to be unconstitutional, such decision shall not affect the validity of the remaining portions of this code.

❖ Once the code is adopted, only a court can set aside any provisions of the code. This is essential to safeguard the application of the code text if a provision of the code is declared illegal or unconstitutional. This section would preserve the legislative action that put the legal provisions in place.

SECTION 102
APPLICABILITY

[A] 102.1 General. Where there is a conflict between a general requirement and a specific requirement, the specific requirement shall govern. Where, in a specific case, different sections of this code specify different materials, methods of construction or other requirements, the most restrictive shall govern.

❖ Bear in mind that conflicts within the code rarely, if ever, occur, but if they do, this section applies. Specific requirements of the code override or take precedence over general requirements. For example, in the 2006 edition of the code, Section 401.4 required a 10-foot (3048 mm) separation between all exhaust openings and property lines, whereas Section 501.2.1, Item 3 required a 3-foot (914.4 mm) separation for a specific type of exhaust. In this case, Section 501.2.1 would overrule. This conflict was resolved in the 2009 edition of the code.

[A] 102.2 Existing installations. Except as otherwise provided for in this chapter, a provision in this code shall not require the removal, *alteration* or abandonment of, nor prevent the continued utilization and maintenance of, a mechanical system lawfully in existence at the time of the adoption of this code.

❖ An existing mechanical system is generally considered to be "grandfathered" in with code adoption if the system meets a minimum level of safety. Frequently, the criteria for this level are the regulations (or code) under which the existing building was originally constructed. If there are no previous code criteria to apply, the code official is to apply those provisions of the code that are reasonably applicable to existing buildings. A specific level of safety is dictated by provisions dealing with hazard abatement in existing buildings and maintenance provisions, as contained in this code, the *International Property Maintenance Code*® (IPMC®) and the *International Fire Code*® (IFC®)

[A] 102.3 Maintenance. Mechanical systems, both existing and new, and parts thereof shall be maintained in proper operating condition in accordance with the original design and in a safe and sanitary condition. Devices or safeguards which are required by this code shall be maintained in compliance with the code edition under which they were installed. The owner or the owner's designated agent shall be responsible for maintenance of mechanical systems. To determine compliance with this provision, the code official shall have the authority to require a mechanical system to be reinspected. The inspection for maintenance of HVAC systems shall be done in accordance with ASHRAE/ACCA/ANSI Standard 180.

❖ All mechanical systems and equipment are subject to deterioration resulting from aging, wear, accumulation of dirt and debris, corrosion and other factors. Maintenance is necessary to keep mechanical systems and equipment in proper operating condition. All required safety devices and controls must be maintained to continue providing the protection that they afford. Existing equipment and systems could be equipped with safety devices or other measures that were necessary because of the nature of the equipment, and such safeguards may have been required by a code that predates the current code. All safeguards required by previous or present codes must be maintained for the life of the equipment or system served by those safeguards.

Maintenance inspections must be performed in accordance with ASHRAE/ACCA/ANSI Standard 180 so that all components of the mechanical system requiring maintenance are properly addressed.

The maintenance of mechanical systems as prescribed in this section is the responsibility of the property owner. The owner may authorize another party to be responsible for the property, in which case that party is responsible for the maintenance of the mechanical systems involved.

The reinspection authority of the code official is needed to accomplish compliance with the maintenance requirements in this section.

[A] 102.4 Additions, alterations or repairs. Additions, alterations, renovations or repairs to a mechanical system shall conform to that required for a new mechanical system without requiring the existing mechanical system to comply with all of the requirements of this code. Additions, alterations or repairs shall not cause an existing mechanical system to become unsafe, hazardous or overloaded.

Minor additions, alterations, renovations and repairs to existing mechanical systems shall meet the provisions for new construction, unless such work is done in the same manner and arrangement as was in the existing system, is not hazardous and is *approved*.

❖ Simply stated, new work must comply with the current requirements for new work. Any alteration or addition to an existing system involves some extent of new work and that new work is subject to the requirements of the code. Additions or alterations can place additional loads or different demands on an existing system and those loads or demands could necessitate changing all or part of the existing system. Additions and alterations must not cause an existing system to be any less in compliance with the code than it was before the changes.

[A] 102.5 Change in occupancy. It shall be unlawful to make a change in the *occupancy* of any structure which will subject the structure to any special provision of this code applicable to the new *occupancy* without approval. The code official shall certify that such structure meets the intent of the provisions of law governing building construction for the proposed new *occupancy* and that such change of *occupancy* does not result in any hazard to the public health, safety or welfare.

❖ When a building undergoes a change of occupancy, the mechanical system must be evaluated to determine what effect the change of occupancy has on it. If an existing system serves an occupancy that is different from the occupancy it served when the code went into effect, the mechanical system must comply with the applicable code requirements for a mechanical system serving the newer occupancy. Depending on the nature of the previous occupancy, changing a building's occupancy classification could result in a change to the mechanical system. For example, if a mercantile building was converted to a restaurant, additional ventilation would be required for the public based on the increased occupant load.

[A] 102.6 Historic buildings. The provisions of this code relating to the construction, *alteration*, repair, enlargement, restoration, relocation or moving of buildings or structures shall not be mandatory for existing buildings or structures identified and classified by the state or local jurisdiction as historic buildings when such buildings or structures are judged by the code official to be safe and in the public interest of health, safety and welfare regarding any proposed construction, *alteration*, repair, enlargement, restoration, relocation or moving of buildings.

❖ This section gives the code official the widest possible flexibility in enforcing the code when the building in question has historic value. This flexibility, however, is not without conditions. The most important criterion for application of this section is that the building must be specifically classified as being of historic significance by a qualified party or agency. Usually, this is done by a state or local authority after considerable scrutiny of the historical value of the building. Most, if not all, states have such authorities, as do many local jurisdictions. The agencies with such authority typically exist at the state or local government level.

[A] 102.7 Moved buildings. Except as determined by Section 102.2, mechanical systems that are a part of buildings or structures moved into or within the jurisdiction shall comply with the provisions of this code for new installations.

❖ Buildings that have been relocated are subject to the requirements of the code as if they were new construction. Placing a building where one did not previously exist is the same as constructing a new building. This section requires alteration of the existing mechanical systems to the extent necessary to bring them into compliance with the provisions of the code applicable to new construction or that the existing mechanical system comply with Section 102.2.

[A] 102.8 Referenced codes and standards. The codes and standards referenced herein shall be those that are listed in Chapter 15 and such codes and standards shall be considered as part of the requirements of this code to the prescribed extent of each such reference and as further regulated in Sections 102.8.1 and 102.8.2.

> **Exception:** Where enforcement of a code provision would violate the conditions of the listing of the *equipment* or *appliance*, the conditions of the listing and the manufacturer's installation instructions shall apply.

❖ A referenced standard or portion of one is an enforceable extension of the code as if the content of the standard were included in the body of the code. For example, Sections 603.4 and 603.5 reference the Sheet Metal and Air Conditioning Contractors National Association (SMACNA) duct construction standards in their entirety for the construction of metal and nonmetallic ducts. Section 301.7 references NFPA 70 for all electrical installations associated with the equipment and appliances regulated by the code. The use and application of referenced standards are limited to those portions of the standards that are specifically identified. It is the intention of the code to be in harmony with the referenced standards. If conflicts occur because of scope or purpose, the code text governs.

The exception recognizes the extremely unlikely but possible occurrence of the code requiring or allowing something less restrictive or stringent than the product's listing or manufacturer's instructions. The intent is for the highest level of safety to prevail. Thus, this exception allows for the conditions of the listing or the manufacturer's installation instructions to override the code requirement if the listing or installation instructions are more restrictive.

[A] 102.8.1 Conflicts. Where conflicts occur between provisions of this code and the referenced standards, the provisions of this code shall apply.

❖ The code takes precedence when the requirements of the standard conflict with the requirements of the code. Although it is the intention of the code to be in harmony with referenced standards, the code text governs if conflicts occur.

[A] 102.8.2 Provisions in referenced codes and standards. Where the extent of the reference to a referenced code or standard includes subject matter that is within the scope of this code, the provisions of this code, as applicable, shall take precedence over the provisions in the referenced code or standard.

❖ Although a standard or code is referenced, its full scope and content are not necessarily applicable. The standard (or code) is applicable only to the extent indicated in the text in which the standard is specifically referenced. A referenced standard or a portion thereof is an enforceable extension of the code as if the content of the standard were included in the body of the code. The use and application of referenced standards are limited to those portions of the standards that are specifically identified.

[A] 102.9 Requirements not covered by this code. Requirements necessary for the strength, stability or proper operation of an existing or proposed mechanical system, or for the public safety, health and general welfare, not specifically covered by this code, shall be determined by the code official.

❖ New technology will sometimes result in a situation or circumstance not specifically covered by the code. This section of the code gives the code official the authority to decide whether and how the code can be used to cover the new situation. Clearly such a section is needed and the code official's reasonable application of the section is necessary. The purpose of the section, however, is not to impose requirements that may be preferred when the code provides alternative methods or is not silent on the circumstances. Additionally, the section can be used to implement the general performance-oriented language of the code to specific enforcement situations.

[A] 102.10 Other laws. The provisions of this code shall not be deemed to nullify any provisions of local, state or federal law.

❖ Other laws enacted by the local, state or federal government may be applicable to a condition that is also governed by a requirement in the code. In such circumstances, the requirements of the code are in addition to those other laws, even though the building official may not be responsible for the enforcement of those laws. For example, the health department might require the pots and pans sink's waste lines in a restaurant to be indirectly connected, whereas the plumbing code allows these sinks to be either directly or indirectly connected.

[A] 102.11 Application of references. Reference to chapter section numbers, or to provisions not specifically identified by number, shall be construed to refer to such chapter, section or provision of this code.

❖ In a situation where the code may make reference to a chapter, section number or to another code provision without specifically identifying its location in the code, assume that the referenced section, chapter, or provision is in this code and not in a referenced code or standard.

PART 2–ADMINISTRATION AND ENFORCEMENT

SECTION 103
DEPARTMENT OF MECHANICAL INSPECTION

[A] 103.1 General. The department of mechanical inspection is hereby created and the executive official in charge thereof shall be known as the code official.

❖ The executive official in charge of the mechanical department is named the "code official" by this section. In actuality, the person who is in charge of the department may hold a different title, such as building commissioner, mechanical inspector or construction official. For the purpose of the code, that person is referred to as the "code official."

[A] 103.2 Appointment. The code official shall be appointed by the chief appointing authority of the jurisdiction.

❖ This section establishes the code official as an appointed position and gives the circumstances under which the code official can be removed from office.

[A] 103.3 Deputies. In accordance with the prescribed procedures of this jurisdiction and with the concurrence of the appointing authority, the code official shall have the authority to appoint a deputy code official, other related technical officers, inspectors and other employees. Such employees shall have powers as delegated by the code official.

❖ This section provides the code official with the authority to appoint other individuals to assist with the administration and enforcement of the code. These individuals have authority and responsibility as designated by the code official.

[A] 103.4 Liability. The code official, member of the board of appeals or employee charged with the enforcement of this code, while acting for the jurisdiction in good faith and without malice in the discharge of the duties required by this code or other pertinent law or ordinance, shall not thereby be rendered liable personally, and is hereby relieved from personal liability for any damage accruing to persons or property as a

result of an act or by reason of an act or omission in the discharge of official duties.

Any suit instituted against any officer or employee because of an act performed by that officer or employee in the lawful discharge of duties and under the provisions of this code shall be defended by the legal representative of the jurisdiction until the final termination of the proceedings. The code official or any subordinate shall not be liable for costs in an action, suit or proceeding that is instituted in pursuance of the provisions of this code.

❖ This section tells us that a code official or an employee in his or her department should not be held personally liable for damage to persons or property resulting from enforcing code requirements in a lawful and honest way. The appointing authority is responsible for providing lawyers to handle any lawsuits against these employees. The best way to be certain that the code official's action is a "lawful duty" is to always cite the applicable code section on which the enforcement action is based.

SECTION 104
DUTIES AND POWERS OF THE CODE OFFICIAL

[A] 104.1 General. The code official is hereby authorized and directed to enforce the provisions of this code. The code official shall have the authority to render interpretations of this code and to adopt policies and procedures in order to clarify the application of its provisions. Such interpretations, policies and procedures shall be in compliance with the intent and purpose of this code. Such policies and procedures shall not have the effect of waiving requirements specifically provided for in this code.

❖ The duty of the code official is to enforce the code, and he or she is the "authority having jurisdiction" for all matters relating to the code and its enforcement. It is the duty of the code official to interpret the code and to determine compliance. Code compliance will not always be easy to determine and will require judgment and expertise, particularly when enforcing the provisions of Sections 105.1 and 105.2. In exercising this authority, however, the code official cannot set aside or ignore any provision of the code.

[A] 104.2 Applications and permits. The code official shall receive applications, review *construction documents* and issue permits for the installation and *alteration* of mechanical systems, inspect the premises for which such permits have been issued and enforce compliance with the provisions of this code.

❖ The code enforcement process is normally initiated with an application for a permit. The code official is responsible for processing the applications and issuing permits for the installation, replacement, addition to or modification of mechanical systems in accordance with the code.

[A] 104.3 Inspections. The code official shall make all of the required inspections, or shall accept reports of inspection by *approved* agencies or individuals. All reports of such inspections shall be in writing and be certified by a responsible officer of such *approved* agency or by the responsible individual. The code official is authorized to engage such expert opinion as deemed necessary to report upon unusual technical issues that arise, subject to the approval of the appointing authority.

❖ The code official is required to make inspections as necessary to determine compliance with the code or to accept written reports of inspections by an approved agency. The inspection of the work in progress or accomplished is another significant element in determining code compliance. Even though a department does not have the resources to inspect every aspect of all work, the required inspections are those that are dictated by administrative rules and procedures based on many factors, including available inspection resources. In order to expand the resources available for inspections, the code official may approve an inspection agency that, in the code official's opinion, possesses the proper qualifications to perform the inspections. When unusual, extraordinary or complex technical issues arise concerning either mechanical installations or the safety of an existing mechanical system, the code official has the authority to seek the opinion and advice of experts. A technical report from an expert requested by the code official can be used to assist the code official in the approval process.

[A] 104.4 Right of entry. Whenever it is necessary to make an inspection to enforce the provisions of this code, or whenever the code official has reasonable cause to believe that there exists in a building or upon any premises any conditions or violations of this code which make the building or premises unsafe, insanitary, dangerous or hazardous, the code official shall have the authority to enter the building or premises at all reasonable times to inspect or to perform the duties imposed upon the code official by this code. If such building or premises is occupied, the code official shall present credentials to the occupant and request entry. If such building or premises is unoccupied, the code official shall first make a reasonable effort to locate the owner or other person having charge or control of the building or premises and request entry. If entry is refused, the code official has recourse to every remedy provided by law to secure entry.

When the code official has first obtained a proper inspection warrant or other remedy provided by law to secure entry, an owner or occupant or person having charge, care or control of the building or premises shall not fail or neglect, after proper request is made as herein provided, to promptly permit entry therein by the code official for the purpose of inspection and examination pursuant to this code.

❖ The first part of this section establishes the right of the code official to enter the premises to conduct the permit inspections required by Section 107. Permit application forms typically include a statement in the certification signed by the applicant (who is the owner or owner's agent) granting the code official the authority to enter areas covered by the permit, as

needed, to enforce code provisions related to the permit. The right to enter other structures or premises is more limited. First, to protect the right of privacy, the owner or occupant must grant the code official permission before an interior inspection of the property can be conducted. Permission is not required for inspections that can be accomplished from within the public right-of-way. Second, access may be denied by the owner or occupant. Unless the inspector has reasonable cause to believe that a violation of the code exists, access may be unattainable. Third, code officials must present proper identification (see commentary, Section 104.6) and request admittance during reasonable hours—usually the normal business hours of the establishment—to be admitted. Fourth, inspections must be aimed at securing or determining compliance with the provisions and intent of the regulations that are specifically within the established scope of the code official's authority.

Searches to gather information for the purpose of enforcing the other codes, ordinances or regulations are considered unreasonable and are prohibited by the Fourth Amendment to the U.S. Constitution. "Reasonable cause" in the context of this section must be distinguished from "probable cause," which is required to gain access to property in criminal cases. The burden of proof establishing reasonable cause may vary among jurisdictions. Usually, an inspector must show that the property is subject to inspection under the provisions of the code; that the interests of the public health, safety and welfare outweigh the individual's right to maintain privacy; and that the inspection is required solely to determine compliance with the provisions of the code.

Many jurisdictions do not recognize the concept of an administrative warrant and may require the code official to prove probable cause in order to gain access upon refusal. This burden of proof is usually greater, often requiring the code official to state in advance why access is needed (usually access is restricted to gathering evidence for seeking an indictment or making an arrest); what specific items or information is sought; its relevance to the case against the subject; how knowledge of the relevance of the information or items sought was obtained; and how the evidence sought will be used. In all such cases, the right to privacy must always be weighed against the right of the code official to conduct an inspection to verify that the public health, safety and welfare are not in jeopardy. Such important and complex constitutional issues should be discussed with the jurisdiction's legal counsel. Jurisdictions should establish procedures for securing the necessary court orders when an inspection is deemed necessary following a refusal.

The last paragraph in this section requires the owner or occupant to permit entry for inspection, if a proper inspection warrant or other documentation required by law has been obtained.

[A] 104.5 Identification. The code official shall carry proper identification when inspecting structures or premises in the performance of duties under this code.

❖ This section requires the code official (including by definition all authorized designees) to carry identification when conducting the duties of the position. The identification removes any question of the purpose and authority of the inspector.

[A] 104.6 Notices and orders. The code official shall issue all necessary notices or orders to ensure compliance with this code.

❖ An important element of code enforcement is the necessary advisement of deficiencies and corrections, which is accomplished through notices and orders. The code official is required to issue orders to abate illegal or unsafe conditions. Sections 108.7, 108.7.1, 108.7.2 and 108.7.3 contain additional information concerning these notices.

[A] 104.7 Department records. The code official shall keep official records of applications received, permits and certificates issued, fees collected, reports of inspections, and notices and orders issued. Such records shall be retained in the official records for the period required for retention of public records.

❖ In keeping with the need for an efficiently conducted business practice, the code official must keep records pertaining to permit applications, permits, fees collected, inspections, notices and orders issued. Such documentation provides a valuable resource if questions arise regarding the department's actions with respect to a building. This section requires that other documents be kept for the length of time mandated by a jurisdiction's, or its state's, laws or administrative rules for retaining public records.

SECTION 105
APPROVAL

[A] 105.1 Modifications. Whenever there are practical difficulties involved in carrying out the provisions of this code, the code official shall have the authority to grant modifications for individual cases upon application of the owner or owner's representative, provided that the code official shall first find that special individual reason makes the strict letter of this code impractical and the modification is in compliance with the intent and purpose of this code and does not lessen health, life and fire safety requirements. The details of action granting modifications shall be recorded and entered in the files of the mechanical inspection department.

❖ The code official may amend or make exceptions to the code, as needed, where strict compliance is impractical. Only the code official has authority to grant modifications. Consideration of a particular difficulty is to be based on the application of the owner and a demonstration that the intent of the code is accomplished. This section is not intended to permit setting aside or ignoring a code provision; rather, it is intended to provide acceptance of equivalent protec-

tion. Such modifications do not, however, extend to actions that are necessary to correct violations of the code. In other words, a code violation or the expense of correcting one cannot constitute a practical difficulty.

[A] 105.2 Alternative materials, methods, equipment and appliances. The provisions of this code are not intended to prevent the installation of any material or to prohibit any method of construction not specifically prescribed by this code, provided that any such alternative has been *approved*. An alternative material or method of construction shall be *approved* where the code official finds that the proposed design is satisfactory and complies with the intent of the provisions of this code, and that the material, method or work offered is, for the purpose intended, at least the equivalent of that prescribed in this code in quality, strength, effectiveness, fire resistance, durability and safety.

❖ The code is not intended to discourage innovative ideas or technological advances. A comprehensive regulatory document such as a mechanical code, cannot envision and then address all future innovations in the industry. As a result, a performance code must be applicable to and provide a basis for the approval of an increasing number of newly developed, innovative materials, systems and methods for which no code text or referenced standards yet exist. The fact that a material, product or method of construction is not addressed in the code is not an indication that the material, product or method is prohibited. The code official is expected to apply sound technical judgment in accepting materials, systems or methods that, while not anticipated by the drafters of the current code text, can be demonstrated to offer equivalent performance. By virtue of its text, the code regulates new and innovative construction practices while addressing the relative safety of building occupants. The code official is responsible for determining whether a requested alternative provides a level of protection of the public health, safety and welfare equivalent to that required by the code.

[A] 105.2.1 Research reports. Supporting data, where necessary to assist in the approval of materials or assemblies not specifically provided for in this code, shall consist of valid research reports from *approved* sources.

❖ When an alternative material or method is proposed for construction, it is incumbent upon the code official to determine whether this alternative is, in fact, an equivalent to the methods prescribed by the code. Reports providing evidence of this equivalency are required to be supplied by an approved source, meaning a source that the code official finds to be reliable and accurate. The ICC Evaluation Service (ICC-ES) is one example of an agency that provides research reports for alternative materials and methods.

[A] 105.3 Required testing. Whenever there is insufficient evidence of compliance with the provisions of this code, or evidence that a material or method does not conform to the requirements of this code, or in order to substantiate claims for alternative materials or methods, the code official shall have the authority to require tests as evidence of compliance to be made at no expense to the jurisdiction.

❖ Sufficient technical data, test reports and documentation must be submitted for evaluation by the code official to provide the basis on which he or she can make a decision regarding an alternative material or type of equipment. If evidence satisfactory to the code official proves that the alternative equipment, material or construction method is equivalent to that required by the code, he or she is obligated to approve it. Any such approval cannot have the effect of waiving any requirements of the code. The burden of proof of equivalence lies with the applicant who proposes the use of alternative equipment, materials or methods.

[A] 105.3.1 Test methods. Test methods shall be as specified in this code or by other recognized test standards. In the absence of recognized and accepted test methods, the code official shall approve the testing procedures.

❖ The code official must require the submission of any appropriate information and data to assist in the determination of equivalency. This information must be submitted before a permit can be issued. The type of information required includes test data in accordance with the referenced standards, evidence of compliance with the referenced standard specifications and design calculations. A research report issued by an authoritative agency is particularly useful in providing the code official with the technical basis for evaluation and approval of new and innovative materials and components. The use of authoritative research reports can greatly assist the code official by reducing the time-consuming engineering analysis necessary to review materials and products. Failure to substantiate a request for the use of an alternative method is a valid reason for the code official to deny a request.

[A] 105.3.2 Testing agency. All tests shall be performed by an *approved* agency.

❖ The testing agency must be approved by the code official. The testing agency should have technical expertise, test equipment and quality assurance to properly conduct and report the necessary testing.

[A] 105.3.3 Test reports. Reports of tests shall be retained by the code official for the period required for retention of public records.

❖ Test reports for retrieval of substantiation of the modification based on the tests are to be retained in accordance with public record laws. The attorney of the jurisdiction could be asked for the specific time period stated in applicable laws of the locality.

[A] 105.4 Approved materials and equipment. Materials, *equipment* and devices *approved* by the code official shall be constructed and installed in accordance with such approval.

❖ The code is a compilation of criteria with which materials, equipment, devices and systems must comply to be suitable for a particular application. The code official has a duty to evaluate such materials, equipment, devices and systems for code compliance and when compliance is determined, approve the same for use. The materials, equipment, devices and systems must be constructed and installed in compliance with, and all conditions and limitations considered as a basis for, that approval. For example, the manufacturer's instructions and recommendations are to be followed if the approval of the material was based, even in part, on those instructions and recommendations. The approval authority given to the code official is a significant responsibility and is a key to code compliance. The approval process is first technical and then administrative and must be approached as such. For example, if data to determine code compliance is required, such data should be in the form of test reports or engineering analysis and not simply taken from a sales brochure.

[A] 105.5 Material, equipment and appliance reuse. Materials, *equipment*, appliances and devices shall not be reused unless such elements have been reconditioned, tested and placed in good and proper working condition and *approved*.

❖ The code criteria for materials, equipment and appliances have changed over the years. Evaluation of testing and materials technology has permitted the development of new criteria that the old materials may not satisfy. As a result, used materials must be evaluated in the same manner as new materials. Used (previously installed) equipment and appliances must be equivalent to that required by the code if they are to be used again in a new installation.

SECTION 106
PERMITS

[A] 106.1 When required. An owner, authorized agent or contractor who desires to erect, install, enlarge, alter, repair, remove, convert or replace a mechanical system, the installation of which is regulated by this code, or to cause such work to be done, shall first make application to the code official and obtain the required permit for the work.

Exception: Where *equipment* and *appliance* replacements or repairs must be performed in an emergency situation, the permit application shall be submitted within the next working business day of the department of mechanical inspection.

❖ In general, a permit is required for all activities that are regulated by the code, and these activities cannot begin until the permit is issued. A mechanical permit is required for the installation, replacement, alteration or modification of mechanical systems and components that are in the scope of applicability of the code. Replacement of an existing piece of equipment or related piping is treated no differently than a new installation in new building construction. The permit causes the work to be inspected to determine compliance with the intent of the code. The exception provides for prompt permit applications for situations where equipment and appliance replacements and repairs are done to address an emergency situation. This action enables the department of mechanical inspection to promptly inspect the work.

[A] 106.2 Permits not required. Permits shall not be required for the following:

1. Portable heating appliances;

2. Portable ventilation appliances and *equipment*;

3. Portable cooling units;

4. Steam, hot water or chilled water piping within any heating or cooling *equipment* or appliances regulated by this code;

5. The replacement of any minor part that does not alter the approval of *equipment* or an *appliance* or make such *equipment* or *appliance* unsafe;

6. Portable evaporative coolers;

7. Self-contained refrigeration systems that contain 10 pounds (4.5 kg) or less of refrigerant, or that are actuated by motors of 1 horsepower (0.75 kW) or less; and

8. Portable fuel cell appliances that are not connected to a fixed piping system and are not interconnected to a power grid.

Exemption from the permit requirements of this code shall not be deemed to grant authorization for work to be done in violation of the provisions of this code or other laws or ordinances of this jurisdiction.

❖ The mechanical installations intended to be exempt from the requirement for a permit are very limited as evidenced by Items 1 through 8. Items 1, 2, 3 and 6 pertain to appliances and equipment that are temporarily used and are not designed for permanent installation. Examples of portable heating, cooling and ventilating appliances and equipment include space heaters; construction site heaters; window unit air conditioners; ventilating fans and blowers; and cooling and ventilation equipment for localized manufacturing processes.

Item 8 also includes the term "portable," but it is used to differentiate this equipment from the permanently installed fuel cell appliances that are hard-piped to a fuel source, such as a natural gas line and/ or connected to the electrical generation power grid (see commentary, Section 924). Portable fuel cell appliances are fueled by on-board gas cylinders and are used as portable generators for such use as charging the batteries of electric cars.

Item 4 applies to steam and water piping that is contained within a packaged assembly or within the enclosure of HVAC equipment, such as air-handling units, heat pumps, cooling towers, chiller units, fan coil units and similar assemblies. This piping is considered part of the equipment and is, therefore, subject to the requirements for the equipment.

Item 5 applies to the replacements of equipment components and appliance components that are minor. A permit would be required if the component replacement could potentially affect either the safety of the equipment or the conditions of approval of the equipment. For example, replacement of a defective control with a control of the same type and specifications as the original factory-supplied control would not require a permit. Replacement of a burner assembly with a burner assembly having a different input capacity or that is designed to burn a different fuel would require a permit.

Item 7 applies to package-type equipment such as freezers, walk-in and reach-in coolers, refrigerated cabinets and cases and similar equipment in which all components of the refrigeration system are located within a single enclosure. The refrigerant charge and compressor motor horsepower are the thresholds of applicability of this item.

The equipment, piping and installation of the eight described items must comply with the code even though a permit is not required for the minor items.

[A] 106.3 Application for permit. Each application for a permit, with the required fee, shall be filed with the code official on a form furnished for that purpose and shall contain a general description of the proposed work and its location. The application shall be signed by the owner or an authorized agent. The permit application shall indicate the proposed *occupancy* of all parts of the building and of that portion of the site or lot, if any, not covered by the building or structure and shall contain such other information required by the code official.

❖ This section limits permit applicants to the building owner or an authorized agent of the owner. An owner's authorized agent could be anyone who is given written permission to act in the owner's interest to obtain a permit, such as an architect, engineer, contractor, tenant or other. Permit forms will generally have enough space for a very brief description of the work to be accomplished, which is sufficient for small jobs. For larger projects, the description will be contained in construction documents.

[A] 106.3.1 Construction documents. *Construction documents,* engineering calculations, diagrams and other data shall be submitted in two or more sets with each application for a permit. The code official shall require *construction documents,* computations and specifications to be prepared and designed by a *registered design professional* when required by state law. Where special conditions exist, the code official is authorized to require additional *construction documents* to be prepared by a *registered design professional. Construction documents* shall be drawn to scale and shall be of sufficient clarity to indicate the location, nature and extent of the work proposed and show in detail that the work conforms to the provisions of this code. *Construction documents* for buildings more than two stories in height shall indicate where penetrations will be made for mechanical systems, and the materials and methods for maintaining required structural safety, fire-resistance rating and fireblocking.

Exception: The code official shall have the authority to waive the submission of *construction documents,* calculations or other data if the nature of the work applied for is such that reviewing of *construction documents* is not necessary to determine compliance with this code.

❖ When the work is of a "minor nature," either in scope or needed description, the code official may use judgment in determining the need for a detailed description of the work. An example of minor work that may not involve a detailed description is the replacement of an existing piece of equipment in a mechanical system or the replacement or repair of a defective portion of a piping system. These provisions are intended to reflect the minimum scope of information needed to determine code compliance. Complex mechanical systems often contain special conditions that can require the involvement of a registered design professional, such as a registered engineer or architect. Many boilers, high heat appliances and appliances located in hazardous areas can create dangerous situations that require more details to be provided on additional construction documents. A statement on the construction documents, such as "All mechanical work must comply with the 2006 edition of the ICC *International Mechanical Code*," is not an acceptable substitute for showing the required information. This section also requires the code official to determine that state professional registration laws are complied with as they apply to the preparation of construction documents.

[A] 106.3.2 Preliminary inspection. Before a permit is issued, the code official is authorized to inspect and evaluate the systems, *equipment*, buildings, devices, premises and spaces or areas to be used.

❖ Some projects might require a preliminary inspection by the code official prior to a permit being issued. This is especially useful for remodel and addition projects where the conditions of the existing building mechanical systems are unknown or are of questionable condition. This section authorizes the code official to make such inspections.

[A] 106.3.3 Time limitation of application. An application for a permit for any proposed work shall be deemed to have been abandoned 180 days after the date of filing, unless such application has been pursued in good faith or a permit has been issued; except that the code official shall have the authority to grant one or more extensions of time for additional periods not exceeding 180 days each. The extension shall be requested in writing and justifiable cause demonstrated.

❖ Once an application for a permit has been submitted for proposed work, a time limit of 180 days is established for issuance of the permit. This prevents the code official from having to hold onto incomplete or delayed applications for an indefinite amount of time. The code official can grant extensions for this time period if provided with a written request with justifiable reasons for the extension request.

[A] 106.4 Permit issuance. The application, *construction documents* and other data filed by an applicant for a permit

shall be reviewed by the code official. If the code official finds that the proposed work conforms to the requirements of this code and all laws and ordinances applicable thereto, and that the fees specified in Section 106.5 have been paid, a permit shall be issued to the applicant.

❖ This section requires the code official to review all submittals for a permit for compliance with the code and further requires code officials to verify that the project will be carried out in accordance with other applicable laws as well. This may involve interagency communication and cooperation so that all laws are being obeyed. Once the code official finds this to be so, a permit may be issued upon payment of the required fees.

[A] 106.4.1 Approved construction documents. When the code official issues the permit where *construction documents* are required, the *construction documents* shall be endorsed in writing and stamped "APPROVED." Such *approved construction documents* shall not be changed, modified or altered without authorization from the code official. Work shall be done in accordance with the *approved construction documents*.

The code official shall have the authority to issue a permit for the construction of part of a mechanical system before the *construction documents* for the entire system have been submitted or *approved*, provided adequate information and detailed statements have been filed complying with all pertinent requirements of this code. The holder of such permit shall proceed at his or her own risk without assurance that the permit for the entire mechanical system will be granted.

❖ Construction documents that reflect compliance with code requirements form an integral part of the permit process. Successful completion of the work depends on these documents. This section requires the code official to stamp the complying construction documents as being "APPROVED" and fixes the status of the document in time. Approved documents may not be revised without the express authorization of the code official to maintain the code-compliance level of the documents.

[A] 106.4.2 Validity. The issuance of a permit or approval of *construction documents* shall not be construed to be a permit for, or an approval of, any violation of any of the provisions of this code or of other ordinances of the jurisdiction. A permit presuming to give authority to violate or cancel the provisions of this code shall be invalid.

The issuance of a permit based upon *construction documents* and other data shall not prevent the code official from thereafter requiring the correction of errors in said *construction documents* and other data or from preventing building operations from being carried on thereunder when in violation of this code or of other ordinances of this jurisdiction.

❖ This powerful code section states the fundamental premise that the permit is only a license to proceed with the work. It is not a license to violate, cancel or set aside any provisions of the code. This statement is important because it means that despite any errors in the approval process, the permit applicant is responsible for code compliance.

[A] 106.4.3 Expiration. Every permit issued by the code official under the provisions of this code shall expire by limitation and become null and void if the work authorized by such permit is not commenced within 180 days from the date of such permit, or if the work authorized by such permit is suspended or abandoned at any time after the work is commenced for a period of 180 days. Before such work recommences, a new permit shall be first obtained and the fee, therefore, shall be one-half the amount required for a new permit for such work, provided no changes have been made or will be made in the original *construction documents* for such work, and provided further that such suspension or abandonment has not exceeded one year.

❖ The permit becomes invalid under two distinct situations, but both are based on a six-month period. The first situation is when no work has been started six months from issuance of the permit. The second situation is when there is no continuation of authorized work for six months. The person who was issued the permit should be notified, in writing, that the permit is invalid and what steps must be taken to restart the work. This section also provides the administrative authority with a means of offsetting the costs associated with the administration of expired, reissued permits by charging a nominal fee for permit reissuance. If, however, the nature or scope of the work to be resumed is different from that contemplated by the original permit, the permit process essentially starts from "scratch" and full fees are charged. The same procedure would also apply if the work has not commenced within one year of the date of permit issuance or if work has been suspended for a year or more.

[A] 106.4.4 Extensions. A permittee holding an unexpired permit shall have the right to apply for an extension of the time within which the permittee will commence work under that permit when work is unable to be commenced within the time required by this section for good and satisfactory reasons. The code official shall extend the time for action by the permittee for a period not exceeding 180 days if there is reasonable cause. A permit shall not be extended more than once. The fee for an extension shall be one-half the amount required for a new permit for such work.

❖ Although it is customary for a project to begin immediately following issuance of a permit, there may be occasions when an unforeseen delay may occur. This section intends to afford the permit holder an opportunity to apply for and receive a single 180-day extension of time within which to begin a project under a still-valid permit (less than 180 days old). The applicant must, however, give the code official an adequate explanation for the delay in starting a project, which could include such things as the need to obtain approvals or permits for the project from other agencies having jurisdiction. This section requires the code official to determine what constitutes "good and satisfactory" reasons for any delay, and further allows

the jurisdiction to offset its administrative costs for extending the permit by charging one-half the permit fee for the extension.

[A] 106.4.5 Suspension or revocation of permit. The code official shall have the authority to suspend or revoke a permit issued under the provisions of this code wherever the permit is issued in error or on the basis of incorrect, inaccurate or incomplete information, or in violation of any ordinance or regulation or any of the provisions of this code.

❖ A permit is in reality a license to proceed with the work. The code official, however, must revoke all permits shown to be based, all or in part, on any false statement or misrepresentation of fact. An applicant may subsequently reapply for a permit with the appropriate corrections or modifications made to the application and the construction documents.

[A] 106.4.6 Retention of construction documents. One set of *approved construction documents* shall be retained by the code official for a period of not less than 180 days from date of completion of the permitted work, or as required by state or local laws. One set of *approved construction documents* shall be returned to the applicant, and said set shall be kept on the site of the building or job at all times during which the work authorized thereby is in progress.

❖ Once the code official has stamped or endorsed as approved the construction documents on which the permit is based (see commentary, Section 106.4.1), one set of approved construction documents must be kept on the construction site to serve as the basis for all subsequent inspections. To avoid confusion, the construction documents on the site must be precisely the documents that were approved and stamped because inspections are to be based on these approved documents. Additionally, the contractor cannot determine compliance with the approved construction documents unless those documents are readily available. Unless the approved construction documents are available, the inspection should be postponed and work on the project halted.

[A] 106.4.7 Previous approvals. This code shall not require changes in the *construction documents*, construction or designated *occupancy* of a structure for which a lawful permit has been heretofore issued or otherwise lawfully authorized, and the construction of which has been pursued in good faith within 180 days after the effective date of this code and has not been abandoned.

❖ This section provides the code official with a useful tool to protect the continuity of permits issued under previous codes or code editions, as long as such permits are being actively executed subsequent to the effective date of the ordinance adopting the newer code.

[A] 106.4.8 Posting of permit. The permit or a copy shall be kept on the site of the work until the completion of the project.

❖ This section requires the permit (or a copy of the permit) to be on the work site until the project is completed. Having the permit at the job site provides project information and evidence to anyone needing to know if the project has been duly authorized.

[A] 106.5 Fees. A permit shall not be issued until the fees prescribed in Section 106.5.2 have been paid, nor shall an amendment to a permit be released until the additional fee, if any, due to an increase of the mechanical system, has been paid.

❖ All fees are to be paid prior to permit issuance. This requirement establishes that the permit applicant intends to proceed with the work and also facilitates payment.

[A] 106.5.1 Work commencing before permit issuance. Any person who commences work on a mechanical system before obtaining the necessary permits shall be subject to 100 percent of the usual permit fee in addition to the required permit fees.

❖ This section is intended to serve as a deterrent to proceeding with work on a mechanical system without a permit (except as provided in Sections 106.1 and 106.2). As a punitive measure, it doubles the permit fee to be charged. This section does not, however, intend to penalize a contractor called upon to do emergency work after hours if he or she promptly notifies the code official the next business day, obtains the requisite permit for the work done and has the required inspections performed.

[A] 106.5.2 Fee schedule. The fees for mechanical work shall be as indicated in the following schedule.

**[JURISDICTION TO INSERT
APPROPRIATE SCHEDULE]**

❖ A published fee schedule must be established for plans examination, permits and inspections. Ideally, the department should generate revenues that cover operating costs and expenses. The permit fee schedule is an integral part of this process.

[A] 106.5.3 Fee refunds. The code official shall authorize the refunding of fees as follows.

1. The full amount of any fee paid hereunder which was erroneously paid or collected.

2. Not more than **[SPECIFY PERCENTAGE]** percent of the permit fee paid when no work has been done under a permit issued in accordance with this code.

3. Not more than **[SPECIFY PERCENTAGE]** percent of the plan review fee paid when an application for a permit for which a plan review fee has been paid is withdrawn or canceled before any plan review effort has been expended.

The code official shall not authorize the refunding of any fee paid, except upon written application filed by the original permittee not later than 180 days after the date of fee payment.

❖ This section allows for a partial refund of fees resulting from the revocation, abandonment or discontinuance of a mechanical project for which a permit has

been issued and fees have been collected. The incomplete work for which the excess fees are to be refunded refers to the work that would have been required by the department had the permit not been terminated. The refund of fees should be related to the cost of enforcement services not provided because of termination of the project.

SECTION 107
INSPECTIONS AND TESTING

[A] 107.1 General. The code official is authorized to conduct such inspections as are deemed necessary to determine compliance with the provisions of this code. Construction or work for which a permit is required shall be subject to inspection by the code official, and such construction or work shall remain accessible and exposed for inspection purposes until *approved*. Approval as a result of an inspection shall not be construed to be an approval of a violation of the provisions of this code or of other ordinances of the jurisdiction. Inspections presuming to give authority to violate or cancel the provisions of this code or of other ordinances of the jurisdiction shall not be valid.

❖ The inspection function is one of the more important aspects of building department operations. This section authorizes the code official to inspect the work for which a permit has been issued and requires that the work to be inspected remain accessible to the code official until inspected and approved. Any expense incurred in removing or replacing material that conceals an item to be inspected is not the responsibility of the code official or the jurisdiction. As with the issuance of permits (see Section 106.4.2), an approval as a result of an inspection is not a license to violate the code. Any work approved which might contain a violation of the code does not relieve the applicant from complying with the code.

[A] 107.2 Required inspections and testing. The code official, upon notification from the permit holder or the permit holder's agent, shall make the following inspections and other such inspections as necessary, and shall either release that portion of the construction or shall notify the permit holder or the permit holder's agent of violations that must be corrected. The holder of the permit shall be responsible for the scheduling of such inspections.

1. Underground inspection shall be made after trenches or ditches are excavated and bedded, piping installed, and before backfill is put in place. When excavated soil contains rocks, broken concrete, frozen chunks and other rubble that would damage or break the piping or cause corrosive action, clean backfill shall be on the job site.

2. Rough-in inspection shall be made after the roof, framing, fireblocking and bracing are in place and all ducting and other components to be concealed are complete, and prior to the installation of wall or ceiling membranes.

3. Final inspection shall be made upon completion of the mechanical system.

Exception: Ground-source heat pump loop systems tested in accordance with Section 1208.1.1 shall be permitted to be backfilled prior to inspection.

The requirements of this section shall not be considered to prohibit the operation of any heating *equipment* or appliances installed to replace existing heating *equipment* or appliances serving an occupied portion of a structure provided that a request for inspection of such heating *equipment* or appliances has been filed with the department not more than 48 hours after such replacement work is completed, and before any portion of such *equipment* or appliances is concealed by any permanent portion of the structure.

❖ Inspections are necessary to determine that an installation conforms to all code requirements. Because the majority of a mechanical system is hidden within the building enclosure, periodic inspections are necessary before portions of the system are concealed. The code official is required to determine that mechanical systems and equipment are installed in accordance with the approved construction documents and the applicable code requirements. All inspections that are necessary to provide verification must be conducted. Generally, the administrative rules of a department may list the interim inspections to be required. Construction that occurs in steps or phases may necessitate multiple inspections; therefore, an exact number of required inspections cannot be specified. Where violations are noted and corrections are required, reinspections may be necessary. As time permits, frequent inspections of some job sites, especially where the work is complex, can be beneficial if they detect code compliance problems or potential problems before they develop or become more difficult to correct. The contractor, builder, owner or other authorized party is responsible for arranging for the required inspections and coordinating inspections to prevent work from being concealed before it is inspected.

1. Inspection of underground piping is especially important because once it is covered, it is the most challenging part of a mechanical system in which to detect a leak. If repairs are necessary, underground repairs are proportionally more expensive because of the need for heavy equipment and the more labor-intensive nature of working below grade level. To reduce possible damage to pipe from rubble, rocks and other rough materials, excavations must be bedded and backfilled with clean fill materials spread and tamped to provide adequate support and protection for piping.

2. A rough-in inspection is an inspection of all parts of the mechanical system that will eventually be concealed in the building structure. The inspection must be made before any of the system is closed up or hidden from view. To gain

approval, the mechanical systems must pass the required rough-in tests.

A rough-in inspection may be completed all at one time or as a series of inspections. This is administratively determined by the local inspections department and is typically dependent on the size of the job.

3. A final inspection may be done as a series of inspections or all at one time, similar to a rough-in inspection. A final inspection is required prior to approval of mechanical work and installations. For the construction of a new building, final approval is required prior to the issuance of the certificate of occupancy as specified in the building code. To verify that all previously issued correction orders have been complied with and to determine whether subsequent violations exist, a final inspection must be made. All violations observed during the final inspection must be noted and the permit holder advised.

The final inspection is made after the completion of the work or installation. Typically, the final inspection is an inspection of all that was installed after the rough-in inspection and not concealed in the building construction. Subsequent reinspections are necessary if the final inspection has generated a notice of violation.

The exception is in relation to Item 1 in this section, which requires inspection before backfill is put in place. The construction sequence of ground-source heat pump loop systems involves trenching, placement and testing of that portion of the system; backfilling; then trenching for another portion of the system. Thus, the exception is needed so that an inspection is not necessary for each test and prior to the backfilling operation for each portion of the system. The frequency of inspection would otherwise be excessive.

The last paragraph of this section provides for prompt operation of replacement heating equipment or appliances, and this allows the occupied areas of a facility to be heated as soon as the new heating equipment or appliance is installed. Any corrections to the installation that are identified by field inspection are to be done.

[A] 107.2.1 Other inspections. In addition to the inspections specified above, the code official is authorized to make or require other inspections of any construction work to ascertain compliance with the provisions of this code and other laws that are enforced.

❖ Any item regulated by the code is subject to inspection by the code official to determine compliance with the applicable code provision, and no list can include all types of work in a given building. Also, other inspections before, during or after the rough-in could be necessary. This section gives the code official the authority to inspect any regulated work.

[A] 107.2.2 Inspection requests. It shall be the duty of the holder of the permit or their duly authorized agent to notify the code official when work is ready for inspection. It shall be the duty of the permit holder to provide *access* to and means for inspections of such work that are required by this code.

❖ This section clarifies that it is the responsibility of the permit holder to arrange for the required inspections when the work is completed. It also establishes his or her responsibility for keeping the work open for inspection and providing all means needed to accomplish the inspections.

[A] 107.2.3 Approval required. Work shall not be done beyond the point indicated in each successive inspection without first obtaining the approval of the code official. The code official, upon notification, shall make the requested inspections and shall either indicate the portion of the construction that is satisfactory as completed, or notify the permit holder or his or her agent wherein the same fails to comply with this code. Any portions that do not comply shall be corrected and such portion shall not be covered or concealed until authorized by the code official.

❖ This section establishes that work cannot progress beyond the point of a required inspection without the code official's approval. Upon making the inspection, the code official must either approve the completed work or notify the permit holder or other responsible party of that which does not comply with the code. Approvals and notices of noncompliance must be in writing, as required by Section 104.3, to avoid any misunderstanding as to what is required. Any work not approved cannot be concealed until it has been corrected and approved by the code official.

[A] 107.2.4 Approved inspection agencies. The code official is authorized to accept reports of *approved* agencies, provided that such agencies satisfy the requirements as to qualifications and reliability.

❖ The determination as to whether to accept an agency test report rests with the code official and the reporting agency must be acceptable to the code official. Appropriate criteria on which to base approval of an inspection agency can be found in Sections 301.5.2.1 through 301.5.2.3.

[A] 107.2.5 Evaluation and follow-up inspection services. Prior to the approval of a prefabricated construction assembly having concealed mechanical work and the issuance of a mechanical permit, the code official shall require the submittal of an evaluation report on each prefabricated construction assembly, indicating the complete details of the mechanical system, including a description of the system and its components, the basis upon which the system is being evaluated, test results and similar information, and other data as necessary for the code official to determine conformance to this code.

❖ As an alternative to the physical inspection by the code official in the plant or location where prefabri-

cated components are produced (such as modular homes and prefabricated structures), the code official has the option of accepting an evaluation report from an approved agency detailing such inspections. Evaluation reports can serve as the basis from which the code official will determine code compliance.

[A] 107.2.5.1 Evaluation service. The code official shall designate the evaluation service of an *approved* agency as the evaluation agency, and review such agency's evaluation report for adequacy and conformance to this code.

❖ The code official is required to review all submitted reports for conformity to the applicable code requirements. If, in the judgment of the code official, the submitted reports are acceptable, the code official should document the basis for the approval.

[A] 107.2.5.2 Follow-up inspection. Except where ready access is provided to mechanical systems, service *equipment* and accessories for complete inspection at the site without disassembly or dismantling, the code official shall conduct the in-plant inspections as frequently as necessary to ensure conformance to the *approved* evaluation report or shall designate an independent, *approved* inspection agency to conduct such inspections. The inspection agency shall furnish the code official with the follow-up inspection manual and a report of inspections upon request, and the mechanical system shall have an identifying label permanently affixed to the system indicating that factory inspections have been performed.

❖ The owner is required to provide special inspections of fabricated assemblies at the fabrication plant. The code official or an approved inspection agency must conduct periodic in-plant inspections to ensure conformance to the approved evaluation report described in Section 107.2.5. Such inspections would not be required where the mechanical systems can be inspected completely at the job site.

[A] 107.2.5.3 Test and inspection records. Required test and inspection records shall be available to the code official at all times during the fabrication of the mechanical system and the erection of the building; or such records as the code official designates shall be filed.

❖ All testing and inspection records related to a fabricated assembly must be filed with the code official to maintain a complete and legal record of the assembly and erection of the building.

[A] 107.3 Testing. Mechanical systems shall be tested as required in this code and in accordance with Sections 107.3.1 through 107.3.3. Tests shall be made by the permit holder and observed by the code official.

❖ The concept of this section is that the testing of mechanical systems is required where testing is specified in the technical chapters of the code. See the "tests" listing in the index of the code for examples of mechanical systems that require testing.

[A] 107.3.1 New, altered, extended or repaired systems. New mechanical systems and parts of existing systems, which have been altered, extended, renovated or repaired, shall be tested as prescribed herein to disclose leaks and defects.

❖ Testing is necessary to make sure that the system is free from leaks or other defects. Testing is also required, to the extent specified in the technical chapters of the code, for portions of existing systems that have been altered, extended, renovated or repaired.

[A] 107.3.2 Apparatus, material and labor for tests. Apparatus, material and labor required for testing a mechanical system or part thereof shall be furnished by the permit holder.

❖ The permit holder is responsible for performing tests, as well as for supplying all of the labor and apparatus necessary to conduct the tests. The code official observes but never performs the test.

[A] 107.3.3 Reinspection and testing. Where any work or installation does not pass an initial test or inspection, the necessary corrections shall be made so as to achieve compliance with this code. The work or installation shall then be resubmitted to the code official for inspection and testing.

❖ If a system or a portion of a system does not pass the initial test or inspection, violations must be corrected and the system must be reinspected. To encourage code compliance and to cover the expense of the code official's time, many code enforcement jurisdictions charge fees for inspections that are required subsequent to the first reinspection.

[A] 107.4 Approval. After the prescribed tests and inspections indicate that the work complies in all respects with this code, a notice of approval shall be issued by the code official.

❖ After the code official has performed the required inspections and observed any required equipment and system tests (or has received written reports of the results of such tests), he or she must determine whether the installation or work is in compliance with all applicable sections of the code. The code official must issue a written notice of approval if the subject work or installation is in apparent compliance with the code. The notice of approval is given to the permit holder and a copy of the notice is retained on file by the code official.

[A] 107.4.1 Revocation. The code official is authorized to, in writing, suspend or revoke a notice of approval issued under the provisions of this code wherever the notice is issued in error, on the basis of incorrect information supplied, or where it is determined that the building or structure, premise or portion thereof is in violation of any ordinance or regulation or any of the provisions of this code.

❖ This section is needed to give the code official the authority to revoke a notice of approval for the reasons indicated in the code text. The code official can suspend the notice until all of the code violations are corrected.

[A] 107.5 Temporary connection. The code official shall have the authority to authorize the temporary connection of a mechanical system to the sources of energy for the purpose of testing mechanical systems or for use under a temporary certificate of *occupancy*.

❖ Typical procedure for a local jurisdiction is to withhold the issuance of the certificate of occupancy until approvals have been received from each code official responsible for inspection of the structure. The code official is permitted to issue a temporary authorization to make connections to the public utility system prior to the completion of all work. The certification is intended to acknowledge that, because of seasonal limitations, time constraints, the need for testing or partial operation of equipment, some building systems may be connected even though the building is not suitable for final occupancy. The intent of this section is that a request for temporary occupancy or the connection and use of mechanical equipment or systems should be granted when the requesting permit holder has demonstrated to the code official's satisfaction that the public health, safety and welfare will not be endangered. The code official should view the issuance of a "temporary authorization or certificate of occupancy" as substantial an act as the issuance of the final certificate. Indeed, the issuance of a temporary certificate of occupancy offers a greater potential for conflict because once the building or structure is occupied, it is very difficult to remove the occupants through legal means.

[A] 107.6 Connection of service utilities. No person shall make connections from a utility, source of energy, fuel or power to any building or system that is regulated by this code for which a permit is required, until authorized by the code official.

❖ This section establishes the authority of the code official to approve utility connections to a building, such as water, sewer, electricity, gas and steam, and to require their disconnection when such approval has not been granted. For the protection of building occupants, including workers, such systems must have had final inspection approvals, except as allowed by Section 110.3 for temporary connections.

SECTION 108
VIOLATIONS

[A] 108.1 Unlawful acts. It shall be unlawful for a person, firm or corporation to erect, construct, alter, repair, remove, demolish or utilize a mechanical system, or cause same to be done, in conflict with or in violation of any of the provisions of this code.

❖ Violations of the code are prohibited. This is the basis for all citations and correction notices related to violations of the code.

[A] 108.2 Notice of violation. The code official shall serve a notice of violation or order to the person responsible for the erection, installation, *alteration*, extension, repair, removal or demolition of mechanical work in violation of the provisions of this code, or in violation of a detail statement or the *approved construction documents* thereunder, or in violation of a permit or certificate issued under the provisions of this code. Such order shall direct the discontinuance of the illegal action or condition and the abatement of the violation.

❖ The code official is required to notify the person responsible for the erection or use of a building found to be in violation of the code. The section that is allegedly being violated must be cited so that the responsible party can respond to the notice.

[A] 108.3 Prosecution of violation. If the notice of violation is not complied with promptly, the code official shall request the legal counsel of the jurisdiction to institute the appropriate proceeding at law or in equity to restrain, correct or abate such violation, or to require the removal or termination of the unlawful *occupancy* of the structure in violation of the provisions of this code or of the order or direction made pursuant thereto.

❖ The code official must pursue, through the use of legal counsel of the jurisdiction, legal means to correct the violation. This is not optional. Any extensions of time so that the violations may be corrected voluntarily must be for a reasonable, bona fide cause or the code official may be subject to criticism for "arbitrary and capricious" actions. In general, it is better to have a standard time limitation for correction of violations. Departures from this standard must be for a clear and reasonable purpose, usually stated in writing by the violator.

[A] 108.4 Violation penalties. Persons who shall violate a provision of this code or shall fail to comply with any of the requirements thereof or who shall erect, install, alter or repair mechanical work in violation of the *approved construction documents* or directive of the code official, or of a permit or certificate issued under the provisions of this code, shall be guilty of a **[SPECIFY OFFENSE]**, punishable by a fine of not more than **[AMOUNT]** dollars or by imprisonment not exceeding **[NUMBER OF DAYS]**, or both such fine and imprisonment. Each day that a violation continues after due notice has been served shall be deemed a separate offense.

❖ A standard fine or other penalty as deemed appropriate by the jurisdiction is prescribed in this section. Additionally, this section identifies a principle that "each day that a violation continues shall be deemed a separate offense" for the purpose of applying the prescribed penalty in order to facilitate prompt resolution.

[A] 108.5 Stop work orders. Upon notice from the code official that mechanical work is being done contrary to the provisions of this code or in a dangerous or unsafe manner, such work shall immediately cease. Such notice shall be in writing and shall be given to the owner of the property, or to the owner's agent, or to the person doing the work. The notice shall state the conditions under which work is authorized to resume. Where an emergency exists, the code official shall not be required to give a written notice prior to stopping the work. Any person who shall continue any work on the system after having been served with a stop work order, except such work as that person is directed to perform to remove a violation or unsafe condition, shall be liable for a fine of not less than **[AMOUNT]** dollars or more than **[AMOUNT]** dollars.

❖ Upon receipt of a violation notice from the code official, the contractor must immediately stop construction activities identified in the notice, except as expressly permitted to correct the violation. A stop work order can result in inconvenience and monetary loss to the affected parties; therefore, justification must be evident and judgement must be exercised before a stop work order is issued. A stop work order can prevent a violation from becoming worse and more difficult or expensive to correct. A stop work order may be issued where work is proceeding without a permit to perform the work. Hazardous conditions could develop where the code official is unaware of the nature of the work and a permit for the work has not been issued. The issuance of a stop work order on a mechanical system may result from work done by the mechanical contractor that affects a nonmechanical component. For example, if a mechanical contractor cuts a structural element to install piping, the structure may be weakened enough to cause a partial or complete structural failure. As determined by the adopting jurisdiction, a penalty may be assessed for failure to comply with this section and the dollar amount is to be inserted in the blanks provided.

[A] 108.6 Abatement of violation. The imposition of the penalties herein prescribed shall not preclude the legal officer of the jurisdiction from instituting appropriate action to prevent unlawful construction or to restrain, correct or abate a violation, or to prevent illegal *occupancy* of a building, structure or premises, or to stop an illegal act, conduct, business or utilization of the mechanical system on or about any premises.

❖ Despite the assessment of a penalty in the form of a fine or imprisonment against a violator, the violation itself must still be corrected. Failure to make the necessary corrections will result in the violator being subject to additional penalties as described in the preceding section.

[A] 108.7 Unsafe mechanical systems. A mechanical system that is unsafe, constitutes a fire or health hazard, or is otherwise dangerous to human life, as regulated by this code, is hereby declared as an unsafe mechanical system. Use of a mechanical system regulated by this code constituting a hazard to health, safety or welfare by reason of inadequate maintenance, dilapidation, fire hazard, disaster, damage or abandonment is hereby declared an unsafe use. Such unsafe *equipment* and appliances are hereby declared to be a public nuisance and shall be abated by repair, rehabilitation, demolition or removal.

❖ Unsafe conditions include those that constitute a health hazard, fire hazard, explosion hazard, shock hazard, asphyxiation hazard, physical injury hazard or are otherwise dangerous to human life and property. In the course of performing duties, the code official may identify a hazardous condition that must be declared in violation of the code and, therefore, must be abated.

[A] 108.7.1 Authority to condemn mechanical systems. Whenever the code official determines that any mechanical system, or portion thereof, regulated by this code has become hazardous to life, health, property, or has become insanitary, the code official shall order in writing that such system either be removed or restored to a safe condition. A time limit for compliance with such order shall be specified in the written notice. A person shall not use or maintain a defective mechanical system after receiving such notice.

When such mechanical system is to be disconnected, written notice as prescribed in Section 108.2 shall be given. In cases of immediate danger to life or property, such disconnection shall be made immediately without such notice.

❖ When a mechanical system is determined to be unsafe, the code official is required to notify the owner or agent of the building as the first step in correcting the difficulty. The notice is to describe the repairs and improvements necessary to correct the deficiency or require removal or replacement of the unsafe equipment or system. Such notices must specify a time frame in which the corrective actions must occur. Additionally, the notice should require the immediate response of the owner or agent. If the owner or agent is not available, public notice of the declaration should suffice for complying with this section. The code official may also determine that disconnection of the system is necessary to correct an unsafe condition and must give written notice to that effect (see commentary, Section 108.2), unless immediate disconnection is essential for public health and safety reasons (see commentary, Section 108.7.2).

[A] 108.7.2 Authority to order disconnection of energy sources. The code official shall have the authority to order disconnection of energy sources supplied to a building, structure or mechanical system regulated by this code, when it is determined that the mechanical system or any portion thereof has become hazardous or unsafe. Written notice of such order to disconnect service and the causes therefor shall be given within 24 hours to the owner and occupant of such building, structure or premises, provided, however, that in cases of immediate danger to life or property, such disconnection shall be made immediately without such notice. Where energy sources are provided by a public utility, the code official shall immediately notify the serving utility in writing of the issuance of such order to disconnect.

❖ Disconnecting a mechanical system from the energy supply is the most radical method of hazard abatement available to the code official and should be reserved for cases in which all other lesser remedies have proven ineffective. Such an action must be preceded by written notice to the owner and any occupants of the building being ordered to disconnect.

Disconnection must be accomplished within the time frame established by the code official in the written notification to disconnect. When the hazard to the public health and welfare is so imminent as to mandate immediate disconnection, the code official has the authority and even the obligation to cause disconnection without notice.

[A] 108.7.3 Connection after order to disconnect. A person shall not make energy source connections to mechanical systems regulated by this code which have been disconnected or ordered to be disconnected by the code official, or the use of which has been ordered to be discontinued by the code official until the code official authorizes the reconnection and use of such mechanical systems.

When a mechanical system is maintained in violation of this code, and in violation of a notice issued pursuant to the provisions of this section, the code official shall institute appropriate action to prevent, restrain, correct or abate the violation.

❖ When any mechanical system is maintained in violation of the code, and in violation of any notice issued pursuant to the provisions of this section, the code official is to institute appropriate action to prevent, restrain, correct or abate the violation. Once the reason for discontinuation of use or disconnection of the mechanical system no longer exists, only the code official may authorize resumption of use or reconnection of the system after it is demonstrated to the code official's satisfaction that all repairs or other work are in compliance with applicable sections of the code. This section also requires the code official to take action to abate code violations (see commentary, Section 108.2).

SECTION 109
MEANS OF APPEAL

[A] 109.1 Application for appeal. A person shall have the right to appeal a decision of the code official to the board of appeals. An application for appeal shall be based on a claim that the true intent of this code or the rules legally adopted thereunder have been incorrectly interpreted, the provisions of this code do not fully apply, or an equally good or better form of construction is proposed. The application shall be filed on a form obtained from the code official within 20 days after the notice was served.

❖ This section literally allows any person to appeal a decision of the code official. In practice, this section has been interpreted to permit appeals only by those aggrieved parties with a material or definitive interest in the decision of the code official. An aggrieved party may not appeal a code requirement per se. The intent of the appeal process is not to waive or set aside a code requirement; rather, it is intended to provide a means of reviewing a code official's decision on an interpretation or application of the code or to review the equivalency of protection to the code requirements.

[A] 109.1.1 Limitation of authority. The board of appeals shall have no authority relative to interpretation of the administration of this code nor shall such board be empowered to waive requirements of this code.

❖ This section establishes limits on the board of appeals. The board may not interpret the administration provisions of the code but does have interpretation authority regarding the technical requirements of the code. The board is not allowed to set aside any of the technical requirements of the code; however, it is allowed to consider alternative methods of compliance with the technical requirements.

[A] 109.2 Membership of board. The board of appeals shall consist of five members appointed by the chief appointing authority as follows: one for five years; one for four years; one for three years; one for two years; and one for one year. Thereafter, each new member shall serve for five years or until a successor has been appointed.

❖ The board of appeals is to consist of five members appointed on a rotating basis by the "chief appointing authority"; typically, the mayor or city manager. This method of appointment allows for a smooth transition of board of appeals members, thus ensuring continuity of action over the years.

[A] 109.2.1 Qualifications. The board of appeals shall consist of five individuals, one from each of the following professions or disciplines.

1. *Registered design professional* who is a registered architect; or a builder or superintendent of building construction with at least 10 years' experience, five of which shall have been in responsible charge of work.

2. *Registered design professional* with structural engineering or architectural experience.

3. *Registered design professional* with mechanical and plumbing engineering experience; or a mechanical contractor with at least 10 years' experience, five of which shall have been in responsible charge of work.

4. *Registered design professional* with electrical engineering experience; or an electrical contractor with at least 10 years' experience, five of which shall have been in responsible charge of work.

5. *Registered design professional* with fire protection engineering experience; or a fire protection contractor with at least 10 years' experience, five of which shall have been in responsible charge of work.

❖ The board of appeals consists of five persons with the qualifications and experience indicated in this section. One must be a registered design professional (see Item 2) with structural or architectural experience. The others must be registered design professionals, construction superintendents or contractors with experience in the various areas of building construction. These requirements are important in that technical people rule on technical matters. The board of appeals is not the place for policy or political deliberations. It is intended that these matters be decided purely on their technical merits, with due regard for state-of-the-art construction technology.

[A] 109.2.2 Alternate members. The chief appointing authority shall appoint two alternate members who shall be called by the board chairman to hear appeals during the absence or disqualification of a member. Alternate members shall possess the qualifications required for board membership and shall be appointed for five years, or until a successor has been appointed.

❖ This section authorizes the chief appointing authority to appoint two alternate members who are to be available if the principal members of the board are absent or disqualified. Alternate members must possess the same qualifications as the principal members and are appointed for a term of five years, or until such time that a successor is appointed.

[A] 109.2.3 Chairman. The board shall annually select one of its members to serve as chairman.

❖ It is customary to determine chairmanship annually so that a regular opportunity is available to evaluate and either reappoint the current chairman or appoint a new one.

[A] 109.2.4 Disqualification of member. A member shall not hear an appeal in which that member has a personal, professional or financial interest.

❖ All members must disqualify themselves from any appeal in which they have a personal, professional or financial interest.

[A] 109.2.5 Secretary. The chief administrative officer shall designate a qualified clerk to serve as secretary to the board. The secretary shall file a detailed record of all proceedings in the office of the chief administrative officer.

❖ The chief administrative officer is to designate a qualified clerk to serve as secretary to the board. The secretary is required to file a detailed record of all proceedings in the office of the chief administrative officer.

[A] 109.2.6 Compensation of members. Compensation of members shall be determined by law.

❖ Members of the board of appeals need not be compensated unless required by the local municipality or jurisdiction.

[A] 109.3 Notice of meeting. The board shall meet upon notice from the chairman, within ten days of the filing of an appeal, or at stated periodic meetings.

❖ The board must meet within 10 days of the filing of an appeal, or at regularly scheduled meetings.

[A] 109.4 Open hearing. All hearings before the board shall be open to the public. The appellant, the appellant's representative, the code official and any person whose interests are affected shall be given an opportunity to be heard.

❖ All hearings before the board must be open to the public. The person who filed the appeal, his or her representative, the code official and any person whose interests are affected must be heard.

[A] 109.4.1 Procedure. The board shall adopt and make available to the public through the secretary procedures under which a hearing will be conducted. The procedures shall not require compliance with strict rules of evidence, but shall mandate that only relevant information be received.

❖ The board is required to establish and make available to the public written procedures detailing how hearings are to be conducted. Additionally, this section provides that although strict rules of evidence are not applicable, the information presented must be relevant to the appeal.

[A] 109.5 Postponed hearing. When five members are not present to hear an appeal, either the appellant or the appellant's representative shall have the right to request a postponement of the hearing.

❖ When all five members of the board are not present, either the person making the appeal or his or her representative may request a postponement of the hearing.

[A] 109.6 Board decision. The board shall modify or reverse the decision of the code official by a concurring vote of three members.

❖ A concurring vote of three members of the board is needed to modify or reverse the decision of the code official.

[A] 109.6.1 Resolution. The decision of the board shall be by resolution. Certified copies shall be furnished to the appellant and to the code official.

❖ A formal decision in the form of a resolution is required to provide an official record. Copies of this resolution are to be furnished to both the person making the appeal and the code official. The code official is bound by the action of the board of appeals, unless it is the opinion of the code official that the board of appeals has acted improperly. In such cases, relief through the court having jurisdiction may be sought by corporate council.

[A] 109.6.2 Administration. The code official shall take immediate action in accordance with the decision of the board.

❖ To avoid any undue delay in the progress of construction, the code official is required to act quickly on the board's decision. This action may be to enforce the decision or to seek legislative relief if the board's action can be demonstrated to be inappropriate.

[A] 109.7 Court review. Any person, whether or not a previous party of the appeal, shall have the right to apply to the appropriate court for a writ of certiorari to correct errors of law. Application for review shall be made in the manner and time required by law following the filing of the decision in the office of the chief administrative officer.

❖ This section allows any person to request a review by the court of jurisdiction if that person believes errors of law have occurred. Application for the review must be made after the decision of the board is filed with the chief administrative officer. This helps to establish the observance of due process for all concerned.

SECTION 110
TEMPORARY EQUIPMENT, SYSTEMS AND USES

[A] 110.1 General. The code official is authorized to issue a permit for temporary *equipment*, systems and uses. Such permits shall be limited as to time of service, but shall not be permitted for more than 180 days. The code official is authorized to grant extensions for demonstrated cause.

❖ The code official is permitted to issue temporary authorization to make connections to a public utility system prior to completion of all work. This acknowledges that, because of seasonal limitations, time constraints, or the need for testing or partial operations of equipment, some building systems may be safely connected even though the building is not suitable for final occupancy. The temporary connection and utilization of connected equipment should be approved when the requesting permit holder has demonstrated to the code official's satisfaction that public health, safety and welfare will not be endangered.

[A] 110.2 Conformance. Temporary *equipment*, systems and uses shall conform to the structural strength, fire safety, means of egress, accessibility, light, ventilation and sanitary requirements of this code as necessary to ensure the public health, safety and general welfare.

❖ Even though a utility connection may be temporary, the only way to make sure that the public health, safety, and general welfare are protected is for those temporary connections to comply with the code.

[A] 110.3 Temporary utilities. The code official is authorized to give permission to temporarily supply utilities before an installation has been fully completed and the final certificate of completion has been issued. The part covered by the temporary certificate shall comply with the requirements specified for temporary lighting, heat or power in the code.

❖ Commonly, the utilities on many construction sites are installed and energized long before all aspects of the system are completed. This section would allow such temporary systems to continue, provided that they comply with the applicable safety provisions of the code.

[A] 110.4 Termination of approval. The code official is authorized to terminate such permit for temporary *equipment*, systems or uses and to order the temporary *equipment*, systems or uses to be discontinued.

❖ This section provides the code official with the necessary authority to terminate the permit for temporary equipment, systems, and uses if conditions of the permit have been violated or if temporary equipment or systems pose an imminent hazard to the public. This enables the code official to act quickly when time is of the essence in order to protect public health, safety and welfare.

Bibliography

The following resource materials were used in the preparation of the commentary for this chapter of the code:

"Building Valuation Data." *Building Safety Journal*. Whittier, CA: International Code Council, Semiannually in May/June and November/December Issues.

IFC-12, *International Fire Code*. Washington, DC: International Code Council, Inc., 2011.

IFGC-12, *International Fuel Gas Code*. Washington, DC: International Code Council, Inc., 2011.

IMC-12, *International Mechanical Code*. Washington, DC: International Code Council, Inc., 2011.

IPMC-12, *International Property Maintenance Code*. Washington, DC: International Code Council, Inc., 2011.

IRC-12, *International Residential Code*. Washington, DC: International Code Council, Inc., 2011.

Legal Aspects of Code Administration. Washington, DC: International Code Council, 2002.

NFPA 70-11, *National Electrical Code*. Quincy, MA: National Fire Protection Association, 2008.

Chapter 2:
Definitions

General Comments

Chapter 2 establishes the meanings of key words and terms used in the code.

Section 201 addresses the practical concerns encountered in writing a technical document as they relate to the use of gender, tense and singular versus plural. This section also provides the code official with guidance for finding definitions of those words or terms not defined herein.

Section 202 is an alphabetical listing of the terms commonly used throughout the code and that are required for the effective application of code requirements.

Purpose

Codes, by their very nature, are technical documents. Literally every word, term and punctuation mark can add to or change the meaning of the intended result. This is even more so with performance code text where the desired result often takes on more importance than the specific words.

Furthermore, the code, with its broad scope of applicability, includes terms used in a variety of construction disciplines. These terms can often have multiple meanings depending on the context or discipline being used at the time.

For these reasons, a consensus on the specific meaning of terms contained in the code must be maintained. Chapter 2 performs this function by stating clearly what specific terms mean for the purpose of the code.

SECTION 201
GENERAL

201.1 Scope. Unless otherwise expressly stated, the following words and terms shall, for the purposes of this code, have the meanings indicated in this chapter.

❖ This chapter contains definitions of terms that are associated with the subject matter of this code. Definitions of terms are necessary for the understanding and application of the code requirements.

201.2 Interchangeability. Words used in the present tense include the future; words in the masculine gender include the feminine and neuter; the singular number includes the plural and the plural, the singular.

❖ Although the definitions contained in Chapter 2 are to be taken literally, gender and tense are considered to be interchangeable; thus, any grammatical inconsistencies within the code text will not hinder the understanding or enforcement of the requirements.

201.3 Terms defined in other codes. Where terms are not defined in this code and are defined in the *International Building Code, International Fire Code, International Fuel Gas Code* or the *International Plumbing Code*, such terms shall have meanings ascribed to them as in those codes.

❖ When a word or term that is not defined in this chapter appears in the code, other references may be used to find its definition, such as the *International Building Code®* (IBC®), the *International Fire Code®* (IFC®), the *International Fuel Gas Code®* (IFGC®) or

the *International Plumbing Code®* (IPC®). These codes contain additional definitions (some parallel and duplicative) that may be used in the enforcement of this code or in the enforcement of the other codes by reference.

201.4 Terms not defined. Where terms are not defined through the methods authorized by this section, such terms shall have ordinarily accepted meanings such as the context implies.

❖ Another resource for defining words or terms not defined in this chapter or in other codes is their "ordinarily accepted meanings." The intent of this statement is that a dictionary definition may suffice, if the definition is in context.

In some cases, construction terms used throughout the code may not be defined in Chapter 2 or in a dictionary. In such case, one would first turn to the definitions contained in the referenced standards (see Chapter 15) and then to published textbooks on the subject in question.

SECTION 202
GENERAL DEFINITIONS

❖ This portion of the commentary addresses only those terms whose definitions appear in Chapter 2. The commentary for definitions that are located elsewhere in the code can be found in the indicated sections that contain those definitions.

ABRASIVE MATERIALS. Moderately abrasive particulate in high concentrations, and highly abrasive particulate in moderate and high concentrations, such as alumina, bauxite, iron silicate, sand and slag.

❖ Abrasive materials flowing in a duct can cause erosion damage to the duct system and air-moving equipment.

ABSORPTION SYSTEM. A refrigerating system in which refrigerant is pressurized by pumping a chemical solution of refrigerant in absorbent, and then separated by the addition of heat in a generator, condensed (to reject heat), expanded, evaporated (to provide refrigeration), and reabsorbed in an absorber to repeat the cycle; the system may be single or multiple effect, the latter using multiple stages or internally cascaded use of heat to improve efficiency.

❖ Absorption refrigeration systems use two fluids (an absorbent and a refrigerant) and a heat source to remove heat through evaporation at a lower pressure and to reject the heat through condensation at a higher pressure. Typical absorption systems use ammonia as the refrigerant and water as the absorbent, or lithium bromide as the absorbent and water as the refrigerant.

ACCESS (TO). That which enables a device, *appliance* or *equipment* to be reached by ready access or by a means that first requires the removal or movement of a panel, door or similar obstruction [see also "Ready access (to)"].

❖ Providing access to mechanical equipment and appliances is necessary to facilitate inspection, observation, maintenance, adjustment, repair or replacement. Access to equipment means that the equipment can be physically reached without having to remove a permanent portion of the structure. It is acceptable, for example, to install equipment in an interstitial space that would require lay-in suspended ceiling panels to be removed to gain access. Mechanical equipment would not be considered as being provided with access if it were necessary to remove or open any portion of a structure other than panels, doors, covers or similar obstructions intended to be removed or opened. Also, see the definition of "Ready access (to)."

Access can be described as the capability of being reached or approached for the purpose of inspection, observation, maintenance, adjustment, repair or replacement. Achieving access may first require the removal or opening of a panel, door or similar obstruction and may require the overcoming of an obstacle such as elevation.

AIR. All air supplied to mechanical *equipment* and appliances for *combustion*, ventilation, cooling, etc. Standard air is air at standard temperature and pressure, namely, 70°F (21°C) and 29.92 inches of mercury (101.3 kPa).

❖ The term "air," for the purposes of the code, includes air used, moved or conditioned by the mechanical systems that are regulated by the code.

AIR CONDITIONING. The treatment of air so as to control simultaneously the temperature, humidity, cleanness and distribution of the air to meet the requirements of a conditioned space.

❖ Air conditioning is commonly referred to only in the context of cooling and dehumidifying air; however, the definition also indicates a much broader scope. In essence, the process of providing ventilation air to a space constitutes air conditioning because the introduction of any ventilation air is an attempt to control the indoor environment.

AIR-CONDITIONING SYSTEM. A system that consists of heat exchangers, blowers, filters, supply, exhaust and return ducts, and shall include any apparatus installed in connection therewith.

❖ This definition is limited to the components commonly used in a mechanical air-conditioning system. Additional apparatus considered as part of the air-conditioning system would include thermostats, humidistats, dampers and any other controls needed for the system to operate properly.

AIR DISPERSION SYSTEM. Any diffuser system designed to both convey air within a room, space or area and diffuse air into that space while operating under positive pressure. Systems are commonly constructed of, but not limited to, fabric or plastic film.

❖ A simple example of an air dispersion system is a tube or hose with holes in it that connects to the supply air duct. This tube extends into the conditioned space and distributes the air more uniformly throughout the space.

Air dispersion systems usually connect to the supplying air duct at a sidewall. The supplying air duct conveys air from the air-handling unit to the destination room, space or area. At this point, a sidewall grille or other type of diffuser could be used to diffuse the air into the space. This diffuser would rely on the velocity of the exiting air and its direction to meet requirements of the space. For this example, the air dispersion system would be mounted in place of the sidewall grille. By the air dispersion system being physically longer, the velocity of air exiting the system is more uniformly distributed throughout the space.

This technology has been used for over 50 years in the United States, and longer in Europe. The concept, in the United States, originated in the agricultural industry, and through innovative fabric technology and proven performance, has evolved into an attractive means to diffuse air within various spaces. These spaces include where food is processed (refrigeration), industrial, warehouses, retail, convention centers, offices, athletic, and laboratory environments. These systems are tested and listed in accordance with UL 2518.

AIR DISTRIBUTION SYSTEM. Any system of ducts, plenums and air-handling *equipment* that circulates air within a space or spaces and includes systems made up of one or more air-handling units.

❖ An air distribution system consists of air-moving equipment; intakes and outlets; supply and return openings and the interconnecting ductwork, plenums and conduit necessary to conduct airflow to and from the inlets and outlets. The primary characteristic is the ability of the system to distribute and circulate air throughout one or more spaces within a building. Air distribution systems in the context of the code are environmental air-conditioning systems that heat, cool and ventilate the occupied spaces of a building. Exhaust equipment, makeup air supply units, rooftop units, turnover units and unit heaters are examples of equipment commonly installed without distribution ductwork. Such installations would not be considered as air distribution systems. Air distribution systems are capable of supplying air to and removing air from spaces and recirculating all or a portion of the air handled.

For the purpose of applying Section 606, an air distribution system that employs multiple air handlers operating in parallel would be considered a single system with an airflow capacity equal to the sum of the individual air handlers. Individual air handlers that operate independently and do not share common supply return ducts would be considered as separate systems (see commentary, Section 606).

AIR, EXHAUST. Air being removed from any space, *appliance* or piece of *equipment* and conveyed directly to the atmosphere by means of openings or ducts.

❖ Exhaust air may be from a space, an appliance or a piece of equipment. Exhaust air may or may not contain contaminants from the space being exhausted.

Exhaust air systems are terminated outside the building, in some cases after the exhaust air has been treated to remove any harmful emissions. Exhaust air is not recirculated.

AIR-HANDLING UNIT. A blower or fan used for the purpose of distributing supply air to a room, space or area.

❖ In addition to blowers, air-handling units may contain heat exchangers, filters and means to control air volume.

AIR, MAKEUP. Air that is provided to replace air being exhausted.

❖ Makeup air is not to be confused with combustion air. Makeup air replaces the air being exhausted through such systems as bathroom and toilet exhausts, kitchen exhaust hoods, hazardous exhaust systems and clothes dryer exhaust systems (refer to Chapter 5 for specific requirements for makeup air). Exhaust systems cannot function at design capacity without

adequate volumes of makeup air to replace the air being exhausted.

[A] ALTERATION. A change in a mechanical system that involves an extension, addition or change to the arrangement, type or purpose of the original installation.

❖ An alteration is any modification or change made to an existing installation. For example, changing refrigerant types or heat transfer fluids in a system would be considered alterations.

APPLIANCE. A device or apparatus that is manufactured and designed to utilize energy and for which this code provides specific requirements.

❖ An appliance is a manufactured component or assembly of components that converts one form of energy into a different form of energy to serve a specific purpose. The term "appliance" generally refers to residential- and commercial-type equipment that is manufactured in standardized sizes or types. The term "appliance" is generally not associated with industrial-type equipment. For the application of the code provisions, the terms "appliance" and "equipment" are mutually exclusive.

Examples of appliances include furnaces; boilers; water heaters; room heaters; refrigeration units; cooking equipment; clothes dryers; wood stoves; pool, spa and hot tub heaters; unit heaters; ovens and similar fuel-fired or electrically operated appliances (see the definition of "Equipment").

APPLIANCE, EXISTING. Any *appliance* regulated by this code which was legally installed prior to the effective date of this code, or for which a permit to install has been issued.

❖ The definition creates a distinction between legally existing appliances and illegally existing appliances. Any appliance that is installed without a permit, if a permit was required at the time of installation, is not a legally existing appliance. An appliance that was illegally installed prior to the effective date of the code is not considered existing, but is "new," and therefore subject to enforcement of the requirements for new installations. This definition is important in the application of Section 102.2 of the code.

APPLIANCE TYPE.

High-heat appliance. Any *appliance* in which the products of *combustion* at the point of entrance to the flue under normal operating conditions have a temperature greater than 2,000°F (1093°C).

❖ A high-heat appliance is one in which the temperature of the products of combustion, measured at the point of entry to the flue under normal operating conditions, exceeds 2,000°F (1093°C). High-heat appliances include industrial furnaces, retorts and kilns.

Low-heat appliance (residential appliance). Any *appliance* in which the products of *combustion* at the point of entrance to the flue under normal operating conditions have a temperature of 1,000°F (538°C) or less.

❖ Residential appliances, including solid-fuel appliances, are classified in this category.

Medium-heat appliance. Any *appliance* in which the products of *combustion* at the point of entrance to the flue under normal operating conditions have a temperature of more than 1,000°F (538°C), but not greater than 2,000°F (1093°C).

❖ This classification includes industrial-type equipment, such as furnaces, kilns, ovens, dryers and incinerators.

APPLIANCE, VENTED. An *appliance* designed and installed in such a manner that all of the products of *combustion* are conveyed directly from the *appliance* to the outdoor atmosphere through an *approved chimney* or vent system.

❖ The majority of fuel-fired appliances are designed to vent the products of combustion to the outdoors through one or more specific types of vent or chimney (see Section 801.2).

[A] APPROVED. Acceptable to the code official or other authority having jurisdiction.

❖ As related to the process of acceptance of mechanical installations, including materials, equipment and construction systems, this definition identifies where ultimate authority rests. Whenever this term is used, it means that only the enforcing authority can accept a specific installation or component as complying with the code. The research reports prepared and published by the member organizations of the International Code Council® (ICC®) may be used by code officials to aid in their review and approval of the material or method described in the report. Publishing a report does not indicate any form of approval for the material or method described in the report.

[A] APPROVED AGENCY. An established and recognized agency that is *approved* by the code official and regularly engaged in conducting tests or furnishing inspection services.

❖ The word "approved" means "as approved by the code official." The basis for approval of an agency for a particular activity may include the capacity and capability of the agency to perform the work in accordance with Sections 301.8 through 301.8.2.3.

AUTOMATIC BOILER. Any class of boiler that is equipped with the controls and limit devices specified in Chapter 10.

❖ As opposed to manually operated boilers, automatic boilers are designed to operate safely with only periodic human attention and supervision.

BATHROOM. A room containing a bathtub, shower, spa or similar bathing fixture.

❖ A bathroom is a room or space that contains bathing fixtures with or without other plumbing fixtures. Examples of bathing fixtures include bathtubs, showers and spas.

BOILER. A closed heating *appliance* intended to supply hot water or steam for space heating, processing or power purposes. Low-pressure boilers operate at pressures less than or equal to 15 pounds per square inch (psi) (103 kPa) for steam and 160 psi (1103 kPa) for water. High-pressure boilers operate at pressures exceeding those pressures.

❖ Boilers are usually manufactured of steel, cast iron or copper and are used to transfer heat, from the combustion of a fuel or from an electric-resistance element, to water to make steam or pressurized hot water for heating or other process or power purposes.

Boilers are usually installed in closed systems where the heat transfer medium is recirculated and retained within the system. Hot water supply boilers are normally part of open systems where the heated water is supplied and used externally to the boiler. Large domestic (potable) water heating systems often use hot water supply boilers.

Boilers must be labeled and installed in accordance with the manufacturer's installation instructions and the applicable sections of the code. Boilers are rated in accordance with standards published by the American Society of Mechanical Engineers (ASME), the Hydronics Institute, the Steel Boiler Institute (SBI), the Canadian Standards Association (CSA) and the American Boiler Manufacturers Association (ABMA). Boilers can be classified by: working temperature; working pressure; type of fuel used (or electric boilers); materials of construction; heat exchanger configuration and whether or not the heat transfer medium changes phase from a liquid to a vapor.

BOILER ROOM. A room primarily utilized for the installation of a boiler.

❖ A room or space that contains a boiler is not necessarily a boiler room unless that room or space is used primarily for that purpose. For example, a basement floor area in which a boiler is located, but which also contains storage areas, electrical, plumbing and mechanical equipment, does not constitute a boiler room because the primary purpose of that space is not to house the boiler. However, if partitions are erected around the boiler to isolate it from other areas of the floor, a boiler room has been created.

BRAZED JOINT. A gas-tight joint obtained by the joining of metal parts with metallic mixtures or alloys which melt at a temperature above 1,000°F (538°C), but lower than the melting temperature of the parts to be joined.

❖ Although both are made with filler metals, brazing must not be confused with soldering. Brazed joints are associated with copper and copper alloy piping materials in the context of the code.

BRAZING. A metal joining process wherein coalescence is produced by the use of a nonferrous filler metal having a melting point above 1,000°F (538°C), but lower than that of the base metal being joined. The filler material is distributed between the closely fitted surfaces of the joint by capillary attraction.

❖ Brazing is the act of producing a brazed joint. Brazing is often referred to as silver soldering. Silver soldering is more accurately described as silver brazing and employs high-silver-bearing alloys primarily composed of silver, copper and zinc. Silver soldering (brazing) typically requires temperatures in excess of 1,000°F (538°C) and the solders are classified as "hard" solders (see the commentary for "Brazed joint").

Confusion has always been present with respect to the distinction between "silver solder" and "silver-bearing" solder. Silver solders are unique and can be further subdivided into soft and hard categories, which are determined by the percentages of silver and the other component elements of the particular alloy. The distinction is that silver-bearing solders [melting point less than 600°F (316°C)] are used in soft-soldered joints, and silver solders [melting point greater than 1,000°F (538°C)] are used in silver-brazed joints.

Brazed joints have traditionally been required for systems containing hazardous refrigerants and for piping and components that are located in environmental air ducts. Brazed joints are considered to have superior strength and stress resistance and, because of the high melting point, are less likely to fail when exposed to fire.

Brazing with a filler metal conforming to AWS A5.8 produces a strong joint that will perform under extreme service conditions. The surfaces to be brazed must be cleaned to be free from oxides and impurities. Flux should be applied as soon as possible after the surfaces have been cleaned. Flux helps to remove residual traces of oxides, to promote wetting and to protect the surfaces from oxidation during heating. Care should be taken to prevent flux from entering the piping system during the brazing operation because flux that remains may corrode the pipe or contaminate the system.

Air and residual refrigerants should be removed from the pipe being brazed by purging the piping with a nonflammable gas such as carbon dioxide or nitrogen. Purging the system has several benefits, such as preventing oxidation from occurring on the inside of the pipe and preventing the creation of toxic gases that can result from the chemical breakdown of some refrigerants. Additionally, purging will eliminate the possibility of an explosion from flammable refrigerants or vaporized oil in the pipe, which could ignite when mixed with air. In a system that has been charged with refrigerant, purging the system eliminates the possibility of the torch igniting the refrigerant and creating highly toxic phosgene gas.

BREATHING ZONE. The region within an occupied space between planes 3 and 72 inches (76 and 1829 mm) above the floor and more than 2 feet (610 mm) from the walls of the space or from fixed air-conditioning *equipment*.

❖ This is the portion of an occupied space that is considered to be the location of the occupants. People will generally not be close to the floor or the walls. The amount of ventilation air required by Chapter 4 to be provided to the occupied space must be distributed throughout the breathing zone to minimize stagnant pockets and stratified layers of air [see Commentary Figure 202(1)]. In previous editions of the code, this was called the occupied zone.

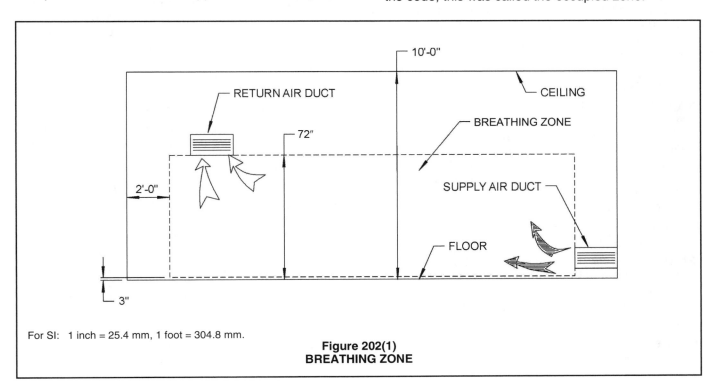

For SI: 1 inch = 25.4 mm, 1 foot = 304.8 mm.

Figure 202(1)
BREATHING ZONE

BTU. Abbreviation for British thermal unit, which is the quantity of heat required to raise the temperature of 1 pound (454 g) of water 1°F (0.56°C) (1 Btu = 1055 J).

❖ The Btu (J) is a unit of energy measurement. Fuel-fired appliances and equipment are rated based upon their Btu/h (W) input or output.

[A] BUILDING. Any structure occupied or intended for supporting or sheltering any *occupancy*.

❖ This definition indicates that where this term is used in the code, it means a structure erected to provide shelter or support for some activity or occupancy.

[B] CEILING RADIATION DAMPER. A *listed* device installed in a ceiling membrane of a fire-resistance-rated floor/ceiling or roof/ceiling assembly to limit automatically the radiative heat transfer through an air inlet/outlet opening.

❖ Habitable spaces that require a ceiling assembly having a fire-resistance rating will often have either heating and cooling supply outlets or return air intakes in the ceiling assembly. Because such penetrations in a fire-resistance-rated assembly present locations where heat and smoke could bypass the rated assembly, devices are required to be installed in the outlet and intake openings to automatically block flow of the heat and smoke when significant heat is sensed at that location. Typical ceiling radiation dampers are shown in Commentary Figure 607.1(4).

CHIMNEY. A primarily vertical structure containing one or more flues, for the purpose of carrying gaseous products of *combustion* and air from a fuel-burning *appliance* to the outdoor atmosphere.

❖ Chimneys differ from vents in their materials of construction and the type of appliance they are designed to serve. Chimneys are capable of venting much higher temperature flue gases than vents.

   Factory-built chimney. A *listed* and *labeled chimney* composed of factory-made components, assembled in the field in accordance with manufacturer's instructions and the conditions of the listing.

❖ A factory-built chimney is a manufactured listed and labeled chimney that has been tested by an approved agency to determine its performance characteristics. Factory built chimneys are manufactured in two basic designs: either a double-wall insulated design or a triple-wall air-cooled design. Both designs use stainless steel inner liners to resist the corrosive effects of combustion products.

   Masonry chimney. A field-constructed *chimney* composed of solid masonry units, bricks, stones or concrete.

❖ Masonry chimneys can have one or more flues within them, and are field constructed of brick, stone, concrete and fire-clay materials. Masonry chimneys can stand alone or be part of a masonry fireplace.

   Metal chimney. A field-constructed *chimney* of metal.

❖ A metal chimney is an unlisted chimney constructed and installed in accordance with NFPA 211 and is sometimes referred to as a "smokestack." Metal chimneys are typically field constructed and installed in industrial structures.

CHIMNEY CONNECTOR. A pipe that connects a fuel-burning *appliance* to a *chimney*.

❖ Chimney connectors are sections of pipe used to convey combustion products from an appliance flue outlet to a chimney inlet. Factory-built chimneys can directly connect to some appliances without the need for a connector; however, masonry chimneys cannot directly connect to an appliance because of the chimney's weight (see the commentary for "Vent connector").

CLEARANCE. The minimum distance through air measured between the heat-producing surface of the mechanical *appliance*, device or *equipment* and the surface of the combustible material or assembly.

❖ Clearances between sources of heat and combustibles are always airspace clearances.

CLOSED COMBUSTION SOLID-FUEL-BURNING APPLIANCE. A heat-producing *appliance* that employs a *combustion* chamber that has no openings other than the flue collar, fuel charging door and adjustable openings provided to control the amount of *combustion air* that enters the *combustion* chamber.

❖ These appliances produce higher flue gas temperatures, compared to open stoves and fireplaces, because of the lack of dilution air entering the firebox.

CLOTHES DRYER. An *appliance* used to dry wet laundry by means of heat.

❖ Although the term "clothes dryer" is well-recognized in modern societies, the term might not have the same meaning in areas where hanging clothes in open air for drying is common practice. This definition clarifies that it is an appliance that uses heat to dry clothes.

[A] CODE. These regulations, subsequent amendments thereto, or any emergency rule or regulation that the administrative authority having jurisdiction has lawfully adopted.

❖ The adopted regulations are generally referred to as "the code" and include not only the *International Mechanical Code®* (IMC®) but any adopted modifications to the code and all other related rules and regulations promulgated and enacted by the jurisdiction.

[A] CODE OFFICIAL. The officer or other designated authority charged with the administration and enforcement of this code, or a duly authorized representative.

❖ The statutory power to enforce the code is normally vested in a building department (or the like) of a state, county or municipality whose designated enforcement officer is termed the "code official" (see commentary, Sections 103 and 104).

[B] COMBINATION FIRE/SMOKE DAMPER. A *listed* device installed in ducts and air transfer openings designed to close automatically upon the detection of heat and resist the passage of flame and smoke. The device is installed to operate automatically, be controlled by a smoke detection system, and where required, is capable of being positioned from a fire command center.

❖ This definition was added to the 2009 edition of the code because the IBC has included this defined term for several editions but the IMC was silent. This type of damper combines the functions of a fire damper and a smoke damper all in one unit for compactness and economy. The damper is automatically controlled by a smoke detection system and, where required, can also have the opening position controlled from a remote command station. Operation from a remote location could be necessary for control and evacuation of smoke during fire-fighting activities.

COMBUSTIBLE ASSEMBLY. Wall, floor, ceiling or other assembly constructed of one or more component materials that are not defined as noncombustible.

❖ A combustible assembly is composed of one or more combustible components that do not comply with the requirements of ASTM E 136 as a noncombustible material (see the definition of "Noncombustible").

[F] COMBUSTIBLE LIQUID. A liquid having a closed cup flash point at or above 100°F (38°C). Combustible liquids shall be subdivided as follows:

Class II. Liquids having a closed cup flash point at or above 100°F (38°C) and below 140°F (60°C).

Class IIIA. Liquids having a closed cup flash point at or above 140°F (60°C) and below 200°F (93°C).

Class IIIB. Liquids having a closed cup flash point at or above 200°F (93°C).

The category of combustible liquids does not include compressed gases or cryogenic fluids.

❖ Combustible liquids differ from flammable liquids in that the closed cup flash point of combustible liquids is at or above 100°F (38°C) (see the definition of "Flash point"). The range of flash point dictates the class of combustible liquid. The flash point range of 100°F to 140°F (38°C to 60°C) for Class II liquids was based on a possible indoor ambient temperature exceeding 100°F (38°C). Only a moderate degree of heating would be required to bring the liquid to its flash point in this type of condition. Class III liquids, which have flash points higher than 140°F (60°C), would require a significant heat source in addition to ambient temperature conditions to reach their flash point (see the definition of "Flammable liquids"). Class III liquids are further divided into Class IIIA and Class IIIB liquids to indicate the two closed cup flash point ranges above 140°F.

COMBUSTIBLE MATERIAL. Any material not defined as noncombustible.

❖ All materials are considered combustible unless they pass the ASTM E 136 test referenced in the code. Combustible materials are those materials that are capable of burning, generally in air, under normal conditions of ambient temperature and pressure. Note that gypsum board (drywall, sheetrock) is considered to be a combustible material in the code.

The IBC considers gypsum board to be a noncombustible material on the basis of Section 703.4.2 of the IBC. The code does not have similar text.

COMBUSTION. In the context of this code, refers to the rapid oxidation of fuel accompanied by the production of heat or heat and light.

❖ The primary components of combustion are fuel, oxygen and heat. The code regulates many aspects of combustion technology, including the process of combustion and providing sufficient air; the use of energy produced from combustion; the safe venting of the products of combustion; combustion efficiency; the containment and control of combustion and the fuel supplies for combustion equipment and appliances.

COMBUSTION AIR. Air necessary for complete *combustion* of a fuel, including *theoretical air* and excess air.

❖ The process of combustion requires a specific amount of oxygen to initiate and sustain the combustion reaction. Combustion air includes primary air, secondary air, draft hood dilution air and excess air. Combustion air is the amount of atmospheric air required for complete combustion of a fuel and is related to the molecular composition of the fuel being burned, the design of the fuel-burning equipment and the percentage of oxygen in the combustion air. Too little combustion air will result in incomplete combustion of a fuel and the possible formation of carbon deposits (soot), carbon monoxide, toxic alcohols, ketones, aldehydes, nitrous oxides and other byproducts. The required amount of combustion air is usually stated in terms of cubic feet per minute (m³/s) or pounds per hour (kg/h).

COMBUSTION CHAMBER. The portion of an *appliance* within which *combustion* occurs.

❖ Combustion chambers are either open to the atmosphere at the burners or are isolated (sealed) from the atmosphere at the burners. For solid fuels, combustion chambers are often referred to as "fireboxes."

COMBUSTION PRODUCTS. Constituents resulting from the *combustion* of a fuel with the oxygen of the air, including the inert gases, but excluding excess air.

❖ Combustion products include water, carbon dioxide, carbon monoxide, nitrogen oxides and various trace compounds.

COMMERCIAL COOKING APPLIANCES. Appliances used in a commercial food service establishment for heating or cooking food and which produce grease vapors, steam, fumes, smoke or odors that are required to be removed through a local exhaust ventilation system. Such appliances include deep fat fryers; upright broilers; griddles; broilers; steam-jacketed kettles; hot-top ranges; under-fired broilers (charbroilers); ovens; barbecues; rotisseries; and similar appliances. For the purpose of this definition, a food service establishment shall include any building or a portion thereof used for the preparation and serving of food.

❖ Recirculating exhaust systems are also referred to as "ductless hoods" because they discharge into the room in which the appliance is located instead of through a duct system to the outdoors. The requirements for these systems are found in UL 197, Supplement SB. Because the emissions (effluent) treatment components cannot remove products of combustion of a fuel, those systems are limited to electrically heated cooking appliances. UL 197 has the same basic requirements as UL 710 for ducted hoods, but UL 197 requires additional emissions tests and the fire-extinguishing system is tested as an integral part of the hood.

COMMERCIAL COOKING RECIRCULATING SYSTEM. Self-contained system consisting of the exhaust hood, the cooking *equipment*, the filters and the fire suppression system. The system is designed to capture cooking vapors and residues generated from commercial cooking *equipment*. The system removes contaminants from the *exhaust air* and recirculates the air to the space from which it was withdrawn.

❖ This definition is important in the application of Section 507, which requires a commercial kitchen hood above commercial cooking equipment. A definition of "Food service establishment" is included within this definition. "Food service" includes operations, such as preparing, handling, cleaning, cooking and packaging food items of any kind (see commentary, Section 507.2).

COMMERCIAL KITCHEN HOODS.

❖ All of the commercial hoods defined below are included in the sizing tables in Section 507.13. The configurations of the various types of hoods require different volumes of exhaust air to properly capture the cooking effluents from the cooking appliances.

Backshelf hood. A backshelf hood is also referred to as a low-proximity hood, or as a sidewall hood where wall mounted. Its front lower lip is low over the *appliance*(s) and is "set back" from the front of the appliance(s). It is always closed to the rear of the appliances by a panel where free-standing, or by a panel or wall where wall mounted, and its height above the cooking surface varies. (This style of hood can be constructed with partial end panels to increase its effectiveness in capturing the effluent generated by the cooking operation).

❖ See Commentary Figures 202(2) and 507.14.

Double island canopy hood. A double island canopy hood is placed over back-to-back appliances or *appliance* lines. It is open on all sides and overhangs both fronts and the sides of the *appliance*(s). It could have a wall panel between the backs of the appliances. (The fact that *exhaust air* is drawn from both sides of the double canopy to meet in the center causes each side of this hood to emulate a wall canopy hood, and thus it functions much the same with or without an actual wall panel between the backs of the appliances).

❖ See Commentary Figures 202(2) and 507.12(4).

Eyebrow hood. An eyebrow hood is mounted directly to the face of an *appliance*, such as an oven and dishwasher, above the opening(s) or door(s) from which effluent is emitted, extending past the sides and overhanging the front of the opening to capture the effluent.

❖ See Commentary Figure 202(2).

Pass-over hood. A pass-over hood is a free-standing form of a backshelf hood constructed low enough to pass food over the top.

❖ See Commentary Figure 202(2).

Single island canopy hood. A single island canopy hood is placed over a single *appliance* or *appliance* line. It is open on all sides and overhangs the front, rear and sides of the *appliance*(s). A single island canopy is more suscepti-

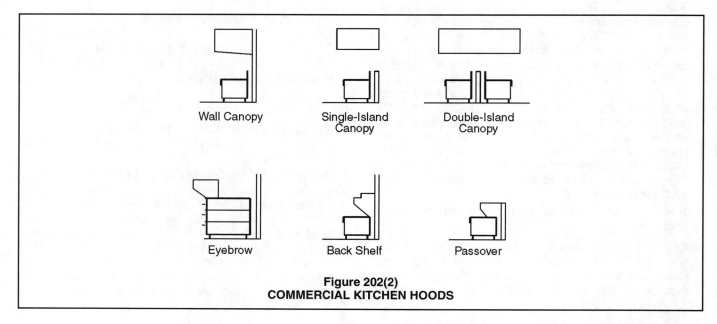

Figure 202(2)
COMMERCIAL KITCHEN HOODS

ble to cross drafts and requires a greater *exhaust air* flow than an equivalent sized wall-mounted canopy to capture and contain effluent generated by the cooking operation(s).

❖ See Commentary Figures 202(2) and 507.12(3).

Wall canopy hood. A wall canopy exhaust hood is mounted against a wall above a single *appliance* or line of *appliance*(s), or it could be free-standing with a back panel from the rear of the appliances to the hood. It overhangs the front and sides of the *appliance*(s) on all open sides.

The wall acts as a back panel, forcing the *makeup air* to be drawn across the front of the cooking *equipment*, thus increasing the effectiveness of the hood to capture and contain effluent generated by the cooking operation(s).

❖ See Commentary Figures 202(2) and 507.12(2).

COMPENSATING HOODS. *Compensating hoods* are those having integral (built-in) *makeup air* supply. The *makeup air* supply for such hoods is generally supplied from: short-circuit flow from inside the hood, air curtain flow from the bottom of the front face, and front face discharge from the outside front wall of the hood. The compensating makeup airflow can also be supplied from the rear or side of the hood, or the rear, front or sides of the cooking *equipment*. The makeup airflow can be one or a combination of methods.

❖ A "short circuit" hood is a type of compensating hood that is designed to provide makeup air directly into the hood intake area. A label, as required in Section 508.2, specifies a minimum total exhaust flow and a maximum makeup air flow that is to be supplied to the compensating hood to ensure capture and containment of the exhaust effluent (see commentary, Section 508.2).

COMPRESSOR. A specific machine, with or without accessories, for compressing a gas.

❖ A compressor is the heart of mechanical refrigeration systems. It is used in a vapor refrigeration cycle to raise the pressure and enthalpy of the refrigerant into the superheated vapor state, at which point the refrigerant vapor enters the condenser and transfers heat energy to a cooler medium. There are two basic types of compressors: positive displacement and dynamic. Positive-displacement compressors increase the pressure of the refrigerant vapor by reducing the volume of the compressor chamber, such as in a reciprocating piston, screw, scroll or rotary compressor. Dynamic compressors, such as centrifugal compressors, increase the pressure of the refrigerant vapor by a continuous transfer of angular momentum from the rotating member to the vapor.

COMPRESSOR, POSITIVE DISPLACEMENT. A compressor in which increase in pressure is attained by changing the internal volume of the compression chamber.

❖ See the commentary for "Compressor."

COMPRESSOR UNIT. A compressor with its prime mover and accessories.

❖ A compressor unit includes the compressor, the motor or engine that drives it and related controls and hardware.

CONCEALED LOCATION. A location that cannot be accessed without damaging permanent parts of the building structure or finish surface. Spaces above, below or behind readily removable panels or doors shall not be considered as concealed.

❖ For example, the space above a "drop-in" tile suspended ceiling system would not be considered as a concealed location.

CONDENSATE. The liquid that condenses from a gas (including flue gas) caused by a reduction in temperature.

❖ Condensate forms when the temperature of a vapor is lowered to its dew point temperature. Air-conditioning systems produce condensate when an airstream contacts cooling coils. The moisture in the air condenses on the cold surface of the coils and the air is "dehumidified." High-efficiency (84 percent and up) fuel-burning appliances produce condensate from the combustion gases. Condensate also forms within improperly designed chimneys and vents when the products of combustion (which contain water vapor) contact the colder walls of the flue. If the temperature of the products of combustion is lowered to the dew point temperature of the water vapor, condensate will form on the inside walls of the vent or chimney. Condensed steam in hydronic systems is also referred to as "condensate."

CONDENSER. A heat exchanger designed to liquefy refrigerant vapor by removal of heat.

❖ In a refrigeration system, the condenser is a heat exchanger that passes (rejects) heat from the system to a cooler medium, such as air or water. As it passes through the condenser, the refrigerant condenses from a vapor into a liquid at the same temperature and pressure. The liquid refrigerant is usually subcooled in or after the condenser before it is passed through the expansion valve, orifice or other restrictor. To maximize heat transfer, condenser design usually includes low thermal-resistance tubes that isolate the refrigerant from the cooling fluid. The condenser rejects the heat of vaporization, compressor heat and any superheat that was absorbed in the system evaporator section.

CONDENSING UNIT. A specific refrigerating machine combination for a given refrigerant, consisting of one or more power-driven compressors, condensers, liquid receivers (when required) and the regularly furnished accessories.

❖ Condensing units always include compressors and condenser heat exchangers and convert refrigerant vapor to liquid.

CONDITIONED SPACE. An area, room or space being heated or cooled by any *equipment* or *appliance*.

❖ Conditioned spaces are intentionally heated or cooled by the expenditure of energy. The term typically refers to conditioning for human comfort.

[A] CONSTRUCTION DOCUMENTS. All of the written, graphic and pictorial documents prepared or assembled for describing the design, location and physical characteristics of the elements of the project necessary for obtaining a building permit. The construction drawings shall be drawn to an appropriate scale.

❖ To determine that proposed construction is in compliance with code requirements, sufficient information must be submitted to the code official for review. This typically consists of the drawings (floor plans, elevations, sections, details, etc.), specifications and product information describing the proposed work.

CONTROL. A manual or automatic device designed to regulate the gas, air, water or electrical supply to, or operation of, a mechanical system.

❖ A control is a device designed to respond to changes in temperature, pressure, liquid or gas flow rates, current, voltage, resistance, humidity and/or liquid levels.

CONVERSION BURNER. A burner designed to supply gaseous fuel to an *appliance* originally designed to utilize another fuel.

❖ Typical conversion burners are designed to convert an appliance such as a boiler from solid fuel fired to oil fired.

COOKING APPLIANCE. See *"Commercial cooking appliances."*

❖ See the commentary for "Commercial cooking appliances."

DAMPER. A manually or automatically controlled device to regulate draft or the rate of flow of air or *combustion* gases.

❖ A damper is used in an air distribution system as a restrictor to regulate airflow through ductwork. When used in flues venting combustion gases, a damper is used to regulate draft (see commentary, Sections 803.5 and 803.6).

   Volume damper. A device that, when installed, will restrict, retard or direct the flow of air in a duct, or the products of *combustion* in a heat-producing *appliance*, its vent connector, vent or *chimney* therefrom.

❖ See the commentary for "Damper."

[B] DESIGN FLOOD ELEVATION. The elevation of the "design flood," including wave height, relative to the datum specified on the community's legally designated flood hazard area map.

❖ The design flood elevation is the height to which the floodwaters are predicted to rise during passage or occurrence of the design flood. The datum specified on the flood hazard map is important because it may differ from that used locally for other purposes. Communities adopt the Flood Insurance Rate Maps (FIRMS) prepared by the Federal Emergency Management Agency (FEMA) or another flood hazard map that shows at least the same flood hazard areas. FEMA uses commonly accepted computer models that estimate hydrologic and hydraulic conditions to determine the 1-percent annual chance (base) flood. Along rivers and streams, statistical methods and computer models may have been used to estimate the runoff and to develop flood elevations. The models take into consideration watershed characteristics and the shape and nature of the floodplain, including natural ground contours and the presence of buildings, bridges and culverts. Along coastal areas, base flood elevations may be developed using models that take into account offshore bathymetry, historical storms and typical wind patterns. In many coastal areas, the base flood elevation includes wave heights.

DESIGN WORKING PRESSURE. The maximum allowable working pressure for which a specific part of a system is designed.

❖ Piping, vessels, ductwork systems and related components are designed to operate within specific pressure ranges. Operation at pressures outside these ranges could result in the failure of one or more system components. The design working pressure is established by the manufacturer of the material or system and is based on material strengths and safety factors.

DIRECT REFRIGERATION SYSTEM. A system in which the evaporator or condenser of the refrigerating system is in direct contact with the air or other substances to be cooled or heated.

❖ In a direct system, the air or substance to be conditioned is passed over or is otherwise in contact with the condenser or evaporator heat exchangers. Heat exchanger failure would allow refrigerants to enter the air or substance being conditioned.

DIRECT-VENT APPLIANCES. Appliances that are constructed and installed so that all air for *combustion* is derived from the outdoor atmosphere and all flue gases are discharged to the outdoor atmosphere.

❖ Such appliances are built with independent exhaust and intake pipes, or have concentric pipes that vent through the inner pipe and convey combustion air in the annular space between pipe walls (see commentary, Sections 304.1 and 804.1).

DRAFT. The pressure difference existing between the *appliance* or any component part and the atmosphere, that causes a continuous flow of air and products of *combustion* through the gas passages of the *appliance* to the atmosphere.

❖ Draft is the negative static pressure, measured relative to atmospheric pressure, that is developed in chimneys and vents and in the flue-ways of fuel-burning appliances. Draft can be produced by hot flue-gas buoyancy ("stack effect") or mechanically by fans and exhausters or by a combination of both natural and mechanical means.

Induced draft. The pressure difference created by the action of a fan, blower or ejector, that is located between the *appliance* and the *chimney* or vent termination.

❖ Induced draft systems use a fan or blower to boost or "induce" draft in a venting system that produces insufficient natural draft. Draft inducers are separate field-installed units that are located between an appliance and its venting system. Draft inducers are used with natural draft venting systems to overcome the resistance of vent or chimney connectors and to compensate for the inability of the chimney or vent to produce sufficient and reliable draft.

Draft induction is also a design principle used in many fan-assisted appliances. Draft inducer fans or blowers that are integral with fuel-fired appliances are necessary to overcome the internal resistance of the heat exchanger flue passageways. To attain higher thermal transfer efficiencies, some heat exchanger designs use flue passageways that retain the combustion gases longer over a greater surface area, thus extracting more heat. Such heat exchanger designs cannot rely upon natural (gravity) venting to overcome the higher resistance to flow and, therefore, must rely on mechanical means to pull the combustion gases through the heat exchanger.

Induced draft systems are not to be confused with systems using power exhausters or other self-venting equipment.

Natural draft. The pressure difference created by a vent or *chimney* because of its height, and the temperature difference between the flue gases and the atmosphere.

❖ Natural draft systems do not use mechanical devices such as fans or blowers, but instead rely on the principle of buoyancy to carry the products of combustion to the atmosphere. Because of the difference in temperature and the resultant difference in density between the hot products of combustion and the ambient atmosphere, the gases within the chimney or flue will rise, creating a buoyant "draft." The phenomenon of natural draft is sometimes referred to as "stack effect" and is measured in inches of water column (kPa). The amount of draft is affected by the height of the chimney or vent and also by the ability of the chimney or vent to maintain the temperature differential between the combustion gases and the ambient air.

DRIP. The container placed at a low point in a system of piping to collect condensate and from which the condensate is removable.

❖ Drip reservoirs, also referred to as "drip legs," are made up of pipe and fittings and are intended to col-

lect liquid in piping systems where condensables are possible. A "drip leg" is distinct from a "sediment trap," even though they may be constructed identically.

DRY CLEANING SYSTEMS. Dry cleaning plants or systems are classified as follows:

Type I. Those systems using Class I flammable liquid solvents having a flash point below 100°F (38°C).

Type II. Those systems using Class II combustible liquid solvents having a flash point at or above 100°F (38°C) and below 140°F (60°C).

Type III. Those systems using Class III combustible liquid solvents having a flash point at or above 140°F (60°C).

Types IV and V. Those systems using Class IV nonflammable liquid solvents.

❖ Five dry cleaning types are defined. The classification is dependent on the flash point of the flammable or combustible liquid solvent being used.

DUCT. A tube or conduit utilized for conveying air. The air passages of self-contained systems are not to be construed as air ducts.

❖ Ducts can be factory manufactured or field constructed of sheet metal, gypsum board, fibrous glass board or other approved material. Ducts are used in air distribution systems, exhaust systems, smoke control systems and combustion air supply systems.

Air passageways that are integral parts of an air handler, packaged air-conditioning unit or similar piece of self-contained, factory-built equipment are not considered as ducts in the context of the code.

DUCT FURNACE. A warm-air furnace normally installed in an air distribution duct to supply warm air for heating. This definition shall apply only to a warm-air heating *appliance* that, for air circulation, depends on a blower not furnished as part of the furnace.

❖ Duct furnaces are vented appliances (if fuel fired) and consist of a heat source, heat exchangers and related controls. Duct furnaces depend on an external air handler or blower.

DUCT SYSTEM. A continuous passageway for the transmission of air that, in addition to ducts, includes duct fittings, dampers, plenums, fans and accessory air-handling *equipment* and appliances.

❖ Duct systems are part of an air distribution system and include supply, return and relief/exhaust air systems.

[B] DWELLING. A building or portion thereof that contains not more than two *dwelling* units.

❖ A dwelling can be a single unit or a two-family unit, often referred to as a "duplex" or a "two-flat" (see the definition of "Dwelling unit").

[B] DWELLING UNIT. A single unit providing complete, independent living facilities for one or more persons, including permanent provisions for living, sleeping, eating, cooking and sanitation.

❖ Many code provisions address dwelling units specifically. All of the elements listed in the definition must be present for a dwelling unit to exist. See the IBC for additional information on dwelling units.

ELECTRIC HEATING APPLIANCE. An *appliance* that produces heat energy to create a warm environment by the application of electric power to resistance elements, refrigerant compressors or dissimilar material junctions.

❖ Such appliances are not fuel fired and include heat pumps and electric resistance units.

ENERGY RECOVERY VENTILATION SYSTEM. Systems that employ air-to-air heat exchangers to recover energy from or reject energy to *exhaust air* for the purpose of preheating, pre-cooling, humidifying or dehumidifying outdoor *ventilation air* prior to supplying such air to a space, either directly or as part of an HVAC system.

❖ An energy recovery ventilation system uses what would otherwise be wasted energy in exhaust air to precondition the incoming ventilation air stream. This preconditioning conserves energy that would otherwise be required by the heating, ventilating and air-conditioning (HVAC) system to heat, cool, humidify or dehumidify ventilation air once it entered the building, room or space for which it was intended (see commentary, Section 514).

ENVIRONMENTAL AIR. Air that is conveyed to or from occupied areas through ducts which are not part of the heating or air-conditioning system, such as ventilation for human usage, domestic kitchen range exhaust, bathroom exhaust, domestic clothes dryer exhaust and parking garage exhaust.

❖ The types of air considered to be environmental air are included in this definition. These include typical building ventilation air, as required in Chapter 4, domestic kitchen and clothes dryer exhaust and exhaust air from both domestic and commercial bathrooms. Commercial cooking exhaust systems are not considered to be environmental air. Hazardous exhaust, such as flammable and explosive vapors; fumes; smoke; spray residues and mists; vapors from corrosive liquids; noxious and toxic gases and other items listed in Section 502 would not be considered environmental air. Dust, stock and refuse systems as detailed in Section 511 are also not considered to be environmental air.

EQUIPMENT. All piping, ducts, vents, control devices and other components of systems other than appliances which are permanently installed and integrated to provide control of environmental conditions for buildings. This definition shall also include other systems specifically regulated in this code.

❖ In the code, appliances are not referred to as equipment and vice versa. Throughout the code, the terms "equipment" and "appliance" have been used as nec-

essary to match the terms with the intent and context of the code text. Traditionally, the term "equipment" has referred to large machinery and specialized hardware not thought of as an "appliance." See the definition of "Appliance" and the commentary for Section 301.4. Note that in the IFGC, the terms "appliance" and "equipment" are synonymous.

EQUIPMENT, EXISTING. Any *equipment* regulated by this code which was legally installed prior to the effective date of this code, or for which a permit to install has been issued.

❖ Although the term "equipment" does not include appliances, it is appropriate to apply this definition to both equipment and appliances. Mechanical equipment that has been illegally installed prior to the effective date of the code is not considered as existing, and is subject to all of the requirements of the code for new installations. Sections that address existing mechanical systems and installations include Sections 102 and 108.

EVAPORATIVE COOLER. A device used for reducing the sensible heat of air for cooling by the process of evaporation of water into an airstream.

❖ Also known as "swamp coolers," such units are used in arid climates and use water as a refrigerant. These units substantially increase the humidity of the air being conditioned.

EVAPORATIVE COOLING SYSTEM. The *equipment* and appliances intended or installed for the purpose of environmental cooling by an evaporative cooler from which the conditioned air is distributed through ducts or plenums to the conditioned area.

❖ In applications using an evaporative cooler attached to an air distribution system, special consideration must be given to the type of materials used because of the high water vapor content of the air being distributed (see Section 603.5.1). Manufacturer's instructions must be consulted concerning approval of the materials to be used. Materials include duct liners, fibrous glass duct, flexible air duct and connectors, and duct tape.

EVAPORATOR. That part of the system in which liquid refrigerant is vaporized to produce refrigeration.

❖ A refrigeration system uses an evaporator to transfer heat from the refrigeration load to the refrigerant fluid. The evaporator is a heat-absorbing heat exchanger. The evaporation process takes place in the evaporator at a constant pressure and temperature, and the refrigerant will change from a liquid state to a saturated vapor as it absorbs heat from the refrigerated space or substance. Heat absorbed by the refrigerant vapor is referred to as "superheat" (or heat in excess of the heat of vaporization). Evaporators are generally constructed of metallic tubes with either plate fins or spine fins attached to enhance the heat transfer capabilities.

EXCESS AIR. The amount of air provided in addition to *theoretical air* to achieve complete *combustion* of a fuel, thereby preventing the formation of dangerous products of *combustion*.

❖ The introduction and efficient intermixing of primary and secondary combustion air into natural-draft fuel-fired appliances is not precise and is coupled with the induced air inflow caused by the internal draft. To achieve complete oxidation (combustion), more air (excess) will be introduced into the appliances than is theoretically necessary for the complete combustion of the fuel.

EXHAUST SYSTEM. An assembly of connected ducts, plenums, fittings, registers, grilles and hoods through which air is conducted from the space or spaces and exhausted to the outdoor atmosphere.

❖ In addition to those components already mentioned, air-handling equipment blowers and fans are also included as part of an exhaust air system.

EXTRA-HEAVY-DUTY COOKING APPLIANCE. Extra-heavy-duty cooking *appliances* include appliances utilizing solid fuel such as wood, charcoal, briquettes, and mesquite to provide all or part of the heat source for cooking.

❖ The increased amount of smoke and grease-laden vapor produced by wood-fired grills and barbecue pits is reflected in the larger required amount of airflow in Section 507.13.1 as compared to the other appliance duties. The phrase "...or part of the heat source..." can be interpreted to mean that any appliance that uses wood or wood chips for flavoring purposes would be considered to be an extra-heavy-duty appliance.

[B] FIRE DAMPER. A *listed* device installed in ducts and air transfer openings designed to close automatically upon detection of heat and to restrict the passage of flame. Fire dampers are classified for use in either static systems that will automatically shut down in the event of a fire, or in dynamic systems that continue to operate during a fire. A dynamic fire damper is tested and rated for closure under elevated temperature airflow.

❖ This definition was added to the 2009 edition of the code because the IBC has included this defined term for several editions but the IMC was silent. Ductwork for HVAC and exhaust systems often passes through fire-resistance-rated walls, ceilings or floors. The open duct presents a potential bypass for fire to breach the fire-resistance-rated assembly. A fire damper senses the heat of a fire and automatically closes to maintain the required fire resistance rating of the rated assembly. The design of fire dampers is specific as to the duct orientation (e.g., vertical or horizontal), as well as the capacity to close against a moving airstream or not. As the air in the duct and the duct itself could become very hot before the damper actuates, the design of the damper and its testing and rating must account for these elevated temperatures.

FIREPLACE. An assembly consisting of a hearth and fire chamber of noncombustible material and provided with a *chimney*, for use with solid fuels.

❖ Fireplaces burn solid fuels (wood, coal, etc.) and are not referred to as appliances in the code. Construction of masonry fireplaces is covered by the IBC.

    Factory-built fireplace. A *listed* and *labeled* fireplace and *chimney* system composed of factory-made components, and assembled in the field in accordance with manufacturer's instructions and the conditions of the listing.

❖ Factory-built fireplaces are solid-fuel-burning units having a fire chamber that is intended to be either open to the room or, if equipped with doors, operated with the doors either open or closed. Figures 903.1(1) and 903.1(2) show two common styles of factory-built fireplaces: a recessed wall fireplace and a free-standing fireplace stove. Note that fireplaces are not referred to as appliances. The term "fireplace" describes a complete assembly, which includes the hearth, the fire chamber and a chimney. A factory-built fireplace is composed of factory-built components representative of the prototypes tested, and is installed in accordance with the manufacturer's installation instructions to form the completed fireplace (see commentary, Section 903).

    Masonry fireplace. A field-constructed fireplace composed of solid masonry units, bricks, stones or concrete.

❖ Masonry fireplaces must be constructed in accordance with the requirements found in the IBC. These specific requirements are based on tradition and field experience and describe the conventional fireplace that has proven to be reliable where properly constructed, used and maintained.

FIREPLACE STOVE. A free-standing chimney-connected solid-fuel-burning heater, designed to be operated with the fire chamber doors in either the open or closed position.

❖ Fireplace stoves are generally of the free-standing type and heat a space by direct radiation. Installation of the various types of fireplace stoves must be in compliance with the listing of the stove (see commentary, Section 308.7 and Chapter 8).

FLAME SAFEGUARD. A device that will automatically shut off the fuel supply to a main burner or group of burners when the means of ignition of such burners becomes inoperative, and when flame failure occurs on the burner or group of burners.

❖ These devices are primary safety controls and are installed on all automatically operated fuel-fired appliances (see IFGC, Section 602.2).

[B] FLAME SPREAD INDEX. The numerical value assigned to a material tested in accordance with ASTM E 84 or UL 723.

❖ The ASTM E 84 test method includes measurements of surface flame spread index (and smoke density index) in comparison with test results obtained by using select red oak as a control material. Red oak is used as a control material for furnace calibration because it is a fairly uniform grade of lumber that is readily available nationally, is uniform in thickness and moisture content and generally gives consistent and reproducible results.

FLAMMABILITY CLASSIFICATION. Refrigerants shall be assigned to one of the three classes—1, 2 or 3—in accordance with ASHRAE 34. For Classes 2 and 3, the heat of *combustion* shall be calculated assuming that *combustion* products are in the gas phase and in their most stable state.

> **Class 1.** Refrigerants that do not show flame propagation when tested in air at 14.7 psia (101 kPa) and 70°F (21°C).

> **Class 2.** Refrigerants having a lower flammability limit (LFL) of more than 0.00625 pound per cubic foot (0.10 kg/m^3) at 70°F (21°C) and 14.7 psia (101 kPa) and a heat of *combustion* of less than 8,174 Btu/lb (19 000 kJ/kg).

> **Class 3.** Refrigerants that are highly flammable, having a LFL of less than or equal to 0.00625 pound per cubic foot (0.10 kg/m^3) at 70°F (21°C) and 14.7 psia (101 kPa) or a heat of *combustion* greater than or equal to 8,174 Btu/lb (19 000 kJ/kg).

❖ All refrigerants are classified relative to flammability and health effects (see commentary, Section 1104).

[F] FLAMMABLE LIQUIDS. Any liquid that has a flash point below 100°F (38°C), and has a vapor pressure not exceeding 40 psia (276 kPa) at 100°F (38°C). Flammable liquids shall be known as Class I liquids and shall be divided into the following classifications:

> **Class IA.** Liquids having a flash point below 73°F (23°C) and a boiling point below 100°F (38°C).

> **Class IB.** Liquids having a flash point below 73°F (23°C) and a boiling point at or above 100°F (38°C).

> **Class IC.** Liquids having a flash point at or above 73°F (23°C) and below 100°F (38°C).

❖ Although all flammable liquids have a flash point less than 100°F (38°C) (see the definition of "Flammable liquids"), the further classification of the Class I liquid is dependent on the boiling point (the temperature at which the vapor pressure of the liquid is equal to the external pressure on the liquid). The flash point limitation of 100°F (38°C) for flammable liquids assumes possible indoor ambient temperature conditions of 100°F (38°C). For the purpose of classifying a material as a flammable or combustible liquid, a vapor pressure (the pressure exerted by the vapor of a material that is in equilibrium with its solid or liquid phase) limitation of 40 psia (276 kPa absolute) at 100°F (38°C) is used as the defining criterion. A liquid is further defined as having a fluidity greater than 300 penetration asphalt when tested in accordance with ASTM D 5. Three hundred penetration asphalt is the most fluid grade of paving asphalt recognized by ASTM D 946. Liquids not meeting these characteristics would not be classified as a flammable or com-

bustible liquid (see the definition of "Combustible liquids").

[F] FLAMMABLE VAPOR OR FUMES. Mixtures of gases in air at concentrations equal to or greater than the LFL and less than or equal to the upper flammability limit (UFL).

❖ Flammable vapors include those generated from flammable liquids, such as paint thinners, gasoline and other flammable solvents. Many flammable vapors are heavier than air. Material Safety Data Sheets (MSDS) provide information on the flammability of a substance.

[F] FLASH POINT. The minimum temperature corrected to a pressure of 14.7 psia (101 kPa) at which the application of a test flame causes the vapors of a portion of the sample to ignite under the conditions specified by the test procedures and apparatus. The flash point of a liquid shall be determined in accordance with ASTM D 56, ASTM D 93 or ASTM D 3278.

❖ The flash point is the characteristic used in the classification of flammable and combustible liquids. The Tag Closed Tester (ASTM D 56) and the Pensky-Martens Closed Tester (ASTM D 93) are the referenced test procedures for determining the flash points of liquids. The applicability of the respective test method is dependent on the viscosity of the test liquid and the expected flash point.

FLOOR AREA, NET. The actual occupied area, not including unoccupied accessory areas or thicknesses of walls.

❖ Section 401.3 requires that occupied spaces be ventilated. This definition defines those areas considered as being occupied and to correlate with the same definition in the IBC. The IBC definition also includes corridors, stairways and closets as unoccupied spaces (see commentary, Section 403.3).

FLOOR FURNACE. A completely self-contained furnace suspended from the floor of the space being heated, taking air for *combustion* from outside such space and with means for observing flames and lighting the *appliance* from such space.

❖ Floor furnaces supply heat through a floor grille placed directly over the unit's heat exchanger. Typically floor furnaces are classified as gravity-type furnaces. Air is circulated from a gravity-type floor furnace by convection. Floor furnaces may be equipped with manufacturer-provided fans to circulate the air.

FLUE. A passageway within a *chimney* or vent through which gaseous *combustion* products pass.

❖ This term is typically associated with chimneys. A flue may be constructed of masonry, metal or plastic products that are approved for the particular application for which the flue is used. The flue passageway includes all the components that make up the chimney or vent from the point of attachment to the appliance or equipment to the point of termination.

FLUE CONNECTION (BREECHING). A passage for conducting the products of *combustion* from a fuel-fired *appliance* to the vent or *chimney* (see also "Chimney connector" and "Vent connector").

❖ The term is typically associated with chimneys. This term is synonymous with "chimney connector" and is typically associated with industrial and commercial equipment and incinerators. Dampers, draft controls and sensors are commonly located on the breeching.

FLUE GASES. Products of *combustion* and excess air.

❖ The exact composition of flue gases will depend on the materials being burned. The primary components of flue gases are nitrogen, carbon dioxide, water vapor, particulates and a myriad of compounds and trace elements that vary with the nature of the fuel. Carbon monoxide is also a component of flue gas.

FLUE LINER (LINING). A system or material used to form the inside surface of a flue in a *chimney* or vent, for the purpose of protecting the surrounding structure from the effects of *combustion* products and conveying *combustion* products without leakage to the atmosphere.

❖ Flue liners must be resistant to heat and the corrosive action of the products of combustion. Flue liners provide insulation to retard the transfer of heat to the chimney structure and limit exposure of the chimney structure to the harmful effects of combustion products. Flue liners are generally made of fire-clay tile, refractory brick, poured-in-place refractory materials and stainless steel alloys for solid fuels. Flue liners used with gas-fired appliances are typically made of stainless steel or aluminum.

FUEL GAS. A natural gas, manufactured gas, liquefied petroleum gas or a mixture of these.

❖ The nature of fuel gases makes proper design, installation and selection of materials and devices necessary to minimize the possibility of fire or explosion. Bringing fuel gases into a building is in itself a risk. The provisions of the code are intended to reduce that risk to a level comparable to that associated with other energy sources such as electricity.

The two most commonly used fuel gases are natural gas and liquefied petroleum gas (LP-gas or LPG). These fuel gases have the following characteristics or properties.

FUEL OIL. Kerosene or any hydrocarbon oil having a flash point not less than 100°F (38°C).

❖ Fuel oils are classified by grades, with a No. 2 grade being the most common used for residential and light commercial heating applications. Kerosene is No. 1 grade fuel oil.

FUEL-OIL PIPING SYSTEM. A closed piping system that connects a combustible liquid from a source of supply to a fuel-oil-burning *appliance*.

❖ A fuel-oil piping system includes the piping, valves and fittings. Also included in the fuel-oil piping system, if they are installed, are pumps, reservoirs, regulators, strainers, filters, relief valves, oil preheaters, controls and gauges. The piping materials and the method of joining sections of pipe must be approved. All other components included in the piping system should be labeled and installed to comply with the manufacturer's instructions.

FURNACE. A completely self-contained heating unit that is designed to supply heated air to spaces remote from or adjacent to the *appliance* location.

❖ The single most distinguishing characteristic of furnaces is that they use air as the heat transfer medium. Furnaces can be fueled by gas, oil, solid fuel or electricity, and can use fans, blowers and gravity (convection) to circulate the heated air to and from the unit. In the context of the code, the primary usage of the term "furnace" refers to heating appliance units that combine a combustion chamber with related components, one or more heat exchangers and an air-handling system.

FURNACE ROOM. A room primarily utilized for the installation of fuel-burning, space-heating and water-heating appliances other than boilers (see also "Boiler room").

❖ A room or space that contains a furnace or water heater is not necessarily a furnace room unless that room or space is used primarily for that purpose. For example, a basement space in which a furnace is located, but which also contains storage areas and electrical, mechanical and plumbing equipment, does not constitute a furnace room because the primary purpose of the space is not to house the furnace. However, if partitions are erected that serve to enclose and isolate the furnace or water heater from other areas of the floor, a furnace room has been created.

FUSIBLE PLUG. A device arranged to relieve pressure by operation of a fusible member at a predetermined temperature.

❖ A fusible plug is designed to open at a set temperature when the fusible element melts and allows the pressure of the system to force the plug out of the relief opening. A fusible plug is a "one-time" device and cannot be reused.

GROUND SOURCE HEAT PUMP LOOP SYSTEM. Piping buried in horizontal or vertical excavations or placed in a body of water for the purpose of transporting heat transfer liquid to and from a heat pump. Included in this definition are closed loop systems in which the liquid is recirculated and open loop systems in which the liquid is drawn from a well or other source.

❖ Closed loop systems consist of loops of piping that act as a heat exchanger. A heat transfer fluid, typically water or a water/antifreeze mixture, is circulated through the piping for the purpose of extracting heat from or rejecting heat to the earth or a body of water. The ground source loop system serves either the condensing or evaporation side of a refrigerant cycle heat pump system. Such systems are part of a geothermal heating and cooling (HVAC) system.

HAZARDOUS LOCATION. Any location considered to be a fire hazard for flammable vapors, dust, combustible fibers or other highly combustible substances. The location is not necessarily categorized in the *International Building Code* as a high-hazard use group classification.

❖ The environment in which mechanical equipment and appliances operate plays a significant role in the safe performance of the equipment installation. Locations that may contain ignitable or explosive atmospheres are classified as hazardous locations for the installation of mechanical equipment and appliances. For example, repair garages can be classified as hazardous locations. Repair garages can contain gasoline vapors from vehicles stored within them as well as other volatile chemicals. As evidenced by Sections 304.3, 304.4 and 304.5, public and private garages can be considered hazardous locations because of the presence of motor vehicles and because they may be used for storing paint, varnish, thinners, lawn and home maintenance products and other chemicals.

HEAT EXCHANGER. A device that transfers heat from one medium to another.

❖ All vented fuel-fired appliances employ heat exchangers of various configurations to capture heat from the combustion gases and transfer that heat to the heating medium. Heat exchangers also transfer heat from one medium to another, such as air-to-air, water-to-water, water-to-air, steam-to-water, steam-to-air, refrigerant-to-water, refrigerant-to-air and other combinations.

Heat exchanger designs are diverse, but they all rely on the same basic concept of using a heat-conducting material as the interface between the media where the exchange of heat is to occur.

HEAT PUMP. A refrigeration system that extracts heat from one substance and transfers it to another portion of the same substance or to a second substance at a higher temperature for a beneficial purpose.

❖ Heat pumps are referred to as reverse cycle refrigeration systems. Special controls and valves allow the system to be used as a comfort heating system, comfort cooling system and potable water heating system. Auxiliary heat is often installed in heat pumps in the form of electric resistance heat or fuel-fired furnaces. In most cases, the design and installation of heat pumps are more critical than other types of comfort air-conditioning systems because of the heat transfer process and the amount of air that must be circulated. Heat pump types include water source (hydronic loop and wells), ground source (earth loop) and air source.

HEAT TRANSFER LIQUID. The operating or thermal storage liquid in a mechanical system, including water or other liquid base, and additives at the concentration present under operating conditions used to move heat from one location to another. Refrigerants are not included as heat transfer liquids.

❖ Heat transfer liquid is the working fluid in hydronic and solar heating or cooling systems. The liquid is pumped from a heat exchanger or collector, where heat is absorbed by the liquid, to the heat exchanger, where the liquid releases the heat energy to another fluid, airstream or some form of thermal storage mass. Some of the commonly used heat transfer fluids are: water, water/ethylene glycol, water/propylene glycol, silicone oils, paraffinic oils and aromatic oils.

HEAVY-DUTY COOKING APPLIANCE. Heavy-duty cooking *appliances* include electric under-fired broilers, electric chain (conveyor) broilers, gas under-fired broilers, gas chain (conveyor) broilers, gas open-burner ranges (with or without oven), electric and gas wok ranges, and electric and gas over-fired (upright) broilers and salamanders.

❖ The list of appliances included in the definition is fairly comprehensive and is intended to represent most of the common appliances that would be classified as "heavy-duty." This classification of appliances produces the most heat, smoke and grease-laden vapors where the cooking fuel is gas or electric (see the definition of "Extra-heavy duty cooking appliances"). The higher airflow rate reflected in Section 507.13.2 is required for the capture and containment of the cooking effluent.

HIGH-PROBABILITY SYSTEMS. A refrigeration system in which the basic design or the location of components is such that a leakage of refrigerant from a failed connection, seal or component will enter an *occupancy* classified area, other than the *machinery room*.

❖ Chapter 11 and ASHRAE 15 classify refrigeration systems into either high-probability or low-probability systems. A high-probability system exists when the components of the system are located where a refrigerant leak could cause refrigerant to enter the occupied space. A direct system is a high-probability system. A low-probability system exists when the components of the system are located where a refrigerant leak will not discharge refrigerant into the occupied space. An indirect system is a low-probability system.

HIGH-SIDE PRESSURE. The parts of a refrigerating system subject to condenser pressure.

❖ The high-side pressure of the refrigeration system is the pressure of the refrigerant in the portion of the system located between the outlet of the compressor and the inlet to the expansion valve, capillary tube or restrictor device, which also includes the condenser.

HOOD. An air intake device used to capture by entrapment, impingement, adhesion or similar means, grease, moisture, heat and similar contaminants before they enter a duct system.

❖ A kitchen exhaust system that includes the hood serving a commercial cooking appliance is a specialized exhaust system. A commercial cooking appliance can generate significant heat as well as large quantities of air contaminants, such as grease vapors, moisture, smoke and combustion byproducts.

Type I. A kitchen hood for collecting and removing grease vapors and smoke. Such hoods are equipped with a fire suppression system.

❖ A Type I exhaust system is required for cooking appliances that are used for commercial purposes and that produce grease-laden vapors and/or smoke. Because of the high potential for grease fires at the cooking appliance and in the hood and duct system, Type I hoods are required to be fitted with a fire suppression system to extinguish an out-of-control fire event at the appliance.

Type II. A general kitchen hood for collecting and removing steam, vapor, heat, odors and products of *combustion*.

❖ A Type II exhaust hood is considered a light-duty hood that would typically be installed over steam kettles, conventional ovens, food warmers, some types of enclosed pizza ovens, steam tables and dishwashing machines. A Type II hood is not intended for grease or smoke removal. The primary purpose of a Type II hood is to capture and remove water vapor, waste heat and any products of combustion that might be associated with the heating of the appliance, such as from fuel gas combustion.

[FG] HYDROGEN GENERATING APPLIANCE. A self-contained package or factory-matched packages of integrated systems for generating gaseous hydrogen. Hydrogen generating appliances utilize electrolysis, reformation, chemical, or other processes to generate hydrogen.

❖ Hydrogen generators use water or hydrocarbon fuels as a feedstock for generation of pure hydrogen or a hydrogen-rich gas. Hydrogen generating appliances based on chemical reformers separate out the hydrogen from fossil fuels, such as natural gas, propane, gasoline, etc. This is the same high-temperature chemical process used at large oil-refineries to produce hydrogen. By generating the hydrogen on-site at the fueling station or customer's facility, these hydrogen generating appliances avoid the high cost of either liquefying hydrogen and delivering it by cryogenic tanker truck, or installing a national hydrogen pipeline system that could cost many tens of billions of dollars. In effect, these on-site hydrogen generating appliances take advantage of one of two existing energy infrastructures: either the natural gas distribution system or the electrical grid. Hydrogen is being generated to fuel a new line of developmental vehicles and to generate electric power in fuel cell power systems.

Installation requirements for these appliances are found in Chapter 7 of the IFGC and ventilation requirements for spaces containing the appliances are found in Section 304.4 of this code.

IGNITION SOURCE. A flame, spark or hot surface capable of igniting flammable vapors or fumes. Such sources include *appliance* burners, burner ignitors and electrical switching devices.

❖ This definition is important in the application of Section 304.3 regarding the elevation of ignition sources. By means of this definition, Section 304.3 applies to unintentional ignition sources such as electrical switching devices, as well as intentional ignition sources such as pilot lights, spark ignitors and hot surface ignitors.

In the context of Section 304.3, an ignition source is something capable of igniting flammable vapors that are present in the atmosphere in the locations listed in Section 304.3.

[F] IMMEDIATELY DANGEROUS TO LIFE OR HEALTH (IDLH). The concentration of airborne contaminants that poses a threat of death, immediate or delayed permanent adverse health effects, or effects that could prevent escape from such an environment. This contaminant concentration level is established by the National Institute of Occupational Safety and Health (NIOSH) based on both toxicity and flammability. It is generally expressed in parts per million by volume (ppm v/v) or milligrams per cubic meter (mg/m^3).

❖ Where the necessary IDLH data are not available, an industrial hygienist, chemist, toxicologist or other approved agency personnel should be consulted for proper determination of the nature of the hazard involved.

INDIRECT REFRIGERATION SYSTEM. A system in which a secondary coolant cooled or heated by the refrigerating system is circulated to the air or other substance to be cooled or heated. Indirect systems are distinguished by the method of application shown below:

❖ Such systems are low-probability systems.

Closed system. A system in which a secondary fluid is either cooled or heated by the refrigerating system and then circulated within a closed circuit in indirect contact with the air or other substance to be cooled or heated.

❖ Closed indirect systems involve heat exchangers between the refrigerant and the substance to be heated or cooled.

Double-indirect open-spray system. A system in which the secondary substance for an indirect open-spray system is heated or cooled by an intermediate coolant circulated from a second enclosure.

❖ Such systems involve multiple heat exchangers and heat transfer fluids.

Open-spray system. A system in which a secondary coolant is cooled or heated by the refrigerating system and then circulated in direct contact with the air or other substance to be cooled or heated.

❖ Open-spray systems typically use brine as the heat transfer medium.

Vented closed system. A system in which a secondary coolant is cooled or heated by the refrigerating system and then passed through a closed circuit in the air or other substance to be cooled or heated, except that the evaporator or condenser is placed in an open or appropriately vented tank.

❖ These systems are closed systems except that the secondary heat transfer medium is open to the atmosphere.

INTERLOCK. A device actuated by another device with which it is directly associated, to govern succeeding operations of the same or allied devices. A circuit in which a given action cannot occur until after one or more other actions have taken place.

❖ The term "interlock" was used indiscriminately in the code, and in some cases it meant one thing and in others it had a different meaning. Adding a definition for the term and deleting of the term in some locations made the code clear on what type of control arrangement was actually intended by the code. For example, Section 508.1 was changed to require simultaneous (parallel) starting of kitchen exhaust fans and makeup air fans instead of referring to an interlock control arrangement. Obviously, a true interlock control arrangement is more complex than simply wiring for parallel starting and involves feedback between related components of a system so as to dictate a controlled sequence of actions. See Sections 504.7 and 507.2.1.1 which use the term interlock and also Section 804.3.2 which does not use the term, but nonetheless requires an interlock. The term is also used in the IFGC, Sections 304.9.2, 304.10 and 505.1.1 and is implied in Sections 503.3.3, 503.3.4.

A good example of an interlock is the control arrangement between a venting system power exhauster and a gas-fired boiler. The gas-fired boiler depends upon the exhauster to convey the vent gases to the outdoors, therefore, the boiler must know the operational status of the exhauster to prevent a very unsafe condition. The exhauster is monitored by supervisory controls (typically pressure switches) that provide feedback to the boiler to prevent the boiler from firing if the exhauster is not producing the required draft. The sequence of operation is as follows: a call for heat will start the exhauster and after the exhauster is proven to be operating properly, a supervisory control will signal the boiler to fire. If the exhauster fails at any time, the boiler will be automatically shut down.

JOINT, FLANGED. A joint made by bolting together a pair of flanged ends.

❖ Flanges are commonly used with large piping at locations where the piping must be capable of being disassembled periodically and at connections to valves, regulators, devices and equipment. Full-face gaskets must be used with bronze and cast-iron flange fittings. Typically, materials for flange gaskets include metal or metal-jacketed asbestos; asbestos and aluminum "O" rings; or spiral-wound metal gaskets. Gaskets must be replaced whenever a joint is opened.

The gasket material must be compatible with the piping contents to prevent a chemical reaction.

JOINT, FLARED. A metal-to-metal compression joint in which a conical spread is made on the end of a tube that is compressed by a flare nut against a mating flare.

❖ Because the pipe end is expanded in a flared joint, only annealed and bending-tempered soft-drawn copper tubing may be flared. Commonly used flaring tools use a screw yoke and block assembly or an expander tool that is driven into the tube with a hammer. The flared tubing end is compressed between a fitting seat and a threaded nut to form a metal-to-metal seal.

JOINT, PLASTIC ADHESIVE. A joint made in thermoset plastic piping by the use of an adhesive substance which forms a continuous bond between the mating surfaces without dissolving either one of them.

❖ Unlike solvent-welded joints, adhesive (glue) joints are surface bonded.

JOINT, PLASTIC HEAT FUSION. A joint made in thermoplastic piping by heating the parts sufficiently to permit fusion of the materials when the parts are pressed together.

❖ Heat-fusion joints for plastic pipe are analogous to the welding of steel pipe. Only polyethylene and polybutylene pipe may be joined by heat fusion, and only in accordance with the pipe manufacturer's instructions. The process involves heating the pipe and fittings with a special iron. When the parts to be joined reach their melting points, they are assembled and allowed to fuse together.

JOINT, PLASTIC SOLVENT CEMENT. A joint made in thermoplastic piping by the use of a solvent or solvent cement which forms a continuous bond between the mating surfaces.

❖ It is important to identify the materials being joined because some plastics cannot be joined by the solvent-cementing process, often referred to as "solvent welding." A solvent-cemented joint is equivalent to chemically welding the fitting and the piping material, resulting in a bond of homogeneous, continuous material. Polyethylene and polybutylene pipe are chemically inert with respect to the solvents in solvent cements and, therefore, cannot be joined by this method. Solvent cements contain a chemical solvent for the plastic material and the plastic raw material (resin). The solvent dissolves the plastic and allows the plastic resin to solidify in the void between the joined parts. Primers are necessary to prepare the surface of the fittings and pipe by softening and dissolving the outer finish surface.

JOINT, SOLDERED. A gas-tight joint obtained by the joining of metal parts with metallic mixtures of alloys which melt at temperatures between 400°F (204°C) and 1,000°F (538°C).

❖ A soldered joint is the most common method of joining copper pipe and tubing. Pipe must be cut square to provide proper alignment, adequate surface area for joining and an interior free from obstructions.

When the pipe is cut, an edge (burr) is left protruding into the pipe. The pipe must be reamed to properly remove the burr. Chamfering is required to bevel the outer edge of a pipe cut end. Undercutting (excessive reaming) can reduce pipe wall thickness. ASTM B 32 covers many grades of solder, including tin-silver solders.

ASTM B 828 governs the procedures for making capillary joints by soldering of copper and copper alloy tubing and fittings. The joint surfaces for a soldered joint must be cleaned to expose untarnished metal, typically accomplished with emery cloth and specially designed socket brushes. Once the copper is sanded and/or brushed, the joint surfaces should not be touched by the human hand since the natural body oils on the hand will affect the joining process.

Flux must be applied to all joint surfaces. Flux is a chemically active material that removes and excludes oxides from the joint area during heating and allows the melted solder to spread out on the surfaces to be joined. The solder will flow by capillary action toward the heat. The temperatures of the joint surfaces, therefore, play a major role in making a properly soldered joint. The fitting and the pipe should be approximately the same temperature when the solder is applied. Heating the copper causes the copper atoms to move farther apart from each other and the solder atoms enter the spaces between the copper atoms, creating a strong bond when the solder solidifies.

If the temperatures are either too high or uneven, the solder will run down the inside of the pipe and fitting or on the outside of the pipe. The result is a solder-lined pipe. Overheating can also nullify the effect of the flux and oxides can form on the copper surfaces, preventing the solder from penetrating.

JOINT, WELDED. A gas-tight joint obtained by the joining of metal parts in molten state.

❖ A welded joint is similar to a brazed joint. The primary differences are the temperature at which the joint is made, the type of filler metals used and the fact that welding reaches the melting point of the base metal, whereas brazing temperatures are well below the melting point of the pipe and fittings.

[A] LABELED. *Equipment*, materials or products to which have been affixed a label, seal, symbol or other identifying mark of a nationally recognized testing laboratory, inspection agency or other organization concerned with product evaluation that maintains periodic inspection of the production of the above-labeled items and whose labeling indicates either that the *equipment*, material or product meets identified standards or has been tested and found suitable for a specified purpose.

❖ When a product is labeled, the label indicates that the material has been tested for conformance to an applicable standard and that the component is subject to third-party inspection to verify that the minimum level of quality required by the standard is maintained.

Labeling provides a readily available source of information that is useful for field inspection of installed products. The label identifies the product or material and provides other information that can be further investigated if there is question concerning the suitability of the product or material for the specific installation. The labeling agency performing the third-party inspection must be approved by the code official. The basis for this approval may include, but is not necessarily limited to, the capacity and capability of the agency to perform the specific testing and inspection.

The referenced standard often states the minimum identifying information that must be on a label. The data contained on a label typically includes, but is not necessarily limited to, name of the manufacturer; product name or serial number; installation specifications; applicable tests and standards; and the approved testing and labeling agency.

This definition was modified for the 2009 edition to correlate with the definitions provided in other *International Codes®* (I-Codes®).

LIGHT-DUTY COOKING APPLIANCE. Light-duty cooking *appliances* include gas and electric ovens (including standard, bake, roasting, revolving, retherm, convection, combination convection/steamer, countertop conveyorized baking/finishing, deck and pastry), electric and gas steam-jacketed kettles, electric and gas pasta cookers, electric and gas compartment steamers (both pressure and atmospheric) and electric and gas cheesemelters.

❖ The list of appliances included in this definition is fairly comprehensive and is intended to represent most of the common appliances that would be classified as "light duty." This classification of appliances produces the smallest amounts of heat, smoke and grease-laden vapors as indicated by the lower airflow requirements in Section 507.13.4. For the 2009 edition of the code, pasta cookers were removed from the medium-duty cooking appliance definition and added to this definition because otherwise pasta cookers would require a Type I hood having a fire suppression system. A pasta cooker does not generate any significant amount of grease-laden vapors.

[FG] LIMIT CONTROL. A device responsive to changes in pressure, temperature or level for turning on, shutting off or throttling the gas supply to an *appliance*.

❖ Limit controls are safety devices used to protect equipment, appliances, property and persons. These controls act at their setpoint to limit a condition such as temperature or pressure and include high and low limits. This definition applies to controls for all types of fuel: gas, liquid and solid.

LIMITED CHARGE SYSTEM. A system in which, with the compressor idle, the design pressure will not be exceeded when the refrigerant charge has completely evaporated.

❖ Limited charge systems are designed so that the maximum system design pressure will not be exceeded if all of the refrigerant in the system is converted to vapor. Very high vapor pressures can develop when liquid refrigerant in a closed system is exposed to heat in standby storage or in a fire.

[A] LISTED. *Equipment,* materials, products or services included in a list published by an organization acceptable to the code official and concerned with evaluation of products or services that maintains periodic inspection of production of *listed equipment* or materials or periodic evaluation of services and whose listing states either that the *equipment*, material, product or service meets identified standards or has been tested and found suitable for a specified purpose.

❖ This definition was modified in the 2009 edition to correlate with the definition of "Listed" in other I-codes. When a product is listed, it indicates that it has been tested for conformance to an applicable standard and is subject to a third-party inspection quality assurance (QA) program. The QA verifies that the minimum level of quality required by the appropriate standard is maintained. The agency performing the third-party inspection must be approved by the code official, and the basis for this approval may include, but is not limited to, the capacity and capability of the agency to perform the specified testing and inspection.

LIVING SPACE. Space within a *dwelling unit* utilized for living, sleeping, eating, cooking, bathing, washing and sanitation purposes.

❖ This definition clarifies the meaning of the term as it is used in Section 304.3. It is not the intent to require elevating appliances in laundry, utility and toilet rooms that open to the habitable spaces of a dwelling unit and that also have a door that opens to a garage.

LOWER EXPLOSIVE LIMIT (LEL). See "LFL."

❖ See the commentary for "LFL."

LOWER FLAMMABLE LIMIT (REFRIGERANT) (LFL). The minimum concentration of refrigerant that is capable of propagating a flame through a homogeneous mixture of refrigerant and air.

❖ Substances that are flammable or explosive when mixed with air have concentration limits within which a mixture will propagate a flame or become explosive. These limits can be determined from product information such as that found in an MSDS furnished by the manufacturer.

[F] LOWER FLAMMABLE LIMIT (LFL). The minimum concentration of vapor in air at which propagation of flame will occur in the presence of an ignition source. The LFL is sometimes referred to as LEL or lower explosive limit.

❖ Substances that are flammable or explosive when mixed with air have concentration limits within which a mixture will propagate a flame or become explosive. These limits can be determined from product information such as that found in material safety data sheets (MSDS) furnished by the manufacturer.

LOW-PRESSURE HOT-WATER-HEATING BOILER. A boiler furnishing hot water at pressures not exceeding 160 psi (1103 kPa) and at temperatures not exceeding 250°F (121°C).

❖ See the commentary for "Boiler."

LOW-PRESSURE STEAM-HEATING BOILER. A boiler furnishing steam at pressures not exceeding 15 psi (103 kPa).

❖ See the commentary for "Boiler."

LOW-PROBABILITY SYSTEMS. A refrigeration system in which the basic design or the location of components is such that a leakage of refrigerant from a failed connection, seal or component will not enter an occupancy-classified area, other than the *machinery room.*

❖ See the commentary for "High-probability systems."

LOW-SIDE PRESSURE. The parts of a refrigerating system subject to evaporator pressure.

❖ The low-side pressure of the refrigeration system is the pressure of the refrigerant in the portion of the system located between the outlet of the expansion or restrictor device and the inlet to the compressor, which includes the evaporator.

MACHINERY ROOM. A room meeting prescribed safety requirements and in which refrigeration systems or components thereof are located (see Sections 1105 and 1106).

❖ Machinery rooms serve to accommodate refrigeration equipment but may also be used to house other equipment and appliances in addition to the refrigeration equipment. ASHRAE 15 defines a machinery room as "a space that is designed to safely house compressors and pressure vessels." Machinery rooms are considered special because of the potential hazard associated with leaking refrigerant equipment. Some refrigerants pose flammability and toxicity hazards to the occupants of the structure, and specially designed rooms are necessary to address those hazards. Sections 1105 and 1106 state the requirements for the construction of machinery rooms.

MECHANICAL DRAFT SYSTEM. A venting system designed to remove flue or vent gases by mechanical means, that consists of an induced-draft portion under nonpositive static pressure or a forced-draft portion under positive static pressure.

❖ Such systems do not depend upon draft; they use fans or blowers.

   Forced-draft venting system. A portion of a venting system using a fan or other mechanical means to cause the removal of flue or vent gases under positive static pressure.

❖ Power exhausters and some power burner systems are examples of forced-draft systems. Vents and chimneys must be listed for positive pressure applications where used with forced-draft systems.

Induced-draft venting system. A portion of a venting system using a fan or other mechanical means to cause the removal of flue or vent gases under nonpositive static vent pressure.

❖ Induced-draft venting is commonly accomplished with field-installed inducer fans designed to supplement natural draft chimneys or vents.

Power venting system. A portion of a venting system using a fan or other mechanical means to cause the removal of flue or vent gases under positive static vent pressure.

❖ See the commentary for "Forced-draft venting system."

MECHANICAL EQUIPMENT/APPLIANCE ROOM. A room or space in which nonfuel-fired mechanical *equipment* and *appliances* are located.

❖ Mechanical equipment rooms typically contain fans, blowers, air-handling units, filters, pneumatic compressors, pumps, refrigeration equipment and other related equipment. The room may or may not be used for purposes other than housing mechanical equipment. Mechanical equipment rooms do not house gas, liquid- or solid-fuel-burning equipment or appliances. A mechanical equipment/appliance room is distinct from a machinery room.

MECHANICAL EXHAUST SYSTEM. A system for removing air from a room or space by mechanical means.

❖ A mechanical exhaust system uses a fan or other air-handling equipment to exhaust air to the outdoors. Mechanical exhaust systems include those used for hazardous exhaust and commercial kitchen exhaust. Mechanical exhaust systems may or may not include ductwork as part of the system and may include air cleaning or filtering equipment and fire suppression equipment.

MECHANICAL JOINT.

1. A connection between pipes, fittings, or pipes and fittings that is not welded, brazed, caulked, soldered or solvent cemented.

2. A general form of gas or liquid-tight connections obtained by the joining of parts through a positive holding mechanical construction such as, but not limited to, flanged, screwed, clamped or flared connections.

❖ There are many different configurations of mechanical joints with some designs being proprietary.

MECHANICAL SYSTEM. A system specifically addressed and regulated in this code and composed of components, devices, *appliances* and *equipment*.

❖ Mechanical systems include, among others, refrigeration systems, air-conditioning systems, exhaust systems, piping systems, duct systems, venting systems, hydronic systems and ventilation systems. A mechanical system may be considered any of the above systems or incorporate one or more of the systems. The system includes all of the equipment and

appliances required to perform the function for which the system is designed.

MEDIUM-DUTY COOKING APPLIANCE. Medium-duty cooking *appliances* include electric discrete element ranges (with or without oven), electric and gas hot-top ranges, electric and gas griddles, electric and gas double-sided griddles, electric and gas fryers (including open deep fat fryers, donut fryers, kettle fryers and pressure fryers), electric and gas conveyor pizza ovens, electric and gas tilting skillets (braising pans) and electric and gas rotisseries.

❖ The list of appliances included in the definition is fairly comprehensive and is intended to represent most of the common appliances that would be classified as "medium duty." The airflow rates required in Section 507.13.3 are less than those required for the heavy-duty cooking appliances but high enough to adequately capture and contain the cooking effluents. For the 2009 edition of the code, electric and gas pasta cookers were removed from this definition and placed under the definition for "Light-duty cooking appliance."

MODULAR BOILER. A steam or hot-water-heating assembly consisting of a group of individual boilers called modules intended to be installed as a unit with no intervening stop valves. Modules are under one jacket or are individually jacketed. The individual modules shall be limited to a maximum input rating of 400,000 Btu/h (117 228 W) gas, 3 gallons per hour (gph) (11.4 L/h) oil, or 115 kW (electric).

❖ Modular units are designed to permit close matching of boiler capacity to building loads and to allow for future expansion of capacity by the addition of modules to a boiler "package" or battery. Modules are, by definition, a number of units that are operated as a single boiler and, therefore, cannot have isolating valves at each module (see commentary, Section 1005.1).

NATURAL DRAFT SYSTEM. A venting system designed to remove flue or vent gases under nonpositive static vent pressure entirely by natural draft.

❖ Natural draft chimneys and vents do not rely upon any mechanical means to convey combustion products to the outdoors. Draft is produced by the temperature difference between the combustion gases (flue gases) and the ambient atmosphere. Hotter gases are less dense and more buoyant than surrounding cooler gases; therefore, they rise, producing a draft.

NATURAL VENTILATION. The movement of air into and out of a space through intentionally provided openings, such as windows and doors, or through nonpowered ventilators.

❖ Natural ventilation relies on pressure differences created by wind or convective air currents to induce airflow through openings in the building envelope.

NET OCCUPIABLE FLOOR AREA. The floor area of an *occupiable space* defined by the inside surfaces of its walls but excluding shafts, column enclosures and other permanently enclosed, inaccessible and unoccupiable areas.

Obstructions in the space such as furnishings, display or storage racks and other obstructions, whether temporary or permanent, shall not be deducted from the space area.

❖ This is the measurable floor area used to calculate the breathing zone outdoor air requirements in Chapter 4. Obstructions, such as display shelving, furniture and racks, are not deducted from the gross floor area.

NONABRASIVE/ABRASIVE MATERIALS. Nonabrasive particulate in high concentrations, moderately abrasive particulate in low and moderate concentrations, and highly abrasive particulate in low concentrations, such as alfalfa, asphalt, plaster, gypsum and salt.

❖ When combined, abrasive and nonabrasive materials have to be addressed as a mixture if they are exhausted through the same ductwork. The ratios of abrasive to nonabrasive materials are related to the nature of the abrasive material.

NONCOMBUSTIBLE MATERIALS. Materials that, when tested in accordance with ASTM E 136, have at least three of four specimens tested meeting all of the following criteria:

1. The recorded temperature of the surface and interior thermocouples shall not at any time during the test rise more than 54°F (30°C) above the furnace temperature at the beginning of the test.

2. There shall not be flaming from the specimen after the first 30 seconds.

3. If the weight loss of the specimen during testing exceeds 50 percent, the recorded temperature of the surface and interior thermocouples shall not at any time during the test rise above the furnace air temperature at the beginning of the test, and there shall not be flaming of the specimen.

❖ A material is defined as a noncombustible material when three of four tested specimens have passed the test method prescribed in ASTM E 136. The use of the test standard is limited to elementary materials and excludes laminated and coated materials because of the uncertainties associated with more complex materials and with products that cannot be tested in a realistic configuration. The test standard is also limited to solid materials and does not measure the self-heating tendencies of large masses of materials, such as resin-impregnated mineral fiber insulation.

The criterion requiring four test specimens recognizes the variable nature of the measurements and the fact that there are difficulties in observing the presence and duration of flaming.

The need to measure the duration of flaming and the rise in temperature recognizes that a brief period of flaming and a small amount of heating are not considered serious limitations on the use of building materials. Test results have shown that such criteria limit the combustible portion of noncombustible materials to a maximum of 3 percent. The 50-percent weight-loss limitation precludes the possibility that combustion of low-density materials will occur so rapidly that the recorded temperature rise and the measured flaming duration will be less than the prescribed limitations. The 50-percent limitation is considered appropriate for materials that contain appreciable quantities of combined water or gaseous components. Note that the provision in the IBC for classifying a composite material is not applicable.

[A] OCCUPANCY. The purpose for which a building, or portion thereof, is utilized or occupied.

❖ The occupancy classification of a building is an indication of the level of hazard to which the occupants are exposed as a function of the actual building use. Occupancy in terms of a group classification is one of the primary considerations in the development and application of many code requirements that are designed to offset the hazards specific to each group designation.

OCCUPIABLE SPACE. An enclosed space intended for human activities, excluding those spaces intended primarily for other purposes, such as storage rooms and *equipment* rooms, that are only intended to be occupied occasionally and for short periods of time.

❖ This is generally all of the spaces within a structure that can be used by people for daily activities. Crawl spaces, attics, closets and shafts are examples of spaces that are not considered to be occupiable even though they can be occupied for brief periods, such as when servicing mechanical equipment.

OFFSET (VENT). A combination of *approved* bends that make two changes in direction bringing one section of the vent out of line but into a line parallel with the other section.

❖ Offsets are used to route vents around obstructions or to direct the location of vent penetrations through construction assemblies. Any limitations of vent capacity as a result of offsets will be specified by the vent and appliance manufacturer's installation instructions.

OUTDOOR AIR. Air taken from the outdoors, and therefore not previously circulated through the system.

❖ Outdoor air, commonly referred to as "fresh air," is mused for ventilation, makeup air and combustion air.

OUTDOOR OPENING. A door, window, louver or skylight openable to the outdoor atmosphere.

❖ Outdoor openings include those used for mechanical and gravity air movement and allow the exchange of air between a building interior and the outside atmosphere.

OUTLET. A threaded connection or bolted flange in a piping system to which a gas-burning *appliance* is attached.

❖ A gas outlet is analogous to an electrical receptacle outlet. See Section 404.13 of the IFGC for requirements regarding the location of gas piping outlets.

PANEL HEATING. A method of radiant space heating in which heat is supplied by large heated areas of room surfaces.

The heating element usually consists of warm water piping, warm air ducts, or electrical resistance elements embedded in or located behind ceiling, wall or floor surfaces.

❖ Panel heating involves the low-temperature heating of large wall, ceiling or floor surfaces as opposed to higher-temperature heating of small radiating elements, such as a steam radiator.

PELLET FUEL-BURNING APPLIANCE. A closed-combustion, vented *appliance* equipped with a fuel-feed mechanism for burning processed pellets of solid fuel of a specified size and composition.

❖ This term refers to an appliance that uses solid fuel in the form of pellets in a closed combustion system. The pellets consist of a blend of wood waste products (sawdust and chips), mixed with resin (binder) and compressed into small pellets. The pellets are fed into the appliance at a controlled rate, resulting in consistent burning.

PIPING. Where used in this code, "piping" refers to either pipe or tubing, or both.

❖ Piping includes tubing and pipe used to convey, among other substances, fuels, water, heat transfer fluids, fire suppression agents, steam and refrigerants.

Pipe. A rigid conduit of iron, steel, copper, brass or plastic.

❖ Commonly used pipes include those made of steel, copper, brass and plastic. Pipe is a rigid material.

Tubing. Semirigid conduit of copper, aluminum, plastic or steel.

❖ Commonly used tubing is made of copper, aluminum, plastic or steel. Tubing is typically semirigid.

PLASTIC, THERMOPLASTIC. A plastic that is capable of being repeatedly softened by increase of temperature and hardened by decrease of temperature.

❖ Polyvinyl chloride (PVC) is a type of thermoplastic. Thermoplastics can be solvent welded and fusion welded.

PLASTIC, THERMOSETTING. A plastic that is capable of being changed into a substantially infusible or insoluble product when cured under application of heat or chemical means.

❖ Polyethylene (PE) and polybutylene (PB) are types of thermosetting plastics. Thermosetting plastics cannot be solvent welded.

PLENUM. An enclosed portion of the building structure, other than an *occupiable space* being conditioned, that is designed to allow air movement, and thereby serve as part of an air distribution system.

❖ A plenum is part of an air distribution system and is usually concealed within the building construction. Plenums can be used for supply, return, exhaust, relief and ventilation air, and can occur in ceiling, attic or under floor spaces; mechanical equipment rooms (air handler rooms) and stud and joist cavities. The

definition clarifies that plenums are uninhabitable, unoccupiable cavities and interstitial spaces only; an unoccupiable room or space is not a plenum (see commentary for Section 602 for restrictions on the use of plenums).

PORTABLE FUEL CELL APPLIANCE. A fuel cell generator of electricity, which is not fixed in place. A portable fuel cell *appliance* utilizes a cord and plug connection to a grid-isolated load and has an integral fuel supply.

❖ These appliances are a smaller version of the stationary fuel cell power plants addressed in Section 924 of this code and defined in this section. Connecting the appliance to the local electrical power grid or hard-piping fuel to the appliance would constitute a permanent installation and require permitting in accordance with Section 106 (see commentary, Section 924 and Section 202, "Hydrogen generating appliances").

POWER BOILER. See "Boiler."

❖ See the commentary for "Boiler."

[A] PREMISES. A lot, plot or parcel of land, including any structure thereon.

❖ When this term is used in the code, the entire lot and all of the structures on it should be included within the scope of that code requirement.

PRESS JOINT. A permanent mechanical joint incorporating an elastomeric seal or an elastomeric seal and corrosion-resistant grip ring. The joint is made with a pressing tool and jaw or ring approved by the fitting manufacturer.

❖ This type of joint is permanent once installed. The joint relies upon an elastomeric seal for water or gas tightness. A special pressing tool is required to make the joint.

PRESSURE, FIELD TEST. A test performed in the field to prove system tightness.

❖ Installed systems of piping and pressure vessels are tested on the job site to verify that the system is free of leaks.

PRESSURE-LIMITING DEVICE. A pressure-responsive mechanism designed to stop automatically the operation of the pressure-imposing element at a predetermined pressure.

❖ In the context of refrigeration systems, the function of a pressure-limiting device is to shut off the compressor when the system pressure reaches a predetermined maximum setting.

PRESSURE RELIEF DEVICE. A pressure-actuated valve or rupture member designed to relieve excessive pressure automatically.

❖ Pressure relief devices are used as safety devices to prevent pressure in a refrigeration system, or other closed system, from exceeding the maximum design pressure of the system or equipment. These devices may be self-closing relief valves or one-time, nonreusable, rupture-type devices or combinations of both.

PRESSURE RELIEF VALVE. A pressure-actuated valve held closed by a spring or other means and designed to relieve pressure automatically in excess of the device's setting.

❖ Pressure relief valves are intended to prevent harm to life and property. These valves are safety devices used to relieve abnormal pressures and to prevent the overpressurization of the vessel or system to which the valves are connected. Pressure relief valves are designed to operate (discharge) only when an abnormal pressure exists in a system. They are not to be used as regulating valves, nor are they intended to control or regulate flow or pressure. Pressure relief valves are set at the factory to begin opening at a predetermined pressure, and are rated according to the maximum energy discharge per unit of time, usually in units of British thermal units per hour (Btu/h)(W).

Pressure relief valves must be properly sized and rated for the boiler, pressure vessel or system served; otherwise, a system malfunction could cause the pressure within the system to continue to rise, thereby creating a hazardous condition. Pressure relief valves do not open fully when they reach the factory-preset pressure; rather, the valves open an amount proportional to the forces produced by the pressure in the system. The valves open fully at a certain percentage above the preset pressure, then close when the pressure drops below the preset pressure. Pressure relief valves are sometimes combined in the same body with temperature relief valves, creating combination temperature and pressure relief valves.

PRESSURE VESSELS. Closed containers, tanks or vessels that are designed to contain liquids or gases, or both, under pressure.

❖ These vessels include boilers, hot water storage tanks, pneumatic tanks, hydropneumatic tanks and pressurized fuel tanks.

PRESSURE VESSELS-REFRIGERANT. Any refrigerant-containing receptacle in a refrigerating system. This does not include evaporators where each separate section does not exceed 0.5 cubic foot (0.014 m³) of refrigerant-containing volume, regardless of the maximum inside dimensions, evaporator coils, controls, headers, pumps and piping.

❖ Containers and reservoirs in refrigeration systems are pressure vessels and they include receivers, oil separators and storage containers.

PROTECTIVE ASSEMBLY (REDUCED CLEARANCE). Any noncombustible assembly that is *labeled* or constructed in accordance with Table 308.6 and is placed between combustible materials or assemblies and mechanical appliances, devices or *equipment*, for the purpose of reducing required airspace clearances. Protective assemblies attached directly to a combustible assembly shall not be considered as part of that combustible assembly.

❖ Such assemblies are used to allow reduction of the required clearance to combustibles based on the protection afforded by the assembly (see commentary, Section 308).

PURGE. To clear of air, water or other foreign substances.

❖ Piping systems are purged (flushed) to remove gaseous, liquid or solid contaminants that could be harmful to the piping contents, the piping system or the system components, or that could create a fire or explosion hazard.

PUSH-FIT JOINTS. A type of mechanical joint consisting of elastomeric seals and corrosion-resistant tube grippers. Such joints are permanent or removable depending on the design.

❖ These joints are a relatively new technology intended primarily for use with copper pipe and tubing. They are considered to be a type of mechanical joint (see the definition for "Mechanical joint").

QUICK-OPENING VALVE. A valve that opens completely by fast action, either manually or automatically controlled. A valve requiring one-quarter round turn or less is considered to be quick opening.

❖ Quick-opening valves are fast-acting valves designed to open fully with a rotor or stem rotation of 90 degrees (1.6 rad) (quarter turn) or less. Ball valves and gas cocks are examples.

RADIANT HEATER. A heater designed to transfer heat primarily by direct radiation.

❖ Radiant heaters use heated metallic or ceramic elements and reflectors to produce radiant heat energy output. Radiant heaters are used where heating of an entire space, such as an automotive repair shop, is either not desirable or not practical.

READY ACCESS (TO). That which enables a device, *appliance* or *equipment* to be directly reached, without requiring the removal or movement of any panel, door or similar obstruction [see "Access (to)"].

❖ Ready access can be described as the capability of being quickly reached or approached for the purpose of operation, inspection, observation or emergency action. Ready access means that nothing must be moved or removed (such as doors or panels), that there are no physical obstructions and that there is no change in elevation to reach the required location.

RECEIVER, LIQUID. A vessel permanently connected to a refrigeration system by inlet and outlet pipes for storage of liquid refrigerant.

❖ Receivers are located between the condenser and the expansion valve or restrictor device and serve to collect and temporarily store the liquid refrigerant output of the condenser.

RECIRCULATED AIR. Air removed from a conditioned space and intended for reuse as supply air.

❖ Conditioned air is recirculated in building spaces by air handlers and other HVAC system components. Only air not used as ventilation air may be recirculated.

RECLAIMED REFRIGERANTS. Refrigerants reprocessed to the same specifications as for new refrigerants by means including distillation. Such refrigerants have been chemically analyzed to verify that the specifications have been met. Reclaiming usually implies the use of processes or procedures that are available only at a reprocessing or manufacturing facility.

❖ As required by federal law, unused refrigerants are recovered and reused or are taken to a reclamation facility to be reconditioned. Section 1102.2.2.3 describes limitations on the reuse of reclaimed refrigerants.

RECOVERED REFRIGERANTS. Refrigerants removed from a system in any condition without necessarily testing or processing them.

❖ Because it is illegal and harmful to the environment to release ozone-depleting refrigerants into the atmosphere, used refrigerants are recovered (captured and contained) from systems being repaired, relocated or retired.

RECYCLED REFRIGERANTS. Refrigerants from which contaminants have been reduced by oil separation, removal of noncondensable gases, and single or multiple passes through devices that reduce moisture, acidity and particulate matter, such as replaceable core filter driers. These procedures usually are performed at the field job site or in a local service shop.

❖ The cleanup process for used refrigerants can occur at the job site where the refrigerants were recovered or at a remote shop location. Note the distinction between recycled and reclaimed refrigerants.

REFRIGERANT. A substance utilized to produce refrigeration by its expansion or vaporization.

❖ The refrigerant is the working fluid in refrigeration and air-conditioning systems. In vapor refrigeration cycles, refrigerants absorb heat from the load side at the evaporator and reject heat at the condenser. In addition to suitable thermodynamic properties, the selection of a refrigerant must also take into consideration chemical stability, flammability, toxicity and environmental compatibility. Refrigeration is a result of the physical laws of vaporization (evaporation) of liquids. Basically, evaporation of liquid refrigerant is an endothermic process and condensing of vapors is an exothermic process.

REFRIGERANT SAFETY CLASSIFICATIONS. Groupings that indicate the toxicity and flammability classes in accordance with Section 1103.1. The classification group is made up of a letter (A or B) that indicates the toxicity class, followed by a number (1, 2 or 3) that indicates the flammability class. Refrigerant blends are similarly classified, based on the compositions at their worst cases of fractionation, as separately determined for toxicity and flammability. In some cases, the worst case of fractionation is the original formulation.

Flammability. Class 1 indicates refrigerants that do not show flame propagation in air when tested by prescribed methods at specified conditions. Classes 2 and 3 signify

refrigerants with "lower flammability" and "higher flammability," respectively; the distinction depends on both the LFL and heat of *combustion*.

Toxicity. Classes A and B signify refrigerants with "lower toxicity" and "higher toxicity," respectively, based on prescribed measures of chronic (long-term, repeated exposures) toxicity.

❖ The toxicity and flammability of refrigerants are classified in order to apply appropriate safety requirements. The classification scheme is based on ASHRAE 34.

REFRIGERATED ROOM OR SPACE. A room or space in which an evaporator or brine coil is located for the purpose of reducing or controlling the temperature within the room or space to below 68°F (20°C).

❖ These spaces are cooled by direct or indirect systems to any temperature below human comfort levels.

REFRIGERATING SYSTEM. A combination of interconnected refrigerant-containing parts constituting one closed refrigerant circuit in which a refrigerant is circulated for the purpose of extracting heat.

❖ Refrigerating systems include at minimum a pressure-imposing element or generator, an evaporator, a condenser and interconnecting piping. A single piece of equipment can contain multiple refrigeration systems (circuits).

REFRIGERATION CAPACITY RATING. Expressed as 1 horsepower (0.75 kW), 1 ton or 12,000 Btu/h (3.5 kW), shall all mean the same quantity.

❖ The terms "1 ton," "1 horsepower" and "12,000 Btu/h" are equivalent capacities and are industry vernacular referring to an equivalent quantity of refrigerating effect.

REFRIGERATION MACHINERY ROOM. See "*Machinery room.*"

❖ See the commentary for "Machinery room."

REFRIGERATION SYSTEM, ABSORPTION. A heat-operated, closed-refrigeration cycle in which a secondary fluid (the absorbent) absorbs a primary fluid (the refrigerant) that has been vaporized in the evaporator.

❖ Absorption systems are powered by a heat source instead of a pressure-imposing element (compressor). Ammonia (R-717) is commonly the refrigerant in absorption systems.

Direct system. A system in which the evaporator is in direct contact with the material or space refrigerated, or is located in air-circulating passages communicating with such spaces.

❖ See the commentary for "Direct refrigeration system."

Indirect system. A system in which a brine coil cooled by the refrigerant is circulated to the material or space refrigerated, or is utilized to cool the air so circulated. Indirect systems are distinguished by the type or method of application.

❖ An indirect system is classified as a low-probability system. See the commentary for "High probability systems."

REFRIGERATION SYSTEM CLASSIFICATION. Refrigeration systems are classified according to the degree of probability that leaked refrigerant from a failed connection, seal or component will enter an occupied area. The distinction is based on the basic design or location of the components.

❖ Refrigerant systems are classified as either high-probability or low-probability systems based on the potential for leakage into an occupied space. These classifications are used to apply the appropriate safety requirements for a given system.

REFRIGERATION SYSTEM, MECHANICAL. A combination of interconnected refrigeration-containing parts constituting one closed refrigerant circuit in which a refrigerant is circulated for the purpose of extracting heat and in which a compressor is used for compressing the refrigerant vapor.

❖ As opposed to heat-operated chemical systems, mechanical systems use compressors.

REFRIGERATION SYSTEM, SELF-CONTAINED. A complete factory-assembled and tested system that is shipped in one or more sections and has no refrigerant-containing parts that are joined in the field by other than companion or block valves.

❖ These systems, as opposed to split systems, contain all components in a single package and include types such as through-the-wall units and roof-top units.

[A] REGISTERED DESIGN PROFESSIONAL. An individual who is registered or licensed to practice their respective design profession as defined by the statutory requirements of the professional registration laws of the state or jurisdiction in which the project is to be constructed.

❖ Legal qualifications for engineers and architects are established by each state. Engineers and architects are licensed and registered using written or oral examinations offered by states or by reciprocity (licensing in other states).

RETURN AIR. Air removed from an *approved* conditioned space or location and recirculated or exhausted.

❖ Return air is air that is being returned to the air handler. Note that only air in excess of the required ventilation air can be recirculated (see commentary, Section 403.2.1).

RETURN AIR SYSTEM. An assembly of connected ducts, plenums, fittings, registers and grilles through which air from the space or spaces to be heated or cooled is conducted back to the supply unit (see also "Supply air system").

❖ In addition to those components already mentioned, air-handling equipment, dampers, controls and conditioning equipment are also included as part of the return air system.

ROOM HEATER VENTED. A free-standing heating unit burning solid or liquid fuel for direct heating of the space in and adjacent to that in which the unit is located.

❖ These heaters are space heating appliances that vent the products of combustion to the outdoors.

SAFETY VALVE. A valve that relieves pressure in a steam boiler by opening fully at the rated discharge pressure. The valve is of the spring-pop type.

❖ Safety valves, sometimes referred to as "pop-off valves," are used with steam systems and have an adjustable or fixed spring that regulates the popping action and the opening pressure setting. Generally, safety valves discharge to the outdoors through piping or discharge to the surrounding atmosphere through openings in the valve housing. Safety valves "pop" wide open at the preset pressure, remaining in that position until the pressure in the vessel has dropped below the preset pressure level, at which time the valve returns to the closed position.

SELF-CONTAINED EQUIPMENT. Complete, factory-assembled and tested, heating, air-conditioning or refrigeration *equipment* installed as a single unit, and having all working parts, complete with motive power, in an enclosed unit of said machinery.

❖ This type of equipment does not require any additional components when installed and is factory assembled and tested. Window air conditioners and packaged terminal units are examples of self-contained units.

[B] SHAFT. An enclosed space extending through one or more stories of a building, connecting vertical openings in successive floors, or floors and the roof.

❖ A shaft is an enclosed, vertical passage that passes through stories and floor levels within a building. Shafts are sometimes referred to as "chases" (see the commentary for "Shaft enclosure").

[B] SHAFT ENCLOSURE. The walls or construction forming the boundaries of a shaft.

❖ Shaft enclosures must be constructed and fire-resistance rated to comply with the IBC.

[B] SLEEPING UNIT. A room or space in which people sleep, which can also include permanent provisions for living, eating, and either sanitation or kitchen facilities but not both. Such rooms and spaces that are also part of a *dwelling unit* are not sleeping units.

❖ This definition is included to coordinate the Fair Housing Act Guidelines with the code. The definition for "Sleeping unit" is needed to clarify the differences between sleeping units and dwelling units. Some examples of sleeping units would be a hotel guestroom, a dormitory room, a boarding house, etc. Another example would be an addition to a studio apartment with a kitchenette (i.e., microwave, sink, refrigerator). Since the cooking arrangements are not permanent, this configuration would be considered a sleeping unit, not a dwelling unit. As already defined in the code, a dwelling unit must contain permanent facilities for living, sleeping, eating, cooking and sanitation.

[B] SMOKE DAMPER. A *listed* device installed in ducts and air transfer openings designed to resist the passage of smoke. The device is installed to operate automatically, controlled by a smoke detection system, and where required, is capable of being positioned from a fire command center.

❖ Similar to a fire damper, smoke dampers are intended to restrict the passage of smoke through ducts or openings in structural assemblies such as smoke barriers and corridor walls. The smoke leakage rates of these devices are used to classify them in accordance with UL 555S.

[B] SMOKE-DEVELOPED INDEX. A numerical value assigned to a material tested in accordance with ASTM E 84.

❖ Materials are tested in accordance with ASTM E 84 to determine the flame spread and smoke-developed indexes of the material. By limiting these parameters, the material's contribution to smoke and spread of flame is lessened. The smoke-developed index is a factor determined by comparison to the smoke-developed index of untreated red oak wood.

SOLID FUEL (COOKING APPLICATIONS). Applicable to commercial food service operations only, solid fuel is any bulk material such as hardwood, mesquite, charcoal or briquettes that is combusted to produce heat for cooking operations.

❖ This definition is needed for the application of Sections 507.2.4 and 507.13.1. The key to the definition is that the solid fuel is combusted to produce heat for cooking, as opposed to solid fuel that is combusted solely to impart flavor to the food being cooked. Many variations of cooking appliances exist, including those that use solid fuel only and those that use solid fuel in conjunction with gas or electric heat.

SOURCE CAPTURE SYSTEM. A mechanical exhaust system designed and constructed to capture air contaminants at their source and to exhaust such contaminants to the outdoor atmosphere.

❖ These systems include hose-type exhaust systems for direct attachment to motor vehicle engine exhaust pipes and hood systems for individual contaminant producing machines. Such systems use hoods or direct attachments to equipment to capture contaminants to prevent their escape into the atmosphere in which the contaminant source is located.

[FG] STATIONARY FUEL CELL POWER PLANT. A self-contained package or factory-matched packages which constitute an automatically operated assembly of integrated systems for generating useful electrical energy and recoverable thermal energy that is permanently connected and fixed in place.

❖ Stationary fuel cell appliances cannot exceed 1,000 kW of power output, listed and tested in accordance with ANSI Z21.83 and installed in accordance with

NFPA 853 and the manufacturer's instructions (see commentary, Section 924 and Section 202, "Hydrogen generating appliances"). These appliances may be independent of or connected to the local electrical power grid and may be fueled by fuel tanks or permanent piping systems.

STEAM-HEATING BOILER. A boiler operated at pressures not exceeding 15 psi (103 kPa) for steam.

❖ Boilers of this type produce steam for comfort heating.

STOP VALVE. A shutoff valve for controlling the flow of liquid or gases.

❖ Stop valves are used to shut off the flow within a system to isolate portions of the system for repairs, replacement or maintenance.

[B] STORY. That portion of a building included between the upper surface of a floor and the upper surface of the floor next above, except that the topmost story shall be that portion of a building included between the upper surface of the topmost floor and the ceiling or roof above.

❖ The definition is required for the application of certain sections of the code, including Sections 506.3.6, 606.2.3, 607.6.1, 607.6.3 and 1104.4.1.

STRENGTH, ULTIMATE. The highest stress level that the component will tolerate without rupture.

❖ The ultimate strength of a material or component is used with safety factors to determine the maximum allowable working pressure or stress for that material or component.

SUPPLY AIR. That air delivered to each or any space supplied by the air distribution system or the total air delivered to all spaces supplied by the air distribution system, which is provided for ventilating, heating, cooling, humidification, dehumidification and other similar purposes.

❖ As opposed to return air, supply air is delivered to the conditioned space by an air handler and may or may not be returned to the air handler. Supply air can include ventilation air.

SUPPLY AIR SYSTEM. An assembly of connected ducts, plenums, fittings, registers and grilles through which air, heated or cooled, is conducted from the supply unit to the space or spaces to be heated or cooled (see also "Return air system").

❖ In addition to those components already mentioned, air-handling equipment, dampers, controls and conditioning equipment are also included as part of the supply air system.

THEORETICAL AIR. The exact amount of air required to supply oxygen for complete *combustion* of a given quantity of a specific fuel.

❖ The calculation of theoretical air is based on ideal conditions that do not often occur in actual appliance operation (see the commentary for "Combustion air" and "Excess air").

THERMAL RESISTANCE (R). A measure of the ability to retard the flow of heat. The R-value is the reciprocal of thermal conductance.

❖ Insulation marketed and sold in the United States is rated with an R-value, which is the commonly accepted unit of measurement for thermal resistance. As the R-value of an element or assembly increases, the transmission of heat energy through the element or assembly decreases [see Commentary Figure 202(3)].

[P] THIRD-PARTY CERTIFICATION AGENCY. An approved agency operating a product or material certification system that incorporates initial product testing, assessment and surveillance of a manufacturer's quality control system.

❖ These agencies evaluate products, verify compliance with specific standards or criteria and conduct periodic inspections of the manufacturer's facilities to determine ongoing compliance with manufacturing processes, standards or criteria.

[P] THIRD-PARTY CERTIFIED. Certification obtained by the manufacturer indicating that the function and performance characteristics of a product or material have been determined by testing and ongoing surveillance by an approved third-party certification agency. Assertion of certification is in the form of identification in accordance with the requirements of the third-party certification agency.

❖ A product that is certified by a third-party certification agency (see commentary, "Third-party certification agency") to comply with requisite standards or criteria, usually identified by a certification mark attached to or permanently marked on it. Additionally, the agency usually provides a list of the products it certifies, which provides information relative to the certification.

[P] THIRD-PARTY TESTED. Procedure by which an approved testing laboratory provides documentation that a product, material or system conforms to specified requirements.

❖ A product that has been tested by an approved agency (see the definition of and commentary for "Approved agency") to determine compliance with specific product standards or criteria.

TLV-TWA (THRESHOLD LIMIT VALUE-TIME-WEIGHTED AVERAGE). The time-weighted average concentration of a refrigerant or other chemical in air for a normal 8-hour workday and a 40-hour workweek, to which nearly all workers are repeatedly exposed, day after day,

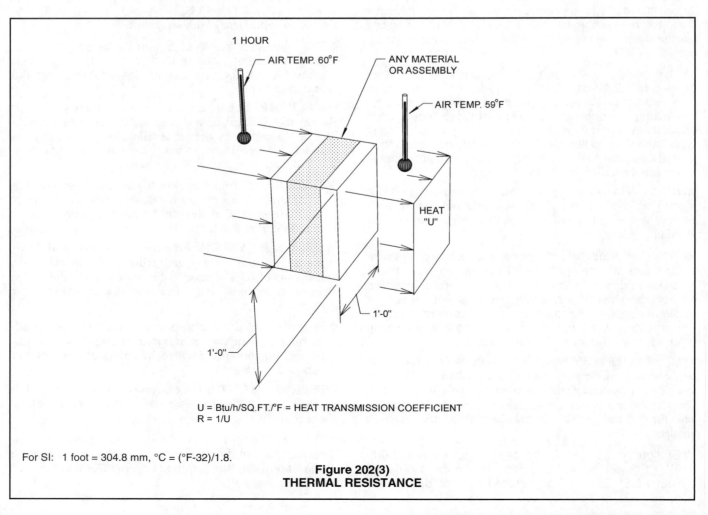

For SI: 1 foot = 304.8 mm, °C = (°F-32)/1.8.

Figure 202(3)
THERMAL RESISTANCE

without adverse effects, as adopted by the American Conference of Government Industrial Hygienists (ACGIH).

❖ These values are based on experimental test data compiled for toxicity to humans and animals. This definition relates to Table 1103.1. Where no value is stated in the table, no value has been determined for that refrigerant.

TOILET ROOM. A room containing a water closet and, frequently, a lavatory, but not a bathtub, shower, spa or similar bathing fixture.

❖ As opposed to bathrooms, toilet room plumbing fixtures include only human waste elimination fixtures and hand-washing fixtures.

TOXICITY CLASSIFICATION. Refrigerants shall be classified for toxicity to one of two classes in accordance with ASHRAE 34:

Class A. Refrigerants for which toxicity has not been identified at concentrations less than or equal to 400 parts per million (ppm), based on data used to determine Threshold Limit Value-Time-Weighted Average (TLV-TWA) or consistent indices.

Class B. Refrigerants for which there is evidence of toxicity at concentrations below 400 ppm, based on data used to determine TLV-TWA or consistent indices.

❖ The classification scheme is used to apply appropriate safety requirements for any given refrigerant.

TRANSITION FITTINGS, PLASTIC TO STEEL. An adapter for joining plastic pipe to steel pipe. The purpose of this fitting is to provide a permanent, pressure-tight connection between two materials which cannot be joined directly one to another.

❖ The joining of piping of dissimilar materials, such as plastic and steel, requires specialized transition fittings. For example, such fittings are used with underground plastic fuel-gas piping and steel meter-setting risers.

UNIT HEATER. A self-contained *appliance* of the fan type, designed for the delivery of warm air directly into the space in which the *appliance* is located.

❖ Unit heaters are similar to warm-air furnaces except that they are not designed for use with ductwork. These heaters are typically suspended from ceilings or roof structures.

VENT. A pipe or other conduit composed of factory-made components, containing a passageway for conveying *combustion* products and air to the atmosphere, *listed* and *labeled* for use with a specific type or class of *appliance*.

❖ In code terminology, vents are distinguished from chimneys and usually are constructed of factory-made listed and labeled components intended to function as a system. Type B and BW vents are constructed of galvanized steel and aluminum sheet metal, and are double wall and air insulated. Such vents are designed to vent gas-fired appliances and equipment that are equipped with draft hoods or are specifically listed (labeled) for use with Type B or BW vents. Type L vents are typically constructed of sheet steel and stainless steel. They are double wall and air insulated, and are designed to vent gas- and oil-fired appliances and equipment. Some appliances are designed for use with corrosion-resistant vents, such as those made of plastic pipe and special alloys of stainless steel.

Pellet vent. A vent *listed* and *labeled* for use with *listed* pellet-fuel-burning appliances.

❖ Pellet vents (PL) are specialized vents similar in design and construction to Type L vents. Pellet vents must not be used with any appliances other than pellet-burning appliances.

Type L vent. A vent *listed* and *labeled* for use with the following:

1. Oil-burning appliances that are *listed* for use with Type L vents.

2. Gas-fired appliances that are *listed* for use with Type B vents.

❖ See the commentary for "Vent."

VENT CONNECTOR. The pipe that connects an *approved* fuel-fired *appliance* to a vent.

❖ In most cases, appliances are not located directly in line with the vertically rising chimney or vent; therefore, a vent connector is necessary to connect the appliance flue outlet to the vent. Vent connectors can be single- or double-wall pipes and are usually made from galvanized steel, stainless steel or aluminum sheet metal. In many installations, the vent connectors must be constructed of the same material as the vent, as is typically done with Type B vent systems.

VENT DAMPER DEVICE, AUTOMATIC. A device intended for installation in the venting system, in the outlet of an individual automatically operated fuel-burning *appliance* that is designed to open the venting system automatically when the *appliance* is in operation and to close off the venting system automatically when the *appliance* is in a standby or shutdown condition.

❖ These devices conserve energy by preventing airflow in the venting system during the appliance "OFF" cycle. Conditioned air is conveyed to the outdoors through the vent and unconditioned outdoor air replaces it by infiltration. Vent dampers also conserve energy by limiting standby losses (stack losses) for appliances, such as boilers and storage type water heaters. Interrupting the draft in the appliance flueways will help retain the residual and stored heat energy in the heat exchangers, heat transfer fluids and stored hot water.

VENTILATION. The natural or mechanical process of supplying conditioned or unconditioned air to, or removing such air from, any space.

❖ Ventilation can be used for comfort cooling, control of air contaminants, equipment cooling and replenishing oxygen levels.

VENTILATION AIR. That portion of supply air that comes from the outside (outdoors), plus any recirculated air that has been treated to maintain the desired quality of air within a designated space.

❖ Ventilation air is supplied to remove or dilute indoor air contaminants. In the context of Chapter 4, ventilation air is 100-percent outdoor air that is not recirculated.

VENTING SYSTEM. A continuous open passageway from the flue collar of an *appliance* to the outside atmosphere for the purpose of removing flue or vent gases. A venting system is usually composed of a vent or a *chimney* and vent connector, if used, assembled to form the open passageway.

❖ Venting systems operate by either gravity (natural draft) or mechanical means. Venting systems include both vents and chimneys.

WATER HEATER. Any heating *appliance* or *equipment* that heats potable water and supplies such water to the potable hot water distribution system.

❖ A water heater is a closed pressure vessel or heat exchanger that is provided with a heat source and that supplies potable (drinkable) water to the building's hot water distribution system. Large commercial water heaters are sometimes referred to as "hot water supply boilers." Water heaters can be of the storage type with integral storage vessel, circulating type for use with an external storage vessel, tankless instantaneous type without storage capacity and point-of-use type with or without storage capacity. Water heaters are regulated by this code and the IPC because they have elements and installation requirements for both mechanical and plumbing systems.

ZONE. One *occupiable space* or several occupiable spaces with similar *occupancy* classification (see Table 403.3), occupant density, zone air distribution effectiveness and zone primary airflow rate per unit area.

❖ One occupiable space can be considered to be a single zone, or multiple zones. More than one occupiable space could be treated as multiple zones or as a single zone. It all depends upon the similarity of the spaces with regard to occupancy classification, population density and ventilation system design. Simply put, a zone is an occupiable space or spaces that have the same characteristics and demands upon the HVAC systems.

Bibliography

The following resource materials were used in the preparation of the commentary for this chapter of the code:

IBC-12, *International Building Code*. Washington, DC: International Code Council, 2011.

IFC-12, *International Fire Code*. Washington, DC: International Code Council, 2011.

IFGC-12, *International Fuel Gas Code*. Washington, DC: International Code Council, 2011.

IPC-12, *International Plumbing Code*. Washington, DC: International Code Council, 2011.

NFPA 211-03, *Standard for Chimneys, Fireplaces, Vents and Solid Fuel Burning Appliances*. Quincy, MA: National Fire Protection Association, 2003.

Chapter 3:
General Regulations

General Comments

A fundamental principle of the code is its dependence on the listing and labeling method of approval for appliances and equipment. Section 301.4 prohibits the installation of unlisted appliances except where approved in accordance with Section 105.

Purpose

Chapter 3 contains requirements for the safe and proper installation of mechanical equipment and appliances to ensure protection of life and property

SECTION 301
GENERAL

301.1 Scope. This chapter shall govern the approval and installation of all *equipment* and appliances that comprise parts of the building mechanical systems regulated by this code in accordance with Section 101.2.

❖ This section states that this chapter governs the approval and installation of all mechanical equipment and appliances that are regulated by the code. Section 101.2 establishes the scope of application of the code (see commentary, Section 101.2).

301.2 Energy utilization. Heating, ventilating and air-conditioning systems of all structures shall be designed and installed for efficient utilization of energy in accordance with the *International Energy Conservation Code.*

❖ Appliances and equipment that use depletable energy sources must be designed and installed to use energy efficiently. The *International Energy Conservation Code®* (IECC®) is the applicable document for regulating the efficiency and performance of the appliances and heating, ventilating and air-conditioning (HVAC) systems. Special applications such as process heating or cooling should be designed for the maximum energy efficiency attainable.

301.3 Identification. Each length of pipe and tubing and each pipe fitting utilized in a mechanical system shall bear the identification of the manufacturer.

❖ The manufacturer is given the option of determining the type of marking for the material. If there is no applicable standard or the applicable standard does not require that a material be identified, identification of the manufacturer is still required by the code. Where the code indicates compliance with an approved standard, the manufacturer must comply with the requirements for marking in accordance with the applicable standard.

301.4 Plastic pipe, fittings and components. Plastic pipe, fittings and components shall be *third-party certified* as conforming to NSF 14.

❖ Plastic piping, fittings and plastic pipe-related components, including solvent cements, primers, tapes, lubricants and seals used in mechanical systems, must be tested and certified as conforming to NSF 14. This includes all piping and fittings and plastic piping system components, including but not limited to pipes, fittings, valves, joining materials, gaskets and appurtenances. This section does not apply to components that only include plastic parts such as brass valves with a plastic stem.

301.5 Third-party testing and certification. Piping, tubing and fittings shall comply with the applicable referenced standards, specifications and performance criteria of this code and shall be identified in accordance with Section 301.3. Piping, tubing and fittings shall either be tested by an approved third-party testing agency or certified by an approved *third-party certification agency.*

❖ The term "third party" refers to an outside organization with no financial or other interest in the outcome. The term "tested" means that the product or material was initially tested, a report or documentation was developed, but retesting at a later date is not performed. The term "certified" means that the product or material was initially tested and a program of periodic testing ensures that the product or material continues to meet the specified requirements.

301.6 Fuel gas appliances and equipment. The approval and installation of fuel gas distribution piping and *equipment*, fuel gas-fired appliances and fuel gas-fired *appliance* venting systems shall be in accordance with the *International Fuel Gas Code.*

❖ Editions of the code prior to the year 2000 addressed installation requirements that were exclusively fuel-gas-related. Starting with the 2000 edition of the code, the provisions that exclusively addressed fuel-gas installations were deleted from the code as the first edition of the *International Fuel Gas Code®* (IFGC®) was developed in 1997. The creation of the IFGC was the result of an agreement between the International Code Council® (ICC®) and the American Gas Association (AGA) to develop and maintain a stand-alone fuel gas code for the family of *International Codes®* (I-Codes®). With the support and input of the fuel gas industry, the IFGC is kept current with new developments and technology in the fuel gas industry. Any code provision that addressed only a fuel-gas-related subject was removed from the code. Code provisions that could apply to other types of fuel (oil, kerosene, wood, coal, pellets, etc.) as well as fuel gas were modified to no longer refer to fuel-gas-related subjects. The IFGC is an inseparable partner to the IMC. Together, they cover all currently used fuels with the most up-to-date text possible.

301.7 Listed and labeled. Appliances regulated by this code shall be *listed* and *labeled* for the application in which they are installed and used, unless otherwise *approved* in accordance with Section 105.

> **Exception:** Listing and labeling of *equipment* and appliances used for refrigeration shall be in accordance with Section 1101.2.

❖ Mechanical appliances must be listed and labeled by an approved agency to show that they comply with the applicable national standards. The code requires listing and labeling for appliances such as boilers, furnaces, space heaters, direct-fired heaters, cooking appliances, clothes dryers, rooftop HVAC units, etc. The code also requires listing for system components as specifically stated in the text addressing those components. The label is the primary, if not the only, assurance to the installer, the inspector and the end-user that a similar appliance has been tested and evaluated by an approved agency and has been determined to perform safely and efficiently when installed and operated in compliance with its listing.

The presence of a label is part of the information that the code official is to consider in the approval of appliances. The only exception to the labeling requirement is when the code official approves the use of a specific appliance with the authority granted in Section 105. The requirement that appliances are to be used only in accordance with their listing is intended to prevent the use of products that have a listing for a particular application but are being used in an application for which they have not been tested. An example would be a fan that is listed for use as a bathroom exhaust fan but is installed for use as an in-line restaurant power ventilator. Another potential misapplication could be duct wrap that has been tested and listed for small ducts but is installed on a much larger diameter duct. Such misapplication has the potential to create hazardous life safety situations.

Caution should be exercised when considering the approval of unlisted and unlabeled appliances.

Approval of unlabeled appliances must be based on documentation that demonstrates compliance with applicable standards or, where no product standards exist, the appliance is appropriate for the intended use and will provide the same level of performance as would listed and labeled appliances. A fundamental principle of the code is the reliance on the listing and labeling process to ensure appliance performance; approvals granted in accordance with Section 105 must be well justified with supporting documentation. To the code official, the installer and the end-user, very little is known about the performance of an appliance that is not tested and built to an appliance standard.

301.8 Labeling. Labeling shall be in accordance with the procedures set forth in Sections 301.8.1 through 301.8.2.3.

❖ As the commentary for Section 301.4 states, the product label is the primary, if not the only, assurance to the code official that the appliance is safe for installation. The labeling of an appliance ensures that testing in compliance with an applicable standard has been performed and that the product will perform acceptably when installed and operated in accordance with the appliance's listing. Before an appliance or other component can be labeled, the code requires specific actions by qualified agencies and personnel. Sections 301.8.1 through 301.8.2.3 describe the requirements that must be complied with before a label can be issued for the appliance or equipment.

301.8.1 Testing. An *approved* agency shall test a representative sample of the mechanical *equipment* and appliances being *labeled* to the relevant standard or standards. The *approved* agency shall maintain a record of all of the tests performed. The record shall provide sufficient detail to verify compliance with the test standard.

❖ An approved agency is one that complies with the requirements of Sections 301.8.2.1 through 301.8.2.3 and is approved by the code official (see commentary, Section 301.8.2.1). The only way that an approved agency can verify that equipment and appliances meet the requirements of the relevant standard(s) is by testing of the appliance or equipment under controlled conditions in a testing laboratory. For mass-produced identical products, the approved agency rarely tests each product. Typically, a representative random sample of a "production run" of products is tested. For example, a test protocol might require that three units out of 1,000 units produced be tested. As long as the design and manufacturing processes for identically produced products do not change, the established sampling and testing frequency provides a high level of assurance that each produced product would pass the test, if actually tested.

The approved agency is responsible for maintaining a record of specific information concerning the product tested, as well as the results of the tests performed. The test standards detail what information is important to record. The records provide proof that the testing was actually performed and that appliance or equipment met or exceeded the minimum requirements of the applicable product standards.

There are numerous standards, not all of which are specifically referenced in the code, applicable to various appliances and equipment. For this reason, the approved agency determines the standards to be used for testing and then, in turn, as the basis for labeling. Each standard contains safety requirements for a given appliance or piece of equipment and specifies tests that must be performed. The labeling agency must maintain sufficient detailed documentation to demonstrate compliance with the test standard. The code official may require that copies of the test reports be submitted to determine the validity of the label.

Examples of standards that are used as a basis for testing and labeling include:

- UL 641—Low-Temperature Venting Systems, Type L;
- UL 727—Oil-Fired Central Furnaces; and
- UL 1482—Room Heaters, Solid-Fuel Type.

The basis for a label is the requirement for testing a representative, perhaps identical, sample of the appliance to indicate conformance to a required standard. For this reason, the appliance must meet the requirements of the standard (see commentary, Section 304.1).

301.8.2 Inspection and identification. The *approved* agency shall periodically perform an inspection, which shall be in-plant if necessary, of the mechanical *equipment* and appliances to be *labeled*. The inspection shall verify that the *labeled* mechanical *equipment* and appliances are representative of the mechanical *equipment* and appliances tested.

❖ The approved agency whose identification insignia appears on the label is required to perform periodic in-plant inspections to verify that the manufactured product is equivalent to the sample that was tested. Because the label is good only for the products that were tested, the in-plant inspections are intended to discover any design changes or production quality control problems. If any discrepancies are found, the labeling agency would discontinue labeling of the particular product and the manufacturer would be required to resolve the problem and, if necessary, have the revised product retested before the labeling process is resumed. The code official may require copies of the periodic inspection reports to determine that the in-plant inspections are being performed in compliance with the requirements for a labeled product. Because appliances and equipment are tested under specific conditions of installation and operation in accordance with the manufacturer's instructions, the issuance of a label requires that these instructions be provided to the installer and end-user to ensure that the product is not misapplied or improperly installed. Because the code requires that the labeled appliances and equipment be installed and operated in accordance with the manufacturer's instructions, the instructions must be attached to or shipped with each appliance. In-plant inspections by the approved agency ensures that the instructions are being shipped with the product, that the design of the product has not substantially changed and that any change in manufacturing processes will not require a change in the testing protocol.

301.8.2.1 Independent. The agency to be *approved* shall be objective and competent. To confirm its objectivity, the agency shall disclose all possible conflicts of interest.

❖ As a part of the basis for a code official's approval of a particular labeling agency, the agency must demonstrate its independence from the manufacturer of the product as well as competence to perform the required tests. The judgment of objectivity is linked to the financial and fiduciary independence of the agency. The competence of the agency is judged by its experience and organization, and the experience of its personnel. As a hypothetical example, the Acme Inspection Agency is testing oil-fired furnaces for the Real Hot Furnace Company. After some investigation, it is discovered that both Acme and Real Hot are subsidiaries of the same parent company. The inspection agency and manufacturer clearly have a relationship that is inappropriate from the standpoint of conflict of interest, and the objectivity of the inspection agency is sufficiently questionable for the code official to justify not approving Acme as a testing and labeling agency for equipment produced by the Real Hot Furnace Company.

While code officials could do their own investigations of testing agencies, many rely upon accredited third-party evaluation services to perform such investigations. One such service is the International Code Council's Evaluation Service (ICC-ES). ICC-ES evaluation reports are public documents, available free of charge on the worldwide Web, not only to building regulators and manufacturers, but also to contractors, specifiers, architects, engineers, and anyone else with an interest in the building industry.

301.8.2.2 Equipment. An *approved* agency shall have adequate *equipment* to perform all required tests. The *equipment* shall be periodically calibrated.

❖ An agency must have proper equipment to perform the specific tests and inspections as required by the product and test standards. Referring to the example in the commentary for Section 301.8.2.1, if the Acme Inspection Agency had the facilities to test only fire doors, they would not be the appropriate agency for testing of an oil-fired furnace. Although this example is oversimplified, the point is that the inspection agency must have all of the necessary equipment to perform the testing required by the applicable standard.

The agency must also keep records of maintenance and calibration of their test and inspection equipment to demonstrate that the equipment can be relied upon to produce accurate, consistent and reproducible results. Often testing apparatus, instruments and equipment must be capable of measurements using very small units of measure within a specified tolerance. To produce accurate, dependable readings and reliable test results, testing apparatus, many pieces of equipment and instruments must be routinely calibrated to established references, such as those maintained by the National Institute of Standards and Technology (NIST).

While code officials could question specific testing agencies with respect to their testing equipment, many rely upon accredited third-party evaluation services to perform such verifications. One such service is the ICC-ES.

301.8.2.3 Personnel. An *approved* agency shall employ experienced personnel educated in conducting, supervising and evaluating tests.

❖ The competence of an inspection agency is based on the agency having the proper equipment to perform the test, as stated in Section 301.8.2.2, and also on the experience and abilities of its personnel. The best calibrated equipment can produce accurate results only when operated by experienced personnel who are trained to conduct, supervise and evaluate tests. For example, consider a newly formed agency that has employed individuals who do not have experience related to the testing to be conducted and have not been adequately trained. The code official may require information that demonstrates the agency personnel have the capability to properly perform the tests. The capabilities and experience of supervisory personnel overseeing their work are also important.

While code officials could question specific testing agencies with respect to their testing personnel, many rely upon accredited third-party evaluation services to perform such verifications. One such service is the ICC-ES.

301.9 Label information. A permanent factory-applied nameplate(s) shall be affixed to appliances on which shall appear in legible lettering, the manufacturer's name or trademark, the model number, serial number and the seal or mark of the *approved* agency. A label shall also include the following:

1. Electrical *equipment* and appliances: Electrical rating in volts, amperes and motor phase; identification of individual electrical components in volts, amperes or watts, motor phase; Btu/h (W) output; and required clearances.

2. Absorption units: Hourly rating in Btu/h (W); minimum hourly rating for units having step or automatic modulating controls; type of fuel; type of refrigerant; cooling capacity in Btu/h (W); and required clearances.

3. Fuel-burning units: Hourly rating in Btu/h (W); type of fuel *approved* for use with the *appliance*; and required clearances.

4. Electric comfort heating appliances: electric rating in volts, amperes and phase; Btu/h (W) output rating; individual marking for each electrical component in amperes or watts, volts and phase; and required clearances from combustibles.

❖ This section requires that the label be a plate, tag or other item made and printed of materials that will have a permanence to last the intended life of the product. In general, label materials other than metal tags or plates usually consist of material that is similar in appearance to a decal, and the label, its adhesive and the printed information are all durable and water resistant. Because of the important information given by a label, the intent is that the label be permanent, not susceptible to damage and legible for the life of the appliance or equipment to which it is attached. The standards that appliances are tested against usually specify the required label performance criteria, the method of attachment and required label information. The code requires that the label be affixed permanently and intends that the label be in a prominent location on the appliance or equipment. Although this section specifies the information that must appear on the label, relevant product standards might require additional information or the manufacturer might choose to provide additional information on the label. Commentary Figure 301.9 shows a typical appliance label.

301.10 Electrical. Electrical wiring, controls and connections to *equipment* and appliances regulated by this code shall be in accordance with NFPA 70.

❖ Field-installed power wiring and control wiring for appliances and equipment must be installed in accordance with NFPA 70.

The power wiring includes all the wiring, disconnects, overcurrent protection devices, starters and related hardware used to supply electrical power to the appliance or equipment. The control wiring includes all the wiring, devices and related hardware that connect the main unit to all external controls and accessories, such as temperature and pressure sensors, thermostats, exhausters, equipment contactors, interlock controls and remote damper motors. The internal factory wiring of appliances and equipment is not covered by this section unless it is specifically addressed in NFPA 70; however, such wiring is covered by the testing and review performed by an approved agency as part of the labeling process.

The mechanical or electrical code official responsible for the inspection of appliances and equipment must be familiar with the applicable sections of NFPA 70.

301.11 Plumbing connections. Potable water supply and building drainage system connections to *equipment* and appliances regulated by this code shall be in accordance with the *International Plumbing Code*.

❖ Plumbing connections to appliances and equipment regulated by the code must be in accordance with the *International Plumbing Code*® (IPC®).

Hydronic systems normally require a means of supplying fill and makeup water to replace any water lost to evaporation, leakage or intentional draining. Where direct connections are made to the potable water supply, the connections must be isolated from the potable water source. This provision is intended to protect the potable water system from contamination by backflow when a direct connection is made to a hydronic system. Hydronic systems are normally pressurized, contain nonpotable water and fluids, and can contain conditioning chemicals or antifreeze solutions. Low-temperature hydronic fluids and cooling towers have also been associated with disease-causing organisms such as the Legionnaires' disease bacterium. The potable water system must be protected from potential contamination resulting from connection to hydronic systems, water-wash filter systems, cooling towers, solar systems, water-cooled heat exchangers, cooking appliances, ice makers, humidifiers, evaporative coolers, etc.

In addition, water heaters must also be considered as both mechanical appliances (equipment) and plumbing appliances and, therefore, must comply with both this code and the IPC.

A water heater installation is complex in that it has a fuel or power supply; a chimney or vent connection, if fuel fired; a combustion air supply, if fuel fired; connections to the plumbing potable water distribution system; and controls and devices to prevent a multitude of potential hazards from conditions such as excessively high temperatures, pressures and ignition failure.

It is not uncommon for jurisdictions to issue both plumbing and mechanical permits for water heater installations, or to require that the installer be licensed in both the plumbing and mechanical trades when performing such installations (see commentary, Section 1002). There is no way to avoid the "code crossover" for water heaters because they are clearly plumbing and mechanically related and under the purview of both plumbing and mechanical codes. This section also triggers the IPC for the drainage associated with mechanical appliances and equipment, such as those addressed in Section 307.

AMERICAN STANDARD INC.
THE TRANE COMPANY
TRENTON, N.J. 08619   MADE IN U.S.A.

FORCED AIR FURNACE   CATEGORY I
ANS Z21.47 - 1990 CENTRAL FURN
FOR INDOOR INSTALLATION IN A BUILDING
CONSTRUCTED ON SITE.   **NRTL**

MODEL NO.	SERIAL NO.	EQUIPPED FOR
TUD080R936A1	G36519785	NAT. GAS
INPUT	LIMIT SETTING	
80,000 BTU/HR.	190 °F	MFRD. 09/92
TEMP. RISE °F	MAX. EXT. STATIC PRESS	MAX. DESIGN AIR
FROM 30 TO 60	.50 INCHES WATER	TEMP. 160 °F
VOLTS/PHASE/HERTZ	TOTAL AMPS	SERVICE CODE
115/1/60	8.5	1

MANIFOLD PRESSURE
(IN INCHES OF WATER)
NAT. 3.5 L P 10.5
SUPPLY PRESSURE
(IN INCHES OF WATER)
MAX. NAT. 10.5, L P 13.0
MIN. NAT. 4.5 L P 11.0 FOR PURPOSE OF INPUT ADJUSTMENT.

FLAME ROLLOUT SWITCH - REPLACE
IF BLOWN WITH CATALOG NO.
WG09X033 (333 F CUTOFF TEMP.)
ONE TIME THERMAL FUSE.

LOW INPUT 52,000 BTU/HR

MINIMUM CLEARANCE COMBUSTIBLE MATERIALS:

FOR	CLOSET	INSTALLATION AS FOLLOWS:			
SIDES	0	IN. W/SINGLE WALL VENT			
FLUE	6	IN. W/SINGLE WALL VENT	1 IN. W/TYPE B-1 VENT		
FRONT	6	IN.	BACK 0 IN.	TOP 1 IN.	

UPFLOW UNITS. FOR INSTALLATION COMBUSTIBLE FLOORING.   21D340159 P01

Figure 301.9
TYPICAL LABEL FOR A CATEGORY I GAS-FIRED FURNACE
(Courtesy of Trane and American-Standard Company)

301.12 Fuel types. Fuel-fired appliances shall be designed for use with the type of fuel to which they will be connected and the altitude at which they are installed. Appliances that comprise parts of the building mechanical system shall not be converted for the usage of a different fuel, except where *approved* and converted in accordance with the manufacturer's instructions. The fuel input rate shall not be increased or decreased beyond the limit rating for the altitude at which the *appliance* is installed.

❖ Mechanical appliances are usually designed by the manufacturer to operate on one specifically designated type of fuel. An element of information used for the approval of appliances is the label, which ensures that the appliance has been tested in accordance with a valid standard and determined to perform acceptably when installed and operated in accordance with the appliance listing (see commentary, Section 301.8). The fuel used in the appliance test must be based on the type of fuel specified by the manufacturer. When an appliance is converted to a different type of fuel, the original label that appears on the appliance is no longer valid. Because the original approval of the appliance is based in part on the label, the appliance is no longer approved for use.

Requiring approval of fuel conversions by the code official and that they be in compliance with the manufacturer's installation instructions is the only way to ensure that the conversion will allow for the safe operation of the appliance. Fuel conversions that are not performed correctly can adversely affect the performance of burners, the venting of combustion gases and the proper clearance to combustibles.

Before a fuel conversion is performed, the manufacturer must be contacted to obtain installation instructions outlining the procedures to follow to ensure proper operation of the appliance. In most cases, conversion kits are available from the manufacturer along with the installation instructions. Once a conversion has been completed, a supplemental label must be installed to update the information contained on the original label, thereby alerting service personnel to the modifications.

All fuel-fired appliances are designed to operate with a maximum and minimum heat energy input capacity. This capacity is field adjusted to suit the elevation because of the change in air density at different elevations. Alteration of heat energy input beyond the allowable limits can result in hazardous overfiring or underfiring. Either condition can cause operation problems that include overheating, vent failure, corrosion, poor draft and poor combustion.

301.13 Vibration isolation. Where vibration isolation of *equipment* and appliances is employed, an *approved* means of supplemental restraint shall be used to accomplish the support and restraint.

❖ Where vibration isolation connections are used in ducts and piping and where equipment is mounted with vibration dampers, support is required for the ducts, piping and equipment to maintain positioning and alignment and to prevent stress and strain on the vibration connectors and dampers.

301.14 Repair. Defective material or parts shall be replaced or repaired in such a manner so as to preserve the original approval or listing.

❖ Repair work must not alter the nature of appliances and equipment in a way that would invalidate the listing or conditions of approval. For example, replacement of safety control devices with different devices could alter the design and operation of an appliance from that intended by the manufacturer and the listing agency.

301.15 Wind resistance. Mechanical *equipment*, appliances and supports that are exposed to wind shall be designed and installed to resist the wind pressures determined in accordance with the *International Building Code*.

❖ Installations of mechanical equipment and appliances that are subject to wind forces must be designed to resist those forces. The wind pressures must be based on the wind provisions in the *International Building Code®* (IBC®) for the site. The requirements in the IBC are based on ASCE 7. The wind pressure requirements are based on the exposure of the building and wind speeds for that region.

[B] 301.16 Flood hazard. For structures located in flood hazard areas, mechanical systems, equipment and appliances shall be located at or above the elevation required by Section 1612 of the *International Building Code* for utilities and attendant equipment.

> **Exception:** Mechanical systems, equipment and appliances are permitted to be located below the elevation required by Section 1612 of the of the *International Building Code* for utilities and attendant equipment provided that they are designed and installed to prevent water from entering or accumulating within the components and to resist hydrostatic and hydrodynamic loads and stresses, including the effects of buoyancy, during the occurrence of flooding up to such elevation.

❖ In areas designated as flood hazard areas in accordance with the requirements of Section 1612 of the IBC, mechanical systems, equipment and appliances must be elevated above the elevation specified in the IBC. Exposure to water can damage most mechanical system components as well as cause serious appliance and equipment malfunctions. For example, the majority of appliance manufacturers require the replacement of safety controls or entire appliances that have been submerged in flood waters. See FEMA 348 for additional guidance.

The exception to this section provides criteria for placing specific equipment below the required elevation. To do so, the equipment must be designed to prevent the entry or accumulation of water. Standard equipment that typically is installed at-grade is not designed to withstand the entry of water and would not meet the requirements of this exception.

[B] 301.16.1 High-velocity wave action. In flood hazard areas subject to high-velocity wave action, mechanical systems and *equipment* shall not be mounted on or penetrate walls intended to break away under flood loads.

❖ Breakaway walls are designed and constructed to fail under flood loads in order to avoid transferring their loads and damaging the primary structural support of the building. When mechanical or plumbing system components penetrate or are attached to the breakaway walls, they can prevent the wall from breaking away, thus transferring additional load to the building. Also, damage to the mechanical systems is certain to occur when the walls fail as intended.

301.17 Rodentproofing. Buildings or structures and the walls enclosing habitable or occupiable rooms and spaces in which persons live, sleep or work, or in which feed, food or foodstuffs are stored, prepared, processed, served or sold, shall be constructed to protect against the entrance of rodents in accordance with the *International Building Code.*

❖ This section references the IBC for the requirements to prevent rodent infestation of a building. Appendix F of the IBC addresses two mechanical issues. The first is the requirement to protect the annular spaces around openings in exterior walls or interior walls enclosing rooms where persons live or where food is prepared or stored. The annular spaces can be sealed by any effective method, but the primary methods used are to seal off the annular space with a hard material such as mortar, or to place a metal collar around the penetrating pipe or duct. The collar material must be durable for the weather exposure and strong enough to prevent the rodents from chewing through and entering the building.

The second IBC requirement is to cover ventilation or combustion air openings with wire cloth of at least 0.035-inch (0.61 mm) wire to prevent the entry of rodents through the opening (see Section 401.5). Instead of the wire cloth, the code allows the openings to be protected by solid sheet metal guards that must be arranged to prevent rodent access.The guard material must be resistant to exposure to the elements.

301.18 Seismic resistance. When earthquake loads are applicable in accordance with the *International Building Code*, mechanical system supports shall be designed and installed for the seismic forces in accordance with the *International Building Code.*

❖ Building mechanical systems are required by the IBC to be designed for specified seismic forces. This section references the detailed mechanical equipment and component seismic support requirements contained in the IBC to bring these requirements to the attention of the design professional and the permit applicant.

Large-diameter piping, mechanical equipment and large HVAC ducts must be braced for earthquake loads as stated in the IBC. The failure of the supports for these components has been shown to be a threat to health and safety in geographical areas where moderate- to high-magnitude earthquakes are likely to occur. The geographical locations where the earthquake design is required for certain piping, mechanical equipment and HVAC ducts, and the size of the components that must be braced are specified in the IBC.

SECTION 302
PROTECTION OF STRUCTURE

302.1 Structural safety. The building or structure shall not be weakened by the installation of mechanical systems. Where floors, walls, ceilings or any other portion of the building or structure are required to be altered or replaced in the process of installing or repairing any system, the building or structure shall be left in a safe structural condition in accordance with the *International Building Code.*

❖ The installation of mechanical systems must not adversely affect the structural integrity of the building components. The IBC dictates the structural safety requirements that must be applied to any structural portion of the building that is penetrated, altered or removed during the installation, replacement or repair of a mechanical system (see commentary, Sections 302.3 through 302.5.3).

302.2 Penetrations of floor/ceiling assemblies and fire-resistance-rated assemblies. Penetrations of floor/ceiling assemblies and assemblies required to have a fire-resistance rating shall be protected in accordance with Chapter 7 of the *International Building Code.*

❖ Penetrations of fire-resistance-rated assemblies can diminish the integrity of the assembly, allowing smoke and fire to pass through and cause it to fail prematurely. For example, the IBC and this code allow the penetration of a fire-resistance-rated floor/ceiling assembly by an air duct if the penetration is protected with a fire damper installed at the floor line and the air duct does not connect more than two stories. If the air duct penetrates more than one floor/ceiling assembly, the penetrations must be protected by a shaft enclosure (see Section 607.6).

[B] 302.3 Cutting, notching and boring in wood framing. The cutting, notching and boring of wood framing members shall comply with Sections 302.3.1 through 302.3.4.

❖ The sections referenced contain prescriptive sizes and locations of acceptable cuts, notches and bored holes in wood framing members. Ideally, framing members should not be altered in any way, but because this is not always practical, the code permits alterations that do not significantly weaken the members.

[B] 302.3.1 Joist notching. Notches on the ends of joists shall not exceed one-fourth the joist depth. Holes bored in joists shall not be within 2 inches (51 mm) of the top or bottom of the joist, and the diameter of any such hole shall not exceed one-third the depth of the joist. Notches in the top or bottom of joists shall not exceed one-sixth the depth and shall not be located in the middle third of the span.

❖ The code recognizes that at times floor and ceiling joists must be cut, notched or bored. The provisions are based on location and size limitations to ensure that the structural member can support the load (see Commentary Figure 302.3.1).

A notch or cut will decrease the allowable stress of the member. A notch or cut may place a lumber defect that was previously in a low-stress location in the interior of the board, to a high-stress area at the edge of the notch or cut. Cutting and notching also cause shear stress concentrations in the member at the corners of the notch or cut. Because of these shear stress concentrations, the full extent of the damage to the member cannot be calculated. In consideration of the detrimental effects of cutting and notching, it is a good practice to avoid cutting, notching or boring adjacent to areas of the member that contain knots and, where possible, use rounded or sloped edged notches to reduce the stress concentrations at the corners.

Notches are not to exceed those permitted in this section unless a structural analysis is completed by a qualified design professional.

Holes cut and bored in joists and studs have the same effect as notches. Holes reduce the strength of the member. This section permits drilling holes in wood structural members within certain limitations. The limitations given in this section are not to be exceeded unless a structural analysis is completed by a qualified design professional.

[B] 302.3.2 Stud cutting and notching. In exterior walls and bearing partitions, any wood stud is permitted to be cut or notched not to exceed 25 percent of its depth. Cutting or notching of studs not greater than 40 percent of their depth is permitted in nonbearing partitions supporting no loads other than the weight of the partition.

❖ The load-sharing characteristics of a wall with studs, sheathing and plates allow for the stated limitation for notches in load-bearing studs. The maximum notch in a stud is not to exceed 25 percent of its depth unless the stud is reinforced to resist the anticipated load (see Commentary Figure 302.3.2). See the commentary to Section 302.3.1 for further discussion of the effects of notching and cutting.

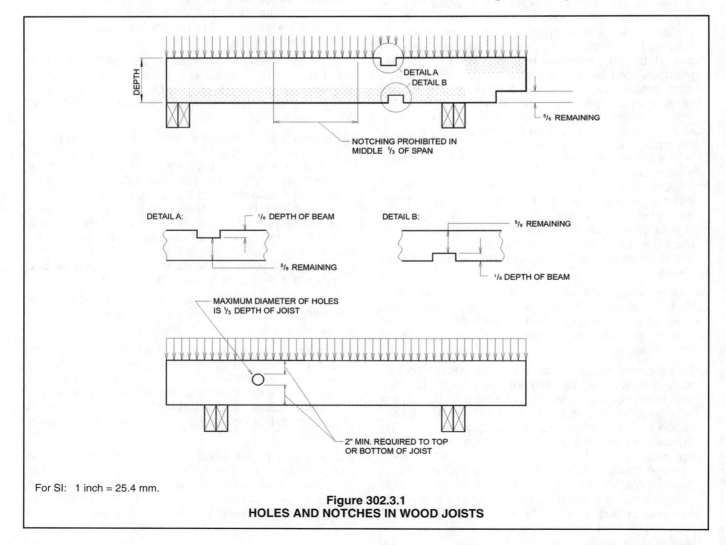

For SI: 1 inch = 25.4 mm.

Figure 302.3.1
HOLES AND NOTCHES IN WOOD JOISTS

[B] 302.3.3 Bored holes. A hole not greater in diameter than 40 percent of the stud depth is permitted to be bored in any wood stud. Bored holes not greater than 60 percent of the depth of the stud are permitted in nonbearing partitions or in any wall where each bored stud is doubled, provided not more than two such successive doubled studs are so bored. In no case shall the edge of the bored hole be nearer than 0.625 inch (15.9 mm) to the edge of the stud. Bored holes shall not be located at the same section of stud as a cut or notch.

❖ This section specifies the limits on bored holes in wood studs (see Commentary Figure 302.3.3). The 40-percent limit is intended for studs in exterior walls and for load-bearing walls. The 60-percent limit for doubled studs is also intended to apply to studs in exterior walls and other load-bearing walls. Bored holes not greater than 60-percent are allowed in any nonload-bearing stud.

The bored-hole edge location limitations maintain the load-carrying capacity of the stud. These limits are intended to apply to load-bearing and nonload-bearing partitions and studs in exterior walls. A bored hole in the same cross section as a cut or notch could cause the stud to fail.

[B] 302.3.4 Engineered wood products. Cuts, notches and holes bored in trusses, structural composite veneer lumber, structural glue-laminated members and I-joists are prohibited except where permitted by the manufacturer's recommendations or where the effects of such alterations are specifically considered in the design of the member.

❖ The structural elements of engineered wood systems, such as metal-plate-connected parallel-chord wood trusses, premanufactured I-joists and metal-plate connected wood roof trusses, must not be altered in anyway unless the manufacturer's installation instructions specifically address the allowable alterations considered in the design of the product. As an alternative, a structural analysis by a qualified design professional could be accepted by the code official.

[B] 302.4 Alterations to trusses. Truss members and components shall not be cut, drilled, notched, spliced or otherwise altered in any way without written concurrence and approval of a *registered design professional*. Alterations resulting in the addition of loads to any member, such as HVAC *equipment* and water heaters, shall not be permitted without verification that the truss is capable of supporting such additional loading.

❖ Trusses are designed and constructed to support specific loads and must not be altered without being evaluated by a registered design professional. The allowable field alterations described in Section 302.3 for joists, rafters and studs cannot be applied to designed trusses. Adding the weight of mechanical equipment such as HVAC units to the truss load is also prohibited without an evaluation of the effects.

[B] 302.5 Cutting, notching and boring in steel framing. The cutting, notching and boring of steel framing members shall comply with Sections 302.5.1 through 302.5.3.

❖ The sections referenced contain prescriptive sizes and locations of acceptable cuts, notches and bored holes in steel framing members.

[B] 302.5.1 Cutting, notching and boring holes in structural steel framing. The cutting, notching and boring of holes in structural steel framing members shall be as prescribed by the *registered design professional*.

❖ This section does not allow cutting, notching or boring of structural steel framing members without approval by a registered design professional. Unlike Section 302.3, which allows field cutting, notching and boring of wood studs and joists within the prescribed limits, this section requires that alterations to steel structural framing members be allowed only as dictated by a design professional prior to the alteration. The code official should not approve any alterations to structural members unless a signed drawing or revision is on the job site showing the allowed size and location of the cut, notch or bored hole.

[B] 302.5.2 Cutting, notching and boring holes in cold-formed steel framing. Flanges and lips of load-bearing cold-formed steel framing members shall not be cut or notched.

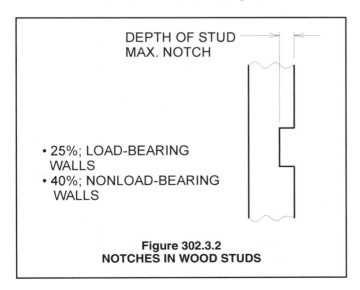

DEPTH OF STUD
MAX. NOTCH

• 25%; LOAD-BEARING WALLS
• 40%; NONLOAD-BEARING WALLS

Figure 302.3.2
NOTCHES IN WOOD STUDS

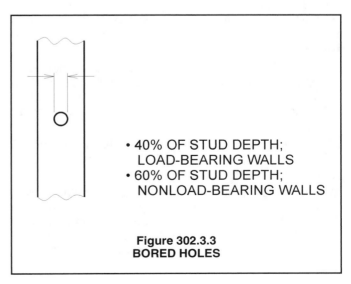

• 40% OF STUD DEPTH; LOAD-BEARING WALLS
• 60% OF STUD DEPTH; NONLOAD-BEARING WALLS

Figure 302.3.3
BORED HOLES

Holes in webs of load-bearing cold-formed steel framing members shall be permitted along the centerline of the web of the framing member and shall not exceed the dimensional limitations, penetration spacing or minimum hole edge distance as prescribed by the *registered design professional*. Cutting, notching and boring holes of steel floor/roof decking shall be as prescribed by the *registered design professional*.

❖ This section does not allow any cutting or notching of the flanges and lips of any load-bearing, cold-formed steel-framing member because these operations would substantially weaken the member. Bored holes in the web of steel-framing members are allowed only as directed by a registered design professional. Cutting, notching and boring of steel floor/roof decking is allowed only as directed by a registered design professional (see commentary, Section 302.5.1). Where possible, the holes that are prebored at the factory in steel-framing members should be used instead of making new holes on the job site.

[B] 302.5.3 Cutting, notching and boring holes in nonstructural cold-formed steel wall framing. Flanges and lips of nonstructural cold-formed steel wall studs shall not be cut or notched. Holes in webs of nonstructural cold-formed steel wall studs shall be permitted along the centerline of the web of the framing member, shall not exceed $1^1/_2$ inches (38 mm) in width or 4 inches (102 mm) in length, and shall not be spaced less than 24 inches (610 mm) center to center from another hole or less than 10 inches (254 mm) from the bearing end.

❖ Similar to Section 302.5.2, this section does not allow any cutting or notching of the flanges and lips of any cold-formed steel-framing member. However, holes may be made in the web of nonstructural cold-formed steel wall studs without the involvement of a registered design professional if the prescribed size, location and spacing requirements are met.

SECTION 303
EQUIPMENT AND APPLIANCE LOCATION

303.1 General. *Equipment* and appliances shall be located as required by this section, specific requirements elsewhere in this code and the conditions of the *equipment* and *appliance* listing.

❖ Section 303 is a consolidation of the code's generally applicable location requirements and limitations. The listing for an appliance or equipment will often contain location requirements parallel with or in addition to these sections.

303.2 Hazardous locations. Appliances shall not be located in a *hazardous location* unless *listed* and *approved* for the specific installation.

❖ Mechanical appliances that are to be installed in a hazardous location must be approved and listed for use in that location. A hazardous location is defined as any location considered to be a potential fire hazard from flammable vapors, dust, combustible fibers

or other highly combustible substances. This location is not necessarily classified in the IBC or NFPA 70 as a high-hazard occupancy classification. Appliances located in a hazardous location require special design, construction and installation because of the potential for an explosion or fire caused by the presence of dust, flammable vapors and highly combustible materials.

This section does not specify the type of appliance to be used in any particular hazardous location; therefore, the code official must evaluate information submitted by the designer on the intended occupancy of a space, the specification for the appliance and the location of the appliance in the space to determine the appropriate requirements. For example, these requirements may include spark-resistant materials for moving parts, such as fan blades; explosionproof construction for electrical switches, starters and motors; or sealed combustion chambers for fuel-burning appliances.

Examples of appliances installed in a hazardous location are a fuel-burning water heater in an area where fuel is dispensed, an industrial baking facility where fine flour dust is present around a fuel-burning furnace or a paint shop where paint fumes and vapors are present around an electric unit heater.

303.3 Prohibited locations. Fuel-fired appliances shall not be located in, or obtain *combustion* air from, any of the following rooms or spaces:

1. Sleeping rooms.

2. Bathrooms.

3. Toilet rooms.

4. Storage closets.

5. Surgical rooms.

Exception: This section shall not apply to the following appliances:

1. *Direct-vent appliances* that obtain all *combustion air* directly from the outdoors.

2. Solid fuel-fired appliances, provided that the room is not a confined space and the building is not of unusually tight construction.

3. Appliances installed in a dedicated enclosure in which all *combustion* air is taken directly from the outdoors, in accordance with Chapter 7. *Access to* such enclosure shall be through a solid door, weather-stripped in accordance with the exterior door air leakage requirements of the *International Energy Conservation Code* and equipped with an *approved* self-closing device.

❖ The intent of this section is to prevent fuel-fired appliances from being installed in rooms and spaces where the combustion process could pose a threat to the occupants. Potential threats include depleted oxygen levels and elevated levels of carbon dioxide, nitrous oxides and carbon monoxide and other combustion gases.

In small rooms, such as bedrooms and bathrooms, the doors are typically closed when the room is occupied, which could allow combustion gases to build up to life-threatening levels. In bedrooms, occupants would not be alert or aware of impending danger.

In surgical rooms, oxygen and flammable anesthetic gases are used.

If an appliance obtains combustion air from a room or space, it communicates with the atmosphere in that room or space whether or not it is considered to be installed. An appliance can be in a room, closet or alcove and obtain combustion air from an adjacent room, which is why Section 303.3 is worded to address both the actual location of an appliance and its source of combustion air. In other words, an appliance in a closet accessed from a bedroom is no different from an appliance located within the bedroom.

Exception 1 recognizes that direct-vent appliances have sealed combustion chambers and obtain all combustion air directly from the outdoors.

Exception 2 exempts solid-fuel-fired appliances if the room construction and volume are such that combustion air is supplied in the manner specified in Chapter 7. If a solid-fuel-fired appliance was not venting properly, smoke would enter the room and alert the occupants.

Exception 3 would allow installation of fuel-fired appliances within a separate dedicated space that is accessed from the rooms and spaces listed in this section. A separated space containing the appliance must be open to the outdoors in accordance with Chapter 7, and the access door to the space must be sealed in accordance with IECC requirements for exterior doors to prevent communication between atmospheres in the separated spaces. The door must also be self-closing and not rely on occupants to keep it closed. The enclosure should not be used for storage or any other purpose.

303.4 Protection from damage. Appliances shall not be installed in a location where subject to mechanical damage unless protected by *approved* barriers.

❖ Appliances must be installed with consideration for potential damage from occupant activities. Damage results in property loss and can cause hazardous malfunctions, fires or explosions. Potential damage includes impact from occupants, vehicles, stored materials and operations that occur in an occupancy. Barriers must be approved and strong enough to resist the type of impact anticipated.

303.5 Indoor locations. Fuel-fired furnaces, water heaters and boilers installed in closets and alcoves shall be *listed* for such installation. For purposes of this section, a closet or alcove shall be defined as a room or space having a volume less than 12 times the total volume of fuel-fired appliances other than boilers and less than 16 times the total volume of boilers. Room volume shall be computed using the gross floor area and the actual ceiling height up to a maximum computation height of 8 feet (2438 mm).

❖ An appliance must be listed for closet/alcove installation if it is installed in a space defined in this section as a closet or alcove. This section defines a closet or alcove by establishing minimum volume criteria. Simply stated, if the room or space volume is less than the stated minimum volume, that room or space is considered to be a closet or alcove.

These volume criteria have traditionally been used to determine whether or not the required clearances to combustibles can be reduced by protection methods such as those found in Section 308 of the code. Fuel-fired appliances installed in small rooms or spaces can pose a potential fire hazard because of the heat emanating from the appliance and the inadequate ventilation afforded by the boundaries of small enclosures.

For appliances installed in closets or alcoves, the clearances stated in the listing (label) are necessary for ventilation cooling of the appliance and must not be reduced as provided for in Section 308 (see Section 308.2). Water heaters were added to the list of appliances that must be listed for installation in a closet or alcove. The concern for adequate room or space volume is no less important for a fuel-fired water heater than it is for fuel-fired furnaces or boilers.

303.6 Outdoor locations. Appliances installed in other than indoor locations shall be *listed* and *labeled* for outdoor installation.

❖ Appliances installed outdoors must be specifically listed for outdoor installation. If an appliance is not labeled for outdoor installation, it must not be installed outdoors because of the risk of malfunction and accelerated deterioration from weather and ambient temperatures. The manufacturer's instructions and listing must be consulted to determine whether a particular appliance is designed for or can be made suitable for outdoor installation. For example, furnaces cannot be installed outdoors in cold climates regardless of any weatherproof enclosure, unless the heat exchanger, burner assemblies and venting system are designed for exposure to temperatures below normal indoor temperatures. Cold ambient temperatures can cause harmful condensation to occur on heat exchanger surfaces of fuel-fired appliances.

Additionally, there may be other local ordinances that govern the outdoor installation of appliances. Before installing an appliance outdoors, consult local zoning regulations, ordinances and subdivision covenants. Many of these regulations strictly limit the location of outdoor mechanical equipment and appliances. Also, for roof installations, the roof structure must be shown to support all imposed static and dynamic structural loads. Substantiating data on the structural adequacy of the entire installation must be submitted and approved.

303.7 Pit locations. Appliances installed in pits or excavations shall not come in direct contact with the surrounding soil. The sides of the pit or excavation shall be held back a minimum of 12 inches (305 mm) from the *appliance*. Where the depth exceeds 12 inches (305 mm) below adjoining grade, the walls of the pit or excavation shall be lined with concrete or masonry. Such concrete or masonry shall extend a minimum of 4 inches (102 mm) above adjoining grade and shall have sufficient lateral load-bearing capacity to resist collapse. The *appliance* shall be protected from flooding in an *approved* manner.

❖ Where installed in a depression, appliances could be damaged by corrosion or a malfunction could be caused by blockage of an opening caused by entry of soil. Because a depression could hold water, protection from flooding is also necessary.

[B] 303.8 Elevator shafts. Mechanical systems shall not be located in an elevator shaft.

❖ The code views an elevator shaft as an unnecessarily, risky location for mechanical system components. There is a potential for leaks to occur in mechanical system equipment and piping that could release refrigerants, products of combustion, heat transfer fluids and other substances into the shaft, which could damage the elevator and harm its passengers. Because the elevator is often relied upon for fire-fighting access, as well as egress for those with disabilities, the shaft must be maintained free from contaminants and potential hazards that could render the elevator inoperative or unsafe for the occupants. The location of mechanical equipment in an elevator shaft could also create access problems for service of the mechanical equipment and the elevator equipment.

SECTION 304
INSTALLATION

304.1 General. *Equipment* and appliances shall be installed as required by the terms of their approval, in accordance with the conditions of the listing, the manufacturer's installation instructions and this code. Manufacturer's installation instructions shall be available on the job site at the time of inspection.

❖ Manufacturer's installation instructions are thoroughly evaluated by the listing agency, verifying that a safe installation is prescribed. The listing agency can require that the manufacturer alter, delete or add information to the instructions, as necessary, to achieve compliance with applicable standards and code requirements. Manufacturer's installation instructions are an enforceable extension of the code and must be in the hands of the code official when an inspection takes place. Without access to the instruc-

tions, the code official would be unable to complete an inspection.

When an appliance is tested to obtain a listing and label, the approved agency installs the appliance in accordance with the manufacturer's installation instructions. The appliance is tested under these conditions; thus, the installation instructions become an integral part of the labeling process (see Commentary Figure 304.1).

The listing and labeling process ensures that the appliance and its installation instructions are in compliance with applicable standards. Therefore, an installation in accordance with the manufacturer's instructions is required, except where the code requirements are more stringent. An inspector must carefully and completely read and comprehend the manufacturer's instructions to properly perform an installation inspection.

In some cases, the code will specifically address an installation requirement that is also addressed in the manufacturer's installation instructions. The code requirements may be the same or may exceed the requirements in the manufacturer's installation instructions or the manufacturer's installation instructions could contain requirements that exceed those in the code. In all cases, the more restrictive requirements would apply (see commentary, Section 304.2).

In addition to the installation requirements for the appliance or equipment itself, this section also regulates the connections to appliances and equipment by requiring compliance with other applicable sections of the code. These connections include, but are not limited to, fuel and electrical supplies; control wiring; hydronic piping; chimneys; vents and ductwork. Some overlap or coincidence of connection requirements may occur in the code and the manufacturer's installation instructions. Where differences or conflicts occur, the requirements that provide the greatest level of safety apply.

Even if an installation appears to be in compliance with the manufacturer's instructions, the installation cannot be complete or approved until all associated components, connections and systems that serve the appliance or equipment are also in compliance with the applicable provisions of the code. For example, an oil-fired boiler installation must not be approved if the boiler is connected to a deteriorated, undersized or otherwise unsafe chimney or vent. Likewise, the same installation must not be approved if the existing oil piping is in poor condition or if the electrical supply circuit is inadequate or unsafe.

In the case of replacement installations, the intent of this section is to require new work associated with the installation to comply with the code without necessarily requiring full compliance for the existing, unchanged portions of the related ductwork, piping, electrical, venting and similar mechanical systems. For example, if a furnace is replaced in an existing building, the new work and connections involved with the replacement would be treated as new construction and the existing unaltered system components would be considered as existing mechanical systems. Existing mechanical systems are accepted on the basis that they are free from hazard although not necessarily compliant with current codes. The code is not retroactive except where specifically stated that it applies to existing systems.

Manufacturer's installation instructions are often updated and changed for various reasons, such as changes in the appliance or equipment design, revi-

sions to the product standard and as a result of field experience related to existing installations. The code official should stay abreast of any changes by reviewing the manufacturer's instructions for every installation.

Equipment and appliances must be installed in accordance with the manufacturer's installation instructions. The manufacturer's label and installation instructions must be consulted in determining whether or not an appliance or piece of equipment

ACCESSORIES

If installing furnace on combustible floor, sub-base (BAYBASE201 or BAYBASE202) must be used depending on furnace width. Installation instructions are packed with the accessories. If a 30″ wide coil is to be used with a 24″ wide furnace, sub-base BAYBASE203 must be used.

In year-round air conditioning installations where the coil enclosure is used with the downflow furnace — a sub-base is not required beneath the coil enclosure on combustible flooring.

DUCT CONNECTIONS

Air duct systems should be installed in accordance with standards for air conditioning systems, National Fire Protection Association Pamphlet No. 90. They should be sized in accordance with ACCA Manual D or whichever is applicable. Check on controls to make certain they are correct for the electrical supply.

Central furnaces, when used with cooling units, shall be installed in parallel or on the upstream side of the cooling units to avoid condensation in the heating element. This applies

unless the furnace has been specifically approved for downstream installation. With a parallel flow arrangement, the dampers or other means used to control air flow shall be adequate to prevent chilled air from entering the furnace. If manually operated, it must be equipped with a means to prevent operation of either unit unless the damper is in full heat or cool position.

The installer must make the plenum chamber at least 8 inches in depth and sufficiently wide to take the largest duct to be attached to it.

Though these units have been specifically designed for quiet, vibration free operation, air ducts can act as sounding boards. If poorly installed, air ducts can amplify the slightest vibration to the annoyance level. If this is the case, a flex connector of nonflammable material may be installed in the return and supply ducts.

Where the furnace is located adjacent to the living area, the system should be carefully designed with returns to minimize noise transmission through the return air grille. Although these winter air conditioners are designed with large blowers operating at moderate speeds, any blower moving a high volume of

Figure 304.1
EXCERPTS FROM TYPICAL MANUFACTURER'S INSTALLATION MANUALS
(Courtesy of Trane and American-Standard Company)

(continued)

can be installed and operated in a particular hazardous location. The manufacturer's installation instructions must be available on the job site when the equipment is being inspected.

304.2 Conflicts. Where conflicts between this code and the conditions of listing or the manufacturer's installation instructions occur, the provisions of this code shall apply.

Exception: Where a code provision is less restrictive than the conditions of the listing of the *equipment* or *appliance*

or the manufacturer's installation instructions, the conditions of the listing and the manufacturer's installation instructions shall apply.

❖ Although very rare, there are some code provisions that intentionally differ or conflict with the manufacturer's installation instructions. For example, Section 903.3 places a restriction on unvented gas log heaters that may or may not agree with the manufacturer's instructions; however, this restriction is

---

air will produce audible noise. This noise could be objectionable when unit is located very close to a living area. It is often advisable to carry the return air ducts under the floor or through the attic. Such design permits the installation of air return remote from the living area (i.e., central hall).

When furnace is installed so supply ducts carry air circulated by the furnace to areas outside the space containing the furnace, air shall also be handled by a duct(s) sealed to the furnace and terminating outside the space containing the furnace.

WHERE THERE IS NO COMPLETE RETURN DUCT SYSTEM, THE RETURN CONNECTION MUST BE RUN FULL SIZE FROM THE FURNACE TO A LOCATION OUTSIDE THE UTILITY ROOM, BASEMENT, ATTIC OR CRAWL SPACE.

IMPORTANT: Do not take return air through back of furnace cabinet.

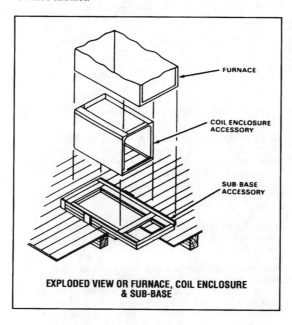

EXPLODED VIEW OR FURNACE, COIL ENCLOSURE & SUB-BASE

GENERAL INFORMATION NOTES

The TUX and TDX series central furnaces have been classified as Category IV furnaces in accordance with ANSI Z21.64-"latest edition" standards. Category IV furnaces operate with

a positive vent pressure and with a vent gas temperature less than 140° F. above its dewpoint. These conditions require special venting systems, which must be gas tight and water tight. The special venting system requirements for these Gas Furnaces are outlined in INSTALLER'S GUIDE (Pub. No. 18-CH15D3), Venting and Condensate Drain Instructions.

GENERAL INSTALLATION INSTRUCTIONS

The manufacturer assumes no responsibility for equipment installed in violation of any code or regulation. The User's Information Manual will instruct you in the service and care of your unit. Have your installer go over the operational portion of that manual with you so you fully understand how your warm air furnace functions.

It is recommended that Manual J of the Air Conditioning Contractors Association (ACCA) or A.R.I. 230 be followed in estimating heating requirements. When estimating heating requirements for installation at altitudes above 2000 ft. remember the gas input must be reduced (See GAS INPUT ADJUSTMENT).

Locate the furnace as close to the flue as permissible and as centralized about the air distribution system as practical.

A gas-fired furnace for installation in a residential garage must be installed so the burner(s) and the ignition source are located not less than 18 inches above the floor. The furnace must be located or protected to avoid physical damage by vehicles.

Consider all clearances that are required:

A. Clearance from combustible materials and for service accessibility (See INSTALLATION CLEARANCES).

B. Clearance for attaching duct work (See DUCT CONNECTIONS).

C. Clearance for proper installation of venting and for running condensate drain to the floor drain (See VENTING AND CONDENSATE DRAIN INSTRUCTIONS, Pub. No. 18-CH15D3).

D. Clearances for any external filter racks or sub-bases to be used.

E. Clearances for gas piping (See GAS PIPING).

The installation must conform with local gas company regulations, national and local electrical codes, local building, plumbing codes.

Figure 304.1—continued
EXCERPTS FROM TYPICAL MANUFACTURER'S INSTALLATION MANUALS
(Courtesy of Trane and American-Standard Company)

intentional and overrides the manufacturer's instructions.

Section 303.3 is another example of code text that intends to override any manufacturer's instructions to the contrary. It is not the intent of this section to cause the manufacturer's requirements that are more stringent or restrictive than the code to be ignored. In the context of this section, conflict refers to instances in which the code intends to require a greater level of safety.

The code official must evaluate each circumstance of perceived conflict and secure the requirements that provide the greatest protection of life and property. At the time of this publication, there are no known instances where the code would require something less stringent than the manufacturer's installation instructions, and this is extremely unlikely to occur in any case. The exception recognizes that the code depends on the listing/labeling process and the manufacturer's instructions for the majority of installation requirements for mechanical equipment and appliances. The conditions of the listing cannot be violated so that a lesser degree of safety would occur.

304.3 Elevation of ignition source. Equipment and appliances having an *ignition source* and located in hazardous locations and public garages, private garages, repair garages, automotive motor fuel-dispensing facilities and parking garages shall be elevated such that the source of ignition is not less than 18 inches (457 mm) above the floor surface on which the *equipment* or *appliance* rests. For the purpose of this section, rooms or spaces that are not part of the living space of a *dwelling unit* and that communicate directly with a private garage through openings shall be considered to be part of the private garage.

> **Exception:** Elevation of the ignition source is not required for appliances that are listed as flammable vapor ignition resistant.

❖ To reduce the fire/explosion hazard in hazardous areas, the potential sources of ignition must be elevated above the surface supporting the equipment or appliance. It is not the intent to measure the elevation from the surface of a dedicated appliance stand or other structure built to support the appliance where that stand or structure does not afford room for the storage of flammable liquids. Some flammable and combustible liquids typically associated with hazardous locations give off vapors that are more dense than air and tend to collect near the floor. The 18-inch (457 mm) height requirement is intended to reduce the possibility of fire or explosion by keeping ignition sources elevated above the anticipated level of accumulated vapors. The 18-inch (457 mm) value is a minimum requirement and must be increased when required by the manufacturer's installation instructions. This section will effectively prohibit the installation of most furnaces, boilers, space heaters, clothes dryers and water heaters directly on the floor of residential garages and other occupancies where flammable or combustible vapors may be present.

The accumulation of flammable vapors more than 18 inches (457 mm) deep is unlikely in most ventilated locations; therefore, maintaining all possible sources of ignition at least 18 inches (457 mm) above the floor will substantially reduce the risk of explosion and fire (see Commentary Figure 304.3).

In the context of this section, a source of ignition could be a pilot flame, burner, burner igniter or electrical component capable of producing a spark. The term "ignition source" is defined and can be interpreted as meaning an intentional source of ignition such as for a burner, or an unintentional source of ignition for any flammable vapors that may be present (see the definition of "Ignition source").

An appliance installed in a closet or room that is accessible only from the garage must be considered as part of the garage for application of this section. Even though the room may be separated from the garage by walls and a door, there is no practical means of making the door vapor tight or any assurance that the door will remain closed during normal use. An appliance room that is accessed only from the outdoors or from the living space would not be considered as part of the garage. Rooms such as utility rooms or laundry rooms that communicate both with the garage and the living space are considered as part of the living space and not part of the garage (see the definition of "Living space").

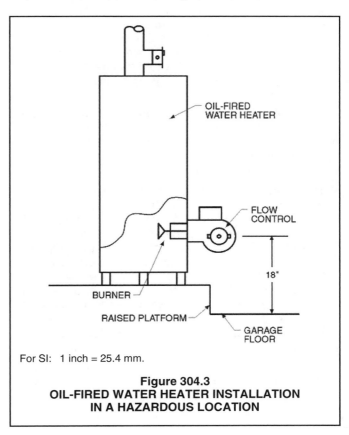

For SI: 1 inch = 25.4 mm.

Figure 304.3
OIL-FIRED WATER HEATER INSTALLATION
IN A HAZARDOUS LOCATION

The exception acknowledges and allows the installation of appliances that are flammable vapor ignition resistant (FVIR) without elevation above the garage floor. FVIR-type appliances are designed to not ignite flammable vapors outside of the appliance combustion chamber. Oil, solid-fuel and electric FVIR appliances are not known by the author to exist.

304.3.1 Parking garages. Connection of a parking garage with any room in which there is a fuel-fired *appliance* shall be by means of a vestibule providing a two-doorway separation, except that a single door is permitted where the sources of ignition in the *appliance* are elevated in accordance with Section 304.3.

> **Exception:** This section shall not apply to *appliance* installations complying with Section 304.6.

❖ In order to minimize the possibility of any spilled flammable liquids and the resulting vapors from coming in contact with the ignition source of a fuel-fired appliance in a parking garage, such equipment must be separated from the areas where the vehicles are parked. Doors connecting the equipment rooms and the main parking area must be arranged with a vestibule or air-lock so that one must pass through two doors prior to entering the other room. An allowance is made where the appliance is elevated at least 18 inches (457 mm) above the floor of the equipment room in accordance with Section 304.3. In such cases, a single door can separate the spaces. If this allowance is used, care must be taken by the code and fire officials that these special stipulations are part of the certificate of occupancy.

The exception recognizes that appliances installed at least 8 feet (2438 mm) above the floor in accordance with Section 304.6 would require neither a two-doorway separation nor, obviously, the 18-inch (457 mm) elevation.

304.4 Prohibited equipment and appliance location. Equipment and appliances having an *ignition source* shall not be installed in Group H occupancies or control areas where open use, handling or dispensing of combustible, flammable or explosive materials occurs.

❖ The installation of appliances and equipment having an ignition source within Group H occupancies or control areas where combustible, flammable or explosive materials are used, handled or dispensed is prohibited regardless of elevation of the ignition source. Although this section is new for the 2009 edition of the code, the requirements were previously included in Section 304.3. Because of the importance of not placing ignition sources in Group H occupancies, and because it was a distinct requirement, this requirement was placed in this independent section.

[FG] 304.5 Hydrogen-generating and refueling operations. Hydrogen-generating and refueling appliances shall be installed and located in accordance with their listing and the manufacturer's instructions. Ventilation shall be required in accordance with Section 304.5.1, 304.5.2 or 304.5.3 in public garages, private garages, repair garages, automotive motor fuel-dispensing facilities and parking garages that contain hydrogen-generating appliances or refueling systems. For the purpose of this section, rooms or spaces that are not part of the living space of a *dwelling unit* and that communicate directly with a private garage through openings shall be considered to be part of the private garage.

❖ The use of hydrogen to fuel vehicles and generate electricity in fuel cell appliances to replace petroleum-based fuels is a rapidly developing technology. Key factors in the increased use of hydrogen are the reduced atmospheric emissions associated with hydrogen and the nation's shift to renewable sources of energy.

Typically, the code official will encounter two classes of equipment—those that generate hydrogen for use by other equipment, such as vehicles, and those that use hydrogen as their energy input, such as fuel cell appliances (see commentary, Section 924.1).

The equipment must be located and installed in accordance with the listing of the equipment and the manufacturer's instructions.

This section intends to minimize the potential for explosions by limiting the source of hydrogen gas and by requiring sufficient ventilation to dissipate any leakage.

[FG] 304.5.1 Natural ventilation. Indoor locations intended for hydrogen-generating or refueling operations shall be limited to a maximum floor area of 850 square feet (79 m²) and shall communicate with the outdoors in accordance with Sections 304.5.1.1 and 304.5.1.2. The maximum rated output capacity of hydrogen generating appliances shall not exceed 4 standard cubic feet per minute (0.00189 m³/s) of hydrogen for each 250 square feet (23.2 m²) of floor area in such spaces. The minimum cross-sectional dimension of air openings shall be 3 inches (76 mm). Where ducts are used, they shall be of the same cross-sectional area as the free area of the openings to which they connect. In such locations, *equipment* and appliances having an *ignition source* shall be located such that the source of ignition is not within 12 inches (305 mm) of the ceiling.

❖ The 850-square-foot (79 m²) maximum floor space used for hydrogen generation or refueling and the maximum rated output capacity of such appliances are intended to limit the amount of hydrogen gas that can accumulate in the space, thus minimizing the potential for explosions.

The location of an ignition source parallels the intent of Section 304.3, but with respect to the ceiling instead of the floor. This section does not apply to spaces that only house vehicles and that do not contain hydrogen-generating or refueling operations. Because it is more buoyant, hydrogen will dissipate more quickly than natural gas, and much more quickly than either propane or gasoline, both of which have vapors that are heavier than air and will linger at an accident site. However, hydrogen and natural gas can both accumulate in unventilated pockets at the top of indoor structures and could represent a risk in such situations.

Similarly, gasoline fumes can accumulate at the floor level in unventilated spaces, posing a different risk. Thus, ignition sources must be avoided at the top of any unventilated spaces for hydrogen gas. Also, hydrogen is odorless, colorless and burns with a flame that is not generally visible to the human eye. This means that it is unlikely that people will be able to detect unsafe conditions without appropriate instrumentation. This is similar to a situation where carbon monoxide (CO) accumulates in a structure.

[FG] 304.5.1.1 Two openings. Two permanent openings shall be provided within the garage. The upper opening shall be located entirely within 12 inches (305 mm) of the ceiling of the garage. The lower opening shall be located entirely within 12 inches (305 mm) of the floor of the garage. Both openings shall be provided in the same exterior wall. The openings shall communicate directly with the outdoors and shall have a minimum free area of $^1/_2$ square foot per 1,000 cubic feet (1 m²/610 m³) of garage volume.

❖ The location requirement will prevent the openings from being more than 12 inches (305 mm) tall because the required openings must be entirely within the 12 inches (305 mm) of wall space measured down from the ceiling and up from the floor (see Commentary Figure 304.5.1.1). The openings must be in the same wall to help create a gravity flow of gases driven by the natural buoyancy of hydrogen gas. The bottom opening is an air inlet and the top is an air outlet.

[FG] 304.5.1.2 Louvers and grilles. In calculating free area required by Section 304.5.1, the required size of openings shall be based on the net free area of each opening. If the free area through a design of louver or grille is known, it shall be used in calculating the size opening required to provide the free area specified. If the design and free area are not known, it shall be assumed that wood louvers will have 25 percent free area and metal louvers and grilles will have 75 percent free area. Louvers and grilles shall be fixed in the open position.

❖ This section recognizes that louvers and grilles are usually installed over air inlets and outlets to prevent rain, snow and animals from entering the building. When louvers or grilles are used, the solid portion of the louver or grille must be considered when determining the unobstructed (net clear) area of the opening.

Air openings are sized based on there being a free, unobstructed area for the passage of air into the space. Louvers or grilles placed over these openings reduce the area of the openings because of the area occupied by the solid portions of the grille or louver. The reduction in area must be considered because only the unobstructed area can be credited toward the required opening size.

The reduction in opening area caused by the presence of grilles or louvers will always require openings to be larger than determined from the sizing ratios of this chapter and larger than any duct of the minimum required size that might connect to these openings.

[FG] 304.5.2 Mechanical ventilation. Indoor locations intended for hydrogen-generating or refueling operations shall be ventilated in accordance with Section 502.16. In such locations, *equipment* and appliances having an *ignition source* shall be located such that the source of ignition is below the mechanical ventilation outlet(s).

❖ Section 502.16 contains criteria for the design and operation of a mechanical ventilation system in repair garages for natural-gas and hydrogen-fueled vehicles. This section makes Section 502.16 applicable to

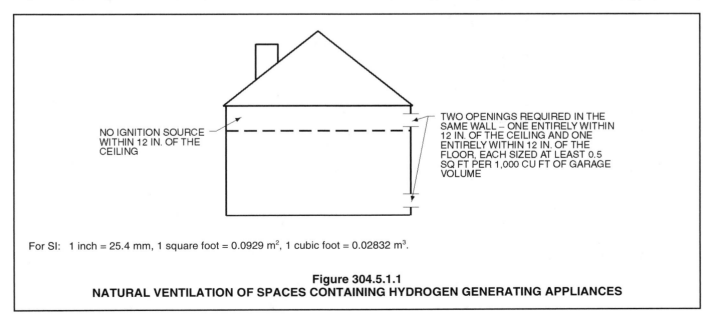

NO IGNITION SOURCE WITHIN 12 IN. OF THE CEILING

TWO OPENINGS REQUIRED IN THE SAME WALL – ONE ENTIRELY WITHIN 12 IN. OF THE CEILING AND ONE ENTIRELY WITHIN 12 IN. OF THE FLOOR, EACH SIZED AT LEAST 0.5 SQ FT PER 1,000 CU FT OF GARAGE VOLUME

For SI: 1 inch = 25.4 mm, 1 square foot = 0.0929 m², 1 cubic foot = 0.02832 m³.

Figure 304.5.1.1
NATURAL VENTILATION OF SPACES CONTAINING HYDROGEN GENERATING APPLIANCES

indoor spaces containing hydrogen-generating and/or refueling operations. Although Section 502.16 does not specify that the outlet must be within 12 inches (305 mm) of the ceiling, as is required for natural ventilation, it does require the opening to be located at the high point of the space. This section requires that any ignition source must be located below the mechanical exhaust outlet opening. It also does not require the ignition source to be located more than 12 inches (305 mm) below the ceiling as is required for natural ventilation.

[FG] 304.5.3 Specially engineered installations. As an alternative to the provisions of Sections 304.5.1 and 304.5.2 the necessary supply of air for ventilation and dilution of flammable gases shall be provided by an *approved* engineered system.

❖ The code is not intended to inhibit innovative ideas or technological advances. A comprehensive regulatory document, such as a mechanical code, cannot envision and then address all future innovations in the industry. As a result, a performance code must be applicable to and provide a basis for the approval of an increasing number of newly developed, innovative materials, systems and methods for which no code text or referenced standards yet exist. The fact that a material, product or method of construction is not addressed in the code is not an indication that prohibition of the material, product or method is intended. The code official is expected to apply sound technical judgment in accepting materials, systems or methods that, while not anticipated by the drafters of the current code text, can be demonstrated to offer equivalent performance. By virtue of its text, the code regulates new and innovative construction practices while addressing the relative safety of building occupants. The code official is responsible for determining whether a requested alternative provides a level of

protection of the public health, safety and welfare as required by the code.

304.6 Public garages. Appliances located in public garages, motor fueling-dispensing facilities, repair garages or other areas frequented by motor vehicles, shall be installed a minimum of 8 feet (2438 mm) above the floor. Where motor vehicles are capable of passing under an *appliance*, the *appliance* shall be installed at the clearances required by the *appliance* manufacturer and not less than 1 foot (305 mm) higher than the tallest vehicle garage door opening.

Exception: The requirements of this section shall not apply where the appliances are protected from motor vehicle impact and installed in accordance with Section 304.3 and NFPA 30A.

❖ Appliances located within building areas where the general public—including employees—operate, repair or fuel motor vehicles must be installed in a manner that provides protection from vehicle impact and limits the possibility of the equipment becoming an ignition source for any flammable/explosive vapors which could collect near the floor. Appliance protection is necessary because the resulting damage from vehicle impact could cause an explosion or fire.

The requirement that appliances be located a minimum of 8 feet (2438 mm) above the floor is intended to provide sufficient clearance for motor vehicles to pass under the appliance without impact. Although 8 feet (2438 mm) is the minimum height requirement, a clearance of at least 1 foot (305 mm) must always be maintained between the top of the highest vehicle entry opening and the bottom of any appliance that is installed where vehicles can pass underneath [see Commentary Figure 304.6(1)]. If the tallest vehicle entry opening is 7 feet (2134 mm) or less in height, the height requirement of 8 feet (2438 mm) will result in a clearance of at least 1 foot (305 mm). If the tallest

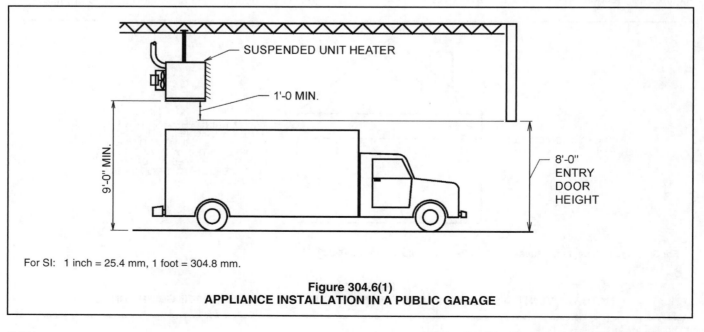

For SI: 1 inch = 25.4 mm, 1 foot = 304.8 mm.

Figure 304.6(1)
APPLIANCE INSTALLATION IN A PUBLIC GARAGE

vehicle entry opening is greater than 7 feet (2134 mm) in height, the minimum installation height requirement must be increased to provide a clearance of not less than 1 foot (305 mm). In buildings such as warehouses, where forklifts or other similar elevating vehicles are used, consideration should also be given to the potential for impact from such vehicles. In vehicle repair garages where vehicles are put on lifts, suspended appliances should not be located over the lift area unless the bottom of the suspended heater is at least as high as the tallest vehicle that could enter the garage plus the highest lift position above the floor. Although this section does not specifically discuss the elevation of suspended appliances over lifts, it would be prudent to add 1 foot (305 mm) to the highest elevation of the vehicle on the lift to ensure a reasonable allowance for construction tolerances.

Except for suspended unit heaters, elevating of appliances above motor vehicles is often impractical, difficult to accomplish, and creates access problems. As such, the exception in this section allows appliances to not be elevated, provided that the appliances are installed in accordance with Section 304.3, protected from vehicle impact and where repair garages or fuel-dispensing facilities are concerned, in accordance with NFPA 30A. NFPA 30A introduces additional criteria and stipulations for nonresidential structures, such as separation and protection of heating appliances. For some applications, NFPA 30A requires continuous mechanical ventilation and could prohibit any operations involving the dispensing or transferring of Class I or II flammable and combustible liquids or liquefied petroleum gas (LP-gas). Even though this section has the exception for nonelevation of appliances, the requirements of NFPA 30A could still require appliance elevation under certain circumstances.

Where appliances are not elevated, the code does not specify how an appliance must be protected from vehicle impact. The most apparent method would be to locate the appliance where it could not reasonably be struck by a vehicle. Another method of protection would be to place one or more barriers in the path of vehicle travel. Commentary Figure 304.6(2) illustrates several methods that could be used to provide a degree of protection for appliances in the path of vehicular travel. The barriers shown in the commentary figure will not eliminate all possibility of a motor vehicle contacting the appliances but will offer a reasonable warning to a driver who is slowly navigating near the appliances. Obviously, if the motor vehicle is narrow enough (e.g., a motorcycle) or a vehicle is traveling at significant speed, such barriers will offer little protection for the appliances. The space required to service, repair and replace an appliance must be taken into account when considering the location and type of barriers. Section 306.1 has requirements for working space in front of the appliance and prohibits the removal of permanent construction (e.g., a pipe embedded in concrete) for service, repair or replacement of the appliance. While a wheel barrier could be located so that the front end of nearly all types of motor vehicles would not contact the appliance, the backing in of a motor vehicle might require the wheel barrier to be located much further away from the appliance. A wheel barrier could also be a pedestrian trip hazard if located in an intended walkway. Where large motor vehicles such as busses and semi-tractor vehicles will be maneuvered near appliances and equipment to be protected, consideration should be given to the size and strength of barriers provided. For example, while the operator of a passenger car will probably be adequately warned by a standard sized concrete wheel barrier or a 3-foot (914 mm) tall, 2-inch (51mm) pipe bollard, operators of larger vehicles might not see such obstructions much less notice impacting such small barriers. Although this section does not specifically require the impact protection provided to stop any type of vehicle at any speed, the intent is for the impact protection to cause the driver to want to stop vehicle movement out of concern for damage that could be occurring. The choice of the type, structural capacity and the location of barriers is the responsibility of the designer.

304.7 Private garages. Appliances located in private garages and carports shall be installed with a minimum clearance of 6 feet (1829 mm) above the floor.

> **Exception:** The requirements of this section shall not apply where the appliances are protected from motor vehicle impact and installed in accordance with Section 304.3.

❖ Appliances located in a private garage or carport must be protected from vehicle impact. This section is applicable to appliances located in an area where motor vehicles can be operated, and includes appliances under which a vehicle can pass and those located anywhere in a vehicle's path where impact is possible. The 6-foot (1829 mm) minimum height requirement is intended to provide adequate clearance above the typical automobile (see Commentary Figure 304.7). With the popularity of conversion vans and recreational vehicles that can be much higher than other automobiles, the 6-foot (1829 mm) minimum installation height above the floor may not provide adequate clearance; additional height might be necessary. The height of the vehicle entry opening of the garage or carport can be used as a guide in determining how tall of a vehicle could be driven into the garage or carport.

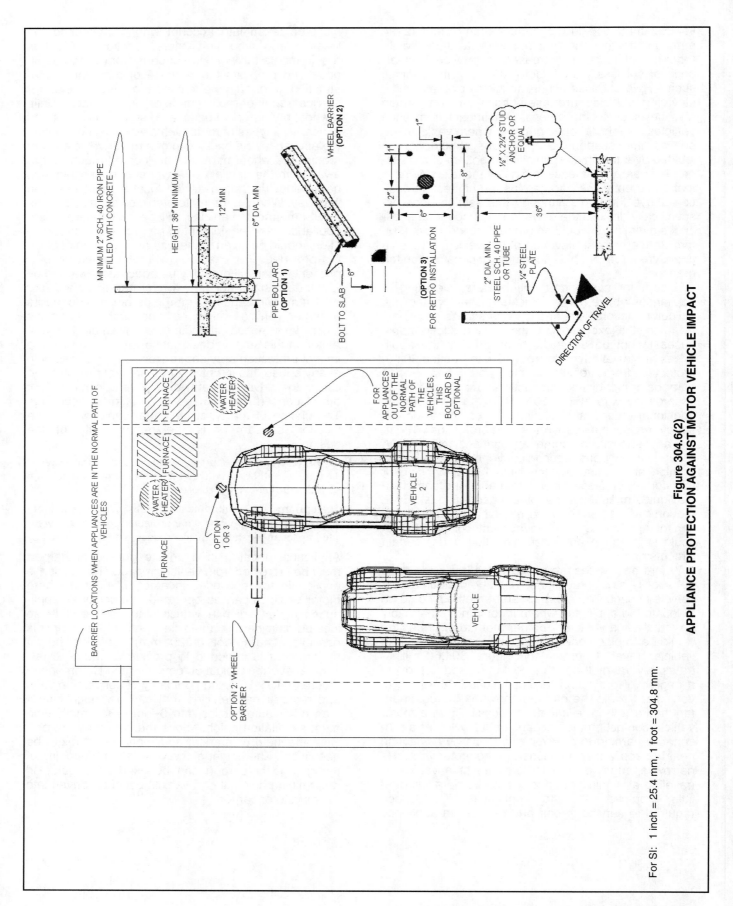

Figure 304.6(2)
APPLIANCE PROTECTION AGAINST MOTOR VEHICLE IMPACT

For SI: 1 inch = 25.4 mm, 1 foot = 304.8 mm.

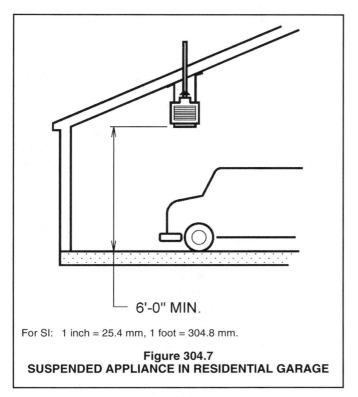

For SI: 1 inch = 25.4 mm, 1 foot = 304.8 mm.

Figure 304.7
SUSPENDED APPLIANCE IN RESIDENTIAL GARAGE

304.8 Construction and protection. Boiler rooms and furnace rooms shall be protected as required by the *International Building Code.*

❖ A "Boiler room" is defined as a room primarily used for the installation of a boiler. A "Furnace room" is defined as a room primarily used for the installation of fuel-fired space heating and water heating appliances other than boilers. Even though the code defines these terms, it is not the intent to require fuel-burning appliances to be enclosed or installed in rooms used primarily for that purpose. A room housing a boiler or a furnace does not necessarily constitute a boiler or furnace room. The key to whether or not a room is a boiler or furnace room can be found in the definitions of these terms. Both definitions state that the room is primarily used for the installation of these appliances. When the room or space is large in comparison to the space occupied by the appliance and has other uses besides housing the boiler, water heater or furnace, it should not be considered as a boiler or furnace room. For example, a boiler or furnace can be installed in a finished basement recreation room without being enclosed and separated from the living space. The basement in which the appliance is located would not be considered a boiler or furnace room. Similarly, an appliance located in an attic space would not cause the attic to be considered a furnace room.

When a boiler or furnace is installed in a room that is used primarily for housing the boiler or furnace, the resulting boiler or furnace room is regulated by the IBC. Depending on the size of the appliances therein, Table 508.2.5 of the IBC requires such rooms to either be enclosed with fire-resistance-rated separa-

tion assemblies or protected with an automatic fire-extinguishing system.

Either separation or protection is required because a boiler, furnace or similar fuel-fired appliance installed in an enclosed space is considered to be a greater hazard than if the appliance or equipment were out in the open and unenclosed. This is based on the fact that furnace or boiler rooms are very often used for unapproved storage and a fire hazard is created by combustible materials stored close to fuel-fired appliances. A small enclosed space typically invites the storage of combustibles and, because of the nature of the typical storage, it is difficult, if not impossible, to maintain the required clearances from the heat-producing appliance. It is not uncommon to find highly combustible materials such as paper, cardboard, rags, cleaning chemicals and similar materials placed directly against vent or chimney connectors, draft hoods, combustion chambers and other components having a required minimum clearance to combustibles.

304.9 Clearances to combustible construction. Heat-producing *equipment* and *appliances* shall be installed to maintain the required *clearances* to combustible construction as specified in the listing and manufacturer's instructions. Such clearances shall be reduced only in accordance with Section 308. *Clearances* to combustibles shall include such considerations as door swing, drawer pull, overhead projections or shelving and window swing, shutters, coverings and drapes. Devices such as doorstops or limits, closers, drapery ties or guards shall not be used to provide the required *clearances.*

❖ Requirements for clearances to combustibles are emphasized because of the potential fire hazard posed when those clearances are not observed. Maintaining an appropriate distance from the outer surfaces of an appliance or piece of equipment to combustible materials reduces the possibility of ignition of combustible materials.

The minimum clearances to combustibles are specified in the manufacturer's installation instructions for a labeled appliance. Because an approved agency tests appliances in accordance with these instructions, the clearances required are necessary for correct installation and operation of the appliance.

Reduction of the required clearances to combustibles is allowed only when the combustibles are protected by one of the methods outlined in Section 308 and clearance reduction is not prohibited by Section 308 or the appliance and equipment listing (see commentary, Section 308). Note that the manufacturer's specified minimum clearances and the clearances specified in Section 308 are all airspace clearances, and those spaces cannot be filled with insulation or any other material, even if the material is noncombustible. In some cases, the manufacturer's installation instructions will specify absolute minimum clearances that may not be reduced by any clearance reduction method.

The most common wall covering material—gypsum wallboard—is a combustible finish material for the purpose of the code. As a result, gypsum wallboard as well as all other combustible wall finishes must be separated from an appliance or equipment in compliance with the prescribed clearance to combustibles (see Section 507.9 for an exception specific to gypsum wallboard). Clearances to combustibles also apply to furnishings, window treatments and moveable items that can be placed within the required clearance range of appliances and equipment.

304.10 Clearances from grade. Equipment and *appliances* installed at grade level shall be supported on a level concrete slab or other *approved* material extending not less than 3 inches (76 mm) above adjoining grade or shall be suspended not less than 6 inches (152 mm) above adjoining grade. Such support shall be in accordance with the manufacturer's installation instructions.

❖ For support and protection from physical damage and contact with soil or water for appliances and equipment installed at grade level indoors or outdoors, the appliances and equipment must be supported on a level concrete slab or other material accepted by the code official, or must be suspended above exposed earth. These slabs must rise at least 3 inches (76 mm) above the surrounding earth. Premanufactured "pads" or support devices could have special requirements for support or connections to support structures. The manufacturer's instructions must be followed.

[B] 304.11 Guards. Guards shall be provided where appliances, *equipment*, fans or other components that require service and roof hatch openings are located within 10 feet (3048 mm) of a roof edge or open side of a walking surface and such edge or open side is located more than 30 inches (762 mm) above the floor, roof or grade below. The guard shall extend not less than 30 inches (762 mm) beyond each end of such appliances, *equipment*, fans, components and roof hatch openings and the top of the guard shall be located not less than 42 inches (1067 mm) above the elevated surface adjacent to the guard. The guard shall be constructed so as to prevent the passage of a 21-inch-diameter (533 mm) sphere and shall comply with the loading requirements for guards specified in the *International Building Code*.

❖ Mechanical equipment and appliances require routine inspection, maintenance and repair. Where the units requiring service or the roof hatch openings are located within 10 feet (3048 mm) of a roof edge or open side with a drop greater than 30 inches (762 mm), a possibility exists for the service person or inspector to be injured by a fall. Appliances installed on roofs are occasionally accessed during inclement weather, emergency situations and darkness. The requirements of this section are intended to protect personnel from the hazard of falls during such conditions. This section pertains to mechanical equipment and appliances, exhaust fans and system components regulated by the code that require routine inspection, maintenance and repair.

The guard serves as a warning and as a protective barrier, and must be 42 inches (1067 mm) high and

constructed to prevent the passage of a 21-inch (533 mm) sphere (see Commentary Figure 304.11). The requirement for the guard to extend 30 inches (762 mm) beyond each end of the equipment, appliance or roof hatch opening provides added protection for the installer or service technician by enlarging the protected area beyond the immediate area of the equipment, appliance or roof hatch. The guardrail must be constructed to resist the imposed loading conditions. It should be noted that the guard requirement only applies to roof hatch openings and not doors that open to the roof.

304.12 Area served. Appliances serving different areas of a building other than where they are installed shall be permanently marked in an *approved* manner that uniquely identifies the *appliance* and the area it serves.

❖ In the event of an emergency or the need for repair or servicing, the area served by the appliances must be easily identifiable. The need for identification is apparent, for example, when a service person looks across the flat roof of a large expansive building and sees a "sea" of rooftop units, each serving some unknown space or tenant below.

SECTION 305
PIPING SUPPORT

305.1 General. All mechanical system piping shall be supported in accordance with this section.

❖ Piping systems, including hydronic, refrigerant, condensate and fuel oil, must be supported as required by this section.

305.2 Materials. Pipe hangers and supports shall have sufficient strength to withstand all anticipated static and specified dynamic loading conditions associated with the intended use. Pipe hangers and supports that are in direct contact with piping shall be of *approved* materials that are compatible with the piping and that will not promote galvanic action.

❖ As with all piping systems, the support of the system is as important as any other part of the overall design. Proper supports are necessary to maintain piping alignment and slope, to support the weight of the pipe and its contents, to control movement and to resist dynamic loads such as thrust. Inadequate support can cause piping to fail under its own weight, resulting in fire, explosion or property damage. Building design must take into consideration the structural loads created by the support of piping systems.

Hangers or supports must not react with or be detrimental to the pipe they support. Hangers or supports for metallic pipe must be of a material that is compatible with the pipe to prevent any corrosive action. For example, copper, copper-clad or specially coated hangers are required if the piping system is constructed of copper tubing. Hangers and supports are typically constructed of noncombustible materials to prevent premature failure in a fire.

305.3 Structural attachment. Hangers and anchors shall be attached to the building construction in an *approved* manner.

❖ A support or hanger can be no stronger than its attachment; therefore, the proper fasteners must be used and installed with the necessary workmanship. Supports for piping systems must also be attached to building construction so that they do not adversely affect the structural integrity of the building.

305.4 Interval of support. Piping shall be supported at distances not exceeding the spacing specified in Table 305.4, or in accordance with MSS SP-69.

❖ The Manufacturers Standardization Society of the Valve & Fittings Industry, Inc.'s (MSS) Pipe Hangers and Supports Standard SP-69 may be used to determine hanger spacing as an alternative to Table 305.4. The MSS standard recognizes the different support needs of piping relative to pipe diameter and specifies a greater interval between supports as the pipe increases in diameter. As the pipe diameter is increased, the material has a greater inherent (beam) strength (see Commentary Figure 305.4).

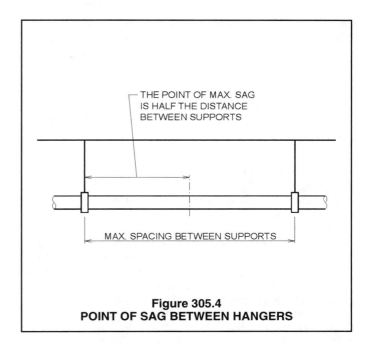

Figure 305.4
POINT OF SAG BETWEEN HANGERS

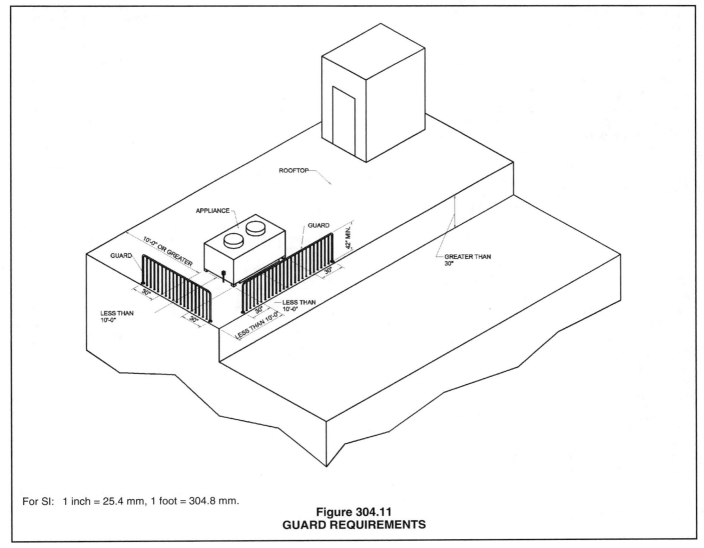

For SI: 1 inch = 25.4 mm, 1 foot = 304.8 mm.

Figure 304.11
GUARD REQUIREMENTS

305.5 Protection against physical damage. In concealed locations where piping, other than cast-iron or steel, is installed through holes or notches in studs, joists, rafters or similar members less than $1^1/_2$ inches (38 mm) from the nearest edge of the member, the pipe shall be protected by shield plates. Protective steel shield plates having a minimum thickness of 0.0575 inch (1.463 mm) (No. 16 gage) shall cover the area of the pipe where the member is notched or bored, and shall extend a minimum of 2 inches (51 mm) above sole plates and below top plates.

❖ This section is intended to minimize the possibility of damage to refrigerant piping and other mechanical piping from nails, screws or other fasteners. Because nails and screws sometimes miss the stud, rafter joist or top or sole plate, the shield must protect the pipe through the full width of the member and must extend not less than 2 inches (51 mm) above or below the sole or top plates, respectively. Commentary Figure 305.5 shows typical shield plates. Cast-iron and galvanized steel pipe have wall thicknesses greater than the required thickness of the shield plate, which makes them inherently resistant to nail and screw penetrations. For the 2009 edition of the code, the thickness dimension for the steel shield plate was changed to reflect the low-end tolerance thickness for 0.0625-inch nominal thickness steel.

TABLE 305.4. See next column.

❖ The support spacing listed in the table is intended to reduce sag and stress in the piping system. The maximum amount of sag occurs at a point halfway between supports and will increase over time if not properly supported. The mid-story guides required by Note c are necessary to prevent excess lateral motion in vertical installations of plastic pipe and tubing. Such lateral motion can result in damage to the pipe and joints. For the 2009 edition of the code, polypropylene piping was added to the table.

SECTION 306
ACCESS AND SERVICE SPACE

306.1 Access for maintenance and replacement. Appliances shall be accessible for inspection, service, repair and replacement without disabling the function of a fire-resistance-rated assembly or removing permanent construction, other appliances, venting systems or any other piping or ducts not connected to the *appliance* being inspected, serviced, repaired or replaced. A level working space at least 30 inches deep and 30 inches wide (762 mm by 762 mm) shall be provided in front of the control side to service an *appliance*.

❖ Because appliances require routine maintenance, repairs and possible replacement, access to appliances is required. The code defines "Access" as being reachable, but which might first require the removal of a panel, door or similar obstruction. An appliance or piece of equipment is not accessible if any portion of the structure's permanent finish materials, such as drywall, plaster, paneling, built-in furni-

ture or cabinets, or any other similar permanently affixed building component must be removed. The access to the appliance must also not require removal or disconnection of other appliances and the ducts, piping or vent systems associated with other appliances. The section is not intended to lessen or negate the access recommendations or requirements stated in the appliance manufacturer's installation instructions but to establish minimum access requirements. As such, the provisions stated here are intended to supplement the manufacturer's installation instructions.

306.1.1 Central furnaces. Central furnaces within compartments or alcoves shall have a minimum working space *clearance* of 3 inches (76 mm) along the sides, back and top with a total width of the enclosing space being at least 12 inches (305 mm) wider than the furnace. Furnaces having a firebox open to the atmosphere shall have at least 6 inches (152 mm) working space along the front *combustion* chamber side.

TABLE 305.4
PIPING SUPPORT SPACING[a]

PIPING MATERIAL	MAXIMUM HORIZONTAL SPACING (feet)	MAXIMUM VERTICAL SPACING (feet)
ABS pipe	4	10[c]
Aluminum pipe and tubing	10	15
Brass pipe	10	10
Brass tubing, $1^1/_4$-inch diameter and smaller	6	10
Brass tubing, $1^1/_2$-inch diameter and larger	10	10
Cast-iron pipe[b]	5	15
Copper or copper-alloy pipe	12	10
Copper or copper-alloy tubing, $1^1/_4$-inch diameter and smaller	6	10
Copper or copper-alloy tubing, $1^1/_2$-inch diameter and larger	10	10
CPVC pipe or tubing, 1 inch and smaller	3	10[c]
CPVC pipe or tubing, $1^1/_4$-inch and larger	4	10[c]
Lead pipe	Continuous	4
PB pipe or tubing	$2^2/_3$ (32 inches)	4
PEX tubing	$2^2/_3$ (32 inches)	10[c]
Polypropylene (PP) pipe or tubing, 1 inch or smaller	$2^2/_3$ (32 inches)	10[c]
Polypropylene (PP) pipe or tubing, $1^1/_4$ inches or larger	4	10[c]
PVC pipe	4	10[c]
Steel tubing	8	10
Steel pipe	12	15

For SI: 1 inch = 25.4 mm, 1 foot = 304.8 mm.

a. See Section 301.18.

b. The maximum horizontal spacing of cast-iron pipe hangers shall be increased to 10 feet where 10-foot lengths of pipe are installed.

c. Mid-story guide.

Combustion air openings at the rear or side of the compartment shall comply with the requirements of Chapter 7.

Exception: This section shall not apply to replacement appliances installed in existing compartments and alcoves where the working space clearances are in accordance with the *equipment* or *appliance* manufacturer's installation instructions.

❖ Central furnaces that are installed in compartments or alcoves must have clearances from the enclosure so that the furnace can be removed, maintained or repaired as necessary. Chapter 7 regulates combustion air ducts and openings for free movement of air.

The minimum clearances specified in the code apply even though the manufacturer's instructions might permit a lesser clearance. Clearances provide access, ventilation cooling of the appliance and equipment and protection for surrounding combustibles. The front (fire box) clearance helps protect combustibles against flame roll-out and allows free movement of combustion air. The exception exempts replacement appliances and equipment installed in existing compartments or alcoves if the installation complies with the manufacturer's instructions.

306.2 Appliances in rooms. Rooms containing appliances shall be provided with a door and an unobstructed passageway measuring not less than 36 inches (914 mm) wide and 80 inches (2032 mm) high.

Exception: Within a *dwelling unit*, appliances installed in a compartment, alcove, basement or similar space shall be accessed by an opening or door and an unobstructed passageway measuring not less than 24 inches (610 mm) wide and large enough to allow removal of the largest *appliance* in the space, provided that a level service space of not less than 30 inches (762 mm) deep and the height of the *appliance*, but not less than 30 inches (762 mm), is present at the front or service side of the *appliance* with the door open.

❖ Access opening and passageway dimensions are specified to afford service personnel reasonable access to appliances and to allow for the passage of system components. Quite often appliances, such as furnaces, boilers and water heaters are installed in spaces with little or no forethought about future access for maintenance or replacement.

306.3 Appliances in attics. Attics containing appliances shall be provided with an opening and unobstructed passageway large enough to allow removal of the largest *appliance*. The passageway shall not be less than 30 inches (762 mm) high and 22 inches (559 mm) wide and not more than 20 feet (6096 mm) in length measured along the centerline of the

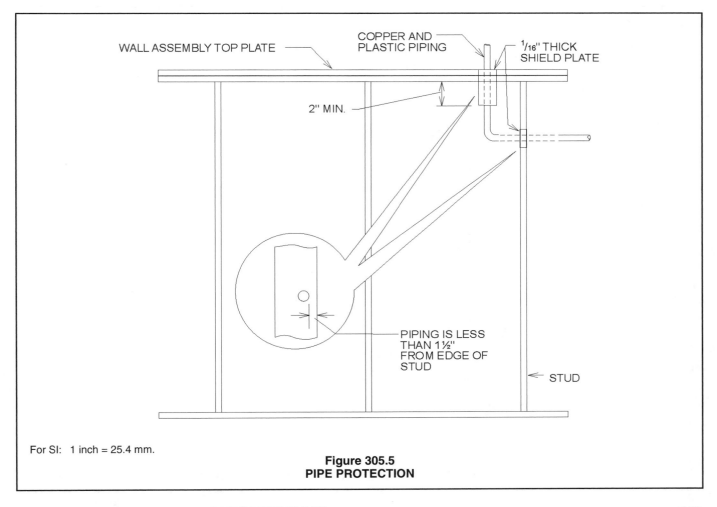

For SI: 1 inch = 25.4 mm.

Figure 305.5
PIPE PROTECTION

passageway from the opening to the *appliance*. The passageway shall have continuous solid flooring not less than 24 inches (610 mm) wide. A level service space not less than 30 inches (762 mm) deep and 30 inches (762 mm) wide shall be present at the front or service side of the *appliance*. The clear access opening dimensions shall be a minimum of 20 inches by 30 inches (508 mm by 762 mm), and large enough to allow removal of the largest *appliance*.

Exceptions:

1. The passageway and level service space are not required where the *appliance* is capable of being serviced and removed through the required opening.

2. Where the passageway is unobstructed and not less than 6 feet (1829 mm) high and 22 inches (559 mm) wide for its entire length, the passageway shall be not greater than 50 feet (15 250 mm) in length.

❖ There is not always sufficient room for mechanical equipment and appliances to be installed in spaces such as basements, alcoves, utility rooms and furnace rooms. In an effort to save floor space or simplify an installation, appliances and mechanical equipment are often installed on roofs, in attics or in similar remote locations. Access to appliances and equipment could be difficult because of roof slope,

stone roof ballasts or the lack of a walking surface, such as might occur in an attic or similar space with exposed ceiling joists. The intent of this section is to require a suitable access opening, passageway and work space that will allow reasonably easy access without endangering the service person (see Commentary Figure 306.3). The longer the attic passageway, the more the service person will be exposed to extreme temperatures and risk of injury.

Exception 1 allows the passageway and level service space to be eliminated if the technician can reach the appliance through the access opening without having to step into the attic. Exception 2 allows the length of the passageway to be extended to 50 feet (15 250 mm) if there is at least 6 feet (1829 mm) of clear headroom for the entire length of the passageway. This is allowed because there is less danger of lengthy exposure to extreme temperatures if the service personnel can walk erect and unimpeded to the equipment rather than crawling.

Note that some appliances might not be listed for attic installation or might otherwise be unsuitable for such conditions.

306.3.1 Electrical requirements. A luminaire controlled by a switch located at the required passageway opening and a

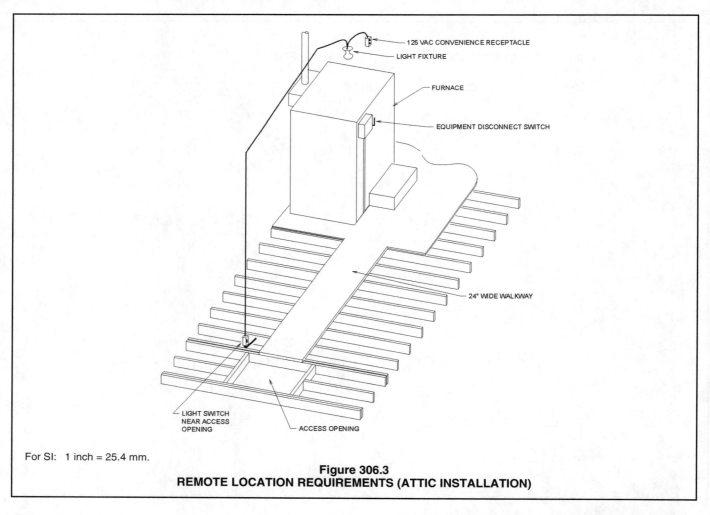

For SI: 1 inch = 25.4 mm.

Figure 306.3
REMOTE LOCATION REQUIREMENTS (ATTIC INSTALLATION)

receptacle outlet shall be provided at or near the *appliance* location in accordance with NFPA 70.

❖ A lighting outlet and receptacle outlet will encourage and facilitate appliance maintenance. The receptacle will accommodate power tools, drop lights and diagnostic instruments. Also, these outlets will negate the need for extension cords, which can be hazardous to service personnel. Exposed-lamp lighting fixtures should be located or have guards installed to avoid impact by personnel.

306.4 Appliances under floors. Underfloor spaces containing appliances shall be provided with an access opening and unobstructed passageway large enough to remove the largest *appliance*. The passageway shall not be less than 30 inches (762 mm) high and 22 inches (559 mm) wide, nor more than 20 feet (6096 mm) in length measured along the centerline of the passageway from the opening to the *appliance*. A level service space not less than 30 inches (762 mm) deep and 30 inches (762 mm) wide shall be present at the front or service side of the *appliance*. If the depth of the passageway or the service space exceeds 12 inches (305 mm) below the adjoining grade, the walls of the passageway shall be lined with concrete or masonry. Such concrete or masonry shall extend a minimum of 4 inches (102 mm) above the adjoining grade and shall have sufficient lateral-bearing capacity to resist collapse. The clear access opening dimensions shall be a minimum of 22 inches by 30 inches (559 mm by 762 mm), and large enough to allow removal of the largest *appliance*.

Exceptions:

1. The passageway is not required where the level service space is present when the access is open and the *appliance* is capable of being serviced and removed through the required opening.

2. Where the passageway is unobstructed and not less than 6 feet high (1929 mm) and 22 inches (559 mm) wide for its entire length, the passageway shall not be limited in length.

❖ The intent of this section, which applies to crawl spaces, is the same as for Section 306.3. The more difficult access to appliances and equipment is, the less likely it is that the appliance or equipment will be inspected and serviced. Attic and crawl space installations suffer from the "out-of-sight, out-of-mind" syndrome (see commentary, Section 306.3).

Exception 1 has the same intent as the first exception of Section 306.3. Exception 2 allows unlimited length of the passageway if there is at least 6 feet (1829 mm) of clear headroom for the entire length of the passageway (see commentary, Section 306.3).

306.4.1 Electrical requirements. A luminaire controlled by a switch located at the required passageway opening and a receptacle outlet shall be provided at or near the *appliance* location in accordance with NFPA 70.

❖ Heating, air-conditioning and refrigeration equipment and appliances located in attics and under-floor spaces are generally not easy to access. The light switch, lighting fixture and receptacle outlet required

by this section provide safety and protection, as well as convenience, for servicers of the equipment and appliances. The receptacle will accommodate power tools, drop lights and diagnostic instruments. Also, these provisions will negate the need for extension cords, which can be hazardous to service personnel. Note that NFPA 70 requires ground fault circuit interrupter protection for receptacles located in a crawl space.

306.5 Equipment and appliances on roofs or elevated structures. Where *equipment* requiring access or appliances are located on an elevated structure or the roof of a building such that personnel will have to climb higher than 16 feet (4877 mm) above grade to access such equipment or appliances, an interior or exterior means of access shall be provided. Such access shall not require climbing over obstructions greater than 30 inches (762 mm) in height or walking on roofs having a slope greater than 4 units vertical in 12 units horizontal (33-percent slope). Such access shall not require the use of portable ladders. Where access involves climbing over parapet walls, the height shall be measured to the top of the parapet wall.

Permanent ladders installed to provide the required access shall comply with the following minimum design criteria:

1. The side railing shall extend above the parapet or roof edge not less than 30 inches (762 mm).

2. Ladders shall have rung spacing not to exceed 14 inches (356 mm) on center. The uppermost rung shall be a maximum of 24 inches (610 mm) below the upper edge of the roof hatch, roof or parapet, as applicable.

3. Ladders shall have a toe spacing not less than 6 inches (152 mm) deep.

4. There shall be a minimum of 18 inches (457 mm) between rails.

5. Rungs shall have a minimum 0.75-inch (19 mm) diameter and be capable of withstanding a 300-pound (136.1kg) load.

6. Ladders over 30 feet (9144 mm) in height shall be provided with offset sections and landings capable of withstanding 100 pounds per square foot (488.2 kg/m^2). Landing dimensions shall be not less than 18 inches (457 mm) and not less than the width of the ladder served. A guard rail shall be provided on all open sides of the landing.

7. Climbing clearance. The distance from the centerline of the rungs to the nearest permanent object on the climbing side of the ladder shall be a minimum of 30 inches (762 mm) measured perpendicular to the rungs. This distance shall be maintained from the point of ladder access to the bottom of the roof hatch. A minimum clear width of 15-inches (381 mm) shall be provided on both sides of the ladder measured from the midpoint of and parallel with the rungs except where cages or wells are installed.

8. Landing required. The ladder shall be provided with a clear and unobstructed bottom landing area having a

minimum dimension of 30 inches (762 mm) by 30 inches (762 mm) centered in front of the ladder.

9. Ladders shall be protected against corrosion by *approved* means.

10. Access to ladders shall be provided at all times.

Catwalks installed to provide the required access shall be not less than 24 inches (610 mm) wide and shall have railings as required for service platforms.

Exception: This section shall not apply to Group R-3 occupancies.

❖ The requirements of this section concern the safety of maintenance and inspection personnel where appliances and equipment are located on roofs or on top of structures. The first sentence establishes 16 feet above grade as the maximum height that personnel will have to climb on a portable ladder. If the climbing height exceeds 16 feet, then a permanent means of access, either from the interior of the building or exterior of the building, must be provided. Note that the last sentence of the first paragraph requires that parapet wall height be included in the climbing height. It is dangerous for maintenance personnel, often carrying tools, parts and diagnostic equipment, to have to negotiate lengthy extension ladders to gain access to the work area [see Commentary Figures 306.5(1) and 306.5(2)].

What the 16-foot maximum climbing height means is that buildings (with appliances or equipment on the roof) that are over one story in height will need permanent ladders, stairways or a combination thereof, installed.

Access to the equipment or appliances cannot require climbing over obstructions greater than 30 inches high. For example, a 36-inch duct running across the roof would require permanent steps or ladders to climb over the duct. Fire walls that extend up from the roof surface more than 30 inches would also require permanent steps or ladders.

Access to the equipment or appliances must not require walking on a roof slope greater than 4 units vertical in 12 units horizontal. Roofs with steep slopes are very difficult to traverse and maintain secure footing. Also, the potential for dropped items to roll off the roof and injure a passerby is great.

Item 1 provides for safe handholds to be able to get on and off the ladder. Item 2 assures that the rung spacing is not so great that climbing would be difficult and that the top rung is not too far below the walking surface that it could not be easily reached. Item 3 provides for adequate space behind the ladder so that a foot can be solidly placed on the rungs. Item 4 provides for an adequate width for even the largest of personnel. Item 5 makes sure that the rungs have sufficient strength.

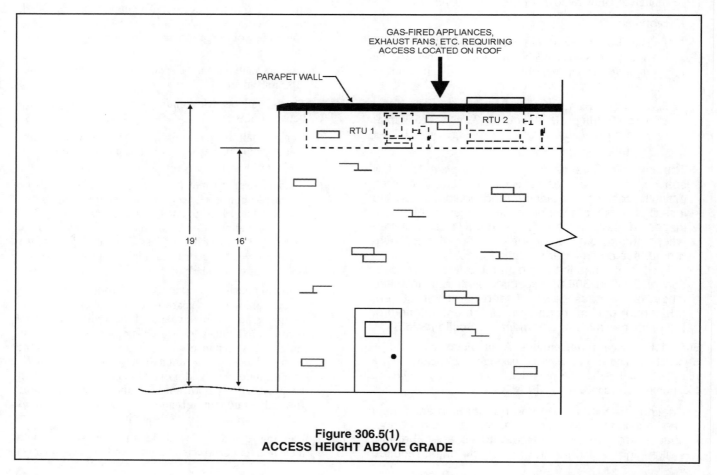

Figure 306.5(1)
ACCESS HEIGHT ABOVE GRADE

Item 6 provides a safe resting point for climbs that are over 30 feet. Item 7 provides for adequate clearance between the climbing side of the ladder and any obstruction. Item 7 also provides for adequate width beyond the sides of the ladder. Item 8 requires that the base of the ladder have a clear area for enabling proper access to the ladder.

Item 9 requires that the ladder be protected from corrosion by an approved means. Typically, steel ladders are hot dipped galvanized. Item 10 requires that access to ladders be provided at all times so that alternative, perhaps unsafe, access means are not attempted because the ladder is blocked by machinery or permanent construction.

The catwalk requirements for a minimum width of 24 inches (610 mm) and installation of guards meeting the requirements of Section 304.11 result in a minimum level of safety for anyone using a catwalk to access mechanical equipment.

The exception exempts one- and two-family dwellings from the access requirements primarily because permanent stairs and ladders are aesthetically unappealing and cost prohibitive for such dwellings.

306.5.1 Sloped roofs. Where appliances, *equipment*, fans or other components that require service are installed on a roof having a slope of three units vertical in 12 units horizontal (25-percent slope) or greater and having an edge more than 30 inches (762 mm) above grade at such edge, a level platform shall be provided on each side of the *appliance* or *equipment* to which access is required for service, repair or maintenance. The platform shall be not less than 30 inches (762 mm) in any dimension and shall be provided with guards. The guards shall extend not less than 42 inches (1067 mm) above the platform, shall be constructed so as to prevent the passage of a 21-inch diameter (533 mm) sphere and shall comply with the loading requirements for guards specified in the *International Building Code*. Access shall not require walking on roofs having a slope greater than four units vertical in 12 units horizontal (33-percent slope). Where access involves obstructions greater than 30 inches (762 mm) in height, such obstructions shall be provided with ladders installed in accordance with Section 306.5 or stairs installed in accordance with the requirements specified in the *International Building Code* in the path of travel to and from appliances, fans or *equipment* requiring service.

❖ A work space platform with guards will provide protection for service personnel and will facilitate inspection, servicing and repairing appliances, equipment or other components. Working on a sloped surface is difficult, dangerous and results in tools and materials sliding off roofs, possibly endangering persons below. Because of the additional expense and unattractive appearance of roof platforms, this section may have the effect of discouraging the installation of appli-

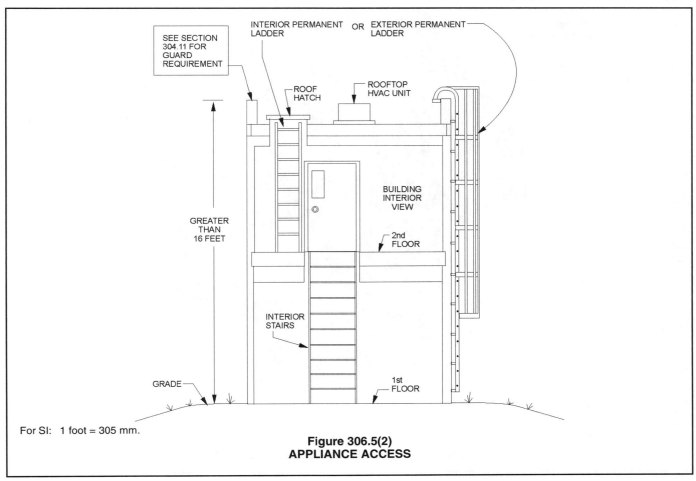

For SI: 1 foot = 305 mm.

Figure 306.5(2)
APPLIANCE ACCESS

ances on sloped roofs. This section, like Section 304.11, is intended to include equipment such as kitchen exhaust and makeup air fans that require routine servicing (see Commentary Figure 306.5.1).

The guard serves as a warning and as a protective barrier, and must be 42 inches (1067 mm) high and constructed to prevent the passage of a 21-inch (533 mm) sphere (see Commentary Figure 304.11). The requirement for the guard to extend 30 inches (762 mm) beyond each end of the equipment, appliance or roof hatch opening provides added protection for the installer or service technician by enlarging the protected area beyond the immediate area of the equipment, appliance or roof hatch. The guardrail must be constructed to resist the imposed loading conditions.

Many installations on roofs are located 20, 50 or even as much as 100 feet (30 480 mm) from the roof's edge. It can be very dangerous to traverse those distances on a sloped roof, especially when carrying tools and repair parts. For this reason, the code prohibits installations requiring walking on a roof with a slope greater than 4 units vertical in 12 units horizontal (33 percent). Similarly, it can be difficult and dangerous to climb over obstructions greater than 30 inches (762 mm) when carrying tools and

repair parts. Such obstacles require either permanent stairs or ladders to assist in crossing over the obstacle.

306.5.2 Electrical requirements. A receptacle outlet shall be provided at or near the *equipment* location in accordance with NFPA 70.

❖ The receptacle outlet required by this section provides convenience for servicers of the equipment and appliances on the roof. The receptacle will accommodate power tools, drop lights and diagnostic instruments. Also, these provisions will negate the need for extension cords, which can be hazardous to service personnel.

SECTION 307
CONDENSATE DISPOSAL

307.1 Fuel-burning appliances. Liquid *combustion* by-products of condensing appliances shall be collected and discharged to an *approved* plumbing fixture or disposal area in accordance with the manufacturer's installation instructions. Condensate piping shall be of *approved* corrosion-resistant material and shall not be smaller than the drain connection on the appliance. Such piping shall maintain a minimum hori-

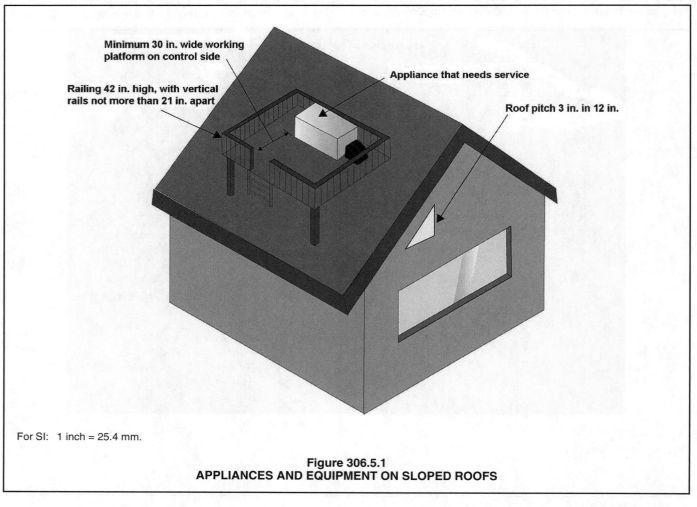

For SI: 1 inch = 25.4 mm.

Figure 306.5.1
APPLIANCES AND EQUIPMENT ON SLOPED ROOFS

zontal slope in the direction of discharge of not less than one-eighth unit vertical in 12 units horizontal (1-percent slope).

❖ This section contains detailed requirements for the disposal of condensate from appliances and equipment such as evaporators, cooling coils and condensing fuel-fired appliances. Condensation must be collected from appliances and equipment in compliance with the manufacturer's installation instructions and this section. High-efficiency condensing-type appliances and some mid-efficiency appliances produce water as a combustion byproduct. This condensate is collected at various points in the appliance heat exchangers and venting system and must be disposed of. Because of impurities in the fuel and combustion air, the condensate can be acidic and thus corrosive to many materials. For example, condensate can contain hydrochloric and sulfuric acids. The slope requirement promotes drainage and eliminates low points (dips) in the piping that would trap water and air that would impede flow.

307.2 Evaporators and cooling coils. Condensate drain systems shall be provided for *equipment* and appliances containing evaporators or cooling coils. Condensate drain systems shall be designed, constructed and installed in accordance with Sections 307.2.1 through 307.2.4.

❖ Appliances and equipment containing evaporators or cooling coils, including refrigeration, dehumidification and comfort cooling equipment, can produce condensate from the water vapor in the atmosphere. A drainage system is necessary to dispose of the condensate and prevent damage to the structure.

307.2.1 Condensate disposal. Condensate from all cooling coils and evaporators shall be conveyed from the drain pan outlet to an *approved* place of disposal. Such piping shall maintain a minimum horizontal slope in the direction of discharge of not less than one-eighth unit vertical in 12 units horizontal (1-percent slope). Condensate shall not discharge into a street, alley or other areas so as to cause a nuisance.

❖ Condensate disposal is a local issue dependent on specific local conditions, such as soil conditions, contour of the area, sewer loading and water treatment plant capacity. Because of this, the approval of the disposal place is the responsibility of the local jurisdiction. Some of the locations that could be accepted are storm or sanitary sewers, rooftops, French drains, drainage ditches, collection ponds or simply into the yard, as is typical for most homes.

Some water treatment facilities operate near maximum capacity and do not allow condensate disposal into the sanitary sewer because condensate does not require treatment. If connected to the sanitary sewer, the connection must be indirect in compliance with the IPC to prevent sewer gases from entering the equipment or system. Disposal to a rooftop will allow the condensate to evaporate on the roof surface or be directed to the roof drain system or gutters. French drains, or seepage pits, are effective depend-

ing on the permeability of the soil, the rate of discharge and the size of the pit.

The piping conveying the condensate to its disposal point must be properly sloped to ensure drainage just the same as any drainage system. Sagging pipe could trap water and air that would impede the flow of the condensate.

307.2.2 Drain pipe materials and sizes. Components of the condensate disposal system shall be cast iron, galvanized steel, copper, cross-linked polyethylene, polybutylene, polyethylene, ABS, CPVC or PVC pipe or tubing. All components shall be selected for the pressure and temperature rating of the installation. Joints and connections shall be made in accordance with the applicable provisions of Chapter 7 of the *International Plumbing Code* relative to the material type. Condensate waste and drain line size shall be not less than $^3/_4$-inch (19 mm) internal diameter and shall not decrease in size from the drain pan connection to the place of condensate disposal. Where the drain pipes from more than one unit are manifolded together for condensate drainage, the pipe or tubing shall be sized in accordance with Table 307.2.2.

❖ Condensate drains must be constructed of a material listed in this section. Note that the materials listed are corrosion resistant. The $^3/_4$-inch (19 mm) minimum pipe size, as well as the pipe sizes required by Table 307.2.2, will ensure that the condensate will flow properly. When drains are merged, the piping must be sized for the aggregate flow by using Table 307.2.2.

TABLE 307.2.2
CONDENSATE DRAIN SIZING

EQUIPMENT CAPACITY	MINIMUM CONDENSATE PIPE DIAMETER
Up to 20 tons of refrigeration	$^3/_4$ inch
Over 20 tons to 40 tons of refrigeration	1 inch
Over 40 tons to 90 tons of refrigeration	$1^1/_4$ inch
Over 90 tons to 125 tons of refrigeration	$1^1/_2$ inch
Over 125 tons to 250 tons of refrigeration	2 inch

1 inch = 25.4 mm, 1 ton = 3.517 kW.

❖ The condensate pipe size for an individual unit must not be less than that specified by the equipment manufacturer. Where the condensate lines from multiple units are manifolded together, Table 307.2.2 must be used for sizing the combined condensate drain pipe.

307.2.3 Auxiliary and secondary drain systems. In addition to the requirements of Section 307.2.1, where damage to any building components could occur as a result of overflow from the *equipment* primary condensate removal system, one of the following auxiliary protection methods shall be provided for each cooling coil or fuel-fired *appliance* that produces condensate:

1. An auxiliary drain pan with a separate drain shall be provided under the coils on which condensation will occur. The auxiliary pan drain shall discharge to a conspicuous point of disposal to alert occupants in the event of a stoppage of the primary drain. The pan shall

have a minimum depth of $1^1/_2$ inches (38 mm), shall not be less than 3 inches (76 mm) larger than the unit or the coil dimensions in width and length and shall be constructed of corrosion-resistant material. Galvanized sheet steel pans shall have a minimum thickness of not less than 0.0236 inch (0.6010 mm) (No. 24 gage). Non-metallic pans shall have a minimum thickness of not less than 0.0625 inch (1.6 mm).

2. A separate overflow drain line shall be connected to the drain pan provided with the *equipment*. Such overflow drain shall discharge to a conspicuous point of disposal to alert occupants in the event of a stoppage of the primary drain. The overflow drain line shall connect to the drain pan at a higher level than the primary drain connection.

3. An auxiliary drain pan without a separate drain line shall be provided under the coils on which condensate will occur. Such pan shall be equipped with a water-level detection device conforming to UL 508 that will shut off the *equipment* served prior to overflow of the pan. The auxiliary drain pan shall be constructed in accordance with Item 1 of this section.

4. A water-level detection device conforming to UL 508 shall be provided that will shut off the *equipment* served in the event that the primary drain is blocked. The device shall be installed in the primary drain line, the overflow drain line, or in the equipment-supplied drain pan, located at a point higher than the primary drain line connection and below the overflow rim of such pan.

Exception: Fuel-fired appliances that automatically shut down operation in the event of a stoppage in the condensate drainage system.

❖ An auxiliary (redundant) drain pan or a secondary drain is required for equipment locations where condensate overflow would cause damage to a building or its contents. The purpose of the auxiliary drain pan and secondary drain is to catch condensate spilling from the primary condensate removal system in the equipment. This "back-up" protects the building from structural and finish damage.

Condensate drains are notorious for clogging because of debris (lint, dust) from air-handling systems and the natural affinity to produce slime growths in drain pans and pipes. It is relatively common for condensate overflows to cause damage to buildings. This section lists four options for preventing damage where the equipment is located in spaces, such as attics, above suspended ceilings and furred spaces and locations on upper stories. One of the four methods must be used.

Method 1 uses an auxiliary drain pan below the coils on which condensation will occur with an independent drain line that discharges to a location that is easily observable to notify the building occupants that a problem with the primary pan exists. The code prescribes the depth of the pan and specific material thicknesses to ensure that the pan will be corrosion

resistant and will have sufficient holding capacity and capture ability. For the 2009 edition of the code, the thickness of metallic material was changed to reflect the low end tolerance for No. 24 gage galvanized sheet metal.

Method 2 uses an independent overflow drain line connected to the primary drain pan at a point higher than the primary drain line. Most evaporator coil pans are factory provided with an overflow drain tap that can be used for this purpose. As in Method 1, the point of discharge must be easily observable.

Method 3 uses a water-level detection device, usually a float switch or electronic sensor that must conform to the requirements of UL 508, located in the auxiliary drain pan. These detection devices will shut down the equipment before the pan overflows. There is no requirement for a separate drain line in this method. Commentary Figure 307.2.3(1) shows a typical float switch for a drain pan.

Method 4 also uses a water-level detection device located in the drain line from the primary drain pan or the overflow line from the primary drain pan rather than the secondary drain pan in Method 3. See Commentary Figures 307.2.3(2) and 307.2.3(3).

Both Methods 3 and 4 will notify the building occupants that a blockage has occurred because the cooling system will cease to function.

The exception recognizes that some fuel-fired appliances that produce condensate have a built-in method of shutting down when a blockage occurs.

Figure 307.2.3(1)
PRIMARY AND AUXILIARY
DRAIN PAN FLOAT SWITCH
(Courtesy of SMD Research, Inc.)

307.2.3.1 Water-level monitoring devices. On downflow units and all other coils that do not have a secondary drain or provisions to install a secondary or auxiliary drain pan, a water-level monitoring device shall be installed inside the primary drain pan. This device shall shut off the *equipment* served in the event that the primary drain becomes restricted. Devices installed in the drain line shall not be permitted.

❖ The intent of this section is to provide adequate over-flow protection on all coils that do not have a second-ary drain and have no provisions for a secondary or auxiliary drain pan. A water-level monitoring device, like the one shown in Commentary Figure 307.2.3(1), is required to be installed in the primary drain pan. An in-line detection device is not allowed by this section because a blockage inside the pan at the drain hole would not be detected by the in-line device because the water would overflow from the pan and never reach the in-line detector. When the water overflows

from the pan, it typically runs into the duct, causing mold and mildew problems. It will eventually leak out through the joints or seams and cause damage to the building structure or its contents.

307.2.3.2 Appliance, equipment and insulation in pans. Where appliances, *equipment* or insulation are subject to water damage when auxiliary drain pans fill, that portion of the *appliance*, *equipment* and insulation shall be installed above the rim of the pan. Supports located inside of the pan to support the *appliance* or *equipment* shall be water resistant and *approved*.

❖ Where appliances, such as upflow furnaces and air handlers are installed such that supports are resting in the bottom of the auxiliary drain pan, all portions that are subject to water damage must be installed above the flood level rim of the pan (see Commentary Figure 307.2.3.2). Electrical components, metal items

Figure 307.2.3(2)
CONDENSATE OVERFLOW SWITCH
(Courtesy of SMD Research, Inc.)

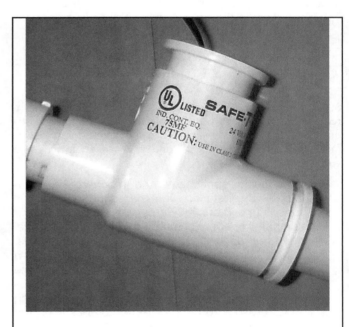

Figure 307.2.3(3)
IN-LINE CONDENSATE OVERFLOW SWITCH
(Courtesy of SMD Research, Inc.)

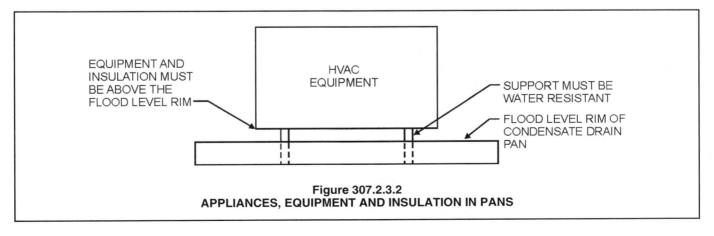

Figure 307.2.3.2
APPLIANCES, EQUIPMENT AND INSULATION IN PANS

subject to rust and insulation within the appliance are examples of items that are subject to damage when submerged in water.

307.2.4 Traps. Condensate drains shall be trapped as required by the *equipment* or *appliance* manufacturer.

❖ The appliance or equipment manufacturer determines the need for a trap and often specifies the depth and configuration of the trap. The traps addressed in this section are unrelated to plumbing traps and serve a different purpose. Condensate drain traps do not directly connect in any way to the plumbing drain or the waste and vent system of a building. Condensate drain traps are installed to prevent air from being pushed or pulled through the drain piping. Airflow can impede condensate flow, causing overflow or abnormal water depth in drain pans. Some drain pans, such as those under pull-through cooling coils, may not drain at all without a trap to block airflow in the drain piping. Airflow in a condensate drain also wastes energy.

SECTION 308
CLEARANCE REDUCTION

308.1 Scope. This section shall govern the reduction in required *clearances* to combustible materials and combustible assemblies for *chimneys*, vents, kitchen exhaust equipment, mechanical appliances, and mechanical devices and *equipment*.

❖ Heat-producing appliances and mechanical equipment must be installed with the required minimum clearances to combustible materials indicated by their listing label. It is not uncommon to encounter practical or structural difficulties in maintaining clearances. Therefore, clearance-reduction methods have been developed to allow, in some cases, reduction of the minimum prescribed clearance distance while achieving equivalent protection. An important understanding is that all prescribed clearances to combustibles are airspace clearances measured from the heat source to the face of the nearest combustible surface, even if that combustible surface is not visible; for example, a wood stud located behind a metal panel [see Commentary Figure 308.1(1)].

When using the assemblies described in Table 308.6, the clearance is measured from the heat source to the face of the combustible surface. In the case of listed equipment, the required clearances are intended to be clear airspace and, therefore, the space is not to be filled with insulation or any other material, including pipes, ducts and conduits. This is especially important where clearances are required from appliances and equipment that rely on the airspace for convection cooling and to maintain its proper operation.

The requirements in Sections 308.2 through 308.11 are based on the principles of heat transfer. Mechanical equipment or appliances producing heat will, by themselves, become hot, and many appliances have hot exterior surfaces by design. The heat energy is then radiated to objects surrounding the appliances or equipment. When mechanical equipment and appliances are tested, the minimum clearances are established so that radiant and, to a lesser extent, convective heat transfer do not represent an ignition hazard to adjacent surfaces and objects. This distance is called the "required clearance" to combustible materials. Appliance and equipment labels are required to specify minimum clearances in all directions.

This section permits the use of materials and systems to act as radiation shields, allowing a reduction

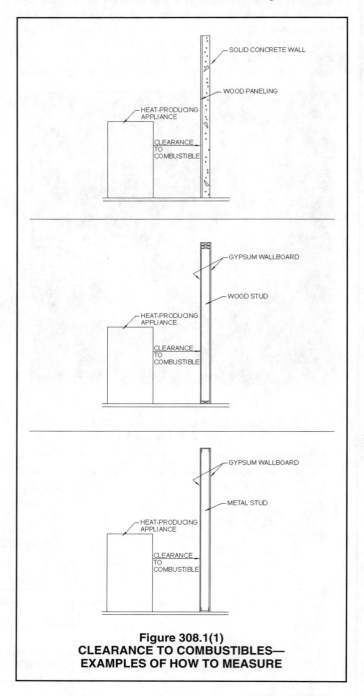

Figure 308.1(1)
CLEARANCE TO COMBUSTIBLES—
EXAMPLES OF HOW TO MEASURE

in the amount of heat energy transferred to surrounding objects and reducing the required clearances between mechanical appliances and equipment and combustible material. Commentary Figures 308.1(2) and 308.1(3) illustrate some methods of clearance reduction.

The definition of "Noncombustible" in the code differs from the definition of "Noncombustible" in the

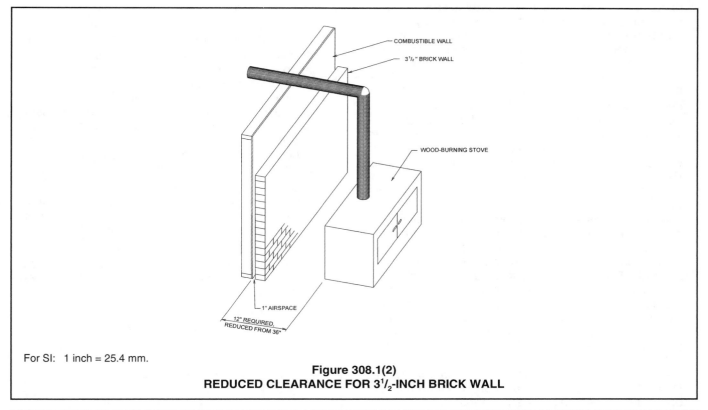

For SI: 1 inch = 25.4 mm.

Figure 308.1(2)
REDUCED CLEARANCE FOR 3¹/₂-INCH BRICK WALL

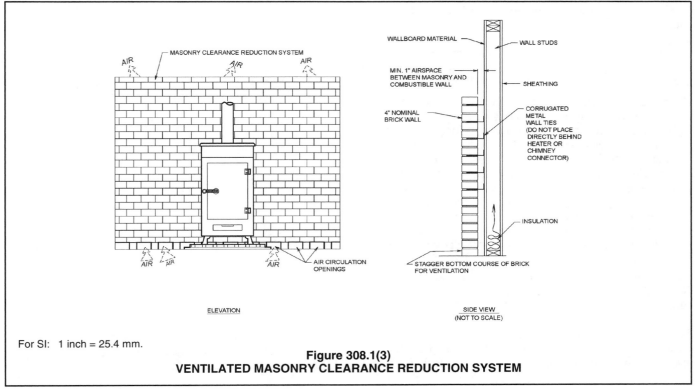

For SI: 1 inch = 25.4 mm.

Figure 308.1(3)
VENTILATED MASONRY CLEARANCE REDUCTION SYSTEM

IBC. The difference is necessary because this code is concerned with the exposure to a continuous heat source, while the IBC emphasizes the behavior of the material under actual fire conditions. An example of this difference is how gypsum wallboard is categorized. In this code, gypsum wallboard is considered to be a combustible material because it is paper faced and breaks down when heated. The IBC classifies it as a noncombustible material because it is used as a thermal barrier. The paper face of gypsum board has a flame spread index that is measurable in the ASTM E 84 test. This alone identifies the need to classify gypsum wallboard as a combustible material for the purpose of requiring a separation from heat-producing equipment and appliances. Thus, with respect to the requirements of this section, applying gypsum wallboard to a wood frame wall does not reduce the required clearance to combustibles and does not change the fact that the adjacent surface is combustible.

Although plaster and gypsum, by themselves, are classified as noncombustible materials, under continued exposure to heat, these materials will gradually decompose as water molecules are driven out of the material. The resulting cracks and spalling of these materials after prolonged exposure to heat makes them unsuitable for protecting underlying combustible material. Therefore, plaster or gypsum products applied to wood lath, plasterboard, and drywall do not protect the combustible materials behind these products.

Section 308.2 and Sections 308.7 through 308.11 specifically prohibit the reduction of clearances in applications described in each of those sections.

308.2 Listed appliances and equipment. The reduction of the required *clearances* to combustibles for *listed* and *labeled* appliances and *equipment* shall be in accordance with the requirements of this section except that such clearances shall not be reduced where reduction is specifically prohibited by the terms of the *appliance* or *equipment* listing.

❖ The clearances specified in the typical appliance or equipment listing are clearances to unprotected combustibles, and unless clearance reduction is prohibited by the listing or label, reduction is allowed in accordance with Section 308 (see Commentary Figure 308.2). The listing of an appliance or piece of equipment could prohibit the reduction of the manufacturer's specified clearances, and in such cases the provisions of Section 308 cannot be applied. Note that it is not the intent to allow clearance reduction for appliances installed in closets and alcoves in accordance with Section 303.5.

308.3 Protective assembly construction and installation. Reduced *clearance* protective assemblies, including structural and support elements, shall be constructed of noncombustible materials. Spacers utilized to maintain an airspace between the protective assembly and the protected material or assembly shall be noncombustible. Where a space between the protective assembly and protected combustible material

or assembly is specified, the same space shall be provided around the edges of the protective assembly and the spacers shall be placed so as to allow air circulation by convection in such space. Protective assemblies shall not be placed less than 1 inch (25 mm) from the mechanical appliances, devices or *equipment*, regardless of the allowable reduced *clearance*.

❖ Spacers are used to support an assembly or material and maintain an airspace between the protection and that which is being protected. The airspace must be created so that unimpeded convection air currents can flow through the airspace. Most of the methods in Table 308.6 depend on convective cooling as an essential part of the system. All protection assemblies must also maintain a minimum clearance from the heat source to reduce the conduction of heat and allow convective cooling on the exposed side of the protective element (see Commentary Figure 308.3).

308.4 Allowable reduction. The reduction of required *clearances* to combustible assemblies or combustible materials shall be based on the utilization of a reduced *clearance* protective assembly in accordance with Section 308.5 or 308.6.

❖ The permissible methods of clearance reduction fall into two categories: factory-built listed and labeled assemblies and field-constructed assemblies specified in Table 308.6.

308.5 Labeled assemblies. The allowable clearance reduction shall be based on an approved reduced clearance protec-

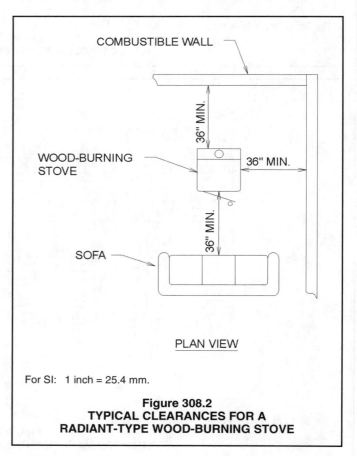

For SI: 1 inch = 25.4 mm.

Figure 308.2
TYPICAL CLEARANCES FOR A
RADIANT-TYPE WOOD-BURNING STOVE

tive assembly that is listed and labeled in accordance with UL 1618.

❖ For a device or system to be labeled as a method for reducing clearances to combustibles, it must be tested. The code requires that the UL 1618 standard be used by the testing agency for determining the performance of materials used for reducing clearances.

The test for floor and wall protection essentially consists of erecting a test enclosure that includes both the heat-producing appliances and the device or system intended to provide the necessary protection. The various tests are performed and temperatures are monitored on both sides of the protection using thermocouples. During the test, if the temperature rise on the surfaces of the protection exposed to the heat source has not exceeded 117°F (47°C), and the temperature rise on the unexposed surfaces of the protection has not exceeded 90°F (32°C), the product passes the test. The product may then be used to reduce the required clearances to combustibles for similar heat-producing appliances.

After a product successfully passes the test, it is eligible for listing and labeling. In the field, a device or assembly that is listed and labeled for use as a method of reducing clearances to combustible materials can be approved only if it is installed in strict compliance with its listing, label and the manufacturer's installation instructions, as required in Section 304.1.

308.6 Reduction table. The allowable *clearance* reduction shall be based on one of the methods specified in Table 308.6. Where required *clearances* are not listed in Table 308.6, the reduced *clearances* shall be determined by linear interpolation between the distances listed in the table. Reduced *clearances* shall not be derived by extrapolation below the range of the table.

❖ Another option for reducing the required clearances to combustible materials is to use one of the on-site field-constructed methods specified in Table 308.6.

TABLE 308.6. See page 3-38.

❖ The bold numbers in the column headings of Table 308.6 represent the clearances to combustible material and assemblies (36, 18, 9, and 6 inches) as required by the appliance or equipment installation instructions where a reduced clearance protection assembly is not provided. The numbers to the right of each method indicate the permissible reduced clearance as measured from the heat-producing appliances to the face of the combustible surface that is behind the protection material or assembly.

The rationale behind the methods of protection listed in Table 308.6 is based on the ability of the protection material or assembly to reduce radiant heat transmission from the appliance and equipment to the combustible material so that the temperature rise of the combustible material will remain below the maximum allowed.

Although the materials referred to in Table 308.6 are common construction materials, confusion over

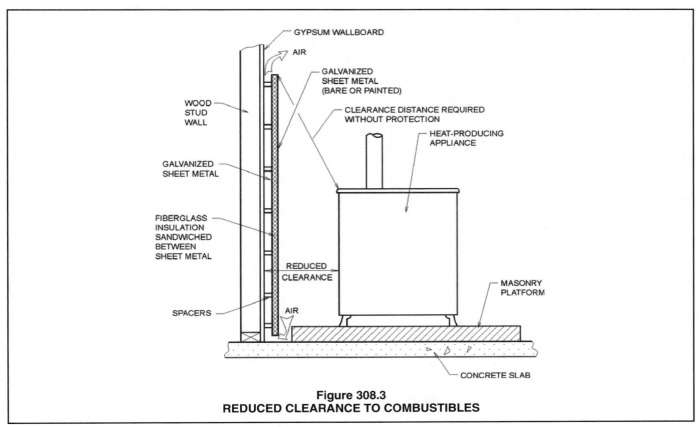

Figure 308.3
REDUCED CLEARANCE TO COMBUSTIBLES

the term "inorganic insulating board" frequently arises. This term refers to noncombustible mineral boards and other noncombustible insulating boards that are not made of carbon-based compounds. Carbon-based compounds are those found in cellulose (wood), plastics and other materials manufactured from raw materials that once existed as living organisms. Note that cement board materials must have a specified maximum C (conductance) value in addition to being noncombustible.

Note a specifies a maximum thermal conductivity value (C) of 1.0 (Btu/h · in)/(ft^2 · h · °F) for insulating board used as a component of clearance reduction methods [conductivity value is the reciprocal of the R-value; C = 1/R]. This translates into a minimum required insulation R-value of 1.0 (ft^2 · h · °F)/Btu/h per inch of insulating material. The methods in Table 308.6 control heat transmission by reflecting heat radiation, retarding thermal conductance and providing convective cooling. Where sheet metal materials or metal plates are specified, the effectiveness of the protection can be enhanced by the reflective surface of the metal. Painting or otherwise covering the surface would reduce the metal's ability to reflect radiant heat and, depending on the color, could increase heat absorption. The airspace between the protected surface and the clearance-reduction method allows convection air currents to cool the protection assembly by carrying away heat that has been conducted through the assembly. Where a clearance-reduction assembly must be spaced 1 inch (25 mm) off the wall, the top, bottom and sides of the assembly must remain open to permit unrestricted airflow (convection currents). If these openings were not there, the air cooling would not take place and the protection assembly would not be as effective in limiting the temperature rise on the protected surfaces. Ideally, the protection assembly should be open on all sides to provide maximum ventilation. Commentary Figures 308.1(3) and 308.6(1) through 308.6(4) show assemblies incorporating an airspace.

Spacers must be noncombustible. Spacers should not be placed directly behind the heat source because the location would increase the amount of heat conduction through the spacer, thus creating a "hot spot." Commentary Figure 308.6(4) specifically shows a noncombustible spacer arrangement.

The performance of a protective assembly when applied to a horizontal surface, such as a ceiling, will differ substantially from the same assembly placed in a vertical plane. Obviously, temperatures at a ceiling surface will be higher because of natural convection

TABLE 308.6
CLEARANCE REDUCTION METHODS[b]

TYPE OF PROTECTIVE ASSEMBLY[a]	REDUCED CLEARANCE WITH PROTECTION (inches)[a]							
	Horizontal combustible assemblies located above the heat source				Horizontal combustible assemblies located beneath the heat source and all vertical combustible assemblies			
	Required clearance to combustibles without protection (inches)[a]				Required clearance to combustibles without protection (inches)			
	36	18	9	6	36	18	9	6
Galvanized sheet steel, having a minimum thickness of 0.0236 inch (0.6010 mm) (No. 24 gage), mounted on 1-inch glass fiber or mineral wool batt reinforced with wire on the back, 1 inch off the combustible assembly	18	9	5	3	12	6	3	3
Galvanized sheet steel, having a minimum thickness of 0.0236 inch (0.6010 mm) (No. 24 gage), spaced 1 inch off the combustible assembly	18	9	5	3	12	6	3	2
Two layers of galvanized sheet steel, having a minimum thickness of 0.0236 inch (0.6010 mm) (No. 24 gage), having a 1-inch airspace between layers, spaced 1 inch off the combustible assembly	18	9	5	3	12	6	3	3
Two layers of galvanized sheet steel, having a minimum thickness of 0.0236 inch (0.6010 mm) (No. 24 gage), having 1 inch of fiberglass insulation between layers, spaced 1 inch off the combustible assembly	18	9	5	3	12	6	3	3
0.5-inch inorganic insulating board, over 1 inch of fiberglass or mineral wool batt, against the combustible assembly	24	12	6	4	18	9	5	3
3$^1/_2$-inch brick wall, spaced 1 inch off the combustible wall	—	—	—	—	12	6	6	6
3$^1/_2$-inch brick wall, against the combustible wall	—	—	—	—	24	12	6	5

For SI: 1 inch = 25.4 mm, °C = [(°F)-32]/1.8, 1 pound per cubic foot = 16.02 kg/m^3, 1.0 Btu · in/ft^2 · h · °F = 0.144 W/m^2 · K.

a. Mineral wool and glass fiber batts (blanket or board) shall have a minimum density of 8 pounds per cubic foot and a minimum melting point of 1,500°F. Insulation material utilized as part of a clearance reduction system shall have a thermal conductivity of 1.0 Btu · in/(ft^2 · h · °F) or less. Insulation board shall be formed of noncombustible material.

b. For limitations on clearance reduction for solid fuel-burning appliances, masonry chimneys, connector pass-throughs, masonry fire places and kitchen ducts, see Sections 308.7 through 308.11.

and because the air circulation between the method of protection and the protected ceiling surface will be substantially reduced or nonexistent. It is for these reasons that Table 308.6 is divided into two application groups.

The manufacturer's instructions or label for many appliances will state an absolute minimum clearance, regardless of any clearance reduction method used. Those clearance requirements take precedence over Table 308.6. For example, a typical wood-burning room heater will require in all cases an airspace clearance of at least 12 inches (305 mm), with no further reduction allowed (see commentary, Section 308.7).

The methods in Table 308.6 are intended to be permanent installations properly supported to prevent displacement or deformation. Movement could adversely affect the performance of the protection method, thus posing a potential fire hazard.

The assemblies in Table 308.6 are the product of experience and testing. To achieve predictable and dependable performance, the components of the various assemblies cannot be mixed, matched, combined or otherwise rearranged to comprise new assemblies, and materials cannot be substituted for those prescribed in the table. Any alterations or substitutions could have an effect on the assembly and its performance would have to be tested and approved.

Table 308.6 does not specify the method of measuring the reduced clearance; however, the intent is to measure the reduced clearance from the heat source to the combustible material, disregarding any intervening protection assembly.

The thicknesses of galvanized sheet metal were changed to reflect the low end of the tolerance for No. 24 gage. This allows the field inspector to verify and accept all thicknesses of material that could be supplied for No. 24 gage. Note b was added to clarify that in some applications the clearance reduction methods are not applicable.

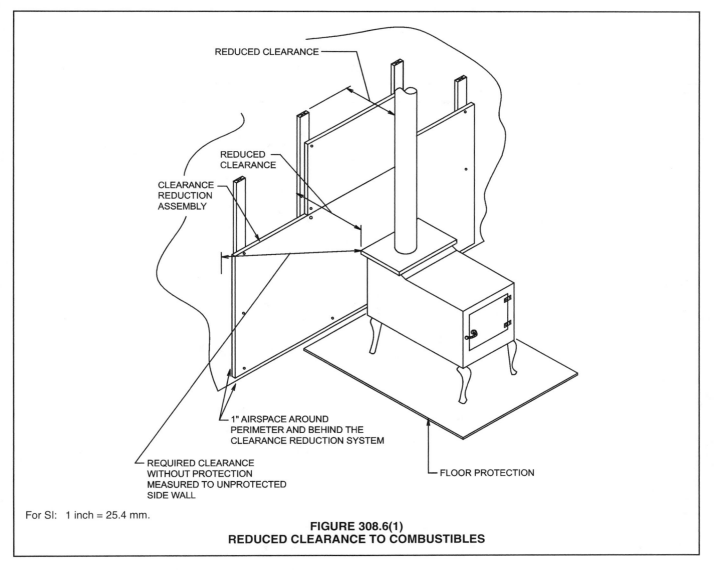

REDUCED CLEARANCE

REDUCED CLEARANCE

CLEARANCE REDUCTION ASSEMBLY

1" AIRSPACE AROUND PERIMETER AND BEHIND THE CLEARANCE REDUCTION SYSTEM

REQUIRED CLEARANCE WITHOUT PROTECTION MEASURED TO UNPROTECTED SIDE WALL

FLOOR PROTECTION

For SI: 1 inch = 25.4 mm.

FIGURE 308.6(1)
REDUCED CLEARANCE TO COMBUSTIBLES

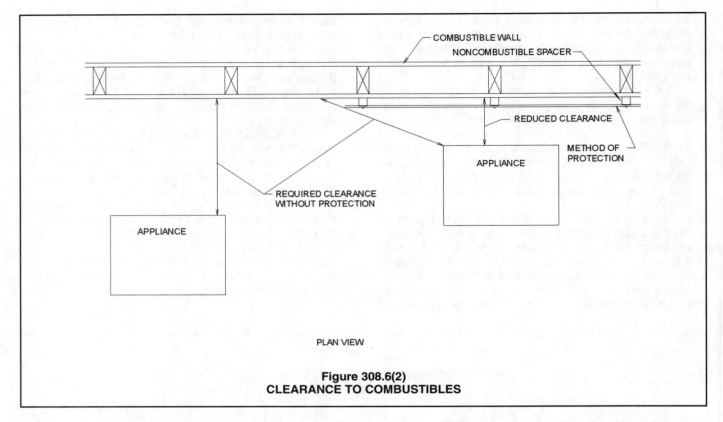

PLAN VIEW

**Figure 308.6(2)
CLEARANCE TO COMBUSTIBLES**

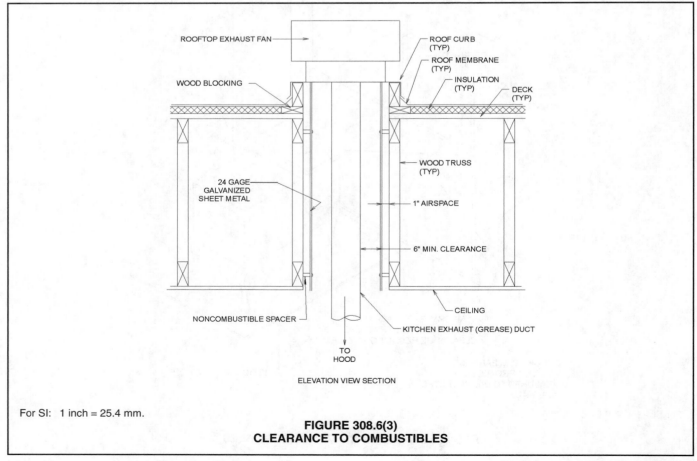

ELEVATION VIEW SECTION

For SI: 1 inch = 25.4 mm.

**FIGURE 308.6(3)
CLEARANCE TO COMBUSTIBLES**

308.7 Solid fuel-burning appliances. The *clearance* reduction methods specified in Table 308.6 shall not be utilized to reduce the *clearance* required for solid fuel-burning appliances that are *labeled* for installation with clearances of 12 inches (305 mm) or less. Where appliances are *labeled* for installation with *clearances* of greater than 12 inches (305 mm), the *clearance* reduction methods of Table 308.6 shall not reduce the *clearance* to less than 12 inches (305 mm).

❖ Section 308.6 does not apply to the installation of solid-fuel-burning appliances listed for clearances of 12 inches (305 mm) or less. If the listed clearance is greater than 12 inches (305 mm), Section 308.6 is applicable if the reduced clearance is not less than 12 inches (305 mm). Solid-fuel-fired appliances can produce high-intensity radiant heat and have wide variations in heat output. With small clearances, the protection method could be inadequate because of the radiation intensity, reradiation from the protection assembly and inability of the protection method to dissipate the heat energy at the rate received from the appliance. A minimum distance is also beneficial to promote convection cooling.

308.8 Masonry chimneys. The *clearance* reduction methods specified in Table 308.6 shall not be utilized to reduce the *clearances* required for masonry *chimneys* as specified in Chapter 8 and the *International Building Code*.

❖ Clearances for masonry chimneys and masonry fireplace's are stated in the IBC. Chapter 8 simply refers to the IBC for masonry chimney construction (see commentary, Section 801.3). The required clearances for low-heat masonry chimneys are already quite small and not within the confines of Table 308.6. Also, the testing and research that led to the

creation of the table involved heat-producing appliances or equipment operating in "free air" in a room or space. The effect of clearance reduction for masonry assemblies could be unpredictable and disastrous.

Masonry chimneys are typically enclosed within building construction and therefore would not be cooled by convection air currents.

308.9 Chimney connector pass-throughs. The *clearance* reduction methods specified in Table 308.6 shall not be utilized to reduce the *clearances* required for *chimney* connector pass-throughs as specified in Section 803.10.4.

❖ The assemblies found in Section 803.10.4 are, in effect, clearance-reduction methods in themselves. These assemblies have been tested and any deviation or further reduction of clearance could change the assembly's performance (see commentary, Section 308.8).

308.10 Masonry fireplaces. The *clearance* reduction methods specified in Table 308.6 shall not be utilized to reduce the *clearances* required for masonry fireplaces as specified in Chapter 8 and the *International Building Code*.

❖ Using the same logic expressed for masonry chimneys, the required clearances to combustibles for masonry fireplaces cannot be reduced (see commentary, Section 308.8).

308.11 Kitchen exhaust ducts. The *clearance* reduction methods specified in Table 308.6 shall not be utilized to reduce the minimum *clearances* required by Section 506.3.11 for kitchen exhaust ducts enclosed in a shaft.

❖ The intent of this section is not to allow the minimum "observation" clearance of Section 506.3.10 to be

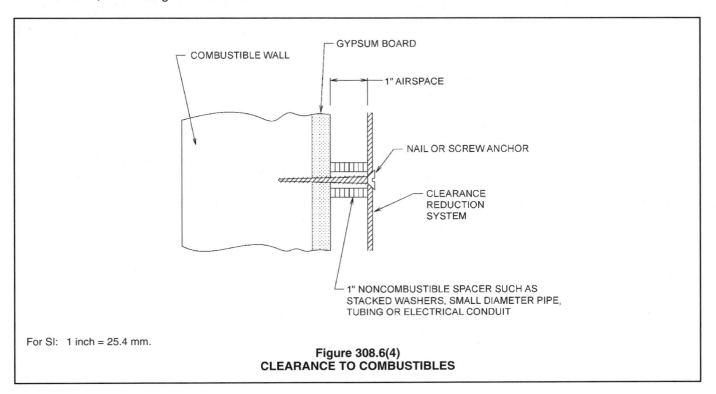

For SI: 1 inch = 25.4 mm.

Figure 308.6(4)
CLEARANCE TO COMBUSTIBLES

reduced. As stated in Section 308.1, Section 308 applies to the clearances required by Sections 506.3.6 and 507.9.

[B] SECTION 309
TEMPERATURE CONTROL

[B] 309.1 Space-heating systems. Interior spaces intended for human occupancy shall be provided with active or passive space-heating systems capable of maintaining a minimum indoor temperature of 68°F (20°C) at a point 3 feet (914 mm) above floor on the design heating day. The installation of portable space heaters shall not be used to achieve compliance with this section.

Exception: Interior spaces where the primary purpose is not associated with human comfort.

❖ The intent of this section is to require some form of space heating system for all interior spaces intended for human occupancy. The heating appliances listed in this code and in the IFGC can be used individually or in combination to provide the heat source to meet this requirement. It should be noted that the IFGC prohibits the use of unvented room heaters as the sole source of heat in a building.

The prescribed temperature of 68°F (20°C) is considered the minimum temperature to maintain an acceptable comfort level for humans. This was the temperature setting for thermostats the federal government recommended for energy savings following the oil shortages of the 1970s. This section does not allow portable space heaters to satisfy the requirement for space heating.

The exception exempts spaces in which human occupancy is not the primary use of the space. An example would be a large storage warehouse.

[F] SECTION 310
EXPLOSION CONTROL

[F] 310.1 Required. Structures occupied for purposes involving explosion hazards shall be provided with explosion control where required by the *International Fire Code*. Explosion control systems shall be designed and installed in accordance with Section 911 of the *International Fire Code*.

❖ The sudden increase in pressure inside a building caused by an explosion can cause massive damage to the building, including raising the roof or leveling the walls. A properly designed explosion control system will limit the damage by allowing the expanding gases to discharge through the vents, thus limiting the internal pressure to a value the structure can withstand.

Section 911 of the *International Fire Code*® (IFC®) contains guidance for the installation and operation of explosion (deflagration) venting in buildings where an explosion hazard exists. Some of the occupancies in the IFC requiring explosion control are facilities that manufacture or store hazardous materials, such as organic coating materials, flammable liquids, oxidizers, flammable gases and others. Note that deflagration venting may not be used as a means to protect buildings from deflagration hazards. An obvious facility that is not included is one that manufacturers or stores fireworks and/or explosives. Table 911.1 of the IFC does not allow explosion (deflagration) venting as a means of explosion control.

Explosion venting is allowed only in exterior walls and roofs, except that specially designed shafts vented to the outside of the building may be used. The size and design of the vents must be sufficient to relieve the sudden increase in internal pressure from an explosion to minimize damage to the structure, contents and occupants. The required vent area is dependent on the expected intensity of the explosion, the strength of the building components (columns, girders, walls, roof, etc.) and the opening response time of the vents.

[F] SECTION 311
SMOKE AND HEAT VENTS

[F] 311.1 Required. *Approved* smoke and heat vents shall be installed in the roofs of one-story buildings where required by the *International Fire Code*. Smoke and heat vents shall be designed and installed in accordance with the *International Fire Code*.

❖ The heated products of combustion, which are less dense than the surrounding air, will rise and accumulate under the roof of a building. If the gases are not removed from the building by mechanical means or through smoke and heat vents, the layer of gases and smoke will build until the entire building is filled. Smoke and heat vents help provide better visibility and additional time for the occupants to exit the building and also help remove some of the toxic by products of the fire that endanger the lives of the occupants and the fire fighters.

Section 910 of the IFC contains guidance for the design and installation of smoke and heat vents. The IFC requires vents in Group F-1 and S-1 buildings larger than 50,000 square feet (4645 m²), certain Group H-1, H-2 and H-3 occupancies, buildings containing high-piled combustibles or rack storage and Group F-1 and S-1 occupancies where the maximum exit travel distance has been increased in accordance with the IBC.

SECTION 312
HEATING AND COOLING LOAD CALCULATIONS

312.1 Load calculations. Heating and cooling system design loads for the purpose of sizing systems, appliances and *equipment* shall be determined in accordance with the procedures described in the ASHRAE/ACCA Standard 183. Alternatively, design loads shall be determined by an *approved* equivalent computation procedure, using the design parameters specified in Chapter 3 of the *International Energy Conservation Code*.

❖ This section was changed to require heating and cooling load calculations to be performed using ASHRAE/ACCA Standard 183. This standard contains all relevant sizing information that is provided in the ASHRAE *Handbook of Fundamentals* but in a more organized and straightforward format. Factors that are involved in calculating loads include types of building materials, building orientation, shading of the building, heat gains and losses from fenestration products (windows, doors, skylights, etc.), heat gains and losses through walls, floors and roof assemblies, duct heat gains and losses and internal heat sources, such as lighting, equipment and occupant load.

Where energy recovery systems are incorporated into an HVAC system, substantial amounts of energy can be recovered or shed to significantly reduce heating and cooling loads, respectively. These load reductions must be factored into sizing the HVAC system so that the system will not be oversized, less efficient and less able to control humidity in the cooling mode.

Design heating and cooling loads can also be computed by other equivalent procedures as approved by the code official. Where equivalent procedures are used, Chapter 3 of the IECC must be consulted for the design parameters.

Bibliography

The following resource materials were used in the preparation of the commentary for this chapter of the code:

IBC-12, *International Building Code*. Washington, DC: International Code Council, 2011.

IECC-12, *International Energy Conservation Code*. Washington, DC: International Code Council, 2011.

IFC-12, *International Fire Code*. Washington, DC: International Code Council, 2011.

IFGC-12, *International Fuel Gas Code*. Washington, DC: International Code Council, 2011.

IMC-12, *International Mechanical Code*. Washington, DC: International Code Council, 2011.

IPC-12, *International Plumbing Code*. Washington, DC: International Code Council, 2011.

Chapter 4:
Ventilation

General Comments

Mechanical ventilation uses fans or blowers to force the movement of air to and from the ventilated spaces. These systems can be dedicated to ventilation or can be part of a heating, cooling and air-conditioning system that serves the space to be ventilated. Spaces not served by an air-handling system, such as those heated and cooled by hot and chilled water, often depend on air-moving equipment that is devoted solely to providing ventilation.

Ventilation air is distinct from combustion air. Ventilation air is required for the occupants of the building; combustion air is necessary for the proper operation of fuel-burning appliances.

The term "occupied" or "occupiable" as used in this chapter also includes those spaces that are inhabited or habitable.

Natural ventilation is dependent on several factors, including: the location of ventilation openings; wind speed and direction; seasonal climate; temperature differences between indoors and outdoors; the building infiltration rate; ventilation opening shape and configuration; barometric pressure; the shape, height and proximity of adjacent structures; ventilation opening size; the number and distribution of openings in the open position and the personal habits and desires of the occupants. Natural ventilation, other than building infiltration and exfiltration through cracks and joints in the building envelope, occurs only when required openings to the outdoors are open.

Natural ventilation is not an exact science. The actual quantity of air movement through windows, doors and other gravity openings cannot be predicted because of the changing variables that affect airflow. The most unpredictable variable affecting natural ventilation is the fact that all such ventilation, other than infiltration and exfiltration, is dependent on one or more manual operations by the occupants of the room or space.

Where contaminants are known to be present in quantities large enough to be irritating or harmful to the occupants' health, naturally ventilated spaces must have mechanical exhaust systems capable of collecting and removing those contaminants. The mechanical exhaust system must comply with this chapter and Chapter 5 which contains exhaust system design criteria. Section 401.6 contains examples of the types of contaminant sources that require a mechanical exhaust system. The application of this section requires the judgment of the designer and the code official on a case-by-case basis. Chapter 5 also prescribes ventilation design for special areas listed there. Spaces designated by Note b of Table 403.3 are examples of spaces requiring mechanical exhaust systems to control contaminants.

In all but mild seasonal temperatures, natural ventilation and human comfort are in direct conflict. It is human nature to avoid opening windows and doors in the winter or summer months when the outdoor conditions are not within the human comfort zone, especially when energy is being expended to heat or cool the building interior.

Section 401 states the scope of the chapter and addresses by what means and when ventilation is to occur. It also covers general installation requirements and exhaust systems for local sources of contamination.

Section 402 contains all the requirements for natural ventilation.

Section 403 addresses mechanical means of ventilation.

Section 404 contains requirements for the ventilation of enclosed parking garages.

Section 405 states the minimum requirements for ventilation system controls.

Section 406 contains requirements for the ventilation of crawl spaces, attic spaces and similar uninhabited spaces.

Purpose

Chapter 4 includes means for protecting building occupant health by controlling the quality of indoor air and protecting property from the effects of inadequate ventilation. In some cases, ventilation is required to prevent or reduce a health hazard by removing contaminants at their source.

Ventilation is the exchange of air from one space to another, usually between an interior space and the outdoors. Ventilation is both necessary and desirable for the control of air contaminants, moisture and temperature. Habitable and occupiable spaces are ventilated to promote a healthy and comfortable environment for the occupants. Uninhabited and unoccupied spaces are ventilated to protect the building structure from the harmful effects of excessive humidity and heat. Ventilation of specific occupancies is necessary to minimize the potential for toxic or otherwise harmful substances to reach dangerously high concentrations in the air.

SECTION 401
GENERAL

401.1 Scope. This chapter shall govern the ventilation of spaces within a building intended to be occupied. Mechanical exhaust systems, including exhaust systems serving clothes dryers and cooking appliances; hazardous exhaust systems; dust, stock and refuse conveyor systems; subslab soil exhaust systems; smoke control systems; energy recovery ventilation systems and other systems specified in Section 502 shall comply with Chapter 5.

❖ This section establishes the scope of the chapter and the basic requirements for where, when and how ventilation is to be provided. This chapter regulates ventilation for rooms and spaces within building interiors that are intended for occupancy or inhabitability. This chapter also includes provisions for the ventilation of unoccupied spaces such as attics and crawl spaces (see commentary, Section 406).

Smoke control systems, smoke venting, mechanical exhaust systems and combustion air supplies are not within the scope of this chapter, but are regulated by other sections of the code, the *International Fuel Gas Code®* (IFGC®) and the *International Building Code®* (IBC®) as follows:

Smoke Control Systems	Section 513
	IBC Section 909
Combustion Air (for appliances not fuel-gas fired)	Chapter 7
Combustion Air (for fuel-gas-fired appliances)	IFGC Section 304

Chapter 5 addresses exhaust systems; however, ventilation is often accomplished using an exhaust-and-makeup-air arrangement such as for toilet rooms, bathrooms, kitchens and specific occupancies denoted by Notes b, g and h in Table 403.3.

401.2 Ventilation required. Every occupied space shall be ventilated by natural means in accordance with Section 402 or by mechanical means in accordance with Section 403. Where the air infiltration rate in a dwelling unit is less than 5 air changes per hour when tested with a blower door at a pressure of 0.2-inch water column (50 Pa) in accordance with Section R402.4.1.2 of the *International Energy Conservation Code*, the dwelling unit shall be ventilated by mechanical means in accordance with Section 403.

❖ Two distinct requirements are established by this section: all occupied spaces must be ventilated and ventilation can by accomplished by either natural (gravity) or mechanical means. The method of ventilation, mechanical or natural, is the choice of the owner or designer except for dwelling units having less than 5 air changes per hour (5 ACH50). If the ACH rate is less than 5 in a dwelling unit, mechanical ventilation is required whether or not natural ventilation criteria have been met. This section is coordinated with Section R303.4 of the *International Residential Code®* (IRC®) and Section 1203.1 of the IBC relative to dwelling unit ventilation. The *Interna-*

tional Energy Conservation Code® (IECC®) mandates that dwelling units be tested for air infiltration to further its goal of creating tighter buildings for the purpose of energy conservation and this same testing is also an indicator of when a building has become too tight to rely on natural ventilation. Testing is performed with an apparatus called a blower door and involves measuring the amount of air that the blower has to push into or pull out of a building in order to maintain a pressure differential between the indoors and outdoors of 0.2 inches of water column (50 Pa). The greater the volumetric flow rate through the blower to maintain the constant pressure differential, the greater the leakage in the building envelope. Multiple studies have shown that natural ventilation alone is not sufficient for dwellings that are tightly sealed such that their infiltration rate is below 5 air changes per hour (ACH). For perspective, 50 Pa = 0.2 inches water column; 1-inch water column = 250 Pa. Traditionally, 0.35 air changes per hour or 15 cfm per occupant has been the required mechanical ventilation rate allowed as an alternative to natural ventilation. An ACH of 0.35 at typical ambient pressure differentials is roughly equivalent to 7 to 10 ACH at a 50 Pa differential, thus the threshold of 5 ACH50 is comparable to the traditional ventilation rate. Note that the IECC intends for the infiltration rate to be 5 ACH50 or less, consistent with the trend for tighter building envelopes. As dwelling envelopes become more air tight, there is evidence that indoor contaminants levels are rising. Poor indoor air quality, the inability to predict ventilation rates from natural ventilation and the decreasing rates of infiltration have all led to this requirement for mechanical ventilation in dwellings. Also, several state codes now mandate mechanical ventilation in dwellings. The requirement for mechanical ventilation applies whether or not the natural ventilation provisions of Section 402 are applied.

The ventilation methodology of this chapter assumes that either the natural or mechanical ventilation method is being used. The code assumes that a building will be in full compliance with one method or the other. There are no provisions in the code for a ventilation system that depends simultaneously on both natural and mechanical ventilation. No criteria are given to evaluate ventilation effectiveness when natural and mechanical ventilation methods are used simultaneously for a room or space. This would be combining apples and oranges because mechanical ventilation is quantifiable and natural ventilation is not. On the other hand, the code does not expressly prohibit the combined use of both natural and mechanical methods. If both natural and mechanical means of ventilation are viable as stand-alone methods, logic would dictate that, for example, 50 percent of the required natural ventilation combined with 50 percent of the required mechanical ventilation would satisfy the intent of the code. Of course, the sum of such fractions (percentages) of the two methods would have to equal or exceed 100 percent. A possible drawback to such a hybrid system would be that the occupants might assume that because some mechanical ventilation is installed, it is therefore unnecessary to open any windows, not realizing that the mechanical ventilation is providing only a portion of the required outdoor air. Note that a system using both natural and mechanical ventilation methods would supply varying and unpredictable ventilation because the natural ventilation component will be unquantifiable.

Of course, naturally ventilated spaces can be served by exhaust systems, such as those prescribed by Section 401.6, and such designs can easily demonstrate compliance with the applicable code provisions.

A building may contain more than one type of ventilation system for different spaces within the same building. For example, an office building with an attached parking garage might use a natural ventilation system for the office structure and mechanical ventilation systems for the parking garage, toilet rooms and smoking lounges.

401.3 When required. Ventilation shall be provided during the periods that the room or space is occupied.

❖ Ventilation must be provided at all times that the room or space is occupied, but can cease when the room or space is unoccupied. This requires a mechanical ventilation system to be designed with controls that provide for continuous ventilation air movement during the entire time that the building is occupied. For example, if a building uses the heating, ventilation and air-conditioning (HVAC) system as the means of providing mechanical ventilation, the HVAC system is not allowed to cycle the air handler off and on. Rather than cycling the blower with the call for heat or cooling, the blower must run continuously while the building is occupied if it is the means of providing mechanical ventilation. This is typically accomplished with timers and energy management control systems (see Section 405.1). Note that even though ventilation must be continuous while the building is occupied, Section 403.5 allows the mechanical ventilation rate to modulate in proportion to the number of occupants in the space being ventilated. Ventilation required by Section 406 is not related to occupancy.

401.4 Intake opening location. Air intake openings shall comply with all of the following:

1. Intake openings shall be located a minimum of 10 feet (3048 mm) from lot lines or buildings on the same lot.

2. Mechanical and gravity outdoor air intake openings shall be located not less than 10 feet (3048 mm) horizontally from any hazardous or noxious contaminant source, such as vents, streets, alleys, parking lots and loading docks, except as specified in Item 3 or Section 501.3.1. Outdoor air intake openings shall be permitted to be located less than 10 feet (3048 mm) horizontally from streets, alleys, parking lots and loading docks provided that the openings are located not less than 25 feet

(7620 mm) vertically above such locations. Where openings front on a street or public way, the distance shall be measured from the closest edge of the street or public way.

3. Intake openings shall be located not less than 3 feet (914 mm) below contaminant sources where such sources are located within 10 feet (3048 mm) of the opening.

4. Intake openings on structures in flood hazard areas shall be at or above the elevation required by Section 1612 of the *International Building Code* for utilities and attendant equipment.

❖ This section addresses intake openings and Section 501.3.1 addresses exhaust openings. These two sections must be applied in harmony because they both can affect the separation between intakes and exhaust openings. To prevent the introduction of contaminants into the ventilation air of a building, Item 1 requires a minimum separation of 10 feet (3048 mm) between outdoor air intake openings and any lot lines or buildings on the same lot. Item 2 addresses powered intakes, (mechanical) and gravity (nonmechanical) intakes and requires separation from potentially harmful contaminant sources including chimneys and vents, plumbing vents and areas where motor vehicles operate. In this item, the 10-foot (3048 mm) distance would be measured from the closest edge of a street, alley, parking lot and loading dock. The last sentence of Item 2 addresses intake openings that face a street or public way [see Commentary Figures 401.4(1) and 401.4(2)].

Item 3 addresses those cases where the required 10-foot (3048 mm) separation cannot be met. For example, if the 10-foot (3048 mm) horizontal separation required in Item 2 cannot be achieved, the intake could be located at least 3 feet (914.4 mm) below the contaminant source. The code assumes that the contaminants likely to be present are buoyant in air because of their temperature or specific gravity and they will rise above and away from the intake opening.

Item 4 intends to prevent floodwaters from entering a building through an air intake opening.

See Chapter 8 and the IFGC for specific regulations for the location of chimney, vent, exhauster, mechanical draft system and appliance vent terminations for fuel-fired appliances. The specific provisions of Chapter 8 and the IFGC take precedence over the general provisions of this section.

See Sections 501.3 and 501.3.1 and consider the fact that an existing intake opening can be impacted by the installation of a new exhaust termination, as well as the fact that an existing exhaust termination can affect the location of a newly installed intake opening.

401.5 Intake opening protection. Air intake openings that terminate outdoors shall be protected with corrosion-resistant screens, louvers or grilles. Openings in louvers, grilles and screens shall be sized in accordance with Table 401.5, and

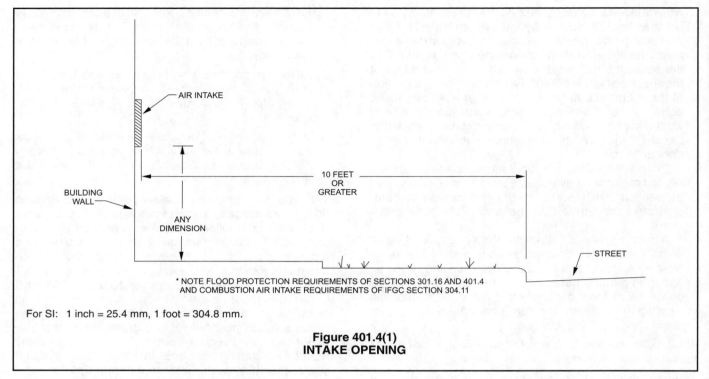

For SI: 1 inch = 25.4 mm, 1 foot = 304.8 mm.

Figure 401.4(1)
INTAKE OPENING

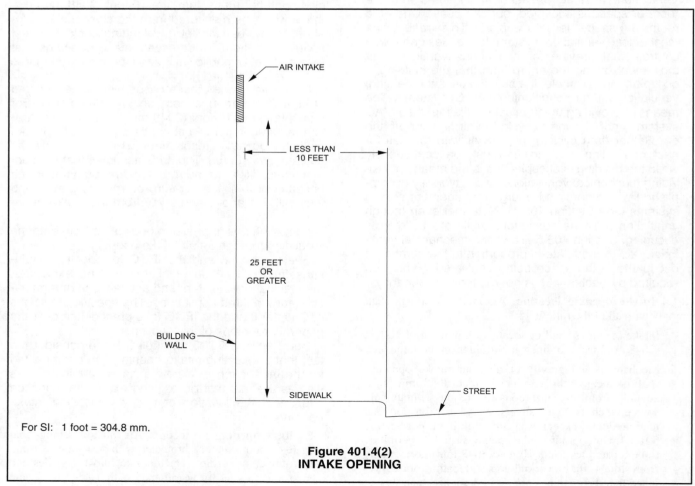

For SI: 1 foot = 304.8 mm.

Figure 401.4(2)
INTAKE OPENING

shall be protected against local weather conditions. Louvers that protect air intake openings in structures located in hurricane-prone regions, as defined in the *International Building Code*, shall comply with AMCA 550. Outdoor air intake openings located in exterior walls shall meet the provisions for exterior wall opening protectives in accordance with the *International Building Code*.

❖ This section aims to protect intake openings, other than windows and doors, from penetration by precipitation, small animals, vermin and debris. The opening sizes for residential intake opening protection range from $\frac{1}{4}$ inch to $\frac{1}{2}$ inch (6.35 mm to 12.7 mm) and the opening sizes for nonresidential intake opening protection are larger than $\frac{1}{4}$ inch (6.35 mm) up to a maximum of 1 inch, as dictated by Table 401.5. The opening size in louvers, grilles and screens must be large enough to inhibit blockage by debris and to prevent significant resistance to airflow, and yet must be small enough to keep out what is intended to be kept out. A screen of the stated mesh sizes would restrict the passage of rodents and large insects and would be resistant to blockage by lint, debris and plant fibers. Louvers in hurricane-prone areas must be resistant to wind-driven rain, hence the requirement to comply with AMCA 500.

TABLE 401.5
OPENING SIZES IN LOUVERS, GRILLES AND SCREENS PROTECTING AIR INTAKE OPENINGS

OUTDOOR OPENING TYPE	MINIMUM AND MAXIMUM OPENING SIZES IN LOUVERS, GRILLES AND SCREENS MEASURED IN ANY DIRECTION
Intake openings in residential occupancies	Not < $\frac{1}{4}$ inch and not > $\frac{1}{2}$ inch
Intake openings in other than residential occupancies	> $\frac{1}{4}$ inch and not > 1 inch

For SI: 1 inch = 25.4 mm.

401.6 Contaminant sources. Stationary local sources producing airborne particulates, heat, odors, fumes, spray, vapors, smoke or gases in such quantities as to be irritating or injurious to health shall be provided with an exhaust system in accordance with Chapter 5 or a means of collection and removal of the contaminants. Such exhaust shall discharge directly to an *approved* location at the exterior of the building.

❖ Stationary local sources include equipment, machinery, processes, chemical handling and storage and special-purpose rooms or areas that produce contaminants in quantities that constitute a health hazard to the building occupants. The contaminants must be removed by an exhaust system or by decontamination equipment designed to collect and remove the contaminants.

This section recognizes that natural and mechanical ventilation methods, in themselves, cannot effectively collect and remove contaminants produced by many types of local sources. Section 402 certainly does not address specific contaminant sources and Section 403 addresses only those contaminant sources listed in Table 403.3. There are many possible sources of contaminants that would be unaddressed if not for Section 401.6. This section is written in performance language and requires a case-by-case evaluation by the designer and the code official to determine the quantity and type of contamination present, whether or not any action needs to be taken and, if so, the method of contaminant control and removal.

The exhaust termination must comply with Sections 501.3, 501.3.1, 501.3.1.1 and 501.3.2.

Attics and crawl spaces are not considered to be outdoors and exhaust ducts cannot terminate in these spaces. Exhaust ducts must connect directly to terminals that pass through the building envelope to the outside atmosphere. Pointing, aiming or similarly directing an exhaust duct at an attic louver, grille, ridge vent, eave vent or soffit vent, for example, in no way ensures that all or any of the exhaust will reach the outdoors. In fact, it is possible that the majority, if not all, of the exhaust vapors and gases will discharge to the attic space rather than to the outdoors.

In the case of a duct that turns down to a soffit vent, the exhaust can rise into the attic as opposed to falling through the perforated soffit. The flow of air through any attic ventilation opening is dependent on attic temperature, wind direction, wind speed and opening configuration and location. In other words, attic air movement is unpredictable and may often be in the opposite direction to that intended. Additionally, grilles and louvers offer resistance and interfere with the exhaust flow directed at them. This may cause deflection of exhaust back into the attic. In cold climates, discharge of exhaust air into an attic space can result in moisture condensation on structural and insulation materials.

SECTION 402
NATURAL VENTILATION

[B] 402.1 Natural ventilation. *Natural ventilation* of an occupied space shall be through windows, doors, louvers or other openings to the outdoors. The operating mechanism for such openings shall be provided with ready access so that the openings are readily controllable by the building occupants.

❖ This section presents the standard of natural ventilation for all occupied spaces. Openings to the outdoor air, such as doors, windows, louvers, etc., provide natural ventilation. The section does not, however, state or intend that the doors, windows or openings actually be constantly open. The intent is that they be maintained in an operable condition so that they are available for use at the discretion of the occupant. Section 402 is consistent with Section 1203.4 of the IBC. See Section 1203.4.2.1 of the IBC, which requires mechanical ventilation in rooms that contain bathing fixtures. Bathtubs, showers, spas and similar bathing fixtures add moisture to the indoor air that must be controlled to prevent unhealthy conditions and damage to building components.

[B] 402.2 Ventilation area required. The minimum openable area to the outdoors shall be 4 percent of the floor area being ventilated.

❖ This section specifies the ratio of openable doors, windows and other openings to the floor space being ventilated, but does not address the distribution around the space or location of these openings. It is the designer's responsibility to distribute openings so as to accomplish the natural ventilation of the space. The placement of ventilation openings should be planned to induce airflow through the space. For example, placing all openings on the same wall will probably not be as effective as placing openings on opposite walls.

When inadequate natural ventilation is provided, mechanical ventilation can supplement any inadequacy (see Section 403). The plan reviewer can determine compliance with this section. For example, in Commentary Figure 402.2, the combined openable area (the net free area of a door, window, louver, vent or skylight, etc., when fully open) of double-hung Windows B and C is equal to 4 percent of the floor area [300 × 0.04 = 12 square feet (1.1 m²)]. Note that only half of the area of the double-hung windows is considered to be openable. The openable area of Window A is not required and need not open onto a court or yard complying with Section 1206 of the IBC.

[B] 402.3 Adjoining spaces. Where rooms and spaces without openings to the outdoors are ventilated through an adjoining room, the opening to the adjoining rooms shall be unobstructed and shall have an area not less than 8 percent of the floor area of the interior room or space, but not less than 25 square feet (2.3 m²). The minimum openable area to the outdoors shall be based on the total floor area being ventilated.

Exception: Exterior openings required for ventilation shall be permitted to open into a thermally isolated sunroom addition or patio cover, provided that the openable area between the sunroom addition or patio cover and the interior room has an area of not less than 8 percent of the floor area of the interior room or space, but not less than 20 square feet (1.86 m²). The minimum openable area to the outdoors shall be based on the total floor area being ventilated.

❖ Adjacent spaces with large connecting openings between them can share sources of ventilation. This section deals with the natural ventilation of connecting interior spaces. The intent is to allow a space without its own ventilation openings to the outdoors to be ventilated by connecting such space to another space that does have openings to the outdoors. It is not the intent to allow any space to be ventilated by another space that does not have openings to the outdoors. In other words, this section applies only to adjacent rooms and spaces, and it is not intended to allow rooms and spaces to be connected in a series of more than two rooms. Imagine Room A connected though an opening to Room B, which is, in turn, connected to yet another Room C and only Room A or C in the series has openings to the outdoors. Clearly the farther the room is from the outdoor openings, the less that ventilation will occur. It is the designer's obligation to locate openings between rooms with exterior openings and connecting spaces without exterior openings to allow for natural ventilation of the connected space.

For purposes of ventilation, this section establishes a minimum openness requirement for the common wall between a room with openings to the outdoors and an interior room without openings to the outdoors. The minimum amount of open area required in that common wall is 8 percent of the floor area of the interior room that has no openings or 25 square feet (2.33 m²), whichever is greater. The openable area of the openings to the outdoors in the "outer" room must

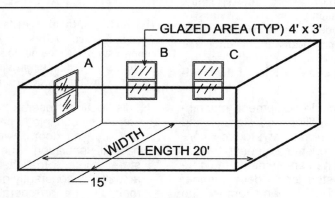

For SI: 1 foot = 304.8 mm, 1 square foot = 0.0929 m².

Figure 402.2
NATURAL LIGHT AND VENTILATION WINDOWS

be equal to or greater than 4 percent (the same as required by Section 402.2) of the total combined floor areas being ventilated. Commentary Figure 402.3 shows a cutaway of an interior room (Room A) adjacent to a room with openings to the outdoors (Room B). The openable area of openings to the outdoors in Space B must be equal to or greater than 0.04 times the area of the entire space (floor area of Space A plus floor area of interior Space B). The opening in the wall between the adjacent spaces must be a minimum of 25 square feet (2.33 m²), but not less than 0.08 times the floor area of interior Space A. Because the opening between the adjacent spaces must be unobstructed to comply with this section, a door cannot be installed in the opening.

[B] 402.4 Openings below grade. Where openings below grade provide required *natural ventilation*, the outside horizontal clear space measured perpendicular to the opening shall be one and one-half times the depth of the opening. The depth of the opening shall be measured from the average adjoining ground level to the bottom of the opening.

❖ This section is applicable whenever occupied spaces below grade depend on natural ventilation through structures like window wells. To provide adequate ventilation, this section sets the minimum horizontal clear space adjacent to the opening used for natural ventilation. Without this minimum horizontal area, air movement through the opening will be inadequate.

As illustrated in Commentary Figure 402.4, the opening area required for the story below grade intended for human occupancy is:

$$A = 0.04(L \times W)$$

The area of the window in the vertical plane (wh) must equal or exceed the required opening area. Additionally, the horizontal dimension from the win-

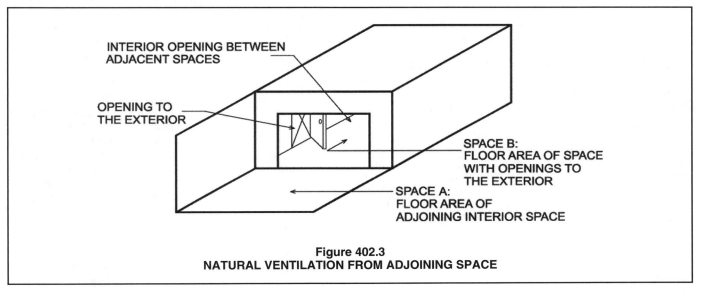

Figure 402.3
NATURAL VENTILATION FROM ADJOINING SPACE

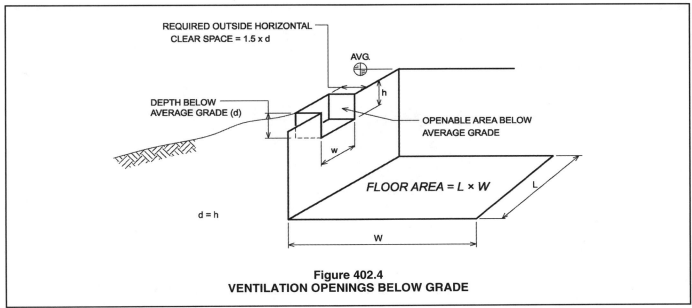

Figure 402.4
VENTILATION OPENINGS BELOW GRADE

dow to the well wall must equal $1^1/_2$ times the depth of the openable portion of the window at the lowest point.

SECTION 403
MECHANICAL VENTILATION

403.1 Ventilation system. Mechanical ventilation shall be provided by a method of supply air and return or *exhaust air*. The amount of supply air shall be approximately equal to the amount of return and *exhaust air*. The system shall not be prohibited from producing negative or positive pressure. The system to convey *ventilation air* shall be designed and installed in accordance with Chapter 6.

❖ This section addresses the mechanical method of ventilation. Mechanical ventilation is the alternative to having natural ventilation. As discussed in Section 401.2, both natural and mechanical ventilation can be provided to a space; however, one method must be in full compliance with the applicable sections of the code or the approved design must use both methods. The code does not state acceptance criteria that can be used to evaluate a ventilation system that depends on both natural and mechanical ventilation components. Mechanical ventilation systems must also comply with the applicable sections of Chapters 3, 5 and 6.

Technological advances are enabling mechanical ventilation to become increasingly economical as a method for ventilating occupied spaces. For example, ventilation systems can be equipped with air-to-air heat exchangers [energy recovery ventilators (ERVs)] that are designed to efficiently extract (reclaim) heat from exhaust air and use that heat to preheat the incoming outdoor makeup air. As a result, air can be exchanged between the indoors and the outdoors with only a slight loss of heat energy. Thus, heat reclamation equipment saves the energy that would otherwise be lost in the exhaust or relief air. ERVs can also be used to extract or reject sensible heat from the outdoor airstream, thus reducing cooling and heating loads. Such equipment is also capable of exchanging water vapor across the exhaust and makeup airstreams, thus retaining the moisture in the indoor space in the heating season or rejecting moisture to the outdoors in the cooling season.

Because certain types of ERVs can allow some degree of cross leakage and carryover between the airstreams, consideration must be given to the type of unit chosen for each application [see Commentary Figure 403.1(1)].

Mechanical ventilation is accomplished by air-handling equipment, fans, blowers and a distribution system that force air to flow through the space being ventilated. The components of a mechanical ventilation system can be dedicated for space ventilation or can be part of an HVAC system that serves the space [see Commentary Figure 403.1(2)].

The air distribution system can involve a duct network or it can be as simple as an exhaust opening with one or more outdoor air openings. A mechanical ventilation system can incorporate a gravity supply, return or exhaust operation within the system. As an example of a gravity supply system, a mechanical ventilation system for an auto repair shop may be equipped with a mechanical exhaust system to remove contaminants within the space. Instead of a blower to draw in supply air, the design relies on louvers placed within the exterior walls to allow outdoor air to enter the space when the exhaust system is operating and thereby creating a negative pressure in the building with respect to the outdoors.

Unlike natural ventilation, mechanical ventilation does not depend on unpredictable air pressure differentials between the indoors and outdoors to create airflow. Mechanical ventilation has the advantage of being both predictable and dependable because it is

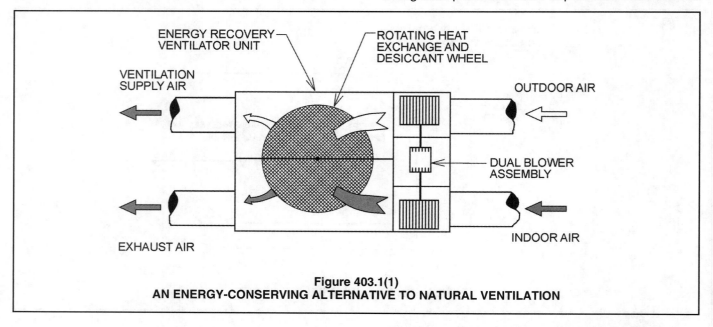

Figure 403.1(1)
AN ENERGY-CONSERVING ALTERNATIVE TO NATURAL VENTILATION

not subject to all of the variables that affect natural ventilation. The volume of air supplied to a space must be approximately equal to the volume of air removed from the space. Otherwise, the space will be either positively or negatively pressurized and the actual ventilation flow rate will be equivalent to the lower rate of either the air supply or air exhaust. To provide the required flow of outdoor air through a space, the rate of air leaving the space must be approximately equal to the rate being supplied. The intent is to make the minimum required airflow rate attainable. Slight negative or positive pressurization of spaces is often desirable to control contaminant or odor migration to or from a space and is, therefore, not prohibited. In the context of the code, negative and positive air pressures refer to pressures that are below or above atmospheric pressure or a reference pressure in another room or space.

Mechanical ventilation systems are air distribution systems, and the components of mechanical ventilation systems are subject to the requirements of Chapter 6, which regulates air distribution systems.

Mechanical ventilation systems involve electrical components and controls that must comply with Section 301.10.

403.2 Outdoor air required. The minimum outdoor airflow rate shall be determined in accordance with Section 403.3. Ventilation supply systems shall be designed to deliver the required rate of outdoor airflow to the *breathing zone* within each *occupiable space*.

Exception: Where the *registered design professional* demonstrates that an engineered ventilation system design

will prevent the maximum concentration of contaminants from exceeding that obtainable by the rate of outdoor air ventilation determined in accordance with Section 403.3, the minimum required rate of outdoor air shall be reduced in accordance with such engineered system design.

❖ To be effective, ventilation air must be delivered to the occupied space at the location of the occupants and in a manner that causes a cross-sectional flow through all portions of the occupied space. Building occupants occupy a zone that extends upward from the floor to approximately 6 feet (1829 mm) in height and that extends to within approximately 2 feet (610 mm) of the enclosing walls of the space (see the definition of "Breathing zone"). The intent is simply to ventilate where the beneficiaries of the ventilation are located. Commentary Figure 403.1(3) shows a dashed line that represents the breathing zone described in this section. Note that it is not the intent of this section to require that outlet diffusers and return air inlets be located within the zone described in this section. If that were so, it would prevent the use of ceiling and floor diffusers. Rather, the intent is to require that the system be designed with the proper airflow characteristics to enable delivery of the required amount of ventilation air to the breathing zone, and be designed to minimize stagnant pockets of air (dead zones) within the breathing zone [see Commentary Figure 403.1(3)].

Section 403 represents an indirect method of controlling air quality by diluting contaminants (ventilation rate procedure) to an acceptable level by introducing outdoor air. Although an engineered ventilation system may be approved by the code official as an alter-

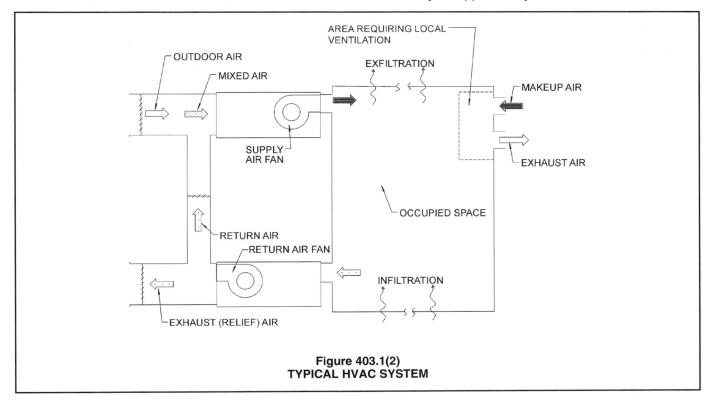

**Figure 403.1(2)
TYPICAL HVAC SYSTEM**

native design in accordance with Section 105, the exception to this section provides a direct reference to such an alternative design in this section. An engineered ventilation system is more of a direct method of controlling air quality and would be classified as an "Indoor Air Quality Procedure" in ASHRAE 62.1. ASHRAE 62.1 is not a referenced standard in the code, but the exception to this section could certainly be viewed as allowing the indoor air quality (IAQ) method of that standard as one of the possible means of complying with the exception. The design professional is responsible for demonstrating to the code official that a proposed engineered system will result in air quality at least equivalent to that achievable by the ventilation rate method of Section 403. A demonstration of equivalence would involve detailed analysis of at least the following: the anticipated contaminants of concern in the space to be ventilated; the anticipated sources and concentrations of the contaminants of concern; the acceptable occupant exposure limits or concentration levels for those contaminants; and the means and methods to control the contaminants. The design documentation should include all criteria and assumptions regarding occupancy conditions, equipment/system performance and contaminants. An engineered ventilation system would be allowed to supply outdoor air at any rate essential to the performance of the design.

403.2.1 Recirculation of air. The outdoor air required by Section 403.3 shall not be recirculated. Air in excess of that required by Section 403.3 shall not be prohibited from being recirculated as a component of supply air to building spaces, except that:

1. Ventilation air shall not be recirculated from one *dwelling* to another or to dissimilar occupancies.

2. Supply air to a swimming pool and associated deck areas shall not be recirculated unless such air is dehumidified to maintain the relative humidity of the area at 60 percent or less. Air from this area shall not be recirculated to other spaces where more than 10 percent of the resulting supply airstream consists of air recirculated from these spaces.

3. Where mechanical exhaust is required by Note b in Table 403.3, recirculation of air from such spaces shall be prohibited. All air supplied to such spaces shall be exhausted, including any air in excess of that required by Table 403.3.

4. Where mechanical exhaust is required by Note g in Table 403.3, mechanical exhaust is required and recirculation is prohibited where more than 10 percent of the resulting supply airstream consists of air recirculated from these spaces.

❖ The amount of ventilation air is always permitted to exceed the minimum quantities specified by Table 403.3. The ventilation air amounts specified by Table 403.3 consist of 100-percent outdoor air, none of which can be recirculated. Only the ventilation air that is in excess of the required amount is permitted to be recirculated, except as prohibited by Items 3 and 4 in conjunction with Notes b and g of Table 403.3. The intent of Section 403 is to bring in outdoor air, force it through the occupied space and then discharge an equal amount of air back to the outdoors. In other words, the inflow rate of outdoor air prescribed by Section 403.3 is always balanced by an outflow rate by means of exhaust or relief. This is what the code means by saying that the required ventilation air cannot be recirculated. Older editions of codes used to require a rate of ventilation air with only a fraction of

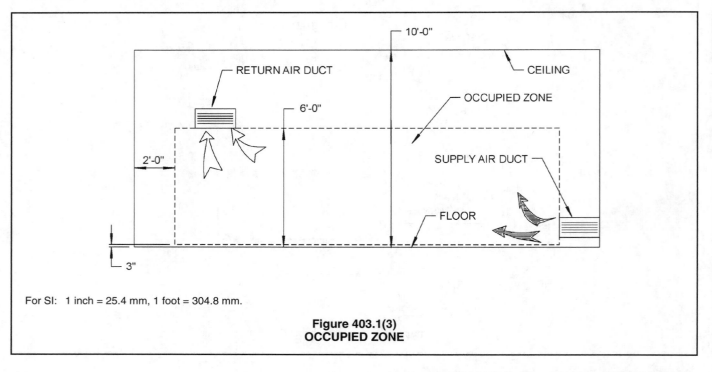

For SI: 1 inch = 25.4 mm, 1 foot = 304.8 mm.

Figure 403.1(3)
OCCUPIED ZONE

that air being outdoor air (i.e., ventilation air was a mix of recirculated air and outdoor air). However, this has not been the case for many years. The ventilation required by Section 403.3 is 100-percent outdoor air, no fraction of which is recirculated within the space served. If more outdoor air is introduced into a space than is required by Section 403.3, that excess air can be recirculated within the space or to other spaces, except as limited by Items 1 through 4.

Any occupancy will have environmental air that contains odors or contaminants that are peculiar or unique to that specific occupancy. For this reason, Item 1 states that ventilation air must not be recirculated to an occupancy of dissimilar use, or from one dwelling unit to another. This requirement is not affected by occupancy classification or tenant space separations. For example, a building classified as a single occupancy, such as a library, may contain a cafeteria. Because the cafeteria and library are of dissimilar occupancies, the ventilation air from one must not be recirculated to the other. Likewise, dissimilar occupancies are not limited to multiple tenancy and can occur in a single tenant space. Recirculating air to another occupancy or area of dissimilar use from which it is taken could subject the occupants to contaminants or odors that are not inherent to that specific occupancy. Such contaminants and odors could be harmful to the occupants or judged to be objectionable.

Mixed occupancy buildings may have different environmental air in each area and recirculation of air between those areas would be prohibited (for example, offices located in a factory). Conversely, mixed occupancy buildings may have similar environmental air in each area and recirculation of air between those areas would be allowed (for example, offices located in a library). Buildings of a single occupancy might contain areas or incidental uses that have different environmental air, therefore, recirculation of air between the areas would be prohibited (for example, a kitchen incidental to an office building or a dressing room incidental to a theater).

Ventilation airflow that is in excess of the required minimum rate is allowed by Item 2 to recirculate in swimming pool areas if the humidity of the area is controlled to maintain a relative humidity of 60 percent or less. Swimming pool areas generate high humidity levels and the air in those spaces must be dehumidified if a portion of it is to be recirculated as permitted by this section. Any recirculation of air from the swimming pool area is restricted to that area and cannot be circulated or transferred to any other area unless the resulting supply airstream consists of not more than 10 percent air from the swimming pool area (i.e., the portion of the mixed supply air that came from the swimming pool space cannot exceed 10 percent of the total). This will allow the designer to take advantage of the energy recovery ventilation technology to capture some of the energy from the humid swimming pool air and reduce the heating or

cooling load for the building (see the commentary, Item 4).

Item 3 relates to Note b of Table 403.3 and is an exception to the recirculation allowance. For example, all air delivered to a public toilet room or a smoking lounge would have to be exhausted, regardless of how much is required exhaust and how much is in excess of the required exhaust rate. No return can be taken directly or indirectly from such spaces. This item does not prohibit space conditioning air from being recirculated within the same space, rather it prohibits air from being recirculated from this space to some other space. For example, consider a beauty and nail salon in a strip mall where such space is served by a dedicated rooftop HVAC unit. The code does not intend to prevent space conditioning air from being turned over (recirculated) within the salon. If this were not the case, such tenants would not be able to afford to heat and cool their space because they would be conditioning 100-percent outdoor air. Besides, the atmosphere within the salon is considered to be healthy for occupancy where the provisions of Section 403.3 have been applied; therefore, there should be no problem with recirculating such "healthy" air within the same space. The intent of Note b is to prevent air from the salon from being circulated to some other tenant space. In other words, all return air taken from this salon must be returned only to the same salon. Imagine an air handler that serves two different spaces and one of the spaces is identified with Note b in Table 403.3. In this case, all supply air delivered by the air handler to the space governed by Note b would have to be exhausted so as to prevent any of such supply air from being returned to the air handler where it would then contaminate the other space served by the system. The key to understanding Item 3 is to keep in mind the intent to prevent contaminated air from spreading beyond the space of origin. As another example, consider smoking lounges and toilet rooms. Air within these spaces can be recirculated within the spaces provided that the required amount of air is exhausted as required by Table 403.3 and provided that no air is circulated to any other space. The second sentence of Item 3 is not an independent requirement and is tied to the first sentence relative to the intent to prevent air from being recirculated from one space or occupancy to another. This sentence has been misinterpreted to prevent recirculation of air within a single space served by a dedicated air handler and this is not the intent. The focus of Item 3 is air handlers that serve multiple zones, spaces or occupancies such that return taken from a space designated with Note b would be distributed to other spaces not so designated.

Item 4 allows the relaxation of the prohibition on recirculation of air from certain rooms and spaces consistent with ASHRAE 62.1. This was added to the code because an absolute ban on recirculation from certain rooms is unfounded and presents a significant barrier to the use of technology that recovers energy from exhaust air to reduce the load on heating and cooling systems from required outdoor ventilation air. The technology and these systems are referred to as ERV systems. See Section 503.2.6 in the IECC regarding the required use of ERV systems. The absolute ban on recirculation presents a significant barrier to the use of energy recovery technology. The lack of opportunity to recover energy from the exhaust air of toilet and locker rooms unnecessarily limits the overall effectiveness and feasibility of ERV technology—the energy in that air is lost to the atmosphere because of the absolute ban on recirculation.

Some of the various types of ERV systems on the market today do not completely eliminate leakage between the exhaust and supply airstreams. Many systems on the market have been engineered to the point that actual leakage is under 5 percent. ERVs designed for sensible heat recovery only are capable of having 0-percent cross-stream leakage, but this is not the case for ERVs that recover latent heat energy (moisture). The most effective technologies have some inherent leakage and even in cases where additional measures, such as purge, could be used to reduce the leakage, the additional equipment and operating costs may not be justified. This section uses 10 percent, not only for consistency with ASHRAE 62.1, but also to provide a level playing field for the entire industry. Ten percent is a limitation that all current suppliers of ERV systems can meet and this section does not create an advantage for one type of system or manufacturer over another.

Note that the intent of Note g is similar to that of Note b in that it intends to prevent air from being circulated to some other space or occupancy and does not prevent recirculation of air within the subject space. Air other than the required outdoor airflow rate required by Table 403.3 can be recirculated within the space designated with Note g, but cannot be circulated outside of such space, except for the 10-percent limit. In other words, mixed air delivered to some other space could contain up to 10-percent air from the space designated with Note g.

403.2.2 Transfer air. Except where recirculation from such spaces is prohibited by Table 403.3, air transferred from occupiable spaces is not prohibited from serving as *makeup air* for required exhaust systems in such spaces as kitchens, baths, toilet rooms, elevators and smoking lounges. The amount of transfer air and *exhaust air* shall be sufficient to provide the flow rates as specified in Section 403.3. The required outdoor airflow rates specified in Table 403.3 shall be introduced directly into such spaces or into the occupied spaces from which air is transferred or a combination of both.

❖ Spaces served by exhaust systems must be supplied with makeup air to replace the exhausted air. This makeup air is commonly supplied in the form of "transfer air" that has been drawn (transferred) from adjoining spaces. For example: toilet, bath and locker rooms often receive makeup air from adjoining assembly areas; kitchens receive makeup air from

dining areas; and elevators and smoking lounges receive makeup air from lobby and open floor areas. Note that the underlying principle is to transfer air only in the direction from "cleaner" to "dirtier" environments. Exhaust with transfer makeup air systems is typically used where control of odor migration is desired. Air in spaces designated by Notes b and g of Table 403.3 must be exhausted to the outdoors and, therefore, cannot be circulated or transferred to any other space. This section lists examples of the types of spaces intended to be covered and, therefore, similar spaces can be considered by the code official.

Ventilation air cannot consist entirely of reused air, meaning that the ventilation air that is transferred into a space cannot consist entirely of the ventilation air that was required to be delivered to an adjoining space. Transfer air will consist of some reused ventilation air and some unused ventilation air, because the space from which the transfer air is taken is, in effect, over ventilated. For example, if the required ventilation air for a kitchen area is not introduced directly into the kitchen, the air would have to be introduced into an adjacent area from which the air is transferred into the kitchen. Consider a large toilet room that needs 700 cubic feet per minute (cfm) of exhaust and is accessed from a lobby. If the required outdoor air is not introduced directly into the toilet room, 700 cfm would have to be introduced into the adjoining lobby in addition to the required ventilation air for the lobby, then air from the lobby could be transferred to the toilet room.

For kitchens, some transfer air is typically taken from dining rooms. The makeup air requirements for the kitchen exhaust hood system will often exceed the required ventilation rate for the kitchen; therefore, the makeup air with the required minimum outdoor air component can serve as the ventilation air for the kitchen. In the case where transfer air is taken from a dining room, the air is actually outdoor air that has been supplied to the dining room by the dining room ventilation system. Any of the required outdoor air for the kitchen that is to be supplied by transfer air from the dining room will have to be added to the required outdoor air rate for the dining room.

403.3 Outdoor airflow rate. Ventilation systems shall be designed to have the capacity to supply the minimum outdoor airflow rate determined in accordance with this section. The occupant load utilized for design of the ventilation system shall not be less than the number determined from the estimated maximum occupant load rate indicated in Table 403.3. Ventilation rates for occupancies not represented in Table 403.3 shall be those for a listed *occupancy* classification that is most similar in terms of occupant density, activities and building construction; or shall be determined by an *approved* engineering analysis. The ventilation system shall be designed to supply the required rate of *ventilation air* continuously during the period the building is occupied, except as otherwise stated in other provisions of the code.

With the exception of smoking lounges, the ventilation rates in Table 403.3 are based on the absence of smoking in occupiable spaces. Where smoking is anticipated in a space other than a smoking lounge, the ventilation system serving the space shall be designed to provide ventilation over and above that required by Table 403.3 in accordance with accepted engineering practice.

Exception: The occupant load is not required to be determined based on the estimated maximum occupant load rate indicated in Table 403.3 where *approved* statistical data document the accuracy of an alternate anticipated occupant density.

❖ Table 403.3 prescribes the amount of ventilation that must be supplied to any ventilated space. The actual amount (rate) supplied must equal or exceed the rate dictated by the table and related equations.

Many users of Section 403.3 have found that in most cases, the ventilation rates have decreased from those required by the 2006 edition of the code. This can be explained by several reasons, including the fact that the later versions of Section 403.3 account for the effectiveness of the air delivery system, thus allowing the economy of doing more with less (or at least the same with less). In other words, the rate can decrease if the ventilation air is delivered so as to be more effective, as opposed to requiring higher ventilation rates in order to compensate for less effectiveness. Also, the rates can decrease because the code now assumes that there is no smoking in all occupancies except smoking lounges.

Two very significant features of Table 403.3 are that it prescribes outdoor air only and the occupant load is determined by the table as opposed to the IBC. Table 403.3 was significantly revised in the 2009 edition and now includes a rate per person and a rate per floor area, both of which apply in most cases. Section 403.3 and the tables therein are based on ASHRAE 62.1, the 2004 edition. It must be understood that Table 403.3 cannot be used without applying the equations in Sections 403.3.1.1 through 403.3.2.3.4. The table does not stand alone. In fact, the application of Table 403.3 has become more complex than ever and requires the application of at least two mathematical equations for single-zone systems and several more for multiple-zone recirculating systems. The increased complexity could encourage code officials to require the design professionals to submit ventilation calculations in accordance with Section 106.3.1.

Another significant change is that Table 403.3 assumes that there is no smoking in all occupancies with the only exception being "smoking lounges." If smoking is allowed by state or local law in an occupancy, the ventilation for that occupancy must be increased over that prescribed by Section 403.3 and, because the code does not state by how much the ventilation must be increased, the system will have to be engineered and approved by the code official. Recall that the 2006 and earlier editions of the code assumed some moderate amount of smoking in occupancies, but were otherwise silent on smoking.

Now the code has clearly stated that it is based on the absence of smoking because there is insufficient data to determine the amount of ventilation for areas in which smoking is permitted and because of the fundamental belief that secondhand smoke is harmful in any concentration.

Although the definition of "Ventilation air" states that it can be a combination of outdoor air and recirculated air, this section makes no allowance for the recirculation of any portion of the required amount of ventilation air. Note that Section 403.2.1 addresses the recirculation of that portion of ventilation air that is in excess of the required amount.

ASHRAE 62.1 permits the use of properly cleaned recirculated air to reduce the outdoor air component of ventilation air if the recirculated air is treated in accordance with the Indoor Air Quality Procedure contained in ASHRAE 62.1. Although the Indoor Air Quality Procedure of ASHRAE 62.1 is not expressly referred to as an option in the code, its use would be permitted if approved by the code official in accordance with Section 105.2 or the exception to Section 403.2. Note that demonstrating compliance with the Indoor Air Quality Procedure of ASHRAE 62.1 is not a simple task and the air treatment involved will likely be much more complex than particulate filtering alone. Sufficient documentation, specifications and analysis must be submitted to the code official for any proposed application of the Indoor Air Quality Procedure of ASHRAE 62.1.

Table 403.3 is used to determine the maximum occupant load of any room or space and it is this number that must be the basis for the design capacity of a mechanical ventilation system when the ventilation rate is based on the number of occupants. Note that the "Occupant Density" column does not specify a "default" occupant density. There is no other occupancy calculation for which this column becomes the "default" and this column always determines the mandatory occupant density. This section and its exception refer to the "estimated maximum occupant load rate" in Table 403.3, but the table refers to this rate as the "occupant density." Note also that the design capacity must be based on the maximum occupant load. The actual quantity of outdoor air supplied can be based on the number of occupants at any given time (see commentary, Section 403.5). In accordance with this section and Section 401.3, the ventilation air must be supplied during the periods that the room or space is occupied. When the room or space is not occupied, the ventilation air can be reduced or shut down completely.

Where a space or occupancy is not represented in the table, the user can base the ventilation rate on the table entry that most closely resembles the subject space or occupancy. Where nothing in the table is truly representative of the subject space or occupancy, an approved engineering analysis must specify the required ventilation rate.

The exception allows the occupant load for ventilation to be based on the actual occupant load for which the space is designed where it can be shown to the code official's satisfaction that the occupant load is realistic. The occupant density entries in Table 403.3 are based on empirical data and are reasonably accurate for the typical cases in each type of occupancy. Sufficient documentation must be submitted to the code official to justify any deviation from the prescribed occupant densities.

The ventilation rates prescribed by this section are based on the Ventilation Rate Procedure of ASHRAE 62.1, which can be described as the dilution rate method. Outdoor air is brought into the space being ventilated in sufficient quantities to dilute the contaminants to an acceptable level and, thus, control the indoor air quality (see commentary, Table 403.3).

TABLE 403.3. See page 4-18.

❖ Table 403.3 is based on ASHRAE 62.1 and prescribes both the use of 100-percent outdoor air as ventilation air and an occupant load calculation method that is independent of the IBC.

Ventilation rates are prescribed based on the actual use (occupancy) of a room or space. The occupancy of the space may or may not coincide with the occupancy of the building or space in accordance with the IBC. If a particular occupancy cannot be found in the table, a similar occupancy classification should be chosen that is closely related to the actual occupancy (e.g., a drug store most closely resembles retail sales). Occupancies not addressed in Table 403.3 are not exempt from ventilation requirements. The code cannot prescribe ventilation rates for all of the possible occupancies; therefore, some judgment has to be made. Ventilation rates for occupancies that are clearly not represented in Table 403.3, such as factory and industrial occupancies, must be determined on a case-by-case basis by an engineering analysis approved by the code official. Note that ASHRAE 62.1 may contain additional guidance for some occupancies that are not listed in Table 403.3. For example, ASHRAE 62.1 references the American Conference on Governmental Industrial Hygienists' (ACGIH) publication titled *Industrial Ventilation—A Manual of Recommended Practice*.

The majority of occupancies listed in Table 403.3 are those in which the primary sources of air contamination are the occupants themselves at various everyday activities and also the contents of the space. Table entries, such as warehouses, shipping/receiving, repair garages and ice arenas obviously address contaminants that are related to processes or material storage. When the parameter listed in the table for out door air is based on cubic feet per minute per square foot only, the anticipated contaminants in the space are not people related and, therefore, the contaminants do not depend on the occupant load. For example, the required outdoor ventilation air for retail stores, shipping and receiving is 0.12 cfm/ft². The anticipated contaminants are not

people related, but are associated with wares, stock, merchandise and storage. Table 403.3 does not attempt to address factory/industrial-type occupancies because the possible number of contaminants would be enormous and the conditions in such occupancies could be unpredictable. Also, for many potential contaminants, the effect on humans is unknown.

Because the occupant load calculation method of the table is not based on egress capacity, the occupant load numbers determined by the table will not always parallel those numbers determined in accordance with the IBC.

The ventilation prescribed by the table is accomplished by bringing in the required amount of outdoor air and expelling, relieving or exhausting an approximately equal quantity of indoor air. Instructions shown in the table and the information given in Notes a through h describe the application and use of the table along with any special requirements applicable to a specific occupancy.

Note a of the table states that the estimated occupant load is based on the net occupiable floor area of the space to be ventilated. The "net occupiable floor area" is very well defined in Chapter 2. Note b in Table 403.3 specifies that some occupancies, such as smoking lounges, beauty and nail salons, and repair garages, must have mechanical exhaust systems installed to exhaust the contaminants produced by that occupancy. The air supplied to these occupancies cannot be circulated to other occupancies (see commentary, Section 403.2.1, Item 3).

Note c intends to exempt from the ventilation requirements spaces in buildings having temperatures well below human comfort levels and that are not continuously occupied. For example, a meat-processing plant, where the employees work in a cold environment, is considered to be continuously occupied and, therefore, must exhibit compliance with the ventilation requirements in the table. The cold meat-storage spaces would not.

Note d provides the option of using the intermittent ventilation design associated with the use of occupant or vehicle operation detectors or CO and NO$_x$ detectors in enclosed parking garages (see commentary, Sections 404.1, 404.2 and 404.3).

Note e explains the intent of the dual rates for exhaust in public toilet rooms, that being, the higher rate applies where the exhaust system operates intermittently and the lower rate applies only where the exhaust system operates continuously while occupied.

Note f explains the intent of the dual rates for exhaust; the higher rate applies when the exhaust is operated on an intermittent basis, such as when tied to the lighting switch or when operated by the occupant by means of a manual switch.

Note that the intent of Note g is similar to that of Note b in that it intends to prevent air from being circulated to some other space or occupancy and does not prevent recirculation of air within the subject space. Air other than the required outdoor airflow rate required by Table 403.3 can be recirculated within the space designated with Note g, but cannot be circulated outside of such space, except for the 10-percent limit. In other words, mixed air delivered to some other space could contain up to 10 percent air taken from the space designated with Note g (see commentary, Section 403.2.1).

Note h addresses nail salons and requires a source capture system for each table/station in addition to the other requirements for ventilation and exhaust in Table 403.3. Of course, it is likely that some beauty salons will contain nail stations which would trigger the requirements of Note f. The intent is to require a moveable hood that can be placed over the nails

(hands or feet) being serviced or a special table with exhaust air intake openings and baffles that will capture the chemical vapors at their source and exhaust them to the outdoors (see Commentary Figure 403.3). See the definitions of "Exhaust system" and "Source capture system." An exhaust fan located in a ceiling or anywhere else other than right next to the chemical source does not meet the intent of the code. Also, an exhaust system that captures the vapors, passes them through a filtering system and discharges the air back into the room does not meet the intent of the code and any such system would have to be approved by the code official under Section 105.2. See the commentary for Item 3 of Section 403.2.1 for discussion of recirculation of space conditioning air.

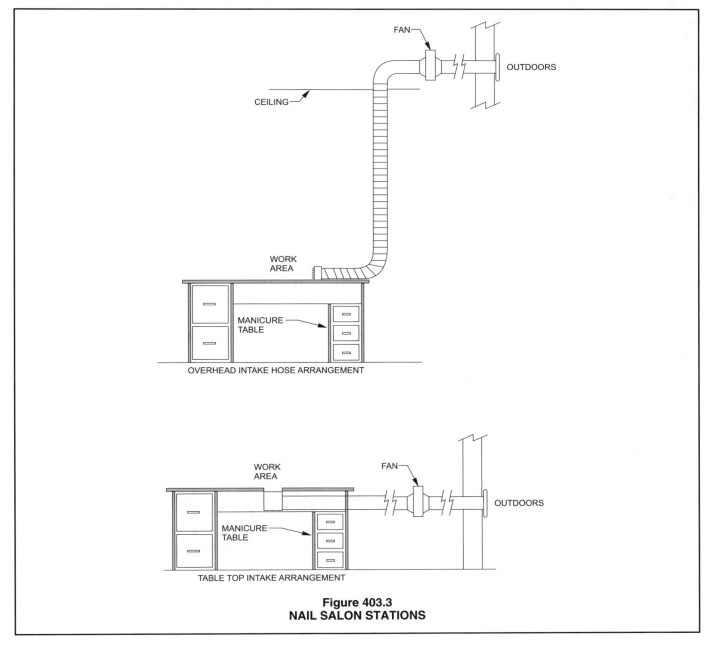

Figure 403.3
NAIL SALON STATIONS

403.3.1 Zone outdoor airflow. The minimum outdoor airflow required to be supplied to each zone shall be determined as a function of *occupancy* classification and space air distribution effectiveness in accordance with Sections 403.3.1.1 through 403.3.1.3.

❖ See the definition of "Zone" in Chapter 2. The zone is where the occupants are located and the required zone outdoor airflow rate is determined (in most cases) by a combined rate per person and rate per area. It is also affected by the configuration of the air distribution system (see Sections 403.3.1.1 and 403.3.1.2).

403.3.1.1 Breathing zone outdoor airflow. The outdoor airflow rate required in the *breathing zone* (V_{bz}) of the *occupiable space* or spaces in a zone shall be determined in accordance with Equation 4-1.

$$V_{bz} = R_p P_z + R_a A_z \qquad \textbf{(Equation 4-1)}$$

where:

A_z = Zone floor area: the *net occupiable floor area* of the space or spaces in the zone.

P_z = Zone population: the number of people in the space or spaces in the zone.

R_p = People outdoor air rate: the outdoor airflow rate required per person from Table 403.3.

R_a = Area outdoor air rate: the outdoor airflow rate required per unit area from Table 403.3.

❖ The equation sums the rate per person times the number of occupants and the rate per area times the square footage of the zone floor. As can be seen from this equation, the code now combines two different airflow requirements that were previously independent in editions of the code prior to 2009. The basis for this approach is that studies have shown that the effects of people-produced contaminants and the effects of building- and building content-produced contaminants are additive in their effect on occupants (see Example 1).

403.3.1.2 Zone air distribution effectiveness. The zone air distribution effectiveness (E_z) shall be determined using Table 403.3.1.2.

❖ The location of supply and return air grilles, diffusers and registers and the temperature of the supply air all affect the movement of air through a conditioned space. Some air distribution configurations are better (more effective) than others at achieving a uniform flow through a space. Uniform flow helps to prevent dead (stagnant) zones and short-circuited flow. Ventilation works best when airflow sweeps through the entire cross section of a space, thereby reaching all occupants and flushing out contaminants. The intent of this section is to rate the air distribution system as

to its ability to produce effective ventilation. Table 403.3.1.2 was developed based on fluid modeling analysis that considers the buoyancy of air masses and predicts air movement, eddy flow, air mixing and flow patterns. The table ratings may not be what one would expect based on the system configuration, but they are logical when the fluid flow analysis is viewed.

The effectiveness number (E_z) is applied in Equation 4-2 of Section 403.3.1.3 and the purpose of that calculation is to adjust the breathing zone airflow rate to account for the effectiveness of the air distribution system in delivering the ventilation air (see commentary Section 403.3.1.3).

TABLE 403.3.1.2
ZONE AIR DISTRIBUTION EFFECTIVENESS[a,b,c,d,e]

Air Distribution Configuration	E_z
Ceiling or floor supply of cool air	1.0[f]
Ceiling or floor supply of warm air and floor return	1.0
Ceiling supply of warm air and ceiling return	0.8[g]
Floor supply of warm air and ceiling return	0.7
Makeup air drawn in on the opposite side of the room from the exhaust and/or return	0.8
Makeup air drawn in near to the exhaust and/or return location	0.5

For SI: 1 foot = 304.8 mm, 1 foot per minute = 0.00506 m/s,
　　°C = [(°F) – 32]/1.8.
a. "Cool air" is air cooler than space temperature.
b. "Warm air" is air warmer than space temperature.
c. "Ceiling" includes any point above the breathing zone.
d. "Floor" includes any point below the breathing zone.
e. "Makeup air" is air supplied or transferred to a zone to replace air removed from the zone by exhaust or return systems.
f. Zone air distribution effectiveness of 1.2 shall be permitted for systems with a floor supply of cool air and ceiling return, provided that low-velocity displacement ventilation achieves unidirectional flow and thermal stratification.
g. Zone air distribution effectiveness of 1.0 shall be permitted for systems with a ceiling supply of warm air, provided that supply air temperature is less than 15°F above space temperature and provided that the 150 foot-per-minute supply air jet reaches to within $4^1/_2$ feet of floor level.

403.3.1.3 Zone outdoor airflow. The zone outdoor airflow rate (V_{oz}), shall be determined in accordance with Equation 4-2.

$$V_{oz} = \frac{V_{bz}}{E_z} \qquad \textbf{(Equation 4-2)}$$

❖ The (E_z) value determined in the previous section is applied in Equation 4-2 to determine the required outdoor airflow rate. The result, (V_{oz}), is adjusted upward in many cases; unaffected in some cases; and even adjusted downward in the case described in Note f of Table 403.3.1.2. For example, if E_z is 1, the rate determined in accordance with Equation 4-1 of Section 403.3.1.1 is unchanged; and if E_z is 0.5, the rate from Section 403.3.1.1 is doubled to account for the poor performance of the air distribution system (see Sections 403.3.2 through 403.3.2.3.4).

403.3.2 System outdoor airflow. The outdoor air required to be supplied by each ventilation system shall be determined in accordance with Sections 403.3.2.1 through 403.3.2.3 as a function of system type and zone outdoor airflow rates.

❖ The actual amount of outdoor air that a ventilation system must deliver is determined in accordance with Section 403.3.2.1, 403.3.2.2 or 403.3.2.3, depending on whether the delivery system serves multiple zones or only one zone, and depending upon whether the system supplies mixed outdoor and recirculated air or only outdoor air.

403.3.2.1 Single zone systems. Where one air handler supplies a mixture of outdoor air and recirculated return air to only one zone, the system outdoor air intake flow rate (V_{ot}) shall be determined in accordance with Equation 4-3.

$$V_{ot} = V_{oz} \qquad \textbf{(Equation 4-3)}$$

❖ This type of system calculation is as simple as it gets. If the ventilation system delivers outdoor air and recirculated air to only one zone (see the definition of "Zone"), the total outdoor air component of the delivered air is the same as calculated in Section 403.3.1.3. In other words, since there is only one zone, the zone outdoor air rate is also the total outdoor air rate (see Example 1).

403.3.2.2 100-percent outdoor air systems. Where one air handler supplies only outdoor air to one or more zones, the system outdoor air intake flow rate (V_{ot}) shall be determined using Equation 4-4.

$$V_{ot} = \Sigma_{all\,zones} V_{oz} \qquad \textbf{(Equation 4-4)}$$

❖ If the ventilation system supplies only outdoor air, with no recirculated air component, and serves multiple zones, the total amount of outdoor air is simply the sum of the zone outdoor air rates. The greek letter, Σ (sigma) in Equation 4-4 means "sum" and the subscript "all zones" that follows Σ mean that the summing is applied to all of the zones in a multiple-zone system. Recall Section 403.3.2.1, which dealt with single-zone systems.

403.3.2.3 Multiple zone recirculating systems. Where one air handler supplies a mixture of outdoor air and recirculated return air to more than one zone, the system outdoor air intake flow rate (V_{ot}) shall be determined in accordance with Sections 403.3.2.3.1 through 403.3.2.3.4.

❖ This section addresses ventilation air systems that supply both outdoor air and recirculated air, and sup-

ply such mixture to multiple zones. It becomes more complicated now because different zones will have different required rates of outdoor air and the typical air-handling system will not be able to deliver a different ratio of outdoor air to recirculated air for each of the zones having different ratio requirements. For example, if an air handler is supplying air that is 25-percent outdoor air and 75-percent recirculated air, how can that same air handler deliver 15-percent outdoor air to one zone, 30-percent outdoor air to another and 25 percent to others? Sections 403.3.2.3.1 through 403.3.2.3.4 address this dilemma by adjusting the total required outdoor air rate to compensate for the inability to deliver different outdoor air ratios to different zones. The theory behind this method is that the spaces having outdoor air rate requirements lower than the "critical" space (Maximum Z_p) will be overventilated and the "unused" (excess) outdoor air will be recirculated to provide a fraction of the outdoor air needed for the "critical" space. The effect of this compensation will be an increase in outdoor air for the "noncritical" (Lower Z_p) spaces and a decrease in outdoor air for the critical space (see Example 2).

403.3.2.3.1 Primary outdoor air fraction. The primary outdoor air fraction (Z_p) shall be determined for each zone in accordance with Equation 4-5.

$$Z_p = \frac{V_{oz}}{V_{pz}} \qquad \textbf{(Equation 4-5)}$$

where:

V_{pz} = Primary airflow: The airflow rate supplied to the zone from the air-handling unit at which the outdoor air intake is located. It includes outdoor intake air and recirculated air from that air-handling unit but does not include air transferred or air recirculated to the zone by other means. For design purposes, V_{pz} shall be the zone design primary airflow rate, except for zones with variable air volume supply and V_{pz} shall be the lowest expected primary airflow rate to the zone when it is fully occupied.

❖ Equation 4-5 is solved solely for the purpose of establishing a ventilation system efficiency value (E_v). That value comes from Table 403.3.2.3.2. In other words, Equation 4-5 yields a value for Z_p which is then entered into Table 403.3.2.3.2 to determine E_v. The outdoor/primary fraction (Z_p) is the ratio of outdoor air to total air delivered by the ventilation system (i.e., the percentage of outdoor air in the total supply air divided by 100). Finally, the value for E_v is entered into Equation 4-8.

TABLE 403.3
MINIMUM VENTILATION RATES

OCCUPANCY CLASSIFICATION	OCCUPANT DENSITY #/1000 FT² ª	PEOPLE OUTDOOR AIRFLOW RATE IN BREATHING ZONE, R_p CFM/PERSON	AREA OUTDOOR AIRFLOW RATE IN BREATHING ZONE, R_a CFM/FT² ª	EXHAUST AIRFLOW RATE CFM/FT² ª
Correctional facilities				
Cells				
without plumbing fixtures	25	5	0.12	—
with plumbing fixtures[g]	25	5	0.12	1.0
Dining halls (see food and beverage service)	—	—	—	—
Guard stations	15	5	0.06	—
Day room	30	5	0.06	—
Booking/waiting	50	7.5	0.06	—
Dry cleaners, laundries				
Coin-operated dry cleaner	20	15	—	—
Coin-operated laundries	20	7.5	0.06	—
Commercial dry cleaner	30	30	—	—
Commercial laundry	10	25	—	—
Storage, pick up	30	7.5	0.12	—
Education				
Auditoriums	150	5	0.06	—
Corridors (see public spaces)	—	—	—	—
Media center	25	10	0.12	—
Sports locker rooms[g]	—	—	—	0.5
Music/theater/dance	35	10	0.06	—
Smoking lounges[b]	70	60		—
Day care (through age 4)	25	10	0.18	—
Classrooms (ages 5-8)	25	10	0.12	—
Classrooms (age 9 plus)	35	10	0.12	—
Lecture classroom	65	7.5	0.06	—
Lecture hall (fixed seats)	150	7.5	0.06	—
Art classroom[g]	20	10	0.18	0.7
Science laboratories[g]	25	10	0.18	1.0
Wood/metal shops[g]	20	10	0.18	0.5
Computer lab	25	10	0.12	—
Multiuse assembly	100	7.5	0.06	—
Locker/dressing rooms[g]	—	—	—	0.25
Food and beverage service				
Bars, cocktail lounges	100	7.5	0.18	—
Cafeteria, fast food	100	7.5	0.18	—
Dining rooms	70	7.5	0.18	—
Kitchens (cooking)[b]	—	—	—	0.7
Hospitals, nursing and convalescent homes				
Autopsy rooms[b]	—	—	—	0.5
Medical procedure rooms	20	15	—	—
Operating rooms	20	30	—	—
Patient rooms	10	25	—	—
Physical therapy	20	15	—	—
Recovery and ICU	20	15	—	—

(continued)

TABLE 403.3—continued
MINIMUM VENTILATION RATES

OCCUPANCY CLASSIFICATION	OCCUPANT DENSITY #/1000 FT²ᵃ	PEOPLE OUTDOOR AIRFLOW RATE IN BREATHING ZONE, R_p CFM/PERSON	AREA OUTDOOR AIRFLOW RATE IN BREATHING ZONE, R_a CFM/FT²ᵃ	EXHAUST AIRFLOW RATE CFM/FT²ᵃ
Hotels, motels, resorts and dormitories				
Multipurpose assembly		5	0.06	—
Bathrooms/toilet—private[g]		—	—	25/50[f]
Bedroom/living room		5	0.06	—
Conference/meeting		5	0.06	—
Dormitory sleeping areas		5	0.06	—
Gambling casinos		7.5	0.18	—
Lobbies/prefunction		7.5	0.06	—
Offices				
Conference rooms	50	5	0.06	—
Office spaces	5	5	0.06	—
Reception areas	30	5	0.06	—
Telephone/data entry	60	5	0.06	—
Main entry lobbies	10	5	0.06	—
Private dwellings, single and multiple				
Garages, common for multiple units[b]	—	—	—	0.75
Garages, separate for each dwelling[b]	—	—	—	100 cfm per car
Kitchens[b]	—	—	—	25/100[f]
Living areas[c]	Based upon number of bedrooms. First bedroom, 2; each additional bedroom, 1	0.35 ACH but not less than 15 cfm/person	—	—
Toilet rooms and bathrooms[g]	—	—	—	20/50[f]
Public spaces				
Corridors	—	—	0.06	—
Elevator car	—	—	—	1.0
Shower room (per shower head)[g]	—	—	—	50/20[f]
Smoking lounges[b]	70	60	—	—
Toilet rooms — public[g]	—	—	—	50/70[e]
Places of religious worship	120	5	0.06	—
Courtrooms	70	5	0.06	—
Legislative chambers	50	5	0.06	—
Libraries	10	5	0.12	—
Museums (children's)	40	7.5	0.12	—
Museums/galleries	40	7.5	0.06	—
Retail stores, sales floors and showroom floors				
Sales (except as below)	15	7.5	0.12	—
Dressing rooms	—	—	—	0.25
Mall common areas	40	7.5	0.06	—
Shipping and receiving	—	—	0.12	—
Smoking lounges[b]	70	60	—	—
Storage rooms	—	—	0.12	—
Warehouses (see storage)	—	—	—	—

(continued)

TABLE 403.3—continued
MINIMUM VENTILATION RATES

OCCUPANCY CLASSIFICATION	OCCUPANT DENSITY #/1000 FT²ᵃ	PEOPLE OUTDOOR AIRFLOW RATE IN BREATHING ZONE, R_p CFM/PERSON	AREA OUTDOOR AIRFLOW RATE IN BREATHING ZONE, R_a CFM/FT²ᵃ	EXHAUST AIRFLOW RATE CFM/FT²ᵃ
Specialty shops				
Automotive motor-fuel dispensing stationsᵇ	—	—	—	1.5
Barber	25	7.5	0.06	0.5
Beauty salonsᵇ	25	20	0.12	0.6
Nail salons ᵇ, ʰ	25	20	0.12	0.6
Embalming roomᵇ	—	—	—	2.0
Pet shops (animal areas)ᵇ	10	7.5	0.18	0.9
Supermarkets	8	7.5	0.06	—
Sports and amusement				
Disco/dance floors	100	20	0.06	—
Bowling alleys (seating areas)	40	10	0.12	—
Game arcades	20	7.5	0.18	—
Ice arenas without combustion engines	—	—	0.30	0.5
Gym, stadium, arena (play area)	—	—	0.30	—
Spectator areas	150	7.5	0.06	—
Swimming pools (pool and deck area)	—	—	0.48	—
Health club/aerobics room	40	20	0.06	—
Health club/weight room	10	20	0.06	—
Storage				
Repair garages, enclosed parking garagesᵇ,ᵈ	—	—	—	0.75
Warehouses	—	—	0.06	—
Theaters				
Auditoriums (see education)	—	—	—	—
Lobbies	150	5	0.06	—
Stages, studios	70	10	0.06	—
Ticket booths	60	5	0.06	—
Transportation				
Platforms	100	7.5	0.06	—
Transportation waiting	100	7.5	0.06	—
Workrooms				
Bank vaults/safe deposit	5	5	0.06	—
Darkrooms	—	—	—	1.0
Copy, printing rooms	4	5	0.06	0.5
Meat processingᶜ	10	15	—	—
Pharmacy (prep. area)	10	5	0.18	—
Photo studios	10	5	0.12	—
Computer (without printing)	4	5	0.06	—

For SI: 1 cubic foot per minute = 0.0004719 m³/s, 1 ton = 908 kg, 1 cubic foot per minute per square foot = 0.00508 m³/(s · m²), °C = [(°F) -32]/1.8, 1 square foot = 0.0929 m².

a. Based upon *net occupiable floor area*.

b. Mechanical exhaust required and the recirculation of air from such spaces is prohibited (see Section 403.2.1, Item 3).

c. Spaces unheated or maintained below 50°F are not covered by these requirements unless the occupancy is continuous.

d. Ventilation systems in enclosed parking garages shall comply with Section 404.

e. Rates are per water closet or urinal. The higher rate shall be provided where the exhaust system is designed to operate intermittently. The lower rate shall be permitted only where the exhaust system is designed to operate continuously while occupied.

f. Rates are per room unless otherwise indicated. The higher rate shall be provided where the exhaust system is designed to operate intermittently. The lower rate shall be permitted only where the exhaust system is designed to operate continuously while occupied.

g. Mechanical exhaust is required and recirculation is prohibited except that recirculation shall be permitted where the resulting supply airstream consists of not more than 10 percent air recirculated from these spaces (see Section 403.2.1, Items 2 and 4).

h. For nail salons, each nail station shall be provided with a *source capture system* capable of exhausting not less than 50 cfm per station.

403.3.2.3.2 System ventilation efficiency. The system ventilation efficiency (E_v) shall be determined using Table 403.3.2.3.2 or Appendix A of ASHRAE 62.1.

❖ The system ventilation efficiency is a measure of how well a system delivers outdoor air to the zone that requires the largest ratio of outdoor air to total supply (primary) air (i.e., Maximum Z_p). As can be seen in the table, the larger the outdoor/primary air ratio, the lower the system efficiency, and as can be seen in Equation 4-8, the lower the system efficiency in the denominator, the greater the amount of outdoor air that the system must supply (see Example 2). Appendix A of ASHRAE 62.1 provides a more detailed and, therefore, more precise method for determining the value of E_v where this is preferred to using Table 403.3.2.3.2. According to ASHRAE 62.1, Table 403.3.2.3.2 (ASHRAE Table 6-3) might yield unrealistically low efficiency values (E_v) where the value of Z_p is greater than 0.15; therefore, a designer may want to use the Appendix A methodology and, possibly, reduce the outdoor air intake flow rate determined by Equation 4-8.

As the outdoor air demand of any one zone becomes significantly greater than the demand of the other zones served by the same air-handling system, it becomes more difficult for the air-handling system to ensure delivery of the outdoor air to the high-demand (critical) zone. Thus, there is the need to establish a system efficiency rating. If the entire system was designed to deliver outdoor air to all zones served at the same rate as the "critical" (Maximum Z_p) zone, the intent of the code would be satisfied; however, the noncritical zones would all be over ventilated. Over-ventilation is not a code violation, but it will cause increased energy consumption from heating, cooling, humidification and dehumidification systems, increased fan power and filter maintenance (see Sections 403.3.2.3.3 and 403.3.2.3.4).

TABLE 403.3.2.3.2
SYSTEM VENTILATION EFFICIENCY[a,b]

Max (Z_p)	E_v
≤ 0.15	1
≤ 0.25	0.9
≤ 0.35	0.8
≤ 0.45	0.7
≤ 0.55	0.6
≤ 0.65	0.5
≤ 0.75	0.4
> 0.75	0.3

a. *Max* (Z_p) is the largest value of Z_p calculated using Equation 4-5 among all the zones served by the system.

b. Interpolating between table values shall be permitted.

403.3.2.3.3 Uncorrected outdoor air intake. The uncorrected outdoor air intake flow rate (V_{ou}) shall be determined in accordance with Equation 4-6.

$$V_{ou} = D\Sigma_{all\,zones}R_pP_z + \Sigma_{all\,zones}R_aA_z \qquad \text{(Equation 4-6)}$$

where:

D = Occupant diversity: the ratio of the system population to the sum of the zone populations, determined in accordance with Equation 4-7.

$$D = \frac{P_s}{\Sigma_{all\,zones}P_z} \qquad \text{(Equation 4-7)}$$

where:

P_s = System population: The total number of occupants in the area served by the system. For design purposes, P_s shall be the maximum number of occupants expected to be concurrently in all zones served by the system.

❖ The uncorrected outdoor air intake is the total outdoor air rate that the ventilation system must deliver if the system efficiency (E_v) is 1 (100 percent). Section 403.3.2.3.4 comes into play where the system efficiency is less than 1 (e.g., if E_v is 1, Equation 4-8 becomes $V_{ot} = V_{ou}$). The purpose of this section is to determine the total amount of outdoor air that must be brought in by the ventilation system where such system is serving multiple zones with differing outdoor air rate demands. Recall that this all started with Section 403.3.2.3 and terminates at Section 403.3.2.3.4 (see Examples 2 and 3). The occupant diversity factor is optional, see Example 2.

403.3.2.3.4 Outdoor air intake flow rate. The outdoor air intake flow rate (V_{ot}) shall be determined in accordance with Equation 4-8.

$$V_{ot} = \frac{V_{ou}}{E_v} \qquad \text{(Equation 4-8)}$$

❖ This is the final step that determines the amount of outdoor air that must be brought into the air-handling system at the outdoor air intake opening. The result of this exercise is that the total amount of required outdoor air ends up being less than if the maximum Z_p was applied to all zones. See Examples 2 and 3 in which it is evident that the outdoor air rate can be reduced in accordance with Section 403.3.2.3, compared to supplying outdoor air to all zones served at the same outdoor/primary air fraction (Z_p) as the critical zone (Maximum Z_p).

Example 1: Single-zone Recirculating System

A 3,000-square-foot dining room is served by a rooftop air-handling unit by means of ducted ceiling supply registers and ceiling return grilles. Determine the system outdoor air intake flow rate (V_{ot}) for the dining room.

Application

This dining room with a single rooftop unit is considered to be a single-zone system. In order to determine the outdoor air intake flow rate (V_{ot}) for a single-zone system using Equation 4-3, the breathing zone airflow rate (V_{bz}) of the occupied space must first be determined using Equation 4-1. Then, the zone airflow effectiveness (E_z) must be determined in accordance with Table 403.3.1.2. Next, the zone outdoor airflow rate (V_{oz}) must be determined using Equation 4-2. The outdoor air intake flow rate for a single-zone system is then simply $V_{ot} = V_{oz}$, Equation 4-3.

Step 1: Determine occupant load (P_z) for the room for use in Equation 4-1:

From Table 403.3 for dining rooms, the occupant density of 70 occupants/1,000 ft² is used:

3,000 ft² × 70

1,000 ft² = 210 occupants = P_z

Step 2: Determine the breathing zone outdoor airflow (V_{bz}) for the room:

First, the outdoor air rates for people (R_p) and area (R_a) must be obtained for the dining room from Table 403.3:

R_p = 7.5 cfm/person and R_a = 0.18 cfm/ft²

Equation 4-1 can now be solved:

$V_{bz} = R_p P_z + R_a A_z$

V_{bz} = (7.5 cfm/person × 210 people) + (0.18 cfm/ft² × 3,000 ft²)

V_{bz} = 2,115 cfm

Step 3: Determine the zone outdoor airflow (V_{oz}):

It is given that the room is served by a rooftop air handling unit by means of ducted ceiling supply registers and ceiling return grilles. Thus, the zone air distribution effectiveness (E_z) can be obtained from Table 403.3.1.2. In cooling mode, E_z = 1.0, and in heating mode, E_z = 0.8. The most restrictive value, E_z = 0.8, must be used (assume that Note g of Table 403.3.1.2 does not apply).

Equation 4-2 can now be solved for the room:

$V_{oz} = \dfrac{V_{bz}}{E_z}$

Cooling Mode: $V_{oz} = \dfrac{2,115}{1.0}$ = 2,115 cfm

Heating Mode: $V_{oz} = \dfrac{2,115}{0.8}$ = 2,644 cfm

As a result, the greater system demand for the dining room requires 2,644 cfm of outdoor air in heating mode. This value of V_{oz} should be used to determine the maximum system requirement for outdoor air intake flow rate (V_{ot}), Equation 4-3:

$V_{ot} = V_{oz}$ = 2,644 cfm

In comparison, the amount of outdoor air required using the 2006 code would be based on the same occupant load. However, the outdoor air ratio is 20 cfm per person. Therefore, the amount of outdoor air is 210 occupants × 20 cfm/occupant = 4,200 cfm of outdoor air required.

Example 2: Multiple-zone Recirculating Systems

A single-story 8,000-square-foot office building consists of 7,250 square feet of general office space and a 500-square-foot conference room.

The building is served by a rooftop unit by means of ducted ceiling supply registers and ceiling return grilles. The primary airflow provided by the rooftop unit is 8,100 cfm for the office space and 780 cfm for the conference room. Determine the outdoor air intake flow rate for the system serving the office area and conference room.

Application

This office area and conference room constitutes two separate zones that are served by a single rooftop unit. In order to determine the outdoor air intake flow rate (V_{ot}) for a multiple-zone system using Equation 4-8, the breathing zone airflow rate (V_{bz}) of the occupied space must first be determined using Equation 4-1. Then, the zone airflow effectiveness (E_z) must be determined in accordance with Table 403.3.1.2. Next, the zone outdoor airflow rate (V_{oz}) must be determined using Equation 4-2. After determining the previously listed variables, the outdoor air intake flow rate for a multiple-zone recirculating system must be determined using Sections 403.3.2.3.1 through 403.3.2.3.4. The primary outdoor air fraction (Z_p) must be determined using Equation 4-5. Then the system ventilation efficiency (E_v) is determined in accordance with Table 403.3.2.3.2 and the uncorrected outdoor air intake flow rate (V_{ou}) is determined using Equations 4-6 and 4-7.

Finally, the outdoor air intake flow rate (V_{ot}) is determined using Equation 4-8.

Step 1: Determine occupant loads (P_z) for each zone:

For the general office space (Zone 1), the occupant density of 5 occupants/1,000 ft² is used (from Table 403.3 for office spaces).

P_z (Zone 1) = 7,250 ft² × 5 occupants/1,000 ft² = 37

For the conference room (Zone 2), the occupant density of 50 occupants/1,000 ft² is used.

P_z (Zone 2) = 500 ft² × 50 occupants/1000 ft² = 25

Step 2: Determine the breathing zone outdoor airflow (V_{bz}) for each zone:

First, the outdoor air rates for people (R_p) and area (R_a) must be obtained for each zone from Table 403.3:

Zone 1 values: R_p (Zone 1) = 5 cfm/person and R_a (Zone 1) = 0.06 cfm/ft²

Zone 2 values: R_p (Zone 2) = 5 cfm/person and R_a (Zone 2) = 0.06 cfm/ft²

Equation 4-1 can now be solved for each zone:

$$V_{bz} = R_p P_z + R_a A_z$$

Zone 1: V_{bz} = (5 cfm/person × 37 people) + (0.06 cfm/ft² × 7,250 ft²)

V_{bz} = 620 cfm

Zone 2: V_{bz} = (5 cfm/person × 25 people) + (0.06 cfm/ft² × 500 ft²)

V_{bz} = 155 cfm

Step 3: Determine the zone outdoor airflow (V_{oz}) for each zone:

It is given that the room is served by a rooftop air-handling unit by means of ducted ceiling supply registers and ceiling return grilles. Thus, the zone air distribution effectiveness (E_z) can be obtained from Table 403.3.1.2. In cooling mode, E_z = 1.0, and in heating mode, E_z = 0.8 (assume that Note g of Table 403.3.1.2 does not apply).

Equation 4-2 can now be solved for each zone:

$$V_{oz} = V_{bz}/E_z$$

$$E_z$$

Zone 1: Cooling Mode: V_{oz} = 620 cfm

1.0 = 620 cfm

Heating Mode: V_{oz} = 620 cfm

0.8 = 775 cfm

Zone 2: Cooling Mode: V_{oz} = 155 cfm

1.0 = 155 cfm

Heating Mode: V_{oz} = 155 cfm

0.8 = 194 cfm

Step 4: Determine the primary outdoor air fraction (Z_p) for each zone:

The primary airflow (V_{pz}) is given:

V_{pz} (Zone 1) = 8,100 cfm and V_{pz} (Zone 2) = 780 cfm.

Equation 4-5 can now be solved for each zone:

$$Z_p = V_{oz}$$

$$V_{pz}$$

Zone 1: Z_p = 775 cfm/8,100 cfm = 0.096

Zone 2: Z_p = 194 cfm/780 cfm = 0.25

Step 5: Determine the system ventilation efficiency (E_v):

The largest value of Z_p among all zones served by the system must be used. Therefore, for Z_p = 0.25, Table 403.3.2.3.2 yields E_v = 0.9.

Step 6: Determine the uncorrected outdoor air intake flow rate (V_{ou}):

Note that the occupant diversity (D) calculation is optional, meaning that if no diversity is desired to be applied or if there is insufficient information for applying Equation 4-7, the designer can simply set D equal to 1 in Equation 4-6. The occupant diversity is used to account for occupants who will be either in one zone or the other at any given time. For this example, it is assumed that D is equal to 1 so as to make the results of Section 403.3.2.3 more obvious, thereby making the example more meaningful. If, however, an occupant diversity (D) was applied in this example it would be calculated as follows:

The design assumption is that the conference room (Zone 2) may at times be occupied by outside visitors to conduct a presentation or meet with the staff with the balance of the conference room occupancy consisting of staff. An assumption is made that, on average, 80 percent of the conference room zone occupancy will be comprised of staff that normally occupies the Zone 1 office area.

This results in a system population (P_s) as follows:

P_s = system population = Zone 1 population + Zone 2 visitors

P_s = 37 people + (25 people × 0.2 visitor rate) = 42 occupants which are expected to be concurrently in all zones served by the system. The denominator of Equation 4-7 is simply the sum of the occupant loads of all zones, that being the sum of Zone 1, 37 occupants and Zone 2, 25 occupants.

Equation 4-7 can now be solved for D, occupant diversity:

$$D = \frac{P_s}{\Sigma_{all\ zones} P_z}$$

D = 42 occupants/(37 occupants + 25 occupants) = 42/62 = 0.68

As stated earlier, D = 1 will be used in the remainder of this example.

Equation 4-6 is solved as follows:

$$V_{ou} = D \times \Sigma_{all\ zones} R_p P_z + \Sigma_{all\ zones} R_a A_z$$

V_{ou} = D {[R_p (Zone 1) × P_z (Zone 1)] + [R_p (Zone 2) × P_z (Zone 2)]} + {[(R_a (Zone 1) × A_z (Zone 1)] + [R_a (Zone 2) × A_z (Zone 2)]}

V_{ou} = 1[(5 cfm/person × 37 people) + (5 cfm/ person × 25 people)] + [(0.06 cfm/ft^2 × 7,250 ft^2) + (0.06 cfm/ft^2 × 500 ft^2)]

V_{ou} = 185 + 125 + 435 + 30 = 775 cfm

Step 7: Determine the outdoor air intake flow rate (V_{ot}):

Equation 4-8 gives the adjusted overall outdoor air flow rate required for the system, using the most restrictive value for system ventilation efficiency (Step 5), E_v:

$$V_{ot} = V_{ou}$$
$$E_v$$

V_{ot} = 775 cfm/0.9 = 861 cfm

This is the overall amount of outdoor air supplied by the rooftop unit to both zones.

Now consider what the overall amount of outdoor air would have been if the Maximum Z_p of 0.25 had been applied to both Zones 1 and 2 instead of working through the process of calculating the "corrected" outdoor air intake flow rate as performed in this example.

For Zone 1, the outdoor/primary air fraction (Z_p) of 0.25 would require that 2,025 cfm be supplied to Zone 1 (2,025/8,100 = 0.25).

For Zone 2, the outdoor/primary air fraction (Z_p) of 0.25 would require the same as before, 194 cfm. (194/780 = 0.25).

Adding both zones: 2,025 + 194 = 2,219 cfm. As can be seen, the process of Section 403.3.2.3 has reduced (corrected) the outdoor rate by 1,358 cfm. If the occupant diversity factor (D) had been applied in this example, the true nature of the "correction" from Equation 4-8 would have been obscured.

In comparison, the amount of outdoor air required using the 2006 code would be the following:

Zone 1:

7,250 ft^2 × 7 occupants

1,000 ft^2 = 51 occupants × 20 cfm/occupant = 1,020 cfm OA

Zone 2:

500 ft^2 × 50 occupants

1,000 ft^2 = 25 occupants × 20 cfm/occupant = 500 cfm OA

Total amount of outdoor air supplied by the rooftop unit is 1,520 cfm under the 2006 code.

Example 3: Multiple-zone Recirculating Systems

A wing of a new high school consists of a 2,200-square-foot art classroom, a 1,400-square-foot science lab, a 1,200-square-foot computer lab and a 750-square-foot corridor. The entire wing is to be served by a single central air-handling unit in an adjoining mechanical room, via ceiling supply and ceiling return grilles and registers.

The primary airflow provided for the building consists of 2,800 cfm for the art classroom, 2,400 cfm for the science lab, 1,800 cfm for the computer lab and 500 cfm for the corridor. Determine the outdoor air intake flow rate for the building area.

Step 1: Determine occupant loads (P_z) for each zone:

For the art classroom (Zone 1), Table 403.3 under the Education occupancy, the occupant density is:

P_z (Zone 1) = (2,200 ft^2/1,000 ft^2) × 20 occupants

P_z (Zone 1) = 44 occupants

For the science lab (Zone 2), the maximum occupancy is given as:

P_z (Zone 2) = (1,400 ft^2/1,000 ft^2) × 25 = 35 occupants)

For the computer lab (Zone 3), the maximum occupancy is:

P_z (Zone 3) = (1,200 ft^2/1,000 ft^2) × 25 occupants

P_z (Zone 3) = 30 occupants

For the corridor (Zone 4), no specific occupancy criterion is given; therefore, from Table 403.3 under the Education occupancy, the table says to go to Public Spaces for corridor requirements. The table entry for corridors does not have a value listed in the Occupant Density or People outdoor air rate columns; therefore, P_z (Zone 4) = 0, and only requires the Area Outdoor Airflow Rate in Breathing Zone component to be considered, which will occur later in the example.

Step 2: Determine the breathing zone outdoor airflow (V_{bz}) for each zone:

First, the outdoor air rates for people (R_p) and area (R_a) must be obtained for each zone from Table 403.3:

Zone 1 values R_p (Zone 1) = 10 cfm/person; R_a (Zone 1) = 0.18 cfm/ft^2

Zone 2 values: R_p (Zone 2) = 10 cfm/person; R_a (Zone 2) = 0.18 cfm/ft^2

Zone 3 values: R_p (Zone 3) = 10 cfm/person; R_a (Zone 3) = 0.12 cfm/ft^2

Zone 4 values: R_p (Zone 4) = 0 cfm/person; R_a (Zone 4) = 0.06 cfm/ft^2

Equation 4-1 can now be solved for each zone:

$$V_{bz} = R_p(Pz) + R_a(A_z)$$

Zone 1: V_{bz} (Zone 1) = 10 cfm/person (44 people) + 0.18 cfm/ft^2 (2,200 ft^2)

V_{bz} (Zone 1) = 836 cfm

Zone 2: V_{bz} (Zone 2) = 10 cfm/person (35 people) + 0.18 cfm/ft^2 (1,400 ft^2)

V_{bz} (Zone 2) = 602 cfm

Zone 3: V_{bz} (Zone 3) = 10 cfm/person (30 people) + 0.12 cfm/sq ft (1,200 ft^2)

V_{bz} (Zone 3) = 444 cfm

Zone 4: V_{bz} (Zone 4) = 0 cfm/person (0 people) + 0.06 cfm/sq ft (750 ft^2)

V_{bz} (Zone 4) = 45 cfm

Step 3: Determine the zone outdoor airflow (V_{oz}) for each zone:

It is given that the building is served by a rooftop air-handling unit via ceiling supply and ceiling return. Thus, the zone air distribution effectiveness (E_z) can be obtained from Table 403.3.1.2; in cooling mode, E_z = 1.0, and in heating mode, E_z = 0.8 (assume that Note g to the Table is not applicable in this example).

Equation 4-2 can now be solved for each zone:

$$V_{oz} = V_{bz}/E_z$$

Zone 1: Cooling Mode V_{oz} (Zone 1) = 836 cfm/1.0 = 836 cfm

Heating Mode: V_{oz} (Zone 1) = 836 cfm/0.8 = 1,045 cfm

Zone 2: Cooling Mode: V_{oz} (Zone 2) = 602 cfm/1.0 = 602 cfm

Heating Mode: V_{oz} (Zone 2) = 602 cfm/0.8 = 753 cfm

Zone 3: Cooling Mode: V_{oz} (Zone 3) = 444 cfm/1.0 = 444 cfm

Heating Mode: V_{oz} (Zone 3) = 444 cfm/0.8 = 555 cfm

Zone 4: Cooling Mode: V_{oz} (Zone 4) = 45 cfm/1.0 = 45 cfm

Heating Mode: V_{oz} (Zone 4) = 45 cfm/0.8 = 57 cfm

Step 4: Determine the primary outdoor air fraction (Z_p) for each zone:

The primary airflow (V_{pz}) is given for each Zone: V_{pz} (Zone 1) = 2,800 cfm; V_{pz} (Zone 2) = 2,400 cfm; V_{pz} (Zone 3) = 1,800 cfm; and V_{pz} (Zone 4) = 500 cfm.

The primary airflow (V_{pz}) is given for each Zone: V_{pz} (Zone 1) = 2,800 cfm; V_{pz} (Zone 2) = 2,400 cfm; V_{pz} (Zone 3) = 1,800 cfm; and V_{pz} (Zone 4) = 500 cfm.

$$Z_p = V_{oz}/V_{pz}$$

Zone 1: Z_p (Zone 1) = 1,045 cfm/2,800 cfm = 0.37

Zone 2: Z_p (Zone 2) = 753 cfm/2,400 cfm = 0.31

Zone 3: Z_p (Zone 3) = 555 cfm/1,800 cfm = 0.31

Zone 4: Z_p (Zone 4) = 57 cfm/500 cfm = 0.12

Step 5: Determine the system ventilation efficiency (E_v):

The largest value of Z_p among all zones served by the system must be used; thus, for Z_p = 0.37 (largest value among all zones), Table 403.3.2.3.2 yields a value of 0.7 for E_v.

Step 6: Determine the uncorrected outdoor air intake (V_{ou}):

For occupant diversity (D), Equation 4-7 is used to account for occupants who will either be in one zone or the other at any given time.

For this example, the school district has noted that the art classroom and computer lab will be normally occupied throughout the school day; however, the school district estimates that the science lab will be used only for two or three periods a day for combined classes and special events; thus, it is only occupied roughly 50 percent of the school day. However, since the science lab will draw its population from other areas of the campus, there is no assumed occupant diversity between the classrooms and labs within our analysis. Additionally, the corridor will be normally unoccupied, and will only have occupants between classes; as such, the corridor does not factor into the occupancy diversity calculation, but is considered only for area outdoor air rate.

Since no one zone will draw occupants away from another zone, the occupant diversity value for our example is D = 1.0.

Equation 4-6 can now be solved:

$$V_{ou} = D [\textstyle\sum_{all\ zones} (R_p P_z)] + [\textstyle\sum_{all\ zones} (R_a A_z)]$$

V_{ou} = D {[R_p (Zone 1) × P_z (Zone 1)] + [R_p (Zone 2) × P_z (Zone 2)] + [R_p (Zone 3) × P_z (Zone 3)] + [R_p (Zone 4) × P_z (Zone 4)]} + {[R_a (Zone 1) × A_z (Zone 1)] + [R_a (Zone 2) × A_z (Zone 2)] + [R_a (Zone 3) × A_z (Zone 3)] + [R_a (Zone 4) × A_z (Zone 4)]}

V_{ou} = 1.0 [(10 cfm/person × 44 people) + (10 cfm/person × 35 people) + (10 cfm/person × 30 people) + (0 cfm/person × 0 people)] + [(0.18 cfm/ft^2 × 2,200 ft^2) + (0.18 cfm/ft^2 × 1,400 ft^2) + (0.12 cfm/ft^2 × 1,200 ft^2) + (0.06 cfm/ft^2 × 750 ft^2)]

V_{ou} = [1,090] + [837]

V_{ou} = 1,927 cfm

Step 7: Determine the outdoor air intake flow rate (V_{ot}):

Equation 4-8 gives the adjusted overall outdoor air flow rate required for the system, using the most restrictive value for system ventilation efficiency, E_v:

$$V_{ot} = V_{ou}/E_v$$

V_{ot} = 1,927 cfm/0.7 = 2,753 cfm

Compare with the 2006 code Table 403.3 method:

Art Classroom = $(2,200 \text{ ft}^2/1,000 \text{ ft}^2) \times 50$ occupants = 110 occupants (110 × 15 cfm/occupant = 1,650 cfm)

Science Lab = $(1,400 \text{ ft}^2/1,000 \text{ ft}^2) \times 30$ occupants = 42 occupants (42 × 20 cfm/occupant = 840 cfm)

Computer Lab = $(1,200 \text{ ft}^2/1,000 \text{ ft}^2) \times 50$ occupants = 60 occupants (60 × 15 cfm/occupant = 900 cfm)

Corridor = $(750 \text{ ft}^2 \times 0.05 \text{ cfm/ft}^2 = 38 \text{ cfm})$

In the 2006 edition of the code, (1,650 + 840 + 900 + 38) = 3,428 cfm would have been required.

Note that the science lab and computer lab both require an exhaust system in accordance with Table 403.3, and such system is in addition to the ventilation calculated in this example (see Section 403.4).

403.4 Exhaust ventilation. Exhaust airflow rate shall be provided in accordance with the requirements in Table 403.3. Exhaust *makeup air* shall be permitted to be any combination of outdoor air, recirculated air and transfer air, except as limited in accordance with Section 403.2.

❖ Table 403.3 has an additional column that was new to the 2009 edition of the code; an exhaust airflow cfm/ft² column. The required exhaust rate airflow must occur in addition to any other ventilation rates prescribed by the table. For example, science labs under "Education" require an exhaust rate, an outdoor air rate per person and an outdoor air rate per square footage. Such cases in the table are associated with extraordinary contaminants thereby justifying an exhaust system that will help maintain a pressure differential to prevent the migration of contaminants to other spaces.

It is important to understand the meaning behind the reference to Section 403.2 and to explain how to apply Table 403.3 where the table specifies both outdoor air rates and exhaust rates for the same occupancy. Where this happens in the table (for example, science labs and barber shops), both the outdoor air and exhaust requirements apply and this could be misinterpreted to mean that the outdoor air needed as makeup air for the exhaust system must be in addition to the outdoor air required based on the occupant load and area of the space. Consider a barber shop of 1000 square feet with a zone air distribution effectiveness (E_z) of one. It will require 500 cfm of exhaust ventilation and 247.5 cfm of outdoor air. Rather than requiring a total outdoor air rate of 747.5 cfm to satisfy both ventilation needs (i.e., 247.5 cfm OA ventilation plus 500 cfm OA as makeup for the exhaust), the code intends to allow the 247.5 cfm of outdoor air to serve as part of the makeup air for the exhaust system with the remaining 252.5 cfm of makeup air being from some other source.

The code intends to make sure that a minimum rate of outdoor air is supplied to the space for occu-

pant well-being in the case where the makeup air for the exhaust system is composed entirely of used ventilation air transferred in from some other space. In most designs, the outdoor air required based on occupant load and area will be used to satisfy a portion of the makeup air required for the exhaust system. The exhaust rate specified in the table will almost always exceed the ventilation outdoor air rate, therefore, the makeup air demand for the exhaust system will have to be satisfied by other sources of air in addition to the outdoor air required based on occupant load and area of the space. The additional makeup air may be provided by air transferred directly from another space, air supplied from return air from another space or outdoor air.

403.5 System operation. The minimum flow rate of outdoor air that the ventilation system must be capable of supplying during its operation shall be permitted to be based on the rate per person indicated in Table 403.3 and the actual number of occupants present.

❖ In accordance with Section 403.3, the ventilation system is required to have the airflow capacity calculated in accordance with Table 403.3 and the related subsections. However, in actual operation, the ventilation system is permitted to supply an airflow quantity based on the actual number of occupants that are present in the space being ventilated. The key words in these sections are system "design" and "operation." The design capacity will not always be the capacity at which the system is operated. Note that this section speaks only of the rate per person and the occupant load, not the rate per floor area; therefore, this section would not apply where Table 403.3 calls for only a rate per floor area in an occupancy. No occupant density is given in such cases. This section applies to the people outdoor air rate (R_a) with no apparent allowance for resetting the area outdoor airflow rate (R_a). ASHRAE 62.1 has a provision (Dynamic reset) which is similar to this section and like this section, it does not indicate how this is to be applied to the area outdoor airflow rate (R_a).

With the exception of those locations in Section 406, ventilation is not required in spaces that are unoccupied; however, ventilation could be necessary during unoccupied periods to prevent the accumulation of contaminants injurious to people, building contents or the structure itself. Therefore, ventilation of a space before and after it is occupied can be necessary to dilute contaminants prior to its next occupied period.

The intent of this section is to allow the rate of ventilation to modulate in proportion to the number of occupants. This can result in significant energy savings. Current technology can permit the design of ventilation systems that are capable of detecting the occupant load of the space and automatically adjusting the ventilation rate accordingly. For example, carbon dioxide (CO_2) detectors can be used to sense the level of CO_2 concentrations that are indicative of the number of occupants. People emit predictable quantities of CO_2 for any given activity, and this knowledge can be used to estimate the occupant load in a space. Occupancies, such as large assembly rooms, could use electronic ingress and egress monitoring technology that keeps an accurate count of the number of occupants at any time.

Note that carbon dioxide (CO_2) concentration monitoring is intended only as a means of determining the approximate number of occupants present at any onetime. Carbon dioxide concentration is not intended to be the sole benchmark or measure of ventilation effectiveness. Carbon dioxide is just one of multiple contaminants of concern. It is not the intent of the code to allow ventilation to be controlled solely by the concentration of CO_2 in the space as the only parameter. For example, setting an arbitrary sensor threshold of 1,000 parts per million (ppm) of CO_2 and allowing the ventilation system to cycle off and on based only on the CO_2 concentration being above or below that setpoint does not meet the intent of the code.

403.6 Variable air volume system control. Variable air volume air distribution systems, other than those designed to supply only 100-percent outdoor air, shall be provided with controls to regulate the flow of outdoor air. Such control system shall be designed to maintain the flow rate of outdoor air at a rate of not less than that required by Section 403.3 over the entire range of supply air operating rates.

❖ Variable air volume HVAC systems continuously vary the supply air delivery rates to a room or space based on the need for heating or cooling within that room or space. Active outdoor air controls are necessary to maintain the required outdoor airflow rate at or above the minimum required for a room or space over the entire range of supply air delivery rates. This is especially critical when several variable air volume (VAV) boxes are served by a single air-handling unit. To accomplish this, the percentage of outdoor air in the mixed air supply to the VAV terminals must be automatically adjusted in response to the output of the VAV terminal units.

Because the required ventilation air is strictly outdoor air, any recirculation of air in a space is solely a function of the heating, cooling and air-conditioning system design. Heating, cooling and air-conditioning system design is basically independent of ventilation system design except for the fact that the heating, cooling and air-conditioning system is typically used to supply the required rate of outdoor airflow, and the HVAC system design must account for the sensible and latent heat of the outdoor air. In HVAC systems, such as VAV systems, it is difficult to control the rate of outdoor air being supplied because the volume of air supplied through VAV systems varies depending on the comfort heating and cooling demand on the system, while the ventilation requirements are constant when the building is occupied and the occupant numbers are constant. In addition, the outdoor air is

normally supplied to the main air handler of the system and the individual VAV units do not control the percentage of outdoor air in the air supply from the main air handler. It is not uncommon for HVAC systems to use an independent outdoor air system to ensure that the outdoor air rate is achieved in zones served by VAV systems. Hardware and controls are available that would allow VAV systems to provide feedback to the main air handler to cause continuous resetting of the outdoor air intake rate, thus overcoming the problem of providing ventilation air through a VAV system. See the following example scenarios for VAV systems.

Example 4:

Scenario 1:

Each of three zones requires 50 cfm (0.024 m³/s) of outdoor air.

Assume that VAV boxes are serving unequal zone loads and are operating as follows:

VAV 1 is operating at full output, 200 cfm (0.094 m³/s).

VAV 2 is operating at half output, 100 cfm (0.047 m³/s).

VAV 3 is operating at minimum output, 50 cfm (0.024 m³/s).

- In this scenario, the supply air to the VAV boxes must be 100-percent outdoor air to satisfy Zone 3 because any recirculated air supplied to the system would dilute the air supplied to Zone 3 to less than 100 percent.
- The required outdoor air rate is 150 cfm (0.071 m³/s) total for Zones 1, 2 and 3. When all three VAVs operate at 100-percent outdoor air, the system would be over-ventilating with a total outdoor air rate of 350 cfm (0.165 m³/s).

Scenario 2:

Assume that the same VAV boxes are operating as follows:

VAV 1 is operating at 100 cfm (0.047 m³/s).

VAV 2 is operating at 150 cfm (0.071 m³/s).

VAV 3 is operating at 200 cfm (0.094 m³/s).

- The required outdoor air rate for the supply air to the VAV boxes must be 50 percent to satisfy Zone 1, and since the outdoor air is supplied by the main air handler, all of the other zones would also receive 50 percent outdoor air.
- The required outdoor air rate is still 150 cfm (0.071 m³/s) total for all zones.
- The system would again be over ventilating with a total outdoor air rate of 225 cfm (0.106 m³/s).

403.7 Balancing. The *ventilation air* distribution system shall be provided with means to adjust the system to achieve at least the minimum ventilation airflow rate as required by Sections 403.3 and 403.4. Ventilation systems shall be balanced by an *approved* method. Such balancing shall verify that the ventilation system is capable of supplying and exhausting the airflow rates required by Sections 403.3 and 403.4.

❖ Balancing of HVAC systems involves the adjustment of the system so that it will perform in the manner intended by the designer. It is an essential part of the final testing and approval of the system and should be performed by experienced and qualified individuals. Balancing of the ventilation system is absolutely necessary to ensure that the system satisfies the requirements of the code. The code does not require balancing of heating and cooling systems where the systems are not supplying ventilation air; however, the IECC does require, in Section 503.2.9.1, that the HVAC system provide means for balancing (see Section 603.17). To ensure that the system can be properly balanced, a sufficient number of dampers and flow-balancing devices must be installed in the system. This section intends to verify that the code-prescribed rate of ventilation is achieved.

SECTION 404
ENCLOSED PARKING GARAGES

404.1 Enclosed parking garages. Mechanical ventilation systems for enclosed parking garages shall be permitted to operate intermittently in accordance with Item 1, Item 2 or both.

1. The system shall be arranged to operate automatically upon detection of vehicle operation or the presence of occupants by approved automatic detection devices.

2. The system shall be arranged to operate automatically by means of carbon monoxide detectors applied in conjunction with nitrogen dioxide detectors. Such detectors shall be installed in accordance with their manufacturers' recommendations.

❖ This section applies to parking garages, not repair garages. As a means of conserving energy, ventilation systems for enclosed public parking garages are allowed to alternate or modulate between full capacity and the minimum capacity specified in Section 404.2. The alternating or modulating operation must be accomplished using approved detection devices that are capable of detecting the operation of vehicles or the presence of occupants in the garage, or by using CO and NO$_x$ detectors in combination. This section is referenced by Note d of Table 403.3 and is intended to apply to parking garages for vehicles such as automobiles, buses and fleet vehicles.

The code does not specify the method of detecting vehicles or occupants. It is left to the designer to select the appropriate detectors for the particular application. Note that the detection system cannot rely solely on measuring carbon monoxide (CO) because other gaseous and particulate contaminants can reach harmful concentrations long before CO levels rise to the detection threshold of the CO detectors. The intent is either to detect the presence of occupants and/or vehicle operation by some means such as infrared motion detectors, or control the exhaust system by means of both CO and NO$_x$ detectors used together.

404.2 Minimum ventilation. Automatic operation of the system shall not reduce the ventilation airflow rate below 0.05 cfm per square foot (0.00025 m^3/s · m^2) of the floor area and the system shall be capable of producing a ventilation airflow rate of 0.75 cfm per square foot (0.0038 m^3/s · m^2) of floor area.

❖ The required ventilation rate for enclosed public parking garages is 0.75 cfm/ft^2 of floor area; however, the rate at which the system operates is permitted to be reduced to not less than 0.05 cfm/ft^2 when the detection systems described in Section 404.1 indicate that there are no occupants or vehicles operating in the garage, or the CO and NO$_x$ concentration levels are within design limits. The exhaust system must be capable of operating automatically at the full rate required by this section and Table 403.3 (see commentary, Section 404.1). This section simply provides an alternative to continuous ventilation at the 0.75-cfm/ft^2 rate by allowing a reduced continuous rate when the detection system says that it is safe to do so.

404.3 Occupied spaces accessory to public garages. Connecting offices, waiting rooms, ticket booths and similar uses that are accessory to a public garage shall be maintained at a positive pressure and shall be provided with ventilation in accordance with Section 403.3.

❖ This section requires that occupied accessory spaces in public garages be adequately separated from the public garage parking area so that a positive pressure can be maintained in the accessory spaces. Maintenance of a positive pressure with respect to the parking areas is intended to prevent contaminants from migrating into spaces that are occupied for extended periods. In addition, compliance with Section 403.3 and Table 403.3 is required. The positive pressure required by this section is to be maintained at all times that the accessory spaces are occupied. Note that the ventilation air for the enclosed accessory spaces must come from the outdoors, not the interior of the parking garage, otherwise, it would make no sense to require positive pressurization to keep out garage contaminants.

SECTION 405
SYSTEMS CONTROL

405.1 General. Mechanical ventilation systems shall be provided with manual or automatic controls that will operate such systems whenever the spaces are occupied. Air-conditioning systems that supply required *ventilation air* shall be provided with controls designed to automatically maintain the required outdoor air supply rate during occupancy.

❖ This section requires that the system that supplies ventilation air, whether it is a dedicated ventilation air system or HVAC system, have controls that will cause ventilation to occur whenever the space is occupied. A dedicated ventilation air system can have manual or automatic controls. If the ventilation air is supplied by an HVAC system, the controls must be automatic and set to supply the required ventilation air at any time the space is occupied. Automatic control is easily accomplished by timers, thermostats with fan programming ability and energy management systems (see commentary, Section 401.3).

SECTION 406
VENTILATION OF UNINHABITED SPACES

406.1 General. Uninhabited spaces, such as crawl spaces and attics, shall be provided with *natural ventilation* openings as required by the *International Building Code* or shall be provided with a mechanical exhaust and supply air system. The mechanical exhaust rate shall be not less than 0.02 cfm per square foot ($0.00001 \text{ m}^3/\text{s} \cdot \text{m}^2$) of horizontal area and shall be automatically controlled to operate when the relative humidity in the space served exceeds 60 percent.

❖ This section addresses spaces that are not intended to be occupied. The ventilation prescribed herein is intended to control temperature, humidity and vapors that emanate from the structure or the earth.

In lieu of natural ventilation required by the IBC, mechanical ventilation can be supplied in accordance with this section.

Without adequate ventilation, such spaces can suffer damage to structural and insulation materials resulting from high temperatures and excessive moisture and molds. These conditions can also affect the performance and the life span of mechanical equipment, appliances and electrical components located in the spaces. This section specifies the required exhaust rate when mechanical ventilation is chosen as the means of ventilation for spaces described herein. Such mechanical systems normally consist of nothing more than an exhaust fan and properly located air inlet openings to the outdoors.

The mechanical ventilation system must be operated automatically by a humidistat control. Operation by a control that senses only relative humidity is appropriate for crawl spaces; however, it may be desirable for attic ventilation systems to also be activated by temperature-sensitive controls for summer seasons. Note that the outdoor air being supplied to such spaces might have a higher relative humidity than the control setpoint; therefore, the system will run continuously and perhaps elevate the relative humidity of the space. In other words, trying to maintain the relative humidity in a space to 60 percent or less will require that the outdoor ventilation air have a relative humidity of 60 percent or less. So called

"smart vents" are available that can recognize when bringing in humid outdoor air will not accomplish the intent of this section.

Bibliography

The following resource materials were used in the preparation of the commentary for this chapter of the code:

IBC-12, *International Building Code.* Washington, DC: International Code Council, 2011.

IECC-12, *International Energy Conservation Code.* Washington, DC: International Code Council, 2011.

IFGC-12, *International Fuel Gas Code.* Washington, DC: International Code Council, 2011.

Chapter 5:
Exhaust Systems

General Comments

Where contaminants are known to be present in quantities that are irritating or harmful to the occupants' health or are hazardous in a fire, both naturally and mechanically ventilated spaces must be equipped with mechanical exhaust systems capable of collecting and removing the contaminants.

As the chapter title implies, Chapter 5 is a compilation of exhaust-system-related code requirements ranging from clothes dryer exhaust and hazardous exhaust to smoke control systems. This chapter addresses exhaust air systems mostly for their impact on the level of fire safety performance provided by the building, and not so much from the standpoint of airflow design or system efficiency. The design, construction, installation, alteration and repair of commercial kitchen exhaust systems are addressed as well.

The code regulates the materials and methods used for constructing and installing ducts, system controls, exhaust, equipment, fire protection systems, smoke control systems and related components. Air brought into the building for ventilation, combustion or makeup is protected from contamination by the outdoor discharge provisions of this chapter. The *International Building Code®* (IBC®) is referenced repeatedly in this chapter for fire protection measures such as fireblocking, draftstopping, fire-resistance-rated assemblies, penetration protection, smoke control systems, and fire protection and suppression systems. A kitchen exhaust system serving commercial cooking appliances is one of the specialized exhaust systems specifically covered in Chapter 5. A commercial cooking appliance can generate large quantities of air contaminants, such as grease vapors, water vapors, smoke, fuel-gas combustion byproducts and waste heat. Without a properly designed kitchen

exhaust system, serious health and fire hazards can develop.

The code official must be aware of the effect that exhaust systems can have on the space conditioning and ventilation systems. For example, improperly designed kitchen exhaust systems can create negative pressure conditions, which can affect the operation of fuel-burning equipment and appliances, the operation of the kitchen and other exhaust systems, and the operation of doors.

Chapter 5 of the code also addresses exhaust systems for some specific occupancies and activities that have the potential for introducing toxic, flammable or otherwise hazardous substances into the indoor environment. The smoke control provisions of Section 513 provide an occupiable route for evacuation or relocation of building occupants in case of a fire. These provisions are not intended to protect contents, enable quick restoration of operations or assist with fire suppression.

Purpose

Chapter 5 provides guidelines for reasonable protection of life, property and health from the hazards associated with exhaust systems, air contaminants and smoke development in the event of a fire. In most cases, these hazards involve materials and gases that are flammable, explosive, toxic or otherwise hazardous. This chapter contains requirements for the installation of exhaust systems, with an emphasis on the structural integrity of the systems and equipment involved and the overall impact of the systems on the fire safety performance of the building. Design considerations, such as energy efficiency, cost effectiveness, system efficiency and convenience, are the responsibility of the design professional.

SECTION 501
GENERAL

501.1 Scope. This chapter shall govern the design, construction and installation of mechanical exhaust systems, including exhaust systems serving clothes dryers and cooking appliances; hazardous exhaust systems; dust, stock and refuse conveyor systems; subslab soil exhaust systems; smoke control systems; energy recovery ventilation systems and other systems specified in Section 502.

❖ This chapter regulates the design, construction and installation of exhaust equipment. The scope of coverage includes the types of exhaust systems specifi-

cally addressed in the chapter. Note that the scope of Chapter 4 is general ventilation and the scope of this chapter is specific mechanical exhaust systems. This chapter applies to systems handling hazardous and nonhazardous exhaust, as well as those exhaust systems necessary for specific equipment, operations and sources of contamination. The definition of "Mechanical exhaust system" in Chapter 2 does not specifically mention specialized exhaust systems such as refuse conveyors, commercial kitchen systems and those smoke control systems that include mechanical exhaust systems. This section states that Chapter 5 covers these systems.

501.2 Independent system required. Single or combined mechanical exhaust systems for environmental air shall be independent of all other exhaust systems. Dryer exhaust shall be independent of all other systems. Type I exhaust systems shall be independent of all other exhaust systems except as provided in Section 506.3.5. Single or combined Type II exhaust systems for food-processing operations shall be independent of all other exhaust systems. Kitchen exhaust systems shall be constructed in accordance with Section 505 for domestic equipment and Sections 506 through 509 for commercial equipment.

❖ To minimize the potential for spreading contaminants, hazardous exhaust, fire and smoke to other parts of the building, most exhaust systems are prohibited from connecting with dissimilar exhaust systems serving the building. Without complete isolation, the fire, health and explosion hazards inherent in some exhaust systems cannot be confined to those systems, would jeopardize other systems and parts of the building. For example, a Type I exhaust system conveying commercial kitchen exhaust must not share ducts or exhaust equipment with any other exhaust system, including other commercial kitchen exhaust systems, except as provided for in Section 506.3.5 (see commentary, Section 506.3.5). This is because of the potential for spreading smoke, grease-laden air and fire from the kitchen exhaust system to other parts of the building. Although grease is not involved, there is the potential for fire and smoke to spread from the kitchen to other parts of the building; therefore, exhaust ducts serving Type II exhaust systems are prohibited from interconnecting with any other exhaust ducts other than Type II exhaust ducts.

Hazardous exhaust systems must be independent in accordance with Section 510.4 (see commentary, Section 510.4). This section does not prohibit exhaust systems exhausting environmental air from sharing common ducts, risers, fans and other exhaust system components. Environmental air is defined as air that is conveyed from occupied spaces such as ventilation for human usage, domestic kitchen range exhaust, bathroom exhaust and domestic dryer exhaust. Note that while domestic dryers are included in the definition of "Environmental air," this section specifically prohibits clothes dryers from being connected to any other exhaust system. Clothes dryer exhaust systems can only be combined where serving clothes dryers located in a multistory building in accordance with Section 504.8.

501.3 Exhaust discharge. The air removed by every mechanical exhaust system shall be discharged outdoors at a point where it will not cause a nuisance and not less than the distances specified in Section 501.3.1. The air shall be discharged to a location from which it cannot again be readily drawn in by a ventilating system. Air shall not be exhausted into an attic or crawl space.

Exceptions:

1. Whole-house ventilation-type attic fans shall be permitted to discharge into the attic space of *dwelling units* having private attics.

2. Commercial cooking recirculating systems.

❖ The primary intent of this section is to avoid exhausting contaminants into areas that may be occupied by people or into other buildings. The term "nuisance" is a legal term commonly thought of as that which is dangerous to human life or detrimental to health. Although there are certainly situations in which a health hazard is clearly present, other "nuisance" conditions that represent an unreasonable interference with the enjoyment of life and property require a subjective decision by the code official. Throughout this subjective evaluation, the code official should consider a "nuisance" as something that is dangerous to human life or detrimental to health, or renders the air or human food, drink or water supply unwholesome.

A nuisance is much more than or much worse than simply an annoyance or an inconvenience. The scope of the term is rather broad in intent, and allows the code official to decide what may or may not constitute a nuisance. Unfortunately, it is not easy to determine whether or not a nuisance will be present. The conditions under which an exhaust system performs vary considerably with the change of seasons, ambient temperatures and prevailing winds. The code official should gather as much information as possible regarding the installation to evaluate the hypothetical worst case. This would include the characteristics and geometry of the installation, as well as the local ambient conditions, so that an educated decision can be made to determine the "nuisance effect" of an exhaust outlet.

Similarly, to prevent the introduction of contaminants into the ventilation air of a building, exhaust opening(s) must not direct exhaust so that it could "...be readily drawn in by a ventilating system." In this situation, the code official must determine an appropriate separation or location for the placement of intake and exhaust openings. The evaluation of each installation should consider the orientation of the exhaust or intake opening and its location relative to a source of contaminants or adjacent intake opening, as well as to the direction of the prevailing winds at the location.

Bear in mind that the air exhausts discharging from a dwelling unit (clothes dryer, kitchen and bathroom) are generally not considered to be significantly hazardous and are of low volume. Furthermore, placing the source of contamination above an air intake takes advantage of the fact that when sources of contamination are lighter (less dense) than the surrounding air they will rise above the vicinity of an air intake located below. The requirements for the termination of commercial kitchen exhaust systems, refuse conveyor systems, subslab exhaust systems and smoke control systems are located in Sections 506.3.13, 506.4.2, 506.5.5, 511.2, 512.4 and 513.10.3, respectively, while requirements for termination of vents and chimneys are in Chapter 8 and the *International Fuel Gas Code*® (IFGC®).

Attics and crawl spaces are not considered to be outdoors. Exhaust ducts cannot terminate in these spaces. Exhaust ducts must connect directly to terminals that pass through the building envelope to the outside atmosphere. Pointing, aiming or similarly directing an exhaust pipe or duct at an opening in the envelope of the building (that is, attic louver, grille, ridge vent, eave vent or soffit vent) in no way ensures that all or any of the exhaust will reach the outdoors. In fact, it is possible that the majority, if not all, of the exhaust vapors and gases will discharge to the attic space rather than to the outdoors.

In the case of a duct that terminates in a ventilated soffit, the exhaust can rise into the attic as opposed to falling through the perforated soffit. The flow of air through an opening for attic ventilation is dependent on attic temperature, wind direction, wind speed and opening configuration and location, making attic air movement unpredictable and often in a direction not intended. Additionally, grilles and louvers offer resistance and interfere with the exhaust flow directed at them. This may cause deflection of exhaust back into the attic.

Exception 1 to this section addresses a mechanical alternative to the natural ventilation of a dwelling unit. A whole-house ventilation-type attic fan is less source specific than a dryer, bathroom or kitchen exhaust system. Whole-house ventilation systems are installed to provide a fresh airflow for the occupants either continuously or at timed intervals. The exception is specific to whole-house ventilation-type attic fans that create an airflow through open windows. Source-specific fans (bathroom exhaust or kitchen exhaust) would not qualify for the exception because their primary function is to exhaust moisture, odors, fumes or products of combustion. The contaminants associated with these functions might be harmful to the occupants, detrimental to the structural integrity of the building or even impact the fire safety performance of the building, and must discharge directly outdoors. The exception applies only to whole-house fans dedicated solely to ventilation and comfort cooling. Within the context of the code, a "private attic" is one in which air discharged from one tenant space cannot enter another tenant's attic space.

Exception 2 addresses recirculating systems for commercial cooking appliances. The cooking effluent is treated to remove grease and smoke particles, and is then returned to the kitchen area rather than discharging to the outdoors. The system must be listed and designed for the cooking appliances being served (see Section 507.1).

501.3.1 Location of exhaust outlets. The termination point of exhaust outlets and ducts discharging to the outdoors shall be located with the following minimum distances:

1. For ducts conveying explosive or flammable vapors, fumes or dusts: 30 feet (9144 mm) from property lines; 10 feet (3048 mm) from operable openings into buildings; 6 feet (1829 mm) from exterior walls and roofs; 30 feet (9144 mm) from combustible walls and operable openings into buildings which are in the direction of the exhaust discharge; 10 feet (3048 mm) above adjoining grade.

2. For other product-conveying outlets: 10 feet (3048 mm) from the property lines; 3 feet (914 mm) from exterior walls and roofs; 10 feet (3048 mm) from operable openings into buildings; 10 feet (3048 mm) above adjoining grade.

3. For all *environmental air* exhaust: 3 feet (914 mm) from property lines; 3 feet (914 mm) from operable openings into buildings for all occupancies other than Group U, and 10 feet (3048 mm) from mechanical air intakes. Such exhaust shall not be considered hazardous or noxious.

4. Exhaust outlets serving structures in flood hazard areas shall be installed at or above the elevation required by Section 1612 of the *International Building Code* for utilities and attendant equipment.

5. For specific systems see the following sections:

   5.1. Clothes dryer exhaust, Section 504.4.

   5.2. Kitchen hoods and other kitchen exhaust *equipment*, Sections 506.3.13, 506.4 and 506.5.

   5.3. Dust stock and refuse conveying systems, Section 511.2.

   5.4. Subslab soil exhaust systems, Section 512.4.

   5.5. Smoke control systems, Section 513.10.3.

   5.6. Refrigerant discharge, Section 1105.7.

   5.7. Machinery room discharge, Section 1105.6.1.

❖ This section details the requirements for the termination points of exhaust ducts. Except for a few notable exceptions (see commentary, Section 501.2), the requirements for termination points for all exhaust ducts are conveniently located in this section. This section gives distances that must be maintained, depending on the type of exhaust, and is more specific than the general requirement that the discharge of exhaust must not create a nuisance. In addition, see Chapter 8 and the IFGC for specific regulations for the location of exhaust vent terminations for fuel-fired appliances.

Item 1 details the requirements for termination points for exhaust ducts that convey explosive or flammable vapors, fumes or dusts, like those exhaust systems that serve operations involving the application of flammable finishes (see Section 502.7), hazardous exhaust systems (see Section 510), and dust, stock and refuse conveyor systems (see Section 511). The intent of this section is to reduce the exposure from the dangerous vapors in the exhaust. This is done to:

1. Protect other parts of the building;

2. Protect other buildings;

3. Reduce a potential reaction from materials that may be incompatible; and

4. Reduce the severity of a fire, in case of an ignition.

Flammable finishes, as well as vapors that are considered flammable, including dusts, have more restrictive termination requirements than other vapors due to the potential for ignition. To avoid recirculation of flammable vapors, fumes or dusts back into the building, the exhaust outlets and ducts must be designed and located to reduce such exposures. This may be achieved by separating the exhaust outlet from openings in the building, walls and roof where sources of ignition or incompatible materials may be present [see Commentary Figures 501.3.1(1) and 501.3.1(2)].

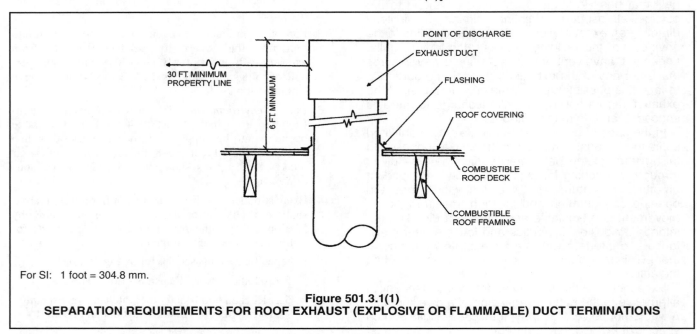

For SI: 1 foot = 304.8 mm.

Figure 501.3.1(1)
SEPARATION REQUIREMENTS FOR ROOF EXHAUST (EXPLOSIVE OR FLAMMABLE) DUCT TERMINATIONS

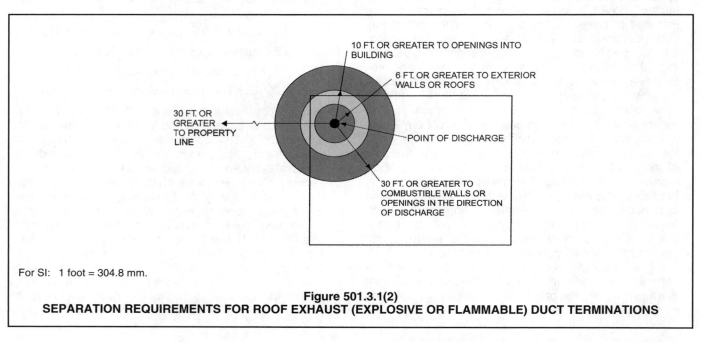

For SI: 1 foot = 304.8 mm.

Figure 501.3.1(2)
SEPARATION REQUIREMENTS FOR ROOF EXHAUST (EXPLOSIVE OR FLAMMABLE) DUCT TERMINATIONS

For health and safety reasons, hazardous exhaust cannot be directed onto adjacent property. Maintaining the required distance allows the hazardous contents of the exhaust to disperse into the atmosphere, thereby minimizing the exposure of the adjoining property to the potential ignition hazard of a burning ember or spark, or the hazardous, noxious and objectionable odors emitted from such systems. Wind and wind-induced eddy currents can react with building structural surfaces to create air pressure zones that can diminish exhaust flow or redirect exhaust into nearby building openings, such as fresh air or combustion air intakes and operable windows.

Exhaust systems sometimes incorporate rotating hoods over the discharge opening to prevent high winds from restricting the flow of exhaust gases out of the system. The hoods align themselves with the direction of the wind to allow the unimpeded and, sometimes, induced discharge from the exhaust outlet. The termination height specified for combustible walls is more restrictive to allow the concentration of explosive or flammable constituents in the exhaust to diminish before landing or accumulating on a combustible wall. The code official must consider prevailing wind conditions in locating hazardous exhaust outlets with respect to other building openings.

Item 2 details the provisions regarding the termination points of other product-conveying ducts, such as those exhausting nonflammable and nonexplosive dusts and waste products. This item states the location of exhaust outlets to be well into the undisturbed wind stream and away from the cavity and wake (eddy) zones around the building. This counteracts the negative effects of wind-induced conditions and also prevents the reentry of exhaust products into the building through openings and fresh air intakes.

Item 3 details the provisions for termination points for exhaust ducts that convey environmental air. "Environmental air" is defined in Section 202 as "air that is conveyed to or from occupied areas through ducts which are not part of the heating or air-conditioning system, such as ventilation for human usage, domestic kitchen range exhaust, bathroom exhaust and domestic clothes dryer exhaust." In general, air that is exhausted from the occupied space, where the occupied space is not subject to any contaminants from a process or operation, would have to meet the termination requirements in Item 3. Note that since the definition of "Environmental air" includes domestic kitchen range exhaust, bathroom exhaust and domestic clothes dryer exhaust, these exhaust terminations would only be required to be 3 feet (914 mm) from operable openings into a building (i.e., windows and doors). Since these three exhaust sources are common in multiple-family residential buildings where the dwelling units are naturally ventilated, the exhaust terminations that penetrate through the exterior wall will only need to be located 3 feet (914 mm) from windows that are providing natural ventilation to the dwelling unit.

Considering that the exhaust from a dwelling unit is not considered to be hazardous or noxious and is of low volume, the 3-foot (914 mm) separation from windows seems reasonable. In accordance with Section 401.4, Item 2, mechanical and gravity outdoor air intakes must be located not less than 10 feet (3048 mm) from any hazardous or noxious contaminant, however, environmental air exhaust is not considered to be hazardous or noxious. Therefore, a 3-foot (914 mm) separation from gravity openings and a 10-foot (3048 mm) separation from mechanical air intakes would be required for exhaust from a bathroom or kitchen and domestic clothes dryer exhaust in a residential dwelling unit.

Based on the definition of environmental air, mechanical exhaust from a parking garage can also be considered environmental air since the parking deck is intended to be occupied by the people parking their cars and the exhaust system is not part of the heating or air-conditioning system. While the vehicle exhaust would be considered a contaminant, the exhaust rate of 0.75 cubic foot per minute per square foot (cfm/ft^2) for a parking garage is considered to be sufficient to provide the occupants with a safe environment. Therefore, the exhaust from a parking garage would then be treated the same as any other environmental air exhaust.

Item 4, for obvious reasons, requires termination points to be located at or above the design flood level where the building is located in a flood hazard area.

Item 5 directs the user to other sections of the code for specific exhaust systems. Note that while users are directed to Section 504.4 for clothes dryer exhaust, the requirements in this section are for general termination and installation. As discussed above for Item 3, clothes dryer exhaust must meet the specific termination requirements for environmental air ducts. In addition, the reference to Section 511 for dust, stock and refuse conveying systems is for outlets for exhaust that exceed 600°F (315°C) that must be designed as a chimney in accordance with Table 511.2. Outlets for exhaust that do not exceed 600°F (315°C) must be in compliance with the termination requirements in Item 1.

501.3.1.1 Exhaust discharge. *Exhaust air* shall not be directed onto walkways.

❖ Exhaust air terminations must not be directed onto walkways. For obvious health reasons, exhaust air cannot be directed onto walkways in such a manner that the users of the walkway are subjected to the exhaust airstream.

501.3.2 Exhaust opening protection. Exhaust openings that terminate outdoors shall be protected with corrosion-resistant screens, louvers or grilles. Openings in screens, louvers and grilles shall be sized not less than $^1/_4$ inch (6 mm) and not larger than $^1/_2$ inch (13 mm). Openings shall be protected against local weather conditions. Louvers that protect exhaust openings in structures located in hurricane-prone regions, as defined in the *International Building Code*, shall comply with

AMCA Standard 550. Outdoor openings located in exterior walls shall meet the provisions for exterior wall opening protectives in accordance with the *International Building Code*.

❖ Exhaust air terminations must be equipped with corrosion-resistant screens, grilles or louvers to prevent foreign objects (such as insects or debris) from entering the system or the building. Also, the openings must be protected against the entry of falling or wind-driven water, snow and ice. Additionally, in hurricane-prone areas, louvers must be tested to determine their water rejection capabilities. Tests conducted in accordance with the standard are intended to show that the louver will limit water infiltration during high-velocity wind conditions. This allows a facility, either essential or nonessential, to continue to operate with the exhaust system in operation.

Exhaust systems sometimes incorporate rotating hoods over the opening to prevent high winds from restricting the flow of gases out of the system. The hoods align themselves with the direction of the wind to allow unimpeded and, actually induced, discharge from the exhaust outlet. Rotating turbines are also used to serve as both weather protection and as a means of inducing airflow.

The opening size in screens, louvers and grilles must be within the specified range. The opening size must be large enough to inhibit blockage by debris and to prevent significant resistance to airflow, and yet must be small enough to keep out what is intended to be kept out. A screen of the stated size would restrict the passage of rodents and large insects, and would be resistant to blockage by lint, debris and plant fibers.

501.4 Pressure equalization. Mechanical exhaust systems shall be sized to remove the quantity of air required by this chapter to be exhausted. The system shall operate when air is required to be exhausted. Where mechanical exhaust is required in a room or space in other than occupancies in R-3 and *dwelling units* in R-2, such space shall be maintained with a neutral or negative pressure. If a greater quantity of air is supplied by a mechanical ventilating supply system than is removed by a mechanical exhaust for a room, adequate means shall be provided for the natural or mechanical exhaust of the excess air supplied. If only a mechanical exhaust system is installed for a room or if a greater quantity of air is removed by a mechanical exhaust system than is supplied by a mechanical ventilating supply system for a room, adequate *makeup air* consisting of supply air, transfer air or outdoor air shall be provided to satisfy the deficiency. The calculated building infiltration rate shall not be used to satisfy the requirements of this section.

❖ Unlike natural ventilation, mechanical ventilation does not depend on unknown air pressure differentials between the indoors and outdoors to create airflow. Mechanical ventilation has the advantage of being both predictable and dependable because it is not subject to all of the variables that affect natural ventilation. The volume of air supplied to a space and the volume of air removed from a space must be approximately equal. Otherwise, the space will be either positively or negatively pressurized and the actual flow rate will be equivalent to the lower rate of either the air supply or the air exhaust. The minimum required airflow rate must be attainable and unaffected by the building's exhaust system(s). Slight negative or positive pressurization of spaces is often desirable to control contaminant or odor migration to or from a space and is, therefore, not prohibited. In the context of the code, negative and positive air pressures refer to pressures that are below or above atmospheric pressure or a reference pressure in another room or space.

Additionally, spaces served by exhaust systems must be supplied with makeup air to replace the air exhausted. The code is clear that the makeup air must be a reliable source, such as supply air or transfer air, rather than relying on the infiltration rate of the building. Relying on the infiltration rate is not acceptable since variables, such as air pressure differential and wind velocity, can result in inadequate makeup air being supplied. Makeup air is commonly provided in the form of "transfer air" that has been drawn (transferred) from adjoining spaces or directly from the outdoors. Exhaust systems with transfer makeup air are typically used where control of odor migration is desired. For example, toilet, bath and locker rooms often receive makeup air from adjoining assembly areas; kitchens receive makeup air from dining areas; and elevators and smoking lounges receive makeup air from lobby and open-floor areas. Conversely, where additional air is supplied to a space above and beyond the air that is exhausted from that space, means must be provided to discharge the excess supply air, either by natural or mechanical means. Gravity relief dampers may be used in such circumstances to dissipate excess air to the outdoors when a designated setpoint pressure is reached to maintain a space at neutral pressure. Air in spaces designated by Notes b and g of Table 403.3 must be exhausted to the outdoors and, therefore, cannot be recirculated or transferred to any other space. Exhaust systems can negatively affect other building systems and appliances; therefore, any such effects must be considered in the design, plan review and inspection phases of construction.

501.5 Ducts. Where exhaust duct construction is not specified in this chapter, such construction shall comply with Chapter 6.

❖ The function of any duct is to convey air between specific points in an air distribution or exhaust system. Chapter 6 contains requirements for the construction of ducts used for general exhaust and makeup air. The construction of clothes dryer exhaust systems; commercial kitchen exhaust systems; hazardous exhaust systems; dust, stock and refuse conveying systems; subslab soil exhaust systems; and smoke control systems are also subject to the provisions of Chapter 6, as well as Sections 504,

506, 510, 511, 512 and 513, respectively. As referenced by this section, Chapter 6 addresses additional performance aspects of ductwork relating to construction, installation, dimensional and structural stability, leakage control, thermal performance, durability, support, vapor permeance and airflow resistance. This section recognizes that Chapter 5 contains duct construction requirements that are specific to certain systems and, therefore, such specific requirements override the general requirements of Chapter 6. The duct systems addressed in Chapter 5 are subject to any requirements of Chapter 6 that are not overridden by a requirement in Chapter 5. Because of this section, no exhaust duct construction is left unregulated.

SECTION 502
REQUIRED SYSTEMS

502.1 General. An exhaust system shall be provided, maintained and operated as specifically required by this section and for all occupied areas where machines, vats, tanks, furnaces, forges, salamanders and other *appliances, equipment* and processes in such areas produce or throw off dust or particles sufficiently light to float in the air, or which emit heat, odors, fumes, spray, gas or smoke, in such quantities so as to be irritating or injurious to health or safety.

❖ Stationary local sources, in addition to those specifically identified, include equipment, machinery, processes, chemical handling and storage, and special-purpose rooms or areas that produce contaminants in quantities that are a health hazard to the building occupants. Exhaust systems must be designed to collect and remove the contaminants. This section addresses specific sources of contaminants and the requirements for exhaust systems to control them. The exhaust systems required by this section are independent of and in addition to any natural or mechanical ventilation required by Chapter 4.

The code recognizes that natural and mechanical ventilation methods, in themselves, cannot effectively collect and remove contaminants produced by local sources. This section is written in performance language and requires a case-by-case evaluation by the designer and the code official to determine the quantity and type of contamination present, whether any action needs to be taken, and, if so, the method of contaminant control and removal. As with all exhaust systems addressed in the code, the exhaust must extend to and terminate to the outdoors. The exhaust termination must be approved and comply with Section 501.3.

Spaces served or affected by exhaust systems must be supplied with makeup air to replace the air exhausted (see commentary, Section 501.4). The volume of makeup air introduced to the space(s) must be approximately equal to the volume of air exhausted.

502.1.1 Exhaust location. The inlet to an exhaust system shall be located in the area of heaviest concentration of contaminants.

❖ To be effective, an exhaust system intake must be located where the exhaust airflow will pick up, entrain and transport the contaminant. The exhaust inlet should be as close as possible to the source of contamination and should be designed to capture the contaminants before they can disperse into the space.

[F] 502.1.2 Fuel-dispensing areas. The bottom of an air inlet or exhaust opening in fuel-dispensing areas shall be located not more than 18 inches (457 mm) above the floor.

❖ If fuel is dispensed within a building, the required exhaust system must be designed with intake openings located within 18 inches (457 mm) of the floor so that any accumulated, heavier-than-air vapors will be pulled into the exhaust flow. Airflow must be designed to circulate across the entire exposed floor area to dilute and exhaust vapors within 18 inches (457 mm) of the floor and to reduce the potential for accumulation of flammable vapor-air mixtures (see commentary, Section 304.3).

502.1.3 Equipment, appliance and service rooms. *Equipment, appliance* and system service rooms that house sources of odors, fumes, noxious gases, smoke, steam, dust, spray or other contaminants shall be designed and constructed so as to prevent spreading of such contaminants to other occupied parts of the building.

❖ This section parallels Section 502.1 with the intent to control the spread of offensive or harmful contaminants to other parts of the building. This section is stated in performance language and does not specify that the means of contaminant control must incorporate an exhaust system.

[F] 502.1.4 Hazardous exhaust. The mechanical exhaust of high concentrations of dust or hazardous vapors shall conform to the requirements of Section 510.

❖ Hazardous exhaust systems are addressed in Section 510. Where a fire or explosion hazard exists as defined in Section 510.2, a mechanical exhaust system would be subject to the requirements for hazardous exhaust systems (see commentary, Section 510).

[F] 502.2 Aircraft fueling and defueling. Compartments housing piping, pumps, air eliminators, water separators, hose reels and similar *equipment* used in aircraft fueling and defueling operations shall be adequately ventilated at floor level or within the floor itself.

❖ The vapor from aircraft fuel is highly flammable and is heavier than air. As a result, the vapors accumulate near the floor, requiring the exhaust system to be at floor level or in the floor with ducts running under the slab. Venting of compartments housing this equipment provides both dilution air to keep possible air-vapor mixtures below the flammable range and airflow for dissipation of any vapors present.

[F] 502.3 Battery-charging areas for powered industrial trucks and equipment. Ventilation shall be provided in an *approved* manner in battery-charging areas for powered industrial trucks and *equipment* to prevent a dangerous accumulation of flammable gases.

❖ Charging lead-acid or nickel-iron batteries is a process of electrolysis in which oxides created by the operation of the battery are reduced to metal and redeposited on the electrode plates. The process results in the rejuvenation of the electrolyte in the battery and the emission of both oxygen and hydrogen gases, as well as corrosive fumes. Hydrogen is highly flammable and if these gases are allowed to accumulate in an enclosed space, they could eventually reach an ignitable or detonable level. The charging area must be ventilated to carry off and dilute the concentrations of hazardous gases. This section is stated in performance language and does not provide a specific exhaust rate that will prevent the accumulation of an ignitable or detonable gas (see commentary, Section 502.4).

[F] 502.4 Stationary storage battery systems. Stationary storage battery systems, as regulated by Section 608 of the *International Fire Code*, shall be provided with ventilation in accordance with this chapter and Section 502.4.1 or 502.4.2.

> **Exception:** Lithium-ion batteries shall not require ventilation.

❖ Section 608 of the *International Fire Code*® (IFC®) applies to certain sizes of stationary storage battery systems that are used for standby power, emergency power or uninterruptible power systems (UPS). It does not apply to individual stationary storage batteries such as those used in vehicles, personal computer UPS devices, emergency lights, fire alarm panels and similar installations. This section is intended to apply to battery charging systems regulated by Section 608 of the IFC. The requirements in Section 608 of the IFC apply only to nonrecombinant (flooded) batteries having an electrolyte capacity of more than 50 gallons (189 L) and to recombinant [valve-regulated lead-acid (VRLA) and lithium-ion] batteries having an electrolyte capacity of more than 1,000 pounds (454 kg). For a more detailed discussion of the different types of stationary lead-acid battery systems, see the IFC commentary to Section 608.

Generally, with lead-acid battery systems, the main concern is the production of hydrogen and oxygen within an enclosed space. With good air circulation around a battery, hydrogen accumulation is normally not a problem. Hydrogen has a wide flammability range and is the lightest element on the Periodic Table of Elements, having an atomic number of 1. The gas mixture is explosive when the amount of hydrogen in air exceeds 4 percent by volume. To address the concern of hydrogen generation and containment in small areas, a minimum ventilation

criterion is set. This section provides two methods of compliance in Sections 502.4.1 and 502.4.2 from which the designer may choose. The exception recognizes the reduced fugitive gas hazard of sealed batteries that contain higher viscosity electrolytes.

[F] 502.4.1 Hydrogen limit in rooms. For flooded lead acid, flooded nickel cadmium and VRLA batteries, the ventilation system shall be designed to limit the maximum concentration of hydrogen to 1.0 percent of the total volume of the room.

❖ This method of compliance is a performance-based approach that sets the maximum concentration of hydrogen at 1.0 percent of the total volume of the room and leaves it to the designer to design the system to maintain that level of concentration or less. If the exhaust fans do not run continuously, they must be able to be turned on automatically when hydrogen accumulation exceeds 1.0 percent of the total volume of the room. This method requires an analysis of plausible failure scenarios to justify the ventilation rate ultimately chosen.

[F] 502.4.2 Ventilation rate in rooms. Continuous ventilation shall be provided at a rate of not less than 1 cubic foot per minute per square foot (cfm/ft^2) [0.00508 m^3/(s · m^2)] of floor area of the room.

❖ This method of compliance is a prescriptive method that establishes a minimum required continuous ventilation rate of 1 cfm/ft^2 [0.00508 m^3/ (s · m^2)] of room area. If the hydrogen limit in Section 502.4.1 cannot be maintained at this exhaust rate, a higher rate would be required.

502.4.3 Supervision. Mechanical ventilation systems required by Section 502.4 shall be supervised by an approved central, proprietary or remote station service or shall initiate an audible and visual signal at a constantly attended on-site location.

❖ The ventilation system in Section 502.4 is required to reduce the likelihood that the concentration of hydrogen will reach 1 percent or present an explosion hazard. Without a supervised system or a signal at a constantly attended location, the required ventilation system could fail without warning, allowing hydrogen concentrations to build to hazardous levels.

[F] 502.5 Valve-regulated lead-acid batteries in cabinets. Valve-regulated lead-acid (VRLA) batteries installed in cabinets, as regulated by Section 608.6.2 of the *International Fire Code*, shall be provided with ventilation in accordance with Section 502.5.1 or 502.5.2.

❖ VRLA batteries differ substantially from the stationary lead-acid batteries addressed in Section 502.4 in design, operation and, especially, potential hazard. A VRLA battery consists of sealed cells furnished with a valve that opens to vent the battery whenever the internal pressure exceeds the ambient pressure by a predetermined amount. This increases the possibility of an explosive amount of hydrogen gas accumulating in the battery cabinet. Compliance with this section can be accomplished by either of two methods: limiting hydrogen concentrations (see Section 502.5.1) or providing a minimum ventilation rate (see Section 502.5.2).

Section 608.1 of the IFC limits the scope of VRLA batteries covered by this section to large systems having an electrolyte capacity greater than 50 gallons (189 L), such as those commonly used in cellular telephone sites. Personal computer UPS devices, emergency lights, fire alarm panels and similar installations are intended to be excluded from the provisions of this section.

[F] 502.5.1 Hydrogen limit in cabinets. The cabinet ventilation system shall be designed to limit the maximum concentration of hydrogen to 1.0 percent of the total volume of the cabinet during the worst-case event of simultaneous boost charging of all batteries in the cabinet.

❖ VRLA batteries are commonly placed in battery cabinets that are exhausted to the outdoors. This method requires that the hydrogen-generation calculations be performed for the worst-case scenario of all batteries in the room being charged simultaneously. The limit for hydrogen concentration and the methods of achieving that limit are the same as those discussed in Section 502.4.1.

[F] 502.5.2 Ventilation rate in cabinets. Continuous cabinet ventilation shall be provided at a rate of not less than 1 cubic foot per minute per square foot (cfm/ft^2) [0.00508 m^3/(s · m^2)] of the floor area covered by the cabinet. The room in which the cabinet is installed shall also be ventilated as required by Section 502.4.1 or 502.4.2.

❖ The formula for determining the exhaust rate for the ventilation of the battery cabinet is the same as that discussed in Section 502.4.2. Note that the ventilation rate is based on the floor area covered by the cabinet and not the area within the cabinet. This section also requires that the room in which the battery cabinet is located be ventilated by one of the methods in Section 502.4.1 or 502.4.2, in addition to the required ventilation for the cabinet. This provides a back-up ventilation system in case the cabinet leaks or the cabinet ventilation system fails.

502.5.3 Supervision. Mechanical ventilation systems required by Section 502.5 shall be supervised by an approved central, proprietary or remote station service or shall initiate an audible and visual signal at a constantly attended on-site location.

❖ The ventilation system in Section 502.5 is required to reduce the likelihood that the concentration of hydrogen will reach 1 percent or present an explosion hazard. Without a supervised system or a signal at a constantly attended location, the required ventilation system could fail without warning, allowing hydrogen concentrations to build to hazardous levels

[F] 502.6 Dry cleaning plants. Ventilation in dry cleaning plants shall be adequate to protect employees and the public

in accordance with this section and DOL 29 CFR Part 1910.1000, where applicable.

❖ Dry cleaning refers to the cleaning of clothing, wearing apparel, fabrics, textiles, drapes, rugs and similar materials using chemical solvents. These solvents can be flammable or combustible and toxic. The intent of this section is to provide a level of ventilation in dry cleaning plants that is adequate to protect the plant employees and the public from the hazards associated with dry cleaning operations using any of the various classes of dry cleaning solvents, both in approved dry cleaning machines, and in spotting and pretreating operations in the open. To achieve this objective, this section requires that mechanical ventilation systems comply with the provisions of Sections 502.6.1 through 502.6.3. Compliance with the applicable provisions of the Occupational Safety and Health Administration (OSHA) workplace regulations pertaining to air contaminants contained in DOL 29 CFR, Part 1910.1000 is also required.

[F] 502.6.1 Type II systems. Type II dry cleaning systems shall be provided with a mechanical ventilation system that is designed to exhaust 1 cubic foot of air per minute for each square foot of floor area (1 cfm/ft^2) [0.00508 m^3/(s · m^2)] in dry cleaning rooms and in drying rooms. The ventilation system shall operate automatically when the dry cleaning *equipment* is in operation and shall have manual controls at an *approved* location.

❖ Type I dry cleaning systems use Class I flammable liquid solvents [flash points less than 73°F (23°C)] and are prohibited by the IFC because of the dangers involved with Class I solvents (see IFC commentary, Chapter 12). Type II systems use Class II combustible liquid solvents [flash points 100°F (38°C) to 140°F (60°C)], such as Stoddard solvent. These solvents are generally safer than Class I solvents and are typically used in closed systems. Nevertheless, because the potential exists for fugitive vapors, NFPA 70 classifies the entire dry cleaning room of a Type II plant as a Class I, Division 2, hazardous (classified) location.

This section intends to prevent solvent vapors from Type II dry cleaning systems from accumulating to an ignitable concentration in the room in which such appliances are located by providing adequate mechanical ventilation. The prescribed rate of mechanical ventilation will also help prevent the ambient temperature in the room from exceeding the flash point of the solvent being used in the cleaning process. It is not the intent that this system be classified as a "hazardous exhaust system" as addressed in Section 510. However, due to the nature of the vapors being exhausted this system should be independent of all other exhaust and ventilation systems.

The required ventilation rate of 1 cfm/ft^2 [0.00508 m^3/(s · m^2)] of floor area is typical of that required in hazardous-materials-related areas where fugitive flammable vapors must be diluted or removed. Although its exact technical origins are not known, this requirement may have been derived from a rule of thumb (possibly from the insurance industry) that established a ventilation requirement in flammable and combustible liquid use areas of six air changes per hour and that has proven to be effective over the years. Though providing effective ventilation and hazard mitigation, the rate of airflow in industrial or storage buildings with larger-than-average floor-to-floor and floor-to-roof dimensions would require large, costly mechanical equipment installations and create concerns over energy conservation. To deal with those concerns, a design ceiling height of 10 feet (3048 mm) was assumed in recognition of the fact that solvent vapors, being heavier than air, will gather at the lowest point in the room. Each square foot of building area, then, would represent 10 cubic feet (0.28 m^3) and, at the rate of six air changes per hour, 60 cubic feet (1.7 m^3) of air per hour would be moved, which yields 1 cfm/ft^2 [0.00508 m^3/(s · m^2)] per square foot of room area.

To be effective, the exhaust equipment must be interlocked with the dry cleaning equipment to be in simultaneous operation with it. Manual controls are also required for additional flexibility and reliability in the event that mechanical system run-on after the dry cleaning system is shut down is desired or needed. The location of the controls is to be approved by the code official.

[F] 502.6.2 Type IV and V systems. Type IV and V dry cleaning systems shall be provided with an automatically activated exhaust ventilation system to maintain a minimum of 100 feet per minute (0.51 m/s) air velocity through the loading door when the door is opened.

> **Exception:** Dry cleaning units are not required to be provided with exhaust ventilation where an exhaust hood is installed immediately outside of and above the loading door which operates at an airflow rate as follows:

$$Q = 100 \times A_{LD} \qquad \text{(Equation 5-1)}$$

where:

Q = Flow rate exhausted through the hood, cubic feet per minute.

A_{LD} = Area of the loading door, square feet.

❖ This section intends to prevent solvent vapors from Type IV and V dry cleaning systems from escaping into the room in which the appliances are located by drawing the required exhaust airflow into the unit through the open door at the minimum velocity of 100 feet per minute (0.51 m/s). In this way, exposure of employees (in Type IV systems) or the public (in Type V systems) to potentially harmful solvent vapors is minimized. The exhaust capability contemplated by this section is integral with the dry cleaning unit. Its operation must be interlocked with the unit door to automatically start the required exhaust airflow as soon as the unit door is opened. It is not the intent

that this system be classified as a hazardous exhaust system as addressed in Section 510. However, due to the nature of the vapors being exhausted, this system should be independent of all other exhaust and ventilation systems.

The exception provides a design alternative to achieve the goal of solvent vapor capture at the door opening through the installation of, what is often referred to as, an "eyebrow hood" located immediately above the dry cleaning unit's door opening so that the exhaust airflow will sweep across the loading door opening of the machine and capture any escaping solvent vapors. This kind of hood may be integral with the unit or may be an after-market-installed accessory to the unit. In either case, it must be either interlocked with the unit door to automatically start the required exhaust airflow as soon as the unit door is opened or operate continuously whenever the dry cleaning machine is in operation.

Exhaust hoods are not designed for a constant air velocity in all parts of the hood, such as can be achieved by drawing air into a dry cleaning machine through a machine-loading door opening. Therefore, the exhaust capability for this alternative design calculated in accordance with Equation 5-1 will be expressed as a flow rate (cfm) rather than a fixed velocity. The fan will be drawing in a large volume of environmental air to achieve the prescribed exhaust airflow across the entire loading door opening. For example, applying the formula to a dry cleaning unit that has a 2-foot-diameter (610 mm) circular loading/unloading door, the exhaust flow rate (Q) would be calculated as follows:

$$Q = 100 \times A_{LD}$$
$$A_{LD} = \pi\, r^2$$
where = π = 3.14 and r = 1, therefore
$$Q = 100 \times (3.14 \times 12)$$
$$Q = 100 \times 3.14$$
$$Q = 314 \text{ cfm } (0.148 \text{ m}^3/\text{s})$$

It is not the intent that this system be classified as a "hazardous exhaust system" as addressed in Section 510. However, due to the nature of the vapors being exhausted, this system should be independent of all other exhaust and ventilation systems.

[F] 502.6.3 Spotting and pretreating. Scrubbing tubs, scouring, brushing or spotting operations shall be located such that solvent vapors are captured and exhausted by the ventilating system.

❖ Spotting and pretreating operations consist of soaking or direct local application of cleaning solvents to articles about to be dry cleaned. These operations are performed to concentrate the solvent's cleaning power on the removal of more stubborn stains, such as those from grease, oils, make-up, paint, dirt or petroleum products, such as tar, asphalt sealer, etc. The operations typically take place on spotting tables or in scrubbing tubs in the open in the dry cleaning plant. The IFC limits solvent use for spotting or pre-treating to Class II or III solvents and, by exception, allows up to 1 gallon (3.8 L) of strictly controlled Class I solvent and limits these operations to Type II, III and IV dry cleaning plants. When locating the spotting or pretreating operation within the plant, care must be taken to choose a location that will allow adequate airflow on all sides of the spotting or pretreating equipment to maximize the effectiveness of the exhaust system (required by Section 502.6.1) in preventing the accumulation of hazardous vapor concentrations.

[F] 502.7 Application of flammable finishes. Mechanical exhaust as required by this section shall be provided for operations involving the application of flammable finishes.

❖ This section is focused on the hazards posed by the vapors and particulates generated when flammable or combustible materials are used as finishes. These provisions come from the IFC and are found primarily in Chapter 15. Flammable finishes can include spray finishing, floor finishing, dip tanks and automobile undercoating.

[F] 502.7.1 During construction. Ventilation shall be provided for operations involving the application of materials containing flammable solvents in the course of construction, *alteration* or demolition of a structure.

❖ Fires are more likely to occur during construction because there are typically more combustible materials present in the form of construction materials and debris. Additionally, when buildings are being constructed, altered or demolished, there is a tendency to forget that the same hazards or increased fire hazards exist, especially since fire protection systems may not yet be operational. Therefore, this section requires specifically that ventilation be provided during those operations.

[F] 502.7.2 Limited spraying spaces. Positive mechanical ventilation which provides a minimum of six complete air changes per hour shall be installed in limited spraying spaces. Such system shall meet the requirements of the *International Fire Code* for handling flammable vapors. Explosion venting is not required.

❖ A limited spraying space is a small area in which spot painting and touch up can occur. It is limited, in the IFC, to 9 square feet (0.83 m²) but can occur in more areas than where typical spray finishing would be allowed. One of the criteria to allow this activity is that the ventilation system must be sized to provide six air changes per hour. However, if the quantity of hazardous material in the room or control area in which the limited spray area is located exceeds the allowable quantities for hazardous materials given in Chapter 50 or 57 of the IFC, both exhaust ventilation requirements for flammable vapors in those chapters and the requirement in this section have to be met. A typical example would be a plating shop where allowable quantities of corrosives and toxins are exceeded (Group H-4) and where exhaust ventilation at a rate

of 1 cfm/ft² [0.00508 m³/(s · m²)] of floor area is required. If a limited spray area is located in such a room, the most restrictive requirement has to be met; that is, the greater of 1 cfm/ft² [0.00508 m³/(s · m²)] or six air changes per hour. The vapor density of the material should be taken into account (whether vapors are heavier or lighter than air) when considering the location of exhaust inlets. Additionally, conformance to the general requirements of Section 502 is recommended, especially Section 502.1.1 ("inlet to an exhaust system shall be located in the area of heaviest concentration of contaminants"). The IFC has additional restrictions, such as frequency of spraying and requirements for classified electrical areas.

[F] 502.7.3 Flammable vapor areas. Mechanical ventilation of flammable vapor areas shall be provided in accordance with Sections 502.7.3.1 through 502.7.3.6.

❖ The following sections are focused on the ventilation of areas. Flammable vapor areas are any areas in which flammable and combustible liquids are found in vapor or atomized form. "Flammable vapor areas" are defined in Section 202 of the IFC and include, but are not limited to, the interior of spray booths, spray rooms, ducts or any portion of the building that is likely to contain these vapors. The terms "spray booth" and "spray room" are also defined in the IFC. Flammable vapor areas may include areas just outside of the openings of a spray booth because those areas may contain overspray. The proper design and installation of exhaust ventilation systems in flammable vapor areas is critical because of the potential production of large amounts of flammable vapors in the processes. Exhaust systems in spray finishing areas must also comply with Section 510, which regulates hazardous exhaust systems. Additionally, compliance with general Section 502, especially Section 502.1.1, is required. The location of the heaviest concentration resulting from the physical characteristics of a material (for example, vapor density) or the process (such as atomization of the material) must be considered when inlets to exhaust systems are designed.

[F] 502.7.3.1 Operation. Mechanical ventilation shall be kept in operation at all times while spraying operations are being conducted and for a sufficient time thereafter to allow vapors from drying coated articles and finishing material residue to be exhausted. Spraying *equipment* shall be interlocked with the ventilation of the flammable vapor area such that spraying operations cannot be conducted unless the ventilation system is in operation.

❖ For effective ventilation, the system must run continuously during the spraying operation phase and the drying phase of a process when vapors are generated. The interlock between the ventilation and spraying equipment will reduce the likelihood of human error, such as the operator's failure to activate the ventilation system prior to the use of the spray equipment. Operating spray equipment without having the exhaust system in operation could allow fugitive flammable vapors to migrate to areas that may contain unprotected ignition sources and ignite. Continuous operation of ventilation systems could also be an integral component of the design of certain alternative fire-extinguishing systems, such as dry chemical systems.

[F] 502.7.3.2 Recirculation. Air exhausted from spraying operations shall not be recirculated.

Exceptions:

1. Air exhausted from spraying operations shall be permitted to be recirculated as *makeup air* for unmanned spray operations provided that:

    1.1. The solid particulate has been removed.

    1.2. The vapor concentration is less than 25 percent of the lower flammable limit (LFL).

    1.3. *Approved equipment* is used to monitor the vapor concentration.

    1.4. An alarm is sounded and spray operations are automatically shut down if the vapor concentration exceeds 25 percent of the LFL.

    1.5. In the event of shutdown of the vapor concentration monitor, 100 percent of the air volume specified in Section 510 is automatically exhausted.

2. Air exhausted from spraying operations is allowed to be recirculated as *makeup air* to manned spraying operations where all of the conditions provided in Exception 1 are included in the installation and documents have been prepared to show that the installation does not pose a life safety hazard to personnel inside the spray booth, spraying space or spray room.

❖ The prohibition of recirculation is specifically directed to the concern that exhaust system makeup air not be added to the fire hazard of the spray area. The exceptions are fairly specific and focus on the lower flammable limit (LFL) and the monitoring of that limit. The exceptions are based on the environmental community's concerns about the volume of emissions generated by spray finishing. The permissible exposure limit (PEL) is generally more restrictive than the LFL.

The recirculation of exhausted air containing flammable vapors could help spread the hazard of flash fires from the area of vapor generation to other parts of the building. At best, it would adversely affect only the area of vapor generation in that it could render the exhaust ventilation useless. It may also give the operators of the facility a false sense that the level of ventilation is safe for the system.

There are two exceptions, one for unmanned spray operations and the other for manned spray operations. The requirements are the same except that for manned spray operations there is an increased concern over how the recirculated air will affect personnel involved in the spraying operations. Therefore,

specific evidence must be available to demonstrate that recirculation will not be harmful to the operators.

Exception 1 lists five conditions that must be met for the recirculation of exhausted air in unmanned spray operations. If solid particulates (dusts) are removed from the exhausted air, the exhausted air is no longer considered a fire hazard because the potential for a dust explosion is eliminated. "Flammable vapors" are defined by Section 202 in the IFC as flammable constituents in air that exceed 25 percent of the LFL. By reducing the flammable vapor concentrations to less than 25 percent of the LFL, the exhausted air is no longer considered a flammable vapor. To increase the likelihood that the concentrations remain at less than 25 percent of the LFL, flammable vapor detection systems must automatically shut down the operations, set off an alarm and exhaust 100 percent of the air. This would allow for energy conservation in unmanned operations, without compromising the safety features typically associated with such ventilation systems.

Exception 2 is applicable to manned operations where additional documentation is need to reduce the risk to personnel, such as people operating the spray equipment, from an injury or life safety standpoint. This documentation may include a risk analysis of fire and health hazards associated with the operation of this equipment when some portion of the exhausted air is recirculated.

In general, fire safety efforts have received some assistance from the environmental regulations that have placed restrictions on the amount of volatile organic compounds (VOCs) emitted to the atmosphere. This, in turn, limits the types of materials used and lowers the flammability characteristics through the use of less volatile liquids.

[F] 502.7.3.3 Air velocity. Ventilation systems shall be designed, installed and maintained such that the average air velocity over the open face of the booth, or booth cross section in the direction of airflow during spraying operations, is not less than 100 feet per minute (0.51 m/s).

❖ To make sure that flammable vapors are kept within a designated spray area and that the amount of overspray is limited, the code requires that the exhaust system be sized to maintain an average velocity over the open face of the booth or booth cross section of not less than 100 linear feet per minute (0.51 m/s), which is the minimum velocity to capture particulate spray material. Velocities exceeding 200 linear feet per minute (1.02 m/s) have been determined to be too great for this purpose. To determine the minimum ventilation/exhaust capacity in cfm, multiply the booth width (feet) by booth height (feet) by 100 (linear feet per minute).

[F] 502.7.3.4 Ventilation obstruction. Articles being sprayed shall be positioned in a manner that does not obstruct collection of overspray.

❖ When ventilation systems for spray operations are designed and installed, the configuration and position of the object being sprayed must be considered because it might disrupt the ventilation pattern in both cross-draft and down-draft ventilation spray booths, thereby compromising the effectiveness of the ventilation system.

[F] 502.7.3.5 Independent ducts. Each spray booth and spray room shall have an independent exhaust duct system discharging to the outdoors.

Exceptions:

1. Multiple spray booths having a combined frontal area of 18 square feet (1.67 m²) or less are allowed to have a common exhaust where identical spray-finishing material is used in each booth. If more than one fan serves one booth, such fans shall be interconnected so that all fans operate simultaneously.

2. Where treatment of exhaust is necessary for air pollution control or energy conservation, ducts shall be allowed to be manifolded if all of the following conditions are met:

    2.1. The sprayed materials used are compatible and will not react or cause ignition of the residue in the ducts.

    2.2. Nitrocellulose-based finishing material shall not be used.

    2.3. A filtering system shall be provided to reduce the amount of overspray carried into the duct manifold.

    2.4. Automatic sprinkler protection shall be provided at the junction of each booth exhaust with the manifold, in addition to the protection required by this chapter.

❖ This section requires independent duct exhaust of residue from spray-finishing operations. These ducts must be routed directly to the exterior of the building to reduce the likelihood of incompatible materials reacting or a fire or explosion in one booth or duct spreading to other booths or ducts.

Exception 1 applies to very small spray booths where the vapor area is very small compared to the area of standard spray booths. Because these individual smaller spray booths are considered as one fire area from a ventilation standpoint, all identical materials are to be used when this exception applies. This will ensure that incompatible materials are not used in booths with a common exhaust system.

Exception 2 notes that because the exhausted air is at times treated, it can be manifolded. However, special hazards must be avoided to ensure fire protection safety. Incompatible materials must be separated in case of a reaction within the ducts.

Nitrocellulose and nitrocellulose-based products are unstable materials that can easily be ignited and, once ignited, need large quantities of water for suppression. Additionally, nitrocellulose is incompatible with many materials (alkalis, amines, etc.). The cleaning products used in other booths may even ignite the nitrocellulose-based products in a mani-

folded exhaust system. Therefore, the exhaust of nitrocellulose-based products is considered an exception to this section.

Additional protection, such as filtering and sprinklers at the junction of each booth exhaust, is also required.

[F] 502.7.3.6 Fan motors and belts. Electric motors driving exhaust fans shall not be placed inside booths or ducts. Fan rotating elements shall be nonferrous or nonsparking or the casing shall consist of, or be lined with, such material. Belts shall not enter the duct or booth unless the belt and pulley within the duct are tightly enclosed.

❖ This requirement within the ventilation section is intended to reduce sources of ignition from spark-producing elements. This section prohibits electric motors that drive exhaust fans from being placed within booths or ducts. This would increase the likelihood that overspray in the booth or duct cannot accumulate on the motor housing, which could ultimately cause the motor to overheat. Products that are subject to sparking should be avoided within spray areas. Belts that drive exhaust fans are not permitted within the spraying area because they are a potential ignition hazard, unless the belts and pulleys are tightly enclosed to prevent solvents in exhaust air from degrading the belt materials and causing a failure of the ventilation system.

Fans used for exhaust systems must also be of nonsparking materials or, alternatively, the surface of the fan housing must be of a nonsparking material. This requirement prevents a fan from causing sparks if it strikes the casing. Note, however, that the term "nonsparking" is somewhat inaccurate. Parts made of brass and similar nonsparking materials produce sparks with ignition energies too low to ignite flammable vapors. Nevertheless, such parts should be designed carefully to avoid producing sufficient frictional heat to cause an ignition.

[F] 502.7.4 Dipping operations. Flammable vapor areas of dip tank operations shall be provided with mechanical ventilation adequate to prevent the dangerous accumulation of vapors. Required ventilation systems shall be so arranged that the failure of any ventilating fan will automatically stop the dipping conveyor system.

❖ This section includes performance-based language that requires ventilation to prevent the dangerous accumulation of vapors. Dipping operations, by nature, have a large surface area of the dipping material and product exposed to the atmosphere. It is important to note that this section does not dictate a particular air change rate because the ventilation system design depends on the operations being conducted and the type of liquid being used. Some materials are more volatile than others; i.e., more readily emit vapors to the atmosphere. Additionally, conveyor systems used with dipping operations need to be interlocked with the ventilation system to avoid continuation of operations in the event of a ventilation system failure.

[F] 502.7.5 Electrostatic apparatus. The flammable vapor area in spray-finishing operations involving electrostatic apparatus and devices shall be ventilated in accordance with Section 502.7.3.

❖ Electrostatic spray-finishing and powder-coating operations electrically charge the materials, which helps to attract the coating material to the object being painted or coated. This generally reduces the amount of overspray generated, which is both more fire safe and more efficient. Electrostatic operations, while producing less overspray, would still require ventilation in accordance with the requirements for normal spray operations as described in Section 502.7.3. This includes the interlocking of the operations with the ventilation systems.

[F] 502.7.6 Powder coating. Exhaust ventilation for powder-coating operations shall be sufficient to maintain the atmosphere below one-half of the minimum explosive concentration for the material being applied. Nondeposited, air-suspended powders shall be removed through exhaust ducts to the powder recovery system.

❖ As with dipping operations, the purpose of this section is to keep the atmosphere below an explosive concentration. Powder would be dealt with as a combustible dust. Powder coating creates explosive atmospheres because of the large surface areas of the particles when dispersed in the air. The explosive limit or concentration, just as with flammable vapors, will depend on the type of material being used. For example, Factory Mutual tested several powder-coating materials and found a range of lower explosive limit (LEL) of 0.026 to 0.097 ounce per square foot (oz/ft^2) and autoignition temperatures between 790°F to 1,039°F (412°C to 559°C). Therefore, ventilation system requirements may vary from one type of coating to another. Additionally, a collection system is required to collect any unused powder.

This requirement is similar to the ventilation requirements for flammable vapors in its intent to limit the amount of vapors or dusts to a concentration that would not support ignition. Therefore, "one-half the minimum explosive concentration" for powder coating is similar in intent to the "25 percent of the lower flammable limit (LFL)" in Section 502.7.3.2. The safety factor in the combustible dust is 2, while the safety factor for the LFL is 4. This is partly the result of the lower ignitability of a dust compared to a vapor.

[F] 502.7.7 Floor resurfacing operations. To prevent the accumulation of flammable vapors during floor resurfacing operations, mechanical ventilation at a minimum rate of 1 cfm/ft^2 [0.00508 m^3/(s · m^2)] of area being finished shall be provided. Such exhaust shall be by *approved* temporary or portable means. Vapors shall be exhausted to the exterior of the building.

❖ An issue that has not been properly addressed in past codes is the concept of ventilation of areas where floors are being refinished. In some of the older codes, this subject was addressed only for bowling alley floor resurfacing. These activities are now addressed explicitly by the *International Codes®* (I-Codes®) and would require a minimum ventilation rate of 1 cfm/ft^2 [0.00508 m^3/(s · m^2)] of space being refinished even in the most remote areas or corners where finishing and surfacing takes place. This criterion is similar to that used for other sections of the I-Codes, such as for the ventilation of rooms housing lead-acid battery systems. This particular ventilation can be either temporary or permanent; i.e., the normal building ventilation system, if adequate, or perhaps the use of several fans and building openings. Please note that floor surfacing and finishing do not occur often in a building. Because outside companies are typically hired to do such surfacing, protection systems, such as exhaust, are temporary and can therefore be portable. To prevent recirculation of exhaust air that may contain flammable vapors, the exhausted air must be discharged to the exterior of the building. Note that the code's scope excludes portable equipment and this requirement would likely be enforced by a fire code official because a mechanical code official would typically have no reason to inspect a maintenance procedure such as floor resurfacing. Section 2410 in the IFC has similar requirements.

[F] 502.8 Hazardous materials—general requirements. Exhaust ventilation systems for structures containing hazardous materials shall be provided as required in Sections 502.8.1 through 502.8.5.

❖ Chapter 50 of the IFC contains guidance for the storage, dispensing, use and handling of hazardous materials. Section 5004.3 of the IFC contains the same ventilation requirements as Sections 502.8.1 and 502.8.1.1 of the code.

[F] 502.8.1 Storage in excess of the maximum allowable quantities. Indoor storage areas and storage buildings for hazardous materials in amounts exceeding the maximum allowable quantity per control area shall be provided with mechanical exhaust ventilation or *natural ventilation* where *natural ventilation* can be shown to be acceptable for the materials as stored.

Exceptions:

1. Storage areas for flammable solids complying with Section 5904 of the *International Fire Code*.

2. Storage areas and storage buildings for fireworks and explosives complying with Chapter 56 of the *International Fire Code*.

❖ Indoor storage areas and buildings must be equipped with either mechanical exhaust ventilation or natural ventilation to maintain the level of vapors below the LFL or the PEL. Keeping the area/building at these levels provides a level of safety for the area. The first exception exempts flammable solids that meet the requirements of the IFC and the second exempts fireworks and explosives that meet the requirements of the IFC.

[F] 502.8.1.1 System requirements. Exhaust ventilation systems shall comply with all of the following:

1. The installation shall be in accordance with this code.

2. Mechanical ventilation shall be provided at a rate of not less than 1 cfm per square foot [0.00508 m³/(s · m²)] of floor area over the storage area.

3. The systems shall operate continuously unless alternate designs are *approved*.

4. A manual shutoff control shall be provided outside of the room in a position adjacent to the access door to the room or in another *approved* location. The switch shall be a break-glass or other *approved* type and shall be *labeled*: VENTILATION SYSTEM EMERGENCY SHUTOFF.

5. The exhaust ventilation shall be designed to consider the density of the potential fumes or vapors released. For fumes or vapors that are heavier than air, exhaust shall be taken from a point within 12 inches (305 mm) of the floor. For fumes or vapors that are lighter than air, exhaust shall be taken from a point within 12 inches (305 mm) of the highest point of the room.

6. The location of both the exhaust and inlet air openings shall be designed to provide air movement across all portions of the floor or room to prevent the accumulation of vapors.

7. The *exhaust air* shall not be recirculated to occupied areas if the materials stored are capable of emitting hazardous vapors and contaminants have not been removed. Air contaminated with explosive or flammable vapors, fumes or dusts; flammable, highly toxic or toxic gases; or radioactive materials shall not be recirculated.

❖ The exhaust ventilation requirement must comply with all of the seven elements:

Element 1: Systems must be installed in accordance with the code.

Element 2: The minimum rate for mechanical ventilation is listed here; however, it is important to note that the Material Safety Data Sheets (MSDS) must be reviewed because the ventilation needed to maintain a safe environment could require a much higher airflow rate for certain materials.

Element 3: Unless an alternative scheme is approved, the exhaust system must provide continuous ventilation in the area.

Element 4: An emergency shutoff must be installed at the entry door into the area and can be used to disable the ventilation system in case of a fire.

Element 5: Because the density of vapor depends on the hazardous material in each case, the system must be designed to consider the vapor density of the chemicals being stored. The vapor density for the chemicals can be found in the MSDS.

Element 6: The system must move air across all of the area being protected to prevent the accumulation of vapors in isolated "dead" spots in the room.

Element 7: Hazardous vapors must not be recirculated back into the storage room or to any other portion of the building. This requirement prevents buildup of vapors over time to hazardous levels and also prevents contamination of other areas.

[F] 502.8.2 Gas rooms, exhausted enclosures and gas cabinets. The ventilation system for gas rooms, exhausted enclosures and gas cabinets for any quantity of hazardous material shall be designed to operate at a negative pressure in relation to the surrounding area. Highly toxic and toxic gases shall also comply with Sections 502.9.7.1, 502.9.7.2 and 502.9.8.4.

❖ The room or enclosure containing the hazardous materials must be designed to operate at a negative pressure to prevent the vapors from escaping into other areas in the event of a leak. The sections listed also have specific ventilation requirements for highly toxic and toxic gases.

[F] 502.8.3 Indoor dispensing and use. Indoor dispensing and use areas for hazardous materials in amounts exceeding the maximum allowable quantity per control area shall be provided with exhaust ventilation in accordance with Section 502.8.1.

Exception: Ventilation is not required for dispensing and use of flammable solids other than finely divided particles.

❖ Exhaust ventilation meeting the requirements of Section 502.8.1 must be installed in indoor use and dispensing areas so that the level of vapors is maintained below the LFL or the PEL. Keeping the area/building at these levels provides a level of safety for the area (see commentary, Section 502.8.1).

[F] 502.8.4 Indoor dispensing and use—point sources. Where gases, liquids or solids in amounts exceeding the maximum allowable quantity per control area and having a hazard ranking of 3 or 4 in accordance with NFPA 704 are dispensed or used, mechanical exhaust ventilation shall be provided to capture gases, fumes, mists or vapors at the point of generation.

Exception: Where it can be demonstrated that the gases, liquids or solids do not create harmful gases, fumes, mists or vapors.

❖ The intent of this section is to ensure that hazardous vapors and fumes are captured and exhausted at the point where the materials are being dispensed or used rather than allowing them to disperse into the room where occupants could be exposed to the harmful effects. This type of system is required only where the materials have an NFPA 704 hazard ranking of 3 or 4 and also exceed the maximum allowable amount for the control area.

[F] 502.8.5 Closed systems. Where closed systems for the use of hazardous materials in amounts exceeding the maximum allowable quantity per control area are designed to be opened as part of normal operations, ventilation shall be provided in accordance with Section 502.8.4.

❖ Where a closed system is designed to be opened during any part of the normal operation cycle, the criteria for designing the ventilation system are the same as for indoor dispensing and use in Section 502.8.4.

[F] 502.9 Hazardous materials—requirements for specific materials. Exhaust ventilation systems for specific hazardous materials shall be provided as required in Section 502.8 and Sections 502.9.1 through 502.9.11.

❖ This section states ventilation requirements for specific hazardous materials that are in addition to the general requirements of Section 502.8.

[F] 502.9.1 Compressed gases—medical gas systems. Rooms for the storage of compressed medical gases in amounts exceeding the permit amounts for compressed gases in the *International Fire Code*, and that do not have an exterior wall, shall be exhausted through a duct to the exterior of the building. Both separate airstreams shall be enclosed in a 1-hour-rated shaft enclosure from the room to the exterior. *Approved* mechanical ventilation shall be provided at a minimum rate of 1 cfm/ft² [0.00508 m³/(s · m²)] of the area of the room.

Gas cabinets for the storage of compressed medical gases in amounts exceeding the permit amounts for compressed gases in the *International Fire Code* shall be connected to an exhaust system. The average velocity of ventilation at the face of access ports or windows shall be not less than 200 feet per minute (1.02 m/s) with a minimum velocity of 150 feet per minute (0.76 m/s) at any point at the access port or window.

❖ Rooms used for the storage of compressed medical gases, such as oxygen, nitrogen, nitrous oxide and carbon dioxide, must have exhaust systems if the amount stored exceeds the permit amount for compressed gases as indicated in Table 105.6.8 in the 2012 IFC. The exhaust must discharge to the exterior of the building at the specified exhaust rate and, if the storage room is not on an exterior wall, a 1-hour-rated shaft enclosure must be constructed to separate the exhaust and makeup air ducts from the occupied space.

The gas cabinet exhaust does not require a 1-hour enclosure but must be connected to an exhaust system. The average velocity of the exhaust stream cannot be less than 200 feet per minute (1.02 m/s) and the minimum velocity at any access port or window cannot be less than 150 feet per minute (0.76 m/s).

[F] 502.9.2 Corrosives. Where corrosive materials in amounts exceeding the maximum allowable quantity per control area are dispensed or used, mechanical exhaust ventilation in accordance with Section 502.8.4 shall be provided.

❖ This section requires mechanical exhaust ventilation for all storage areas containing corrosive liquids with positive vapor pressures, which, if exposed under standard room temperature and atmospheric pressure, give off hazardous fumes and vapors. Adequate mechanical ventilation will reduce the chance for the accumulation of hazardous concentration levels of toxic fumes and vapors. Corrosive liquids without a positive vapor pressure do not readily give off vapors at hazardous concentration levels under normal conditions and, therefore, do not require mechanical exhaust ventilation.

[F] 502.9.3 Cryogenics. Storage areas for stationary or portable containers of cryogenic fluids in any quantity shall be ventilated in accordance with Section 502.8. Indoor areas where cryogenic fluids in any quantity are dispensed shall be ventilated in accordance with the requirements of Section 502.8.4 in a manner that captures any vapor at the point of generation.

Exception: Ventilation for indoor dispensing areas is not required where it can be demonstrated that the cryogenic fluids do not create harmful vapors.

❖ The ventilation of areas where cryogenic fluids are stored, used and dispensed is very important for a number of reasons. First, these materials might be flammable, explosive or reactive. Second, the materials are gases at normal temperature and pressure. Even if these materials are inert, they still have the potential to release large amounts of gas that can asphyxiate the building occupants. Areas where cryogenic liquids are stored must be ventilated in accordance with Section 502.8, regardless of the amount of cryogenics present. Note that these ventilation requirements appear to be for normal operation rather than emergency operation.

These requirements focus on a minimum ventilation rate, recirculation, even air movement and manual shutoff controls. Ventilation is required at a minimum rate of 1 cfm/ft² [0.00508 m³/(s · m²)] and is to operate continuously. In areas where cryogenics are used or dispensed, additional ventilation requirements that are specifically focused on the source of a leak or failure may apply. More specifically, when a cryogenic fluid has a hazard classification of 3 or 4 in accordance with NFPA 704, the ventilation system must be able to capture any gases at the point of discharge or generation. This requirement provides redundancy by addressing the more likely modes of failure.

There is an exception to the ventilation requirements for the use and dispensing of Class 3 and 4 materials that can be shown to not create harmful vapors.

For instance, some materials disperse very quickly

and may become sufficiently diluted without the need for ventilation.

[F] 502.9.4 Explosives. Squirrel cage blowers shall not be used for exhausting hazardous fumes, vapors or gases in operating buildings and rooms for the manufacture, assembly or testing of explosives. Only nonferrous fan blades shall be used for fans located within the ductwork and through which hazardous materials are exhausted. Motors shall be located outside the duct.

❖ These requirements are intended to prevent exhaust systems from contributing to hazards associated with the exhaust of hazardous fumes, vapors or gases in buildings and rooms that are involved in the manufacture, assembly or testing of explosives. Squirrel cage blowers are considered unsatisfactory because the blower wheel could contact portions of the cage housing if parts are bent or otherwise improperly aligned. Ferrous metal fan blades mounted in ductwork represent a similar hazard because a loose or misaligned blade could strike its housing. In both cases, contact between ferrous metal blower or fan parts and housings could cause sparks that could lead to fires or explosions.

[F] 502.9.5 Flammable and combustible liquids. Exhaust ventilation systems shall be provided as required by Sections 502.9.5.1 through 502.9.5.5 for the storage, use, dispensing, mixing and handling of flammable and combustible liquids. Unless otherwise specified, this section shall apply to any quantity of flammable and combustible liquids.

> **Exception:** This section shall not apply to flammable and combustible liquids that are exempt from the *International Fire Code*.

❖ Minimum requirements for mechanical exhaust ventilation for inside areas in which flammable or combustible liquids are stored are contained in Section 5004.3 of the IFC and Section 502.8.1.1 of the code. The use of natural ventilation as an alternative to mechanical ventilation is subject to approval by the code official (see commentary, Sections 502.9.5.1 through 502.9.5.5).

[F] 502.9.5.1 Vaults. Vaults that contain tanks of Class I liquids shall be provided with continuous ventilation at a rate of not less than 1 cfm/ft^2 of floor area [0.00508 m^3/(s · m^2)], but not less than 150 cfm (4 m^3/min). Failure of the exhaust airflow shall automatically shut down the dispensing system. The exhaust system shall be designed to provide air movement across all parts of the vault floor. Supply and exhaust ducts shall extend to a point not greater than 12 inches (305 mm) and not less than 3 inches (76 mm) above the floor. The exhaust system shall be installed in accordance with the provisions of NFPA 91. Means shall be provided to automatically detect any flammable vapors and to automatically shut down the dispensing system upon detection of such flammable vapors in the exhaust duct at a concentration of 25 percent of the LFL.

❖ The removal of Class I vapors from a vault requires that a ventilation system be installed according to

Section 502.8.1 and the IFC. The ventilation system must have supply and exhaust ducts within 3 inches (76 mm) to 12 inches (305 mm) of the vault floor. These ducts are to provide ventilation across the entire vault floor to remove the vapors and provide breathable air. The vapor detection system is to shut off dispensing pumps when the detection system senses a vapor concentration of 25 percent or greater of the flammable or combustible LFL. NFPA 91 contains design and construction requirements for exhaust systems. The standard covers duct material, access, hangers and supports. Additional requirements for vaults can be found in Section 5704.2.8 of the IFC (see commentary, Section 502.8.1).

[F] 502.9.5.2 Storage rooms and warehouses. Liquid storage rooms and liquid storage warehouses for quantities of liquids exceeding those specified in the *International Fire Code* shall be ventilated in accordance with Section 502.8.1.

❖ Ventilation to eliminate flammable vapors may be either natural or mechanical. Although natural ventilation, which depends on many conditions including temperature and wind, has the advantage of not being dependent on either manual starting or power supply, it is not as easily controlled as mechanical ventilation. Mechanical ventilation should be used whenever there are extensive indoor operations involving flammable and combustible liquids.

[F] 502.9.5.3 Cleaning machines. Areas containing machines used for parts cleaning in accordance with the *International Fire Code* shall be adequately ventilated to prevent accumulation of vapors.

❖ The use of parts-washing machines must be restricted to areas that are ventilated in accordance with Section 502.8.1 of the code to reduce the likelihood of vapor buildup from solvent exposure and any solvent spill that might occur to concentrations above the LFL.

[F] 502.9.5.4 Use, dispensing and mixing. Continuous mechanical ventilation shall be provided for the use, dispensing and mixing of flammable and combustible liquids in open or closed systems in amounts exceeding the maximum allowable quantity per control area and for bulk transfer and process transfer operations. The ventilation rate shall be not less than 1 cfm/ft^2 [0.00508 m^3/(s · m^2)] of floor area over the design area. Provisions shall be made for the introduction of *makeup air* in a manner that will include all floor areas or pits where vapors can collect. Local or spot ventilation shall be provided where needed to prevent the accumulation of hazardous vapors.

> **Exception:** Where *natural ventilation* can be shown to be effective for the materials used, dispensed or mixed.

❖ The requirements of this section for a mechanical ventilation system are identical to those in Section 502.8.1. The mechanical ventilation system must remove the vapors to prevent accumulation in concentrations in the flammable range for the flammable and combustible liquids being used. The ventilation system must be designed to sweep all areas involved. Exhaust and makeup air openings must be located to create the intended airflow in areas where the vapors collect. The exception allows use of natural ventilation complying with Section 502.8.1 that can be demonstrated to be effective for the flammable and combustible liquids being used.

[F] 502.9.5.5 Bulk plants or terminals. Ventilation shall be provided for portions of properties where flammable and combustible liquids are received by tank vessels, pipelines, tank cars or tank vehicles and which are stored or blended in bulk for the purpose of distributing such liquids by tank vessels, pipelines, tank cars, tank vehicles or containers as required by Sections 502.9.5.5.1 through 502.9.5.5.3.

❖ Bulk plants and terminals are used for the blending or transfer of large volumes of flammable and combustible liquids. Besides normally involving large quantities of flammable and combustible liquids, these facilities can create numerous opportunities for leaks, spills and the escape of vapors. The safeguards and practices described in Sections 502.9.5.5.1 through 502.9.5.5.3 are intended to reduce the likelihood of flammable vapors reaching hazardous concentrations.

[F] 502.9.5.5.1 General. Ventilation shall be provided for rooms, buildings and enclosures in which Class I liquids are pumped, used or transferred. Design of ventilation systems shall consider the relatively high specific gravity of the vapors. Where *natural ventilation* is used, adequate openings in outside walls at floor level, unobstructed except by louvers or coarse screens, shall be provided. Where *natural ventilation* is inadequate, mechanical ventilation shall be provided.

❖ The low flash point for Class I liquids requires that where a Class I liquid is present the area be ventilated to prevent the accumulation of a vapor-air mixture above the LFL. Ventilation can be either mechanical or natural. Because the vapors from Class I liquids are heavier than air, the ventilation must remove the air at the floor level. Natural ventilation must be to the outdoors. It is imperative that vapors be drawn from the rooms or equipment in which they are generated and taken directly to the outside of the building.

[F] 502.9.5.5.2 Basements and pits. Class I liquids shall not be stored or used within a building having a basement or pit into which flammable vapors can travel, unless such area is provided with ventilation designed to prevent the accumulation of flammable vapors therein.

❖ The vapors from Class I liquids are heavier than air, so these vapors can settle from the upper floors and accumulate in a basement or pit. In a building with a basement or pit that has any connection between the basement or pit and upper floors, the basement or pit is not to be used for the storage or mixing of Class I liquids. The vapors from Class I liquids may settle in the basement or pit through openings for mechanical systems or that are part of an egress path, creating a

hazard. Unless there is a mechanical or natural ventilation system that is adequate for removing accumulated vapors from Class I liquids from the basement or pit, the building cannot be used for the storage or mixing of Class I liquids.

[F] 502.9.5.5.3 Dispensing of Class I liquids. Containers of Class I liquids shall not be drawn from or filled within buildings unless a provision is made to prevent the accumulation of flammable vapors in hazardous concentrations. Where mechanical ventilation is required, it shall be kept in operation while flammable vapors could be present.

❖ Containers of Class I liquids cannot be dispensed or filled in a building unless ventilation is installed to remove accumulated vapors. To make certain that vapors do not accumulate, any mechanical ventilation system must operate continuously or whenever containers of Class I liquids are open.

[F] 502.9.6 Highly toxic and toxic liquids. Ventilation exhaust shall be provided for highly toxic and toxic liquids as required by Sections 502.9.6.1 and 502.9.6.2.

❖ The following two sections state the requirements for treatment systems and exhaust ventilation when highly toxic and toxic liquids are being used. Generally, the requirements for liquids are less restrictive than those for gases because a liquid is much easier to control and emits smaller amounts of hazardous vapor that could be harmful to the occupants. The following requirements apply only when the maximum allowable quantities per control area for such liquids have been exceeded.

[F] 502.9.6.1 Treatment system. This provision shall apply to indoor and outdoor storage and use of highly toxic and toxic liquids in amounts exceeding the maximum allowable quantities per control area. Exhaust scrubbers or other systems for processing vapors of highly toxic liquids shall be provided where a spill or accidental release of such liquids can be expected to release highly toxic vapors at normal temperature and pressure.

❖ Because of the dangers associated with the potential vapors created by an accidental release or spill of highly toxic liquids in amounts exceeding the maximum allowable quantities per control area, the IFC and the code require that storage and use areas, both indoors and outdoors, have an exhaust scrubber or similar system installed to process released vapors. Such systems are termed treatment systems and are to be designed, constructed and installed in accordance with the IFC, specifically Section 6004.2.2.7. This requirement is specific to highly toxic liquids and would require a treatment system to collect and process any vapors that might escape if a spill should occur at "normal temperature and pressure." In other words, if at normal temperature and pressure conditions vapors would not be highly toxic, a treatment system would not be required. The focus of this section is on the inhalation hazards associated with highly toxic materials. A material may be considered highly toxic by skin contact or ingestion, but not create an inhalation hazard because of the low volatility of the liquid.

[F] 502.9.6.2 Open and closed systems. Mechanical exhaust ventilation shall be provided for highly toxic and toxic liquids used in open systems in accordance with Section 502.8.4. Mechanical exhaust ventilation shall be provided for highly toxic and toxic liquids used in closed systems in accordance with Section 502.8.5.

Exception: Liquids or solids that do not generate highly toxic or toxic fumes, mists or vapors.

❖ When highly toxic or toxic liquids in excess of the maximum allowable quantities per control area are used in open systems, there is a higher likelihood that harmful vapors will escape to the atmosphere. This section specifically references Section 502.8.4, which requires that any gases, liquids or solids having a hazard ranking of 3 or 4 be used with a ventilation system that captures the vapors at the point of generation. The method of capturing the vapors at the point of generation or release will vary depending on the specific operation and configuration of the operation.

Closed systems are those systems not normally exposed to the atmosphere and, thus, are less hazardous than open systems. In certain instances, systems that are generally considered closed, but on a somewhat regular basis are open to the atmosphere, would require mechanical ventilation in accordance with Section 502.8.5. Section 502.8.5 requires ventilation in accordance with Section 502.8.4 if the closed system is designed to be opened during normal operations. Section 502.8.4 is for open systems using materials with a hazard ranking of 3 or 4. This section would not require mechanical ventilation where only occasional maintenance requires opening of the system. The requirements for closed systems would apply only when the maximum allowable quantities per control area have been exceeded.

This section does not require a treatment system for the vapors collected by such ventilation systems. The approach is to remove the vapors from the area and simply dilute the materials through the ventilation system. This ventilation is specifically aimed at the normal operation of the systems and not necessarily at a failure or emergency. A failure, with respect to highly toxic liquids, would be addressed by Section 502.9.6.1, which would require a treatment system. Generally, liquids are a much lower hazard because they do not readily give off dangerous concentrations of vapors to the atmosphere as a gaseous version of such material does.

The exception states that liquids that do not produce hazardous vapors, mists or fumes do not require compliance with these ventilation requirements. Much of this will depend on the volatility of the liquid, the degree of hazard of the liquid and how it is used.

[F] 502.9.7 Highly toxic and toxic compressed gases—any quantity. Ventilation exhaust shall be provided for highly toxic and toxic compressed gases in any quantity as required by Sections 502.9.7.1 and 502.9.7.2.

❖ This section describes the requirements for gas cabinets and exhausted enclosures containing toxic and highly toxic gases. Gas cabinets and exhausted enclosures are required for all amounts of highly toxic gases. Toxic gases must be stored only in such cabinets when the maximum allowable quantities have been exceeded. Section 502.9.8.5 requires that all exhaust ventilation from gas cabinets, exhausted enclosures and gas rooms be directed to a treatment system. Connection to a treatment system would be required only in cases where the maximum allowable quantities per control area have been exceeded.

[F] 502.9.7.1 Gas cabinets. Gas cabinets containing highly toxic or toxic compressed gases in any quantity shall comply with Section 502.8.2 and the following requirements:

1. The average ventilation velocity at the face of gas cabinet access ports or windows shall be not less than 200 feet per minute (1.02 m/s) with a minimum velocity of 150 feet per minute (0.76 m/s) at any point at the access port or window.

2. Gas cabinets shall be connected to an exhaust system.

3. Gas cabinets shall not be used as the sole means of exhaust for any room or area.

❖ Gas cabinets that contain toxic and highly toxic gases must comply with both the general requirements found in Chapter 50 of the IFC and the specific requirements found in this section. Chapter 50 of the IFC provides the basic construction requirements and states that the cabinet must be maintained at a negative pressure relative to the surrounding environment. Because of the hazards presented by toxic and highly toxic gases, this section includes additional requirements that are the same ventilation exhaust requirements found in Section 6004.1.2 of the IFC. These include specific velocity minimums for ventilation at the face of the cabinets, a connection to an exhaust system and prohibition of the use of gas cabinets as the sole method of ventilation in a room or area of a building. Additionally, Section 6004.1.2 of the IFC requires that gas cabinets containing toxic and highly toxic gases specifically be protected with an automatic sprinkler system complying with NFPA 13. Other types of extinguishing systems are prohibited.

[F] 502.9.7.2 Exhausted enclosures. Exhausted enclosures containing highly toxic or toxic compressed gases in any quantity shall comply with Section 502.8.2 and the following requirements:

1. The average ventilation velocity at the face of the enclosure shall be not less than 200 feet per minute (1.02 m/s) with a minimum velocity of 150 feet per minute (0.76 m/s).

2. Exhausted enclosures shall be connected to an exhaust system.

3. Exhausted enclosures shall not be used as the sole means of exhaust for any room or area.

❖ Exhausted enclosures must comply with both the general requirements found in Chapter 50 of the IFC and the specific requirements found in this section. Chapter 50 of the IFC contains the basic construction requirements, requires that the exhausted enclosure beat a negative pressure relative to the surrounding environment and requires an approved automatic extinguishing system. Because of the hazards presented by toxic and highly toxic gases, this section includes additional requirements that are the same ventilation exhaust requirements found in Section 6004.1.3 of the IFC. These include specific velocity minimums for ventilation at the face of the enclosure, a connection to an exhaust system and restrictions on the use of exhausted enclosures as the sole method of ventilation in a room or any area of a building. Additionally, Section 6004.1.3 of the IFC requires that exhausted enclosures be protected with an automatic sprinkler system complying with NFPA 13. Other types of extinguishing systems are prohibited.

[F] 502.9.8 Highly toxic and toxic compressed gases—quantities exceeding the maximum allowable quantity per control area. Ventilation exhaust shall be provided for highly toxic and toxic compressed gases in amounts exceeding the maximum allowable quantities per control area as required by Sections 502.9.8.1 through 502.9.8.6.

❖ The following sections are the ventilation requirements for areas where highly toxic and toxic gases exceed the maximum allowable quantities. More specifically, these provisions focus on both stationary and portable containers and require that the exhaust collected from these areas, whether through gas cabinets, exhausted enclosures, gas rooms or local exhaust system, be directed to a treatment system to avoid the release of harmful amounts of toxic and highly toxic gas to the atmosphere. The IFC does allow, under certain conditions, the use of containment vessels or containment systems instead of the use of gas cabinets and exhausted enclosures.

[F] 502.9.8.1 Ventilated areas. The room or area in which indoor gas cabinets or exhausted enclosures are located shall be provided with exhaust ventilation. Gas cabinets or exhausted enclosures shall not be used as the sole means of exhaust for any room or area.

❖ This section requires that rooms or spaces containing gas cabinets and exhausted enclosures must be connected to an exhaust system. Although gas cabinets and exhausted enclosures provide a mechanism to exhaust hazardous gases, they are not allowed to be the only method of exhaust for the room or area of the building in which they are located. The ventilation for the room or area should be in accordance with

Section 502.8. Section 502.9.8.5 requires that gas cabinets and exhausted enclosures be directed through a treatment system when mandated by the IFC.

[F] 502.9.8.2 Local exhaust for portable tanks. A means of local exhaust shall be provided to capture leakage from indoor and outdoor portable tanks. The local exhaust shall consist of portable ducts or collection systems designed to be applied to the site of a leak in a valve or fitting on the tank. The local exhaust system shall be located in a gas room. Exhaust shall be directed to a treatment system where required by the *International Fire Code*.

❖ This section requires that portable tanks have some means available to collect gases that escape through either a valve or a fitting. These gases must be collected through either ducts or some type of collection system. Because these tanks are portable, they may require a system in which the point of collection is flexible to address the different types and configurations of tanks being used. The local exhaust system used to collect these gases must be located in a gas room connected to a treatment system. This section applies to both indoor and outdoor situations.

[F] 502.9.8.3 Piping and controls—stationary tanks. Filling or dispensing connections on indoor stationary tanks shall be provided with a means of local exhaust. Such exhaust shall be designed to capture fumes and vapors. The exhaust shall be directed to a treatment system where required by the *International Fire Code*.

❖ Stationary tanks installed indoors and containing highly toxic or toxic gases must be equipped with an exhaust system that is specifically located at the filling and dispensing portions of the tank. The most likely locations for a release of gases is at an opening to the tank where connections are made and broken. Therefore, air from those areas must be exhausted and directed to a treatment system.

[F] 502.9.8.4 Gas rooms. The ventilation system for gas rooms shall be designed to operate at a negative pressure in relation to the surrounding area. The exhaust ventilation from gas rooms shall be directed to an exhaust system.

❖ Chapter 50 of the IFC has general requirements for gas rooms. These requirements focus on construction and basic ventilation requirements. More specifically, gas rooms must be constructed so that they are separated from the remainder of the building. The separation should be based on the primary occupancy classification of the building. This section requires that the room be kept at a negative pressure relative to the surroundings and that it be connected directly to an exhaust system. Again, this exhaust system must be directed to a treatment system in accordance with Section 502.9.8.5.

[F] 502.9.8.5 Treatment system. The exhaust ventilation from gas cabinets, exhausted enclosures and gas rooms, and local exhaust systems required in Sections 502.9.8.2 and

502.9.8.3 shall be directed to a treatment system where required by the *International Fire Code*.

❖ When highly toxic and toxic gases are stored in gas cabinets, exhausted enclosures and gas rooms, the exhaust is not to be released directly to the atmosphere. Instead all exhaust must go through a treatment system that complies with the code and Sections 6004.2.2.7.1 through 6004.2.2.7.5 of the IFC.

There are two exceptions to the requirements for treatment systems in the IFC that are related to highly toxic and toxic gases in storage when several specific controls have been installed. These controls include redundancy with the use of caps or plugs on valve outlets; handwheel-operated valves to prevent movement and the use of containment vessels or containment systems in accordance with the IFC. Exception 2 is specific to toxic gases in use where a gas detection system is installed that has a sensing interval not exceeding 5 minutes and also an approved fail-safe automatic-closing valve located next to the cylinder valves. The fail-safe valve is to close when gas is detected at or above the PEL.

The rest of the IFC requirements provide guidance on the intended treatment of the gases, which could be through dilution, absorption, containment, neutralization, burning or any other method as long as the largest single vessel is accounted for. Generally, the criterion is to reduce the gas to one-half of the immediate dangerous-to-life-and-health (IDLH) level. The IFC further discusses sizing of the system based on the rate of release and whether the container is stationary or portable.

[F] 502.9.8.6 Process equipment. Effluent from indoor and outdoor process *equipment* containing highly toxic or toxic compressed gases which could be discharged to the atmosphere shall be processed through an exhaust scrubber or other processing system. Such systems shall be in accordance with the *International Fire Code*.

❖ Industrial processes that use toxic or highly toxic compressed gases have the potential for accidently discharging these gases to the atmosphere. This section requires installation of a system to capture, exhaust and process the effluent through a scrubber or other approved processing system before it is exhausted to the atmosphere. The processing system should absorb, burn, neutralize, dilute or otherwise contain the toxic gases so that they are not at hazardous levels when released to the atmosphere. Section 6004.2.2.7 of the IFC includes requirements for treatment systems.

[F] 502.9.9 Ozone gas generators. Ozone cabinets and ozone gas-generator rooms for systems having a maximum ozone-generating capacity of $^1/_2$ pound (0.23 kg) or more over a 24-hour period shall be mechanically ventilated at a rate of not less than six air changes per hour. For cabinets, the average velocity of ventilation at *makeup air* openings with cabi-

net doors closed shall be not less than 200 feet per minute (1.02 m/s).

❖ The primary requirements for ozone gas generators are located in Section 6005 of the IFC. Ozone is considered, among other hazards, a highly toxic gas. It is dealt with separately because it is not simply a matter of storage and use. The generation of ozone is being regulated. This section is applicable if the ozone generator system produces more than 2 pounds (0.9 kg) of ozone within a 24-hour period. Section 502.9.9 of the code discusses only the ventilation requirements. Section 6005 of the IFC has more specific construction-related requirements for generating systems. Section 502.9.9 requires that ozone cabinets and ozone-gas-generator rooms be ventilated at a rate not less than six air changes per hour. The cabinet ventilation system must produce an average velocity of 200 feet per minute (1.02 m/s) at makeup air openings with the cabinet doors closed. This criteria results in a negative pressure in the cabinet with appropriate intake air sizing.

[F] 502.9.10 LP-gas distribution facilities. LP-gas distribution facilities shall be ventilated in accordance with NFPA 58.

❖ Ventilation systems in buildings in which liquefied petroleum gas (LP-gas) products, such as propane and butane, are handled or stored must be arranged and sized in accordance with NFPA 58. LP-gases are heavier than air and, therefore, will accumulate near the floor or at the lowest level of any enclosure. Because of the danger associated with flammable gas, the building must have a means for purging any accumulated gas that may occur as a result of leakage or carelessness. NFPA 58 is referenced for the design, installation and operation of ventilation systems for these facilities.

[F] 502.9.10.1 Portable container use. Above-grade underfloor spaces or basements in which portable LP-gas containers are used or are stored awaiting use or resale shall be provided with an *approved* means of ventilation.

Exception: Department of Transportation (DOT) specification cylinders with a maximum water capacity of 2.5 pounds (1 kg) for use in completely self-contained hand torches and similar applications. The quantity of LP-gas shall not exceed 20 pounds (9 kg).

❖ This section is specific to portable LP-gas containers such as those used for gas grills or travel trailers. Enclosed spaces, such as basements or under-floor spaces, used for the storage of portable containers must be properly ventilated to prevent the buildup of gas fumes in the enclosed space. This would not apply to a garage or storage room subject to normal infiltration of outside air caused by loose construction methods. The exception exempts small storage cylinders of less than 2.5 pounds (1 kg) capacity as long as the aggregate amount stored does not exceed 20 pounds (9 kg).

[F] 502.9.11 Silane gas. Exhausted enclosures and gas cabinets for the indoor storage of silane gas in amounts exceeding the maximum allowable quantities per control area shall comply with Chapter 64 of the *International Fire Code.*

❖ Silane gas (SiH₄) is a pyrophoric material capable of spontaneous ignition when exposed to air at low, normal or slightly elevated temperatures, even in very small quantities. Silane is used in the manufacturing processes of energy-efficient glass coatings, transistors, photosensitive copier drums and photovoltaics, and in the semiconductor industry for several film manufacturing process steps. This section refers users to Chapter 64 of the IFC for the exhaust requirements when the indoor storage of silane gas exceeds the maximum allowable quantities per control area.

[F] 502.10 Hazardous production materials (HPM). Exhaust ventilation systems and materials for ducts utilized for the exhaust of HPM shall comply with this section, other applicable provisions of this code, the *International Building Code* and the *International Fire Code.*

❖ This section is a summary of the ventilation requirements for semiconductor facilities and references the IBC and IFC for further requirements. Parallel requirements may be found in Section 2703.14 of the IFC.

[F] 502.10.1 Where required. Exhaust ventilation systems shall be provided in the following locations in accordance with the requirements of this section and the *International Building Code.*

1. Fabrication areas: Exhaust ventilation for fabrication areas shall comply with the *International Building Code.* Additional manual control switches shall be provided where required by the code official.

2. Workstations: A ventilation system shall be provided to capture and exhaust gases, fumes and vapors at workstations.

3. Liquid storage rooms: Exhaust ventilation for liquid storage rooms shall comply with Section 502.8.1.1 and the *International Building Code.*

4. HPM rooms: Exhaust ventilation for HPM rooms shall comply with Section 502.8.1.1 and the *International Building Code.*

5. Gas cabinets: Exhaust ventilation for gas cabinets shall comply with Section 502.8.2. The gas cabinet ventilation system is allowed to connect to a workstation ventilation system. Exhaust ventilation for gas cabinets containing highly toxic or toxic gases shall also comply with Sections 502.9.7 and 502.9.8.

6. Exhausted enclosures: Exhaust ventilation for exhausted enclosures shall comply with Section 502.8.2. Exhaust ventilation for exhausted enclosures containing highly toxic or toxic gases shall also comply with Sections 502.9.7 and 502.9.8.

7. Gas rooms: Exhaust ventilation for gas rooms shall comply with Section 502.8.2. Exhaust ventilation for

gas rooms containing highly toxic or toxic gases shall also comply with Sections 502.9.7 and 502.9.8.

8. Cabinets containing pyrophoric liquids or Class 3 water-reactive liquids: Exhaust ventilation for cabinets in fabrication areas containing pyrophoric liquids shall be as required in Section 2705.2.3.4 of the *International Fire Code.*

❖ Items 1 through 8 list the specific ventilation requirements for semiconductor facilities based on the use of the particular area. Many of these requirements are references to other sections. These are essentially the same requirements found in Section 2703.14 of the IFC.

1. Fabrication areas must be ventilated in accordance with Section 415.10.1.6 of the IBC. This section also gives the code official the authority to ask for additional manual control switches where facility arrangements make the additional switches necessary.

2. Workstations, typically found within each fabrication area, must have individual exhaust systems to collect any exhaust fumes and vapors as close to the point of generation as possible.

3. Liquid storage rooms within semiconductor facilities are to be treated the same as other liquid storage rooms and should be ventilated according to the general requirements for hazardous materials found in Section 502.8.1.1.

4. Hazardous production materials (HPM) rooms, which are essentially a Group H-2, H-3 or H-4 occupancy, must be ventilated in the same way as any another Group H-2, H-3 and H-4 occupancy would be ventilated in accordance with Section 502.8.1.1.

5. Gas cabinets used in semiconductor facilities should be treated the same as any other use of gas cabinets with hazardous materials. Gas cabinets can share the exhaust system of an individual workstation. Also, semiconductor facilities tend to make use of both toxic and highly toxic gases; therefore, additional requirements are specifically highlighted for these applications.

6. Exhausted enclosures are also regulated just as they are for other applications with hazardous materials. As with gas cabinets, semiconductor facilities tend to make use of both toxic and highly toxic gases; therefore, additional requirements are specifically highlighted for those applications.

7. Gas rooms must also be treated the same as any other gas rooms used with hazardous materials.

8. Cabinets containing pyrophoric liquids or Class 3 water-reactive liquids: Exhaust ventilation for

cabinets in fabrication areas containing pyrophoric liquids or Class 3 water-reactive liquids shall be as required in Section 2705.2.3.4 of the IFC.

[F] 502.10.2 Penetrations. Exhaust ducts penetrating fire barriers constructed in accordance with Section 707 of the *International Building Code* or horizontal assemblies constructed in accordance with Section 711 of the *International Building Code* shall be contained in a shaft of equivalent fire-resistance-rated construction. Exhaust ducts shall not penetrate fire walls. Fire dampers shall not be installed in exhaust ducts.

❖ Semiconductor facilities are essentially a combination of many activities involving the storage and use of hazardous materials. These facilities are large and generally complex, and exhaust systems may lead through several areas of a building before the exhaust is processed or released to the atmosphere. This section requires that the protection surrounding an exhaust system is at least equivalent to the fire-resistance rating of the fire barriers and horizontal assemblies being penetrated.

Ventilation should not be interrupted by a fire damper when a fire or other emergency occurs involving a workstation. This helps reduce the likelihood that hazardous combustion byproducts or hazardous concentrations of HPM will be forced back into the workstation or cleanroom. Continuous ventilation through a duct enclosed in a fire-resistance-rated shaft is one alternative to where a fire damper would normally be required at a duct penetration of a fire-resistance-rated assembly. Fire walls define separations between buildings. Ducts must never penetrate a barrier common to another building or occupancy. This reduces the likelihood of tampering with or interrupting the duct integrity.

[F] 502.10.3 Treatment systems. Treatment systems for highly toxic and toxic gases shall comply with the *International Fire Code*.

❖ This is simply a general reference back to the IFC for treatment systems when toxic or highly toxic gases are being used.

502.11 Motion picture projectors. Motion picture projectors shall be exhausted in accordance with Section 502.11.1 or 502.11.2.

❖ This section contains criteria for exhausting hot gases and smoke from motion picture projectors.

502.11.1 Projectors with an exhaust discharge. Projectors equipped with an exhaust discharge shall be directly connected to a mechanical exhaust system. The exhaust system shall operate at an exhaust rate as indicated by the manufacturer's installation instructions.

❖ This section requires a dedicated exhaust system for projection equipment that is designed to connect to an exhaust system. Projection equipment using an electric arc (carbon arc) as the light source produces hot gases and smoke that must be exhausted to the

outdoors. Other light sources, such as incandescent and xenon lamps, produce intense heat and must be air cooled.

Projection equipment that has an exhaust connection and is intended by the manufacturer to be connected to an exhaust system depends on the external exhaust system to cool the equipment and remove waste heat and any smoke and gases produced by the light source. The projection equipment manufacturer will specify the required rate of exhaust for the specific equipment, and the exhaust system must be designed to provide that capacity. Article 540 of NFPA 70 addresses motion picture projectors.

Projectors are installed in specially constructed projection equipment rooms that contain one or more projectors and the associated hardware and controls. Projector exhaust systems may serve a single projector or multiple units if the system has the necessary capacity for all units served. Projection rooms normally have a general room exhaust system to control the room's ambient temperature.

502.11.2 Projectors without exhaust connection. Projectors without an exhaust connection shall have contaminants exhausted through a mechanical exhaust system. The exhaust rate for electric arc projectors shall be a minimum of 200 cubic feet per minute (cfm) (0.09 m³/s) per lamp. The exhaust rate for xenon projectors shall be a minimum of 300 cfm (0.14 m³/s) per lamp. Xenon projector exhaust shall be at a rate such that the exterior temperature of the lamp housing does not exceed 130°F (54°C). The lamp and projection room exhaust systems, whether combined or independent, shall not be interconnected with any other exhaust or return system within the building.

❖ If the projectors do not connect to an exhaust system, the projection room must have an exhaust system with the capacity specified in this section. The required system must exhaust the waste heat, smoke and gases produced by the projection equipment. The exhaust rate specified for xenon lamp projectors is intended to control the equipment surface temperatures to reduce the fire hazard and protect personnel from injury. The projector lamp and general room exhaust systems can be combined or they can be independent systems; however, these systems must be independent of all other building exhaust, ventilation and air-conditioning systems. Makeup air for these exhaust systems can be outdoor air or transfer air from adjoining spaces, such as the auditorium.

[F] 502.12 Organic coating processes. Enclosed structures involving organic coating processes in which Class I liquids are processed or handled shall be ventilated at a rate of not less than 1 cfm/ft² [0.00508 m³/(s · m²)] of solid floor area. Ventilation shall be accomplished by exhaust fans that intake at floor levels and discharge to a safe location outside the structure. Noncontaminated intake air shall be introduced in such a manner that all portions of solid floor areas are provided with continuous uniformly distributed air movement.

❖ Ventilation systems must be designed to prevent the accumulation of vapors within the building where

Class I flammable liquids are processed or handled. Uncontaminated air, such as outdoor air or transfer air, must be distributed to dilute and disperse vapors over the entire solid floor area and then discharge them to a safe location outside the building. A ventilation rate of 1 cfm/ft^2 [0.00508 m^3/(s · m^2)] is required.

502.13 Public garages. Mechanical exhaust systems for public garages, as required in Chapter 4, shall operate continuously or in accordance with Section 404.

❖ This section applies to parking garages and does not apply to repair garages. As a means of conserving energy, ventilation systems for enclosed public parking garages are allowed to modulate between full capacity and the minimum capacity specified in Section 404.2. This modulating operation must use approved detection devices that are capable of detecting the operation of vehicles in the garage. This section is intended to apply to parking garages for vehicles such as automobiles, buses and fleet vehicles (see commentary, Section 404).

502.14 Motor vehicle operation. In areas where motor vehicles operate, mechanical ventilation shall be provided in accordance with Section 403. Additionally, areas in which stationary motor vehicles are operated shall be provided with a *source capture system* that connects directly to the motor vehicle exhaust systems.

Exceptions:

1. This section shall not apply where the motor vehicles being operated or repaired are electrically powered.

2. This section shall not apply to one- and two-family dwellings.

3. This section shall not apply to motor vehicle service areas where engines are operated inside the building only for the duration necessary to move the motor vehicles in and out of the building.

❖ In repair garages, parking garages, warehouses, vocational shops and similar locations where internal combustion engines are operated, engine exhaust contamination can accumulate. To prevent the accumulation of harmful contaminants, these spaces must have mechanical ventilation as prescribed in Section 403. For example, Table 403.3 specifies an exhaust rate of 0.75 cfm/ft^2 [0.00381 m^3/(s · m^2)] for repair garages. This section requires mechanical ventilation to be provided and Note b of Table 403.3 specifically requires mechanical exhaust where it is indicated for an occupancy in Table 403.3. The intent is to contain the contaminants in that occupancy, thus preventing the contaminants from migrating or being transferred to other occupancies.

It should be noted that Note b does not permit the recirculation of the minimum required amount of ventilation air to any other occupancies (see commentary, Section 403.2.1). For example, where an air-handling unit serves both a repair garage and the sales floor of a car dealership, recirculation of air from the repair garage to the sales floor is not permitted. Note that if the air-handling unit served only the repair garage, then air could be recirculated through the unit and back to the repair garage for the purpose of conditioning the space.

In addition to mechanical ventilation being required for the space, a source capture system is required where stationary motor vehicles operate. A source capture system is defined as a mechanical system designed to capture the contaminats at their source and discharge them to the outdoor atmosphere. The source capture system must connect directly to the vehicle's exhaust system.

A number of repair garages have large overhead doors to permit the movement of vehicles in and out of the service area. Because of the size of the overhead doors, these spaces usually meet the requirements for natural ventilation in Section 402.2 of the code. Note that this section, in conjunction with Section 401.6 of the code and Section 1203.4.2 of the IBC, requires installation of a mechanical exhaust system in spaces where motor vehicles operate. Therefore, a repair garage must be mechanically ventilated; providing natural ventilation is not an acceptable option. Note that Exception 3 would not require mechanical ventilation where the engines are operated inside the building only for the duration necessary to move the vehicles in and out (see commentary for Exception 3).

The requirement for repair garages to have mechanical exhaust also requires an approximately equal amount of makeup air to be provided (see commentary, Section 501.3). It is not uncommon for repair garages to have exhaust fans in the exterior wall of a garage. When the outdoor temperature permits, this exhaust is usually intended to provide ventilation by exhausting the air in the garage while the outdoor makeup air enters through the open overhead doors or motorized wall louvers that simultaneously operate with the exhaust fans. It should be noted that while this system might be sized to provide the required 0.75 cfm/ft^2 [0.00381 m^3/(s · m^2)] of exhaust when the outdoor temperature is moderate, this system may see little use during colder temperatures, especially in more northern climates. For this system to be acceptable, the method to heat the incoming makeup air must be indicated (i.e., the heating system must have sufficient capacity to perform such a task).

Motor vehicle operation would also include forklifts, with an internal combustion engine, in a warehouse. Note that the code does not have any specific requirements that address a method to deal with the exhaust from forklifts operating in a warehouse. Lacking any specific requirements, the mechanical exhaust system required for a public parking garage is the closest occupancy in Table 403.3. A public parking garage requires 0.75 cfm/ft^2 [0.00381 m^3/(s · m^2)] of floor area of outdoor and exhaust air, which would be a substantial amount of air in a large warehouse. In lieu of providing 0.75 cfm/ft^2 [0.00381 m^3/(s · m^2)] of outdoor and exhaust air, the design professional must propose a system to address the exhaust from the forklifts and submit it to the code official for approval. Note that OSHA, the American Society of Heating, Refrigerating and Air-Conditioning Engineers (ASHRAE) or the American Conference on Governmental Industrial Hygienists (ACGIH) may have information that could be useful when dealing with exhaust from forklifts in a building.

As stated in Exception 3, those areas where motor vehicles operate include repair garages where the vehicle is driven into a service bay for routine maintenance and repair services that do not require running the engine. Spaces for routine maintenance work that involves the exchange of parts or fluids, such as oil, batteries and tires, and do not require the vehicles to operate as part of the maintenance work would not be required to provide mechanical ventilation or to have a source capture system. For example, quick oil-change facilities where the vehicle is pulled in and shut off while servicing would not be required to have mechanical ventilation or a source capture system. Where internal combustion engines must be run during repair and maintenance, mechanical ventilation is required based on Section 403, and an independent source capture exhaust system must be installed that will attach directly to the motor vehicle exhaust system's tailpipes and discharge directly outdoors [see Commentary Figures 502.14(1) and (2)]. This tailpipe-connected system is in addition to the mechanical ventilation exhaust system required by this section. In the previous example, if the same quick oil-change facility performed other services that require the

engine to be running (e.g., radiator flushing, full transmission fluid flushing and fuel injection cleaning), mechanical ventilation is required along with a source capture exhaust system directly attached to the motor vehicle exhaust system's tailpipes.

A number of fire stations have source capture systems for the fire trucks located in the apparatus bays. These systems usually connect to the exhaust pipes of the trucks with a flexible hose that will automatically disconnect when the truck is driven out of the bay. Note that there is no specific code requirement for having a source capture system connected to the fire trucks. If the trucks are driven in and shut off and then started and driven out, a source capture system would not be required. However, if it is common practice for the fire department to run the engines while the trucks are stationary in the apparatus bay, then a source capture system would be required.

The provisions of this section do not apply to one- and two-family dwellings, and where the motor vehicles being operated, stored or repaired are electrically powered and, therefore, do not produce combustion byproducts.

[F] 502.15 Repair garages. Where Class I liquids or LP-gas are stored or used within a building having a basement or pit wherein flammable vapors could accumulate, the basement or pit shall be provided with ventilation designed to prevent the accumulation of flammable vapors therein.

❖ A ventilation system must be installed in any below-grade area where flammable and combustible vapors might accumulate. The ventilation system must be in operation any time the repair garage is open for busi-

Figure 502.14(1)
OVERHEAD SOURCE CAPTURE SYSTEM

ness and any time there is a chance that vapors may accumulate in the below-grade area.

[F] 502.16 Repair garages for natural gas- and hydrogen-fueled vehicles. Repair garages used for the repair of natural gas- or hydrogen-fueled vehicles shall be provided with an *approved* mechanical ventilation system. The mechanical ventilation system shall be in accordance with Sections 502.16.1 and 502.16.2.

> **Exception:** Where *approved* by the code official, *natural ventilation* shall be permitted in lieu of mechanical ventilation.

❖ Repair garages that install and repair compressed natural gas (CNG) or hydrogen motor fuels must be equipped with ventilation and gas detection systems in accordance with Sections 502.16.1 and 502.16.2. The intent of this section is to prevent the accumulation of lighter-than-air flammable and combustible gases inside the repair garage.

An example of natural ventilation that a code official may approve at his or her discretion is a repair garage with at least two opposite sides open all the way to the ceiling. The two opposite sides would allow for cross ventilation. Having the walls open to the ceiling would prevent lighter-than-air gases from accumulating at the ceiling level. The ceiling would have to be sealed to prevent gases from entering the attic space; otherwise, mechanical ventilation would be required.

[F] 502.16.1 Design. Indoor locations shall be ventilated utilizing air supply inlets and exhaust outlets arranged to pro-vide uniform air movement to the extent practical. Inlets shall be uniformly arranged on exterior walls near floor level. Outlets shall be located at the high point of the room in exterior walls or the roof.

Ventilation shall be by a continuous mechanical ventilation system or by a mechanical ventilation system activated by a continuously monitoring natural gas detection system, or for hydrogen, a continuously monitoring flammable gas detection system, each activating at a gas concentration of 25 percent of the lower flammable limit (LFL). In all cases, the system shall shut down the fueling system in the event of failure of the ventilation system.

The ventilation rate shall be at least 1 cubic foot per minute per 12 cubic feet [0.00138 m^3/(s · m^3)] of room volume.

❖ The intent of this requirement is to provide uniform ventilation throughout the garage area that will exchange at least 1 cubic foot of air for every 12 cubic feet of room volume every minute. A detection system must be able to recognize hydrogen where hydrogen-fueled vehicles are involved.

[F] 502.16.2 Operation. The mechanical ventilation system shall operate continuously.

Exceptions:

1. Mechanical ventilation systems that are interlocked with a gas detection system designed in accordance with the *International Fire Code*.

2. Mechanical ventilation systems in garages that are used only for the repair of vehicles fueled by liquid fuels or odorized gases, such as CNG, where the

Figure 502.14(2)
WALL-MOUNTED SOURCE CAPTURE SYSTEM

ventilation system is electrically interlocked with the lighting circuit.

❖ The intent of this section is to prevent the accumulation of lighter-than-air gases inside vehicle repair garages by requiring the ventilation system to operate continuously. Exceptions allow intermittent operation of the ventilation system where it is interlocked with one of the systems specified. Exception 2 does not apply to hydrogen.

502.17 Tire rebuilding or recapping. Each room where rubber cement is used or mixed, or where flammable or combustible solvents are applied, shall be ventilated in accordance with the applicable provisions of NFPA 91.

❖ NFPA 91 requires that the exhaust flow within the duct system be maintained at a vapor concentration no greater than 25 percent of the LFL.

502.17.1 Buffing machines. Each buffing machine shall be connected to a dust-collecting system that prevents the accumulation of the dust produced by the buffing process.

❖ The dust produced by the buffing process is a potentially explosive mixture of loose rubber particles, cement, and solvent vapors and fumes. Accordingly, each buffing machine must be equipped with a dust-collecting system that will confine and evacuate both the vapor and solid particles created by the buffing process. Besides cloth and paper-type filters, other filtration devices, such as scrubbers, cyclone collectors, dust collectors and reduced transport velocity collectors, may be used in the collection process.

502.18 Specific rooms. Specific rooms, including bathrooms, locker rooms, smoking lounges and toilet rooms, shall be exhausted in accordance with the ventilation requirements of Chapter 4.

❖ This section recognizes that besides this chapter, Chapter 4 also contains provisions that mandate mechanical exhaust. This section specifies that some occupancies, such as smoking lounges, require mechanical exhaust systems to exhaust the contaminants produced by that occupancy. Note b in Table 403.3 applies to these occupancies. Where Note b appears in Table 403.3, the ventilation air, both the required air and any excess air quantities, must not be recirculated to any other occupancy. This may require 100 percent exhaust and 100 percent supply with no return air in rooms that are served by air-handling units that also serve other rooms. In these cases, the recirculation provision of Section 403.2.1 is not an option. As with all exhaust systems addressed in the code, the exhaust must extend to and terminate at the outdoors. The exhaust termination must be approved and comply with Section 501.2.

502.19 Indoor firing ranges. Ventilation shall be provided in an *approved* manner in areas utilized as indoor firing ranges. Ventilation shall be designed to protect employees and the public in accordance with DOL 29 CFR 1910.1025 where applicable.

❖ This section requires a ventilation system to be designed to protect employees and the public by referencing the DOL standard. This DOL standard covers lead and air-borne particulate matter in the workplace and is applicable to this type of facility. The use of this standard along with guidance from the National Rifle Association will provide information to the design professional to help design and the code official to help approve a ventilation system for indoor firing ranges.

SECTION 503
MOTORS AND FANS

503.1 General. Motors and fans shall be sized to provide the required air movement. Motors in areas that contain flammable vapors or dusts shall be of a type *approved* for such environments. A manually operated remote control installed at an *approved* location shall be provided to shut off fans or blowers in flammable vapor or dust systems. Electrical *equipment* and appliances used in operations that generate explosive or flammable vapors, fumes or dusts shall be interlocked with the ventilation system so that the *equipment* and appliances cannot be operated unless the ventilation fans are in operation. Motors for fans used to convey flammable vapors or dusts shall be located outside the duct or shall be protected with *approved* shields and dustproofing. Motors and fans shall be provided with a means of access for servicing and maintenance.

❖ Exhaust fans, motors and other related prime moving components must be of adequate size. The designer must select the air volume, fan static pressure, transport velocity and pressure drop across all components in order to move the contaminant(s) of concern from the source to the point of termination. It is important to note, however, that the code does not dictate the sizing methodology for such equipment. Furthermore, exhaust fans, motors and other related prime moving components must be specifically designed for the application, be compatible with the contaminants being exhausted and be approved for installation in those environments. For example, exhaust systems handling corrosive vapors must be constructed of materials that will not be damaged by the vapors.

Components of exhaust systems conveying flammable vapors or dusts, such as fans, blowers, motors, shaft bearings, collectors, filters and volume controls, must be located out of the exhaust flow or must be constructed of materials that are impervious to the exhaust flow.

Protection from abrasion may be necessary for systems handling dusts and stock. Abrasive exhaust materials, such as those produced by cutting, grinding, milling and forming stock materials, can erode the components of an exhaust system.

The type of air-moving equipment must be carefully considered when designing an exhaust system to handle flammable or explosive substances (see commentary, Section 510). Access must be provided for servicing, maintenance and inspection of the exhaust systems prime moving equipment.

Electrical equipment involved in operations generating explosive or flammable vapors, fumes or dusts must be interlocked to the ventilating or exhaust system to achieve simultaneous operation of the equipment and the exhaust system. When ventilating or exhaust equipment is installed, it becomes an essential part of the appliance or equipment it serves.

The equipment involved in operations generating explosive or flammable vapors, fumes or dusts relies on the ventilation or exhaust equipment for proper removal of these flammable or explosive contaminants from the building. The interlocking controls will ensure that the equipment involved in operations generating explosive or flammable vapors, fumes or dusts will not function if there is insufficient ventilation or exhaust to evacuate these contaminants. Requirements for electrical installations in hazardous exhaust systems must conform to NFPA 70. Certain materials can generate a static charge when conveyed by the system. Locations where static arcing occurs must be grounded.

503.2 Fans. Parts of fans in contact with explosive or flammable vapors, fumes or dusts shall be of nonferrous or nonsparking materials, or their casing shall be lined or constructed of such material. When the size and hardness of materials passing through a fan are capable of producing a spark, both the fan and the casing shall be of nonsparking materials. When fans are required to be spark resistant, their bearings shall not be within the airstream, and all parts of the fan shall be grounded. Fans in systems-handling materials that are capable of clogging the blades, and fans in buffing or woodworking exhaust systems, shall be of the radial-blade or tube-axial type.

❖ Of significant concern is the possibility of mechanical exhaust equipment being an ignition source in systems handling explosive or flammable materials. Additionally, abrasive exhaust materials, such as those produced by cutting, grinding, milling and forming stock materials, can produce sparks while passing through the exhaust system. Nonferrous, nonspark-producing fans and fan housings must be used to reduce the potential for ignition in explosive or flammable material-handling systems. Bronze and aluminum are commonly used where nonferrous parts are specified.

Fans required to be spark resistant must have all components grounded in accordance with NFPA 70. The bearings of these fans must not be located in the airstream, because these moving parts typically reach high temperatures during operation and could create a spark when hit by flying debris. For guidance, the Air Movement and Control Association

(AMCA), in its standard AMCA 99-0401, outlines three types of spark-resistant construction:

1. Type A construction requires that all parts of the fan in contact with the air or gas being handled be made of nonferrous metal.

2. Type B construction requires that the fan have an entirely nonferrous wheel and a nonferrous ring about the opening through which the shaft passes.

3. Type C construction specifies that the fan must be constructed so that a shift of the wheel or shaft will not permit two ferrous parts of the fan to rub or strike.

Fans in systems-handling materials must be a type that is not subject to clogging. Forward-curved squirrel-cage centrifugal fans are a type of fan subject to clogging. Radial blade fans are less susceptible to loading with debris.

503.3 Equipment and appliance identification plate. *Equipment* and appliances used to exhaust explosive or flammable vapors, fumes or dusts shall bear an identification plate stating the ventilation rate for which the system was designed.

❖ The exhaust flow rate is critical to the proper capture, containment, dilution and transport of contaminants. Equipment is often repaired, adjusted or modified in its life span, and the flow rate identification plate will provide valuable information to help maintain the original design performance of the system. For example, if a fume hood system is to be modified or a fan is to be replaced, the original design flow rate information for the fume hoods must be known so that system performance will not be adversely affected.

503.4 Corrosion-resistant fans. Fans located in systems conveying corrosives shall be of materials that are resistant to the corrosive or shall be coated with corrosion-resistant materials.

❖ Corrosive materials conveyed in exhaust systems may exhibit a wide range of characteristics. Any mechanical equipment located in the exhaust airstream must be compatible with and unaffected by the exhausted material. For example, a laboratory fume hood exhaust system may handle corrosive materials, in which case the steel fan and housing assembly should be constructed of stainless steel or protected by a corrosion-resistant coating. The intent is to provide a particular material or coating to protect the fan from attack by corrosive contaminants. Ultimately, when corrosion does take place, two types of attack may be involved: chemical and electrochemical. Direct chemical attack is generally limited to high temperatures, highly corrosive environments or both. The rapid scaling of steel at temperatures above 900°F (482°C) and the effect of concentrated acids or alkalies are examples.

Electrochemical attack is much more common. Such a reaction requires discrete anodic and cathodic regions connected by solid material and submerged in an electrolyte. In any case, preventive maintenance in the form of frequent cleaning will remove deposits and thus extend the equipment's useful life.

In choosing a particular material or coating to protect a fan from attack by a corrosive contaminant, the design professional must consider the effects of temperature, local concentration, velocity, impurities and fabrication in terms of actual system performance. Various metals and alloys have been used to fabricate corrosion-resistant fans, such as aluminum-based alloys at room temperature for protection against most gases in the absence of moisture; copper at room temperature for protection against dry halogen gases in the absence of moisture; and copper-zinc, copper-nickel, copper-tin and copper-silicon alloys for protection against most dry gases at ordinary temperatures. In all cases, the formation of galvanic couples by joining dissimilar metals should be avoided. The protection provided by other coatings, such as lead, rubber or plastics, is primarily a result of their inert properties. The optimum thickness of inert coating varies with the corrosive medium.

In addition to corrosion resistance, protection from abrasion may be necessary for systems handling dusts and stock.

SECTION 504
CLOTHES DRYER EXHAUST

504.1 Installation. Clothes dryers shall be exhausted in accordance with the manufacturer's instructions. Dryer exhaust systems shall be independent of all other systems and shall convey the moisture and any products of *combustion* to the outside of the building.

> **Exception:** This section shall not apply to *listed* and *labeled* condensing (ductless) clothes dryers.

❖ Clothes dryers must be exhausted in compliance with the dryer manufacturer's installation instructions. The manufacturer's installation instructions apply and where the code provisions are more strict, the code applies. Note that some of the provisions in the code are redundant with the appliance installation instructions. Dryers are designed and built to meet industry safety standards. The manufacturer's installation instructions are evaluated by the agency responsible for testing the appliance and, therefore, those instructions will prescribe an installation that is consistent with the appliance installation that was tested. This requirement is consistent with Section 304.1, which addresses the installation of mechanical equipment.

Because clothes dryer exhaust contains high concentrations of combustible lint, debris and water vapor, dryer exhaust systems must be independent of all other systems. This requirement prevents the fire hazards associated with that exhaust system

from extending into or affecting other systems or other areas in the building. Additionally, this section intends to prevent toxic products of combustion from entering the building through other systems.

Dryer exhaust ducts must be independent of other dryer exhaust ducts unless connected to an engineered exhaust system specifically designed to serve multiple dryers or where connected to a multiple-story system in accordance with Section 504.8.

Clothes dryer exhaust systems must convey the moisture and any products of combustion directly to the exterior of the building (outdoors); therefore, any devices that allow all or part of the exhaust to discharge to the indoors are in violation of the code. The requirement to convey the exhaust to the outdoors applies to both fuel-fired and nonfuel-fired (i.e., electric) appliances. Clothes dryer exhaust systems cannot terminate in or discharge to any enclosed space, such as an attic or crawl space, regardless of whether or not the space is ventilated through openings to the outdoors. The high levels of moisture in the exhaust air can cause condensation to form on exposed surfaces or in insulation materials. Water vapor condensation can cause structural damage, deterioration of building materials and contribute to the growth of mold and fungus. Clothes dryer exhausts that discharge to enclosed spaces will also cause an accumulation of combustible lint and debris, creating a significant fire hazard. An improperly installed clothes dryer exhaust system not only reduces dryer efficiency and increases running time, but can also cause a significant increase in exhaust temperature, causing the dryer to cycle on its high limit control, which is an unsafe operating condition.

Clothes dryer exhaust ducts must be installed to comply with the dryer manufacturer's installation instructions and the requirements of this section. The clothes dryer manufacturer's installation instructions control the type of exhaust duct material allowed and the method of installation. For example, typical dryer installation instructions will require metallic duct materials and will impose more stringent length limitations for corrugated ducts than for rigid ducts because of the poorer flow characteristics of corrugated duct materials.

The exception specifically excludes condensing clothes dryers from the requirement to exhaust to the outdoors. Condensing dryers are electrically heated and recirculate the same air through the clothes drum. An air-to-air heat exchanger condenses the water vapor in the clothes drum air by using room air as the condensing medium. The condensate is collected in a reservoir that is either manually emptied or automatically emptied by an integral condensate pump. These dryers do not discharge heated air to the outdoors and have very little effect on the indoor temperature; thus, they are energy efficient.

These "ductless" dryers must be listed for this application and installed to comply with the listing. They are commonly installed in applications where a conventional dryer exhaust duct cannot be used because of distance, aesthetics or problems with the penetration of rated assemblies.

504.2 Exhaust penetrations. Where a clothes dryer exhaust duct penetrates a wall or ceiling membrane, the annular space shall be sealed with noncombustible material, *approved* fire caulking or a noncombustible dryer exhaust duct wall receptacle. Ducts that exhaust clothes dryers shall not penetrate or be located within any fireblocking, draftstopping or any wall, floor/ceiling or other assembly required by the *International Building Code* to be fire-resistance rated, unless such duct is constructed of galvanized steel or aluminum of the thickness specified in Section 603.4 and the fire-resistance rating is maintained in accordance with the *International Building Code*. Fire dampers, combination fire/smoke dampers and any similar devices that will obstruct the exhaust flow shall be prohibited in clothes dryer exhaust ducts.

❖ Rigid clothes dryer exhaust ducts are permitted to penetrate assemblies that are not fire-resistance-rated and building elements not used as fireblocking or draftstopping. In such cases, the annular space around the dryer exhaust duct penetration must be protected with a noncombustible seal, fire caulking or a noncombustible dryer receptacle (see Commentary Figure 504.2). The penetrations must be properly protected to help prevent a fire that is associated with a dryer from spreading into the wall or ceiling cavity. In all other cases, ducts must be constructed of galvanized steel or aluminum of the thickness specified in Section 603.4, and the penetration must be protected to maintain the fire-resistance rating and integrity of the assembly or element being penetrated. Because of the strength and rigidity differences between steel and aluminum, aluminum ducts generally must be of a heavier (thicker) gage than steel ducts for a given application.

The metal thickness requirements of Section 603.4 practically necessitate rigid pipe and all but rule out the use of flexible duct where the duct must penetrate fireblocking, draftstopping or a fire-resistance-rated assembly. Note that Section 504.6.1 states that the duct must be constructed of metal and must have a smooth interior finish. This requires rigid duct and thereby disallows flexible duct for domestic clothes dryer exhaust. Where penetrating fireblocking or draftstopping, the exhaust duct must be constructed of galvanized steel or aluminum in accordance with Section 603.4, and the annular space around the duct must be fireblocked in accordance with the IBC.

It is important to note that a clothes dryer is exhausted through a duct and not a vent. As such, clothes dryer exhaust ducts must follow the same requirements as any other ducts with respect to where the ducts penetrate fire-resistance-rated assemblies. Therefore, where penetrating a fire-resistance-rated assembly, the penetration must be protected in accordance with Sections 302.2 and 607 that will, in most cases, require a fire damper and, in some cases, require a smoke damper. However, this section specifically prohibits the installation of fire and smoke dampers in a clothes dryer exhaust. The moisture, heat, lint and debris could damage, impair or obstruct fire and smoke dampers and, if the dampers

Figure 504.2
NONCOMBUSTIBLE WALL RECEPTACLE FOR CLOTHES DRYER EXHAUST DUCT

were to close, a hazardous condition could result from the continued operation of the dryer. These facts certainly suggest that dryer exhaust ducts cannot penetrate any fire-resistance-rated assemblies unless the installation satisfies one of the limited fire and smoke damper exceptions in the IBC.

Where penetrating fire-resistance-rated floor/ceiling or roof/ceiling assemblies, dryer exhaust duct penetrations are subject to more stringent penetration protection requirements that, in practically all cases, will necessitate a shaft enclosure for the dryer exhaust ducts. Note that where a duct penetrates a shaft, a fire and smoke damper is required to be installed. As previously stated, a fire and smoke damper cannot be installed in a dryer exhaust duct. There are, however, provisions in Section 504.8 for conveying exhaust from multiple clothes dryers through multiple stories (see commentary, Section 504.8).

Note that there is an exception in Section 607.6.1 of the code that would apply to domestic clothes dryer exhaust ducts. This exception permits a duct to penetrate three floors (i.e., connecting four stories) without a fire damper at each floor level and without using a shaft. To qualify for this exception the duct must be located within the cavity of a wall; be constructed of steel not less than 0.0187 inches (0.4712 mm); be continuous from one dwelling unit to the exterior; not exceed 4 inches (102 mm) in diameter; the total area of such ducts must not exceed 100 square inches (0.065 m^2) in any 100 square feet (9.3 m^2) of floor area and the annular space around the duct is protected with materials that prevent the passage of flame and hot gases sufficient to ignite cotton waste.

The difficulty and expense of maintaining the fire-resistance rating of assemblies penetrated by clothes dryer exhaust ducts are an obvious justification for avoiding these penetrations wherever possible. The requirements of this section, combined with Section 504.6, make a compelling case for always placing clothes dryers against outside walls to avoid long duct runs and penetrations of other than exterior walls.

504.3 Cleanout. Each vertical riser shall be provided with a means for cleanout.

❖ Lint and debris carried in the dryer exhaust will settle in the lowest point of any vertical riser in the system; therefore, an accessible means for removing accumulations in the system must be provided to prevent duct blockages and to eliminate the fire hazard of such combustible accumulations. Because of the difficulty of transporting suspended solids vertically against gravity, a vertical section of exhaust duct might eventually require maintenance. The exhaust duct connection to an individual dryer outlet is typically considered as a cleanout because the code requires a "means for cleanout" without specifying what that "means" can or cannot be. Where exhaust ducts can be accessed and readily disassembled, the intent of this section has been met. This section does not require that a tee and cap or similar arrangement be installed where ducts can otherwise be accessed

for cleaning. Note that tee fittings can offer significant flow resistance in an exhaust duct.

504.4 Exhaust installation. Dryer exhaust ducts for clothes dryers shall terminate on the outside of the building and shall be equipped with a backdraft damper. Screens shall not be installed at the duct termination. Ducts shall not be connected or installed with sheet metal screws or other fasteners that will obstruct the exhaust flow. Clothes dryer exhaust ducts shall not be connected to a vent connector, vent or *chimney*. Clothes dryer exhaust ducts shall not extend into or through ducts or plenums.

❖ Exhaust ducts must connect directly to terminals that pass through the building envelope to the outdoor atmosphere. Clothes dryer exhaust must terminate outdoors because of the high levels of moisture, combustible lint and, for gas-fired dryers, combustion products in the exhaust. If discharged indoors, such exhaust could present a health and fire hazard, could cause structural damage and deterioration of building materials, and could contribute to the growth of mold and fungus. It should be noted that clothes dryer exhaust is considered to be environmental air and, as such, has less stringent termination requirements in Section 501.3.1. Attics and crawl spaces are not considered to be outdoors, and exhaust ducts cannot terminate in those spaces (see commentary, Section 501.3). Backdraft dampers must be installed in dryer exhaust ducts to avoid outdoor air infiltration during periods when the dryer is not operating and to prevent the entry of animals. These dampers should be designed and installed to provide an adequate seal when in a closed position to minimize air leakage (infiltration). Backdraft dampers are typically of the gravity type, which are opened by the energy of the exhaust discharge. Some dryer manufacturers prohibit the use of magnetic backdraft dampers because of the extra resistance that the exhaust flow must overcome.

Exhaust terminal opening size is also governed by the dryer manufacturer's instructions. Full-opening terminals present less resistance to flow and might be mandated by the dryer manufacturer. A "full opening" is considered to be an opening having no dimension less than the diameter of the exhaust duct. Dryer exhaust flow must not be restricted by screens or fastening devices, such as sheet metal screws. Any type of screen would become completely blocked with fibers in a very short time. Consider that the filter screen integral with the appliance becomes restricted with lint in each cycle of operation. These restrictions and projections will promote the accumulation of combustible lint and debris in the exhaust duct, thereby creating a potential fire hazard and causing flow resistance. Duct tape should not be relied on as the sole means of joining dryer exhaust ducts because adhesives can deteriorate with age and when exposed to high temperatures, causing joints to separate. However, sealing joints is still desirable to limit the leakage of lint, fibers, moisture vapor and combustion products into the occupied space. Note that this section and Section 504.6.2 both address duct joint fasteners, but do so differently. This section prohibits fasteners from obstructing flow thereby implying that limited protrusions into the duct can be allowed; however, Section 504.6.2 prohibits fastener protrusion into the duct without regard for the degree of obstruction created (see commentary, Section 504.6.2).

504.5 Makeup air. Installations exhausting more than 200 cfm (0.09m³/s) shall be provided with *makeup air*. Where a closet is designed for the installation of a clothes dryer, an opening having an area of not less than 100 square inches (0.0645 m²) shall be provided in the closet enclosure or *makeup air* shall be provided by other *approved* means.

❖ Makeup air must be supplied to compensate for the air exhausted by the dryer exhaust system where the amount exhausted is more than 200 cfm (0.09 m³/s). A typical domestic clothes dryer will exhaust less than 200 cfm (0.09 m³/s). For closet installations, an opening with a minimum area of 100 square inches (0.0645 m²) must be cut in the closet door or in the closet enclosure. Where louvers or grilles are used, the solid portion of the louver or grille should be evaluated in accordance with Section 304.10 of the IFGC. Makeup air is necessary to prevent the room or space housing the dryer(s) from developing a negative pressure with respect to adjacent spaces or to the outdoors, which could result in the improper and dangerous operation of the dryer and other fuel-burning appliances. The required amount of makeup air should be approximately equal to the amount exhausted, and is normally supplied by the infiltration of air from outdoors or through openings to the outdoors.

The makeup air not only supplies the air that is to be exhausted from the dryer, but also supplies combustion air. Commercial dryer manufacturers' installation instructions will prescribe opening requirements for supplying the necessary makeup and combustion air.

To simplify the installation, commercial and multiple-dryer installations commonly have transfer or ducted openings to the outdoors that introduce outdoor air into the immediate vicinity of the appliances. Commercial units typically obtain makeup air from an enclosed accessible space (plenum) behind the units. This enclosure provides maintenance access; houses the unit's power, piping and duct connections; and isolates the makeup air from the conditioned occupiable spaces. Makeup air is rarely supplied by the building's heating, ventilating and air-conditioning (HVAC) system because if it were, the HVAC system would have to be interlocked with the dryers to avoid starving any one dryer of makeup air if the HVAC system was not operating. Also, supplying makeup air through the HVAC system would waste energy because large volumes of conditioned air would be exhausted through the dryers. Additionally, cold outdoor air will normally have low moisture content and, thus, serves well as makeup air for drying operations. (The warmed outdoor air will have low relative humidity.)

Note that clothes dryers installed in closets should be listed for that application.

504.6 Domestic clothes dryer ducts. Exhaust ducts for domestic clothes dryers shall conform to the requirements of Sections 504.6.1 through 504.6.7.

❖ See commentary for subsequent sections.

504.6.1 Material and size. Exhaust ducts shall have a smooth interior finish and shall be constructed of metal a minimum 0.016 inch (0.4 mm) thick. The exhaust duct size shall be 4 inches (102 mm) nominal in diameter.

❖ This section requires that the diameter of the exhaust duct be 4 inches, no larger and no smaller. A 4-inch duct is the basis for the design of the appliance exhaust system. Ducts that are too small would restrict air movement and would cause the dryer to consume more energy because of the longer drying time. They could also cause appliance overheating. Ducts that are too large will cause a decrease in the flow velocity, causing lint and debris to drop out of the exhaust air stream. For example, increasing the duct size from 4 inches to 5 inches (101.6 to 127 mm) will result in a reduction of duct velocity of approximately 37 percent for the typical dryer. The exhaust duct material must be smooth-wall metal with the specified minimum wall thickness. This translates to rigid metal duct of not less than 28 gage or heavier. Plastic ducts, corrugated ducts and flexible ducts are all prohibited with the exception of listed and labeled transition ducts in accordance with Section 504.6.3.

504.6.2 Duct installation. Exhaust ducts shall be supported at 4-foot (1219 mm) intervals and secured in place. The insert end of the duct shall extend into the adjoining duct or fitting in the direction of airflow. Ducts shall not be joined with screws or similar fasteners that protrude into the inside of the duct.

❖ Clothes dryer ducts must be well supported to prevent deformation and joint separation. The male end of the duct and fittings must point in the direction of airflow to lessen flow resistance, joint leakage and collection of lint. This is common practice for all ducts, regardless of application. The last sentence of this section presents somewhat of a dilemma because it says that duct fasteners cannot protrude into the inside of the ducts. All ducts have been required by tradition, codes or installation standards to be mechanically fastened at all joints (see Section 603.9 applicable to HVAC ducts); however, this section departs from that practice by specifically prohibiting the only means of joining such ducts other than spot welding (i.e., screws and rivets). Joint sealing is not equivalent to or a substitute for mechanical fastening; yet, this section has, from a practical standpoint, forced the installation of dryer exhaust ducts that rely solely on joint sealants to hold them together. Dryer exhaust ducts must be exceptionally well-supported and sealed so that the joints will not separate. This is especially important where the ducts are to be concealed within the building construction and where the ducts will be mechanically cleaned by duct cleaning apparatus. See Section 504.4 which also addresses duct fasteners, but does not expressly prohibit fasteners from protruding into the duct.

It should be noted that Section M1502.4.2 in the 2012 *International Residential Code*® (IRC®) permits dryer exhaust ducts to be joined with screws and similar fasteners that protrude not more than $^1/_8$ inch into the inside of the duct.

504.6.3 Transition ducts. Transition ducts used to connect the dryer to the exhaust duct system shall be a single length that is *listed* and *labeled* in accordance with UL 2158A. Transition ducts shall be a maximum of 8 feet (2438 mm) in length and shall not be concealed within construction.

❖ Within the context of this section, a transition duct is a flexible connector used as a transition between the dryer outlet and the connection point to the exhaust duct system. Transition duct connectors must be listed and labeled as transition ducts for clothes dryer application. Transition ducts are currently listed to comply with UL 2158A. Such transition ducts are not considered to be flexible air ducts or flexible air connectors and are not subject to the material requirements of Section 603.6.

Transition ducts are flexible ducts constructed of a metalized (foil) laminated fabric supported on a spiral wire frame. They are more fire resistant than the typical plastic spiral duct. Transition duct connectors are necessary for domestic dryers because of appliance movement, vibration and outlet location. In many cases, connecting a domestic clothes dryer directly to rigid duct would be difficult.

Transition duct connectors are limited to 8 feet (1829 mm) in length and must be installed in accordance with their listing and the manufacturer's instructions. These duct connectors must not be concealed by any portion of the structure's permanent finish materials, such as drywall, plaster, paneling, built-in furniture or cabinets, or any other similar permanently affixed building component; they must remain entirely within the room in which the appliance is installed. Transition duct connectors cannot be joined to extend beyond the 8-foot (1829 mm) maximum length limit. Transition ducts are to be cut to length as needed to avoid excess duct and unnecessary bends. Note that in the application of Section 504.6.4.1, the length of the transition duct is not counted in the overall duct system length, despite the fact that the transition duct does contribute considerable airflow resistance to the exhaust duct system.

504.6.4 Duct length. The maximum allowable exhaust duct length shall be determined by one of the methods specified in Section 504.6.4.1 or 504.6.4.2.

❖ The duct length, the number and angle of fittings and the smoothness of the duct interior all contribute to the total friction loss of the airflow. When the friction loss is excessive, a reduction of the air velocity occurs, allowing lint and debris to accumulate in the duct and thus creating a fire hazard. Impeded airflow will also increase drying time, decrease appliance efficiency and raise system temperatures. The manufacturers' instructions and the code limitations concerning duct length, the number of elbows installed, and the type of termination fittings must be adhered to help prevent a fire hazard and allow proper appliance operation.

Note that this section introduces two distinct length choices in its two subsections.

504.6.4.1 Specified length. The maximum length of the exhaust duct shall be 35 feet (10 668 mm) from the connection to the transition duct from the dryer to the outlet terminal. Where fittings are used, the maximum length of the exhaust duct shall be reduced in accordance with Table 504.6.4.1.

❖ The maximum exhaust duct length of 35 feet (10 668 mm) is the requirement for domestic clothes dryers. The 35-foot (10 688 mm) limit is based on the worst-case scenario where the dryer is rated for a maximum duct length of 35 feet (10 688 mm). The intent was to ensure that the least capable dryer available would be compatible with a 35-foot (10 688 mm) duct system. If the exhaust duct system is designed for worst case, it should work properly in all cases. It should be noted that the code does not recognize the use of booster fans with clothes dryer exhaust systems. At this time, there is not a consensus standard to list booster fans for use in clothes dryer exhaust systems. A UL standard for dryer exhaust duct power ventilators is currently being developed. The use of a booster fan would have to be approved by the code official under the provisions of Section 105.2.

Because the maximum length is based on equivalent length, all fittings must be accounted for in accordance with Table 504.6.4.1. The length is measured from the point where the transition duct (if used) connects to the rigid exhaust duct system to the terminal outdoors (see Commentary Figure 504.6.4.1). The transition duct is excluded from the calculation of equivalent length of the duct system.

Table 504.6.4.1 assigns an equivalent length to the various types of elbow fittings that are available. The first two rows of the table cover the traditional elbows that the code and dryer manufacturer's installation instructions have always addressed. The other rows in the table cover newer high-performance elbows that have longer turning radii and smoother walls (i.e., no mitered joints); thus, the equivalent length is substantially reduced. When an installer discovers that the length of an exhaust duct will exceed the limits of the code, he or she may find that using higher performance elbows will keep the length within the allowable limits. Note that there may be some dryers on the market that specify a duct length of less than 35 feet (10 668 mm), and the manufacturer's instructions would prevail in such case.

TABLE 504.6.4.1
DRYER EXHAUST DUCT FITTING EQUIVALENT LENGTH

DRYER EXHAUST DUCT FITTING TYPE	EQUIVALENT LENGTH
4″ radius mitered 45-degree elbow	2 feet 6 inches
4″ radius mitered 90-degree elbow	5 feet
6″ radius smooth 45-degree elbow	1 foot
6″ radius smooth 90-degree elbow	1 foot 9 inches
8″ radius smooth 45-degree elbow	1 foot
8″ radius smooth 90-degree elbow	1 foot 7 inches
10″ radius smooth 45-degree elbow	9 inches
10″ radius smooth 90-degree elbow	1 foot 6 inches

For SI: 1 inch = 25.4 mm, 1 foot = 304.8 mm, 1 degree = 0.0175 rad.

504.6.4.2 Manufacturer's instructions. The maximum length of the exhaust duct shall be determined by the dryer manufacturer's installation instructions. The code official shall be provided with a copy of the installation instructions for the make and model of the dryer. Where the exhaust duct is to be concealed, the installation instructions shall be provided to the code official prior to the concealment inspection. In the absence of fitting equivalent length calculations from the clothes dryer manufacturer, Table 504.6.4.1 shall be used.

❖ This section allows the 35-foot (10 668 mm) limit to be exceeded where longer exhaust duct lengths are allowed by the appliance manufacturer's instructions.

Today's appliances often, but not always, permit longer distances, therefore this section allows the installer to take advantage of those longer distances where specified by the manufacturer. The make and model of the dryer must be provided to the code official, along with the respective installation instructions, to permit the code official to inspect the duct installation based on the manufacturer's instructions. Because the installation is specific to and dependent upon a certain appliance make and model, the code official must follow through to make sure that the appliance did get installed. Remember that the prime objective is to prevent a dangerous mismatch between a dryer and its exhaust system.

Where not otherwise specified by the dryer manufacturer, the flow resistance contributed by fittings installed in the duct system must be in accordance with Table 504.6.4.1. Each type of fitting and its angle is equated to a certain length of straight duct and this "equivalent length" is added to the actual straight duct length in the system to obtain the total length. The table recognizes newer fitting designs that have smooth bore interiors and much larger turning radii than traditionally used fittings. Such designs greatly reduce friction loss as can be seen by their much smaller equivalent lengths. If an installer is unable to stay within the allowable length on a particular job, he/she may find that substituting better fittings will correct the problem.

504.6.5 Length identification. Where the exhaust duct is concealed within the building construction, the equivalent length of the exhaust duct shall be identified on a permanent

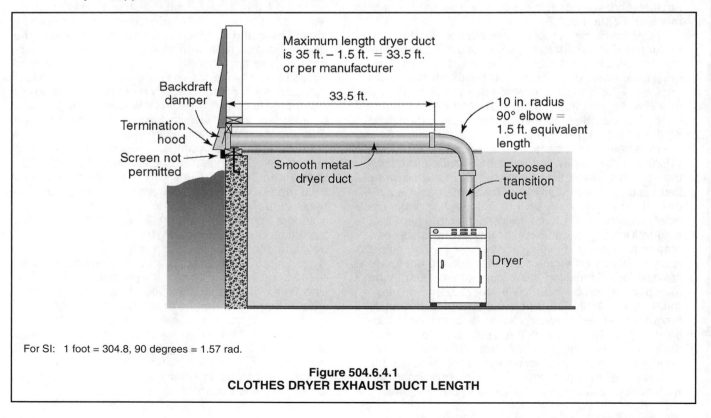

For SI: 1 foot = 304.8, 90 degrees = 1.57 rad.

Figure 504.6.4.1
CLOTHES DRYER EXHAUST DUCT LENGTH

label or tag. The label or tag shall be located within 6 feet (1829 mm) of the exhaust duct connection.

❖ A point to consider is the situation that will exist when the original dryer is replaced or the original occupant moves out and the new occupant installs a different dryer. In either case, the replacement dryer might not be compatible with the existing duct length. If the duct system equivalent length is not identified in a conspicuous manner, the new resident will be unaware of the potentially hazardous situation that has been created by the dryer and exhaust mismatch (see Commentary Figure 504.6.5). It is assumed that the installer will be aware of the equivalent length of exhaust ducts that are entirely exposed and observable.

504.6.6 Exhaust duct required. Where space for a clothes dryer is provided, an exhaust duct system shall be installed. Where the clothes dryer is not installed at the time of occupancy, the exhaust duct shall be capped at the location of the future dryer.

Exception: Where a *listed* condensing clothes dryer is installed prior to occupancy of structure.

❖ Where a space is provided for the future installation of a clothes dryer, an exhaust duct system must be installed so that such system can be inspected and approved (see Commentary Figure 504.6.6). Otherwise, it is likely that the dryer and exhaust duct system will be installed later without the code official's knowledge, and it will not be inspected for proper installation. Note that Section 504.6.4.2 does not appear to be an option when an exhaust duct system

is "roughed-in" for future use because the make, model and exhaust capability of the dryer will be unknown. If the equivalent length of the exhaust duct system does exceed the maximum length of 35 feet (10 668 mm) and the model and manufacturer of the dryer are unknown, then the code official could approve identifying the equivalent length in accordance with Section 504.6.5 under the alternative approval provisions of Section 105.2.

The requirement to cap the exhaust duct where a dryer is not installed at the time of occupancy is intended to keep the duct from becoming a conduit for air to travel between the inside of the structure and the outdoors. While a backdraft damper is required at the outdoor termination, capping the duct further reduces the chance of air movement. The capping will also prevent the entry of debris and foreign objects. The cap is required inside at the future dryer's location so that it will be readily apparent when the future dryer is installed. Requiring that a listed condensing dryer be installed prior to occupancy will prevent someone from taking advantage of this exception to avoid installing an exhaust duct system by indicating that a future condensing dryer will be installed.

504.6.7 Protection required. Protective shield plates shall be placed where nails or screws from finish or other work are likely to penetrate the clothes dryer exhaust duct. Shield plates shall be placed on the finished face of all framing members where there is less than $1\frac{1}{4}$ inches (32 mm) between the duct and the finished face of the framing member. Protective shield plates shall be constructed of steel, have

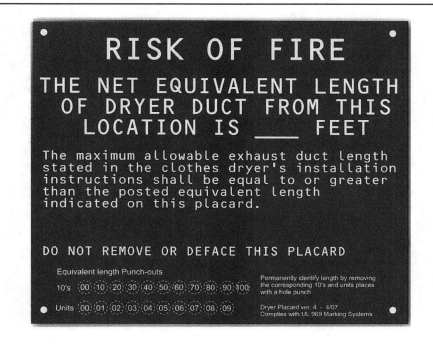

Figure 504.6.5
EXAMPLE OF PERMANENT LABEL OF EQUIVALENT LENGTH

a thickness of 0.062 inch (1.6 mm) and extend a minimum of 2 inches (51 mm) above sole plates and below top plates.

❖ Protection from nail and screw penetration has been required for wiring, vent systems, plumbing piping, fuel piping, hydronic piping and refrigerant piping for some time in the codes and now it is also required for dryer exhaust ducts (see Commentary Figure 504.6.7). Depending upon the depth of penetration, a nail or screw could create a serious obstruction problem and potential fire hazard.

504.7 Commercial clothes dryers. The installation of dryer exhaust ducts serving commercial clothes dryers shall comply with the *appliance* manufacturer's installation instructions. Exhaust fan motors installed in exhaust systems shall be located outside of the airstream. In multiple installations, the fan shall operate continuously or be interlocked to operate when any individual unit is operating. Ducts shall have a minimum *clearance* of 6 inches (152 mm) to combustible materials. Clothes dryer transition ducts used to connect the *appliance* to the exhaust duct system shall be limited to single lengths not to exceed 8 feet (2438 mm) in length and shall be *listed* and *labeled* for the application. Transition ducts shall not be concealed within construction.

❖ The code classifies clothes dryers into two categories: domestic and commercial. Domestic clothes dryers are primarily used in the family living environment and do not have the capacity to generate the amount of heat necessary to dry large volumes of clothing. The scope of this section does not regulate domestic clothes dryers.

Commercial clothes drying appliances and exhaust ducts are designed for the higher volumes, tempera-

tures and frequency of use found in public access and commercial laundries. Commercial dryers must be installed in accordance with the manufacturer's instructions. These dryers are tested to the appropriate safety standard for the appliance, and the manufacturer's installation instructions convey the information needed to duplicate the installation configuration that was tested and found to meet the requirements of the safety standard. The manufac-

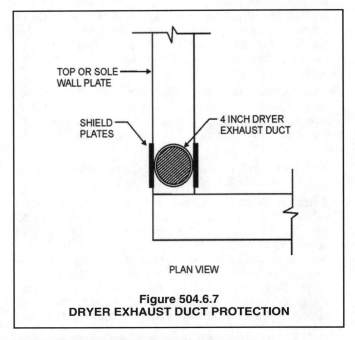

Figure 504.6.7
DRYER EXHAUST DUCT PROTECTION

Figure 504.6.6
DOMESTIC CLOTHES DRYER ROUGH-IN

turer's installation instructions are evaluated by the agency responsible for testing the appliance and, therefore, the instructions will prescribe an installation that is consistent with Section 304.1, which addresses the installation of all mechanical equipment. Common exhaust systems for multiple dryers must be designed and engineered in accordance with the equipment and the appliance manufacturer's installation instructions (see Commentary Figure 504.7).

The potential fire hazards associated with commercial clothes dryers are similar to those of domestic clothes dryers (combustible lint, debris and high temperatures). However, the complex exhaust systems, intense heat and larger volume of combustible lint and combustion byproducts associated with commercial dryers call for additional safety requirements.

Commercial and multiple dryer installations can rely on a single exhaust fan to prevent backpressure, overcome duct friction and induce the proper flow within the common exhaust system. Commentary Figure 504.7 shows a four-unit installation with the exhaust fan located at the roof level. This section requires that exhaust fans for commercial clothes dryer installations be operated continuously or be interlocked to operate when any individual unit is operating.

It is possible to exhaust multiple commercial dryers through a common duct system, such as a common riser or common manifold serving groups of adjacent units. This configuration will usually require an exhaust fan to prevent backpressure and to overcome static pressure in the duct system. In such installations, the common duct system will be connected to the inlet side of an exhaust fan and the fan will be located at the duct system termination to the outdoors.

The motor of the exhaust fan must be positioned outside of the exhaust airstream so that lint, debris and condensation do not accumulate in or on the motor. This requirement is generally satisfied by belt-driven fans with externally mounted motors. Location in the exhaust stream could result in deterioration and failure of the motor, as well as a fire hazard associated with the accumulation of combustible fibers in contact with a motor that may be hot enough to cause ignition. Failure of the exhaust fan motor could result in an unsafe operating condition for the dryers served by the fan.

Where multiple dryer units vent to a common venting system using a common fan, the common exhaust fan must operate when any one or more of the individual dryer unit(s) are operating. This can be achieved in two ways.

The first, continuous exhaust fan operation, is not desirable and is rarely done because of increased energy usage and the unreliability of the design. Continuous operation without interlocks (fan supervision)

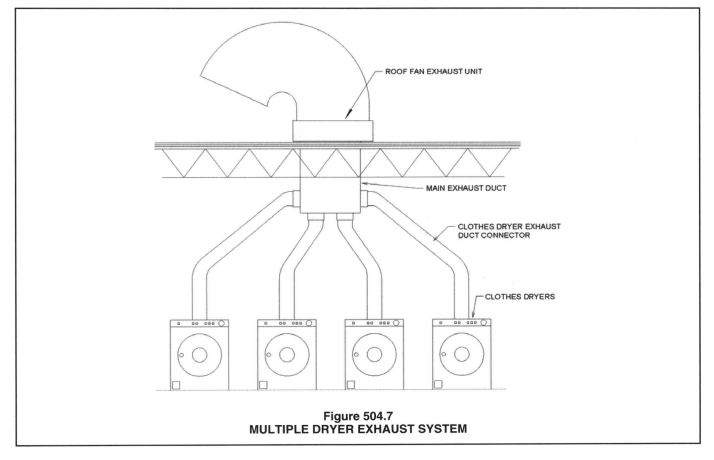

Figure 504.7
MULTIPLE DRYER EXHAUST SYSTEM

is not reliable because the dryer units would not respond to or be "aware" of an exhaust fan failure. Any mechanical failure of the exhaust fan or opening of an overcurrent protection device could create an unsafe condition. The second and preferred method is to install exhaust fan supervision interlocked with each dryer unit so that the dryers would be stopped or prevented from starting in the event of exhaust fan failure. The supervision/interlock design should first prove exhaust fan operation by sensing duct pressures before enabling dryer operation. The supervision/interlock logic is identical to that discussed in the commentary to Section 804.3.2.

Transition ducts are listed factory-built connectors designed to connect an appliance discharge outlet to the exhaust duct system.

504.8 Common exhaust systems for clothes dryers located in multistory structures. Where a common multistory duct system is designed and installed to convey exhaust from multiple clothes dryers, the construction of the system shall be in accordance with all of the following:

1. The shaft in which the duct is installed shall be constructed and fire-resistance rated as required by the *International Building Code.*

2. Dampers shall be prohibited in the exhaust duct. Penetrations of the shaft and ductwork shall be protected in accordance with Section 607.5.5, Exception 2.

3. Rigid metal ductwork shall be installed within the shaft to convey the exhaust. The ductwork shall be constructed of sheet steel having a minimum thickness of 0.0187 inch (0.4712 mm) (No. 26 gage) and in accordance with SMACNA *Duct Construction Standards.*

4. The ductwork within the shaft shall be designed and installed without offsets.

5. The exhaust fan motor design shall be in accordance with Section 503.2.

6. The exhaust fan motor shall be located outside of the airstream.

7. The exhaust fan shall run continuously, and shall be connected to a standby power source.

8. Exhaust fan operation shall be monitored in an *approved* location and shall initiate an audible or visual signal when the fan is not in operation.

9. Makeup air shall be provided for the exhaust system.

10. A cleanout opening shall be located at the base of the shaft to provide *access* to the duct to allow for cleaning and inspection. The finished opening shall be not less than 12 inches by 12 inches (305 mm by 305 mm).

11. Screens shall not be installed at the termination.

12. The common multistory duct system shall serve only clothes dryers and shall be independent of other exhaust systems.

❖ As noted in the commentary to Section 504.2, a clothes dryer exhaust duct is considered to be a duct and must follow the same requirements as any other ducts with respect to where the ducts penetrate fire-resistance-rated and nonrated floor/ceiling assemblies. Therefore, where penetrating a floor/ceiling assembly, the penetration must be protected in accordance with Sections 302.2 and 607 that will, in most cases, require a fire damper and, in some cases, require fire and smoke dampers along with a fire-resistance-rated shaft. However, Section 504.2 specifically prohibits the installation of fire and smoke dampers in a clothes dryer exhaust duct (see commentary, Section 504.2).

The intent behind this section is to provide prescriptive requirements on how to accomplish the exhausting of clothes dryers on multiple floors in a building such as a multiple-family apartment building. A common shaft to exhaust multiple clothes dryers on multiple floors has been used for years, although the code has been silent on this issue. While this practice is not new, the model codes have never contained any requirements, and there are no industry standards for this application other than recommendations by some appliance manufacturer's.

This section is based on a concept that has been in the model codes for many years, which is the use of subducts and continuous vertical airflow within the shaft in lieu of providing fire and smoke dampers where the ducts penetrate the shaft enclosure. This section expands this concept by requiring compliance with 12 different items. The first item requires the shaft to be constructed and fire rated in accordance with the IBC, and the second item requires that penetrations be protected in accordance with Section 607.5.5, Exception 2. This exception has the limitations that it is only applicable in Group B and R occupancies that are equipped throughout with an NFPA 13 automatic sprinkler system. As this exception only eliminates the requirement for the smoke damper, Section 607.5.5, Exception 1, Item 1.1 is used to eliminate the requirement for the fire damper.

Item 3 requires the shaft to be lined with rigid metal ductwork. The moisture from the dryer exhaust may be detrimental to the materials that make up the construction of the shaft. Gypsum board is commonly used to construct fire-resistance-rated shafts and requiring metal ductwork prevents the gypsum enclosure itself from serving as the exhaust passageway. Item 4 helps to reduce the effects of friction that may occur at offsets that, in turn, may reduce the airflow velocity within the shaft. Additionally, offsets could become a collection point for lint accumulation.

Because of the combustible nature of the lint produced by clothes dryers, Item 5 requires the construction of the fan to be in accordance with Section 503.2 (see commentary, Section 503.2). This is intended to ensure that the proper type of fan is utilized for this application. Item 6 requires the fan to be located outside of the airstream which is a current requirement for commercial dryers (see commentary, Section 504.7). Since the use of subducts requires continuous vertical airflow within the shaft, Item 7 requires the exhaust fan to run continuously and to be connected to a standby power source. This requirement is redundant with Section 607.5.5, Exception 2 which requires the exhaust fan to be powered continuously in accordance with the provisions of Section 909.11 of the IBC, which requires an approved standby power system.

The exhaust fan is required by Item 8 to be monitored to make sure that the fan is operating. The reason that fire and smoke dampers can be omitted with the use of subducts is because of the continuous vertical airflow within the shaft. If the fan is not running, there is no continuous airflow and excess lint will accumulate within the shaft. Additionally, without upward airflow, any flames or hot gases could make their way back down the subduct and into the occupied space. Therefore, an audible or visual signal must indicate when the fan is not in operation. The signal should be located such that it alerts building supervisory personnel so that they can take appropriate action.

Item 9 requires makeup air to be provided into the shaft for the exhaust system. Failure to provide makeup air could result in air being taken through the dryer exhaust system from the occupied spaces when the dryers are not operating. This could create an air balance problem within the occupied spaces that would have to be addressed and could also increase energy consumption in the spaces. Item 10 requires a cleanout at the base of the shaft. The current duct cleaning industry has the technology to be able to clean the shaft with a cleanout located at the base and through the top where the fan is located. Item 11 states a common requirement that screens are not permitted in clothes dryer exhaust systems because screens will become obstructed by lint. Item 12 is consistent with Section 501.2 which requires dryer exhaust to be independent of all other exhaust systems.

One item not addressed by this section is the amount of continuous upward airflow that is required. In the case of dryer exhaust and because of the difficulty of transporting suspended solids vertically against gravity, the velocity of the airflow within the shaft must be such that the clothes fibers (lint) would remain in the air stream and exit the top of the shaft. If the lint falls out of the air stream, it will collect on the interior of the shaft creating maintenance issues and

a potential fire hazard. The code leaves this issue up to the design professional since the amount of airflow would have to take into account variables, such as the number of dryers, the number of stories, the fan size, the required transport velocity and the size of the shaft.

SECTION 505
DOMESTIC KITCHEN EXHAUST EQUIPMENT

505.1 Domestic systems. Where domestic range hoods and domestic appliances equipped with downdraft exhaust are located within dwelling units, such hoods and appliances shall discharge to the outdoors through sheet metal ducts constructed of galvanized steel, stainless steel, aluminum or copper. Such ducts shall have smooth inner walls, shall be air tight, shall be equipped with a backdraft damper, and shall be independent of all other exhaust systems.

Exceptions:

1. Where installed in accordance with the manufacturer's installation instructions and where mechanical or *natural ventilation* is otherwise provided in accordance with Chapter 4, *listed* and *labeled* ductless range hoods shall not be required to discharge to the outdoors.

2. Ducts for domestic kitchen cooking appliances equipped with downdraft exhaust systems shall be permitted to be constructed of Schedule 40 PVC pipe and fittings provided that the installation complies with all of the following:

   2.1. The duct shall be installed under a concrete slab poured on grade.

   2.2. The underfloor trench in which the duct is installed shall be completely backfilled with sand or gravel.

   2.3. The PVC duct shall extend not more than 1 inch (25 mm) above the indoor concrete floor surface.

   2.4. The PVC duct shall extend not more than 1 inch (25 mm) above grade outside of the building.

   2.5. The PVC ducts shall be solvent cemented.

❖ A domestic kitchen exhaust system is one that serves appliances typically found in residential occupancies such as within dwelling units. When compared to commercial cooking operations, residential cooking operations are far less frequent, of shorter duration, have lower heat output and produce fewer grease-laden vapors. However, air-borne contaminant control may be even more important in residential cooking operations because of the lower or nonexistent ventilation rates typical of dwelling units.

Note that domestic cooking appliances are installed in a number of occupancies that are not dwelling units. In a number of these installations a domestic kitchen exhaust system could be installed over these appliances. For a discussion on where a domestic range hood versus a commercial kitchen hood is required to be installed, see the commentary to Section 507.2.1.

Ducts used for dwelling unit kitchen exhaust must be constructed of sheet metal to reduce the potential fire hazard associated with the collection of grease or the absorption of grease into a more porous duct material. This section requires that domestic kitchen exhaust ducts be constructed of galvanized or stainless steel, aluminum or copper. The ducts must be sealed air tight to prevent leakage of air and grease into wall and ceiling cavities. Ducts should be sealed with a material that is suitable for long-term exposure to elevated temperatures. Backdraft dampers prevent the infiltration of outdoor air when the exhaust system is not operating. The hood manufacturer's instructions may require that the ducts be installed with a minimum clearance to combustibles (see commentary, Section 304.1).

In multistory buildings, it is not uncommon to see a shaft used to route the domestic kitchen exhaust duct to the exterior of the building. The code is silent on such installation of kitchen exhaust shafts. Note that where a duct penetrates a shaft, a fire and smoke damper is required to be installed. There are, however, exceptions to Section 607.5.5 that would not require fire and smoke dampers to be installed where 22-inch (559 mm) subducts and continuous airflow upward to the outdoors is provided in the shaft (see the commentary for Exceptions 1 and 2 of Section 607.5.5 for the specific requirements). Since this section requires a domestic kitchen exhaust system to exhaust through ducts made of sheet metal, one can interpret this as meaning that the inside of the shaft must be constructed of metal with a smooth interior finish.

By requiring the ducts to have smooth inner walls, this section prohibits the use of flexible and semirigid corrugated ducts for range hood exhaust and for the exhausting of cooking appliances having integral downdraft exhaust fan systems. Domestic kitchen exhaust must be independent of all other exhaust systems which is consistent with Section 501.2.

Exception 1 to this section allows the use of ductless (recirculating) range hoods that have no means for discharge to the outdoors. Note that exhaust to the outdoors would be required if natural or mechanical ventilation was not provided for the kitchen (see Chapter 4).

If natural or mechanical ventilation is otherwise provided for a space, note that this section does not require the installation of a domestic range hood over a domestic appliance.

Exception 2 allows the use of plastic duct to exhaust the downdraft-type domestic kitchen cooking appliances when all five of the specified installation criteria are met. Plastic ducts are not allowed in above-ground applications because the plastic cannot contain a fire within the confines of the duct. This exception allows plastic for the duct material because the duct must be below a slab on grade and the trench for the duct must be completely backfilled with sand or gravel to: (1) limit the amount of combustion air available to sustain a duct fire and (2) prevent a fire within the duct from spreading to the rest of the building. Additionally, the fumes and smoke from the burning plastic will be exhausted to the outdoors, and not introduced into the occupied space. The plastic material also has the advantage of corrosion resistance. Metal ducts below a slab are susceptible to deterioration.

505.2 Makeup air required. Exhaust hood systems capable of exhausting in excess of 400 cfm (0.19 m³/s) shall be provided with *makeup air* at a rate approximately equal to the *exhaust air* rate. Such *makeup air* systems shall be equipped with a means of closure and shall be automatically controlled to start and operate simultaneously with the exhaust system.

❖ It is becoming more common for residential kitchens to resemble commercial kitchens, both aesthetically and functionally. As a result, much larger capacity range hoods are being installed which aggravate the already existing problem of a lack of adequate makeup air for all exhaust systems in the dwelling. Homes typically suffer multiple ills because of indoor negative pressures caused by an imbalance of exhaust air to makeup air. This section requires a makeup air supply system at the prescribed threshold and further requires it to be tied to the exhaust system controls such that both systems operate simultaneously. The intent is to not allow the homeowner the option of operating the exhaust fan without also operating the makeup air system.

While this section does not specifically state whether the makeup air system can be passive or active, it is the author's opinion that the intent of the section is for an active system with a fan to bring in the makeup air. The makeup air is supposed to be provided "at a rate approximately equal to the exhaust air rate," which implies an active makeup air system that utilizes a fan. In addition, the makeup air system must "be automatically controlled to start and operate simultaneously," which also implies an active system where a fan will operate. Certainly, opening a window or normal infiltration is not acceptable.

SECTION 506
COMMERCIAL KITCHEN HOOD VENTILATION SYSTEM DUCTS AND EXHAUST EQUIPMENT

506.1 General. Commercial kitchen hood ventilation ducts and exhaust *equipment* shall comply with the requirements of this section. Commercial kitchen grease ducts shall be designed for the type of cooking *appliance* and hood served.

❖ Section 506 addresses ducts serving Type I and II hoods installed over commercial cooking appliances.

506.2 Corrosion protection. Ducts exposed to the outside atmosphere or subject to a corrosive environment shall be protected against corrosion in an *approved* manner.

❖ Grease ducts and other ducts can deteriorate where exposed to the outdoors or to any indoor corrosive atmosphere. For example, ducts on a building rooftop and ducts run up a building exterior wall must be protected by an approved covering or encasement. Such protection must not create a fire hazard because of clearances to combustibles.

506.3 Ducts serving Type I hoods. Type I exhaust ducts shall be independent of all other exhaust systems except as provided in Section 506.3.5. Commercial kitchen duct systems serving Type I hoods shall be designed, constructed and installed in accordance with Sections 506.3.1 through 506.3.13.3.

❖ Because of the potential for spreading smoke, grease-laden air and fire from the kitchen exhaust system to other parts of the building, exhaust ducts serving Type I hoods are prohibited from interconnecting with any other exhaust ducts other than another Type I exhaust duct and then only where all four provisions of Section 506.3.5 are met (see commentary, Section 506.3.5).

Sections 506.3.1 through 506.3.13.3 address ducts used for Type I hoods and Sections 506.4 through 506.4.2 address ducts used for Type II hoods.

506.3.1 Duct materials. Ducts serving Type I hoods shall be constructed of materials in accordance with Sections 506.3.1.1 and 506.3.1.2.

❖ This section addresses the fact that grease ducts and makeup air ducts have different material requirements, which are prescribed in Sections 506.3.1.1 and 506.3.1.2, respectively.

506.3.1.1 Grease duct materials. Grease ducts serving Type I hoods shall be constructed of steel having a minimum thickness of 0.0575 inch (1.463 mm) (No. 16 gage) or stainless steel not less than 0.0450 inch (1.14 mm) (No. 18 gage) in thickness.

Exception: Factory-built commercial kitchen grease ducts *listed* and *labeled* in accordance with UL 1978 and installed in accordance with Section 304.1.

❖ Commercial kitchen hood and duct systems are designed to resist structural failure in the event of a grease fire. Stainless steel and steel of the prescribed thicknesses are the only materials that can be used for field-fabricated, unlabeled grease ducts serving a Type I hood. The thicknesses specified for ducts are greater than those specified for hoods because the

ducts are usually concealed in the building structure and because the likelihood of a fire occurring in a duct is greater than in the hood. Also, the intensity of a fire in a duct will be greater because of the concentrated fuel load and the airflow velocity created by the exhaust fan or the chimney effect of vertical ducts.

The exception allows a factory-built grease duct system that is listed and labeled in accordance with UL 1978 to serve a commercial kitchen exhaust hood instead of a shop or field-fabricated, unlabeled grease duct system. A duct conforming to the requirements in this exception must be tested and labeled by an approved testing agency in accordance with UL 1978 and installed in accordance with the manufacturer's installation instructions. This referenced standard is specifically for factory-built commercial kitchen grease ducts. Note that this standard only tests the performance of the factory-built duct with respect to serving as a grease duct and does not change the need for a grease duct enclosure required by Section 506.3.11 (see commentary, Section 506.3.11). Grease ducts are intended to withstand a grease fire within the duct; therefore, this standard requires a simulated internal fire test to show that the factory-built duct will perform as intended. This standard also determines the clearances required to combustible materials. A duct that is labeled in accordance with this standard may be installed with the clearance to combustibles that is specified in the manufacturer's installation instructions. These clearances may be less than the 18 inches (457 mm) required by Section 506.3.6. All components of the duct system, including the duct, fittings, access doors and joint materials, must be approved for use with the tested and labeled duct system. Each individual grease duct assembly or part must be marked with the following:

- The approved agency's identification;
- The manufacturer's specified minimum clearance (airspace) to combustible materials;
- The manufacturer's or private labeler's name or identifying symbol;
- The part or model number or designation; and
- The requirement to install in accordance with the manufacturer's installation instructions [see Commentary Figure 506.3.1.1(1)].

The manufacturer's installation instructions must also specify the type of exhaust hood and cooking appliance(s) that the grease duct can serve. Factory-built ducts offer installation alternatives to the requirement of Section 506.3.6 and are often used where grease ducts must pass through combustible construction assemblies. Some of these factory-built grease ducts are listed for reduced clearance to combustibles [see Commentary Figure 506.3.1.1(2)].

506.3.1.2 Makeup air ducts. Makeup air ducts connecting to or within 18 inches (457 mm) of a Type I hood shall be constructed and installed in accordance with Sections 603.1, 603.3, 603.4, 603.9, 603.10 and 603.12. Duct insulation installed within 18 inches (457 mm) of a Type I hood shall be noncombustible or shall be *listed* for the application.

❖ Makeup air ducts bring in outdoor air to the kitchen to replace the air exhausted by the hood. If they do not connect to a Type I hood or come within 18 inches (457 mm) of the hood, the ducts may be constructed of any material listed in Section 603 because there is little possibility of excess heat or fire being introduced into the duct. Where the makeup air duct is connected directly to the hood or comes within 18 inches (457 mm) of the hood, the duct materials must be noncombustible or listed for the intended application. Since flexible ducts are combustible, they are not permitted to connect to the hood or come within 18 inches (457 mm) of a Type I hood [see Commentary Figures 506.3.1.2(1) and 506.3.1.2(2)].

Where duct insulation is located within 18 inches (457 mm) of the hood, it has the potential for exposure to excessive heat that could ignite the insulation. For this reason, the insulation must be composed of a noncombustible material or must be listed for the application.

A METALBESTOS.
ZERO CLEARANCE GREASE DUCT
AND FIRE-RATED INTEGRAL CHASE

Figure 506.3.1.1(2)
FACTORY-BUILT GREASE DUCTS
(Photo courtesy of Selkirk, L.L.C.)

c(UL)us
LISTED

CLASSIFIED
(UL)

SELKIRK
METALBESTOS.
Logan, OH   Nampa, ID

GREASE DUCT
A218

GREASE DUCT
FOR USE IN GREASE DUCT ASSEMBLIES
CLASSIFIED IN ACCORDANCE WITH UL 2221
SEE UL FIRE RESISTANCE DIRECTORY
18LS

Model Zero Clear - GREASE DUCT, CLASSIFIED AS AN ALTERNATE TO A 2 HOUR FIRE RESISTIVE SHAFT ENCLOSURE WITH A MINIMUM ZERO CLEARANCE TO COMBUSTIBLES.

Model Zero Clear – GREASE DUCT PART FOR RESTAURANT COOKING APPLIANCE. LISTED IN ACCORDANCE WITH UL1978 - STANDARD FOR GREASE DUCTS.

FLUE DIAMETER: 5" – 36"   CLEARANCE TO COMBUSTIBLES: 0" (ALL SIZES)

FOR GREASE DUCT SYSTEMS INSTALLED WITHOUT A CONTINUOUS FIRE-RATED ENCLOSURE, AN EVALUATED THROUGH-PENETRATION FIRESTOP ASSEMBLY SHALL BE USED. SEE THROUGH PENETRATION FIRESTOP SYSTEM NO. C-AJ-7101 IN THE UNDERWRITERS LABORATORIES, INC. FIRE RESISTANCE DIRECTORY.

INSTALL AND USE ONLY IN ACCORDANCE WITH SELKIRK METALBESTOS GREASE DUCT INSTALLATION AND MAINTENANCE INSTRUCTIONS.

WARNING: RISK OF FIRE. DO NOT FULLY ENCLOSE WITHIN COMBUSTIBLE MATERIAL PLACED AT ZERO CLEARANCE TO ASSEMBLY.

Figure 506.3.1.1(1)
TYPICAL FACTORY-BUILT GREASE DUCT LABEL
(Photo courtesy of Selkirk, L.L.C.)

Figure 506.3.1.2(1)
PROHIBITED INSTALLATION OF FLEXIBLE AIR DUCT CONNECTING
TO A MAKEUP AIR PLENUM ABOVE A TYPE I HOOD

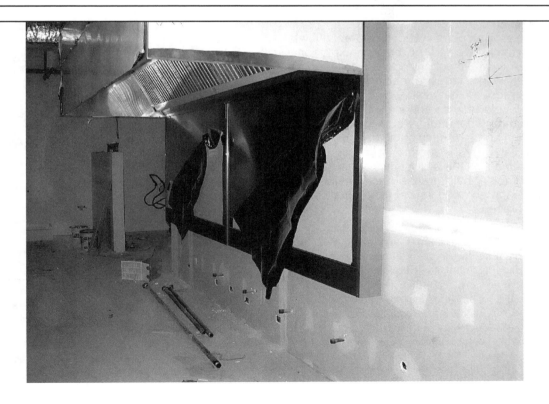

Figure 506.3.1.2(2)
BACKWALL MAKEUP AIR PLENUM

506.3.2 Joints, seams and penetrations of grease ducts. Joints, seams and penetrations of grease ducts shall be made with a continuous liquid-tight weld or braze made on the external surface of the duct system.

Exceptions:

1. Penetrations shall not be required to be welded or brazed where sealed by devices that are *listed* for the application.

2. Internal welding or brazing shall not be prohibited provided that the joint is formed or ground smooth and is provided with ready access for inspection.

3. Factory-built commercial kitchen grease ducts *listed* and *labeled* in accordance with UL 1978 and installed in accordance with Section 304.1.

❖ The joints and connections of an exhaust duct must be welded or brazed liquid tight to prevent grease and residues from leaking from the duct interior. This would prohibit the use of sheet metal locking joints, rivets, screws or any mechanical connectors, except as stated in the exceptions to this section. Joint surfaces must be smooth to facilitate cleaning and to prevent the accumulation of grease. Penetrations of grease ducts would include fire suppression system piping and nozzles.

Exception 1 recognizes the existence of penetration sealing devices that are specifically listed for that application. These devices would have to rely on some type of noncombustible or heat-tolerant gasket or compression seal.

Exception 2 allows internal welded or brazed joints instead of external welded or brazed joints only where those joints would not promote the collection of grease and only where they are readily observable.

Exception 3 requires that factory-built grease ducts be listed and labeled in accordance with UL 1978, a consensus standard for factory-built grease ducts. The construction and assembly of factory-built grease ducts are regulated by the product listing and therefore are not subject to the provisions of this section. For example, factory-built grease duct sections are joined with proprietary mechanical joints and cannot be welded or brazed.

506.3.2.1 Duct joint types. Duct joints shall be butt joints, welded flange joints with a maximum flange depth of $^1/_2$ inch (12.7 mm) or overlapping duct joints of either the telescoping or bell type. Overlapping joints shall be installed to prevent ledges and obstructions from collecting grease or interfering with gravity drainage to the intended collection point. The difference between the inside cross-sectional dimensions of overlapping sections of duct shall not exceed $^1/_4$ inch (6 mm). The length of overlap for overlapping duct joints shall not exceed 2 inches (51 mm).

❖ The provisions of this section intend to achieve liquid-tight joints that will not collect grease. The provision for welded flange joints was added to the code because this type of joint had been used for years under one of the legacy codes. The use of this joint is not only faster and easier to construct, but it yields a better looking joint that is easier to make liquid tight. Use of this type of joint helps to prevent warping of the duct from welding, which can produce puddles in the duct that can lead to grease accumulation. Since a $^1/_2$-inch (12.7 mm) flange depth is the maximum permitted, only a minimal amount of grease could collect in the groove between the flanges. Telescoping overlap joints must be installed so that the grease drainage flows from the duct with the smaller internal diameter to the duct with the larger internal diameter. This will prevent the duct edge from creating a dam in the grease flow, especially in horizontal ducts [see Commentary Figures 506.3.2.1(1) and 506.3.2.1(2)].

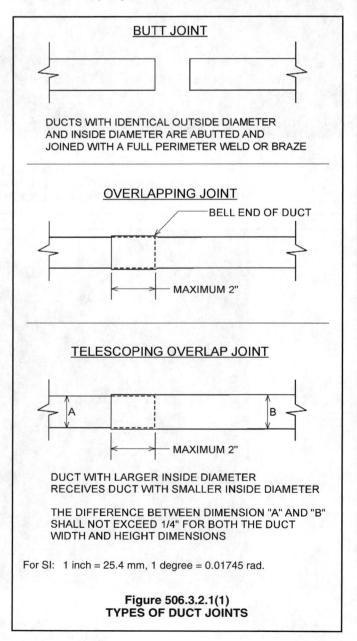

For SI: 1 inch = 25.4 mm, 1 degree = 0.01745 rad.

Figure 506.3.2.1(1)
TYPES OF DUCT JOINTS

506.3.2.2 Duct-to-hood joints. Duct-to-hood joints shall be made with continuous internal or external liquid-tight welded or brazed joints. Such joints shall be smooth, accessible for inspection, and without grease traps.

Exceptions: This section shall not apply to:

1. A vertical duct-to-hood collar connection made in the top plane of the hood in accordance with all of the following:

    1.1. The hood duct opening shall have a 1-inch-deep (25 mm), full perimeter, welded flange turned down into the hood interior at an angle of 90 degrees (1.57 rad) from the plane of the opening.

    1.2. The duct shall have a 1-inch-deep (25 mm) flange made by a 1-inch by 1-inch (25 mm by 25 mm) angle iron welded to the full perimeter of the duct not less than 1 inch (25 mm) above the bottom end of the duct.

    1.3. A gasket rated for use at not less than 1,500°F (815°C) is installed between the duct flange and the top of the hood.

    1.4. The duct-to-hood joint shall be secured by stud bolts not less than $^1/_4$ inch (6.4 mm) in diameter welded to the hood with a spacing not greater than 4 inches (102 mm) on center for the full perimeter of the opening. All bolts and nuts are to be secured with lockwashers.

2. *Listed* and *labeled* duct-to-hood collar connections installed in accordance with Section 304.1.

❖ This section applies to Type I hoods only. Grease ducts and hoods must be liquid tight; therefore, welded or brazed joints are required. It is often difficult to weld or braze the joint between a hood and its duct(s) because of lack of access to the juncture and space constraints caused by structural members. Either the hood or the duct is installed first and the companion element (hood or duct) is then joined to the already installed component. This typical sequence of construction is what makes it difficult to weld or braze hood-to-duct joints and is the reason for the two exceptions to this section. Both exceptions refer to mechanical joints designed to be mechanically strong and liquid tight. These kinds of joints are considered to be functionally equivalent to welded or brazed joints (see Commentary Figure 506.3.2.2).

506.3.2.3 Duct-to-exhaust fan connections. Duct-to-exhaust fan connections shall be flanged and gasketed at the base of the fan for vertical discharge fans; shall be flanged, gasketed and bolted to the inlet of the fan for side-inlet utility fans; and shall be flanged, gasketed and bolted to the inlet and outlet of the fan for in-line fans. Gasket and sealing materials shall be rated for continuous duty at a temperature of not less than 1500°F (816°C).

❖ Joints between grease ducts and exhaust fans must be liquid tight but need not be welded or brazed. The joints must be readily disassembled to allow for main-

Figure 506.3.2.1(2)
WELDED FLANGE JOINT
(Photo courtesy of Guy McMann)

tenance and repair or fan replacement; therefore, mechanical joints consisting of flanges, gaskets and bolts are used. Bolts are not required for flanged joints under vertical discharge (upblast) fans because fan curbs would make the bolts inaccessible. Grease ducts are supported independently of vertical discharge fans and the fan flange rests on the duct flange, relying on the flange gasket and the fan body to maintain a seal. The flange gasket must be rated for a temperature of not less than 1500°F, which is consistent with the temperature requirements for the gasket used for the duct-to-hood joint in Section 506.3.2.2, Exception 1.

506.3.2.4 Vibration isolation. A vibration isolation connector for connecting a duct to a fan shall consist of noncombustible packing in a metal sleeve joint of *approved* design or shall be a coated-fabric flexible duct connector *listed* and *labeled* for the application. Vibration isolation connectors shall be installed only at the connection of a duct to a fan inlet or outlet.

❖ Vibration isolation is not required by the code, but is regulated if installed in an exhaust duct system. To avoid the potentially hazardous accumulation of grease and the installation of ignition-prone materials, the code prescribes the use of an approved metal sleeve joint with noncombustible packing or an approved coated-fabric flexible duct connector for vibration isolation in grease duct systems. These joints must prevent the escape of grease, exhaust flow, smoke and fire. The joints do not resemble the fabric isolation joints typically found in HVAC ductwork. The packed metal sleeve allows motion in only one plane, whereas the flexible fabric connector allows motion in multiple planes.

Vibration isolation joints are intended for use only where necessary to control noise and vibration, which means only at the point of connection to exhaust fans. Vibration isolation joints are not intended for and must not be used for joints in duct runs.

506.3.2.5 Grease duct test. Prior to the use or concealment of any portion of a grease duct system, a leakage test shall be performed. Ducts shall be considered to be concealed where installed in shafts or covered by coatings or wraps that prevent the ductwork from being visually inspected on all sides. The permit holder shall be responsible to provide the neces-

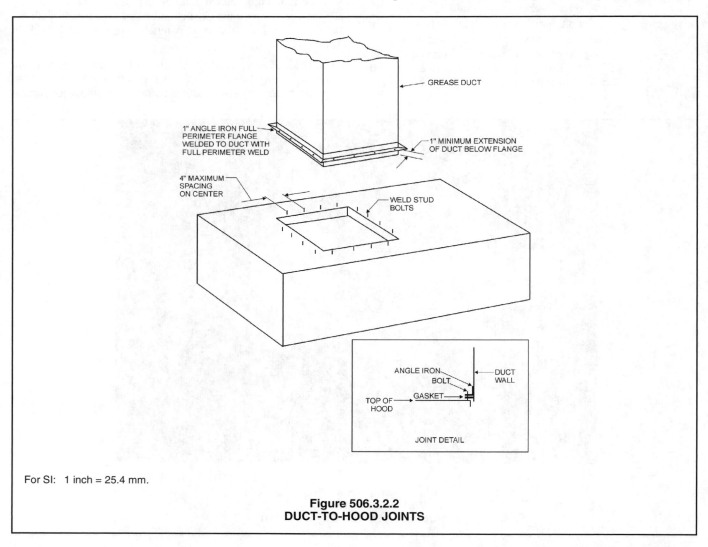

For SI: 1 inch = 25.4 mm.

Figure 506.3.2.2
DUCT-TO-HOOD JOINTS

sary *equipment* and perform the grease duct leakage test. A light test shall be performed to determine that all welded and brazed joints are liquid tight.

A light test shall be performed by passing a lamp having a power rating of not less than 100 watts through the entire section of ductwork to be tested. The lamp shall be open so as to emit light equally in all directions perpendicular to the duct walls. A test shall be performed for the entire duct system, including the hood-to-duct connection. The duct work shall be permitted to be tested in sections, provided that every joint is tested. For *listed* factory-built grease ducts, this test shall be limited to duct joints assembled in the field and shall exclude factory welds.

❖ The intent of this section is to prescribe a method of testing grease ducts to check that all welded and brazed joints and seams are sealed liquid tight. A light test is the most cost-efficient and effective method to locate any defects in welded or brazed grease duct installations and has been successfully used for many years. Leaks in grease ducts can lead to the passage of grease to surrounding areas of the ductwork and the building components, thus creating a potentially dangerous fire condition. This potentially dangerous condition can be eliminated by performing this simple test. Using the light test also allows for the grease duct to be tested as the sections of duct are completed. For factory-built grease ducts, only the field-assembled joints are required to undergo this test. Note that in accordance with Section 107.3, the code official will witness the required test or possibly accept reports from approved third parties.

506.3.3 Grease duct supports. Grease duct bracing and supports shall be of noncombustible material securely attached to the structure and designed to carry gravity and seismic loads within the stress limitations of the *International Building Code*. Bolts, screws, rivets and other mechanical fasteners shall not penetrate duct walls.

❖ Proper support of the duct system requires designing for the weight of the duct, as well as for dynamic loads caused by vibrations. Support is necessary for proper alignment of the duct system and avoidance of stresses that could lead to joint failure. A sagging duct will increase internal resistance to airflow, reduce the efficiency of the system and lead to the hazardous accumulation of grease. As with piping systems, ducts and ductwork systems that are improperly supported will deteriorate structurally and functionally.

Support for grease ducts must be installed in accordance with Section 603.10 and the seismic provisions of the IBC. Support spacing is limited to not more than 12 feet (3048 mm) between supports because of the inability of commonly used duct materials to resist deflection at larger intervals. This spacing criterion is primarily applicable to rigid ducts constructed of sheet metal; some duct configurations and duct materials will require shorter spacing intervals. For example, some duct materials would require support at each joint. The SMACNA HVAC *Duct Con-*

struction Standards—Metal and Flexible contains information for choosing support systems for ducts. Additionally, the IBC could require that the duct support system be designed to resist earthquake loads, depending on the location and use of the building.

If duct hangers and supports are not used and a portion of the building structure is used to support the duct, the support system must be designed by a registered professional engineer or architect to comply with the IBC. Duct hangers must be noncombustible, approved for the application and designed for the loads to be carried. Fasteners must not penetrate ducts because of the liquid-tight design of grease ducts.

506.3.4 Air velocity. Grease duct systems serving a Type I hood shall be designed and installed to provide an air velocity within the duct system of not less than 500 feet per minute (2.5 m/s).

> **Exception:** The velocity limitations shall not apply within duct transitions utilized to connect ducts to differently sized or shaped openings in hoods and fans, provided that such transitions do not exceed 3 feet (914 mm) in length and are designed to prevent the trapping of grease.

❖ The cross-sectional area of a grease duct system serving a Type I hood must be sized so that the exhaust air velocity is maintained at 500 feet per minute (2.5 m/s) or higher. This velocity is based on an ASHRAE research project (RP-1 033) conducted by the University of Minnesota, which found that duct velocities below 1,500 feet per minute (7.6 m/s) did not result in an increase in the rate of grease deposition on the walls of the ductwork. It should be noted that NFPA 96 also limits the minimum duct velocity to 500 feet per minute (2.5 m/s).

The exception allows the velocity of exhaust flow to fall below the specified velocity only in duct transition fittings that are necessary to connect grease ducts to hoods and fans. Because the shape of the transition fitting differs from that of the duct, the flow characteristics in the fitting will be different. This deviation in velocity is tolerable for small distances, especially at hood and fan connections.

506.3.5 Separation of grease duct system. A separate grease duct system shall be provided for each Type I hood. A separate grease duct system is not required where all of the following conditions are met:

1. All interconnected hoods are located within the same story.

2. All interconnected hoods are located within the same room or in adjoining rooms.

3. Interconnecting ducts do not penetrate assemblies required to be fire-resistance rated.

4. The grease duct system does not serve solid-fuel-fired appliances.

❖ A grease duct system serving a Type I hood may interconnect with, or share common components with, another grease duct system serving a Type I hood only as specified in this section. Grease duct systems serving Type I hoods cannot, under any circumstances, interconnect with, or share common components with, any building ventilation or exhaust system other than another grease duct system.

Although Section 506.3.11 limits a shaft to enclosing a single grease exhaust duct system, this section does not prohibit multiple grease exhaust ducts from connecting to a common trunk, riser or system if the additional safety conditions are met. Interconnected hoods must be located within the same story and within the same room or in adjoining rooms, and the interconnecting grease ducts that join hoods to the common exhaust duct must not penetrate fire-resistance-rated assemblies. A grease duct system serving a solid-fuel-fired appliance, such as a barbecue, smoking pit or a wood-fired pizza oven, cannot be interconnected with other grease hood exhausts. The intent of this section is to prevent fire from spreading from one system to another through the common portion shared by the hood systems and to prevent fire from spreading to other stories or fire areas (see Section 507.2.4).

506.3.6 Grease duct clearances. Where enclosures are not required, grease duct systems and exhaust *equipment* serving a Type I hood shall have a *clearance* to combustible construction of not less than 18 inches (457 mm), and shall have a *clearance* to noncombustible construction and gypsum wallboard attached to noncombustible structures of not less than 3 inches (76 mm).

Exceptions:

1. Factory-built commercial kitchen grease ducts *listed* and *labeled* in accordance with UL 1978.

2. *Listed* and *labeled* exhaust *equipment* installed in accordance with Section 304.1.

3. Where commercial kitchen grease ducts are continuously covered on all sides with a *listed* and *labeled* field-applied grease duct enclosure material, system, product or method of construction specifically evaluated for such purpose in accordance with ASTM E 2336, the required *clearance* shall be in accordance with the listing of such material, system, product or method.

❖ The required 18-inch (457 mm) distance from exhaust duct surfaces to combustible materials applies to exposed and concealed ducts (see Commentary Figure 506.3.6) where enclosures are not required in accordance with Section 506.3.11 (see commentary, Section 506.3.11). Although normal operating temperatures in a hood and duct system might be relatively low, especially compared to temperatures encountered in a fire, combustibles near the duct system become more susceptible to ignition because of the long-term exposure to even this moderate heat and can be ignited by the radiant or con-

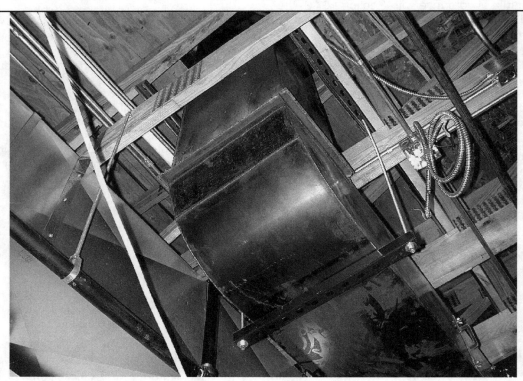

For SI: 1 inch = 25.4 mm.

Figure 506.3.6
PROHIBITED INSTALLATION OF A GREASE DUCT (Less than 18-inch clearance to combustibles)

vective heat given off by the duct system surfaces during "normal" hood and duct system operation. Note that the required clearance to combustibles for a Type I hood is also 18 inches (457 mm); however, this section only applies to the grease duct. The clearance requirements for a Type I hood can be found in Section 507.9 (see commentary, Section 507.9).

The clearances from the ducts to vertical and horizontal surfaces can be reduced in accordance with the provisions of Section 308.

In the context and application of the code in general, a composite material such as gypsum wallboard (drywall, sheetrock, etc.) is considered to be combustible (see the definition of "Noncombustible materials"); however, this section treats gypsum wallboard as a limited combustible material and only requires a 3-inch (76 mm) clearance. Remember that gypsum wallboard installed on a combustible substrate or on wood studs does not cause the wall to be considered as a noncombustible assembly, and the 18-inch (457 mm) minimum clearance still applies. A 3-inch (76 mm) clearance would be allowed for gypsum board attached to metal studs, for example. The classification of combustible and noncombustible materials is not changed by the use of fire-retardant-treated wood products or fire-rated (Type X) gypsum wallboard.

The first exception recognizes that factory-built commercial kitchen grease ducts are required to be listed and labeled in accordance with UL 1978, which tests to establish the required clearance to combustibles for the grease duct. The second exception recognizes that listed and labeled exhaust equipment

that is tested and approved for a specified clearance to combustibles can be installed in accordance with the manufacturer's installation instructions. The third exception is applicable to field-applied grease duct enclosure systems (e.g., duct wrap systems) which have, as a part of their listing, a specific allowable clearance from the outside of the enclosure system to adjacent combustible materials [see Commentary Figures 506.3.1.1(1) and 506.3.1.1(2)].

506.3.7 Prevention of grease accumulation in grease ducts. Duct systems serving a Type I hood shall be constructed and installed so that grease cannot collect in any portion thereof, and the system shall slope not less than one-fourth unit vertical in 12 units horizontal (2-percent slope) toward the hood or toward a grease reservoir designed and installed in accordance with Section 506.3.7.1. Where horizontal ducts exceed 75 feet (22 860 mm) in length, the slope shall be not less than one unit vertical in 12 units horizontal (8.3 percent slope).

❖ Sections of ducts serving Type I hoods must be constructed with the code-prescribed slopes and installed without forming any dips, pockets or low points that are capable of collecting grease or residue. Sloping the duct back toward the exhaust hood or to an approved grease-collection reservoir (see commentary, Section 506.3.7.1) will minimize the retention of grease in the duct system. A greater slope is necessary for long duct runs to encourage grease to flow to the collection points. Without adequate slope, grease could congeal before reaching the collection point, thus forming dams and allowing significant buildup [see Commentary Figure 506.3.7].

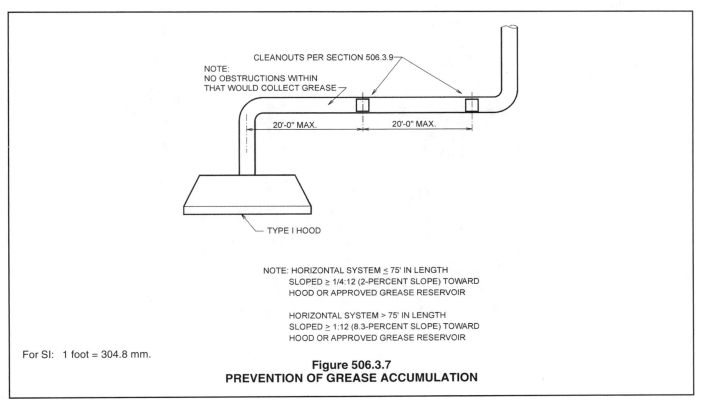

For SI: 1 foot = 304.8 mm.

Figure 506.3.7
PREVENTION OF GREASE ACCUMULATION

506.3.7.1 Grease reservoirs. Grease reservoirs shall:

1. Be constructed as required for the grease duct they serve.

2. Be located on the bottom of the horizontal duct or the bottommost section of the duct riser.

3. Have a length and width of not less than 12 inches (305 mm). Where the grease duct is less than 12 inches (305 mm) in a dimension, the reservoir shall be not more than 2 inches (51 mm) smaller than the duct in that dimension.

4. Have a depth of not less than 1 inch (25.4 mm).

5. Have a bottom that is sloped to a point for drainage.

6. Be provided with a cleanout opening constructed in accordance with Section 506.3.8 and installed to provide direct access to the reservoir. The cleanout opening shall be located on a side or on top of the duct so as to permit cleaning of the reservoir.

7. Be installed in accordance with the manufacturer's instructions where manufactured devices are utilized.

❖ In previous editions of the code, grease reservoirs were required to be approved with no specific guidance on how a reservoir should be constructed. The intent behind this section is to provide prescriptive requirements on how to construct a grease reservoir. The seven listed items provide a commonsense approach to the construction of a grease reservoir.

Considering that a grease duct is designed to resist structural failure in the event of a grease fire, Item 1 requires the grease reservoir to be constructed as required for the grease duct, which ensures that the grease reservoir does not compromise the integrity of the grease duct system (i.e., does not provide a weak link). In Item 2, locating the reservoir in the lowest sections of the grease duct system provides the best chance to collect the grease and retain it until it can be drained and cleaned out. The size of the reservoir with respect to the size of the duct will have an effect on how much grease will be collected. Item 3 requires the reservoir to have a length and width of not less than 12 inches. The best case scenario would be for the reservoir to span across the entire width of the horizontal duct, perpendicular to the direction of the airflow [see Commentary Figures 506.3.7.1(1) and 506.3.7.1(2)]. While this is not specifically stated in the text, this would allow any grease along the bottom of the duct to collect in the reservoir rather than running around the reservoir. In the case where the duct is less than 12 inches, this item does permit the reservoir to be not be less than 2 inches smaller than the duct in that dimension.

A minimum depth of 1 inch is required for the reservoir in Item 4. It should be noted that the deeper the reservoir, the more grease that can be retained between cleanings; therefore, more time between cleanings is allowed. The reservoir is required to be sloped to a drainage point. While not specifically required, a drain can be provided as indicated in Commentary Figure 506.3.7.1(1); however, the usefulness of the drain comes into question when the grease congeals such that it will not flow through the drainage opening. This is why Item 6 requires a cleanout opening to be located so that the reservoir can be reached and cleaned. In the case where a manufactured grease reservoir device is installed, it

Figure 506.3.7.1(1)
GREASE COLLECTION RESERVOIR
(Notice that there is no cleanout)
(Photo courtesy of Guy McMann)

must be installed in accordance with the manufacturer's installation instructions.

506.3.8 Grease duct cleanouts and openings. Grease duct cleanouts and openings shall comply with all of the following:

1. Grease ducts shall not have openings except where required for the operation and maintenance of the system.

2. Sections of grease ducts that are inaccessible from the hood or discharge openings shall be provided with cleanout openings.

3. Cleanouts and openings shall be equipped with tight-fitting doors constructed of steel having a thickness not less than that required for the duct.

4. Cleanout doors shall be installed liquid tight.

5. Door assemblies including any frames and gaskets shall be approved for the application and shall not have fasteners that penetrate the duct.

6. Gasket and sealing materials shall be rated for not less than 1500°F (816°C).

7. Listed door assemblies shall be installed in accordance with the manufacturer's instructions.

❖ To maintain the integrity of grease duct systems, openings into the systems are limited to those necessary for proper operation and maintenance. Operation and maintenance are closely related. Operation creates the need for maintenance, and maintenance allows continued operation. Because grease can and will liquefy on a duct wall, cleanouts are required in any portion of a grease duct system that cannot be reached for cleaning and inspection from the duct entry or discharge. Cleanout doors must be constructed to comply with Section 506.3.1.1, and must have a latching method to hold the door tightly closed. Latching devices must hold cleanout doors reasonably air tight when closed. Air leakage at cleanout openings would reduce exhaust system effectiveness and, in positive-pressure ducts, would allow exhaust leakage to the exterior of the duct. Because ducts are designed to contain a grease fire, cleanout openings must not reduce duct integrity. The code recognizes that cleanout doors may have to be opened by the use of hand tools [see Commentary Figures 506.3.8(1) and 506.3.8(2)].

The gasket and sealing materials for the cleanout must be rated for a temperature of not less than 1500°F which is consistent with the temperature requirements for the gasket used for the duct-to-hood joint in Section 506.3.2.2, Exception 1.

506.3.8.1 Personnel entry. Where ductwork is large enough to allow entry of personnel, not less than one *approved* or *listed* opening having dimensions not less than 22 inches by 20 inches (559 mm by 508 mm) shall be provided in the hori-

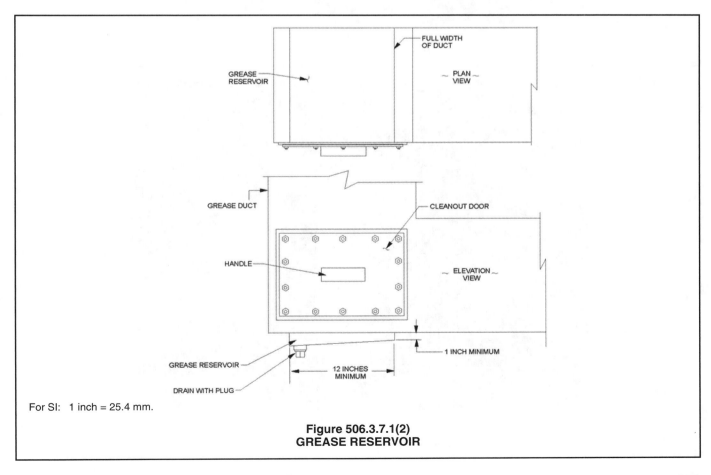

For SI: 1 inch = 25.4 mm.

Figure 506.3.7.1(2)
GREASE RESERVOIR

zontal sections, and in the top of vertical risers. Where such entry is provided, the duct and its supports shall be capable of supporting the additional load, and the cleanouts specified in Section 506.3.8 are not required.

❖ Large grease ducts could be designed to allow maintenance personnel to enter the duct for cleaning. This could make the cleanouts specified in Section 506.3.9 unnecessary. Where such access is intended, the access doors must be of adequate size to allow personnel to easily enter the duct. The duct must also be capable of supporting the additional weight of personnel and the dynamic loading caused by movement of personnel.

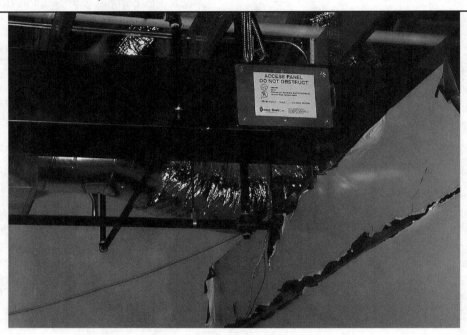

Figure 506.3.8(1)
GREASE DUCT CLEANOUT

Figure 506.3.8(2)
GREASE DUCT CLEANOUT

506.3.8.2 Cleanouts serving in-line fans. A cleanout shall be provided for both the inlet side and outlet side of an in-line fan except where a duct does not connect to the fan. Such cleanouts shall be located within 3 feet (914 mm) of the fan duct connections.

❖ In-line fans present a unique problem when it comes to maintenance and service since they are located within the building and are in-line with the grease duct attached to both the inlet and outlet side of the fan housing. In order to allow for maintenance and cleaning, a cleanout is required on the inlet and outlet sides of the fan within 3 feet (914 mm) of the fan to duct connections.

506.3.9 Grease duct horizontal cleanouts. Cleanouts serving horizontal sections of grease ducts shall:

1. Be spaced not more than 20 feet (6096 mm) apart.

2. Be located not more than 10 feet (3048 mm) from changes in direction that are greater than 45 degrees (0.79 rad).

3. Be located on the bottom only where other locations are not available and shall be provided with internal damming of the opening such that grease will flow past the opening without pooling. Bottom cleanouts and openings shall be approved for the application and installed liquid-tight.

4. Not be closer than 1 inch (25.4 mm) from the edges of the duct.

5. Have opening dimensions of not less than 12 inches by 12 inches (305 mm by 305 mm). Where such dimensions preclude installation, the opening shall be not less than 12 inches (305 mm) on one side and shall be large enough to provide access for cleaning and maintenance.

6. Shall be located at grease reservoirs.

❖ Given the nature of the exhaust from a Type I hood, grease will collect on the sides of the grease duct. Since this cannot be prevented from occurring, the only option is to provide access to the duct for periodic cleaning. Such access is especially important for a horizontal grease duct, where grease will tend to accumulate more and access from the hood or the fan on the roof is not available.

The 20-foot (6096 mm) interval requirement in Item 1 ensures that not more than 10 feet (3048 mm) of duct will extend beyond the access provided by any cleanout (see Commentary Figure 506.3.7). While this is implied by Item 1, Item 2 requires cleanouts to be located no farther than 10 feet (3048 mm) from any change of direction greater than 45 degrees in the duct. Cleanout openings should be located only on the vertical sides and top of a horizontal run to prevent the leakage of grease. Cleanout openings in duct sidewalls should not extend to the bottom of the duct because this would allow grease to leak from the duct. At least 1 inch (25 mm) of the duct sidewall should remain below the bottom edge of the cleanout to prevent grease from running out of the duct.

Cleanouts are allowed in the top and bottom of horizontal ducts only where it is not possible to put the required size cleanout in the vertical sidewalls.

Additional cleanouts should be installed in any portion of a duct system that cannot be reached for cleaning and inspection from the required cleanouts. The cleanout openings must have a minimum dimension of 12 inches (305 mm) to permit access for cleaning of the system. If the side of the duct is not large enough to permit the required cleanout size, the cleanout must be relocated to the top of the duct, or as a last resort, the bottom of the duct. Cleanouts located on the bottom of a duct would promote leakage of grease; therefore, internal barriers (dams) must be installed around the full perimeter of the opening. These barriers/dams must not prevent grease from draining to the intended collection point. If the largest dimension of a duct is less than that necessary to accommodate the required size cleanout, the cleanout would have to be as large as the duct dimensions would permit, on the duct face allowing the largest opening and would have to meet the opening edge requirements of this section.

506.3.10 Underground grease duct installation. Underground grease duct installations shall comply with all of the following:

1. Underground grease ducts shall be constructed of steel having a minimum thickness of 0.0575 inch (1.463 mm) (No. 16 gage) and shall be coated to provide protection from corrosion or shall be constructed of stainless steel having a minimum thickness of 0.0450 inch (1.140 mm) (No. 18 gage).

2. The underground duct system shall be tested and approved in accordance with Section 506.3.2.5 prior to coating or placement in the ground.

3. The underground duct system shall be completely encased in concrete with a minimum thickness of 4 inches (102 mm).

4. Ducts shall slope toward grease reservoirs.

5. A grease reservoir with a cleanout to allow cleaning of the reservoir shall be provided at the base of each vertical duct riser.

6. Cleanouts shall be provided with access to permit cleaning and inspection of the duct in accordance with Section 506.3.

7. Cleanouts in horizontal ducts shall be installed on the topside of the duct.

8. Cleanout locations shall be legibly identified at the point of access from the interior space.

❖ Item 1 requires a steel grease duct to be the same gage as an above-ground grease duct; however, the duct must be coated to provide protection from corrosion. In lieu of using coated steel, a stainless steel grease duct can be used. Due to the underground location of the grease duct, Item 2 requires the duct system to be tested prior to being coated or placed in the ground. Since the duct will be encased in concrete as required by Item 3, testing the duct prior to it being encased or coated is necessary so that any defects can be spotted and repaired. Encasing the duct in concrete provides added protection for the occupants who, in the event of a grease duct fire, could be sitting just above the grease duct (see Commentary Figure 506.3.10).

Running an underground grease duct from a table top cooking appliance creates a "trap" where the duct goes down into the floor, is run under the floor and then goes up through the floor and out through the roof. This trap creates a perfect place for grease to collect. Items 4 and 5 attempt to address this situation by requiring the duct to slope to a grease reservoir which must be located at the base of the vertical duct riser. A cleanout must be provided at this reservoir to facilitate cleaning. In addition to this cleanout, Item 6 requires cleanouts to be provided for this underground duct system just as they are required for above-ground grease ducts. The fact that some of these table top cooking appliances will most likely be located in open area dining rooms means that the run of the underground duct to a surrounding wall where the duct can be concealed and go through the roof could be of significant length. Section 506.3.9 requires horizontal ducts to have cleanouts every 20 feet. This may require cleanouts in the dining room floor unless careful planning is used to arrange the tables such that the duct run remains under the 20-foot maximum. Due to the nature of a horizontal underground grease duct, cleanouts must be located in the topside of the duct as indicated in Item 7. Item 8 requires that the location of all cleanouts be identified at the point of access. This would make it even more desirable to locate the tables such that a cleanout would not be required in the dining room floor. Not many restaurant owners would like to have the cleanout locations identified on the floor for the dining guests to see.

506.3.11 Grease duct enclosures. A grease duct serving a Type I hood that penetrates a ceiling, wall, floor or any concealed spaces shall be enclosed from the point of penetration to the outlet terminal. A duct shall penetrate exterior walls only at locations where unprotected openings are permitted by the *International Building Code*. The duct enclosure shall serve a single grease duct and shall not contain other ducts, piping or wiring systems. Duct enclosures shall be either field-applied or factory-built. Duct enclosures shall have a fire-resistance rating of not less than that of the assembly penetrated and not less than 1 hour. Duct enclosures shall be as prescribed by Section 506.3.11.1, 506.3.11.2 or 506.3.11.3.

❖ An enclosure is required where a grease duct serving a Type I hood penetrates a ceiling, wall or floor whether or not the ceiling, wall or floor is fire-resistance rated. The portion of a grease duct system that is exposed within the same room as the hood it serves would not have to be enclosed if such duct did not pass through any ceiling, wall or floor membrane. Once a duct passes through a ceiling, wall or floor membrane, it must be enclosed from that point on, all the way to the outdoor terminal. The enclosure maintains the integrity of the assembly that the grease duct is penetrating, and reduces the possibility of smoke and fire in the kitchen spreading to other parts of the building. The IBC, in some cases, requires fire-resistance-rated vertical shafts for ducts that penetrate floor/ceiling assemblies. Typically, when one thinks of ducts being in a fire-rated enclosure, it is associated with a vertical passageway through stories of a building. However, the enclosure required by this section requires both vertical and horizontal sections of the grease exhaust duct system to be in a rated enclosure. The enclosure must begin at the first point of penetration and must be continuous to the duct termination. In other words, the enclosure must originate in the room or space containing the hood and must extend to where the duct discharges to the

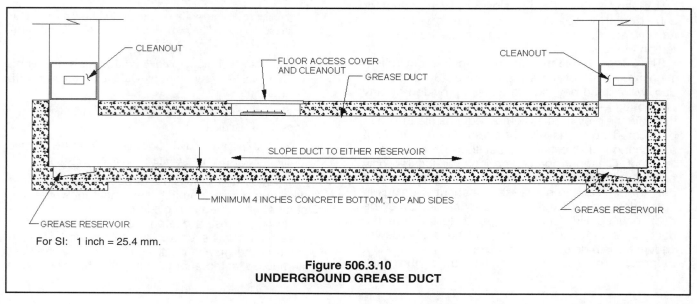

For SI: 1 inch = 25.4 mm.

Figure 506.3.10
UNDERGROUND GREASE DUCT

outdoors. In the case where a grease duct runs horizontally through nonrated walls, a grease duct enclosure that is fire-resistance rated for 1 hour would be required. Panels or doors in any grease duct enclosure providing access to cleanouts must be approved opening protectives that comply with the IBC or with the listed and labeled grease duct enclosure system.

This section permits a grease duct to penetrate an exterior wall where the wall is not required to have protected openings (see Commentary Figure 506.3.11). Depending on the building's proximity to lot lines and other buildings, the IBC might require that openings in exterior walls be protected with fire-resistance-rated opening protectives.

A grease duct enclosure must serve only a single grease duct, and it must not contain any other ducts, piping or wiring. Considering the hazards associated with a grease duct and the expectation that there could be a fire within the grease duct, the grease duct enclosure can serve only a single grease duct. Nothing else is permitted in the enclosure. Duct enclosures are required to have a fire-resistance rating not less than the assembly penetrated and not less than 1 hour. Therefore, this should be applied to any wall or floor assembly penetrated and not less than 1 hour.

This section provides three options for construction of grease duct enclosures. The first option is the traditional method of installing grease ducts in shafts,

the second is field-applied enclosure systems installed in accordance with ASTM E 2336 and the third is factory-built assemblies in accordance with UL 2221.

506.3.11.1 Shaft enclosure. Commercial kitchen grease ducts constructed in accordance with Section 506.3.1 shall be permitted to be enclosed in accordance with the *International Building Code* requirements for shaft construction. Such grease duct systems and exhaust *equipment* shall have a *clearance* to combustible construction of not less than 18 inches (457 mm), and shall have a *clearance* to noncombustible construction and gypsum wallboard attached to noncombustible structures of not less than 6 inches (76 mm). Duct enclosures shall be sealed around the duct at the point of penetration and vented to the outside of the building through the use of weather-protected openings.

❖ One of the options for a grease duct enclosure is providing a fire-resistance-rated shaft. Commentary Figure 506.3.11.1(1) illustrates an exhaust duct enclosed by a vertical shaft. The grease duct enclosure must be constructed to comply with the shaft enclosure requirements in the IBC, including the bottom or origin of the shaft enclosure. The reference to the IBC for shaft construction means that the IBC regulates the fire-resistance rating and construction of the assemblies that form the duct enclosure (shaft), whereas this code determines when and where an enclosure is required. Where the duct penetrates the duct enclosure, it must be sealed in accordance with

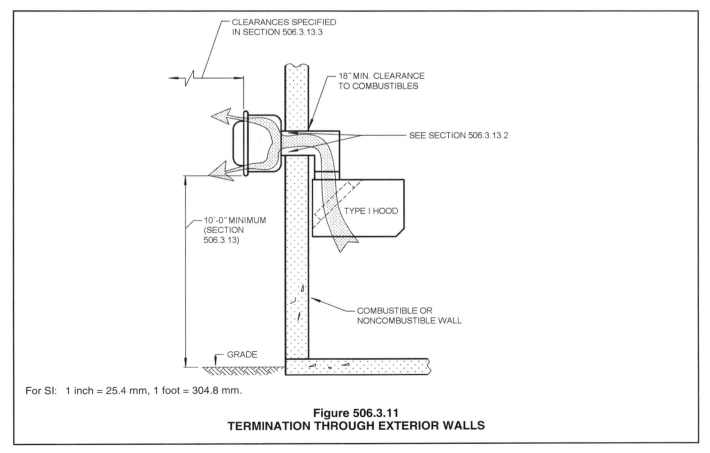

For SI: 1 inch = 25.4 mm, 1 foot = 304.8 mm.

Figure 506.3.11
TERMINATION THROUGH EXTERIOR WALLS

the IBC at the point of penetration with noncombustible materials to prevent fire and hot gases from entering or exiting the shaft enclosure. The duct enclosure must be vented to the exterior to allow heat, hot gases and fire to escape (see Commentary Figure 506.5.3). In the context of this section, shafts can be horizontal as well as vertical.

Commentary Figure 506.3.11.1(2) illustrates an exhaust duct penetrating a wall and extending through an adjacent room before exiting the building. All segments of the duct beyond the wall penetration are protected by the enclosure. The duct termination at the exterior of the building must be protected against damage caused by precipitation, wind or other weather conditions.

The minimum clearance dimensions are required around the grease duct to protect combustibles, allow for some convection cooling and allow for uniform inspection of the outside surface of the duct. In the event that a fire has occurred in the system, the duct system must be inspected to determine whether it has been damaged or is otherwise unfit for continued service. Clearance to combustibles for ducts not enclosed in a shaft is addressed in Section 506.3.6.

The 6-inch (152 mm) minimum clearance is required to allow visual inspection of the duct, especially after a fire event. The shaft is intended to isolate the grease duct from all other ducts, including other grease ducts. In other words, if there are two separate Type I kitchen hoods in the same kitchen with separate grease ducts that are routed to separate exhaust fans on the roof, the grease ducts are not permitted to be located in the same fire-resistance-rated shaft.

If the duct enclosure (shaft) construction is combustible, grease ducts must maintain a minimum clearance to combustibles of 18 inches (457 mm). Typical construction for shafts is gypsum board on metal studs and this kind of assembly is treated as noncombustible under the provisions of this section. An assembly of gypsum board on wood studs is treated as a combustible assembly and the 18-inch (457 mm) clearance would apply.

In the case where an in-line grease fan is installed in the grease duct, in other than the room of duct origin, the in-line fan would have to be enclosed within the shaft enclosure. Note that the manufacturer's installation instructions may require ventilation to help

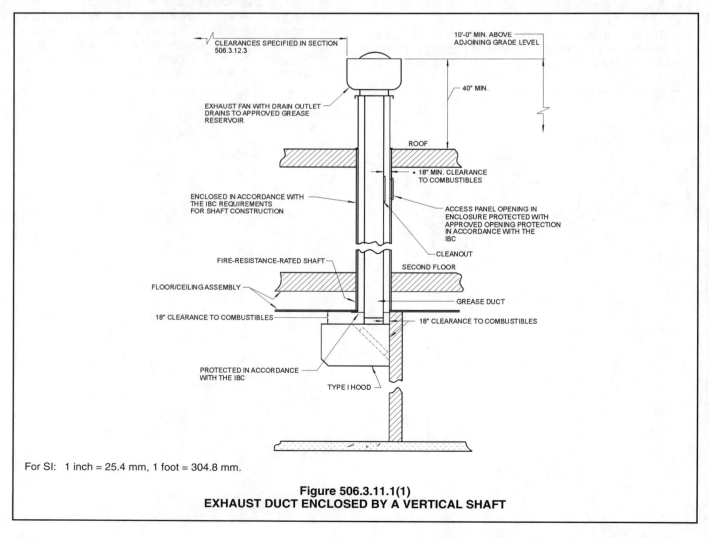

For SI: 1 inch = 25.4 mm, 1 foot = 304.8 mm.

Figure 506.3.11.1(1)
EXHAUST DUCT ENCLOSED BY A VERTICAL SHAFT

cool the fan motor. Also, access doors would be required for servicing the fan.

506.3.11.2 Field-applied grease duct enclosure. Commercial kitchen grease ducts constructed in accordance with Section 506.3.1 shall be enclosed by field-applied grease duct enclosure that is a listed and labeled material, system, product, or method of construction specifically evaluated for such purpose in accordance with ASTM E 2336.

The surface of the duct shall be continuously covered on all sides from the point at which the duct originates to the outlet terminal. Duct penetrations shall be protected with a through-penetration fire-stop system classified in accordance with ASTM E 814 or UL 1497 and having a "F" and "T" rating equal to the fire-resistance rating of the assembly being penetrated. Such systems shall be installed in accordance with the listing and the manufacturer's installation instructions. Partial application of a field-applied grease duct enclosure system shall not be installed for the sole purpose of reducing clearances to combustibles at isolated sections of grease duct. Exposed duct-wrap systems shall be protected where subject to physical damage.

❖ The second option for providing a duct enclosure is a field-applied grease duct enclosure [see Commentary Figure 506.3.11.2(1)]. This field-applied duct enclosure system must be evaluated for the intended pur-

pose in accordance with ASTM E 2336, *Fire Resistive Grease Duct Enclosure Systems*. Such systems include what is commonly referred to as "grease duct fire wrap." The duct covering/enclosure system is evaluated by testing for the following: noncombustibility, fire endurance, durability, internal fire and fire engulfment with a through-penetration firestop.

In addition to being tested as a grease duct enclosure system in accordance with ASTM E 2336, the system is required to be tested as an approved through-penetration firestop system. An approved through-penetration firestop system is one that has been tested in accordance with ASTM E 814 or UL 1479. The test method determines the performance of the protection system when exposed to a standard time-temperature fire test and hose-stream test. The performance of the protection system is dependent on the specific assembly of materials tested, including the number, type and size of penetrations and the type of floor, wall or ceiling in which it is installed. It should also be noted that tests have been conducted at various pressure differentials; however, the current criterion used is 0.01 inch (0.025 kPa) of water gauge, and only tests with this minimum pressure throughout the test period are to be accepted.

In evaluating test reports, the code official must determine that the tested assembly is truly represen-

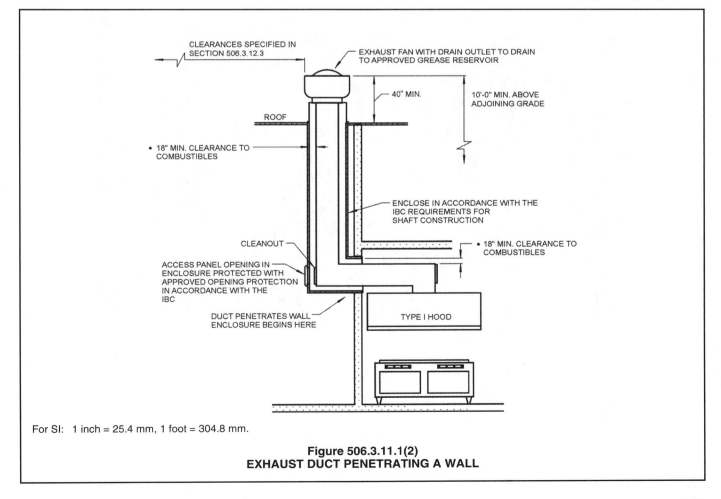

For SI: 1 inch = 25.4 mm, 1 foot = 304.8 mm.

Figure 506.3.11.1(2)
EXHAUST DUCT PENETRATING A WALL

tative of the proposed system installation. The ASTM E 814 test establishes two ratings. These are the F rating, which identifies the ability of the material to resist the passage of flame, and the T rating, which identifies the thermal transmission characteristics of the material or assembly.

A field-applied grease duct enclosure must begin at the first point of penetration of a ceiling, wall or floor and must be continuous to the duct termination. Since the grease duct enclosure is required to be tested and specifically evaluated for such purpose, it is critical that the enclosure be installed in accordance with the listing and the manufacturer's installation instructions. A field-applied duct wrap is usually a foil-encapsulated proprietary insulation material that, depending on its location, could be subject to damage. For this reason, if the duct enclosure is located in an area where it would be subject to physical damage, it must be protected with some type of physical barrier.

Note that in addition to being evaluated as a grease duct enclosure, field-applied duct enclosures also offer reduced clearance to combustibles, with some offering zero clearance to combustibles. The actual clearance to combustibles will be indicated on the label for the particular product and the installation must be in accordance with the manufacturer's instal-lation instructions. The reduced clearances that these products offer make them a possible solution to clearance problems that may be encountered when, for example, a grease duct is routed up through a roof of a building of wood-frame construction (see Commentary Figure 506.3.6). This section prohibits the practice of partially applying the duct wrap to only certain portions of a duct for the purpose of reducing clearance to combustibles. This normally occurs when Section 506.3.11.4 does not require a duct enclosure and at the point where the grease duct goes up through the roof. Where the roof is of combustible construction or the insulation or roof membrane is combustible, the practice of wrapping the duct from a point 18 inches from the bottom of the roof deck and up through the roof curb has been proposed in the past [see Commentary Figure 506.3.11.2(2)]. This, however, is not how these products were intended to be used nor have they been tested and approved for this application. This is because such materials are tested and listed only as a complete system. The duct wrap materials are required to meet all five of the tests that are in ASTM E 2336 including an internal fire test and the external full engulfment test. The duct wrap would never pass these tests under partial application and it has never been tested to reduce clearances in small sections of grease ducts.

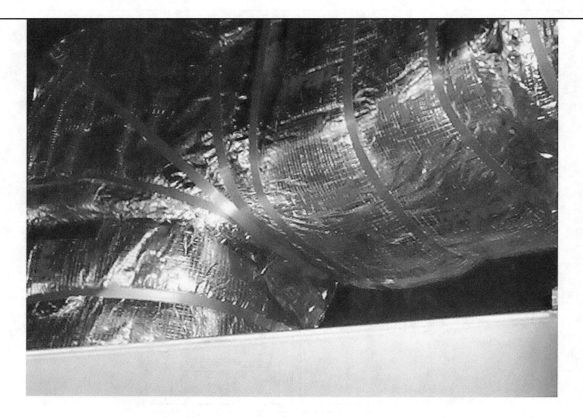

Figure 506.3.11.2(1)
FIELD-APPLIED GREASE DUCT ENCLOSURE
(Photo courtesy of Guy McMann)

Where an in-line grease fan is installed in the grease duct, a fan enclosure must be installed to maintain the integrity of the fire-resistant-rated enclosure provided by the grease duct wrap [see Commentary Figures 506.3.11.2(1) and 506.5.1(2)].

506.3.11.3 Factory-built grease duct assemblies. Factory-built grease duct assemblies incorporating integral enclosure materials shall be *listed* and *labeled* for use as commercial kitchen grease duct assemblies in accordance with UL 2221. Duct penetrations shall be protected with a through-penetration firestop system classified in accordance with ASTM E 814 or UL 1479 and having an "F" and "T" rating equal to the fire-resistance rating of the assembly being penetrated. Such assemblies shall be installed in accordance with the listing and the manufacturer's installation instructions.

❖ The third option for providing a duct enclosure is a factory-built grease duct, which is tested as a through-penetration firestop system in accordance with ASTM E 814 or UL 1479 (see commentary, Section 506.3.11.2), as well as being evaluated in accordance with UL 2221. The prefabricated grease duct enclosure system is evaluated by testing that is very similar to the testing for the grease duct fire wrap: noncombustibility, internal and external fire tests, external fire test with fire engulfment and external fire test with a firestop test.

There are two types of factory-built grease ducts, factory-built grease ducts without an enclosure (see commentary, Section 506.3.1.1) and factory-built grease ducts that are listed and labeled with an enclosure in accordance with UL 2221. In order to comply with UL 2221, the factory-built grease duct must first comply with UL 1978. A factory-built grease duct that is in compliance with UL 1978 must then incorporate insulation materials that are above and beyond those required to meet UL 1978 in order to show compliance with UL 2221.

Note that in addition to being evaluated as a grease duct enclosure, factory-built grease ducts offer reduced clearance to combustibles, with some offering zero clearance to combustibles. The actual clearance to combustibles will be indicated on the label for the particular product, and the installation must be in accordance with the manufacturer's installation instructions. The reduced clearances that these products offer make them a possible solution to clearance problems that may be encountered when, for example, a grease duct is routed up through a roof of a building of wood-frame construction [see Commentary Figures 506.3.1.1(2) and 506.3.6].

Where an in-line grease fan is installed in a factory-built grease duct, a fan enclosure must be provided (see commentary, Section 506.3.10.2).

Figure 506.3.11.2(2)
PARTIAL APPLICATION OF FIELD-APPLIED GREASE DUCT ENCLOSURE
(PROHIBITED)

506.3.11.4 Duct enclosure not required. A duct enclosure shall not be required for a grease duct that penetrates only a nonfire-resistance-rated roof/ceiling assembly.

❖ This section allows penetration of an unrated roof/ceiling assembly without a shaft enclosure. This could allow penetration of a drop-in ceiling in a one-story building or the top story of a building. The intent of this section is duct enclosures to be omitted where they would have little value. The typical single-story restaurant of Type VB construction will have grease ducts penetrating only nonrated roof/ceiling assemblies. Grease ducts in such buildings do not pass through stories and are inherently protected by their inaccessible location in the above ceiling and attic spaces. Note that where the grease duct goes up through a combustible roof, there may be clearance to combustibles issues (see Commentary, Section 506.3.11.2).

506.3.12 Grease duct fire-resistive access opening. Where cleanout openings are located in ducts within a fire-resistance-rated enclosure, access openings shall be provided in the enclosure at each cleanout point. Access openings shall be equipped with tight-fitting sliding or hinged doors that are equal in fire-resistive protection to that of the shaft or enclosure. An *approved* sign shall be placed on access opening panels with wording as follows: "ACCESS PANEL. DO NOT OBSTRUCT."

❖ The requirements of this section relate to the requirements of Sections 506.3.8 and 506.3.11. The construction of the duct enclosure, including opening protectives, must meet IBC requirements. Where a cleanout occurs in a grease duct system that is enclosed as required by Section 506.3.11, an opening must be placed in the enclosure to provide access to the cleanout. To maintain shaft integrity, these opening protectives must afford the same fire protection as that of the duct enclosure.

506.3.13 Exhaust outlets serving Type I hoods. Exhaust outlets for grease ducts serving Type I hoods shall conform to the requirements of Sections 506.3.13.1 through 506.3.13.3.

❖ This section regulates the termination of grease ducts that serve Type I hoods. While Section 506.5.5 references the user back to this section for the termination for exhaust equipment (i.e., exhaust fans), this section is only referring to outlets serving a Type I hood. An example of this would be where an in-line grease fan is installed and only the grease duct outlet extends to the exterior of the building.

506.3.13.1 Termination above the roof. Exhaust outlets that terminate above the roof shall have the discharge opening located not less than 40 inches (1016 mm) above the roof surface.

❖ The requirement for the discharge opening to be at least 40 inches (1016 mm) above the roof surface is intended to prevent the accumulation of grease residue on surrounding surfaces and to allow the diffusion of the exhaust into the surrounding air currents.

The intent is that the discharge be directed away from the roof.

This section addresses only grease duct exhaust outlets serving Type I hoods. Exhaust outlets for Type II hoods must comply with Sections 501.3.1 and 506.4.2.

506.3.13.2 Termination through an exterior wall. Exhaust outlets shall be permitted to terminate through exterior walls where the smoke, grease, gases, vapors and odors in the discharge from such terminations do not create a public nuisance or a fire hazard. Such terminations shall not be located where protected openings are required by the *International Building Code*. Other exterior openings shall not be located within 3 feet (914 mm) of such terminations.

❖ Exhaust outlets can terminate through a wall only if a fire hazard or public nuisance is not created. Because of the many different places the exhaust can terminate and the various types of discharge, exterior wall terminations must be evaluated on a case-by-case basis by the code official (see Commentary Figure 506.3.11). In some locations, a wall termination could create a public nuisance because of odors, smoke, grease discharge and the proximity of other buildings, walkways and other occupied areas. Terminations through the exterior wall must be located so that they do not create a fire hazard. For example, a termination through the exterior wall below a roof soffit should be avoided due to the possibility of grease collecting on the soffit.

This section prohibits wall terminations in walls required to have protected openings. Depending on the building's proximity to lot lines and other buildings, the building code might require that openings in exterior walls be protected with fire-resistance-rated opening protectives. The wall termination must be located at least 3 feet (914 mm) from all openings, including any window (fixed or openable), door, air exhaust or intake opening.

This section addresses only grease duct exhaust outlets serving Type I hoods. Exhaust outlets for Type II hoods must comply with Sections 501.3.1 and 506.4.2.

506.3.13.3 Termination location. Exhaust outlets shall be located not less than 10 feet (3048 mm) horizontally from parts of the same or contiguous buildings, adjacent buildings and adjacent property lines and shall be located not less than 10 feet (3048 mm) above the adjoining grade level. Exhaust outlets shall be located not less than 10 feet (3048 mm) horizontally from or not less than 3 feet (914 mm) above air intake openings into any building.

Exception: Exhaust outlets shall terminate not less than 5 feet (1524 mm) horizontally from parts of the same or contiguous building, an adjacent building, adjacent property line and air intake openings into a building where air from the exhaust outlet discharges away from such locations.

❖ This section addresses only grease duct exhaust outlets serving Type I hoods. Exhaust outlets for Type II hoods must comply with Sections 501.3.1 and 506.4.2.

The discharge of the exhaust system must be located to minimize accumulation of grease on parts of the same, contiguous or adjacent buildings and to prevent the entry of exhaust discharge into any fresh air intake or other opening to any building. The exhaust outlet must be located a minimum distance above grade to protect passersby and to help the dispersion of the exhaust into the atmosphere. Exhaust outlets must be a minimum of 10 feet (3048 mm) horizontally from or 3 feet (914 mm) above any air intakes which is consistent with other sections of the code.

The exception allows reduced horizontal clearance where the exhaust discharge is directed so as not to affect any property or enter any building. The intent of the exception is to grant relief from the 10-foot (3048 mm) clearance requirement where the exhaust is directed away from what the code is trying to protect. Discharging the exhaust "away" from the protected buildings can be interpreted as not directing the exhaust at any angle towards them. The outlet can discharge in a direction parallel to a building wall, for example, but cannot be directed at some angle towards such wall. Good judgment is required for securing the intent of this exception.

506.4 Ducts serving Type II hoods. Commercial kitchen exhaust systems serving Type II hoods shall comply with Sections 506.4.1 and 506.4.2.

❖ Sections 506.4.1 and 506.4.2 address ducts used for Type II hoods. Exhaust ducts serving Type II exhaust systems are prohibited from interconnecting with any other exhaust ducts other than Type II exhaust ducts in accordance with Section 501.2.

506.4.1 Ducts. Ducts and plenums serving Type II hoods shall be constructed of rigid metallic materials. Duct construction, installation, bracing and supports shall comply with Chapter 6. Ducts subject to positive pressure and ducts conveying moisture-laden or waste-heat-laden air shall be constructed, joined and sealed in an *approved* manner.

❖ Nongrease ducts and plenums serving Type II hoods are not subjected to the same hazards as grease ducts serving Type I hoods. Because nongrease duct systems control the waste heat and vapors associated with Type II hoods, they must be constructed of rigid metallic materials, and be braced and supported to comply with Section 603. SMACNA *HVAC Duct Construction Standards—Metal and Flexible* contains information on duct sealing. For positive duct pressure applications, duct joints must be sealed, including longitudinal joints, transverse joints and connections. Ducts must be sealed with approved materials (such as mastic with fibrous backing tape or pressure sensitive tape) consistent with Chapter 6. Type II hood ducts could be subjected to wetting from condensation and, therefore, must be constructed with this in mind.

506.4.2 Type II terminations. Exhaust outlets serving Type II hoods shall terminate in accordance with the hood manufacturer's installation instructions and shall comply with all of the following:

1. Exhaust outlets shall terminate not less than 3 feet (914 mm) in any direction from openings into the building.

2. Outlets shall terminate not less than 10 feet (3048 mm) from property lines or buildings on the same lot.

3. Outlets shall terminate not less than 10 feet (3048 mm) above grade.

4. Outlets that terminate above a roof shall terminate not less than 30 inches (762 mm) above the roof surface.

5. Outlets shall terminate not less than 30 inches (762 mm) from exterior vertical walls

6. Outlets shall be protected against local weather conditions.

7. Outlets shall not be directed onto walkways.

8. Outlets shall meet the provisions for exterior wall opening protectives in accordance with the *International Building Code*.

❖ This section provides prescriptive requirements for the termination of Type II exhaust systems. Listed manufactured hoods must be installed in accordance with manufacturer's installation instructions and any requirements in this section that are more restrictive. Shop-fabricated hoods that are not listed must follow these prescriptive requirements for the termination of the Type II exhaust.

Because grease is not involved, the termination requirements for Type II exhaust systems are not nearly as stringent as those for Type I exhaust systems; however, there are many similarities. Type II exhaust outlets must be a minimum of 10 feet (3048 mm) from lot lines and buildings on the same lot and must be located a minimum of 10 feet (3048 mm) above grade. Unlike Type I systems, a Type II exhaust may terminate through an exterior wall that is required by the IBC to have protected openings. In such cases, the duct termination opening would have to be protected by a closure device such as a fire damper. The 30-inch (762 mm) requirement for the exhaust outlet to be above the roof or from an exterior wall is an attempt to decrease the likelihood that any exhaust air will impinge on the exterior surfaces due.

506.5 Exhaust equipment. Exhaust *equipment*, including fans and grease reservoirs, shall comply with Sections 506.5.1 through 506.5.5 and shall be of an *approved* design or shall be *listed* for the application.

❖ Exhaust fans, motors, grease reservoirs and other related equipment can be subjected to high temperatures and severe duty; therefore, they must be tested and listed for the intended application or specifically approved.

506.5.1 Exhaust fans. Exhaust fan housings serving a Type I hood shall be constructed as required for grease ducts in accordance with Section 506.3.1.1.

> **Exception:** Fans *listed* and *labeled* in accordance with UL 762.

❖ Commercial kitchen exhaust fan housings serving Type I hoods must be constructed of steel and must be capable of handling hot grease-laden air and smoke (see the definition of "Hood, Type I"). The fan must also be designed to collect grease and divert it to a point of collection to prevent a fire hazard and damage to the building surfaces. Fan housings constructed of steel have the integrity to withstand the operating temperatures and conditions related to a duct system fire. There are two types of exhaust fans that serve Type I hoods; the traditional exterior-mounted exhaust fan (i.e., either through the roof or exterior wall) and interior-mounted in-line fans [see Commentary Figures 506.5.1(1) and 506.5.1(2)].

The exception recognizes other fan housing materials such as aluminum alloys. Fans of materials other than steel must be specifically listed for commercial kitchen applications.

506.5.1.1 Fan motor. Exhaust fan motors shall be located outside of the exhaust airstream.

❖ An exhaust fan motor cannot be installed in locations that expose it to the exhaust airflow. Exposure to high temperatures and grease-laden vapors will shorten the life of the motor, and a malfunctioning motor can

Figure 506.5.1(1)
ROOF-MOUNTED EXHAUST FAN FOR A TYPE I HOOD
(Notice that the duct does not extend
18 inches above the roof surface)

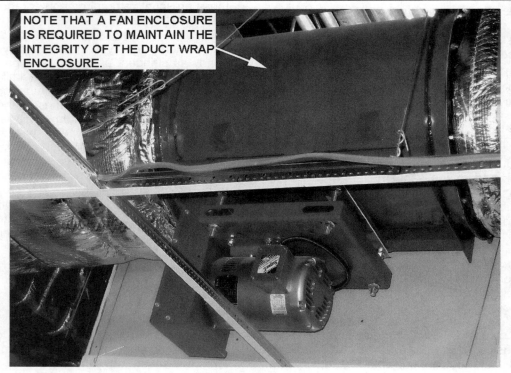

Figure 506.5.1(2)
INLINE EXHAUST FAN FOR A TYPE I HOOD

be a potential ignition source. The exhaust fan would be better located at or as close as possible to the discharge end of the duct to minimize exposure temperatures and the amount of the duct system that is exposed to positive pressures. Typical commercial kitchen exhaust fan designs, both direct drive and belt driven, place motors in shielded locations not subject to the exhaust flow.

506.5.2 Exhaust fan discharge. Exhaust fans shall be positioned so that the discharge will not impinge on the roof, other *equipment* or appliances or parts of the structure. A vertical discharge fan shall be manufactured with an *approved* drain outlet at the lowest point of the housing to permit drainage of grease to an *approved* grease reservoir.

❖ Exhaust must be discharged away from the roof, other equipment and other parts of the structure to eliminate damage or a fire hazard caused by the high-temperature discharge and condensable grease-laden vapor. Vertical discharge (upblast) fans must have an integral grease drainage path leading to an approved grease collection container. Commentary Figure 506.5.1(1) illustrates a grease drain to a reservoir.

506.5.3 Exhaust fan mounting. An upblast fan shall be hinged and supplied with a flexible weatherproof electrical cable to permit inspection and cleaning. The ductwork shall extend a minimum of 18 inches (457 mm) above the roof surface.

❖ Scheduled inspections and maintenance of upblast exhaust fans are not easily performed if a fan has to be lifted off of its roof curb. A hinged base simplifies inspection and maintenance. The flexible electrical

cable provides freedom of movement when opening the hinged fan body without damaging the electrical connection. The 18-inch (457 mm) duct extension is also required by NFPA 96 (see Commentary Figure 506.5.3). Commentary Figure 506.5.1(1) illustrates a hinged upblast fan with a flexible electrical connection and a grease drain that will pull out of the grease reservoir when the fan is tilted.

506.5.4 Clearances. Exhaust *equipment* serving a Type I hood shall have a *clearance* to combustible construction of not less than 18 inches (457 mm).

Exception: Factory-built exhaust *equipment* installed in accordance with Section 304.1 and *listed* for a lesser *clearance*.

❖ Exhaust equipment is subject to the same exposure to grease fires as the ducts themselves. Therefore, the clearance requirements are the same as those in Section 506.3.6 (see commentary, Section 506.3.6).

506.5.5 Termination location. The outlet of exhaust *equipment* serving Type I hoods shall be in accordance with Section 506.3.13.

Exception: The minimum horizontal distance between vertical discharge fans and parapet-type building structures shall be 2 feet (610 mm) provided that such structures are not higher than the top of the fan discharge opening.

❖ Exhaust equipment must terminate in the same manner as specified for Type I duct systems in Section 506.3.13 (see commentary, Section 506.3.13). Note that Section 506.3.13 is for grease duct outlets and that exhaust equipment must terminate in accor-

Figure 506.5.3
KITCHEN EXHAUST FAN MOUNTED 18 INCHES ABOVE THE ROOF SURFACE WITH VENTS FOR THE SHAFT
(Photo courtesy of Guy McMann)

dance with Section 506.3.13 but also at a point where there is access for servicing the equipment.

The exception allows reduced horizontal clearance between vertical discharge fans and parapet walls where the fan discharge opening is as high as or higher than the parapet wall (see Commemtary Figure 506.5.5).

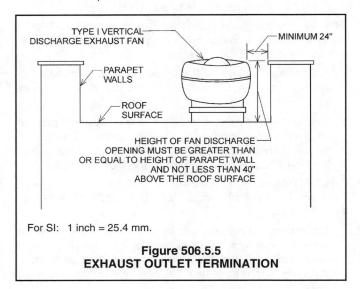

For SI: 1 inch = 25.4 mm.

Figure 506.5.5
EXHAUST OUTLET TERMINATION

SECTION 507
COMMERCIAL KITCHEN HOODS

507.1 General. Commercial kitchen exhaust hoods shall comply with the requirements of this section. Hoods shall be Type I or II and shall be designed to capture and confine cooking vapors and residues. Commercial kitchen exhaust hood systems shall operate during the cooking operation.

Exceptions:

1. Factory-built commercial exhaust hoods that are listed and labeled in accordance with UL 710, and installed in accordance with Section 304.1 shall not be required to comply with Sections 507.4, 507.5, 507.7, 507.11, 507.12, 507.13, 507.14, and 507.15.

2. Factory-built commercial cooking recirculating systems that are listed and labeled in accordance with UL 710B, and installed in accordance with Section 304.1 shall not be required to comply with Sections 507.4, 507.5, 507.7, 507.11, 507.12, 507.13, 507.14, and 507.15. Spaces in which such systems are located shall be considered to be kitchens and shall be ventilated in accordance with Table 403.3. For the purpose of determining the floor area required to be ventilated, each individual *appliance* shall be considered as occupying not less than 100 square feet (9.3 m^2).

3. Net exhaust volumes for hoods shall be permitted to be reduced during part-load cooking conditions, where engineered or *listed* multispeed or variable-speed controls automatically operate the exhaust system to maintain capture and removal of cooking effluents as required by this section. Reduced vol-

umes shall not be below that required to maintain capture and removal of effluents from the idle cooking appliances that are operating in a standby mode.

❖ Type I and II commercial kitchen exhaust hoods must be listed and labeled by an approved agency or must be constructed to conform to the requirements of this chapter. Type I kitchen hoods must be installed above all cooking appliances that produce grease or smoke. Type II kitchen hoods must be installed above cooking or dishwashing appliances that produce heat and moisture and/or products of combustion. Note that the code states that cooking appliances must never operate without the commercial kitchen exhaust hood system also being in operation. Exception 1 states the requirements for factory-built commercial exhaust hoods and specifies which code provisions are not applicable to factory-built hoods. Shop-built and field-constructed hoods are subject to all of the design and fabrication requirements of Section 507.

A factory-built commercial exhaust hood that has been tested in accordance with UL 710 and listed and labeled by an approved agency must be installed in accordance with the manufacturer's instructions [see Commentary Figure 507.1(1)]. The importance of installing the system in strict compliance with the manufacturer's instructions cannot be overemphasized. These instructions contain specific installation requirements that are critical to the proper and efficient operation of the hood.

This section states that listed and labeled hoods have demonstrated compliance with the construction and design requirements of this section and others, such as Sections 507.4, 507.5, 507.7, 507.11, 507.12, 507.13, 507.14 and 507.15. Note that hoods listed and labeled in accordance with UL 710 are not exempt from Section 507.9, therefore, the 18-inch (457 mm) clearance requirement applies regardless of what is stated on the hood labels. The following is a list of some of the information that must be contained within the manufacturer's installation instructions or on the label:

- Minimum and maximum spacing between the front lower edge of the hood and the cooking surface;
- Minimum exhaust airflow quantity;
- Maximum supply airflow if the supply air is directed into the hood;
- Minimum overhangs of the exhaust hood over the cooking surface;
- Maximum allowable surface temperature of the cooking appliance; and
- The specific type of cooking appliance an exhaust hood is intended to serve.

It is also important to determine that all parts and subassemblies of an exhaust hood are a component of the listed and labeled exhaust hood, or that the parts and subassemblies have been evaluated under the same conditions of fire severity as the exhaust hood. Furthermore, the exhaust hood must be compatible with and intended for the type of cooking appliance it will serve. Typically, the label of a factory-built hood tested in accordance with UL 710 will indicate the maximum temperature of cooking appliances that can be located under the hood. It should be noted that the UL 710 standard is being updated and revised and will require the duty classification (i.e., extra-, heavy-, medium- and light-duty) of the appliances the hood is intended to serve to be indicated on the label. This will make it easier for the inspector in the field to determine whether the hood that is installed is suited for the appliances that are being installed under the hood. Note that these changes to UL 710 are still undergoing the review process and that it may take some time before this information appears on the hood label.

It is not the intent of this section to require labeling of exhaust hoods. Unlabeled factory-built hoods and shop/field-constructed hoods are permitted.

A factory-built or field-constructed commercial exhaust hood that has not been listed and labeled in accordance with UL 710 is permitted if it is designed, constructed and installed in accordance with Section 507 and all other applicable requirements of this chapter. This section addresses hood material requirements, hood construction, hood dimensions, exhaust quantities and makeup air requirements. Kitchen exhaust systems must discharge all effluent to the outdoors to comply with Sections 501.2, 506.3.13 and 506.4.

Exception 2 allows installation of factory-built commercial cooking recirculating systems if they have

been listed and labeled in accordance with UL 710B. It is important that recirculating systems be installed in accordance with the manufacturer's installation instructions so that the listing requirements are met. An improper installation could result in hazardous vapors being discharged back into the kitchen.

Factory-built commercial cooking recirculating systems consist of an electric cooking appliance and an integral or matched packaged hood assembly [see Commentary Figures 507.1(2) and 507.1(3)]. The hood assembly consists of a fan, collection hood, grease filter, fire damper, fire-extinguishing system and air filter such as an electrostatic precipitator. These systems are tested for fire safety and emissions. The grease vapor (condensable particulate matter) in the effluent at the system discharge is not allowed to exceed a concentration of 5.0 mg/m³. Recirculating systems are not used with fuel-fired appliances because the filtering systems do not remove products of combustion.

Recirculating systems are becoming more and more popular especially in smaller businesses that serve food, such as a sandwich shop, that wants to expand their menu without becoming a full kitchen. A possible concern with using such a system is that they are tested to prevent particulate matter from being redistributed into the space, but they do not completely remove all contaminants and odors. For this reason, a space where one of these appliances is installed must be considered as a kitchen for the purpose of applying Table 403.3. A space not considered to be a traditional kitchen and containing recirculating systems with no exhaust to the outdoors, would not be ventilated if not for this provision in the code. The

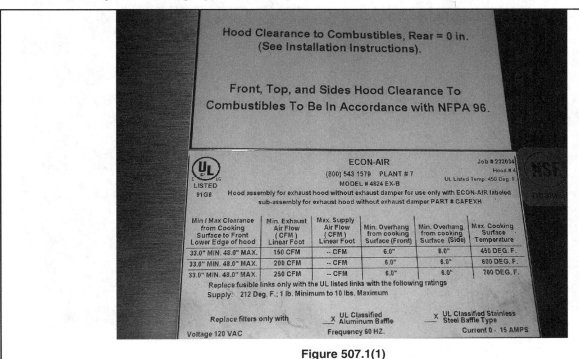

Figure 507.1(1)
TYPICAL LABEL FOR A TYPE I HOOD

ventilation rate for kitchens is 0.7 cfm/ft² of exhaust, so to apply this rate, each individual recirculating appliance is considered as occupying 100 square feet. The purpose of this requirement is to provide a minimal amount of exhaust in the area where one or more of these appliances is installed. If a recirculating appliance is installed in a typical commercial kitchen where the kitchen ventilation rate is being applied to the whole space, then the area of the kitchen would be used to determine the ventilation rate without adding the additional 100 square feet for each recirculating appliance. Note however, if the kitchen is small and the number of recirculating appliances multiplied by 100 square feet is greater than the area of the kitchen, the exhaust rate for the kitchen must be based on the area calculated in accordance with this section.

Exception 3 allows reduction of the exhaust rate of commercial hoods when no cooking is taking place. Net exhaust volumes are based on the assumption that full-load cooking is taking place. In fact, part-load cooking conditions or appliances operating in standby mode represent the majority of time for most food service operations. Although the hoods should obviously be designed for maximum load, this should not prevent operating the hoods at a reduced exhaust rate when the actual load is less than maximum provided that the intent of the code is still satisfied (i.e., capture and removal of effluents). Reducing the exhaust rates

during reduced cooking periods can improve system performance (fire safety, occupant health, energy efficiency and kitchen comfort).

The majority of gas-fired cooking appliance installations use the exhaust hood as a means of venting the combustion byproducts of the cooking appliances. Commercial cooking appliances are either connected to a vent or chimney or the flue outlet discharges into the exhaust hood. The exhaust hood system must be operating when appliances that depend on the exhaust system to vent combustion byproducts are in use. Gas-fired appliances must be interlocked with the hood system if combustion gases are vented by that system (see the definition of "Interlock" in Chapter 2 and Section 505.1.1 of the IFGC).

All hoods, listed and unlisted, must capture and confine cooking vapors within the hood to prevent spillage into the room (see commentary, Section 507.16).

507.2 Where required. A Type I or Type II hood shall be installed at or above all *commercial cooking appliances* in accordance with Sections 507.2.1 and 507.2.2. Where any cooking *appliance* under a single hood requires a Type I hood, a Type I hood shall be installed. Where a Type II hood is required, a Type I or Type II hood shall be installed.

Exception: Where cooking appliances are equipped with integral down-draft exhaust systems and such appliances and exhaust systems are listed and labeled for the applica-

Figure 507.1(2)
ELECTRIC DEEP FAT FRYER
WITH RECIRCULATING HOOD

Figure 507.1(3)
ELECTRIC GRIDDLE WITH RECIRCULATING HOOD

tion in accordance with NFPA 96, a hood shall not be required at or above them.

❖ An exhaust system is required for "Commercial cooking appliances," as defined in Chapter 2. In addition to the specific cooking appliances, which are identified in the definition, further examples of commercial cooking appliances that require a commercial kitchen exhaust system are: griddles (flat or grooved); tilting skillets or woks; braising and frying pans; roasters; pastry ovens; pizza ovens; charbroilers; salamander and upright broilers; infrared broilers; and open-burner stoves and ranges. Furthermore, the definition of "Commercial cooking appliances" defines a food service establishment as "any building or portion thereof used for the preparation and serving of food." Within the context of Section 507, the "preparation and serving of food" includes operations such as preparing, handling, cleaning, cooking and packaging foodstuffs of any sort. The obvious examples of a food service establishment are restaurants and school cafeterias. A less obvious example is a church with a fellowship hall that holds fund-raising events, such as spaghetti dinners, fish fries or pancake breakfasts. Even a child day care facility may be loosely classified as a food service establishment if a hot breakfast or lunch is served to the children as part of their care. For a discussion on where a Type I versus Type II hood is required, see the commentary to Section 507.2.1.

A Type I hood must always be installed above a cooking appliance that produces grease or smoke (see commentary, Section 507.2.1). Grease and smoke go hand-in-hand such that where one is present, the other is likely present. The last sentence of this section simply states that either a Type I or II hood may be installed above a cooking appliance that requires only a Type II hood.

The exception recognizes the hoodless griddle type of cooking appliance that is becoming very popular in restaurants. These types of cooking appliances are often referred to as hibachi tables where food is prepared for the customers in front of them at the table where they are dining [see Commentary Figure 507.2(1)]. These cooking tables have a built-in integral down-draft exhaust system that is designed to capture the cooking vapors by drawing the air across the table into exhaust inlets located at the edge of the cooking surface. The cooking vapors are routed to grease filters located under the cooking surface and then to the grease duct that is under the floor [see Commentary Figure 507.2(2)]. The grease duct must be installed in accordance with Section 506.3.10 (see commentary, Section 506.3.10). These hoods must be listed and labeled for this application. Chapter 15 of NFPA 96 contains specific requirements for integral down-draft exhaust systems. It should be noted that other sections of NFPA 96 are referenced in this chapter and Section 102.8.2 of this code addresses such references.

507.2.1 Type I hoods. Type I hoods shall be installed where cooking *appliances* produce grease or smoke as a result of the cooking process. Type I hoods shall be installed over *medium-duty, heavy-duty* and *extra-heavy-duty cooking appliances*. Type I hoods shall be installed over *light-duty cooking appliances* that produce grease or smoke.

Exception: A Type I hood shall not be required for an electric cooking appliance where an approved testing

Figure 507.2(1)
HIBACHI TABLE WITH INTEGRAL DOWN-DRAFT EXHAUST SYSTEM
(Photo courtesy of Roaster Tech, Inc.)

agency provides documentation that the appliance effluent contains 5 mg/m³ or less of grease when tested at an exhaust flow rate of 500 cfm (0.236 m³/s) in accordance with Section 17 of UL 710B

❖ This section requires Type I hoods for cooking appliances that produce grease or smoke as a result of the cooking process (see definition of "Hood, Type I"). The term "grease" refers to animal and vegetable fats and oils that are used to cook foods or that are a byproduct of cooking foods. Cooking appliances are used for commercial purposes when the appliance is primarily used for the preparation of food for compensation, trade or services rendered. When the nature of the cooking produces grease or smoke then a Type I hood is required. A Type I hood is required where smoke is produced as part of the cooking process. The intent is not to require a Type I hood where there is a possibility of food being burned and producing smoke. For example, smoke that is produced when toast is burned does not mean that a Type I hood is required over a toaster. This section makes it clear that a Type I hood is required over medium-duty, heavy-duty and extra-heavy-duty cooking appliances. If there exists a light-duty cooking appliance that produces grease or smoke, a Type I hood is required for that appliance.

Cooking appliances installed in cafeterias, restaurants, dormitory kitchens, hotels, motels, schools and institutional occupancies are examples of appliances that typically require Type I exhaust hood systems. Some examples of commercial cooking appliances that require a commercial kitchen exhaust system are: deep fat fryers; griddles (flat or grooved); tilting skillets or woks; braising and frying pans; charbroilers; salamander and upright broilers; infrared broilers; open burner stoves and ranges; and barbecue equipment.

A common question that is asked is, what type of hood is required for conveyor and deck-style pizza ovens? Conveyor-type pizza ovens are listed in the definition of "Medium-duty cooking appliances." Type I hoods are required to be installed over medium-duty cooking appliances. Deck-type ovens are listed in the definition of "Light-duty cooking appliances." A Type I hood is required over a light-duty cooking appliance that produces grease or smoke. There is no longer a specific reference to deck-style pizza ovens, just the reference to deck-type ovens. Considering that a deck pizza oven is an enclosed oven and that the primary byproducts given off are heat and moisture, deck pizza ovens have commonly been approved for use under a Type II hood.

Unusual circumstances sometimes arise that may warrant a close evaluation of a cooking appliance or a cooking appliance installation before determining whether a Type I hood is required. For example,

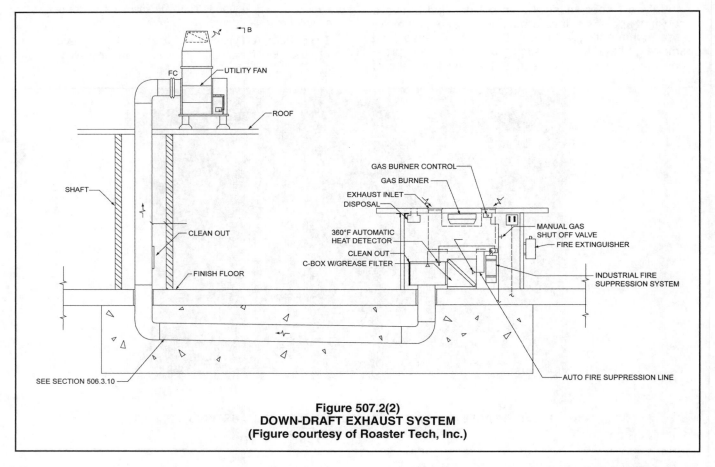

Figure 507.2(2)
DOWN-DRAFT EXHAUST SYSTEM
(Figure courtesy of Roaster Tech, Inc.)

cooking appliances used in a way that does not produce grease or smoke may need to be equipped only with a Type II hood or, depending on the occupancy where the cooking appliance is located, a residential hood or no hood at all. The key issues in making such determinations are the frequency of use and whether grease or smoke is produced by the cooking appliance and the cooking operation. The following are examples of kitchens serving occupancies that, depending on the nature of the cooking and the code official's interpretation of this section, might require only a Type II hood, a residential-type hood or no hood at all for the cooking appliances: church assembly halls; child care facilities; office or factory lunch rooms; employee break rooms; police and fire stations; bed-and-breakfast lodgings; VFW and similar halls; domestic-type kitchens in institutional occupancies; classrooms used to teach cooking; cooking demonstration displays and charity soup kitchens.

The code official should examine the nature of cooking operations before determining whether a Type I or II hood is required for a particular cooking appliance or a cooking appliance installation. Note that this section has been tightened up by stating that a Type I hood must be installed over medium-, heavy- and extra-heavy-duty cooking appliances. Bear in mind the primary purpose of a Type I hood is to control a potential fire hazard associated with grease and the purpose of a Type II hood is to control waste heat and moisture that burden HVAC systems and promote an unhealthy workplace. Excess moisture can deteriorate building components, promote the growth of mold and fungi, and create unhealthy and uncomfortable working conditions for employees.

Some common scenarios that come up are the type of hoods that are required in a life science classroom in a high school (i.e., a classroom used to teach, among other things, cooking to students) and the type of hood required over a cooking appliance(s) in a fire station. In both cases, the type of cooking is the deciding factor on the type of hood required.

Typically, students in a life science class are learning to prepare meals that are the same as those that are prepared for a family in a residential dwelling unit. In most cases, residential-type range/ovens are installed in the classroom. As such, the same byproducts that are produced in a kitchen in a dwelling unit would be produced in the classroom. Based on the residential style of cooking that is being taught, it would seem appropriate that the same type of hood installed in a residential dwelling could be installed over the residential range/ovens used in a classroom. Therefore, a Type I or II hood would not be required and residential kitchen hoods that are ducted to the outdoors could be installed.

Note that if the high school offers a culinary arts class and uses commercial cooking appliances to teach students how to prepare meals that are normally prepared in a restaurant, then the appropriate Type I or II hood could be required based on the type of cooking operations that are performed under the hood.

In the case of a kitchen located in a fire station, once again it depends on the type of cooking and the intended use of the facility. Meals prepared in a kitchen in a fire station that has a residential-type range/oven that is only intended to be used to prepare meals for the fire fighters on that particular shift is similar, if not the same, as those prepared in a home environment. As such, the same byproducts that are produced in a kitchen in a dwelling unit would be produced in the kitchen in the fire station. Based on the residential style of cooking that is being performed, it would seem appropriate that the same type of hood installed in a residential dwelling could be installed or, in a case where the space meets its ventilation requirements in Chapter 4 of the code, no hood at all.

It is not uncommon, however, for fire stations to have a community room with a kitchen used for preparing meals. The community room is often used to hold fund-raising events, such as spaghetti dinners, fish fries or pancake breakfasts, or used by members of the community for special events, such as parties or weddings. The kitchen may or may not have commercial cooking appliances installed. In this case, it would appear that such a situation is intended for the preparation of food for revenue generation. In this case, a Type I or II hood is required based on the cooking operations that are performed under the hood. This would also apply to VFW and other fraternal organizations, church assembly halls and other similar halls.

It is important to note that cooking appliances installed in commercial occupancies do not necessarily require the installation of a Type I or II hood. There are a number of installations in a commercial occupancy where residential-type cooking occurs that would not require a commercial kitchen hood (see the discussion above for school classrooms and fire stations). Lunchrooms and breakrooms in commercial businesses often have residential ranges/ovens installed. In addition, many multiple-family residential buildings (e.g., condominiums and townhomes) have a clubhouse or community room that the residents can reserve for special functions. Typically these are seldom used, and when they are, it is to warm food or bake frozen food like pizza, lasagna or premade appetizers. Based on the residential style of cooking that is performed on these appliances, it would seem appropriate that the same type of hood installed in a residential dwelling could be installed or there may be no hood at all.

If multiple cooking appliances are installed under a single hood and one or more of those appliances requires a Type I hood, a Type I hood would be required to serve the entire appliance line.

With the trend for larger kitchens in new dwelling units, kitchens designed with commercial-type cooking appliances have become more popular. Although these installations would generally not require commercial exhaust hoods, commercial appliances should be carefully evaluated for use in dwellings. Commercial cooking appliances are typically not listed for domestic use and might lack certain safety features that would be required for domestic cooking appliances. Note that Sections 917.2 and 917.3 require appliances in dwelling units to be designed and listed for domestic use (see commentary, Sections 917.2 and 917.3).

This chapter does not require exhaust hoods for cooking equipment or appliances installed outdoors where the grease-laden vapors, etc., discharge directly to the outside atmosphere, nor does this chapter intend to regulate cooking appliances installed in vehicles or towed trailers (see definition of "Commercial cooking appliances"). Note that cooking appliances installed outdoors but located under a roof should be evaluated for installation under a Type I or II hood just as if they were located inside a building having enclosing walls.

The exception recognizes the growing use of small electrical appliances used for cooking, such as in small sandwich shops and convenience stores, where little or no grease is produced. The installation of a Type I hood in these small establishments creates the expense of the hood and the energy costs of running the fan and tempering the makeup air for the owner where grease emissions are minimal or nonexistent. The grease emission threshold requirement is consistent with NFPA 96 and the testing procedure is done in accordance with Section 17 of UL 710B. In order for an appliance to qualify for use without a Type I hood it must be tested by an approved agency and shown that the effluent contains 5 mg/m^3 or less of grease when tested at an exhaust flow rate of 500 cfm. If the appliance is below the grease emission threshold, the provisions of Section 507.2.2 are still applicable and a Type II hood may still be required.

507.2.1.1 Operation. Type I hood systems shall be designed and installed to automatically activate the exhaust fan whenever cooking operations occur. The activation of the exhaust fan shall occur through an interlock with the cooking appliances, by means of heat sensors or by means of other *approved* methods. A method of interlock between an exhaust hood system and appliances equipped with standing pilot burners shall not cause the pilot burners to be extinguished. A method of interlock between an exhaust hood system and cooking appliances shall not involve or depend upon any component of a fire extinguishing system.

❖ This section and Section 507.1 state that the hood system must operate whenever cooking operations are taking place. In order to perform the intended function, a Type I hood is required to automatically operate when cooking operations occur or must be activated in an arrangement that prevents cooking without hood exhaust system operation. There are several methods indicated to achieve this and it is left up to the designer/installer/owner and code official to

determine what they all agree will be necessary to verify that fan operation will occur whenever cooking operations occur.

The activation of the exhaust fan must occur through an interlock with the appliances, by means of heat sensors or other approved methods. It should be noted that an interlock with the cooking appliances is one of the methods to accomplish this, but is not the only method. This text has been misinterpreted as meaning that all appliances must be fitted with controls that would start the hood system. This is not the case. In fact, tampering/altering with listed and labeled appliances may in itself create a code violation. However, if a cooking appliance has provisions incorporated into its listed and labeled design that included some type of interlock option, that would certainly meet the requirements of this text.

It should be pointed out that the text states that "hood systems shall be designed and installed..." and this means that the hood system needs the controls and not necessarily the actual cooking appliances. The hood system must cooperate with appliances by means of heat sensors or other approved methods. All this means is that something needs to activate the exhaust fan when a cooking operation takes place. This can be achieved through the use of controls such as heat sensors/infrared technology, light beam interference detection or through methods such as electric relays that control the branch circuit that the appliances are connected to or, in the case of gas appliances, a solenoid valve in the gas supply piping. This section does not prevent manual starting of the exhaust system, provided that there is a means to prevent cooking appliance operation when the exhaust system is not operating (e.g., hood and appliance interlock).

The part of this code text that says, "or by other approved methods" leaves the door open for many options. This leaves it up to the designer/installer/owner and code official to determine what will be necessary to verify that fan operation will occur whenever cooking operations take place. One way might be to tie the fan to the lighting control serving the kitchen area, assuming that the cooking would not be possible if the lights were off. This option may work very well because of the allowance that permits the use of variable speed exhaust fans (see commentary, Section 507.1). When the lights are turned on the fan might not even be running, but when cooking operations begin the heat created would cause the fan to begin to run on a light load condition. This variable speed technology already has the interlock incorporated into it, which is how the fan knows to automatically change speeds throughout the day. Another "approved" method may be one that some of the chain restaurants use in which the standard operating procedure is that the fan always runs when the building is occupied or upon startup of any cooking appliance. While not stating a specific method to interlock the hood to the operation of the appliances, the code does state that the method used must not cause the pilot burners to be extinguished. For example, if the interlock method uses an electrically actuated (solenoid) main fuel valve, that valve will close when the exhaust fan system is shut down. This, in turn, will extinguish any standing pilots and necessitate the relighting of pilots each day or each time the kitchen is "off line." Besides the obvious inconvenience, there is also a safety concern with having employees and owners routinely relighting pilots that may be difficult to access. For this reason, the code prohibits a method of interlock that would cause standing pilot burners to be extinguished. Although some attempts have been made to circumvent this problem, a practical and safe solution to the standing pilot problem is not known at this time. It should be noted that in Section 505.1.1 of the IFGC, gas piping is specifically prohibited from being installed to bypass the solenoid valve (see IFGC commentary, Section 505.1.1).

The method of interlock must not involve or depend on any component of the fire-extinguishing system. Since fire-extinguishing systems provide a shutoff device for the fuel to the appliances being protected, someone might try to utilize such valve in an effort to comply with this requirement. This could pose a potential problem for the shutoff device, since it is not listed to provide this function, which might also compromise the effectiveness of the fire-extinguishing systems in an emergency situation.

This is an operational requirement to ensure that the ventilation is operational when the cooking appliances are in use. This requirement is not part of the operations that occur when a fire occurs. The shutoff device in the extinguishing system is listed as part of the extinguishing system and is not listed to perform this additional function.

Where solenoid gas valves are used to interlock cooking appliances with an exhaust system, a manual reset device should be considered for inclusion in such arrangements. In the event of a power outage, the solenoid valve will close and because the gas appliances are inoperative, the kitchen personnel might walk away from them leaving manual burner controls in the on position. When power is restored, personnel may not be aware that the solenoid has reopened allowing gas to escape from open burners. A manual reset device will prevent such reopening of the solenoid without a deliberate action by informed personnel.

507.2.1.2 Exhaust flow rate label. Type I hoods shall bear a label indicating the minimum exhaust flow rate in cfm per linear foot (1.55 L/s per linear meter) of hood that provides for capture and containment of the exhaust effluent for the cooking appliances served by the hood, based on the cooking appliance duty classifications defined in this code.

❖ The requirement for Type I hoods to bear a label is for factory-built commercial exhaust hoods that are listed and labeled in accordance with UL 710. Type I

hoods that are not listed and labeled in accordance with UL 710 would have to meet the requirements of Section 507.13. The minimum exhaust flow rate and the cooking appliance duty classification is information that is necessary for the inspector in the field to determine that the hood has the minimum exhaust flow rate for the appliances that are installed beneath the hood. The exhaust hood must be compatible with and intended for the type of cooking appliance it will serve. Typically, the label of a factory-built hood tested in accordance with UL 710 will indicate the cfm per linear foot and the maximum temperature of cooking appliances that can be located under the hood [see Commentary Figure 507.1(1)]. However, the code references duty classifications for cooking appliances and not temperature ratings. This creates enforcement problems for the inspector in the field if he or she does not know the temperature ratings of the appliances installed beneath the hood. Requiring the cfm per linear foot and the duty classifications of the appliances will help the inspector in the field to verify that the hood system is appropriate for the appliances served.

It should be noted that the UL 710 standard is currently being updated and revised and will require that the duty classification (i.e., extra-, heavy-, medium- and light-duty) of the appliances the hood is intended to serve be indicated on the label. Note that these changes to UL 710 are still undergoing a review process and that it may take some time before this information appears on the hood label.

507.2.2 Type II hoods. Type II hoods shall be installed above dishwashers and appliances that produce heat or moisture and do not produce grease or smoke as a result of the cooking process, except where the heat and moisture loads from such appliances are incorporated into the HVAC system design or into the design of a separate removal system. Type II hoods shall be installed above all appliances that produce products of *combustion* and do not produce grease or smoke as a result of the cooking process. Spaces containing cooking appliances that do not require Type II hoods shall be provided with exhaust at a rate of 0.70 cfm per square foot (0.00033 m³/s). For the purpose of determining the floor area required to be exhausted, each individual *appliance* that is not required to be installed under a Type II hood shall be considered as occupying not less than 100 square feet (9.3 m²). Such additional square footage shall be provided with exhaust at a rate of 0.70 cfm per square foot [.00356 m³/(s · m²)].

❖ Type II hoods are required above dishwashers and appliances that produce heat or moisture and do not produce grease or smoke, except where the heat or moisture loads are incorporated into the HVAC system (see Commentary Figure 507.2.2). Where light-duty cooking appliances produce products of combustion and do not produce grease or smoke, they must be located under a Type II hood. Where smoke is produced as part of the cooking process, a Type I hood is required. The intent is not to require a Type I hood where there is a possibility of food being burned and producing smoke. For example, smoke that is produced when toast is burned does not mean that a Type I hood is required over a toaster. Where a dishwasher or appliance has a separate removal system that is specific to that appliance, and it discharges the heat or moisture to the exterior, a Type II hood is not required. Any light-duty cooking appliance that produces grease or smoke must be located under a Type I hood (see commentary, Section 507.2.1).

In previous editions of the code, there were a number of exceptions that did not require a Type II hood over light-duty electric cooking appliances such as convection, bread and microwave ovens; toasters; steam tables; popcorn poppers and coffee makers, as long as the additional heat and moisture loads were accounted for in the design of the HVAC system. This laundry list of exceptions kept growing with every code change cycle until the list of appliances that did not require a hood nearly exceeded the list that required a hood. The exceptions are now gone and replaced with criteria that are twofold; (1) is heat or moisture produced? (2) are the heat and moisture loads accounted for in the design of the HVAC system? If heat and moisture are produced and the loads are incorporated into the HVAC design, then no Type II hood is required. If heat and moisture is produced and the loads are not incorporated into the HVAC design, then a Type II hood would be required. Note that if a Type II hood is not required, there is no limit to the number of electric appliances that can be installed as long as the loads from all the appliances

are accounted for in the HVAC design. The designer should consider if it is more energy efficient to design the HVAC system to handle the heat and moisture loads or if it is more efficient to provide a Type II hood and makeup air. Since the code permits either option, the designer must make the decision on which design is more energy efficient. Outside weather conditions, the number of appliances, the heat and moisture loads generated and hours of operation may all be factors that help decide which option is more energy efficient.

If cooking appliances are provided and a Type II hood is not required, the space where the appliances are located must be provided with exhaust at a rate of 0.7 cfm/ft^2, which is the same rate as kitchens in Table 403.3. To apply this rate, each individual appliance is considered as occupying 100 square feet. The purpose of this requirement is to provide a minimal amount of exhaust in the area where one or more of these appliances are installed. If cooking appliances are installed in a typical commercial kitchen where the kitchen ventilation rate is being applied to the whole space, then the area of the kitchen would be used to determine the ventilation rate without adding the additional 100 square feet for each appliance. Note however, if the kitchen is small and the number of appliances multiplied by 100 square feet is greater than the area of the kitchen, the exhaust rate for the kitchen must be based on the area calculated in accordance with this section.

Figure 507.2.2
TYPE II HOOD ABOVE A DISHWASHER
(Photo courtesy of Guy McMann)

507.2.3 Domestic cooking appliances used for commercial purposes. Domestic cooking appliances utilized for commercial purposes shall be provided with Type I or Type II hoods as required for the type of appliances and processes in accordance with Sections 507.2, 507.2.1 and 507.2.2.

❖ Domestic cooking appliances used for commercial purposes are subject to the same requirements as commercial appliances and are therefore subject to the requirements of Sections 507.1 and 507.2. For example, food catering services that work out of a residential kitchen must have Type I or II hoods. It is important for the code official to examine the nature of the cooking operation(s) before determining whether an exhaust hood is required for a particular kitchen facility. Note that just because a domestic cooking appliance is installed in a commercial building does not mean that the domestic appliance is being used for commercial purposes (see commentary, Section 507.2.1).

507.2.4 Extra-heavy-duty. Type I hoods for use over *extra-heavy-duty cooking appliances* shall not cover *heavy-*, *medium-* or *light-duty appliances*. Such hoods shall discharge to an exhaust system that is independent of other exhaust systems.

❖ "Extra-heavy-duty cooking appliances" are defined as appliances that utilize solid fuel for all or part of the heating source for cooking. The creation of air-borne sparks and embers is potentially hazardous and typical of a solid fuel-burning cooking operation. Oftentimes, Type I hoods serving solid fuel-burning cooking appliances require the installation of spark arrester devices ahead of the grease removal device to minimize the possibility of passing these sparks and embers into the grease removal device, and possibly into the hood and duct system. To minimize the potential for spreading fire, air-borne sources of ignition or grease-laden vapors to other exhaust systems, Type I hoods for use over solid fuel-burning cooking appliances must discharge to independent exhaust systems that do not connect to any other exhaust system.

Without complete isolation, the smoke, fire and ignition hazards of the effluent developed by solid fuel-burning kitchen exhaust systems might not be confined to the kitchen exhaust system and could jeopardize other systems and other parts of the building. The exclusion of other appliances is consistent with NFPA 96 and is intended to prevent hot embers or sparks emitted from extra-heavy-duty cooking appliances from igniting grease or oils present on or in heavy-, medium- or light-duty cooking appliances. For example, a deep fat fryer would not be permitted under the same hood since it is a medium-duty appliance. Two extra-heavy-duty cooking appliances would be permitted to be installed next to each other and under the same Type I hood. NFPA 96 addresses solid fuel-cooking operations in detail, including inspection and cleaning operations, fuel storage and handling and ash removal.

507.3 Fuel-burning appliances. Where vented fuel-burning appliances are located in the same room or space as the hood, provisions shall be made to prevent the hood system from interfering with normal operation of the *appliance* vents.

❖ A significant reduction in building pressure could be created by hood systems, exhaust fans, ventilation systems and similar equipment, which can negatively affect appliance vents and chimneys. Fuel-burning appliances are often in competition with other mechanical equipment or systems for the available combustion air that infiltrates the building envelope or is otherwise introduced into a building, room or space. The competition between powered exhaust equipment and natural-draft fuel-fired appliances is an unfair contest; the powered equipment will starve the natural-draft appliances unless provisions are made to compensate for the effect of the powered exhaust equipment. Natural-draft appliances also compete among themselves for combustion air. The appliance that produces the strongest draft, such as a solid fuel appliance, can cause combustion air shortages for the appliances that produce a weaker draft.

Exhaust fans and similar equipment and appliances can produce significant negative building pressures that can interfere with the operation of vents and chimneys. This interference can cause reverse flow as outdoor air enters the building through the vents and chimneys as a result of the pressure difference. Any such interference with vents or chimneys would cause discharge of combustion products into the building and, therefore, must be avoided. The amount of combustion air provided could be inadequate or nullified where there is no compensation for the effect of appliances or equipment removing air from a room or space. In many cases, additional combustion air or makeup air must be supplied to offset the deficiency. For example, in restaurants it is common to find fuel-fired appliances such as water heaters that are spilling combustion products into the kitchen area because of the negative pressure created by the kitchen exhaust fan (see Commentary Figure 507.3). Direct-vent appliances are a wise choice in commercial kitchens to avoid venting problems.

507.4 Type I materials. Type I hoods shall be constructed of steel having a minimum thickness of 0.0466 inch (1.181 mm) (No. 18 gage) or stainless steel not less than 0.0335 inch [0.8525 mm (No. 20 MSG)] in thickness.

❖ Steel and stainless steel are the only materials permitted for the construction of unlisted Type I hoods. The construction of listed and labeled hoods is regulated by the standard UL 710. Metal thicknesses are one gage lighter for hoods than for ducts. Compared to a hood, a duct has a greater potential for grease buildup and has greater airflow. Therefore, a fire is more likely to occur in a duct and would be more severe; hence the heavier gage requirements for ducts.

Plain steel and galvanized steel are rarely used for Type I or II hood construction because they are not as corrosion resistant or as easily cleaned as stainless steel. This section does not prohibit a hood from being sheathed for aesthetic reasons in a decorative metal, such as copper.

Stainless steel is used for construction of both Type I and II hoods. Because of its superior properties, stainless steel is used almost exclusively for hood construction. Commercial kitchen Type I hood and duct systems are designed to resist structural failure caused by a grease fire. Therefore, any alternative materials approved under the provisions of Section 105.2 must be evaluated for equivalence to steel or stainless steel.

507.5 Type II hood materials. Type II hoods shall be constructed of steel having a minimum thickness of 0.0296 inch (0.7534 mm) (No. 22 gage) or stainless steel not less than 0.0220 inch (0.5550 mm) (No. 24 gage) in thickness, copper sheets weighing not less than 24 ounces per square foot (7.3 kg/m^2) or of other *approved* material and gage.

❖ Type II hoods are not designed to withstand a fire; therefore, the metal gages are lighter than required for Type I hoods. Copper is rarely used for hood construction and is allowed only for Type II hoods.

507.6 Supports. Type I hoods shall be secured in place by non-combustible supports. All Type I and Type II hood supports shall be adequate for the applied load of the hood, the unsupported ductwork, the effluent loading and the possible weight of personnel working in or on the hood.

❖ Because Type I hoods are designed to resist failure from a fire within, noncombustible supports are necessary to prevent duct displacement and stress failures. Both Type I and II hoods must have supports

capable of carrying all potential loads that may reasonably be applied to the hood.

507.7 Hood joints, seams and penetrations. Hood joints, seams and penetrations shall comply with Sections 507.7.1 and 507.7.2.

❖ Sections 507.7.1 and 507.7.2 contain the requirements for sealing the joints and seams of Type I and II hoods, respectively. The nature of the products being exhausted by the two types of hoods require different levels of sealing protection.

507.7.1 Type I hoods. External hood joints, seams and penetrations for Type I hoods shall be made with a continuous external liquid-tight weld or braze to the lowest outermost perimeter of the hood. Internal hood joints, seams, penetrations, filter support frames and other appendages attached inside the hood shall not be required to be welded or brazed but shall be otherwise sealed to be grease tight.

Exceptions:

1. Penetrations shall not be required to be welded or brazed where sealed by devices that are *listed* for the application.

2. Internal welding or brazing of seams, joints and penetrations of the hood shall not be prohibited provided that the joint is formed smooth or ground so as to not trap grease, and is readily cleanable.

❖ A Type I exhaust hood must be designed to contain, without leakage, all liquid residue that accumulates from the grease-laden vapors. Therefore, seams, joints, connections and penetrations within the portions of the hood that direct, capture and collect grease-laden vapors must have a liquid-tight continu-

Figure 507.3
WATER HEATER NEXT TO A TYPE I HOOD
(Photo courtesy of Guy McMann)

ous welded or brazed joint. The intent is to require welded or brazed joints for all seams that form the external shell of the hood to prevent grease from escaping the confines of the hood. Rivets, screws and sheet metal locking joints would not be acceptable means of joining parts of the hood because they cannot create a liquid-tight joint and because they provide surfaces that accumulate grease that cannot be removed without disassembly of the hood. Internal joints not affecting the grease-containment integrity of the hood need not be welded or brazed; however, they must be sealed or otherwise made grease tight. For example, a continuous external weld is not required for the grease filter support frame or appendages contained within the hood shell. Joint surfaces must be smooth to facilitate cleaning and to prevent the accumulation of grease (see commentary, Section 507.8).

The connection between the hood and the exhaust duct must also be welded or brazed liquid tight except as allowed by the exceptions to Section 506.3.2.2 (see commentary, Section 506.3.2.2).

Exception 1 recognizes the use of mechanical sealing devices that are specifically listed for grease duct penetration. Hoods are typically penetrated for the installation of suppression system piping and electrical conduit.

Exception 2 allows internal welds and brazing as opposed to external, if the joints will not harbor grease.

507.7.2 Type II hoods. Joints, seams and penetrations for Type II hoods shall be constructed as set forth in Chapter 6, shall be sealed on the interior of the hood and shall provide a smooth surface that is readily cleanable and water tight.

❖ This section requires that Type II hoods be constructed and sealed to comply with the duct requirements of Chapter 6. Because Type II hoods do not exhaust the grease- and smoke-laden air associated with Type I hoods, the more restrictive liquid-tight welding and brazing of joints and seams is not necessary. The interior of the hood still must be smooth to allow cleaning and to prevent the accumulation of particles in crevices, which could promote the growth of mold and bacteria. Because Type II hoods are likely to be exposed to condensed water vapor, they must be made water tight.

507.8 Cleaning and grease gutters. A hood shall be designed to provide for thorough cleaning of the entire hood. Grease gutters shall drain to an *approved* collection receptacle that is fabricated, designed and installed to allow access for cleaning.

❖ The inside surface of the exhaust hood must be free from protrusions, ridges and ledges that can accumulate grease. This section recognizes that troughs, gutters and receptacles are intentionally designed to collect and drain grease along the lower perimeter (skirt) of the hood and under the grease removal device (filter) supports. These troughs or gutters must drain to an approved grease collection receptacle that can be readily emptied and cleaned (see Commentary Figure 507.8). Joints, seams and ridges must be made smooth enough to allow cleaning and to prevent grease from collecting. The intent is to minimize the likelihood of a fire or health hazard.

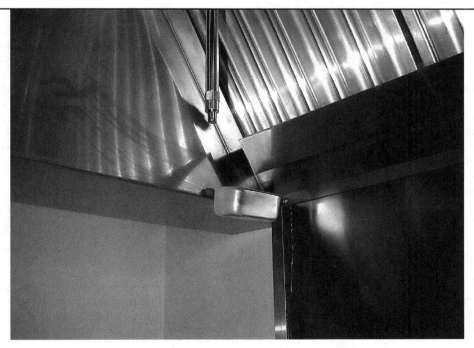

Figure 507.8
GREASE GUTTER WITH RESERVOIR

507.9 Clearances for Type I hood. A Type I hood shall be installed with a *clearance* to combustibles of not less than 18 inches (457 mm).

> **Exception:** *Clearance* shall not be required from gypsum wallboard or $^1/_2$-inch (12.7 mm) or thicker cementitious wallboard attached to noncombustible structures provided that a smooth, cleanable, nonabsorbent and noncombustible material is installed between the hood and the gypsum or cementitious wallboard over an area extending not less than 18 inches (457 mm) in all directions from the hood.

❖ Type I hoods are required to maintain an 18-inch (457 mm) air space clearance to combustible materials. Although normal operating temperatures in a hood might be relatively low, especially compared to temperatures encountered in a fire, combustibles near the hood become more susceptible to ignition because of the long-term exposure to even this moderate heat and can be ignited by the radiant or convective heat given off by the hood surfaces during "normal" hood operation. The clearances to combustibles can be reduced in accordance with the provisions of Section 308.

There are some hoods that have been tested, listed and labeled for "zero clearance" to combustibles for one or more surfaces of the hood. It is important to note that Section 507.1, Exception 1, does not list Section 507.9 as a section that is not applicable when the hood is listed and labeled in accordance with UL 710; therefore, Section 507.9 still applies. It should be noted that UL 710 does not contain specific requirements to test hoods for clearance reduction. Therefore, hoods that are tested in accordance with UL 710 and that are listed for zero clearance have been evaluated by the testing agency for clearance reduction using criteria borrowed from various other standards. Without a specific test standard to evaluate clearance reduction for Type I hoods, the 18-inch (457 mm) clearance would be required unless the code official would approve the zero clearance hood using the provisions of Section 105.2. Such hoods must be installed in accordance with the manufacturer's installation instructions, which may contain specific installation requirements to achieve the zero clearance to combustibles such as by adding insulation. The instructions may also state the minimum clearance between the appliance and the back wall and a maximum temperature of the cooking surface [see Commentary Figure 507.1(1)].

The exception allows gypsum board and $^1/_2$-inch cementitious wallboard to be considered as noncombustible only with respect to hood installations and only where the gypsum or cement wallboard is attached to noncombustible structures (see commentary, Sections 506.3.6 and 506.3.11). Hoods can be installed without clearance to walls constructed of metal studs and gypsum or cementitious wallboard that are protected with noncombustible finish materi-

als. This allowance is very useful in restaurant construction. In the case where a restaurant is of Type VB combustible construction, building a metal stud and gypsum wall next to or in contact with an exterior wall of combustible construction does not negate the 18-inch (457 mm) clearance that is required from the combustible exterior wall. Gypsum board or cementitious wallboard on wood studs is, of course, a combustible wall in any case and the 18-inch (457 mm) clearance of this section still applies. The code considers composite materials such as gypsum wallboard to be combustible, while the IBC considers gypsum wallboard to be noncombustible. It is common practice to install exhaust hoods on walls constructed of gypsum wallboard and metal studs as a means of avoiding the clearance required for combustible walls.

All surfaces within 18 inches (457 mm) of the cooking surface and exhaust hood opening must be smooth, readily cleanable, noncombustible and nonabsorbent, such as those surfaces finished with sheet metal or ceramic tile. In all cases, wall and ceiling surfaces close to the hood opening and cooking surfaces will accumulate grease that must be removed to minimize the fire and health hazard. The exception applies only to Type I hoods and is not applicable to grease ducts (see commentary, Sections 506.3.6 and 506.3.11).

507.10 Hoods penetrating a ceiling. Type I hoods or portions thereof penetrating a ceiling, wall or furred space shall comply with Section 506.3.11. Field-applied grease duct enclosure systems, as addressed in Section 506.3.11.2, shall not be utilized to satisfy the requirements of this section.

❖ A shaft enclosure for grease ducts is required in accordance with Section 506.3.11. Where Type I hoods or portions of them extend into a wall cavity, floor system cavity, interstitial space above a ceiling, furred space or similar space, the portion of the hood extending into that space must be contained within a shaft constructed in accordance with Section 506.3.11 (see commentary, Section 506.3.11). This section applies where any part of the "shell" of the hood structure penetrates a ceiling or wall. The intent is to treat the penetrating hood no differently than a penetrating duct (see commentary, Section 506.3.11.2).

This section prohibits the practice of applying field-applied grease duct enclosure systems to the portion of a hood that penetrates a ceiling, wall or furred space for the purpose of complying with Section 506.3.11 (see Commentary Figure 507.10). Field-applied grease duct enclosure systems have not been tested nor listed for this application. The manufacturers have not had their products tested for this application and their installation instructions do not contain any provisions for installing the duct wrap on the Type I hood.

507.11 Grease filters. Type I hoods shall be equipped with grease filters listed and labeled in accordance with UL 1046 and designed for the specific purpose. Grease-collecting *equipment* shall be provided with access for cleaning. The lowest edge of a grease filter located above the cooking surface shall be not less than the height specified in Table 507.11.

❖ Grease filters are required in Type I hoods to prevent large amounts of grease from collecting in the hood, in exhaust ducts, on fan blades and at the exhaust system termination. The accumulation of grease can cause restrictions in ducts or equipment failure, and can create a fire or health hazard. The grease filters must be listed and labeled in accordance with UL 1046 for the application and must be placed inside the hood so that all air being exhausted by the hood passes through them. This standard has a comprehensive set of performance requirements that are used to evaluate and list grease filters. The most common type of grease filters, often referred to as "extractors," remove grease using baffles that cause abrupt changes in direction of airflow combined with an increase in flow velocity. New types of filters are being developed that use specialized media to trap grease particulates with greater efficiency and over a larger range of particle sizes.

Through direct impingement and centrifugal force, grease is extracted from the airstream and deposited on the baffle surfaces. Grease filters must be protected from radiant heat, hot gases and direct flame impingement that can occur during normal cooking operations. At the same time, the filters have to be located to achieve the maximum extraction of grease from the cooking vapors.

Filter location should be as high above the cooking surface as the hood design will allow and must not be less than the minimum distances specified in Table 507.11 and by the grease filter manufacturer. The primary intent of this section is to prevent high-temperature cooking vapors and open flames from igniting the grease collected on the filters. Also, when the filter is located too close to the cooking surface, the temperature of the cooking vapors passing through it could exceed the acceptable temperature range for maximum grease collection efficiency.

A filter subjected to temperatures in excess of 200°F (93°C) could allow grease-laden vapors to pass straight through. The high temperatures can vaporize any grease accumulated on the filter and also allow that grease to pass through. Also, the high temperatures can approach the ignition temperature of animal fat and vegetable shortening, creating a potential fire hazard.

Appliances must be vented in accordance with the cooking equipment manufacturer's installation instructions, Chapter 8 and the IFGC. The flue gas outlet of many cooking appliances will terminate underneath an exhaust hood, thus relying on the hood for removal of the combustion byproducts. Commentary Figure 507.11 illustrates a cooking appliance equipped with a flue outlet fitting that directs the flue gases into the exhaust hood for

Figure 507.10
GREASE DUCT ENCLOSURE SYSTEM USED ON THE TOP OF A TYPE I HOOD
(Prohibited)

removal. Electrically heated appliances do not produce flue gases and do not have exposed flames; however, a high-temperature resistance-heating element can cause ignition or grease flame flare-ups similar to the ignition potential posed by open-flame appliances.

This section does not prohibit the installation or use of alternative methods for grease extraction or removal. Bear in mind that the components of any alternative grease removal system (other than grease filters) must establish equivalency with the requirements of this section in accordance with Section 105 and as approved by the code official. Grease removal devices range from simple designs, such as a configuration of baffles, to elaborate hot water scrubbers and electrostatic precipitators.

TABLE 507.11
MINIMUM DISTANCE BETWEEN THE LOWEST EDGE OF A GREASE FILTER AND THE COOKING SURFACE OR THE HEATING SURFACE

TYPE OF COOKING APPLIANCES	HEIGHT ABOVE COOKING SURFACE (feet)
Without exposed flame	0.5
Exposed flame and burners	2
Exposed charcoal and charbroil type	3.5

For SI: 1 foot = 304.8 mm.

❖ The table classifies cooking appliances into three categories based on the presence of an exposed fire and the amount of heat produced at the cooking surface. The hazards associated with an appliance without an exposed flame are significantly less than those of appliances that have exposed flames. Without exposed flames, there is no contact between the food and the heat source because they are completely separated by a heat transfer surface, thus reducing the likelihood of grease ignition.

An appliance having a cooking surface equipped with open burners is classified as an "exposed flame" appliance. This category of appliances tends to produce a consistent, predictable and moderate-temperature flame. The food is usually contained within a cooking vessel and the animal fat, vegetable shortening or other oily matter produced by cooking does not have direct contact with the flame under normal cooking operations. These appliances are usually gas fired.

An appliance having a cooking surface that has an uncontrolled exposed flame, and that will allow the food to have direct contact with the flame, is classified as a "charcoal-burning" appliance. There are different types of charcoal-type broilers, including gas- and electrically heated charbroilers. These cooking appliances might also present the same hazards associated with an open solid fuel fire. The cooking surface produces high temperatures, and food and grease have direct contact with the heat source.

The code official must evaluate the cooking equipment arrangement and determine that the location and type of grease filter used in conjunction with these appliances is in compliance with Section 507.11.

Note that factory-built hoods listed in accordance with UL 710 are not required to meet this section. Grease filters for UL 710 tested hoods must be installed in accordance with the manufacturer's installation instructions.

507.11.1 Criteria. Filters shall be of such size, type and arrangement as will permit the required quantity of air to pass through such units at rates not exceeding those for which the filter or unit was designed or *approved*. Filter units shall be installed in frames or holders so as to be readily removable without the use of separate tools, unless designed and installed to be cleaned in place and the system is equipped for such cleaning in place. Removable filter units shall be of a size that will allow them to be cleaned in a dishwashing machine or pot sink. Filter units shall be arranged in place or provided with drip-intercepting devices to prevent grease or other condensate from dripping into food or on food preparation surfaces.

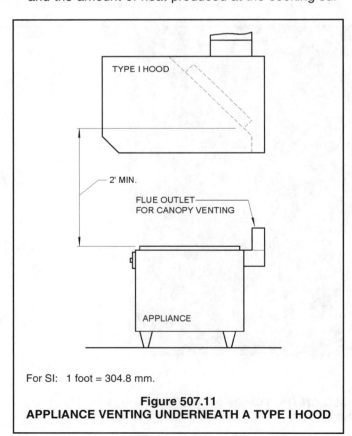

For SI: 1 foot = 304.8 mm.

Figure 507.11
APPLIANCE VENTING UNDERNEATH A TYPE I HOOD

❖ Grease filter surface area is a function of the volume of exhaust air that must pass through the filter. If the filter surface area is too small, it will have insufficient grease-retention capacity and the airflow velocity will exceed the filter design flow. The airflow velocity is directly related to the filter's ability to capture grease by impingement. Because a grease filter is designed to alter the flow of air in order to collect grease, it is essential that the filter be large enough to not create too much resistance and affect the capacity of the exhaust system. Approved filters sized in accordance with the filter manufacturer's specifications and this section must be capable of functioning properly with the airflow rate specified in Section 507.13.

The quantity of airflow is dictated by Section 507.13 and is a fixed quantity; therefore, if the filter area is decreased, the velocity of flow through the filter is increased. Because filters are designed for optimum operational efficiency at specified flow velocities, they must be sized to take advantage of this optimum performance in the approved design.

Filter units must be arranged to direct grease or other condensate to the troughs, gutters and receptacles of the hood, which are intentionally designed to collect and drain these effluents to an approved grease collection receptacle.

507.11.2 Mounting position. Filters shall be installed at an angle of not less than 45 degrees (0.79 rad) from the horizontal and shall be equipped with a drip tray beneath the lower edge of the filters.

❖ Exhaust hoods are designed to achieve the greatest possible surface area of grease filters exposed to the airflow. Also, it is desirable to have all grease-laden vapors contact the filters before contacting any other hood surface. For these reasons, it might appear that placing filters in the horizontal plane would result in the greatest surface area and most effective location; however, filter placement at any angle less than 45 degrees (0.79 rad) from horizontal can interfere with grease retention. The minimum angle is required to prevent grease droplets from dripping back onto the cooking surfaces and to promote proper drainage. Grease filters must be designed or equipped to collect and drain off grease to a channel that drains to an approved grease collection receptacle [see Commentary Figures 507.11.2(1) and 507.11.2(2)].

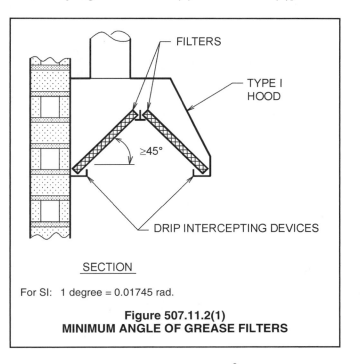

SECTION

For SI: 1 degree = 0.01745 rad.

Figure 507.11.2(1)
MINIMUM ANGLE OF GREASE FILTERS

507.12 Canopy size and location. The inside lower edge of canopy-type Type I and II commercial hoods shall overhang or extend a horizontal distance of not less than 6 inches (152 mm) beyond the edge of the top horizontal surface of the *appliance* on all open sides. The vertical distance between the front lower lip of the hood and such surface shall not exceed 4 feet (1219 mm).

Exception: The hood shall be permitted to be flush with the outer edge of the cooking surface where the hood is closed to the *appliance* side by a noncombustible wall or panel.

❖ A canopy-type hood can be classified into three basic categories: wall, island and back-to-back or double island [see Commentary Figures 507.12(1-5)] and commentary for the definition of "Commercial kitchen hoods," Section 202]. A canopy hood must be designed to cover the entire horizontal surface of the appliance. The canopy overhang must project a minimum horizontal distance of 6 inches (152 mm) beyond the edge of the horizontal surfaces of the appliance on all open sides not bounded by walls or panels. The overhang is intended to improve the vapor-capturing ability of the hood. Note that this is applicable to both Type I and II hoods. For example, a Type II canopy hood would have to extend 6 inches (152 mm) beyond the horizontal surface of a commercial dishwasher just like a Type I canopy hood would have to extend beyond the horizontal cooking surface of a commercial range. The vertical distance between the front lower edge (skirt) of the hood, and the horizontal surface of the appliance must not exceed 4 feet (1219 mm). A maximum vertical separation is specified so that the hood develops an effective capture velocity at or near the horizontal surface of the appliance. This limitation also shields the ascending cooking vapors from the influences of background air currents. The exhaust volume, cooking load, height of hood, hood design and room air currents are all factors that affect the vapor-capturing ability of the hood. It is important to note that, as indicated in Section 507.1, Exception 1, this section is not applicable to factory-built hoods that are tested, listed and labeled in accordance with UL710.

The exception allows elimination of the overhang beyond the horizontal surface of the appliance where the appliance is placed against a noncombustible sidewall or has one or more side panels (skirts) [see Commentary Figures 507.12(6) and 507.12(7)]. All open sides must have the required overhang. The wall and side panels serve as physical barriers to reduce the amount of heat radiated into the kitchen, to confine the vapor plume, to direct the upward draft into the hood and to increase the velocity of incoming air from the remaining open sides. This, in turn, reduces the effects of room cross drafts and increases the efficiency and effectiveness of the hood. Side panels/skirts are sometimes used to eliminate the required overhang because the panels/skirts serve the same purpose as the overhang.

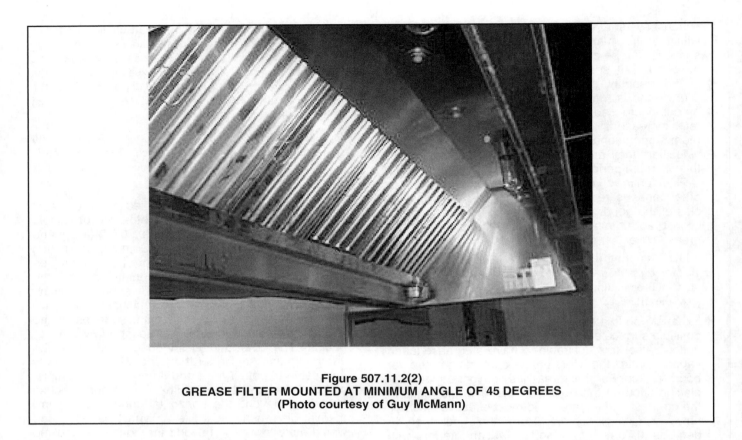

Figure 507.11.2(2)
GREASE FILTER MOUNTED AT MINIMUM ANGLE OF 45 DEGREES
(Photo courtesy of Guy McMann)

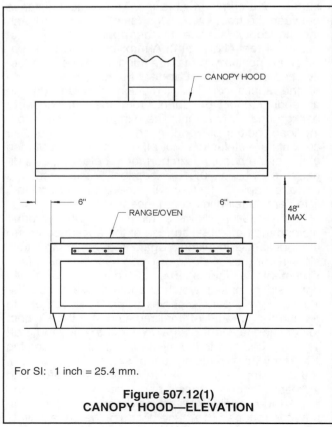

For SI: 1 inch = 25.4 mm.

Figure 507.12(1)
CANOPY HOOD—ELEVATION

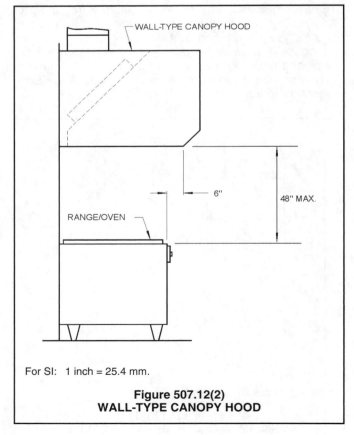

For SI: 1 inch = 25.4 mm.

Figure 507.12(2)
WALL-TYPE CANOPY HOOD

Figure 507.12(3)
WALL-TYPE CANOPY HOOD WITH MAKEUP AIR

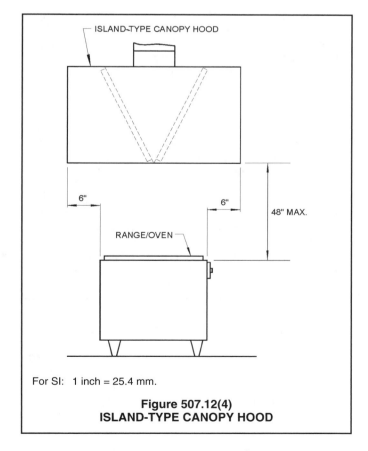

For SI: 1 inch = 25.4 mm.

Figure 507.12(4)
ISLAND-TYPE CANOPY HOOD

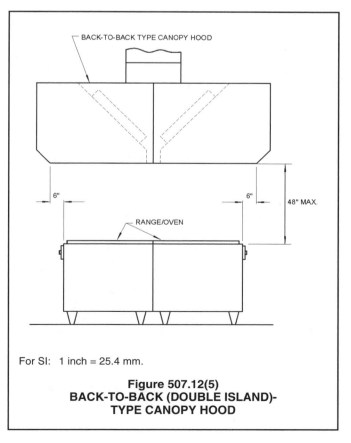

For SI: 1 inch = 25.4 mm.

Figure 507.12(5)
BACK-TO-BACK (DOUBLE ISLAND)-
TYPE CANOPY HOOD

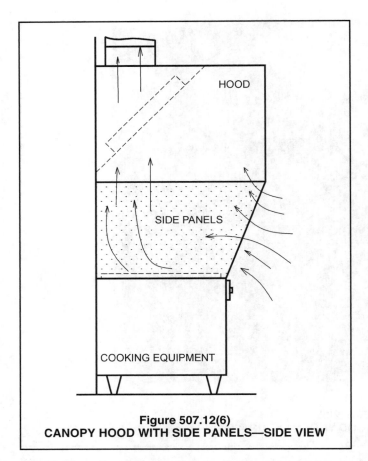

Figure 507.12(6)
CANOPY HOOD WITH SIDE PANELS—SIDE VIEW

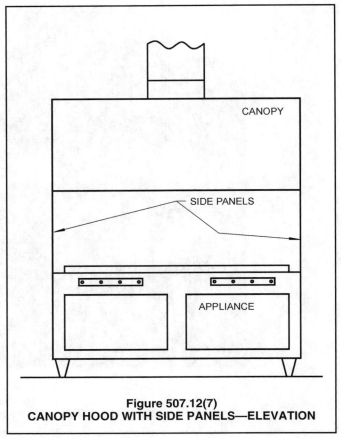

Figure 507.12(7)
CANOPY HOOD WITH SIDE PANELS—ELEVATION

507.13 Capacity of hoods. Commercial food service hoods shall exhaust a minimum net quantity of air determined in accordance with this section and Sections 507.13.1 through 507.13.5. The net quantity of *exhaust air* shall be calculated by subtracting any airflow supplied directly to a hood cavity from the total exhaust flow rate of a hood. Where any combination of *heavy-duty*, *medium-duty* and *light-duty cooking appliances* are utilized under a single hood, the exhaust rate required by this section for the heaviest duty *appliance* covered by the hood shall be used for the entire hood.

❖ The hood must be designed to adequately collect and exhaust smoke and vapors from the area over which it is installed. To accomplish this, the hood must cause an airflow pattern that will sweep and direct the smoke and vapors upward from the cooking surfaces and into the hood inlet.

Where multiple appliances are located under a single hood in any combination of heavy-, medium- and light-duty cooking appliances, the required exhaust rate for the heaviest-duty appliance applies and the highest exhaust rate required by this section must be used for the entire hood [see Commentary Figure 507.13(1)]. This is a conservative design requirement to ensure that all of the cooking effluents are captured by specifying the highest exhaust rate. Note that extra-heavy-duty cooking appliances cannot be served by a hood that also serves any other duty classification of cooking appliance (see commentary, Section 507.2.4).

Although the dimensional and location design parameters of the exhaust hood are regulated by Sections 507.12 and 507.14, this section specifies the minimum quantity of exhaust air necessary for effective removal of cooking vapors and the approximate amount of makeup air necessary for proper operation. The quantity of required exhaust is as much a function of the operational characteristics of the cooking equipment (solid-fuel-burning; low-, medium- or high-heat) as it is a function of the size of the cooking surface or the exhaust hood opening area and the presence of any walls and side panels [see Commentary Figures 507.13(2) and 507.13(3)]. Walls and side panels serve as physical barriers that enhance the capture ability of the hood and increase the velocities of incoming air from the remaining open sides. This, in turn, reduces the effects of room cross drafts and flare-ups, and increases the efficiency and effectiveness of the hood. Island canopies, which are exposed on all four sides, lack the benefits of adjacent walls and, therefore, are required to exhaust greater quantities of air than wall-mounted canopy hoods. Note that this section does not apply to hoods listed in accordance with UL 710.

The formulas for sizing exhaust flow rates in Sections 507.13.1 through 507.13.4 are based on the linear feet of hood, the type of hood (island hood, eyebrow hood, etc.) and the duty rating of the appliance being exhausted (heavy-duty, light-duty, etc.).

The requirement for determining the exhaust rate based on the net exhaust flow is the result of testing that has shown that hood performance is dependent on "net exhaust" (total exhaust flow minus internal makeup air delivered inside the hood). Internal makeup air is also known as short-circuit air. Short-circuit hoods discharge unconditioned makeup air directly into the hood cavity to reduce the amount of conditioned air that is exhausted from the kitchen. However, this air does not assist in capturing the cooking effluents; therefore, the formulas in Sections 507.13.1 through 507.13.4 are based on net exhaust flow. In other words, any "short-circuit" air supplied directly into a hood is not counted toward satisfying the code requirement. Only the exhaust air flow taken from the kitchen can satisfy the code requirement.

Based on the requirements in Section 507.2.1, the first three subsections of this section specify the quantity of required exhaust for Type I hoods, and the last subsection of this section contains criteria for Type I and II hoods. The quantity of exhaust required for Type II hoods must comply with Section 507.13.4 (see commentary, Section 507.13.4). It is important to note that, as indicated in Section 507.1, Exception 1, this section is not applicable to factory-built hoods that are tested, listed and labeled in accordance with UL 710.

New energy-saving technology can allow the exhaust flow rate to modulate in response to the cooking effluent load. Such systems use heat and vapor detectors and variable or multispeed exhaust fans (see Section 507.1, Exception 3).

507.13.1 Extra-heavy-duty cooking appliances. The minimum net airflow for hoods, as determined by Section 507.2, used for *extra-heavy-duty cooking appliances* shall be determined as follows:

Type of Hood	CFM per linear foot of hood
Backshelf/pass-over	Not allowed
Double island canopy (per side)	550
Eyebrow	Not allowed
Single island canopy	700
Wall-mounted canopy	550

For SI: 1 cfm per linear foot = 1.55 L/s per linear meter.

❖ The quantity of exhaust required for Type I hoods serving extra-heavy-duty cooking appliances must be supplied in accordance with this section. Extra-heavy-duty cooking appliances use solid fuel as all or any part of the source of heat for cooking (see commentary for the definition of "Extra-heavy-duty cooking appliances," Section 202). Some examples of extra-heavy-duty cooking appliances are barbecue pits and barbecue cooking appliances; solid fuel-burning stoves and ovens; grease-burning charbroilers and charcoal grills (see commentary, Section 507.2.4).

Based on the definition of "Extra-heavy-duty cooking appliances" in Section 202, appliances that burn wood chips to impart flavor to the food being cooked are considered to be extra-heavy-duty appliances. Various species of wood logs and chips are burned in such cooking appliances as barbecue grills, smokers and broilers. The classification of those appliances as

Figure 507.13(1)
CANOPY HOOD SERVING MULTIPLE APPLIANCES

extra heavy duty is not based on whether the consumption of solid fuel is the primary heat source for cooking. The consumption of solid fuel only for the purpose of flavoring foods, where the primary heat source for cooking is another fuel or electricity, is still viewed as solid fuel burning.

Note that the exhaust rate is based on the linear length of the face of the hood and not the area of the hood opening. The length of the hood is measured along the front side that is parallel with the front line of the cooking appliances. Laboratory testing and empirical data support this methodology, which is

Figure 507.13(2)
BACKSHELF/PASS-OVER HOOD

Figure 507.13(3)
BACKSELF HOOD WITH ONE SIDE PANEL
(Photo courtesy of Guy McMann)

consistent with ASHRAE guidelines. The exhaust rates are not tied to the hood opening area; therefore, the designer can take advantage of the improved capture and containment ability of larger area hoods without having to increase the required exhaust rate because of the larger hood area (see Commentary Figure 507.13.1).

507.13.2 Heavy-duty cooking appliances. The minimum net airflow for hoods, as determined by Section 507.2, used for *heavy-duty cooking appliances* shall be determined as follows:

Type of Hood	CFM per linear foot of hood
Backshelf/pass-over	400
Double island canopy (per side)	400
Eyebrow	Not allowed
Single island canopy	600
Wall-mounted canopy	400

For SI: 1 cfm per linear foot = 1.55 L/s per linear meter.

❖ The quantity of exhaust required for Type I hoods serving heavy-duty cooking appliances, such as under-fired broilers, open-burner ranges and wok ranges, must be in accordance with this section (see commentary for the definition of "Heavy-duty cooking appliances," Section 202). Higher quantities of grease-laden vapors and smoke are associated with these appliances.

Note that the exhaust rate is based on the linear length of the face of the hood and not the area of the hood opening. The length of the hood is measured along the front side that is parallel with the front line of the cooking appliances. Laboratory testing and empirical data support this methodology, which is consistent with ASHRAE guidelines. The exhaust rates are not tied to the hood opening area; therefore, the designer can take advantage of the improved capture and containment ability of larger area hoods

without having to increase the required exhaust rate because of the larger hood area (see Commentary Figure 507.13.1).

507.13.3 Medium-duty cooking appliances. The minimum net airflow for hoods, as determined by Section 507.2, used for *medium-duty cooking appliances* shall be determined as follows:

Type of Hood	CFM per linear foot of hood
Backshelf/pass-over	300
Double island canopy (per side)	300
Eyebrow	250
Single island canopy	500
Wall-mounted canopy	300

For SI: 1 cfm per linear foot = 1.55 L/s per linear meter.

❖ The quantity of exhaust required for Type I hoods serving medium-duty cooking appliances must be in accordance with this section. Some examples of medium-duty cooking appliances are rotisseries, griddles, deep fat fryers and conveyor pizza ovens (see commentary for the definition of "Medium-duty cooking appliances," Section 202).

Note that the exhaust rate is based on the linear length of the face of the hood and not the area of the hood opening. The length of the hood is measured along the front side that is parallel with the front line of the cooking appliances. Laboratory testing and empirical data support this methodology, which is consistent with ASHRAE guidelines. The exhaust rates are not tied to the hood opening area; therefore, the designer can take advantage of the improved capture and containment ability of larger area hoods without having to increase the required exhaust rate because of the larger hood area (see Commentary Figure 507.13.1).

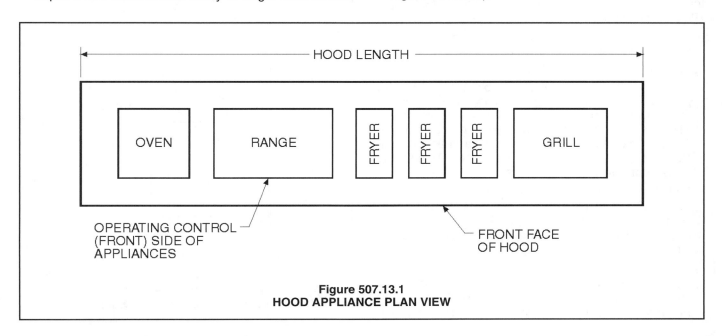

Figure 507.13.1
HOOD APPLIANCE PLAN VIEW

507.13.4 Light-duty cooking appliances. The minimum net airflow for hoods, as determined by Section 507.2, used for *light-duty cooking appliances* and food service preparation shall be determined as follows:

Type of Hood	CFM per linear foot of hood
Backshelf/pass-over	250
Double island canopy (per side)	250
Eyebrow	250
Single island canopy	400
Wall-mounted canopy	200

For SI: 1 cfm per linear foot = 1.55 L/s per linear meter.

❖ The quantity of exhaust required for Type I and II hoods serving light-duty cooking appliances must be in accordance with this section. Some examples of light-duty cooking appliances are ovens, steamers and steam kettles (see commentary for the definition of "Light-duty cooking appliances," Section 202). A Type II hood serving a dishwashing appliance must be in accordance with Section 507.13.5 (see commentary, Section 507.13.5).

Note that the exhaust rate is based on the linear length of the face of the hood and not the area of the hood opening. The length of the hood is measured along the front side that is parallel with the front line of the cooking appliances. Laboratory testing and empirical data support this methodology, which is consistent with ASHRAE guidelines. The exhaust rates are not tied to the hood opening area; therefore, the designer can take advantage of the improved capture and containment ability of larger area hoods without having to increase the required exhaust rate because of the larger hood area (see Commentary Figure 507.13.1).

507.13.5 Dishwashing appliances. The minimum net airflow for Type II hoods used for dishwashing appliances shall be 100 CFM per linear foot of hood length.

Exception: Dishwashing appliances and *equipment* installed in accordance with Section 507.2.2.

❖ The minimum quantity of exhaust required for Type II hoods serving a dishwashing appliance is 100 cfm per linear foot of hood length. This is a specific airflow rate that is for Type II hoods serving dishwashers. The flow rates in Section 507.13.4 of the code are for Type II hoods serving light-duty cooking appliances. Dishwashers are not cooking appliances. This rate is consistent with industry practice.

507.14 Noncanopy size and location. Noncanopy-type hoods shall be located a maximum of 3 feet (914 mm) above the cooking surface. The edge of the hood shall be set back a maximum of 1 foot (305 mm) from the edge of the cooking surface.

❖ A kitchen exhaust hood that does not meet the minimum extension criteria listed in Section 507.12 is classified as a noncanopy hood (see Commentary Figures 507.13(2), 507.13(3) and 507.14). A noncan-

opy hood, often called a "low-wall," "proximity" or "backshelf" hood, relies on the closeness of the hood to the cooking surface for capture and containment of the cooking vapors. Therefore, a noncanopy hood must have the intake located within 3 feet (914 mm), measured vertically, of the cooking surface. Backshelf, pass-over and eyebrow hoods are examples of noncanopy hoods. This section does not apply to hoods listed in accordance with UL 710.

The front edge of the hood must be located not more than 1 foot (305 mm), measured horizontally, from the front edge of the cooking surface to facilitate movement of the cooking vapors into the hood before they can escape upward beyond the hood inlet.

The net exhaust airflow requirements for backshelf, pass-over and eyebrow hoods are prescribed in Sections 507.13.1 through 507.13.4 or are in accordance with the manufacturer's installation instructions for a factory-built hood that is tested, listed and labeled in accordance with UL 710.

Without the capture area characteristics of a canopy hood, a noncanopy hood is usually not capable of handling large surges of grease-laden vapors produced by charbroilers or similar high-heat and grease-producing appliances. To confine vapor effectively, the proper design of noncanopy hoods is especially critical (see commentary, Section 507.13).

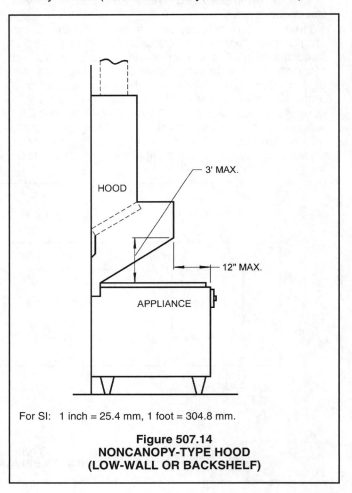

For SI: 1 inch = 25.4 mm, 1 foot = 304.8 mm.

Figure 507.14
NONCANOPY-TYPE HOOD
(LOW-WALL OR BACKSHELF)

A noncanopy hood is required to meet all the provisions specified in this chapter for hoods with the exception of Section 507.12.

507.15 Exhaust outlets. Exhaust outlets located within the hood shall be located so as to optimize the capture of particulate matter. Each outlet shall serve not more than a 12-foot (3658 mm) section of hood.

❖ Exhaust outlets in a hood are the openings to which exhaust ducts connect and, depending on the size of the hood, more than one outlet might be needed. The exhaust outlets are distributed in the hood to result in uniform exhaust flow rates over the entire hood opening area. The 12-foot (3568 mm) limitation is intended to prevent widely varying exhaust flow rates within the hood opening area by requiring multiple outlets on large (long) hoods.

507.16 Performance test. A performance test shall be conducted upon completion and before final approval of the installation of a ventilation system serving *commercial cooking appliances*. The test shall verify the rate of exhaust airflow required by Section 507.13, makeup airflow required by Section 508 and proper operation as specified in this chapter. The permit holder shall furnish the necessary test *equipment* and devices required to perform the tests.

❖ Upon completion of an installation and before final approval is given, the code official must determine that the kitchen exhaust equipment will function properly under the most severe anticipated cooking conditions. The performance test includes the determination that the code-required airflow and containment of all cooking vapors will occur. It is not the intent of this section to require a full-scale test of the automatic fire suppression system.

The performance test is very important because it will determine whether or not the system, including the hood, ducts and fans, is designed and operating properly. For example, a factory-built hood tested and listed in accordance with UL 710 will have the minimum required exhaust rate that is determined in accordance with the testing. However, because the testing occurs in an ideal laboratory setting, that exhaust rate might be inadequate in the environment in which the hood is actually installed. The only way to determine whether the exhaust rate is adequate for capture and containment of the cooking effluents is to test the installed system on the job site. Determining the actual exhaust and makeup airflow rates requires specialized apparatus and instruments, as well as skilled technicians. Such testing should be performed by experienced third-party personnel.

The test must verify that the required exhaust rate and capture and containment of the cooking effluent is achieved. Verification of proper makeup air system operation is also part of the performance test. It is more realistic to conduct a test that actually involves the cooking operations rather than a test that uses only a steam or smoke generator. The rising thermal draft that is generated by the cooking operation will generally aid in the capture and containment of

smoke and vapors. Testing procedures must be acceptable to the code official and are typically designed to simulate the actual worst-case cooking operation that is expected to take place under the hood. In accordance with Section 507.16.1, any test must involve simultaneous use of the appliances served with all of the appliances at their operating temperature.

507.16.1 Capture and containment test. The permit holder shall verify capture and containment performance of the exhaust system. This field test shall be conducted with all appliances under the hood at operating temperatures, with all sources of outdoor air providing *makeup air* for the hood operating and with all sources of recirculated air providing conditioning for the space in which the hood is located operating. Capture and containment shall be verified visually by observing smoke or steam produced by actual or simulated cooking, such as with smoke candles, smoke puffers, etc.

❖ This section contains requirements for performing the capture and containment test that is performed as part of the field performance test required by Section 507.16 (see commentary, Section 507.16). To make the test realistic, the appliances must all be at operating temperatures, as required for actual cooking. The heat from the appliances produces a rising thermal plume that affects the capture and containment ability of the hood. Obviously, the most realistic test would involve actual cooking, but, if that is not practical, visual aids such as smoke generators can be used to simulate cooking.

In addition to the cooking appliances being at operating temperature, the source of makeup air and the recirculated air that provides conditioning for the kitchen space must also be operating at the time of the test. It has been demonstrated that all sources of air introduced around and within the vicinity of the hood can impact the hood's ability to capture and contain. Four-way diffusers used for ventilation air to condition the kitchen space have been shown to have a negative impact on hood performance when located improperly. Transfer air introduced from adjacent spaces has been shown to create turbulence or velocities that can have a negative impact on a hood's performance. Therefore, all sources of airflow in and around the hood must be in operation to demonstrate actual operating conditions during the performance test.

SECTION 508
COMMERCIAL KITCHEN MAKEUP AIR

508.1 Makeup air. *Makeup air* shall be supplied during the operation of commercial kitchen exhaust systems that are provided for *commercial cooking appliances*. The amount of *makeup air* supplied to the building from all sources shall be approximately equal to the amount of *exhaust air* for all exhaust systems for the building. The *makeup air* shall not reduce the effectiveness of the exhaust system. *Makeup air* shall be provided by gravity or mechanical means or both.

Mechanical *makeup air* systems shall be automatically controlled to start and operate simultaneously with the exhaust system. *Makeup air* intake opening locations shall comply with Section 401.4.

❖ The air exhausted from a kitchen exhaust system must be replaced with enough air to allow the required exhaust airflow and to prevent excessive negative pressure in the commercial cooking area. Kitchen exhaust systems are usually designed with the quantity of makeup air being slightly less than that exhausted, thereby creating a slight negative pressure in the kitchen to help confine cooking odors within the kitchen [see Commentary Figure 508.1(1)]. NFPA 96 and some hood manufacturers require that negative pressures not exceed 0.02 inch of water column (0.050 kPa).

The introduction of makeup air is critical to the proper operation of all kitchen exhaust systems and the fuel-burning appliances located in kitchen areas. Too little makeup air will cause excessive negative pressures to develop in the kitchen, thereby reducing the exhaust airflow. Too little makeup air can also affect nearby fuel-burning appliances and cause loss of draft in appliance vents and chimneys, or cause combustion byproducts to discharge into the building. A lack of combustion air and makeup air can result in incomplete fuel combustion, appliance malfunction and flue gas spillage.

Makeup air is not required to be provided through dedicated makeup units. Makeup air can be supplied through the HVAC system in the building that supplies outdoor air for ventilation in accordance with Section 403.2. Rather than having just a dedicated makeup system for the kitchen, makeup air can be supplied from all the sources within the building, including dedicated units, and must balance with all of the exhaust systems in the building. This would allow outdoor air supplied to provide ventilation for a dining room, for example, to be transferred to the kitchen to serve as makeup air. An air balance schedule is usually provided to show all of the sources of makeup air within the building and all of the air that is exhausted from the building. As previously discussed, kitchen exhaust systems are commonly designed to create a slight negative pressure within the kitchen to help confine cooking odors. For this reason, the makeup air rate is allowed to be "approximately equal" to or slightly less than the exhaust air rate. The difference in quantity between the makeup air supplied and the exhaust rate is made up with transfer air from adjacent rooms or spaces.

An exhaust system can be designed to introduce makeup air through the exhaust hood [see Commentary Figures 508.1(2), 508.1(3) and 508.1(4)]. Commentary Figures 508.1(2) and 508.1(3) illustrate an exhaust hood that has makeup air registers located on the face of the hood. Such hoods are referred to as compensating hoods because they compensate for the exhaust flow by supplying makeup air to the space in which the hood is located. Compensating

hoods include short-circuit hoods (see the definition of "Compensating hoods").

Commentary Figure 508.1(4) illustrates an exhaust hood that introduces air at the face of the hood and within the hood cavity. The figure shows a hood that is required by the code to exhaust 5,000 cfm (2.36 m³/s). Because 1,500 cfm (0.71 m³/s) of untempered air is introduced directly into the hood, only 3,500 cfm (1.65 m³/s) of air is exhausted from the kitchen space. The additional makeup air quantity must be approximately 3,500 cfm (1.65 m³/s).

Section 507.13 requires a "net" quantity of exhaust airflow, which is determined by subtracting from the actual flow rate of the hood any airflow supplied directly to the hood cavity. The short-circuit air shown in Commentary Figure 508.1(4) must be subtracted from the total exhaust airflow through the hood exhaust duct to determine the net exhaust flow.

This design is prohibited by the code because only 3,500 cfm (1.65 m³/s) is being exhausted from the kitchen area instead of the required 5,000 cfm (2.36 m³/s). A hood that introduces air directly into the hood cavity is called a short-circuit hood system (see the definition of "Compensating hoods"). A short-circuit hood was originally intended for use only where the code-required exhaust rate exceeds that which is actually necessary for a cooking operation. Using this type of hood could result in energy savings because the supply air brought into the hood does not have to be tempered. Short-circuit hoods are not addressed in the code other than indirectly by the second sentence of Section 507.13 and the third sentence of this section, which says that makeup air must not reduce the effectiveness of the exhaust system. It is important to note that, as indicated in Section 507.1, Exception 1, Section 507.13 is not applicable to factory-built hoods that are tested, listed and labeled in accordance with UL 710. Therefore, a short-circuit hood that is tested in accordance with UL 710 and passes the performance, capture and containment tests required by Section 507.16.1 would be permitted.

Testing has shown that the location of makeup air discharge can negatively affect hood performance, depending on where and at what velocity the makeup air is introduced into the kitchen or hood. Makeup air introduced near or into a hood can cause air currents that will cause cooking effluent to spill from the hood or be drawn away from the hood. The full-scale test in accordance with Section 507.16 is intended to be used by the code official to determine that the exhaust hood will effectively confine and exhaust the cooking vapors. In all cases, the method of introducing makeup air to the kitchen space or hood system must not reduce the effectiveness of the exhaust system.

The mechanical makeup air system and the kitchen exhaust system must operate simultaneously through the use of automatic controls. The fans must be wired for parallel starting and operation. Supervisory controls, like those required for systems that are required to be interlocked, are not required for the operation of these fans. If the HVAC system is supplying all or part of the makeup air, then the HVAC units must be controlled to operate via the automatic controls. There must not be an option to operating the exhaust hood without operating the makeup air supply system.

Makeup air may be supplied by gravity (openings to the outdoors) or mechanical means, or both. All

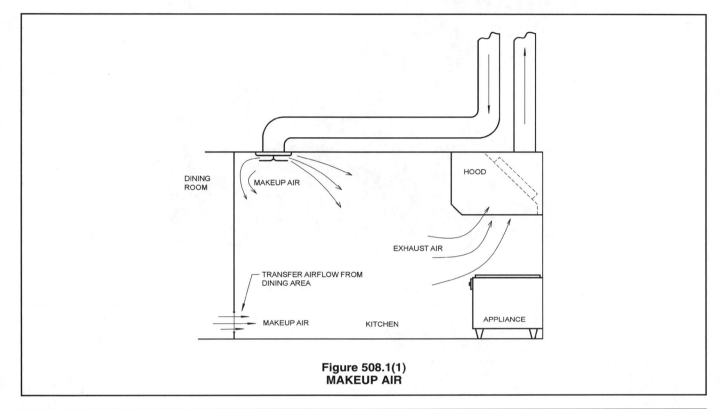

Figure 508.1(1)
MAKEUP AIR

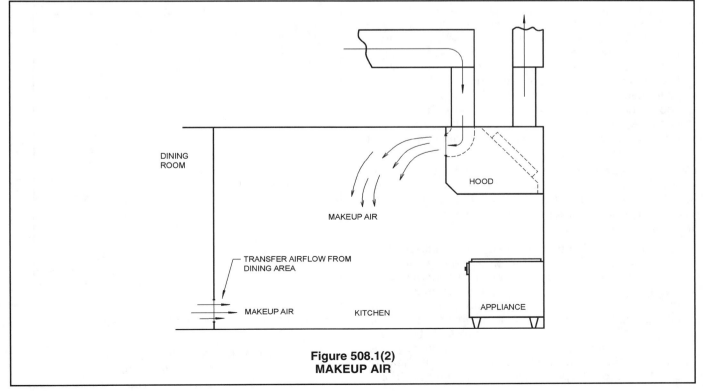

Figure 508.1(2)
MAKEUP AIR

Figure 508.1(3)
MAKEUP AIR SUPPLIED FROM FACE OF HOOD

such mechanical systems and motorized dampers must be tied to the operation of the cooking appliances served. Kitchen hoods within and serving dwelling units are exempt from the code provisions for makeup air if the exhaust hood system does not exhaust more than 400 cfm (see commentary, Section 505.2).

Makeup air is typically supplied mechanically by makeup air supply fans equipped with air filters and air-heating units comparable to duct heaters. Regional and local public health agencies might require filtration of makeup air because it discharges into food-handling and preparation areas. Makeup air distribution systems and HVAC supply terminals should not introduce air too near the exhaust hood because this has been proven to cause interference with the hood's ability to capture exhaust as a result of the turbulence created.

508.1.1 Makeup air temperature. The temperature differential between *makeup air* and the air in the conditioned space shall not exceed 10°F (6°C) except where the added heating and cooling loads of the *makeup air* do not exceed the capacity of the HVAC system.

❖ Makeup air that is not introduced directly into or close to the exhaust hood must be tempered to within 10°F (6°C) of the temperature of the conditioned air within the space. The intent is to prevent the makeup air from causing employee discomfort, which might encourage employees to shut down or restrict the makeup air supply. The temperature limitation does not apply to makeup air that is supplied through the HVAC system since that system will either heat or cool the air to the space conditions that are controlled

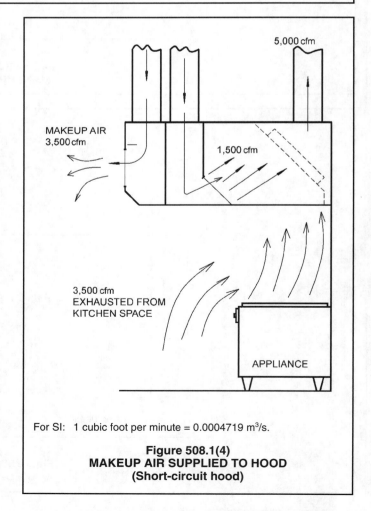

For SI: 1 cubic foot per minute = 0.0004719 m³/s.

Figure 508.1(4)
MAKEUP AIR SUPPLIED TO HOOD
(Short-circuit hood)

by the thermostat within the space. If the makeup air is supplied to a compensating hood or in close proximity to any hood, it is possible that employee comfort conditions will not be affected because a major portion of the makeup air will mix with the cooking effluent and enter the hood before affecting comfort. This may permit providing makeup air that, by design, is not heated or cooled to space conditions. By providing the makeup air near the exhaust hood and directing it towards the exhaust hood the makeup air could be exhausted before imposing significant heating and cooling loads to the space. Additionally, even if the makeup air becomes a heating or cooling load to the HVAC system, comfort conditions will not be affected if the HVAC system capacity is sufficient to ensure that space heating and cooling thermostat setpoints are not exceeded. The ventilation provisions of Chapter 4 also apply to kitchen spaces (see commentary, Table 403.3).

Kitchen ventilation may be satisfied by the outdoor air supplied as makeup air to the exhaust hood system where the exhaust rate of the hood system equals or exceeds the required ventilation air rate. In that case, the makeup air also serves as ventilation air. Also, the quantity of outdoor air supplied as ventilation air may be sufficient to supply the makeup air for exhaust hoods, in which case no separate source of makeup air would be used.

The focus of this section is employee comfort and is typically satisfied by makeup air being heated to a temperature not greater than the heating/cooling balance point of the space. Because makeup air tempering controls may be separate from HVAC controls, it is important to ensure that heating and cooling do not occur simultaneously, which might happen, for example, with a makeup air heat setting higher than the balance point—a temperature that might be 50°F (10°C) or lower. Rarely, if ever, would the need arise to cool makeup air. Systems should be designed to avoid the use of energy to simultaneously heat and cool kitchen air supplies. For example, it is possible that makeup air is being heated while the kitchen space is being cooled.

508.2 Compensating hoods. Manufacturers of compensating hoods shall provide a label indicating minimum exhaust flow and/or maximum makeup airflow that provides capture and containment of the exhaust effluent.

> **Exception:** Compensating hoods with *makeup air* supplied only from the front face discharge and side face discharge openings shall not be required to be labeled with the maximum makeup airflow.

❖ "Compensating hoods" are defined in Section 202 as those hoods having an integral makeup air supply. The methods of providing the compensating air, include:

- From inside the hood (also called short-circuit air);
- Air curtain flow from the bottom of the front face;

- From the front face, rear or side of the hood;
- From the front, rear or sides of the cooking equipment itself; or
- From a combination of the methods above.

The primary reason for using compensating hoods is the energy savings resulting from not having to temper the makeup air. One of the major complaints from kitchen employees is the cold, drafty makeup air flowing across the kitchen during the winter. To avoid this condition, the air must be heated, adding to the energy costs of the building. Makeup air introduced within the boundaries of the hood or cooking equipment may not require tempering because the employees are less likely to be exposed to this air flow.

Improper exhaust and makeup airflow could negatively affect hood performance; therefore, compensating hoods must specify the parameters within which the hood will function (see commentary, Section 508.1).

This section requires the hood to have a label indicating the minimum exhaust flow rate and maximum makeup airflow required to adequately capture, contain and remove the smoke and grease-laden air. These flow rates are established through testing under ideal conditions in the manufacturer's test facilities. To verify the performance of the compensating hood under actual conditions, a full scale test must be performed as required by Section 507.16. The location of doors, windows, HVAC supply and return openings and other exhaust fans might adversely affect the hood's operation.

The exception recognizes that the requirements in UL 710 do not require the evaluation and marking of the maximum makeup air for hoods with front and side face discharge openings. Makeup air discharging from the front and side face of compensating hoods will effectively combine with room supply air to provide for the capture of cooking product effluent.

SECTION 509
FIRE SUPPRESSION SYSTEMS

509.1 Where required. *Commercial cooking appliances* required by Section 507.2.1 to have a Type I hood shall be provided with an *approved* automatic fire suppression system complying with the *International Building Code* and the *International Fire Code.*

❖ Fire is a well-documented problem in restaurant and commercial kitchens. This section requires an effective fire suppression system to combat fire on the cooking surfaces of grease-producing appliances, and within the hood and exhaust system of a commercial kitchen installation. Commentary Figure 509.1 illustrates a typical suppression system serving a commercial exhaust hood and cooking appliance. Automatic fire-extinguishing systems installed in commercial food heat-processing appliances must be installed in accordance with the IBC and IFC, and be approved by the code official.

In accordance with the IBC, the IFC and typical manufacturer's installation instructions, the fire-extinguishing system must protect the cooking surfaces (which are the most likely source of a fire) and must also protect all portions of the exhaust hood system, including the hood, the filters, the extractors and the duct system. The most common type of fire suppression systems used with kitchen equipment include dry-chemical, wet-chemical and water sprinkler systems.

Exhaust systems serving only commercial cooking appliances that produce no grease-laden vapor have no specific fire protection requirements in this section. However, an exhaust system that serves any grease-producing cooking appliance must be treated entirely as a grease exhaust system, even if it also serves grease-free cooking appliances.

SECTION 510
HAZARDOUS EXHAUST SYSTEMS

510.1 General. This section shall govern the design and construction of duct systems for hazardous exhaust and shall determine where such systems are required. Hazardous exhaust systems are systems designed to capture and control hazardous emissions generated from product handling or processes, and convey those emissions to the outdoors. Hazardous emissions include flammable vapors, gases, fumes, mists or dusts, and volatile or airborne materials posing a health hazard, such as toxic or corrosive materials. For the purposes

of this section, the health hazard rating of materials shall be as specified in NFPA 704.

For the purposes of the provisions of Section 510, a laboratory shall be defined as a facility where the use of chemicals is related to testing, analysis, teaching, research or developmental activities. Chemicals are used or synthesized on a nonproduction basis, rather than in a manufacturing process.

❖ Hazardous exhaust systems include, but are not limited to, those conveying flammable vapors, gases, paint residue, corrosive fumes, dust and particulate matter and volatile or air-borne materials posing a health hazard. The intent of this section is to reduce the hazards associated with exhaust systems that convey combustible, flammable, explosive, toxic or corrosive materials in any state (solid, liquid or gaseous). For example, laboratory exhaust hoods or booths are designed to contain and exhaust many different substances that may or may not be flammable, toxic, corrosive or pathogenic. Also, the nature of those types of exhaust systems is dependent on the quantities of materials released within the hoods or booths and the amount of dilution air introduced into the hoods or booths.

NFPA 704 describes a system for the identification of hazards to the life and health of people. The standard contains quantitative guidelines for determining the numerical health-hazard rating of a material based on the physical properties and characteristics of the material that are known or can be determined

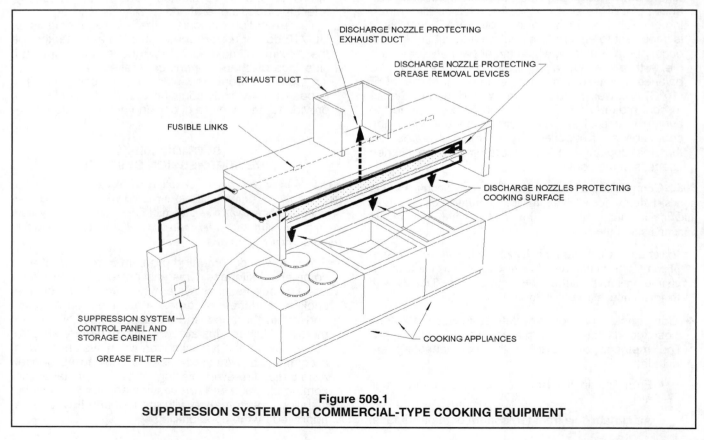

Figure 509.1
SUPPRESSION SYSTEM FOR COMMERCIAL-TYPE COOKING EQUIPMENT

by standard methods. The provisions of Section 510 clearly point to the concern for combustibility, flammability, toxicity, corrosiveness and explosiveness, but the code does not cover microbial, pathogenic and similarly dangerous exhausts. NFPA 45 presents additional guidance for the regulation of laboratory exhaust systems.

The definition of "Laboratories" provided in this section is intended to describe those environments that, unlike manufacturing operations that may use large quantities of chemicals on a constant basis, are characterized by the use of a wide variety of chemicals in very small quantities, and often for very short periods of time or on an infrequent basis. These are operations in which standard laboratory exhaust practices provide significant "in duct" dilution, which prevent in-duct incompatible material reactions and buildup of flammable vapors. This definition is the basis for exceptions found for laboratories in Sections 510.2, 510.4 and 510.7.

510.2 Where required. A hazardous exhaust system shall be required wherever operations involving the handling or processing of hazardous materials, in the absence of such exhaust systems and under normal operating conditions, have the potential to create one of the following conditions:

1. A flammable vapor, gas, fume, mist or dust is present in concentrations exceeding 25 percent of the lower flammability limit of the substance for the expected room temperature.

2. A vapor, gas, fume, mist or dust with a health-hazard rating of 4 is present in any concentration.

3. A vapor, gas, fume, mist or dust with a health-hazard rating of 1, 2 or 3 is present in concentrations exceeding 1 percent of the median lethal concentration of the substance for acute inhalation toxicity.

Exception: Laboratories, as defined in Section 510.1, except where the concentrations listed in Item 1 are exceeded or a vapor, gas, fume, mist or dust with a health-hazard rating of 1, 2, 3 or 4 is present in concentrations exceeding 1 percent of the median lethal concentration of the substance for acute inhalation toxicity.

❖ This section specifies where the installation of an effective hazardous exhaust system is required based on the stated conditions. This section contains the criteria to determine whether a given operation requires a hazardous exhaust system. If, in the absence of an exhaust system, the normal operation would exceed the threshold concentrations indicated, a hazardous exhaust system is required. The code does not indicate how to determine if any of the criteria are triggered. It may be necessary to consult an expert in the field such as a professional toxicologist or an industrial hygienist or contact an organization, such as the Occupational Safety and Health Administration or the National Institute for Occupational Safety and Health. The hazardous exhaust system is intended to keep the threshold levels from being reached, either in the room containing the point

source (for all types of hazardous materials) or in the exhaust duct system (for flammability/explosion hazards). The criteria are based on normal operating conditions, not on accident conditions. It is important to note and emphasize that the threshold concentrations are without the exhaust system operating. Section 510.3 requires the exhaust system to operate below threshold concentrations; therefore, the threshold limits are determined without the exhaust system operating.

Item 1 requires a hazardous exhaust system where flammable vapor, gas, fume, mist or dust reaches concentrations that exceed 25 percent of the lower flammability limit (LFL) of the substance for the range of anticipated space temperature(s). The limits of flammability (high and low) are the extreme concentration limits of a substance. The LFL of a substance is the lowest concentration by volume of a flammable vapor, gas, fume, mist or dust mixed with air (or any oxidant) that will ignite and burn with a flame. For example, hydrogen-air mixtures will propagate flames in concentrations between 4 and 74 percent by volume of hydrogen at 70°F (21°C) and at atmospheric pressure. The smaller value is the lower (lean) limit, and the larger value is the upper (rich) limit of flammability. For purposes of this example, anything greater than a 1-percent concentration by volume of hydrogen in air would exceed 25 percent of hydrogen's LFL at 70°F (21°C) and atmospheric pressure. Keep in mind that when the mixture temperature is increased, the flammability range widens; when the temperature is decreased, the range narrows.

The term "lower flammability limit" describes the minimum concentration by volume of vapor to air that will propagate a flame in the presence of an ignition source. The upper flammability limit (UFL) is the maximum vapor-to-air concentration that will propagate a flame. If a vapor-to-air mixture is below the LFL, it is described as being "too lean" to burn, and if it is above the UFL, it is "too rich" to burn. When the vapor-to-air ratio is somewhere between the LFL and the UFL, fires and explosions can occur upon introduction of an ignition source. The mixture is then said to be within its flammable or explosive range. When the mixture is in the range between the LFL and UFL [synonymous with LEL (lower explosive limit) and UEL (upper explosive limit)], the ignition is more intense and violent than if the mixture were closer to either the upper or lower limits.

Item 2 requires a hazardous exhaust system where the general health hazard posed by a single exposure to a vapor, gas, fume, mist or dust is classified with a health-hazard rating of 4 in accordance with NFPA 704. Health-hazard ratings are quantified on an intensifying numerical scale from zero to 4, with zero posing no health hazard beyond that of ordinary combustible material and 4 being too dangerous to health to expose even the best trained personnel. For a material with a health hazard rating of 4, a few whiffs of the gas could cause death, or the vapor or liquid could be fatal upon penetrating the full protective clothing of a fire fighter. Exposure may vary from a few seconds up to an hour. However, the intent here is to consider hazardous vapors, gases, fumes, mists or dusts that may arise out of an inherent property of the exposed material in the presence or absence of a catalyst; for example, a material (solid, liquid or gas) that when in the presence of a catalyst (water) gives off a gas too dangerous to the health of those exposed to it, such as fire fighters. In this instance, the normal full protective clothing and self-contained breathing apparatus (standard issue to the average fire department) will not provide adequate protection against inhalation or skin contact with these materials.

Item 3 requires a hazardous exhaust system where the concentration of a particular vapor, gas, fume, mist or dust with a health-hazard rating of 1, 2 or 3 is present in excess of 1 percent of the median lethal concentration of the substance for acute inhalation toxicity. Health-hazard rankings are based primarily on criteria detailed in the United Nations (UN) publication, *Recommendations on the Transport of Dangerous Goods*. The UN criterion for inhalation toxicity is based on the LC50 (lower concentration limit fatal to 50 percent of the test population) and saturated vapor concentration of the material. Furthermore, in addition to inhalation toxicity, the UN has established criteria for oral (exposure by ingestion) and dermal (exposure by skin) toxicity, as well as corrosivity; all of which have been considered in developing NFPA 704. Consult NFPA 704 for further information.

The exception for laboratories is to recognize the occasional presence of Class 4 materials in laboratories in such small concentrations to preclude the need for a hazardous exhaust system. This exception does not change the threshold limits for Items 1 and 3. If these limits are exceeded, then a hazardous exhaust system for the laboratory is required. This exception has the affect of making the requirements of Item 2 less stringent. Instead of having the threshold for a vapor, gas, fume, mist or dust classified as health-hazard rating of 4 at any concentration, it will now be treated as hazard ratings 1, 2 or 3 in Item 3. In effect, the exception simply exempts labs that have some chemicals with a health-hazard rating of 4 if the concentrations of such chemicals do not exceed 1 percent of the medium lethal concentration. Minute quantities used in laboratories do not necessarily drive the requirement for a hazardous exhaust system. This exception addresses the need for a hazardous exhaust system when hazardous quantities of Class 1, 2, 3 or 4 materials are in use in any lab environment at concentrations exceeding the prescribed limit.

[F] 510.2.1 Lumber yards and woodworking facilities. *Equipment* or machinery located inside buildings at lumber yards and woodworking facilities which generates or emits combustible dust shall be provided with an *approved* dust-collection and exhaust system installed in conformance with this section and the *International Fire Code*. *Equipment* and systems that are used to collect, process or convey combustible dusts shall be provided with an *approved* explosion-control system.

❖ The basic wood-processing activities related to lumber yards and woodworking facilities generate large volumes of wood particles suspended in the air to create a potentially explosive form of dust. This section requires that combustible dust be controlled by an approved dust-collection and exhaust system designed and installed in accordance with Sections 510 and 511. There are two basic designs of dust collection systems. One is a single-stage system that consists of a single dust collector in the form of a cyclone separator or a combination cyclone/baghouse unit. The other is a two-stage system that consists of a cyclone separator followed by a bag-type filter house.

An approved explosion control system, installed in accordance with Section 911 of the IFC, is also required for the equipment and systems that collect and convey the combustible dust.

[F] 510.2.2 Combustible fibers. *Equipment* or machinery within a building which generates or emits combustible fibers shall be provided with an *approved* dust-collecting and exhaust system. Such systems shall comply with this code and the *International Fire Code*.

❖ Machinery that causes readily combustible fibers, such as cotton, hay or wastepaper to become airborne during operation must be equipped with a dust-collection and exhaust system designed and installed in accordance with Section 511 of this code and Chapter 52 of the IFC. There are two basic designs of dust collection systems. One is a single-stage system that consists of a single dust collector in the form of a cyclone separator or a combination cyclone/baghouse unit. The other is a two-stage system that consists of a cyclone separator followed by a bag-type filter house.

Unlike woodworking facilities, buildings housing operations that cause combustible fibers to become airborne do not need an explosion control system. Although the fibers are highly combustible, the dust created by the particles of combustible fibers does not have the potentially explosive nature of the wood dust.

510.3 Design and operation. The design and operation of the exhaust system shall be such that flammable contaminants are diluted in noncontaminated air to maintain concentrations in the exhaust flow below 25 percent of the contaminant's lower flammability limit.

❖ This section contains performance-based criteria to quantify the amount of exhaust in relation to the control of the contaminant source. In other words, a direct solution is to restrict the concentration of all known contaminants of concern to a specified and acceptable level. That "level" is below 25 percent of the LFL for a particular contaminant. Where the concentration of a flammable contaminant in air is maintained below 25 percent of the lower flammability limit for that contaminant, there is insufficient fuel vapor in the gas phase to sustain homogeneous ignition. Managing the contaminant(s) in a diluted state means there is less chance of catastrophic flame propagation, ignition or explosion. The quantity or volume of exhaust required is source specific.

Because maintaining acceptable contaminant levels is not easily or exactly determined, the services of a registered design professional should be considered. In choosing a particular quantity or rate of exhaust, the design professional must consider the effects of pressure, temperature, local concentration, impurities and background velocities in terms of actual system performance.

Section 510.3 and Item 1 of Section 510.2 appear to be in conflict; however, each addresses a distinct criterion. Item 1 of Section 510.2 establishes a concentration that triggers the requirement for the hazardous exhaust system and Section 510.3 prescribes the operating parameters of the system.

510.4 Independent system. Hazardous exhaust systems shall be independent of other types of exhaust systems. Incompatible materials, as defined in the *International Fire Code*, shall not be exhausted through the same hazardous exhaust system. Hazardous exhaust systems shall not share common shafts with other duct systems, except where such systems are hazardous exhaust systems originating in the same fire area.

Exception: The provision of this section shall not apply to laboratory exhaust systems where all of the following conditions apply:

1. All of the hazardous exhaust ductwork and other laboratory exhaust within both the occupied space and the shafts are under negative pressure while in operation.

2. The hazardous exhaust ductwork manifolded together within the occupied space must originate within the same fire area.

3. Each control branch has a flow regulating device.

4. Perchloric acid hoods and connected exhaust shall be prohibited from manifolding.

5. Radioisotope hoods are equipped with filtration and/or carbon beds where required by the *registered design professional*.

6. Biological safety cabinets are filtered.

7. Provision is made for continuous maintenance of negative static pressure in the ductwork.

Contaminated air shall not be recirculated to occupiable areas. Air containing explosive or flammable vapors, fumes or dusts; flammable, highly toxic or toxic gases; or radioactive material shall be considered to be contaminated.

❖ To minimize the potential for spreading hazardous exhaust to other parts of a building, hazardous exhaust systems must not connect with any other exhaust system of the building. Without complete iso-

lation, the fire, health and explosion hazards inherent in hazardous exhaust systems cannot be confined only to the hazardous system and could jeopardize other systems and other parts of the building. The intent is to prohibit the combining of hazardous and nonhazardous exhaust systems. For example, an exhaust system conveying flammable vapors must not share ducts or exhaust equipment with a toilet room exhaust system. This does not, however, prohibit multiple hazardous exhaust ducts of the same type from connecting to a common trunk, riser or system. It also does not prohibit dissimilar hazardous systems from sharing common ducts if the dissimilar exhausts are compatible, as defined by the IFC, and the intermixing of their exhausts does not increase the overall hazard. When combined, dissimilar exhausts could react to form an exhaust mixture that is more combustible, flammable, explosive or otherwise more hazardous than the individual components.

Ductwork from two or more independent hazardous exhaust systems may be enclosed in the same fire-resistance-rated shaft only if the ducts originate in the same fire area. HVAC ductwork and nonhazardous exhaust ducts cannot be enclosed in the same shaft enclosure with the hazardous exhaust system and ducts, and hazardous exhaust systems originating in different fire areas cannot share common shaft enclosures. See the commentary to Section 602.1 for a description of fire areas.

The exception permits manifolding of laboratory exhaust ducts when all of the conditions are met. Laboratories are characterized by the use of a wide variety of chemicals in very small quantities and often for very short periods of time or on an infrequent basis. These are operations in which standard laboratory exhaust practices provide significant "in-duct" dilution that prevent in-duct incompatible material reactions and buildup of flammable vapors. The benefit of manifolding laboratory exhaust ducts is that with multiple hood systems manifolded together, the duct dilution factor is increased even further. The text highlights those few conditions, referenced in laboratory ventilation standards, that should preclude manifolding. The intent is to keep materials that are incompatible from being part of the manifolded system. The hazardous exhaust ductwork in the occupied space and in the shaft is required to operate under a negative pressure so as to minimize the possibility of any leakage into the occupied space or shaft from the duct. Item 7 specifically addresses the need for continuous maintenance of negative static pressure in the duct, but, does not indicate how this is to be accomplished. In the supporting documentation for the proposal that created Item 7, it was indicated that in a manifolded system with a common duct extending to the terminal, it is simple and economically feasible to include redundant fans and to have emergency power. This will also provide the benefit of being able to inspect and maintain the system without having to shut the system down. Having redundant fans and back-up power would appear to meet the intent of maintaining a continuos negative pressure in the duct system. Any hazardous exhaust ducts that are manifolded together within occupied spaces must originate in the same fire area, thereby further limiting any possible exposure to fire areas that do not contain laboratories.

Note that the code is silent on hazardous exhaust ducts that are manifolded together within the fire-resistance-rated shaft enclosure, but logic dictates that such ducts must also originate in the same fire area.

The final paragraph of this section recognizes that air contaminated with explosive or flammable vapors, fumes or dusts; flammable or toxic gases or radioactive material must be exhausted to the outdoors and, therefore, cannot be recirculated or transferred to any other space. Note that Section 510.1 requires hazardous exhaust to discharge to the outdoors.

510.5 Design. Systems for removal of vapors, gases and smoke shall be designed by the constant velocity or equal friction methods. Systems conveying particulate matter shall be designed employing the constant velocity method.

❖ The primary task of the duct designer is to design duct systems that will fulfill the required exhaust system function in a practical, economical and energy-conserving manner within the prescribed limits of available space, friction loss, velocity, sound level and leakage losses or gains. Whether the duct system is designed manually or by computer, exhaust systems conveying vapors, gases and smoke must be designed using either the constant velocity or the equal friction method. In the equal friction method, ducts are sized for a constant pressure loss per unit length. In the constant velocity method, ducts are sized for a constant velocity per unit length to maintain minimum contaminant transport velocities.

Minimum transport velocity is the velocity required to transport particulates without settling. The American Conference of Governmental Industrial Hygienists' (ACGIH) *Industrial Ventilation—Manual of Recommended Practice* lists some generally accepted transport velocities as a function of the nature of the contaminant. Duct velocities can be higher than the minimum transport velocities but should never be significantly lower.

When the equal friction or constant velocity method is used manually, the time to calculate duct sizes can be shortened by using duct calculators offered by various HVAC equipment manufacturers. Round ducts are preferred because they offer a more uniform air velocity to resist settling of material and can withstand the higher static pressures normally found in exhaust systems. For additional guidance on duct sizing using one of the methods above, consult a standard design handbook such as the American Society of Heating, Refrigerating, and Air-Conditioning Engineers' (ASHRAE) *Handbook of Fundamen-*

tals or the ACGIH *Industrial Ventilation—Manual of Recommended Practice.*

510.5.1 Balancing. Systems conveying explosive or radioactive materials shall be prebalanced by duct sizing. Other systems shall be balanced by duct sizing with balancing devices, such as dampers. Dampers provided to balance airflow shall be provided with securely fixed minimum-position blocking devices to prevent restricting flow below the required volume or velocity.

❖ General procedures for balancing systems are described in this section. Most exhaust systems have more than one hood. If the design pressures are not the same for merging parallel airstreams, the system adjusts to equalize pressure at the common point; however, the flow rates of the two merging airstreams will not necessarily be the same as designed. As a result, the hoods can fail to control the contaminant adequately, exposing occupants to potentially hazardous contaminant concentrations.

Two design methods are used to achieve the required design flow rates at all hoods and inlets. One method uses balancing devices such as blast gates or dampers to obtain design airflow at each hood or inlet. The other approach balances systems by adding resistance to ductwork sections without external aids (that is, changing duct size, selecting different fittings and increasing airflow). This self-balancing method is preferred, especially when the system handles abrasive materials. Where potentially explosive or radioactive materials are conveyed, flow balancing is critical and the self-balancing method is mandatory because contaminants could accumulate or settle at the balancing devices and because manual balancing is subject to human error and tampering.

Where balancing devices are installed in exhaust systems, the design must include a minimum stop or means for preventing restriction of flow below the required volume or velocity. It is obvious that modulating a balancing device to its fully closed position would restrict flow, thereby creating a potentially dangerous situation.

510.5.2 Emission control. The design of the system shall be such that the emissions are confined to the area in which they are generated by air currents, hoods or enclosures and shall be exhausted by a duct system to a safe location or treated by removing contaminants.

❖ Exhaust air systems are either general systems that remove air from large spaces or local systems that capture heat, vapors, gases, fumes, mists or dust at source-specific locations within a room or space. This section requires that hazardous exhaust systems be designed to prevent the spread of contaminants beyond the area of origin. Confinement is accomplished using controlled air currents or barriers such as hoods, booths and similar enclosures. For example, the air movement through a spray-painting or dipping room must create a flow pattern that effectively sweeps the entire cross-sectional area of that room. An enclosure such as a room or booth is typically necessary to contain the paint overspray and prevent the migration of vapors and fumes from the paint application area.

Additional information concerning procedures for evaluating and controlling contaminant levels can be found in ACGIH *Industrial Ventilation—Manual of Recommended Practice* or ACGIH *Threshold Limit Values for Chemical Substances in the Work Environment.*

510.5.3 Hoods required. Hoods or enclosures shall be used where contaminants originate in a limited area of a space. The design of the hood or enclosure shall be such that air currents created by the exhaust systems will capture the contaminants and transport them directly to the exhaust duct.

❖ When sources within the building generate hazardous contaminants (see commentary, Section 510.2), direct exhaust through hoods is more effective than control by general ventilation (dilution). This section requires the installation of an exhaust hood or enclosure at the location of greatest concentration to improve the capture ability of the exhaust system. Most often, this is at the source of contamination such as a particular piece of equipment or appliance or a particular process or operation such as evaporation; plating; container filling; welding; chute loading of conveyors; crushing; cool or hot shakeout processes; grinding; blasting or tumbling. The intent is to maximize capture efficiency by preventing contaminants from spreading beyond the immediate area of the source.

510.5.4 Contaminant capture and dilution. The velocity and circulation of air in work areas shall be such that contaminants are captured by an airstream at the area where the emissions are generated and conveyed into a product-conveying duct system. Contaminated air from work areas where hazardous contaminants are generated shall be diluted below the thresholds specified in Section 510.2 with air that does not contain other hazardous contaminants.

❖ The references to "contaminated air" and "hazardous contaminants" make clear that this section is dealing with mixtures of environmental air plus hazardous contaminants. The section does not apply, for instance, to simply objectionable odor contaminants or to gaseous mixtures that are part of the processes. Rather than addressing only the flammability hazard, this section also addresses the health hazard, to be consistent with Section 510.2, because it applies to (occupied) work areas. Dilution air may contain contaminants, as long as they are not "hazardous" contaminants.

The code recognizes that the minimum transport velocity required for the capture of large particles, such as mists or dusts differs from that required for contaminants, such as vapors, gases or fumes and must be considered in the design of exhaust systems to ensure that capture and direct transport to the exhaust duct system is achieved. Capture velocities

are air velocities at the point of contaminant generation upstream of the hood or inlet. The contaminant enters the airstream at the point of generation and is conducted along with the air into the hood and from there directly to the exhaust duct system.

This section also prescribes the maximum contaminant concentration allowed in occupant work areas where the contaminants are generated. At lower levels, the contaminant is considered diluted or innocuous.

510.5.5 Makeup air. *Makeup air* shall be provided at a rate approximately equal to the rate that air is exhausted by the hazardous exhaust system. *Makeup-air* intakes shall be located so as to avoid recirculation of contaminated air.

❖ Exhaust flow can occur only if air is constantly supplied to replace the air being exhausted. The air exhausted from a hazardous exhaust system must be replaced with air at the required exhaust flow rate. Hazardous exhaust systems are usually designed with the quantity of makeup air being slightly less than that exhausted, thereby creating a slight negative pressure that helps confine contaminants to the area of origin.

The introduction of makeup air is critical to the proper operation of all hazardous exhaust systems and the fuel-burning appliances that may be located in nearby areas (see commentary, Section 508.1). The mechanical makeup air source and the hazardous exhaust system should be electrically interlocked and controlled by a single start switch to make certain makeup air is supplied when the exhaust hood is in operation.

Makeup air intakes must be located at sufficient distances from exhaust outlets to prevent exhaust discharge from contaminating the makeup air.

510.5.6 Clearances. The minimum *clearance* between hoods and combustible construction shall be the *clearance* required by the duct system.

❖ Requirements for clearances between hoods and combustibles are emphasized because of the potential fire hazard posed where those clearances are not observed. Section 510.8.2 prescribes clearances for hazardous exhaust ducts. Greater clearances are required if specified by the duct or hood manufacturer.

Reduction of the required clearances to combustibles is allowed only where the combustibles are protected by one of the methods outlined in Section 308 (see commentary, Section 308).

510.5.7 Ducts. Hazardous exhaust duct systems shall extend directly to the exterior of the building and shall not extend into or through ducts and plenums.

❖ The intent of this section is to minimize the potential for spreading hazardous exhaust to other parts of the building as a result of duct leakage or failure. In the event of a duct fire or explosion, other areas of the building could be jeopardized. The intent is to require routing of hazardous exhaust ducts to the outdoors as directly as practicable, thereby avoiding unnecessary duct lengths and travel through other spaces. In all cases, ducts conveying hazardous exhaust must not extend into or through other ducts or plenum spaces.

510.6 Penetrations. Penetrations of structural elements by a hazardous exhaust system shall conform to Sections 510.6.1 through 510.6.4.

 Exception: Duct penetrations within H-5 occupancies as allowed by the *International Building Code*.

❖ This section contains requirements for the design of duct penetrations through structural elements of a building, including penetrations through floors and walls, and prohibitions for fire walls. The exception refers to Sections 415.10.1.4 and 415.10.1.5 of the IBC. Section 415.10.1.4 allows unprotected openings between floors of a Group H-5 fabrication if the interconnected levels consist of the fabrication area and a mechanical room used solely for mechanical equipment related to the operations on the fabrication floor. Section 415.10.1.5 allows penetrations through no more than two floors of the fabrication area by mechanical duct and piping without requiring a shaft.

510.6.1 Fire dampers and smoke dampers. Fire dampers and smoke dampers are prohibited in hazardous exhaust ducts.

❖ Fire and smoke dampers must not be installed within hazardous exhaust systems because a closed damper will prevent all or part of the exhaust system from functioning and could possibly create a dangerous condition in the room or space where the exhaust system originates. The purpose of the hazardous exhaust system is to dilute and remove the hazardous materials so that they are not a threat to the building occupants. The installation of fire or smoke dampers may adversely affect the operation of the exhaust system, thereby preventing the exhaust system from achieving its intended purpose. Additionally, the materials conveyed in hazardous exhaust ducts could damage, impair or obstruct the dampers, seriously affecting their performance. Hazardous exhaust systems are permitted to penetrate fire-resistance-rated assemblies, other than fire walls, and fire and smoke dampers are not required at those penetrations because protection is provided by the fire-resistance-rated enclosure required by Sections 510.6.2 and 510.6.3.

510.6.2 Floors. Hazardous exhaust systems that penetrate a floor/ceiling assembly shall be enclosed in a fire-resistance-rated shaft constructed in accordance with the *International Building Code*.

❖ To reduce the risk of spreading fire from the hazardous exhaust system to other parts of the building, the hazardous exhaust system must be enclosed in a fire-resistance-rated shaft enclosure from the point where the system penetrates a floor level to the termination outdoors. A minimum 1-hour fire-resistance-

rated shaft enclosure is required even though the floor/ceiling assembly itself may not be fire-resistance-rated. The requirements for constructing the shaft enclosure and for the minimum fire-resistance ratings are given in the IBC. In addition to reducing the risk associated with fire, the rated enclosures will also help protect the duct from physical damage and may provide some protection against the spread of hazardous exhaust in the event of an explosion or duct failure.

510.6.3 Wall assemblies. Hazardous exhaust duct systems that penetrate fire-resistance-rated wall assemblies shall be enclosed in fire-resistance-rated construction from the point of penetration to the outlet terminal, except where the interior of the duct is equipped with an approved automatic fire suppression system. Ducts shall be enclosed in accordance with the *International Building Code* requirements for shaft construction and such enclosure shall have a minimum fire-resistance-rating of not less than the highest fire-resistance-rated wall assembly penetrated.

❖ Hazardous exhaust ducts that penetrate a vertical fire-resistance-rated assembly (wall) must be enclosed in fire-resistance-rated construction, unless the hazardous exhaust duct is protected internally by an approved automatic fire suppression system. The enclosure must be continuous (both horizontally and vertically) from the first fire-resistance-rated assembly penetrated to the termination point of the exhaust. The entire enclosure must have a fire-resistance rating equivalent to the greatest fire rating of any assembly penetrated by the duct. Additionally, if the enclosure also penetrates a floor/ceiling assembly, the fire-resistance rating must not be less than that required by the IBC based on the number of stories connected.

 The enclosure requirements include vertical shafts and what could be viewed as "horizontal shafts" and can be summarized as follows:

- Hazardous exhaust ducts, whether protected by a fire suppression system or not, that penetrate any floor/ceiling assembly (fire-resistance rated or not), must be enclosed in a fire-resistance-rated shaft.

- Hazardous exhaust ducts without fire suppression system protection that penetrate a fire-resistance-rated wall assembly must be enclosed in fire-resistance-rated enclosures from the first penetration to the exhaust termination.

- Hazardous exhaust ducts having fire suppression system protection that penetrate a fire-resistance-rated assembly other than a floor/ceiling assembly need not be enclosed at that penetration.

 One of the purposes for the horizontal enclosure of hazardous exhaust ducts is to compensate for the lack of fire dampers in such systems.

510.6.4 Fire walls. Ducts shall not penetrate a fire wall.

❖ The IBC explains the differences between fire walls, fire barriers and other fire-resistance-rated assemblies. A fire wall creates separate buildings. Hazardous exhaust ducts are not allowed to penetrate fire walls because the penetration of a fire wall by a hazardous exhaust duct would introduce a potential hazard from one building into another.

510.7 Suppression required. Ducts shall be protected with an *approved* automatic fire suppression system installed in accordance with the *International Building Code*.

Exceptions:

1. An approved automatic fire suppression system shall not be required in ducts conveying materials, fumes, mists and vapors that are nonflammable and noncombustible under all conditions and at any concentrations.

2. Automatic fire suppression systems shall not be required in metallic and noncombustible, nonmetallic exhaust ducts in semiconductor fabrication facilities.

3. An *approved* automatic fire suppression system shall not be required in ducts where the largest cross-sectional diameter of the duct is less than 10 inches (254 mm).

4. For laboratories, as defined in Section 510.1, automatic fire protection systems shall not be required in laboratory hoods or exhaust systems

❖ To provide protection against the spread of fire within a hazardous exhaust system and to prevent a duct fire from involving the building, an automatic fire suppression system conforming to the IBC must be installed to protect the exhaust duct system. The fire suppression system need not be a sprinkler system because some materials conveyed in exhaust ducts may not be compatible with water. Other systems that may be more suitable, such as dry chemical and carbon dioxide suppression systems, are described in the IBC. The type of suppression system used is the choice of the designer; however, the design of any fire suppression system must conform to the requirements of the IBC.

Exception 1 recognizes that an automatic fire suppression system would be of little value for an exhaust system that conveys only materials, fumes, vapors and gases that are nonflammable and noncombustible under all conditions and at any concentrations. Remember that Section 510.3 requires hazardous exhaust systems to maintain an exhaust flow mixture that is well below the LFL for the flammable contaminant; thus, the exception can be applicable only to contaminants that are essentially nonflammable and noncombustible. The fire suppression requirement is intended to apply to exhaust systems having an actual fire hazard.

Exception 2 is for metallic and noncombustible, nonmetallic ducts that are located in semiconductor fabrication facilities. This exception is similar to Section 2703.10.4.1 in the IFC which does not require ducts to have a suppression system in H-5 facilities where the ducts are conveying nonflammable gases, vapors or fumes.

Exception 3 recognizes the reduced hazard associated with smaller ducts and the impracticality of installing fire suppression systems in those ducts.

Exception 4 recognizes that a fire suppression system within laboratory hood exhaust ducts increases the potential for workers to be exposed to chemicals. In addition, an uncontrolled flow of water from sprinklers in a chemical hood duct could transport the chemicals out of the hood and duct system and create a very dangerous situation, especially if the hood system contains water-reactive chemicals. The dilution that occurs in laboratory exhaust ducts, and the further dilution that occurs when exhaust ducts are manifolded together, reduce the need for suppression. This, along with the increased potential for worker exposure, provides the necessary justification to omit the automatic fire suppression system. The language used in this section for the laboratory exception is similar to the language used in NFPA 45.

510.8 Duct construction. Ducts used to convey hazardous exhaust shall be constructed of *approved* G90 galvanized sheet steel, with a minimum nominal thickness as specified in Table 510.8. Nonmetallic ducts used in systems exhausting nonflammable corrosive fumes or vapors shall be *listed* and *labeled*.

Nonmetallic ducts shall have a flame spread index of 25 or less and a smoke-developed index of 50 or less, when tested in accordance with ASTM E 84 or UL 723. Ducts shall be *approved* for installation in such an exhaust system. Where the products being exhausted are detrimental to the duct material, the ducts shall be constructed of alternative materials that are compatible with the exhaust.

❖ To provide structural strength and a degree of corrosion resistance, ducts in hazardous exhaust systems must be constructed of G90 galvanized sheet steel. The amount of galvanizing (zinc or zinc-alloy coating) is expressed in terms of weight per unit area, and G90 is an ASTM inch-pound (mm-kg) designation for the weight of coating on a given area of sheet metal. Sheet steel is a noncombustible material, which provides protection against the spread of fire and smoke outside of the duct enclosure. The thickness of the exhaust duct is regulated by Table 510.8. Part of the justification for allowing the omission of fire dampers in hazardous exhaust ducts is the fact that steel ducts possess an inherent resistance to the spread of fire.

Galvanized steel is not an appropriate duct material for all substances handled in the broad category of hazardous exhaust systems. A duct material compatible with the exhaust must be selected and factors such as corrosion resistance, abrasion resistance, chemical resistance and operating temperatures must be taken into account. For example, stainless steel or steel coated with polyvinyl chloride (PVC) might be used in systems where the exhaust is corrosive to G90 galvanized steel. Nonmetallic ducts used to convey nonflammable corrosive fumes or vapors must be listed and labeled for the application. Note that this section implies that nonmetallic ducts are permitted only for conveying nonflammable corrosive contaminants.

To reduce the possible contribution to the spread of fire and smoke, these materials must be tested to ASTM E 84 or UL 723. These materials are limited to a maximum flame spread index of 25 and a smoke-developed index of 50. Information on industrial exhaust systems can be obtained from NFPA 91, SMACNA's *Accepted Industry Practice for Industrial Duct Construction*, ASHRAE's *Handbook of Applications* and ASHRAE's *HVAC Systems and Equipment Handbook*.

TABLE 510.8
MINIMUM DUCT THICKNESS

DIAMETER OF DUCT OR MAXIMUM SIDE DIMENSION	MINIMUM NOMINAL THICKNESS		
	Nonabrasive materials	Nonabrasive/ Abrasive materials	Abrasive materials
0-8 inches	0.028 inch (No. 24 gage)	0.034 inch (No. 22 gage)	0.040 inch (No. 20 gage)
9-18 inches	0.034 inch (No. 22 gage)	0.040 inch (No. 20 gage)	0.052 inch (No. 18 gage)
19-30 inches	0.040 inch (No. 20 gage)	0.052 inch (No. 18 gage)	0.064 inch (No. 16 gage)
Over 30 inches	0.052 inch (No. 18 gage)	0.064 inch (No. 16 gage)	0.079 inch (No. 14 gage)

For SI: 1 inch = 25.4 mm.

❖ Table 510.8 establishes the minimum material thickness for hazardous exhaust duct construction. The material referred to in Table 510.8 is galvanized sheet steel, as prescribed in Section 510.8. However, alternative materials can be approved for use in accordance with Section 105.2 if the code official has evaluated the alternative material for its compatibility with the exhausted material and its suitability for the application. Specifically, the alternative duct material must demonstrate its equivalence to the structural strength, fire resistance, combustibility, corrosion resistance, friction loss and vapor permeability of galvanized steel. One commonly used alternative material is stainless steel.

Table 510.8 shows the minimum duct thickness based on the size of the duct and whether the materials being exhausted are abrasive, nonabrasive or both. The column designated "nonabrasive/abrasive materials" indicates that the duct will be used to exhaust materials that are somewhat abrasive or that are mixtures of nonabrasive and abrasive components. Abrasive particles can erode the duct material;

therefore, the required minimum duct thicknesses for exhaust systems handling abrasive materials are greater than required for ducts handling nonabrasive materials.

An example that demonstrates the use of Table 510.8 would be to determine the minimum thickness of a 10-inch-diameter (254 mm) round duct used to exhaust abrasive materials. The minimum thickness required for the duct is 0.052 inch (1.32 mm) (No. 18 gage).

510.8.1 Duct joints. Ducts shall be made tight with lap joints having a minimum lap of 1 inch (25 mm). Joints used in ANSI/SMACNA Round Industrial Duct Construction Standards and ANSI/SMACNA Rectangular Industrial Duct Construction Standards are also acceptable.

❖ Because of the hazardous nature of the duct contents, prevention of duct leakage is very important. Poor joints in positive-pressure ducts will cause the material being conveyed to leak from the duct. In the case of negative pressure ducts, poor joints allow infiltration, thereby reducing the level of performance of the overall system. Leakage of any kind can be a fire, explosion or health hazard. Duct joints should be lapped so that the male end points in the direction of flow (see Commentary Figure 510.8.1). This orientation will reduce turbulence and friction loss and will help prevent the accumulation of solids under the ledges created by the joints. The types of joints permitted in either of these referenced standards have been used in the industrial exhaust and conveyance systems for years and have provided acceptable alternatives to lap joints.

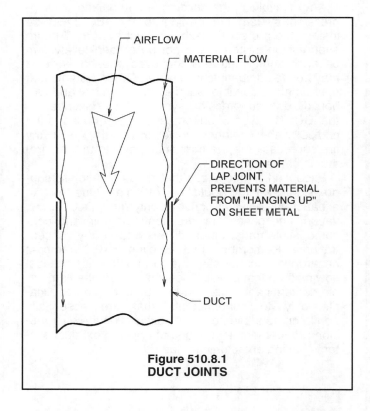

Figure 510.8.1
DUCT JOINTS

510.8.2 Clearance to combustibles. Ducts shall have a *clearance* to combustibles in accordance with Table 510.8.2. Exhaust gases having temperatures in excess of 600°F (316°C) shall be exhausted to a *chimney* in accordance with Section 511.2.

❖ To protect the building and any adjacent buildings from potential fire hazards, high-temperature exhaust [600°F (316°C)] must discharge to the atmosphere through a metal chimney (smokestack) (see commentary, Section 511.2).

TABLE 510.8.2
CLEARANCE TO COMBUSTIBLES

TYPE OF EXHAUST OR TEMPERATURE OF EXHAUST (°F)	CLEARANCE TO COMBUSTIBLES (inches)
Less than 100	1
100-600	12
Flammable vapors	6

For SI: 1 inch = 25.4 mm, °C = [(°F)- 32]/1.8.

❖ Table 510.8.2 contains the minimum required clearance from the hazardous exhaust duct to combustible materials. The clearance is a function of the temperature of the exhausted material and the flammability of the material. If an exhaust falls into more than one category, the more stringent requirements apply. For example, if a hazardous exhaust system is exhausting flammable vapors at a temperature of 70°F (21°C), the minimum clearance to combustible materials must be 6 inches (152 mm) and not 1 inch (25 mm) as stated in the table for exhaust temperatures less than 100°F (38°C) (see commentary, Sections 510.3 and 510.7).

510.8.3 Explosion relief. Systems exhausting potentially explosive mixtures shall be protected with an *approved* explosion relief system or by an *approved* explosion prevention system designed and installed in accordance with NFPA 69. An explosion relief system shall be designed to minimize the structural and mechanical damage resulting from an explosion or deflagration within the exhaust system. An explosion prevention system shall be designed to prevent an explosion or deflagration from occurring.

❖ Hazardous exhaust systems that exhaust potentially explosive mixtures, such as flammable gases and combustible dusts in air, must be protected by an explosion relief system or explosion prevention system designed to comply with NFPA 69. Deflagration is defined as the propagation of a combustion zone at a velocity that is less than the speed of sound in the unreacted medium. The explosion relief system is used to automatically vent the combustion gases and pressures resulting from a deflagration within the hazardous exhaust system so that structural and mechanical damage is minimized.

During an explosion, a sudden release of high-pressure gas occurs and the energy is dissipated in the form of a shock wave. Ensuring that the exhaust system remains intact affords protection for the building and its occupants. Typical venting methods for hazardous exhaust system ducts include displaceable diaphragms, rupture discs, hinged covers and blowout panels or caps.

An explosion prevention system reduces the probability of a deflagration within the hazardous exhaust system by reducing the concentration of the oxidants or combustibles involved. Oxidant concentration is reduced by supplying enough purge gas to dilute the mixture, which creates an oxidant deficiency. Combustible concentrations are reduced by dilution or recirculation of the combustible atmosphere through a catalytic oxidation unit. Significant instrumentation and control interaction is required in explosion prevention systems to make certain that the desired concentration reductions are achieved.

An explosive mixture consists of a fuel in the proper concentration and an oxidant in sufficient quantity to support combustion of the fuel. The fuel may be combustible liquids and dusts, flammable gases, mists and vapors or mixtures of these or other flammable/combustible substances. The oxidant in a hazardous exhaust system is normally, although not always, the oxygen in the air that serves as the transport medium for the substances being exhausted. This section addresses exhaust that has a potential for explosion.

Many factors, such as fuel and oxidant concentrations, flammability limitations, particle size, the presence of moisture and inert material, determine the explosive characteristics of exhaust mixtures. This information may be obtained from a number of sources, including the manufacturer of the material, the United States Department of Transportation (USDOT) and chemical handbooks.

510.9 Supports. Ducts shall be supported at intervals not exceeding 10 feet (3048 mm). Supports shall be constructed of noncombustible material.

❖ Proper support of the duct system requires designing for the weight of the duct and its contents, as well as for dynamic loads, such as those created by exhaust velocity and vibrations. Proper support is necessary to maintain proper alignment of the duct system and to prevent excessive stress on ducts and duct joints. A sagging duct will increase the internal resistance to airflow, reduce the efficiency of the system and cause the accumulation of exhaust products at low points. Duct support spacing is limited to not more than 10 feet (3048 mm) between supports because commonly used duct materials will sag at larger spacing intervals. Supports must be constructed of "Noncombustible materials," as defined in Chapter 2. The use of combustible supports could allow the duct system to fail in the event of a fire in the building, possibly contributing to the fire hazard. Also, combustible supports are obviously not compatible with high-temperature exhaust ducts.

Some duct configurations and duct materials require smaller spacing intervals. The SMACNA publication *HVAC Duct Construction Standards—Metal and Flexible* contains information for designing support systems for ducts. Additionally, the IBC may require designing the duct support system to be designed to resist earthquake loads, depending on the location and use of the building and the duct sizes and duct support configurations.

SECTION 511
DUST, STOCK AND REFUSE CONVEYING SYSTEMS

511.1 Dust, stock and refuse conveying systems. Dust, stock and refuse conveying systems shall comply with the provisions of Section 510 and Sections 511.1.1 through 511.2.

❖ Sections 511.1.1 through 511.2 contain specific fire safety requirements for mechanical equipment, ducts and piping used in hazardous exhaust systems. Dust, stock and refuse conveying systems are hazardous exhaust systems as addressed in Section 510. The requirements of this section are to be applied in conjunction with the provisions of Section 510.

511.1.1 Collectors and separators. Collectors and separators involving such systems as centrifugal separators, bag filter systems and similar devices, and associated supports shall be constructed of noncombustible materials and shall be located on the exterior of the building or structure. A collector or separator shall not be located nearer than 10 feet (3048 mm) to combustible construction or to an unprotected wall or floor opening, unless the collector is provided with a metal vent pipe that extends above the highest part of any roof with a distance of 30 feet (9144 mm).

Exceptions:

1. Collectors such as "Point of Use" collectors, close extraction weld fume collectors, spray finishing booths, stationary grinding tables, sanding booths, and integrated or machine-mounted collectors shall be permitted to be installed indoors provided the installation is in accordance with the *International Fire Code* and NFPA 70.

2. Collectors in independent exhaust systems handling combustible dusts shall be permitted to be installed indoors provided that such collectors are installed in compliance with the *International Fire Code* and NFPA 70.

❖ Dust and particle collectors and separators such as centrifugal separators and bag filter systems are used to separate solids from exhaust air in dust and waste-product conveyor systems serving manufacturing processes. These accumulated solids are highly combustible materials that represent a fire and explosion hazard. Because of the risk of fire or explosion and the subsequent threat to the building(s), collection equipment must be constructed of noncombustible materials and must be located outdoors. The support structures for the equipment must also be noncombustible to help prevent the spread of fire beyond the equipment and to help maintain the ability of the equipment to contain a fire.

As an additional precaution, the equipment must be located at least 10 feet (3048 mm) from any combustible construction or any unprotected opening in a building. Opening protection for windows, doors, vents and other openings is addressed in the IBC. The 10-foot (3048 mm) separation is not required if the equipment discharge vent pipe is metallic and extends above any part of a building within a 30-foot (9144 mm) horizontal distance from the discharge vent pipe. A vertical discharge vent, as described here, would act as a chimney and direct the fire and hot gases upward away from the building.

Exception 1 recognizes that in some buildings, locating the collector or separator outdoors results in having to use larger capacity systems and motors than would ordinarily be necessary, as well as installing long runs of duct that often must pass through fire-resistance-rated construction. This can result in transporting hazardous byproducts throughout different parts of a building, sometimes through other occupancies, in order to reach the exterior of the building. These conditions are neither ideal or safe, nor are they always necessary. This exception is necessary in order to avoid conflicts with the IFC and to apply the appropriate explosion protection standards for specific types of combustible dust-producing operations.

Exception 2 allows combustible dust collectors in independent exhaust systems to be installed indoors.

The reasoning for Exception 2 is the same as the reasoning stated above for Exception 1. This is consistent with the IFC and Exception 1.

511.1.2 Discharge pipe. Discharge piping shall conform to the requirements for ducts, including clearances required for high-heat appliances, as contained in this code. A delivery pipe from a cyclone collector shall not convey refuse directly into the firebox of a boiler, furnace, dutch oven, refuse burner, incinerator or other *appliance*.

❖ This section addresses the discharge pipes or ducts that connect to blowers, fans, collectors and separators used in dust, stock and refuse conveyor systems, which are hazardous exhaust systems that must comply with the provisions of Section 510. The required clearance to combustibles for these pipes or ducts must be as specified for high-heat appliances.

The definition of a "High-heat appliance" and the provisions of Section 510.8.2 make clear that the required clearances to combustibles must be determined by the system designer and approved by the code official. Although the system's normal operating temperature may be lower than that of high-heat appliances (because the discharge pipe or duct could be subjected to an internal fire), the clearances must be based on the temperatures associated with such fires. NFPA 91 contains guidance for duct clearances to combustibles. Because of the possibility of igniting the combustibles within the exhaust system, a centrifugal (cyclone) collector is prohibited from discharging collected waste products directly into the firebox of solid-fuel-burning equipment.

511.1.3 Conveying systems exhaust discharge. An exhaust system shall discharge to the outside of the building either directly by flue or indirectly through the bin or vault into which the system discharges except where the contaminants have been removed. Exhaust system discharge shall be permitted to be recirculated provided that the solid particulate has been removed at a minimum efficiency of 99.9 percent at 10 microns (10.01 mm), vapor concentrations are less than 25 percent of the LFL, and *approved equipment* is used to monitor the vapor concentration.

❖ Because of the potential fire and explosion hazard, dust, stock and refuse conveyor systems must discharge the solids and the transport medium (air) flow to a location on the exterior of the building. Typically, these systems discharge to an outdoor separator or collector system that separates the solids from the exhaust flow and deposits them in a bin, vault or hopper.

If equipment is provided that can reduce the contaminants to an acceptable level and the vapor concentrations are monitored, the exhaust system discharge is permitted to be recirculated. This will allow the designer more flexibility to employ energy recovery methods to save energy and reduce costs. Note that if the contaminants have been removed in accordance with the stated criteria, then the prohibition in Section 514.2 on using dust, stock and refuse systems for energy recovery ventilation systems is no longer applicable.

511.1.4 Spark protection. The outlet of an open-air exhaust terminal shall be protected with an *approved* metal or other noncombustible screen to prevent the entry of sparks.

❖ The discharge termination must be protected by an approved means that will prevent the entry of sparks into the exhaust system. The intent is to help prevent a fire or explosion in the exhaust system by eliminating a possible ignition source.

511.1.5 Explosion relief vents. A safety or explosion relief vent shall be provided on all systems that convey combustible refuse or stock of an explosive nature, in accordance with the requirements of the *International Building Code.*

❖ An explosion relief vent designed in accordance with the IBC is required to prevent or minimize the structural or mechanical damage that could result from an explosion or deflagration within the exhaust system. A properly designed explosion relief system will normally require multiple vents located at various points in the exhaust system. Typical relief vents consist of blowout panels or diaphragms, rupture discs, hinged panels and closure assemblies held in position by friction, magnets, springs or special retainers. NFPA 68 contains guidance for the design of explosion-venting systems.

511.1.5.1 Screens. Where a screen is installed in a safety relief vent, the screen shall be attached so as to permit ready release under the explosion pressure.

❖ Where screens are installed to protect relief vent assemblies or the opening of a relief vent duct, the screen must not be allowed to interfere with the proper release and operation of the explosion vent. Protective screens, as required by the IBC, are used to prevent damage to a relief device or to prevent the entry of debris into a relief vent duct.

511.1.5.2 Hoods. The relief vent shall be provided with an *approved* noncombustible cowl or hood, or with a counterbalanced relief valve or cover arranged to prevent the escape of hazardous materials, gases or liquids.

❖ The intent of this section is to prevent the materials being exhausted or conveyed in the system from escaping through the required relief vent or vents. Relief vents are designed to remain closed under normal operating conditions so as not to interfere with the exhaust/conveyor system's operation. Relief vents must discharge directly to the outdoors or through a relief vent duct that terminates at the building exterior. In the event of an explosion in the exhaust system, the relief vent or vents will discharge hot gases, flames or debris that must be directed to a safe location outside of the building.

511.2 Exhaust outlets. Outlets for exhaust that exceed 600°F (315°C) shall be designed as a *chimney* in accordance with Table 511.2.

❖ Dust, stock and refuse conveyor systems are hazardous exhaust systems as addressed in Section 510. Although the system's normal operating temperature may be lower than that of high-heat appliances, the discharge pipe or duct could be subjected to an internal fire. For this reason, Table 511.2 includes requirements for the materials, liners, termination height and proper clearances to combustibles for the exhaust outlet.

TABLE 511.2. See page 5-104.

❖ This table establishes the requirements for a single-wall metal chimney serving a dust, stock or refuse conveyor system with exhaust exceeding 600°F (316°C) and includes requirements for chimney materials, liners, termination height and proper clearances to combustibles. This table does not preclude the use of a factory-built chimney system listed and labeled for the application.

SECTION 512
SUBSLAB SOIL EXHAUST SYSTEMS

512.1 General. When a subslab soil exhaust system is provided, the duct shall conform to the requirements of this section.

❖ A subslab soil exhaust system is often associated with radon mitigation. Yet, evidence from some studies suggests that contaminants other than radon may enter buildings from the surrounding soil, although the extent of such occurrences is unknown. For example, methane migrating away from landfills through unsaturated soils has reached explosive levels in nearby buildings. In other instances, potentially toxic and carcinogenic volatile organic compounds present in soil have been transported into buildings.

Radon is a colorless, odorless, radioactive gas found in various concentrations in all soils, ground water, stone, concrete and indeed virtually anything that comes out of the ground. Studies have shown it to be harmful when inhaled in high concentrations.

Based on early studies, the Environmental Protection Agency (EPA) estimated that radon causes between 5,000 and 20,000 lung cancer deaths each year; however, the health risk associated with the low levels of radon gas typically found within buildings is the subject of scientific debate. In fact, most recent studies show no correlation between the levels of radon typically encountered in residential construction and increased lung cancer rates. Therefore, the reduction of radon or any other contaminant transported by soil gas is not required by the code. This section is added to provide material guidance where a soil exhaust system is installed.

The concept is simply to install a vent system to serve as an alternative path (of least resistance) for any soil gases to reach the atmosphere above the roof. A recent study has found that under certain circumstances, improper operation of the HVAC system could create negative pressure within a structure, resulting in increased soil-gas migration into the structure. The study also found that when ceiling return air plenums are intersected by masonry block walls that penetrate the floor slab, the negative pressure in the plenum can induce soil-gas migration through the core of the block wall.

Specific guidance for proper system performance, minimum ventilation rates, floor area served per vent, sealing methods, particulate arrestors and fan selection is beyond the current scope of the code. Appendix F of the IRC contains radon control information. Additional design criteria have been developed by the EPA and others for use as guidance by the code official. They are: *Washington State Energy Code—*

Builder's Field Guide; EPA—Citizens' Guide to Radon (EPA 86-004); EPA and Centers for Disease Control—The Influences of HVAC Design and Operation on Radon Mitigation in Existing School Buildings; and ASHRAE—IAQ The Human Equation: Health and Comfort. The basic standards for permissible concentrations of radon and a variety of other soil gases in the air are those of the National Committee on Radiation Protection, published by the National Bureau of Standards (NBS) as *Handbook No. 69.* Industries operating under licenses from the U.S. Nuclear Regulatory Commission (NRC) or state licensing agencies must meet the requirements of DOE 10 CFR, Part 20. Some states have additional requirements.

512.2 Materials. Subslab soil exhaust system duct material shall be air duct material *listed* and *labeled* to the requirements of UL 181 for Class 0 air ducts, or any of the following piping materials that comply with the *International Plumbing Code* as building sanitary drainage and vent pipe: cast iron; galvanized steel; brass or copper pipe; copper tube of a weight not less than that of copper drainage tube, Type DWV; and plastic piping.

❖ The subslab exhaust vent is a pipe or duct that keeps unwanted soil gases out of the structure by conducting them directly to the outdoors. The duct materials allowed include ducts that have been shown to meet the requirements of UL 181 for a Class 0 air duct and are so labeled. Although UL 181 is used primarily to evaluate nonmetallic duct materials, the use of the standard for evaluating metallic ductwork is not precluded by its scope. Class 0 indicates a flame spread index of zero and a smoke-developed index of zero when tested to ASTM E 84. UL 181 tests samples of the duct to determine fire performance characteristics, corrosion and erosion resistance, leakage resistance, mold growth and humidity resistance and structural integrity. Air ducts that conform to the requirements of UL 181 are identified by the manu-

TABLE 511.2
CONSTRUCTION, CLEARANCE AND TERMINATION REQUIREMENTS FOR SINGLE-WALL METAL CHIMNEYS

CHIMNEYS SERVING	MINIMUM THICKNESS		TERMINATION				CLEARANCE			
	Walls (inch)	Lining	Above roof opening (feet)	Above any part of building within (feet)			Combustible construction (inches)		Noncombustible construction	
				10	25	50	Interior inst.	Exterior inst.	Interior inst.	Exterior inst.
High-heat appliances (Over 2,000°F)[a]	0.127 (No. 10 MSG)	4¹/₂″ laid on 4¹/₂″ bed	20	—	—	20	See Note c			
Low-heat appliances (1,000°F normal operation)	0.127 (No. 10 MSG)	none	3	2	—	—	18	6	Up to 18″ diameter, 2″ Over 18″ diameter, 4″	
Medium-heat appliances (2,000°F maximum)[b]	0.127 (No. 10 MSG)	Up to 18″ dia.—2¹/₂″ Over 18″—4¹/₂″ On 4¹/₂″ bed	10	—	10	—	36	24		

For SI: 1 inch = 25.4 mm, 1 foot = 304.8 mm, °C = [(°F)-32]/1.8.

a. Lining shall extend from bottom to top of outlet.

b. Lining shall extend from 24 inches below connector to 24 feet above.

c. Clearance shall be as specified by the design engineer and shall have sufficient clearance from buildings and structures to avoid overheating combustible materials (maximum 160°F).

facturer's or vendor's name, the rated velocity and the rated negative and positive pressures. Additionally, most of the materials that comply with the *International Plumbing Code®* (IPC®) as building sanitary drainage and vent pipe (DWV) may be used.

A subslab exhaust system typically consists of continuous sealed piping that runs from beneath the ground cover to a point outside the building. The vent installation should provide a location for the possible future installation of an in-line fan.

512.3 Grade. Exhaust system ducts shall not be trapped and shall have a minimum slope of one-eighth unit vertical in 12 units horizontal (1-percent slope).

❖ During the summer months (typically June through September), soil gas may contain high amounts of moisture that may condense on cool pipe walls. A minimum slope is required and traps are prohibited in an effort to allow this condensation to return to the soil while maintaining an open path for the soil gas to escape above the roof.

512.4 Termination. Subslab soil exhaust system ducts shall extend through the roof and terminate at least 6 inches (152 mm) above the roof and at least 10 feet (3048 mm) from any operable openings or air intake.

❖ To allow for the proper dispersion of soil gases into the atmosphere and minimize the exposure of the building occupants, the subslab exhaust duct must terminate through the roof to the building exterior at a minimum distance from openings through which radon gas could enter the building.

512.5 Identification. Subslab soil exhaust ducts shall be permanently identified within each floor level by means of a tag, stencil or other *approved* marking.

❖ In an effort to avoid mistaking the subslab exhaust duct for a component of the building's plumbing or mechanical system(s), the radon duct must be permanently identified as such at each floor level. Although the means for permanent marking (tag, stencil, label, stamp, sticker, etc.) is subject to the approval of the code official, the phraseology "radon vent" is most often used to comply with the identification provisions of Section 512.5.

SECTION 513
SMOKE CONTROL SYSTEMS

[F] 513.1 Scope and purpose. This section applies to mechanical and passive smoke control systems that are required by the *International Building Code* or the *International Fire Code*. The purpose of this section is to establish minimum requirements for the design, installation and acceptance testing of smoke control systems that are intended to provide a tenable environment for the evacuation or relocation of occupants. These provisions are not intended for the preservation of contents, the timely restoration of operations, or for assistance in fire suppression or overhaul activities. Smoke control systems regulated by this section serve a different purpose than the smoke- and heat-venting provisions found in Section 910 of the *International Building Code* or the *International Fire Code*.

❖ This section is clarifying the intent of smoke control provisions, which is to provide a tenable environment to occupants during evacuation and relocation and not to protect the contents, enable timely restoration of operations or facilitate fire suppression and overhaul activities. There are provisions for high-rise buildings in Section 403.4.7 of the IBC that are focused upon the removal of smoke for post fire and overhaul operations, which is very different than the smoke control provisions in Section 513. Another element addressed in this section is that smoke control systems serve a different purpose than smoke and heat vents (see Section 910 of the IBC). This eliminates any confusion that smoke and heat vents can be used as a substitution for smoke control. Additionally, a clarification is provided to note that smoke control systems are not considered an exhaust system in accordance with Chapter 5 of the IMC. This is due to the fact that such systems are unique in their operation and are not necessarily designed to exhaust smoke but are focused upon tenability for occupants during egress. It should be noted that the smoke control provisions are duplicated in Section 909 of both the IBC and IFC.

These provisions only apply when smoke control is required by various sections of the IBC. The IBC requires smoke management within atrium spaces (see Section 404.5 of the IBC) and underground buildings (see Section 405.5 of the IBC). High-rise facilities require smokeproof exit enclosures in accordance with Sections 909.20 and 1019.1.8 (see also Section 403.5.4) of the IBC. Also, covered mall buildings that contain atriums that connect more than two stories require smoke control (see Section 402.10 of the IBC). Section 513 focuses primarily on mechanical smoke control systems, but there are many instances within the code where smoke is required to be managed in a passive way through the use of concepts such as smoke compartments. Smoke compartments are formed through the use of smoke barriers in accordance with Section 709 of the IBC. Smoke barriers can be used simply as a passive smoke management system or can be a design component of a mechanical smoke control system in accordance with Section 513. Some examples of occupancies requiring passive systems include hospitals, nursing homes and similar facilities (Group I-2 occupancies) and detention facilities (Group I-3 occupancies) (see Sections 407.5 and 408.6 of the IBC).

In some cases, mechanical smoke control in accordance with Section 513 is allowed as an option for compliance. More specifically, if a Group I-3 occupancy contains windowless areas of the facility, natural or mechanical smoke management is required (see Section 408.9 of the IBC).

In the last several years, smoke control provisions have become more complex. The reason is related to the fact that smoke is a complex problem, while a generic solution of six air changes has repeatedly and scientifically been shown to be inadequate. Six air changes per hour does not take into account factors such as buoyancy; expansion of gases; wind; the geometry of the space and of communicating spaces; the dynamics of the fire, including heat release rate; the production and distribution of smoke and the interaction of the building systems.

Smoke control systems can be either passive or active. Active systems are sometimes referred to as mechanical. Passive smoke control systems take advantage of smoke barriers surrounding the zone in which the fire event occurs or high bay areas that act as reservoirs to control the movement of smoke to other areas of the building. Active systems utilize pressure differences to contain smoke within the event zone or exhaust flow rates sufficient to slow the descent of the upper-level smoke accumulation to some predetermined position above necessary exit paths through the event zone. On rare occasions, there is also a possibility of controlling the movement of smoke horizontally by opposed airflow, but this method requires a specific architectural geometry to function properly that does not create an even greater hazard.

Essentially, there are three methods of mechanical or active smoke control that can be used separately or in combination within a design: pressurization, exhaust and, in rare and very special circumstances, opposed airflow.

Of course, all of these active approaches can be used in combination with the passive method. Typically, the mechanical pressurization method is used in high-rise buildings when pressurizing stairways and for zoned smoke control. Pressurization is not practical in large open spaces such as atriums or malls, since it is difficult to develop the required pressure differences due to the large volume of the space.

The exhaust method is typically used in large open spaces such as atriums and malls. As noted, the pressurization method would not be practical within large spaces. The opposed airflow method, which basically uses a velocity of air horizontally to slow the movement of smoke, is typically applied in combination with either a pressurization method or exhaust method within hallways or openings into atriums and malls.

The application of each of these methods will be dependent on the specifics of the building design. Smoke control within a building is fundamentally an architecturally driven problem. Different architectural geometries first dictate the need or lack thereof for smoke control, and then define the bounds of available solutions to the problem.

[F] 513.2 General design requirements. Buildings, structures, or parts thereof required by the *International Building Code* or the *International Fire Code* to have a smoke control system or systems shall have such systems designed in accor-

dance with the applicable requirements of Section 909 of the *International Building Code* and the generally accepted and well-established principles of engineering relevant to the design. The *construction documents* shall include sufficient information and detail to describe adequately the elements of the design necessary for the proper implementation of the smoke control systems. These documents shall be accompanied with sufficient information and analysis to demonstrate compliance with these provisions.

❖ This section simply states that when smoke control systems are required by the code, the design is required to be in accordance with the provisions of this section. As noted in the commentary to Section 513.1, there are instances within the code that have smoke management systems that are purely passive in nature and do not reference Section 513.

This section stresses that designs in accordance with this section need to follow "generally accepted and well-established principles of engineering relevant to the design," essentially requiring a certain level of qualifications in the applicable areas of engineering to prepare such designs. The primary engineering disciplines tend to be fire engineering and mechanical engineering. It should be noted that each state in the U.S. typically requires minimum qualifications to undertake engineering design. Two important resources when designing smoke control systems are ICC's *A Guide to Smoke Control in the 2006 IBC* and American Society of Heating, Refrigerating and Air-Conditioning Engineers' (ASHRAE) Design of Smoke Management Systems. Additionally, Section 513.8 requires the use of NFPA 92B for the design of smoke control systems using the exhaust method. This standard has many relevant aspects beyond the design that are beneficial. In particular, Annex B provides resources in terms of determination of fire size for design. ICC's *A Guide to Smoke Control in the 2006 IBC* also provides guidance on design fires.

A key element covered in this section is the need for detailed and clear construction documents so that the system is installed correctly. In most complex designs, the key to success is appropriate communication to the contractors as to what needs to be installed. The more complex a design becomes, the more likely there is to be construction errors. Most smoke control systems are complex, which is why special inspections in accordance with Section 513.3 and Chapter 17 of the IBC are critical for smoke control systems. Additionally, in order for the design to be accepted, analyses and justifications need to be provided in enough detail to evaluate for compliance. Adequate documentation is critical to the commissioning, inspection, testing and maintenance of smoke control systems and significantly contributes to the overall reliability and effectiveness of such systems.

[F] 513.3 Special inspection and test requirements. In addition to the ordinary inspection and test requirements which buildings, structures and parts thereof are required to undergo, smoke control systems subject to the provisions of Section 909 of the *International Building Code* shall undergo special inspections and tests sufficient to verify the proper commissioning of the smoke control design in its final installed condition. The design submission accompanying the *construction documents* shall clearly detail procedures and methods to be used and the items subject to such inspections and tests. Such commissioning shall be in accordance with generally accepted engineering practice and, where possible, based on published standards for the particular testing involved. The special inspections and tests required by this section shall be conducted under the same terms as found in Section 1704 of the *International Building Code*.

❖ Due to the complexity and uniqueness of each design, special inspection and testing must be conducted. The designer needs to provide specific recommendations for special inspection and testing within his or her documentation. In fact, the code specifies in Chapter 17 of the IBC that special inspection agencies for smoke control have expertise in fire protection engineering, mechanical engineering and certification as air balancers.

Since the designs are unique to each building, there probably will not be a generic approach available to inspect and test such systems. The designer can and should, however, use any available published standards or guides when developing the special inspection and testing requirements for that particular design. ICC's *A Guide to Smoke Control in the 2006 IBC®* provides some background on such inspections, Also, ASHRAE Guideline 5 is a good starting place, but only as a general outline. In addition, NFPA 92A and NFPA 92B also have extensive testing, documentation and maintenance requirements that may be a good resource. NFPA 92B is referenced in Section 513.8 for the design of smoke control systems using the exhaust method. Each system will require a unique commissioning plan that can be developed only after careful and thoughtful examination of the final design and all of its components and interrelationships. Generally, these provisions may be included in design standards or engineering guides.

[F] 513.4 Analysis. A rational analysis supporting the types of smoke control systems to be employed, their methods of operation, the systems supporting them and the methods of construction to be utilized shall accompany the submitted *construction documents* and shall include, but not be limited to, the items indicated in Sections 513.4.1 through 513.4.6.

❖ This section indicates that simply determining airflow, exhaust rates and pressures to maintain tenable conditions is not adequate. There are many factors that could alter the effectiveness of a smoke control system, including stack effect, temperature effect of fire, wind effect, heating, ventilating and air-conditioning (HVAC) system interaction and climate, as well as the placement, quantity of inlets/outlets and velocity of

supply and exhaust air. These factors are addressed in the sections that follow. Additionally, the duration of operation of any smoke control system is mandated at a minimum of 20 minutes or 1.5 times the egress time, whichever is less. The code cannot reasonably anticipate every conceivable building arrangement and condition the building may be subject to over its life and must depend on such factors being addressed through a rational analysis.

[F] 513.4.1 Stack effect. The system shall be designed such that the maximum probable normal or reverse stack effects will not adversely interfere with the system's capabilities. In determining the maximum probable stack effects, altitude, elevation, weather history and interior temperatures shall be used.

❖ Stack effect is the tendency for air to rise within a heated building when the temperature is colder on the exterior of the building. Reverse stack effect is the tendency for air to flow downward within a building when the interior is cooler than the exterior of the building. This air movement can affect the intended operation of a smoke control system. If stack effect is great enough, it may overcome the pressures determined during the design analyses and allow smoke to enter areas outside the zone of origin (see Commentary Figure 513.4.1).

[F] 513.4.2 Temperature effect of fire. Buoyancy and expansion caused by the design fire in accordance with Section 513.9 shall be analyzed. The system shall be designed such that these effects do not adversely interfere with its capabilities.

❖ This section requires that the design account for the effect temperature may have on the success of the system. When air or any gases are heated they will expand. This expansion makes the gases lighter and, therefore, more buoyant. The buoyancy of hot gases is important when the design is to exhaust such gases from a location in or close to the ceiling; therefore, if sprinklers are part of the design, as required by Section 513, the gases may be significantly cooler than an unsprinklered fire, making it more difficult to remove the smoke and alter the plume dynamics. The fact that air expands when heated needs to be accounted for in the design.

When using the pressurization method, the expansion of hot gases needs to be accounted for, since it will take a larger volume of air to create the necessary pressure differences to maintain the area of fire origin in negative pressure. The expansion of the gases has the effect of pushing the hot gases out of the area of fire origin. Since sprinklers will tend to cool the gases, the effect of expansion is lower. The pressure differences required in Section 513.6.1 are specifically based on a sprinklered building. If the building is nonsprinklered, higher pressure differences may be required. The minimum pressure difference for certain unsprinklered ceiling height buildings is as follows:

Ceiling Height (feet)	Minimum Pressure Difference (inch water gage)
9	0.10
15	0.14
21	0.18

For SI: 1 foot = 304.8 mm, 1 inch water gage = 250 pascal.

This is a very complex issue that needs to be part of the design analysis. It needs to address the type and reaction of the fire protection systems, ceiling heights and the size of the design fire.

[F] 513.4.3 Wind effect. The design shall consider the adverse effects of wind. Such consideration shall be consistent with the wind-loading provisions of the *International Building Code.*

❖ The effect of wind on a smoke control system within a building is very complex. It is generally known that wind exerts a load upon a building. The loads are looked at as windward (positive pressure) and leeward (negative pressure). The velocity of winds will vary based on the terrain and the height above grade; therefore, the height of the building and surrounding obstructions will have an effect on these velocities. These pressures alter the operation of fans, especially propeller fans, thus altering the pressure differences and airflow direction in the building. There is not an easy solution to dealing with these effects. In fact, little research has been done in this area. It should be noted that in larger buildings a wind study is normally undertaken for the structural design. The data from those studies can be used in the analysis of the effects on the pressures and airflow within the building with regard to the performance of the smoke control system.

[F] 513.4.4 HVAC systems. The design shall consider the effects of the heating, ventilating and air-conditioning (HVAC) systems on both smoke and fire transport. The analysis shall include all permutations of systems' status. The design shall consider the effects of fire on the HVAC systems.

❖ If not properly configured to shut down or included as part of the design, the HVAC system can alter the smoke control design. More specifically, if dampers are not provided between smoke zones within the HVAC system ducts, smoke could be transported from one zone to another. Additionally, if the HVAC system places more supply air than assumed for the smoke control system design, the velocity of the air may adversely affect the fire plume or a positive pressure may be created. Generally, an analysis of the smoke control design and the HVAC system in all potential modes should occur and be noted within the design documentation as well as incorporated into inspection, testing and maintenance procedures. This is critical as these systems need to be maintained and tested to help ensure that they operate and shut down systems as required.

[F] 513.4.5 Climate. The design shall consider the effects of low temperatures on systems, property and occupants. Air inlets and exhausts shall be located so as to prevent snow or ice blockage.

❖ This section is focused on properly protecting equipment from weather conditions that may affect the reliability of the design. For instance, extremely cold or hot air may damage critical equipment within the system when pulled directly from the outside. Some listings of duct smoke detectors are for specific temperature ranges; therefore, placing such detectors within areas exposed to extreme temperatures may void the listing. Also, the equipment and air inlets and outlets should be designed and located so

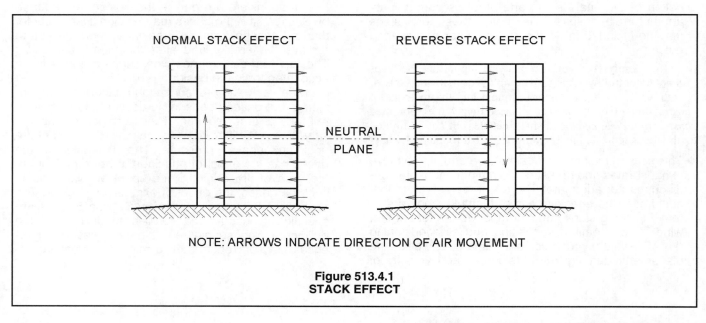

NOTE: ARROWS INDICATE DIRECTION OF AIR MOVEMENT

Figure 513.4.1
STACK EFFECT

EXHAUST SYSTEMS

as to not collect snow and ice that could block air from entering or exiting the building.

[F] 513.4.6 Duration of operation. All portions of active or passive smoke control systems shall be capable of continued operation after detection of the fire event for a period of not less than either 20 minutes or 1.5 times the calculated egress time, whichever is less.

❖ The intent of the smoke control provisions is to provide a tenable environment for occupants to either evacuate or relocate to a safe place. Evacuation and relocation activities include notifying occupants, possible investigation time for the occupants, decision time and the actual travel time. In order to achieve this goal, the code has established 20 minutes or 1.5 times the calculated egress time, whichever is less, as a minimum time for evacuation or relocation. Basically, this allows a designer to undertake an egress analysis to more closely determine the necessary time for egress. The code provides a safety factor of 1.5 times the egress time to account for uncertainty related to human behavior. It is stressed that the 20-minute duration as well as the calculated egress time, whichever approach is chosen, begins after the detection of the fire event and notification to the building occupants to evacuate has occurred, since occupants need to be alerted before evacuation can occur. The calculation of evacuation time needs to include delays with notification and the start of evacuation (i.e., pre-movement time, etc.). It is stressed that the code states 20 minutes or 1.5 times the egress time, whichever is less (i.e., 20 minutes is a maximum). Egress of occupants can be addressed through hand calculations or through the use of computerized egress models. Some of the more advanced models can address a variety of factors, including the building layout, different sizes of people, different movement speeds and different egress paths available. With these types of programs the actual time can be even more precisely calculated. Of course, it is cautioned that in many cases these models provide the optimal time for egress. The safety factor of 1.5 within the code is intended to address many of these uncertainties.

Note that this section applies to all types of smoke control designed in accordance with Section 513. Also, most smoke control systems will typically have the ability to run for longer than the 20-minute maximum as they are on standby power and may be able to continue to achieve the tenability goals. In some cases, even if the system runs longer than 20 minutes, the tenability may not be able to continue. It simply depends on the system design and the fire hazards within the building. System response as required in Section 909.17 of the IFC needs to be accounted for when determining the ability of the smoke control system to keep the smoke layer interface at the appropriate level (see commentary, Section 909.17 of the IFC).

[F] 513.5 Smoke barrier construction. Smoke barriers shall comply with the *International Building Code*. Smoke barriers shall be constructed and sealed to limit leakage areas exclusive of protected openings. The maximum allowable leakage area shall be the aggregate area calculated using the following leakage area ratios:

1. Walls: $A/A_w = 0.00100$
2. Interior exit stairways and ramps and exit passageways: $A/A_w = 0.00035$
3. Enclosed exit access stairways and ramps and all other shafts: $A/A_w = 0.00150$
4. Floors and roofs: $A/A_F = 0.00050$

where:

A = Total leakage area, square feet (m²).

A_F = Unit floor or roof area of barrier, square feet (m²).

A_w = Unit wall area of barrier, square feet (m²).

The leakage area ratios shown do not include openings due to doors, operable windows or similar gaps. These shall be included in calculating the total leakage area.

❖ Part of the strategy of smoke control systems, particularly smoke control systems using the pressurization method (often termed zoned smoke control), is the use of smoke barriers to divide a building into separate smoke zones (or compartments). This strategy is used in both passive and mechanical systems. It should be noted that not all walls, ceilings or floors would be considered smoke barriers. Only walls that designate separate smoke zones within a building need to be constructed as smoke barriers. This section is simply providing requirements for walls, floors and ceilings that are used as smoke barriers. It should be noted that it is possible that a smoke control system utilizing the exhaust method may not need to utilize a smoke barrier to divide the building into separate smoke zones; therefore, the evaluation of barrier construction and leakage area may not be necessary and as noted is primarily focused upon designs using the pressurization method.

In order for smoke to not travel from one smoke zone to another, specific construction requirements are necessary in accordance with the code. It should be noted that openings such as doors and windows are dealt with separately within Section 513.5.2 from openings such as cracks or penetrations.

[F] 513.5.1 Leakage area. Total leakage area of the barrier is the product of the smoke barrier gross area times the allowable leakage area ratio, plus the area of other openings such as gaps and operable windows. Compliance shall be determined by achieving the minimum air pressure difference across the barrier with the system in the smoke control mode for mechanical smoke control systems. Passive smoke control systems tested using other *approved* means such as door fan testing shall be as *approved* by the code official.

❖ It is impossible for walls and floors to be constructed that are completely free from openings that may allow the migration of smoke; therefore, leakage needs to be compensated for within the design by calculating the leakage area of walls, ceilings and floors. The factors provided in Section 513.5, which originated from ASHRAE's provisions on leaky buildings, are used to calculate the total leakage area. The total leakage area is then used in the design process to determine the proper amount of air to create the required pressure differences across these surfaces that form smoke zones. These pressure differences then need to be verified when the system is in smoke control mode.

Additionally, Section 513.5 provides ratios to determine the maximum allowable leakage in walls, interior exit stairways, shafts, floors and roofs. These leakage areas are critical in determining whether the proper pressure differences are provided when utilizing the pressurization method of smoke control. Pressure differences will decrease as the openings get larger.

[F] 513.5.2 Opening protection. Openings in smoke barriers shall be protected by automatic-closing devices actuated by the required controls for the mechanical smoke control system. Door openings shall be protected by door assemblies complying with the requirements of the *International Building Code* for doors in smoke barriers.

Exceptions:

1. Passive smoke control systems with automatic-closing devices actuated by spot-type smoke detectors *listed* for releasing service installed in accordance with the *International Building Code*.

2. Fixed openings between smoke zones which are protected utilizing the airflow method.

3. In Group I-2 where such doors are installed across corridors, a pair of opposite-swinging doors without a center mullion shall be installed having vision panels with *approved* fire-rated glazing materials in *approved* fire-rated frames, the area of which shall not exceed that tested. The doors shall be close-fitting within operational tolerances, and shall not have undercuts, louvers or grilles. The doors shall have head and jamb stops, astragals or rabbets at meeting edges and automatic-closing devices. Positive latching devices are not required.

4. Group I-3.

5. Openings between smoke zones with clear ceiling heights of 14 feet (4267 mm) or greater and bank down capacity of greater than 20 minutes as determined by the design fire size.

❖ Similar to concerns of smoke leakage between smoke zones, openings may compromise the necessary pressure differences between smoke zones. Openings in smoke barriers, such as doors and windows, must be either constantly or automatically closed when the smoke control system is operating.

This section requires that doors be automatically closed through the activation of an automatic closing device linked to the smoke control system. Essentially, when the smoke control system is activated, all openings are automatically closed. This most likely would mean that the mechanism that activates the smoke control system would also automatically close all openings. The smoke control system will be activated by a specifically zoned smoke detection or sprinkler system as required by Sections 513.12.2 and 513.12.3.

In terms of actual opening protection, Section 513.5.2 simply refers the user to Section 716.5.3 for specific construction requirements for doors located in smoke barriers. Note that smoke barriers are different from fire barriers since the intended measure of performance is different. One is focused on fire spread from the perspective of heat, the other from the perspective of smoke passage. Smoke barriers do require a 1-hour fire-resistance rating.

There are several exceptions to this particular section. Exception 1 is specifically for passive systems. Passive systems, as noted, are systems in which there is no use of mechanical systems. Instead, the system operates primarily upon the configuration of barriers and layout of the building to provide smoke control. Passive systems can use spot-type detectors to close doors that constitute portions of a smoke barrier. Essentially, this means a full fire alarm system would not be required. Instead, single station detectors would be allowed to close the doors. Such doors would need to fail in the closed position if power is lost. The specifics as to approved devices are found in NFPA 72.

Exception 2 is based on the fact that some systems take advantage of the opposed airflow method such that smoke is prevented from migrating past the doors. Therefore, since the design already accounts for potential smoke migration at these openings through the use of air movement, it is unnecessary to require the barrier to be closed.

Exception 3 is specifically related to the unique requirements for Group I-2 occupancies. Essentially, a very specific alternative, which meets the functional needs of Group I-2 occupancies, is provided. One aspect of the alternative approach is that doors have vision panels with approved fire protection-rated glazing in fire protection-rated frames of a size that does not exceed the type tested.

Exception 4 allows an exemption from the automatic-closing requirements for all Group I-3 occupancies. This is related to the fact that facilities that have occupants under restraint or with specific security restrictions have unique requirements in accordance with Section 408 of the IBC. These requirements accomplish the intent of providing reliable barriers between each smoke zone since, for the most part, such facilities will have a majority of doors closed and in a locked position due to the nature of the facility. The staff very closely controls these types of facilities.

Exception 5 relates to the behavior of smoke. The assumption is that smoke rises due to the buoyancy of hot gases, and if the ceiling is sufficiently high, the smoke layer will be contained for a longer period of time before it begins to move into the next smoke zone. Therefore, it is not as critical that the doors automatically close. This allowance is dependent on the specific design fire for a building. See Section 513.9 for more information on design fire determination. Different size design fires create different amounts of smoke that, depending on the layout of the building, may migrate in different ways throughout the building. This section mandates that smoke cannot begin to migrate into the next smoke zone for at least 20 minutes. This is consistent with the 20-minute maximum duration of operation of smoke control systems required in Section 513.4.6. It should be noted that a minimum of 14-foot (4267 mm) ceilings are required to take advantage of this exception. This exception would require an engineering analysis.

[F] 513.5.2.1 Ducts and air transfer openings. Ducts and air transfer openings are required to be protected with a minimum Class II, 250°F (121°C) smoke damper complying with the *International Building Code*.

❖ Another factor that adds to the reliability of smoke barriers is the protection of ducts and air transfer openings within smoke barriers. Left open, these openings may allow the transfer of smoke between smoke zones. These ducts and air transfer openings most often are part of the HVAC system. Damper operation and the reaction with the smoke control system will be evaluated during acceptance testing. It should be noted that there are duct systems used within a smoke control design that are controlled by the smoke control system and should not automatically close upon detection of smoke via a smoke damper.

It should be noted that a smoke damper works differently than a fire damper. Fire dampers react to heat via a fusible link, while smoke dampers activate upon the detection of smoke. The smoke dampers used should be rated as Class II, 250°F (121°C). The class of the smoke damper refers to its level of performance relative to leakage. The temperature rating is related to its ability to withstand the heat of smoke resulting from a fire. It should be noted that although smoke barriers are only required to utilize smoke dampers, there may be many instances where a fire damper is also required. For instance, the smoke barrier may also be used as a fire barrier. Also, Section 716.5.3 of the IBC would require penetration of shafts to contain both a smoke and fire damper. Therefore, in some cases both a smoke damper and fire damper would be required. There are listings specific to combination smoke and fire dampers. Note that the exceptions to Section 717.5.3 recognize that smoke

and fire dampers may interfere with a smoke control design.

More specific requirements about dampers can be found in Chapter 7 of the IBC and Chapter 6 of this code.

[F] 513.6 Pressurization method. The primary mechanical means of controlling smoke shall be by pressure differences across smoke barriers. Maintenance of a tenable environment is not required in the smoke control zone of fire origin.

❖ There are several methods or strategies that may be used to control smoke movement. One of these methods is pressurization, wherein the system primarily utilizes pressure differences across smoke barriers to control the movement of smoke. Basically, if the area of fire origin maintains a negative pressure, then the smoke will be contained to that smoke zone. A typical approach used to obtain a negative pressure is to exhaust the fire floor. This is a fairly common practice in high-rise buildings. Interior exit stairways also utilize the concept of pressurization by keeping the interior exit stairways under positive pressure. The pressurization method in large open spaces, such as malls and atria, is impractical since it would take a large quantity of supply air to create the necessary pressure differences. It should be noted that pressurization is mandated as the primary method for mechanical smoke control design but this is related to the primary methods historically used for smoke control in high rise buildings. Currently, high-rise buildings do not require smoke control. Airflow and exhaust methods are only allowed when appropriate. The exhaust method is the most commonly applied method due to the use of the atrium provisions in Section 404.5 of the IBC.

The pressurization method does not require that tenable conditions be maintained in the smoke zone where the fire originates. Maintaining this area tenable would be impossible, based on the fact that pressures from the surrounding smoke zones would be placing a negative pressure within the zone of origin to keep the smoke from migrating.

Pressurization is used often with interior exit stairways. This method provides a positive pressure within the interior exit stairways to resist the passage of smoke. Stair pressurization is one method of compliance for stairways in high-rise or underground buildings where the floor surface is located more than 75 feet (22 860 mm) above the lowest level of fire-department vehicle access or more than 30 feet (9144 mm) below the floor surface of the lowest level of exit discharge. It should be noted that there are two methods found in the IBC that address smoke movement—smokeproof enclosures or pressurized stairs. A smokeproof enclosure requires a certain fire-resistance rating along with access through a ventilated vestibule or an exterior balcony. The vestibule can be ventilated in two ways: using natural ventilation or mechanical ventilation as outlined in Sections

909.20.3 and 909.20.4 of the IBC. The pressurization method requires a sprinklered building and a minimum pressure difference of 0.15 inch (37 Pa) of water and a maximum of 0.35 inch (87 Pa) of water. These pressure differences are to be available with all doors closed under maximum stack pressures (see Sections 909.20 and 1022.9 of the IBC for more details).

As noted, the pressurization method utilizes pressure differences across smoke barriers to achieve control of smoke. Sections 513.6.1 and 513.6.2 provide the criteria for smoke control design in terms of minimum and maximum pressure differences.

In summary, the pressurization method is used in two ways. The first is through the use of smoke zones where the zone of origin is exhausted, creating a negative pressure. The second is stair pressurization that creates a positive pressure within the stair to avoid the penetration of smoke. Note that the code allows the use of a smokeproof enclosure instead of pressurization.

[F] 513.6.1 Minimum pressure difference. The minimum pressure difference across a smoke barrier shall be 0.05-inch water gage (12.4 Pa) in fully sprinklered buildings.

In buildings permitted to be other than fully sprinklered, the smoke control system shall be designed to achieve pressure differences at least two times the maximum calculated pressure difference produced by the design fire.

❖ The minimum pressure difference is established as 0.05-inch water gage (12 Pa) in fully sprinklered buildings. This particular criterion is related to the pressures needed to overcome buoyancy and the pressures generated by the fire, which include expansion. This particular criterion is based upon a sprinklered building. The pressure difference would need to be higher in a building that is not sprinklered. Additionally, the pressure difference needs to be provided based upon the possible stack and wind effects present.

[F] 513.6.2 Maximum pressure difference. The maximum air pressure difference across a smoke barrier shall be determined by required door-opening or closing forces. The actual force required to open exit doors when the system is in the smoke control mode shall be in accordance with the *International Building Code*. Opening and closing forces for other doors shall be determined by standard engineering methods for the resolution of forces and reactions. The calculated force to set a side-hinged, swinging door in motion shall be determined by:

$$F = F_{dc} + K(WA\Delta P)/2(W-d) \qquad \textbf{(Equation 5-2)}$$

where:

A = Door area, square feet (m^2).

d = Distance from door handle to latch edge of door, feet (m).

F = Total door opening force, pounds (N).

F_{dc} = Force required to overcome closing device, pounds (N).

K = Coefficient 5.2 (1.0).

W = Door width, feet (m).

ΔP = Design pressure difference, inches (Pa) water gage.

❖ The maximum pressure difference is based primarily upon the force needed to open and close doors. The code establishes maximum opening forces for doors. This maximum opening force cannot be exceeded, taking into account the pressure differences across a doorway in a pressurized environment. Essentially, based on the opening force requirements of Section 1008.1.3 of the IBC, the maximum pressure difference can be calculated in accordance with Equation 5-2. In accordance with Chapter 10 of the IBC, the maximum opening force of a door has three components, including:

Door latch release: Maximum of 15 pounds (67 N)

Set door in motion: Maximum of 30 pounds (134 N)

Swing to full open
position: Maximum of 15 pounds (67 N)

Equation 5-2 is used to calculate the total force to set the door into motion when in the smoke control mode; therefore, the limiting criteria would be 30 pounds (134 N). It should be noted that although the accessibility requirements related to door opening force are more restrictive in Section 404.2.8 of ICC A117.1, *Accessible and Usable Buildings and Facilities*, fire doors do not require compliance with these requirements.

[F] 513.7 Airflow design method. When *approved* by the code official, smoke migration through openings fixed in a permanently open position, which are located between smoke control zones by the use of the airflow method, shall be permitted. The design airflows shall be in accordance with this section. Airflow shall be directed to limit smoke migration from the fire zone. The geometry of openings shall be considered to prevent flow reversal from turbulent effects.

❖ This method is only allowed when approved by the building official. As the title states, this method utilizes airflow to avoid the migration of smoke across smoke barriers. This has been referred to as opposed airflow. Specifically, this method is suited for the protection of smoke migration through doors and related openings fixed in a permanently open position. This method consists of providing a particular velocity of air based upon the temperature of the smoke and the height of the opening. The temperature of the smoke will depend on the design fire that is established for the particular building. The higher the temperature of the smoke and the larger the opening, the higher the velocity necessary to maintain the smoke from migrating into the smoke zone. It should be noted that the airflow method seldom works for large openings, since the velocity to oppose the smoke becomes too high. This method tends to work better for smaller openings, such as pass-through windows. Equation 5-3 provides the method to calculate the necessary velocity.

[F] 513.7.1 Velocity. The minimum average velocity through a fixed opening shall not be less than:

$$v = 217.2[h(T_f - T_o)/(T_f + 460)]^{1/2} \quad \text{(Equation 5-3)}$$

For SI: $v = 119.9 [h(T_f - T_o)/T_f]^{1/2}$

where:

h = Height of opening, feet (m).

T_f = Temperature of smoke, °F (K).

T_o = Temperature of ambient air, °F (K).

v = Air velocity, feet per minute (m/minute).

❖ This section provides the formula for the minimum average velocity through a fixed opening. The minimum velocity is based on the velocity needed to prevent the smoke from migrating into the smoke zone. Consideration needs to be given to the eventual exhaust of the air introduced for this approach. See commentary, Section 513.7, for further discussion.

[F] 513.7.2 Prohibited conditions. This method shall not be employed where either the quantity of air or the velocity of the airflow will adversely affect other portions of the smoke control system, unduly intensify the fire, disrupt plume dynamics or interfere with exiting. In no case shall airflow toward the fire exceed 200 feet per minute (1.02 m/s). Where the formula in Section 513.7.1 requires airflow to exceed this limit, the airflow method shall not be used.

❖ The airflow method has a limitation on maximum velocity. This limitation is based upon the fact that air may distort the flame and cause additional entrainment and turbulence; therefore, having a high velocity of air entering the zone of fire origin has the potential of increasing the amount of smoke produced. The velocity may also interact with other portions of the smoke control design. For instance, the pressure differences in other areas of the building may be altered, which may exceed the limitations of Sections 513.6.1 and 513.6.2. This section requires that when a velocity of over 200 feet per minute (1.02 m/sec) is calculated, the airflow method is not allowed. The solution may result in requiring a barrier such as a wall or door.

If the airflow design method is chosen to protect areas communicating with an atrium, the air added to the smoke layer needs to be accounted for in the exhaust rate.

[F] 513.8 Exhaust method. When *approved* by the building official, mechanical smoke control for large enclosed volumes, such as in atriums or malls, shall be permitted to utilize the exhaust method. Smoke control systems using the exhaust method shall be designed in accordance with NFPA 92B.

❖ This method is only allowed when approved by the building official. The primary application of the exhaust method is in large spaces, such as atriums and malls, and is the most widely used method in

Section 909 of the IBC and IFC and Section 513 of this code. The strategy of this method is to keep the smoke layer at a certain level within the space. This is primarily accomplished through exhausting smoke. The amount of exhaust depends upon the design fire [see Commentary Figure 513.8(1)]. Essentially, fires produce different amounts and properties of smoke based on the material being burned, size of the fire and the placement of the fire; therefore, NFPA 92B is referenced for the design of such systems. NFPA 92B presents several ways to address the control of smoke, which includes the use of the following tools:

- Scale Modeling (Small scale testing)—Utilizes the concept of scaling to allow small scale tests to be conducted to understand the smoke movement within a space.
  - Benefits—More realistic understanding of smoke movement in spaces with unusual configurations or projections than algebraic calculations.
  - Disadvantages—Expensive and the application of results is limited to the uniqueness of the space being analyzed.
- Algebraic (Calculations—similar to 2003 IBC)—Empirically derived (based upon testing) modeling in its simplest form.
  - Benefits—Simple, cost-effective analysis.
  - Disadvantages—Limited applicability due to the range of values they were derived from, only appropriate with certain types of design fires, typically overconservative outputs that increase equipment needs, equipment costs and can impact aesthetics and architectural design.
- Computer Modeling [Computational fluid dynamics (CFD) or zone models]—Combination of theory and empirical values to determine the smoke movement and fire-induced conditions within a space and effectiveness of the smoke control system.
  - Benefits—More realistic understanding of smoke movement in spaces with unusual configurations or projections and less expensive than scale modeling. Helps significantly in designing smoke control systems tailored to spaces and achieving cost-effective designs, and can help limit the impact to architectural design.
  - Disadvantages—Computing time and cost can be longer than algebraic calculations but benefits typically outweigh this disadvantage. Early planning is important and can limit these adverse impacts.

In terms of computer modeling, there are essentially two methods that include zone models and CFD models. Zone models are based upon the unifying assumption that in any room or space where the effects of the fire are present there are distinct layers (hot upper layer, cool lower layer). In real life such distinct layers do not exist. Some examples of zone models used in such applications include Consolidated Model of Fire Growth And Smoke Transport (C-FAST) and Available Safe Egress Time (ASET) (see Section 3-7 of the *SFPE Handbook of Fire Protection Engineering* for further information). CFD models take this much further and actually divide the space into thousands or millions of interconnected "cells" or "fields." The model then evaluates the fire dynamics and heat and mass in each individual cell and how it interacts with those adjacent to it. The use of such models becomes more accurate with more numerous and smaller cells but the computing power and expertise required is much higher than for zone models. As noted the use of either types of models can be advantageous but such use must be undertaken by someone qualified. Proper review and verification of the input and output is critical. The most popular model in the area of CFD with regard to fire is the Fire Dynamics Simulator (FDS) developed by National Institute for Standards and Technology (NIST). Other models, such as Fluent (Fluent, Inc.), are sometimes used.

Depending upon the space being evaluated, some design strategies may provide a better approach than others. Past editions of the IBC smoke control provisions for the exhaust method mandated the use of the algebraic methods with a steady fire. This, of course, also mandated a mechanical system be used whereas NFPA 92B allows an overall review of smoke layer movement and whether the design goals, which in this case are mandated by the code, can be met. Therefore, if it can be shown that the smoke layer interface can be held at 6 feet (1829 mm) as mandated in Section 513.8.1 for the design operation time required by Section 513.4.6 without mechanical ventilation, then the space would comply with Section 513. NFPA 92B presents several design approaches. This allows more flexibility in design than that found in previous editions of the IBC. NFPA 92B as a standard does not set the minimum smoke layer interface height or duration for system operation. Such criteria is found within Sections 513.8.1 and 513.4.6, respectively. See the commentary for those sections.

If the algebraic approach is used, consideration of three types of fire plumes may be required to determine which one is the most demanding in terms of smoke removal needs based upon the space being assessed. They include:

- Axisymmetric plumes—Smoke rises unimpeded by walls, balconies or similar projections [see Commentary Figure 513.8(2)].

- Balcony spill plumes—Smoke flows under and around edges of a horizontal projection [see Commentary Figure 513.8(3)].

- Window plumes—Smoke flows through an opening into a large-volume space [see Commentary Figure 513.8(4)].

It should be noted that prior to the reference to NFPA 92B in the code, the balcony spill and window plume calculations had been eliminated from the smoke control requirements of the code due to concerns with the applicability of those calculations. The major difference is that NFPA 92B does not mandate the use of such equations as did previous editions of the IBC. The use of such equations will depend upon the design fires agreed upon for the particular design and whether an algebraic approach is chosen. These equations are used to determine a mass flow rate of smoke to ultimately determine the required exhaust volume for that space. If the potential for a balcony or window spill plumes are known to exist within the

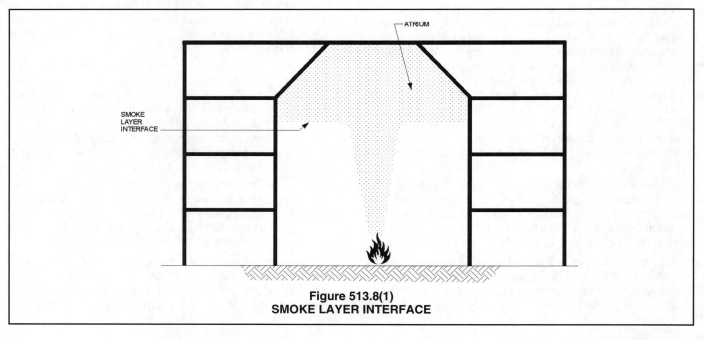

Figure 513.8(1)
SMOKE LAYER INTERFACE

space, then appropriate measures need to be taken to address these, as they typically result in more onerous exhaust and supply requirements. Part of the reason for the initial deletion of these equations was the fact that such scenarios are not as likely or their impact is significantly reduced in sprinklered buildings. There is also some concern with the applicability of the balcony spill plume equation in a variety of applications. These potential fire scenarios and resulting plumes may further the need to undertake a CFD analysis to address such hazards more appropriately and effectively.

Another key aspect that NFPA 92B included within the algebraic methods is equations to determine that a minimum number of exhaust inlets are available to prevent plugholing. Plugholing occurs when air from below the smoke layer is pulled through the smoke layer into the smoke exhaust inlets. As such, if plugholing occurs, some of the fan capacity is used to exhaust air rather than smoke and thus can affect the ability to maintain the smoke layer at or above the design height. Scale modeling and computer fire modeling would demonstrate these potential problems during the testing and analysis, respectively [see Commentary Figure 513.8(5)].

It should be noted that this section specifically references NFPA 92B for the design of smoke control using the exhaust method. Therefore the requirements in NFPA 92B related to testing, documentation and maintenance would not be applicable though they may be a good resource. Equipment and controls would be part of the design; therefore, related provisions of NFPA 92B would apply. Generally the IBC addresses equipment and controls in a similar fashion.

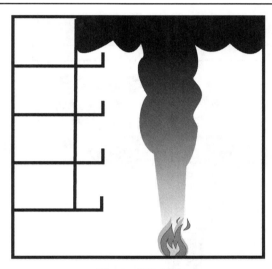

Figure 513.8(2)
AXISYMMETRICAL PLUME

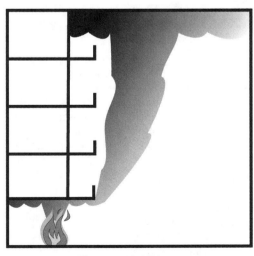

Figure 513.8(3)
BALCONY SPILL PLUME

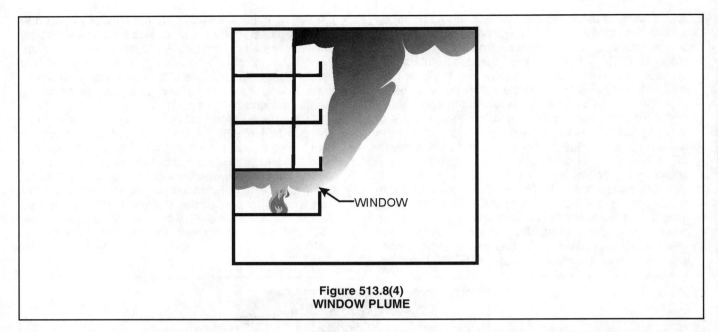

**Figure 513.8(4)
WINDOW PLUME**

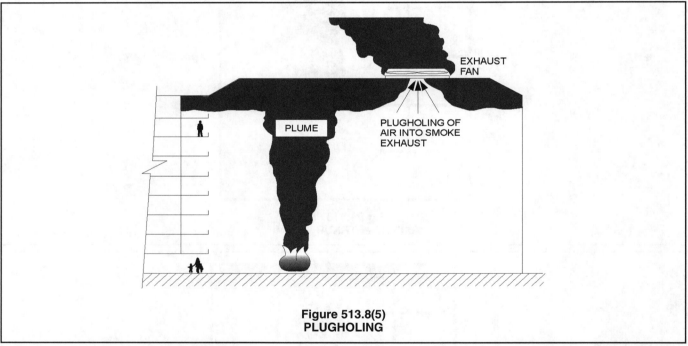

**Figure 513.8(5)
PLUGHOLING**

[F] 513.8.1 Exhaust rate. The height of the lowest horizontal surface of the accumulating smoke layer shall be maintained at least 6 feet (1829 mm) above any walking surface which forms a portion of a required egress system within the smoke zone.

❖ The design criteria to be used when applying NFPA 92B is to maintain the smoke layer interface at least 6 feet (1829 mm) above any walking surface that is considered part of the required egress within the particular smoke zone, such as an atrium, for 20 minutes or 1.5 times the calculated egress time (see Section 513.4.6). Chapter 10 of the IBC considers the majority of occupiable space as part of the means of

egress system. Also, keep in mind that the criteria of 6 feet (1829 mm) does not apply just to the main floor surface of the mall or atrium but to any level where occupants may be exposed (for example, balconies) see Commentary Figure 513.8.1(1).

The code uses the terminology "lowest horizontal surface of the accumulating smoke layer interface." NFPA 92B has several definitions related to smoke layer, which include the following:

Smoke layer. The accumulated thickness of smoke below a physical barrier.

Smoke layer interface. The theoretical boundary between a smoke layer and the smoke-free air.

(Note: This boundary is at the beginning of the transition zone.)

First indication of smoke. The boundary between the transition zone and the smoke-free air.

Transition zone. The layer between the smoke layer interface and the first indication of smoke in which the smoke layer temperature decreases to ambient. The transition zone may be several feet thick (large open space) or may barely exist (small area with intense fire) [see Commentary Figure 513.8.1(2)].

NFPA 92B provides algebraic equations to determine the first indication of smoke but is limited to very specific conditions, such as a uniform cross section, specific aspect ratios, steady or unsteady fires and no smoke exhaust operating. When using algebraic equations for smoke layer interface looking at different types of plumes, the smoke layer interface terminology is used, and the user enters the desired smoke layer interface height. Zone models use simplifying assumptions so the layers are distinct from one another. In contrast, when CFD or scale modeling is used, the data must be analyzed to verify that the smoke layer interface is located at or above the 6 feet (1829 mm) during the event. This is not a simple analysis as CFD and scale modeling provide more detail on actual smoke behavior; therefore, the location of the smoke layer interface may not be initially clear without some level of analysis. Again, it depends on the depth of the transition layer. This may require reviewing tenability within the transition zone. Tenability limits need to be agreed upon by the stakeholders involved. Using CFD or scale modeling would likely need to occur through the alternative methods and materials (see Section 105.2) because of the need to review tenability limits. It should be noted that NFPA 92B Annex A suggests that there are methods to determine where the smoke layer interface and first indication of smoke are located when undertaking CFD and scale modeling using a limited number of point measurements.

Also, Section 513.8.1 specifies a minimum distance for the smoke layer interface from any walking surface whereas Section 4.5.3 of NFPA 92B has provisions that simply allow the analysis to demonstrate tenability regardless of where the layer height is located above the floor. Defining tenability can be more difficult as there is not a standard definition. Any design using that approach would need to be addressed through Section 105.2.

Note that the response time of the system components (detection, activation, ramp up time, shutting down HVAC, opening/closing doors and dampers, etc.) needs to be accounted for when analyzing the location of the smoke layer interface in relation to the duration of operations stated in Section 513.4.6 (see commentary, Section 909.17 of the IFC).

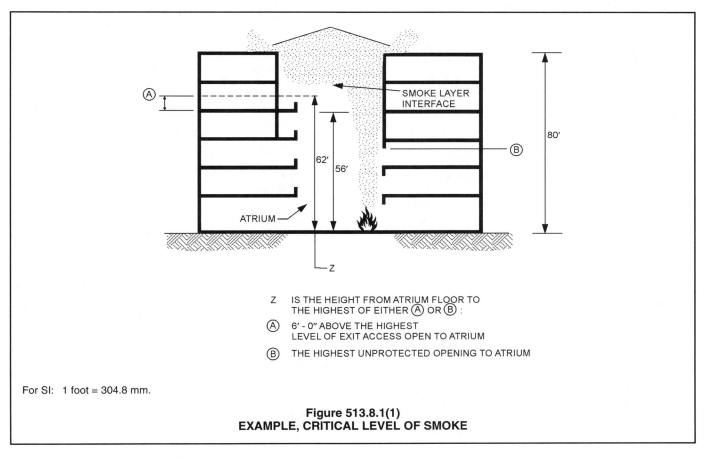

For SI: 1 foot = 304.8 mm.

Figure 513.8.1(1)
EXAMPLE, CRITICAL LEVEL OF SMOKE

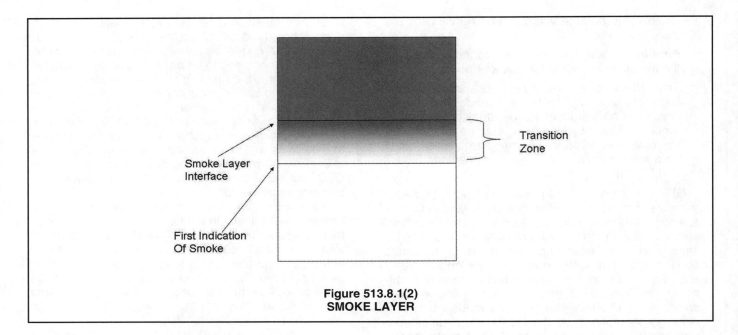

Figure 513.8.1(2)
SMOKE LAYER

[F] 513.9 Design fire. The design fire shall be based on a rational analysis performed by the *registered design professional* and *approved* by the code official. The design fire shall be based on the analysis in accordance with Section 513.4 and this section.

❖ The design fire is the most critical element in the smoke control system design. The fire is what produces the smoke to be controlled by the system; thus, the size of the fire directly impacts the quantity of smoke being produced. This section ensures that the design fire be determined through a rational analysis by a registered design professional with knowledge in this area. Such professionals should have experience in the area of fire dynamics, fire engineering and general building design, including mechanical systems. When determining the design fire the designer should work with various stakeholders to determine the types of hazards and combustible materials (fire scenarios) on a permanent as well as temporary basis (i.e., Christmas/holiday decorative materials or scenery, temporary art exhibits) that may be present throughout the use of the building once occupied. Those hazards then need to be translated to potential design fires to be used when determining the smoke layer interface height for the duration as determined by Section 513.4.6. See the commentary to Section 513.9.3 for potential sources when determining design fires.

This section also does not mandate the type of fire (i.e., steady versus unsteady). A steady fire assumes a constant heat release rate over a period of time, where unsteady fires do not. An unsteady fire includes the growth and decay phases of the fire, as well as the peak heat release rate. An unsteady fire will hit a peak heat release rate when burning in the open, like an axisymmetric fire. An unsteady fire is a more realistic view of how fires actually burn. It

should be noted that fires can be a combination of unsteady and steady fires when the sufficient fuel is available. In other words, the fire initially grows (unsteady) then reaches a steady state and burns for some time at a particular heat release rate before decay occurs.

Design fire information should therefore typically include growth rate, peak heat release rate, duration and decay as well as information related to fire locations and products of combustion yield (CO, smoke, etc.) that are produced by the various design fires that are deemed credible for the space.

To provide an order of magnitude of fire sizes obtained from various combustibles, the following data from fire tests is provided. The following heat release rates, found in Section 3, Chapter 3-1 of the 2nd edition of the *SFPE Handbook of Fire Protection Engineering*, are peak heat release rates:

Plastic trash bags/ paper trash:	114-332 Btu/sec (120-350 kw)
Latex foam pillow:	114 Btu/sec approximately (120 kw)
Dry Christmas tree:	475-618 Btu/sec (500-650 kw)
Sofa:	2,852 Btu/sec approximately (3,000 kw)
Plywood wardrobe:	2,947-6,084 Btu/sec (3,100-6,400 kw)

[F] 513.9.1 Factors considered. The engineering analysis shall include the characteristics of the fuel, fuel load, effects included by the fire and whether the fire is likely to be steady or unsteady.

❖ This section simply provides more detail on the factors that should be taken under consideration when determining the design fire size. To determine the appropriate fire size, an engineering analysis is necessary that takes into account the following elements: fuel (potential burning rates), fuel load (how much), effects included by the fire (smoke particulate size and density), steady or unsteady (burn steadily or simply peak and dissipate) and likelihood of sprinkler activation (based on height and distance from the fire).

[F] 513.9.2 Design fire fuel. Determination of the design fire shall include consideration of the type of fuel, fuel spacing and configuration.

❖ The design fire size may also be affected by surrounding combustibles, which may have the effect of increasing the fire size. More specifically, there is concern that if sufficient separation is not maintained between combustibles, then a larger design fire is likely. The code does not provide extensive detail on this as such determination is left to the rational analysis undertaken by the design professional. NFPA 92B provides one method in which to determine the critical separation distance, *R*. This is based upon fire size and the critical radiant heat flux for nonpiloted ignition. Nonpiloted ignition means the radiated heat from the fire without direct flame contact will ignite adjacent combustibles.

[F] 513.9.3 Heat-release assumptions. The analysis shall make use of the best available data from *approved* sources and shall not be based on excessively stringent limitations of combustible material.

❖ This section is merely stressing the fact that data obtained for use in a rational analysis needs to come from relevant and appropriate sources. Data can be obtained from groups such as the NIST or from Annex B of NFPA 92B. Data from fire tests is available and is a good resource for such analysis. As noted earlier, such data is not prevalent (see also Chapter 8, Analysis of Design Fires, of *A Guide to Smoke Control in the 2006 IBC®* and Section 3, Chapter 3-1 of the *SFPE Handbook of Fire Protection Engineering*).

[F] 513.9.4 Sprinkler effectiveness assumptions. A documented engineering analysis shall be provided for conditions that assume fire growth is halted at the time of sprinkler activation.

❖ This section raises a few questions regarding activation of sprinklers and their impact on a fire both in terms of their ability to "control" as well as "extinguish" a fire. The first is concerning an assumption that sprinklers will immediately control the fire as soon as they are activated (i.e., control results in limiting further growth and maintaining the heat release rate at approximately the same fire size as when the sprinklers activated). This assumption may be true in some cases, but for high ceilings the sprinkler may not activate or may be ineffective. Sprinklers may be ineffective in high spaces, since by the time they are

activated the fire is too large to control. Essentially, the fire plume may push away and evaporate the water before it actually reaches the seat of the fire. In addition the fire may be shielded from sprinkler spray so that insufficient quantities of water reach the fuel. These are common problems with high-piled storage as well as other fires including retail and has been shown in actual tests. Also, if the fire becomes too large before the sprinklers are activated, the available water supply and pressure for the system may be compromised. Additionally, based on the layout of the room and the movement of the fire effluents, the wrong sprinklers could be activated, which leads to a larger fire size and depletion of the available water supply and pressure.

Another issue is whether the sprinklers "control" or "extinguish" the fire. Typical sprinklers are assumed only to control fires as opposed to extinguishing them. Sprinklers may be able to extinguish the fire but it should not automatically be assumed. A fire that is controlled will achieve steady state and maintain a certain fire size, which is very different from a fire that is actually extinguished.

Based upon these concerns, each scenario needs to be looked at individually to determine whether sprinklers would be effective in halting the growth or extinguishing the fire. More specifically, the evaluation should include droplet size, density and area of coverage and should also be based on actual test results.

[F] 513.10 Equipment. *Equipment* such as, but not limited to, fans, ducts, automatic dampers and balance dampers shall be suitable for their intended use, suitable for the probable exposure temperatures that the rational analysis indicates, and as *approved* by the code official.

❖ Section 513.10 and subsequent sections are primarily related to the reliability of the system components to provide a smoke control system that works according to the design. One of the largest concerns when using smoke control provisions is the overall reliability of the system. Such systems have many different components, such as smoke and fire dampers; fans; ducts and controls associated with such components. The more components a system has, the less reliable it becomes. In fact, one approach in providing a higher level of reliability is utilizing the normal building systems such as the HVAC to provide the smoke control system. Basically, systems used every day are more likely to be working appropriately, since they are essentially being tested daily; however, there are many components that are specific to the smoke control system, such as exhaust fans in an atrium or the smoke control panel.

Also, there is not a generic prescriptive set of requirements as to how all smoke control system elements should operate, since each design may be fairly unique. The specifics on operation of such a system need to be included within the design and construction documents. Most components used in smoke control systems are elements used in many other applications such as HVAC systems; therefore, the basic mechanisms of a fan used in a smoke control system may not be different, although they may be applied differently.

[F] 513.10.1 Exhaust fans. Components of exhaust fans shall be rated and certified by the manufacturer for the probable temperature rise to which the components will be exposed. This temperature rise shall be computed by:

$$T_s = (Q_c/mc) + (T_a) \qquad \textbf{(Equation 5-4)}$$

where:

c = Specific heat of smoke at smoke-layer temperature, Btu/lb°F (kJ/kg · K).

m = Exhaust rate, pounds per second (kg/s).

Q_c = Convective heat output of fire, Btu/s (kW).

T_a = Ambient temperature, °F (K).

T_s = Smoke temperature, °F (K).

Exception: Reduced T_s as calculated based on the assurance of adequate dilution air.

❖ Fans used for smoke control systems must be able to tolerate the possible elevated temperatures to which they will be exposed. Again, like many other factors this depends upon the specifics of the design fire. Essentially, Equation 5-4 requires the calculation of the potential temperature rise. The exhaust fans must be specifically rated and certified by the manufacturer to be able to handle these rises in temperature. There is an exception that allows reduction of the temperature if it can be shown that adequate temperature reduction will occur. In many cases, if the exhaust fans are near the ceiling, the smoke will be much cooler than the value resulting from Equation 5-4 since the smoke may cool considerably by the time it reaches the ceiling. Also, sprinkler activation will assist in cooling the smoke further.

[F] 513.10.2 Ducts. Duct materials and joints shall be capable of withstanding the probable temperatures and pressures to which they are exposed as determined in accordance with Section 513.10.1. Ducts shall be constructed and supported in accordance with Chapter 6. Ducts shall be leak tested to 1.5 times the maximum design pressure in accordance with nationally accepted practices. Measured leakage shall not exceed 5 percent of design flow. Results of such testing shall be a part of the documentation procedure. Ducts shall be supported directly from fire-resistance-rated structural elements of the building by substantial, noncombustible supports.

Exception: Flexible connections, for the purpose of vibration isolation, that are constructed of *approved* fire-resistance-rated materials.

❖ The next essential component of a smoke control system is the integrity of the ducts to transport supply and exhaust air. The integrity of ducts is also important for an HVAC system, but is more critical in this

case since it is not simply a comfort issue but one of life safety. The key concern with ducts in smoke control systems is that they can withstand elevated temperatures and that there will be minimal leakage. The concern with leakage is the potential of leaking smoke into another smoke zone or not providing the proper amount of supply air to support the system.

More specifically, all ducts need to be leak tested to 1.5 times the maximum static design pressure. The leakage resulting should be no more than 5 percent of the design flow. For example, a duct that has a design flow of 300 cubic feet per minute (cfm) (0.141 m³/s) would be allowed 15 cfm (0.007 m³/s) of leakage when exposed to a pressure equal to 1.5 times the design pressure for that duct. The tests should be in accordance with nationally accepted practices. This criterion will often limit ductwork for smoke control systems to lined systems, since the amount of leakage in such systems is much less.

As part of the concern for possible exposure to fire and fire products, the ducts are required to be supported by way of substantial noncombustible supports connected to the fire-resistance-rated structural elements of the building. As noted, the system needs to be able to run for 20 minutes starting from the detection of the fire. The exception to this section is really more of an acknowledgement that flexible connections for vibration isolation are acceptable when constructed of approved fire-resistance-rated materials. More specifically, it is often necessary to use such connections for connecting the duct to the fan. These connections cannot necessarily meet the requirements of the main section, but are a minimal part of the ductwork and as long as they perform adequately with regard to fire resistance, they are permitted. Note that the term "approved" is used to determine the required fire resistance, therefore, flexibility is provided. The code does not specifically address this determination but perhaps a relationship to the duration or operation and these flexible connections could be made to determine the necessary performance.

[F] 513.10.3 Equipment, inlets and outlets. *Equipment* shall be located so as to not expose uninvolved portions of the building to an additional fire hazard. Outdoor air inlets shall be located so as to minimize the potential for introducing smoke or flame into the building. Exhaust outlets shall be so located as to minimize reintroduction of smoke into the building and to limit exposure of the building or adjacent buildings to an additional fire hazard.

❖ The intent of this section is to minimize the likelihood of smoke being reintroduced into the building due to poorly placed outdoor air inlets and exhaust air outlets; therefore, placing one right next to another on the exterior of the building would be inappropriate. In addition, wind and other adverse conditions should be considered when choosing locations for these inlets and outlets. Particular attention should be paid to introducing exhausted smoke into another smoke zone. Also, smoke should be exhausted in a direction that will not introduce it into surrounding buildings or facilities. Within the building itself, the supply air and exhaust outlets should also be strategically located. The exhaust inlets and supply air should be evenly distributed to reduce the likelihood of a high velocity of air that may disrupt the fire plume and also push smoke back into occupied areas. See the commentary for Section 513.8 for discussion on avoiding plugholing.

[F] 513.10.4 Automatic dampers. Automatic dampers, regardless of the purpose for which they are installed within the smoke control system, shall be *listed* and conform to the requirements of *approved* recognized standards.

❖ This section addresses the reliability of any dampers used within a smoke control system. This particular provision requires that the dampers be listed and conform to the appropriate recognized standards. More specifically, Section 717 of the IBC and Section 607 of this code contain more detailed information on the specific requirements for smoke and fire dampers. Smoke and fire dampers should be listed in accordance with UL 555S and 555, respectively. Also, remember that each smoke control design is unique and the sequence and methods used to activate the dampers may vary from design to design. This information needs to be addressed in the construction documents.

Another factor to take into account, with regard to timing of the system, is the fact that some dampers react more quickly than others, simply due to the particular smoke damper characteristics. Additionally, during the commissioning of the system, the damper is going to be exposed to many repetitions. These repetitions need to be accounted for in the overall reliability of the system.

[F] 513.10.5 Fans. In addition to other requirements, belt-driven fans shall have 1.5 times the number of belts required for the design duty with the minimum number of belts being two. Fans shall be selected for stable performance based on normal temperature and, where applicable, elevated temperature. Calculations and manufacturer's fan curves shall be part of the documentation procedures. Fans shall be supported and restrained by noncombustible devices in accordance with the structural design requirements of the *International Building Code*. Motors driving fans shall not be operating beyond their nameplate horsepower (kilowatts) as determined from measurement of actual current draw. Motors driving fans shall have a minimum service factor of 1.15.

❖ Part of the overall reliability requires that fans used to provide supply air and exhaust capacity will be functioning when necessary; therefore, a safety factor of 1.5 is placed upon the required belts for fans. All fans used as part of a smoke control system must provide 1.5 times the number of required belts with a minimum of two belts for all fans.

This section also points out that the fan chosen should fit the specific application. It should be able to withstand the temperature rise as calculated in Section 513.10.1 and generally be able to handle typical exposure conditions, such as location and wind. For instance, propeller fans are highly sensitive to the effects of wind. When located on the windward side of a building, wall-mounted, nonhooded propeller fans are not able to compensate for wind effects. Additionally, even hooded propeller fans located on the leeward side of the building may not adequately compensate for the decrease in pressure caused by wind effects. In general, when designing a system, it should be remembered that field conditions might vary from the calculations; therefore, flexibility should be built into the design that would account for things such as variations in wind conditions. Finally, this section stresses that fan motors not be operated beyond their rated horsepower.

[F] 513.11 Power systems. The smoke control system shall be supplied with two sources of power. Primary power shall be the normal building power systems. Secondary power shall be from an *approved* standby source complying with Chapter 27 of the *International Building Code*. The standby power source and its transfer switches shall be in a room separate from the normal power transformers and switch gear and ventilated directly to and from the exterior. The room shall be enclosed with not less than 1-hour fire-resistance-rated fire barriers constructed in accordance with Section 707 of the *International Building Code* or horizontal assemblies constructed in accordance with Section 711 of the *International Building Code*, or both. Power distribution from the two sources shall be by independent routes. Transfer to full standby power shall be automatic and within 60 seconds of failure of the primary power. The systems shall comply with NFPA 70.

❖ As with any life safety system, a level of redundancy with regard to power supply is required to enable the functioning of the system during a fire. The primary source is the building's normal power system. The secondary power system is by means of standby power. One of the key elements is that standby power systems are intended to operate within 60 seconds of loss of primary power. It should be noted that the primary difference between standby power and emergency power is that emergency power must operate within 10 seconds of loss of primary power versus 60 seconds. This section also requires isolation from normal building power systems via a 1-hour fire barrier or 1-hour horizontal assembly or both depending upon the location within the building. This increases the reliability and reduces the likelihood that a single event could remove both power supplies.

[F] 513.11.1 Power sources and power surges. Elements of the smoke management system relying on volatile memories or the like shall be supplied with uninterruptible power sources of sufficient duration to span 15-minute primary power interruption. Elements of the smoke management system susceptible to power surges shall be suitably protected by conditioners, suppressors or other *approved* means.

❖ Smoke control systems have many components, sometimes highly sensitive electronics, that are adversely affected by any interruption in or sudden surges of power. Therefore, Section 513.11.1 requires that any components of a smoke control system, such as volatile memories, be supplied with an uninterruptible power system for the first 15 minutes of loss of primary power. Volatile memory components will lose memory upon any loss of power no matter how short the time period. Once the 15 minutes elapse, these elements can be transitioned to the already operating standby power supply.

With regard to components sensitive to power surges, they need to be provided with surge protection in the form of conditioners, suppressors or other approved means.

[F] 513.12 Detection and control systems. Fire detection systems providing control input or output signals to mechanical smoke control systems or elements thereof shall comply with NFPA 72 and the requirements of Chapter 9 of the *International Building Code* or the *International Fire Code*. Such systems shall be equipped with a control unit complying with UL 864 and listed as smoke control *equipment*.

Control systems for mechanical smoke control systems shall include provisions for verification. Verification shall include positive confirmation of actuation, testing, manual override, the presence of power downstream of all disconnects and, through a preprogrammed weekly test sequence report, abnormal conditions audibly, visually and by printed report.

❖ This section is focused upon two main elements. The first is the proper operation and monitoring of the fire detection system that activates the smoke control system through compliance with Section 907 in the IBC and UL 864. This requires a specific listing as smoke control equipment. UL 864 has a subcategory (UUKL) specific to fire alarm control panels for smoke control system applications.

The second aspect is related to the mechanical elements of the smoke control system once the system is activated. In particular, there is a focus upon verification of activities. Verification would include the following two aspects according to the second paragraph of this section:

1. The system is able to verify actuations, testing, manual overrides and the presence of power downstream. This would require information reported back to the smoke control panel, which can be accomplished via the weekly test sequence or through full electronic monitoring of the system.

2. Conduct a preprogrammed weekly test that simulates an actual (smoke) event to test the

components of the system. These components would include elements such as smoke dampers, fans and doors. Abnormal conditions need to be reported in three ways:

   a. Audibly;

   b. Visually; and

   c. Printed report.

It should be noted that electrical monitoring of the control components is not required (supervision). Such supervision verifies integrity of the conductors from a fire alarm control unit to the control system input. The weekly test is considered sufficient verification of system performance and is often termed end to end verification. In other words, the control system input provides the expected results. Verification can be accomplished through any sensor that can be calibrated to distinguish the difference between proper operation and a fault condition. For fans, proper operation means that the fan is moving air within the intent of its design. Fault conditions include power failure, broken fan belts, adverse wind effects, a locked rotor condition and/or filters or large ducts that are blocked, causing significantly reduced airflow. In addition to differential pressure transmitters and sail switches, this can be accomplished by the present state of the art current sensors. More discussion on verification for elements such as ducts and fire doors is discussed in Chapter 9 of *A Guide to Smoke Control in the 2006 IBC®*.

Also, the fact that a smoke control system is non-dedicated (integrated with an HVAC system) does not mean that it is automatically being tested on a daily basis. It is cautioned that simply depending upon occupant discomfort, for example, is sometimes an insufficient indicator of a fully functioning smoke control system. There may be various modes in which the HVAC system could operate that may not exercise the smoke control features and the sequence in which the system should operate. An example is an air-conditioning system operating only in full recirculating mode versus exhaust mode. This failure will likely not affect occupants and will not exercise the exhaust function. Plus, doors, which may be part of the smoke barrier, may not need to be closed in normal building operations but would need to be closed during smoke control system operation. This is why this section does not necessarily differentiate between dedicated and nondedicated smoke control systems and requires the system components to be tested.

It is important to note that this weekly test sequence is not an actual smoke event and is only intended to activate the system to ensure that the components are working correctly. Although NFPA 92B is only referenced for design, Sections 7.3.1 and 7.3.8 of that standard coordinate with this section.

More specifically, Section 7.3.8 also requires the weekly test but as provided for by the UL 864-UUKL-listed smoke control panel. NFPA 92B Section 7.3.6.2 requires off-normal indication at the smoke control panel within 200 seconds when a positive confirmation is failed to be achieved. Section 513.12 only requires that abnormal conditions be reported weekly.

[F] 513.12.1 Wiring. In addition to meeting the requirements of NFPA 70, all wiring, regardless of voltage, shall be fully enclosed within continuous raceways.

❖ Wiring is required to be placed within continuous raceways, which provides an additional level of reliability for the system. The definition of the term "Raceway" in NFPA 70 lists several acceptable types of complying raceway that can be used, however manufactured cable assemblies such as metal-clad cable (Type MC) or armored cable (Type AC) are not included.

[F] 513.12.2 Activation. Smoke control systems shall be activated in accordance with the *International Building Code* or the *International Fire Code*.

❖ The activation of a smoke control system is dependent on when such a system is required. Mechanical smoke control systems, which could include pressurization, airflow or exhaust methods, require an automatic activation mechanism. When using a passive system, which depends upon compartmentation, spot-type detectors are acceptable for the release of door closers and similar openings. Whereas with more complex mechanical systems, such activation needs to go beyond single station detectors and be part of an automatic coordinated system.

[F] 513.12.3 Automatic control. Where completely automatic control is required or used, the automatic control sequences shall be initiated from an appropriately zoned automatic sprinkler system complying with Section 903.3.1.1 of the *International Fire Code*, from manual controls that are readily accessible to the fire department, and any smoke detectors required by engineering analysis.

❖ When automatic activation is required, it must be accomplished by a properly zoned automatic sprinkler system and, if the engineering analysis requires them, smoke detectors. Manual control for the fire department needs to be provided. An important point with this particular requirement is that smoke control systems are engineered systems and a prescribed smoke detection system may not fit the needs of the specific design. Other types of detectors, such as beam detectors (within an atrium), may be used and could be more useful and more practical from a maintenance standpoint. Also, it may not be practical or appropriate for the building's fire alarm system to activate such systems, as it may alter the effectiveness of the system by pulling smoke through the building versus removing or containing the smoke. For example, a building with an atrium may have several floors below the space. If a fire occurs in one of the floors

not associated with the atrium, the atrium smoke control system could possibly pull smoke throughout the building if the detection is zoned incorrectly.

[F] 513.13 Control-air tubing. Control-air tubing shall be of sufficient size to meet the required response times. Tubing shall be flushed clean and dry prior to final connections. Tubing shall be adequately supported and protected from damage. Tubing passing through concrete or masonry shall be sleeved and protected from abrasion and electrolytic action.

❖ Control tubing is a method that uses pneumatics to operate components such as the opening and closing of dampers. Due to the sophistication of electronic systems today, control tubing is becoming less common.

These particular requirements provide the criteria for properly designing and installing control tubing. Essentially, it is up to the design professional to determine the size requirements and to properly design appropriate supports. This information needs to be detailed within the construction documents. Additionally, due to the effect of moisture and other contaminants on control tubing, it must be flushed clean then dried before installation.

[F] 513.13.1 Materials. Control-air tubing shall be hard-drawn copper, Type L, ACR in accordance with ASTM B 42, ASTM B 43, ASTM B 68, ASTM B 88, ASTM B 251 and ASTM B 280. Fittings shall be wrought copper or brass, solder type in accordance with ASME B 16.18 or ASME B 16.22. Changes in direction shall be made with appropriate tool bends. Brass compression-type fittings shall be used at final connection to devices; other joints shall be brazed using a BCuP5 brazing alloy with solidus above 1,100°F (593°C) and liquids below 1,500°F (816°C). Brazing flux shall be used on copper-to-brass joints only.

Exception: Nonmetallic tubing used within control panels and at the final connection to devices provided all of the following conditions are met:

1. Tubing shall comply with the requirements of Section 602.2.1.3.

2. Tubing and connected device shall be completely enclosed within a galvanized or paint-grade steel enclosure having a minimum thickness of 0.0296 inch (0.7534 mm) (No. 22 gage). Entry to the enclosure shall be by copper tubing with a protective grommet of Neoprene or Teflon or by suitable brass compression to male barbed adapter.

3. Tubing shall be identified by appropriately documented coding.

4. Tubing shall be neatly tied and supported within the enclosure. Tubing bridging cabinets and doors or moveable devices shall be of sufficient length to avoid tension and excessive stress. Tubing shall be protected against abrasion. Tubing serving devices on doors shall be fastened along hinges.

❖ This section addresses the materials allowed for control air tubing along with approved methods of connection. All of this information needs to be documented, as it will be subject to review by the special inspector.

[F] 513.13.2 Isolation from other functions. Control tubing serving other than smoke control functions shall be isolated by automatic isolation valves or shall be an independent system.

❖ This section requires separation of control tubing used for other functions through the use of isolation valves or a completely separate system. This is due to the difference in requirements for control tubing used in a smoke control system versus other building systems. The isolation of the control air tubing for a smoke control system needs to be specifically noted on the construction documents.

[F] 513.13.3 Testing. Test control-air tubing at three times the operating pressure for not less than 30 minutes without any noticeable loss in gauge pressure prior to final connection to devices.

❖ As part of the acceptance testing of the smoke control system, the control air tubing will be pressure tested three times the operating pressure for 30 minutes or more. The performance criteria as to whether the control tubing is considered a failure is when there is any noticeable loss in gauge pressure prior to final connection of devices during the 30-minute duration test.

[F] 513.14 Marking and identification. The detection and control systems shall be clearly marked at all junctions, accesses and terminations.

❖ This section requires that all portions of the fire detection system that activate the smoke control system be marked and identified appropriately. This includes all applicable fire alarm-initiating devices, the respective junction boxes, all data-gathering panels and fire alarm control panels. Additionally, all components of the smoke control system, which are not considered a fire detection system, are required to be properly identified and marked. This would include all applicable junction boxes, control tubing, temperature control modules, relays, damper sensors, automatic door sensors and air movement sensors.

[F] 513.15 Control diagrams. Identical control diagrams shall be provided and maintained as required by the *International Fire Code*.

❖ Control diagrams provide consistent information on the system in several key locations, including the building department, the fire department and the fire command center. If a fire command center is not required or provided the diagrams need to be located such that they can be readily accessed during an emergency. Some possible locations may be the security office, building managers' office or, if possible, within the smoke control panel. This information is intended to assist in the use and operation of the smoke control system. The format of the control diagram must be approved by the fire chief because the fire department is the agency that will be using the system during a fire as well as observing system tests in the future. The more clearly the information is communicated, the more effective the smoke control system will be.

Note that the fire department may want all smoke control systems within a jurisdiction to follow a particular protocol for control diagrams.

Generally, the control diagrams should indicate required reaction of the system in all scenarios. The status or position of every fan and damper in every scenario must be clearly identified.

[F] 513.16 Fire fighter's smoke control panel. A fire fighter's smoke control panel for fire department emergency response purposes only shall be provided in accordance with the *International Fire Code*.

❖ One of the elements that make a smoke control system effective is successful communication of system activity to the fire department. Another is giving the fire department the ability to manually operate the system. Section 909.16 of the IFC outlines requirements for a control panel specifically for smoke control systems located in the fire command center. Two components include the requirements for the display and also for the controls. When a fire command center is not required, the control panel should be located with the fire alarm panel. The specific location will depend on the needs of the fire department in a particular jurisdiction. This control panel will provide an ability to override any other controls within the building as they relate to the smoke control system, whether manual or automatic.

[F] 513.17 System response time. Smoke control system activation shall comply with the *International Fire Code*.

❖ Section 909.17 of the IFC defines when the smoke control system must begin operation. Whether the activation is manual or automatic, the system must be initiated immediately. Also, the IFC requires that components activate in a sequence that will not potentially damage the fans, dampers, ducts and other equipment. Unrealistically accounting for timing of the system has the potential of creating an unsuccessful system. Delays in the system can be seen in slow dampers, fans that ramp up or down, systems that poll slowly and intentional built-in delays. These factors can add significantly to the reaction time of the system and may prevent the system from achieving the design goals.

The key element is that the system be fully operational before the smoke conditions exceed the design parameters. The sequence of events must be justified in the design analysis and described clearly in the construction documents.

[F] 513.18 Acceptance testing. Devices, *equipment*, components and sequences shall be tested in accordance with the *International Fire Code*.

❖ In order to achieve a certain level of performance, the smoke control system needs to be thoroughly tested. Section 909.18 of the IFC requires that all devices, equipment components and sequences be individually tested. The IFC addresses components such as detection devices, smoke barriers and ductwork. In addition, the IFC requires a qualified special inspector for acceptance testing. Acceptance testing is required to be properly documented and reports prepared for the fire code official.

[F] 513.19 System acceptance. Acceptance of the smoke control system shall be in accordance with the *International Fire Code*.

❖ The IFC and IBC stipulate that the certificate of occupancy cannot be issued unless the smoke control system has been accepted. It is essential that the system be inspected and approved since it is a life safety system. There is an exception to these requirements in Section 909.19 of the IFC for buildings that are constructed in phases where a temporary certificate of occupancy is allowed. An example would be a building where the portion that requires smoke control is not yet occupied; therefore, the performance of the smoke control system is not an issue. The exception does require that a significant hazard is not placed upon the occupants due to the unfinished portion of the building.

The IFC also requires a maintenance program for smoke control systems as the long-term success of such systems depends heavily on proper maintenance in addition to rigorous acceptance testing. The code simply provides a reference to that section of the IFC.

SECTION 514
ENERGY RECOVERY VENTILATION SYSTEMS

514.1 General. Energy recovery ventilation systems shall be installed in accordance with this section. Where required for purposes of energy conservation, energy recovery ventilation systems shall also comply with the *International Energy Conservation Code*. Ducted heat recovery ventilators shall be listed and labeled in accordance with UL 1812. Nonducted heat recovery ventilators shall be listed and labeled in accordance with UL 1815.

❖ This section addresses the provisions for installing energy recovery ventilation (ERV) systems and equipment (see definition in Section 202). ERV systems have some unique features and the specific requirements and limitations given in Sections 514.2 and 514.3 are necessary to regulate installation of these systems.

There are several different types of ERV systems. ERVs have tremendous potential for energy savings because they are designed to reduce both heat energy losses in the heating season and heat energy gains in the cooling season. ERVs can also be equipped with heat exchangers that reclaim or reject latent heat by transferring water vapor from the exhaust air stream to the makeup air stream or vice versa. ERVs can significantly lower heating loads by extracting heat from exhaust air, such as toilet room and ventilation air exhaust systems, that would otherwise be lost to the outdoors. In the cooling season, ERVs can significantly reduce cooling loads by rejecting heat and water vapor that would otherwise be brought into the building with the makeup air and ventilation air.

This section works with Item 4 of Section 403.2.1, which prohibits recirculation except as required by Table 403.3 (see Note g), which allows up to 10 percent of exhaust to be recirculated in the resulting supply air stream from the ERV system. The allowance for 10-percent air from such spaces is to allow the use of technology that recovers energy from exhaust air to reduce the load on heating and cooling systems. Air from such spaces designated by Note g in Table 403.3 can be part of the supply air to other spaces as long as it does not exceed 10 percent of the supply air (10-percent exhaust air from occupancies designated with Note g from Table 403.3 plus 90-percent outdoor air). Chapter 5 of the *International Energy Conservation Code*® (IECC®) also requires the use of ERV systems consistent with ASHRAE 90.1 and this section.

The 10-percent allowance removes the absolute ban on recirculation that presented a significant barrier to the use of energy recovery technology. The lack of opportunity to recover energy from the exhaust air of toilet and locker rooms unnecessarily limited the overall effectiveness and feasibility of ERV technology (i.e., the energy in that air is lost to the atmosphere because of the absolute ban on recirculation). Medical and research sectors represented on the ASHRAE 62 committee have long affirmed that there is no health issue related to toilet room exhaust. The issue is one of odor perception. In the broadest sense, odor perception is not a health and safety issue and should therefore not be used as a basis for determining code requirements. Further, if the ventilation systems in which the air being exhausted are a mixture of air from toilet rooms and other spaces where recirculation is allowed, the problem is compounded (i.e., none of that air can be subject to energy recovery since some of the leakage would involve toilet room air). ERV technology is a highly effective and cost beneficial means of energy conservation. ASHRAE 90.1 and the IECC require the use of ERV systems within certain threshold conditions.

Most of the various types of ERV systems on the market today do not completely eliminate leakage between the exhaust and supply airstreams. Many systems on the market have been engineered to the point that actual leakage is under 5 percent. The most effective technologies have some inherent leakage and even in cases where additional measures, such as purge, could be used to reduce the leakage, the additional equipment and operating costs are hard to justify. The 10-percent allowance is consistent with ASHRAE 62 and also provides a level playing field for the entire industry. The 10-percent limitation is one that all current suppliers of ERV systems can meet and will therefore not have the effect of creating an advantage for one type of system or one manufacturer over another. Leakage is a certified value in the *ARI Certified Products Directory for ERVs* and is supported by industry catalogs and software tools.

Exhaust and ventilation have long been at odds with energy conservation because makeup air and ventilation air have to be conditioned nearly year round in most climates. With ERV technology, the energy price tag for exhaust systems and ventilation can be reduced significantly. Some ERV technology can be up to 80-percent efficient in reclaiming or rejecting heat energy:

* Static heat exchanger;
* Rotating (wheel or drum) heat exchanger; and
* Sealed tube heat exchanger containing a refrigerant.

Some static and some rotating heat exchangers use desiccants or molecular sieve technology to selectively transfer certain vapors or gases, while selectively rejecting others.

ERV systems may be voluntarily installed or they may, in certain conditions, be required for purposes of energy conservation. In either case, it is important that these systems be installed to comply with the requirements in Sections 514.2 and 514.3 [see Section 403.1 and Figure 403.1(1)].

514.2 Prohibited applications. Energy recovery ventilation systems shall not be used in the following systems:

1. Hazardous exhaust systems covered in Section 510.
2. Dust, stock and refuse systems that convey explosive or flammable vapors, fumes or dust.
3. Smoke control systems covered in Section 513.
4. Commercial kitchen exhaust systems serving Type I and Type II hoods.
5. Clothes dryer exhaust systems covered in Section 504.

❖ All of the systems addressed by this section have the potential to foul or attack the ERV components because of the particulates and chemicals entrained in the exhaust airflow, such as dust, grease, smoke, lint, vapors, gases, etc. There is also the possibility that a portion of these hazardous contaminants could be recirculated back into the occupied areas of the building. Some ERV designs, such as those using rotating heat exchangers, will allow some amount of cross-leakage (air exchange) between the exhaust air stream and the makeup air stream. This must be avoided where the contaminants in the exhaust air are hazardous.

514.3 Access. A means of access shall be provided to the heat exchanger and other components of the system as required for service, maintenance, repair or replacement.

❖ Access openings must be provided to any components of the ERV system that could require periodic inspection, maintenance or repair (see commentary for the definition of "Access," Section 202). Typical designs use filters and heat exchangers that must be cleaned or replaced periodically

514.4 Recirculated air. Air conveyed within energy recovery systems shall not be considered as recirculated air where the energy recovery ventilation system is constructed to limit cross-leakage between air streams to less than 10 percent of the total airflow design capacity.

❖ This section was created because some people were interpreting the sections that prohibit the recirculation of air as prohibiting the use of energy recovery ventilation systems. As long as the cross contamination between the supply and exhaust air streams is less than 10 percent, the air from an energy recovery ventilation system is not considered recirculated.

Bibliography

The following resource materials were used in the preparation of the commentary for this chapter of the code:

IBC-12, *International Building Code*. Washington, DC: International Code Council, 2011.

IFC-12, *International Fire Code*. Washington, DC: International Code Council, 2011.

IFGC-12, *International Fuel Gas Code*. Washington, DC: International Code Council, 2011.

Klote, J. H. and Milke, J.A., *Design of Smoke Management Systems*. Atlanta, GA: ASHRAE/SFPE, 1992.

NFPA 92A-09, *Smoke Control Systems Utilizing Barriers and Pressure Differences*. Quincy, MA: National Fire Protection Association, 2009.

NFPA 92B-05, *Smoke Management Systems in Malls, Atria and Large Areas*. Quincy, MA: National Fire Protection Association, 2005.

Chapter 6:
Duct Systems

General Comments

Chapter 6 of the code addresses duct systems by instituting requirements for the protection of the occupants of the building and the building itself. This code regulates the materials and methods used for constructing and installing ducts, system controls, exhausts, fire protection systems and related components that affect the overall performance of a building's air distribution system.

Other chapters of the code contain provisions relative to Chapter 6, including Chapters 3, 4, and 5. The *International Building Code®* (IBC®) is referenced repeatedly for requirements pertaining to fire protection measures such as fireblocking, draftstopping, fire-resistance-rated assemblies, penetration protection and fire protection and suppression systems.

Purpose

Chapter 6 contains reasonable protection to life and property from the hazards associated with air-moving equipment and systems. This chapter states requirements for the installation of supply, return and exhaust air systems. Chapter 6 contains certain information on the design of these systems from the standpoint of air movement, and is concerned with the structural integrity of the systems and the overall impact of the systems on the fire-safety performance of the building. Design considerations, such as duct sizing, maximum efficiency, cost effectiveness, occupant comfort and convenience are the responsibility of the design professional.

SECTION 601
GENERAL

601.1 Scope. Duct systems used for the movement of air in air-conditioning, heating, ventilating and exhaust systems shall conform to the provisions of this chapter except as otherwise specified in Chapters 5 and 7.

> **Exception:** Ducts discharging combustible material directly into any *combustion* chamber shall conform to the requirements of NFPA 82.

❖ Chapter 6 governs the construction, installation, alteration, maintenance and repair of duct systems used for the movement of environmental air (see commentary, Sections 501.5, and 1401.5 and the definition of "Duct system"). Duct systems are addressed mostly within the context of their impact on the performance of the building during a fire.

Duct system materials and their installation are regulated to address the hazards associated with their burning characteristics and contribution to the spread of flame and smoke within a building. This chapter also addresses the relationship between an air distribution system and other building requirements, such as floodproofing and structural integrity.

Chapter 5 contains exhaust system requirements. Additional information can be obtained from sources, such as the American Society of Heating, Refrigeration and Air-Conditioning Engineers, Inc. (ASHRAE), the Sheet Metal and Air-Conditioning Contractors National Association, Inc. (SMACNA) and Air Conditioning Contractors of America (ACCA).

The requirements of Chapter 6 apply to new construction, as well as to alterations, additions, repairs and maintenance of existing air distribution systems and ducts.

The exception requires duct systems that discharge combustible materials directly to an incinerator, boiler, furnace or other combustion chamber to be designed, constructed and installed in compliance with the requirements of NFPA 82.

[B] 601.2 Air movement in egress elements. Corridors shall not serve as supply, return, exhaust, relief or *ventilation air* ducts.

> **Exceptions:**
> 1. Use of a corridor as a source of *makeup air* for exhaust systems in rooms that open directly onto such corridors, including toilet rooms, bathrooms, dressing rooms, smoking lounges and janitor closets, shall be permitted, provided that each such corridor is directly supplied with outdoor air at a rate greater than the rate of *makeup air* taken from the corridor.
> 2. Where located within a *dwelling unit*, the use of corridors for conveying return air shall not be prohibited.
> 3. Where located within tenant spaces of 1,000 square feet (93 m²) or less in area, use of corridors for conveying return air is permitted.
> 4. Incidental air movement from pressurized rooms within health care facilities, provided that the corri-

dor is not the primary source of supply or return to the room.

❖ This section prohibits corridors, whether they are fire-resistance rated or not, from being used as air distribution system components because of the potential for spreading smoke and fire into elements of the building's required means of egress. Smoke could make those egress elements unusable. Note that Section 602.1 defines "Plenums" as being unoccupiable spaces and cavities; thus, a corridor is never a plenum.

Corridors are designed to protect the people using them. The intent of this section is to prohibit air movement that would introduce smoke into the corridor. This section does not prohibit the air movement necessary for ventilation and space conditioning of corridors, but does prevent them from serving as conduits for the distribution or transport of air to, or the collection of air from, spaces other than the corridor. Air movement in a corridor must be restricted to that necessary for ventilating, heating, cooling and conditioning (see Section 401.4).

Additionally, opening protectives must be installed in all openings that penetrate assemblies that are fire-resistance rated (see commentary, Section 607). This section would prohibit transfer grilles and transoms over doors that connect adjacent rooms and spaces

with corridors when those openings are used to convey air to and from the adjacent rooms and spaces (see IBC Section 716.5.3).

At times a pressure differential across corridor doors will be required. For example, positive pressures are maintained in hospital and other institutional corridors to prevent the migration of infectious organisms and odors into the corridors, and negative pressures are often maintained in kitchens to prevent odor migration into dining rooms. This section is not intended to restrict those pressure differences; however, it does prohibit the use of the corridor as the source of supply air to, or the means of providing return air from, adjacent spaces.

Exception 1 addresses the common practice of using air from the corridor as makeup air for small exhaust fans in adjacent rooms. When the corridor is supplied directly with outdoor air at a rate greater than the makeup air rate, positive pressure will be created in the corridor with respect to the adjoining rooms and smoke would not be drawn into the corridor [see Commentary Figures 601.2(1) and 601.2(2)].

Exception 2 permits using a corridor within a dwelling unit as a return air duct. Dwelling units may have unprotected openings between floors. Exit access corridors in dwelling units serve small occupant loads, are short in length and need not be fire-resistance rated. For these reasons, the use of the corri-

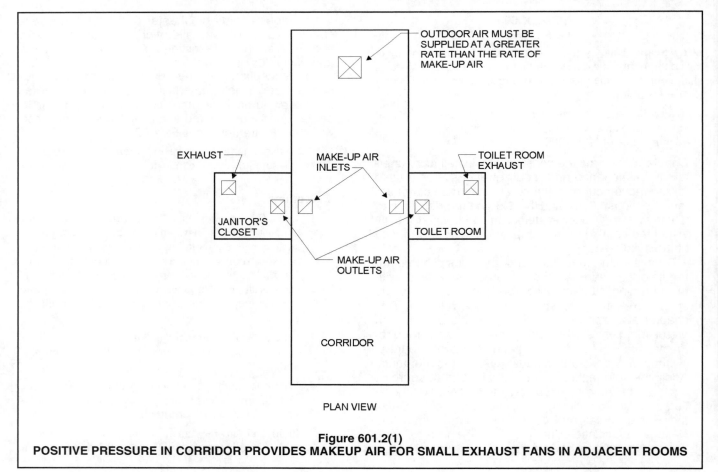

Figure 601.2(1)
POSITIVE PRESSURE IN CORRIDOR PROVIDES MAKEUP AIR FOR SMALL EXHAUST FANS IN ADJACENT ROOMS

dor or the space above a corridor ceiling for return air does not constitute an unacceptable hazard. It is a common practice to locate return air openings in the corridors of dwelling units and draw return air from adjoining spaces through the corridor.

Exception 3 permits use of corridors located in small occupancies for return air based on the small size of the tenant space, the small occupant load and the fact that the corridor length will be relatively short.

Section 602.1 limits plenums to specific uninhabited, unoccupiable spaces that do not include corridors. This is why the exceptions to this section do not refer to corridors as plenums.

Exception 4 recognizes that some rooms in health care facilities require positive pressurization to significantly reduce the spreading of germs and contaminants into the room. While the air moving from the room into the corridor is return air, and the corridor carries that air to the return air intake for the corridor, the amount of air is considered to be incidental and not considered to be a safety hazard.

[B] 601.2.1 Corridor ceiling. Use of the space between the corridor ceiling and the floor or roof structure above as a return air *plenum* is permitted for one or more of the following conditions:

1. The corridor is not required to be of fire-resistance-rated construction;

2. The corridor is separated from the *plenum* by fire-resistance-rated construction;

3. The air-handling system serving the corridor is shut down upon activation of the air-handling unit smoke detectors required by this code;

4. The air-handling system serving the corridor is shut down upon detection of sprinkler waterflow where the building is equipped throughout with an automatic sprinkler system; or

5. The space between the corridor ceiling and the floor or roof structure above the corridor is used as a component of an *approved* engineered smoke control system.

❖ This section permits use of the space between the corridor ceiling and the floor or roof above such ceiling as a return air plenum when one or more of five specified conditions are met. Using the space above a corridor ceiling as a plenum is not specifically prohibited by Section 601.2. However, this section implies that intent by limiting such use to the stated conditions. The plenum is limited to return air applications only because, in this application, the plenum space will be under a negative pressure with respect to the corridor. Thus, the pressure difference will assist in containing smoke and gases within the plenum space. In a supply plenum, however, the air in the plenum will be at a positive pressure with respect to the corridor, thereby increasing the likelihood that smoke and gases will infiltrate the corridor.

Section 1018.1 of the IBC determines where corridors are required to be fire-resistance rated. Where a corridor is not required to be fire-resistance rated, Condition 1 of this section allows for the space above the corridor ceiling to serve as a return air plenum without requiring the plenum to be separated from the corridor by fire-resistance-rated construction [see Commentary Figure 601.2.1(1)]. If the corridor was required to be fire-resistance rated, the space above

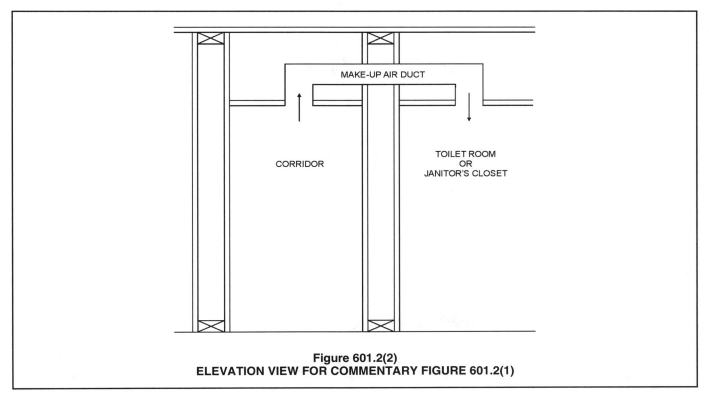

Figure 601.2(2)
ELEVATION VIEW FOR COMMENTARY FIGURE 601.2(1)

the corridor ceiling is prohibited from being used as a plenum [see Commentary Figure 601.2.1(2)].

Condition 2 is applicable to corridors that are enclosed with fire-resistance-rated construction. The plenum must be separated from the corridor by a ceiling membrane of fire-resistance-rated construction equivalent to the required rating of the corridor enclosure itself [see Commentary Figure 601.2.1(3)].

Conditions 3 and 4 recognize that the hazard associated with smoke spread through a plenum is minimized if the air movement is stopped upon receipt of a signal from the smoke detectors or the automatic sprinkler system. The intention of Conditions 3 and 4 is that any air-handling equipment that is moving air across or through a corridor in the space between the corridor ceiling and the floor or roof be shut down upon detection of smoke or fire regardless of the cubic foot per minute (cfm) rating of the air handler [see Commentary Figure 601.2.1(4)].

It is not uncommon for the space above a nonrated ceiling in a fire-resistance-rated corridor to be used as part of an approved engineered smoke removal system. In such cases, Condition 5 allows the space to serve as a plenum [see Commentary Figure 601.2.1(5)].

[B] 601.3 Exits. *Equipment* and ductwork for exit enclosure ventilation shall comply with one of the following items:

1. Such *equipment* and ductwork shall be located exterior to the building and shall be directly connected to the exit enclosure by ductwork enclosed in construction as required by the *International Building Code* for shafts.

2. Where such *equipment* and ductwork is located within the exit enclosure, the intake air shall be taken directly from the outdoors and the *exhaust air* shall be discharged directly to the outdoors, or such air shall be conveyed through ducts enclosed in construction as required by the *International Building Code* for shafts.

3. Where located within the building, such *equipment* and ductwork shall be separated from the remainder of the building, including other mechanical *equipment*, with construction as required by the *International Building Code* for shafts.

In each case, openings into fire-resistance-rated construction shall be limited to those needed for maintenance and operation and shall be protected by self-closing fire-resistance-rated devices in accordance with the *International Building Code* for enclosure wall opening protectives. Exit enclosure ventilation systems shall be independent of other building ventilation systems.

❖ Where exit enclosures are ventilated by mechanical systems, the mechanical equipment and duct systems must be installed as required by this section. These requirements are intended to maintain a separation of exit enclosure ventilation systems from other portions of the building and other building mechanical systems, and are intended to protect the exit enclosures from the spread of smoke from other areas. The key point is that heating, ventilating and air-conditioning (HVAC) systems serving an exit enclosure must not serve any other room or space.

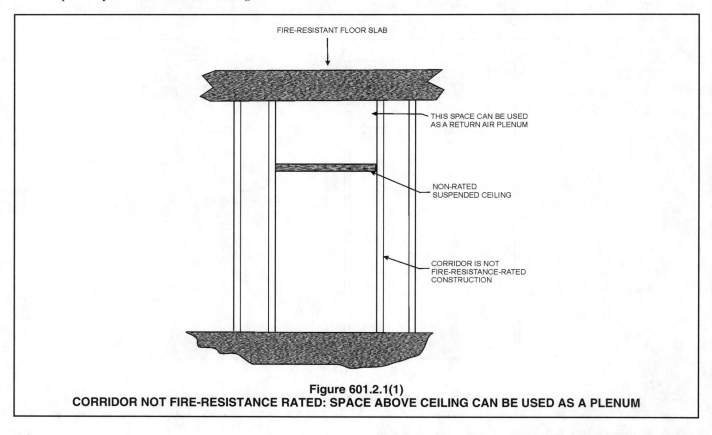

Figure 601.2.1(1)
CORRIDOR NOT FIRE-RESISTANCE RATED: SPACE ABOVE CEILING CAN BE USED AS A PLENUM

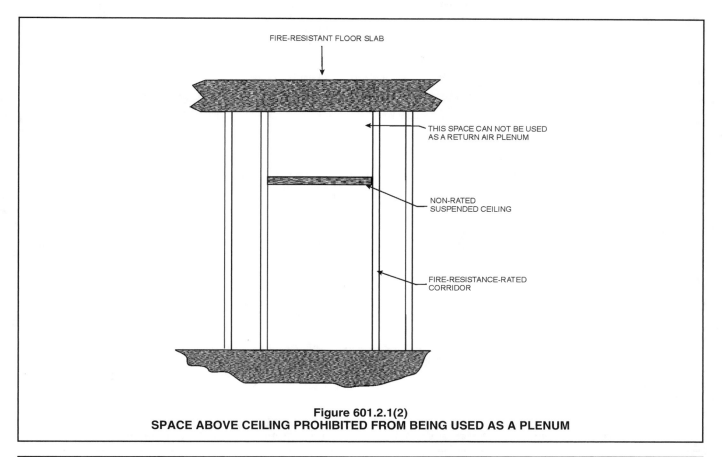

Figure 601.2.1(2)
SPACE ABOVE CEILING PROHIBITED FROM BEING USED AS A PLENUM

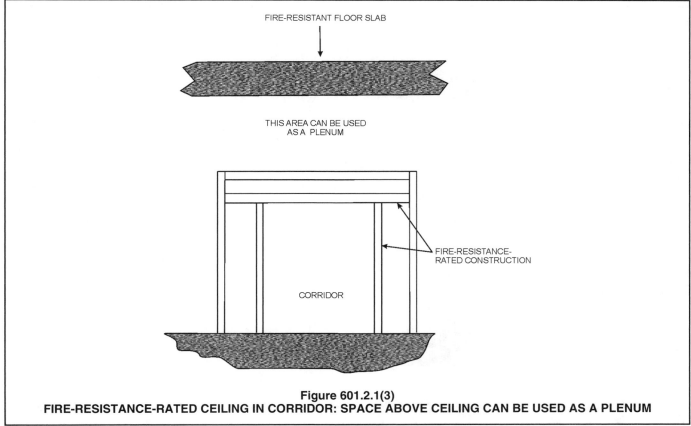

Figure 601.2.1(3)
FIRE-RESISTANCE-RATED CEILING IN CORRIDOR: SPACE ABOVE CEILING CAN BE USED AS A PLENUM

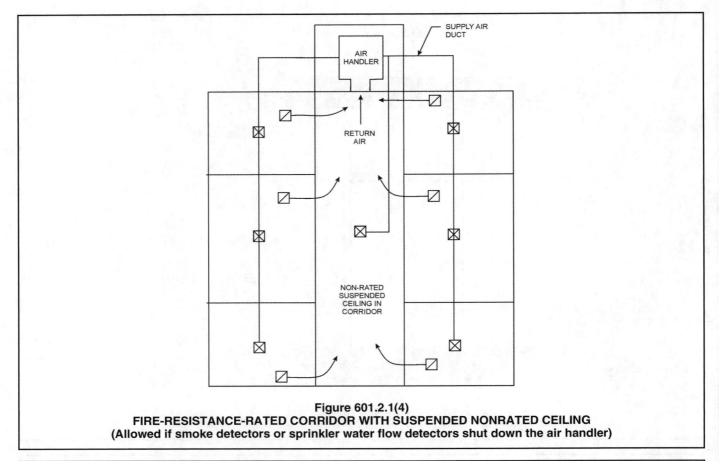

Figure 601.2.1(4)
FIRE-RESISTANCE-RATED CORRIDOR WITH SUSPENDED NONRATED CEILING
(Allowed if smoke detectors or sprinkler water flow detectors shut down the air handler)

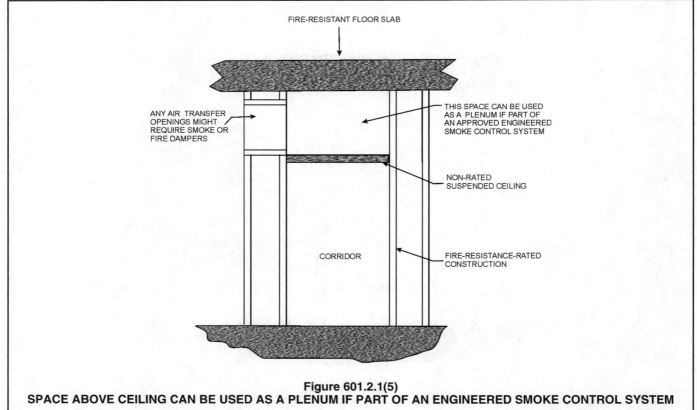

Figure 601.2.1(5)
SPACE ABOVE CEILING CAN BE USED AS A PLENUM IF PART OF AN ENGINEERED SMOKE CONTROL SYSTEM

601.4 Contamination prevention. Exhaust ducts under positive pressure, chimneys and vents shall not extend into or pass through ducts or plenums.

Exceptions:

1. Exhaust systems located in ceiling return air plenums over spaces that are permitted to have 10 percent recirculation in accordance with Section 403.2.1, Item 4. The exhaust duct joints, seams and connections shall comply with Section 603.9.

2. This section shall not apply to chimneys and vents that pass through plenums where such venting systems comply with one of the following requirements:

    2.1. The venting system shall be listed for positive pressure applications and shall be sealed in accordance with the vent manufacturer's instructions.

    2.2. The venting system shall be installed such that fittings and joints between sections are not installed in the above ceiling space.

    2.3. The venting system shall be installed in a conduit or enclosure with sealed joints separating the interior of the conduit or enclosure from the ceiling space.

❖ To prevent cross contamination, exhaust ducts under positive pressure, as well as chimneys and vents, are not permitted to extend into or pass through ducts or plenums. Any exhaust duct, vent or chimney leakage could escape into the air duct because of the pressure differential. Also, a negative duct or plenum pressure could affect draft and/or draw combustion gases from a natural draft chimney or vent (see Section 602.1).

Exception 1 applies only where the exhaust duct is conveying air taken from the space below the plenum. This exception recognizes the specific situation of an exhaust duct located in a ceiling return air plenum over a space that the air is being exhausted from. If the space is allowed to have up to 10 percent air recirculation in accordance with Section 403.2.1, any leakage from the exhaust duct to the plenum would be insignificant considering the recirculation allowance of Section 403.2.1, Item 4, for such spaces.

Exception 2 allows for specific installations of chimneys and vents to pass through a plenum. This exception is based on Section 503.3.6 of the *International Fuel Gas Code®* (IFGC®). One arrangement is where the venting system is listed for positive pressure service and the joints are sealed according to the manufacturer's instructions. A listed positive pressure vent, when installed properly, should have extremely low leakage of gases to the exterior of the vent. Another arrangement would be if the venting system had no joints or fittings within the plenum space. The absence of any joints eliminates potential leakage points and, therefore, no leakage of gases into the plenum would occur. Finally, if the venting

system is installed within a sealed conduit, then any leakage from the venting system would be channeled out of the plenum (see Figure 503.3.6 of the IFGC commentary).

SECTION 602
PLENUMS

602.1 General. Supply, return, exhaust, relief and *ventilation air* plenums shall be limited to uninhabited crawl spaces, areas above a ceiling or below the floor, attic spaces and mechanical *equipment* rooms. Plenums shall be limited to one fire area. Fuel-fired appliances shall not be installed within a *plenum*.

❖ The definition of the term "Plenum" in Section 202 states that a plenum is not an occupiable space being conditioned. This section addresses the use of plenums as part of air distribution systems in buildings. A plenum is an unoccupiable enclosed portion (cavity) of the building structure that is designed to allow the movement of air, thereby forming part of an air distribution system. Plenums can be used for supply, return, exhaust, relief and ventilation air, and can occur in ceiling, attic or under-floor spaces, mechanical equipment rooms (air handler rooms) and in stud and joist cavities.

A mechanical equipment room is occupiable and therefore is not technically a plenum. A mechanical equipment room is unique in that it is an enclosure that can be used as a ductwork plenum and can be occupied for maintenance. The plenums described in this section are commonly used to reduce construction costs by taking advantage of the existing concealed spaces created as part of the building construction. Plenums are permitted in all building construction types. Commentary Figure 602.1(1) shows a typical floor/ceiling or roof/ceiling return air plenum. A mechanical equipment room and exit access corridors in dwelling units and specific small occupancies are the only exceptions where an occupiable room or an inhabited space is permitted to serve as an air conduit [see Commentary Figure 602.1(2) and commentary, Section 601.2]. An occupied room, other than a mechanical equipment room, is never considered to be a plenum in the context of Section 602.

Limiting the location of air distribution system plenums reduces the potential for the spread of fire and smoke through the air distribution system. To prevent the rapid spread of smoke through occupied areas, plenums are restricted to concealed and uninhabited spaces within a building. Plenums are limited to one fire area to allow for the containment of fire and smoke, thereby reducing the risk of it spreading to other areas of the building. A "fire area" is a floor area that is enclosed and bounded by fire walls, fire barriers, fire-resistance-rated horizontal assemblies or exterior walls of the building. Plenums must be designed to maintain the integrity of the fire area enclosure.

Consider return air ceiling plenums over two fire areas that are linked together by transfer openings. Fire dampers are installed in the transfer openings, thereby maintaining the integrity of the fire areas. This means that the two return plenums are discrete and are each limited to a single fire area. However, this begs the question, when would a plenum not be limited to a single fire area? This presents a puzzle in that the code appears to restrict a plenum arrangement that could never exist because of other code provisions that regulate fire areas and fire barriers.

The conservative interpretation of the restriction of plenums to a single fire area is that the code intends to prohibit plenums in different fire areas from being linked through protected transfer openings in fire barriers, thus requiring each fire area plenum to be individually ducted back to the air handler [see Commentary Figure 602.1(3)].

Fuel-fired appliances must not be installed within a plenum because of the potential for toxic products of combustion to be spread throughout the building in the event that the appliance, equipment or venting system malfunctions. Also, the negative or positive pressures developed within a plenum space can seriously affect the operation of fuel-fired appliances located in it. For example, if a natural-draft fuel-fired

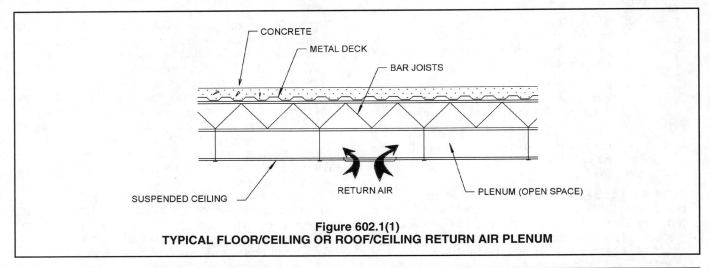

Figure 602.1(1)
TYPICAL FLOOR/CEILING OR ROOF/CEILING RETURN AIR PLENUM

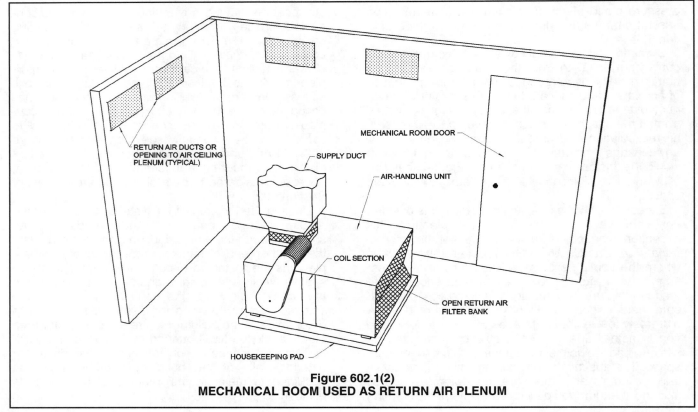

Figure 602.1(2)
MECHANICAL ROOM USED AS RETURN AIR PLENUM

appliance was installed in a return air plenum, the negative pressure of the plenum could overcome the draft of the appliance venting system and flue gases could be drawn into the plenum space. Also, improper operation of the appliance would greatly accelerate the deterioration of the appliance and its vent.

602.2 Construction. *Plenum* enclosures shall be constructed of materials permitted for the type of construction classification of the building.

The use of gypsum boards to form plenums shall be limited to systems where the air temperatures do not exceed 125°F (52°C) and the building and mechanical system design conditions are such that the gypsum board surface temperature will be maintained above the airstream dew-point temperature. Air plenums formed by gypsum boards shall not be incorporated in air-handling systems utilizing evaporative coolers.

❖ A building of noncombustible construction as defined in the IBC is permitted to have only noncombustible plenums constructed of noncombustible materials. The intent of requiring noncombustible plenums in buildings of noncombustible construction is to maintain consistency between the materials used in plenum construction and the materials used in building construction. Materials classified as noncombustible in the IBC are used to construct the plenum. These are materials that have been tested to the requirements of ASTM E 136.

In general, noncombustible plenums are those formed and bounded by noncombustible materials and contain few, if any, combustible materials within the plenum space. These plenums represent a relatively low risk for fire spread and smoke generation because of the limited amount of combustible materials present. Although some combustible materials may be exposed within the plenum space, the types and amounts of these materials are strictly limited by Section 602.2.1. The area of a noncombustible plenum is limited only by the size of the fire area containing it.

A building of combustible construction is permitted by the IBC to use combustible or noncombustible building materials for specified components of the structure. In these buildings, either noncombustible plenums or combustible plenums can be used. Again, the intent is to require that the plenum be constructed of or formed by materials that are consistent with the materials allowed for building construction.

A combustible plenum may be constructed using any approved combustible or noncombustible material that is permitted for combustible construction in accordance with the IBC. The flame spread and

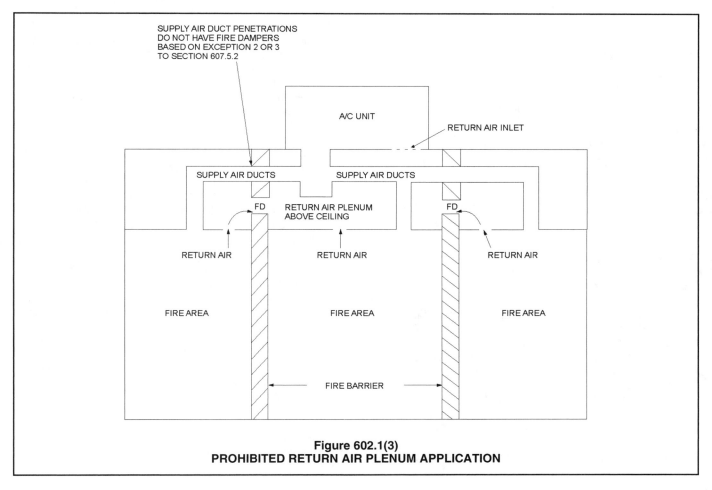

Figure 602.1(3)
PROHIBITED RETURN AIR PLENUM APPLICATION

smoke-developed indexes of all materials within a combustible plenum must also comply with the applicable sections of the IBC. Combustible materials, such as ducts, pipe, wire, tubing and insulation coverings may be exposed within the plenum only in accordance with Section 602.2.1.

Note, however, that in buildings that are not permitted to have concealed spaces, the use of a plenum is not feasible. See the IBC for the definitions of the different construction types.

The airstream and surface temperature restrictions on the use of gypsum boards in plenums are intended to prevent premature deterioration of the gypsum board because of condensation on the plenum walls or excessive drying of the gypsum board (see commentary, Section 603.5.1).

602.2.1 Materials within plenums. Except as required by Sections 602.2.1.1 through 602.2.1.5, materials within plenums shall be noncombustible or shall be listed and labeled as having a flame spread index of not more than 25 and a smoke-developed index of not more than 50 when tested in accordance with ASTM E 84 or UL 723.

Exceptions:

1. Rigid and flexible ducts and connectors shall conform to Section 603.

2. Duct coverings, linings, tape and connectors shall conform to Sections 603 and 604.

3. This section shall not apply to materials exposed within plenums in one- and two-family dwellings.

4. This section shall not apply to smoke detectors.

5. Combustible materials fully enclosed within one of the following:

    5.1. Continuous noncombustible raceways or enclosures.

    5.2. Approved gypsum board assemblies.

    5.3. Materials listed and labeled for installation within a plenum.

6. Materials in Group H, Division 5 fabrication areas and the areas above and below the fabrication area that share a common air recirculation path with the fabrication area.

❖ Materials located within a plenum, regardless of whether the plenum is constructed of or bounded by combustible or noncombustible materials, must be noncombustible or must have a flame spread index of 25 or less and a smoke-developed index of 50 or less when tested in accordance with ASTM E 84. The requirement for listing and labeling provides assurance from a third-party agency that the product has been tested and found to be in compliance with the requirements of the code.

This section addresses those items that are installed within plenums, not the materials that bound and create the plenum space (see Section 602.2).

The code recognizes that the plenum space is often used to accommodate components of other building systems, such as electrical, plumbing, fire protection, communication and mechanical. For this reason, the code permits exposure of limited types of combustible materials used for these systems within the plenum. A flame spread index of 25 or less and a smoke-developed index of 50 or less are viewed as acceptable values for combustibles in a plenum because of the minimal risk associated with those materials. Any hazard associated with smoke production in a plenum is also reduced by the air handler shutdown requirements of Section 606.

Previous editions of the code used the term "exposed within plenums" when addressing this issue. However, some designers and installers used that language to install plastic pipe and other combustible material with some insulation wrapped around it, claiming that the material was no longer exposed. If the wrapping material were to become damaged, loosened or destroyed in a fire, the combustible material could be exposed to the fire and produce hazardous smoke that would be spread to other parts of the building through the plenum. The word "exposed" was deleted in the 2006 edition of the code to close this loophole. Note that Exception 5 still allows electrical wiring and cable to be installed in a plenum when enclosed in noncombustible conduit.

There is no stated limit on the aggregate amount of these combustible materials that may be exposed in a plenum. For example, a test of electrical wiring using the procedures of UL 910 accounts only for the amount of wiring present within that tested assembly. Once installed in the plenum, however, additional combustible materials, such as insulated wire, pneumatic tubing and insulation, may also be present. The aggregate effect of exposing all of these materials within a plenum is usually not known.

Exception 1 permits both rigid and flexible ducts within plenums if the ducts conform to the requirements of Section 603. Metallic ducts are usually noncombustible, and nonmetallic ducts must be tested and classified in accordance with UL 181 to determine their fire and smoke characteristics. Duct materials for rigid and flexible ducts, both metallic and nonmetallic, are limited to Class 0 and Class 1 materials. Class 0 duct materials have a flame spread and a smoke-developed index of zero. Class 1 duct materials have a flame spread index of 25 or less and a smoke-developed index of 50 or less.

Exception 2 permits exposure of ductwork system materials such as vapor barriers, insulation, joint sealers, vibration connector fabrics and acoustical liners in plenums if they comply with the requirements of Sections 603 and 604, which require that all such components have a flame spread index of 25 or less and a smoke-developed index of 50 or less when tested using ASTM E 84. Also, duct coverings and linings must be tested in accordance with ASTM C 411 to demonstrate that they do not flame, glow, smolder or smoke when tested at their rated temperatures or to 250° F (121°C), whichever is higher.

Exception 3 exempts one- and two-family dwellings from the requirements of this section. It is important to note that the definition of a dwelling is a building or portion thereof that contains not more than two dwelling units. The intention of this exception is to exempt one and two unit dwellings from this section. If a building has more than two attached dwelling units, materials that are exposed in plenums will have to be noncombustible or have a flame spread index of not more than 25 and a smoke-developed index of not more than 50 when tested in accordance with ASTM E 84. It should be noted that the construction of a plenum must comply with Section 602.2.

Exception 4 allows installation of combustible smoke detectors in plenums. This section is intended to regulate materials that can be of considerable quantity and surface area, such as ducts, piping, tubing and electrical and optical conductors; however, something as small as a plastic smoke detector housing would literally fall under the limitations of Section 602.2.1. The protection provided by a smoke detector far outweighs any risk of the detector generating additional smoke in a fire situation.

Exception 5 recognizes the practice of enclosing combustible materials, such as plastic piping, in gypsum board assemblies or in insulation materials that are listed and labeled for this specific application. It also allows the installation of electrical wiring and cable in a noncombustible raceway that is completely enclosed and continuous throughout the plenum. If a combustible material is covered by an approved thermal barrier meeting the flame spread and smoke-developed index of this section and the covering will remain intact for the anticipated time for egress, the material would no longer be subject to fire exposure and the intent of the code would be met.

Exception 6 provides an exemption for materials in plenums that circulate air for Hazardous, Division 5 fabrication rooms (commonly referred to as a "clean rooms"). The interstitial spaces above a clean room and below the raised floor of a clean room represent a unique air movement system that uses high-velocity air to quickly sweep any contaminants from the room. Because a Hazardous, Division 5 fabrication area already contains materials that are combustible, the air moving in the plenum spaces represents no less of a hazard than a material in the plenum that is combustible or has flame spread and smoke-developed indexes greater than 25 and 50, respectively. In other words, because the air in the plenum comes from an area that has combustible materials, the plenum space is already contaminated with combustible material such that requiring other materials in the plenum to comply with this section would be senseless.

602.2.1.1 Wiring. Combustible electrical wires and cables and optical fiber cables exposed within a plenum shall be listed as having a maximum peak optical density of 0.50 or less, an average optical density of 0.15 or less, and a maximum flame spread distance of 5 feet (1524 mm) or less when tested in accordance with NFPA 262 or shall be installed in metal raceways or metal sheathed cable. Combustible optical fiber and communication raceways exposed within a plenum shall be listed as having a maximum peak optical density of 0.5 or less, an average optical density of 0.15 or less, and a maximum flame spread distance of 5 feet (1524 mm) or less when tested in accordance with ANSI/UL 2024. Only plenum-rated wires and cables shall be installed in plenum-rated raceways. Electrical wires and cables, optical fiber cables and raceways addressed in this section shall be listed and labeled and shall be installed in accordance with NFPA 70.

❖ Electrical wires, cables and optical fiber cables have components that are combustible such that their installation in a plenum presents a hazard unless these materials are specially manufactured to provide for limited flame spread and low smoke optical density. These materials must be tested in accordance with NFPA 262 and must exhibit smoke and flame characteristics less than or equal to those specified in this section. Wires, cables, and optical fiber cables that exceed the limitations can be installed in plenums only if they are installed in metal raceways or metal sheathed cable.

Raceways of combustible material are frequently used in plenums because they are extremely flexible and provide for an organized installation of wires, cables and optical fiber cables. These combustible raceways present the same hazard in a plenum as do wires, cables and optical fiber cables having combustible components. The raceways must be tested in accordance with UL 2024 and must meet the same criteria that wires, cables and optical fiber cables must meet. Note that wires, cables and optical fiber cables installed in these compliant combustible raceways must meet the requirements in the first sentence of this section. In other words, a combustible raceway offers no protection for the wires, cables and optical fiber cables contained in such raceways.

This section requires that the electrical wires, cables, optical fiber cables and combustible raceways be listed and labeled indicating that a third-party certification agency must verify that the criteria for such products has been met. Installation of these products must be in accordance with NFPA 70.

602.2.1.2 Fire sprinkler piping. Plastic fire sprinkler piping exposed within a *plenum* shall be used only in wet pipe systems and shall have a peak optical density not greater than 0.50, an average optical density not greater than 0.15, and a flame spread of not greater than 5 feet (1524 mm) when tested in accordance with UL 1887. Piping shall be *listed* and *labeled*.

❖ The plastic fire sprinkler piping described in Section 602.2.1.2 has been shown through testing to perform under fire conditions as well as or better than other combustible materials permitted within the plenum.

A wet pipe sprinkler system is one in which the piping is filled with water at all times. The heat sink effect of water-filled piping offers some protection for plastic fire sprinkler piping that is exposed to fire. Test data must be submitted to indicate the performance characteristics of the plastic piping when it is tested in accordance with UL 1887. Peak and average optical density values are used to quantify the amount of smoke produced after the piping is subjected to the test procedures of UL 1887. A flame spread value is a measure of the extent that the flame travels along the length of the pipe being tested. The plastic pipe must also be labeled by a third-party agency to indicate that it has been tested in accordance with the referenced standard (see commentary, Section 301.5). The criteria of this section are considered to be more stringent than the requirements of Section 602.2.1.

602.2.1.3 Pneumatic tubing. Combustible pneumatic tubing exposed within a *plenum* shall have a peak optical density not greater than 0.50, an average optical density not greater than 0.15, and a flame spread of not greater than 5 feet (1524 mm) when tested in accordance with UL 1820. Combustible pneumatic tubing shall be *listed* and *labeled*.

❖ This section contains requirements for plastic tubing used in pneumatic [heating, ventilating and air-conditioning (HVAC)] control systems. UL 1820 is the test standard used to evaluate combustible pneumatic tubing for its fire and smoke characteristics. Exposure of combustible pneumatic tubing is permitted within plenums when it is tested to UL 1820 and the flame spread, the peak optical density and the average optical density are as specified in this section. The pneumatic tubing must also be labeled by an approved agency (see commentary, Section 301.5).

Pneumatic tubing is used to carry pressurized air to and from HVAC system controls, sensors, operators and thermostats. Formerly constructed of only metal pipe and tubing, pneumatic lines are now almost exclusively constructed of plastic tubing.

602.2.1.4 Electrical equipment in plenums. Electrical *equipment* exposed within a *plenum* shall comply with Sections 602.2.1.4.1 and 602.2.1.4.2.

❖ This section refers the reader to compliance requirements in Sections 602.2.1.4.1 and 602.2.1.4.2 for plenum-located electrical equipment having metallic or combustible enclosures that are integral to the equipment.

602.2.1.4.1 Equipment in metallic enclosures. Electrical *equipment* with metallic enclosures exposed within a *plenum* shall be permitted.

❖ Because metallic enclosures are noncombustible, the electrical equipment having an integral metallic enclosure is not required to comply with Section 602.2.1 (see Item No. 5).

602.2.1.4.2 Equipment in combustible enclosures. Electrical *equipment* with combustible enclosures exposed within a *plenum* shall be *listed* and *labeled* for such use in accordance with UL 2043.

❖ This section addresses components, such as heat and smoke detectors, speakers, horns, and other

similar types of equipment that have integral combustible enclosures. The enclosure along with the integral equipment must be listed and labeled in accordance with UL 2043.

602.2.1.5 Foam plastic insulation. Foam plastic insulation used as interior wall or ceiling finish, or as interior trim, in plenums shall exhibit a flame spread index of 75 or less and a smoke-developed index of 450 or less when tested in accordance with ASTM E 84 or UL 723 and shall also comply with one or more of Sections 602.2.1.5.1, 602.2.1.5.2 and 602.2.1.5.3.

❖ Foam plastic insulation and foam plastic trim are combustible and, if used in a plenum space, the flame and smoke of combusting foam plastic could spread to other parts of the building. If these materials are used as a component of the interior walls or ceiling of a plenum space, the materials must be specially manufactured to provide for limited flame spread and smoke-developed indexes. At a minimum, all foam plastic insulation and trim must exhibit smoke and flame characteristics less than or equal to those specified in this section. This is the same requirement found in IBC Section 2603.3. Where used as an interior component of a wall or ceiling surface in a plenum, the foam plastic material must be completely covered by a noncombustible material (see Section 602.2.1.5.1 or 602.2.1.5.3) or the exposed foam plastic material must meet more stringent flame spread and smoke-developed indexes and the testing requirement of Section 602.2.1.5.2.

602.2.1.5.1 Separation required. The foam plastic insulation shall be separated from the plenum by a thermal barrier complying with Section 2603.4 of the *International Building Code* and shall exhibit a flame spread index of 75 or less and a smoke-developed index of 450 or less when tested in accordance with ASTM E 84 or UL 723 at the thickness and density intended for use.

❖ Section 2603.4 of the IBC requires all foam plastic to be separated from the interior of the building by a thermal barrier such as 0.5-inch-thick (12.7 mm) gypsum board or an approved equivalent material (see commentary, IBC Section 2603.4). This section is concerned only with separating the foam plastic insulation from the plenum space.

The IBC sets forth test methods and criteria by which alternative thermal barriers are to be qualified to limit the average temperature rise of the unexposed face to 250°F (121°C) for 15 minutes of fire exposure while complying with the time-temperature conditions of ASTM E 119.

602.2.1.5.2 Approval. The foam plastic insulation shall exhibit a flame spread index of 25 or less and a smoke-developed index of 50 or less when tested in accordance with ASTM E 84 or UL 723 at the thickness and density intended for use and shall meet the acceptance criteria of Section 803.1.2 of the *International Building Code* when tested in accordance with NFPA 286.

The foam plastic insulation shall be approved based on tests conducted in accordance with Section 2603.10 of the *International Building Code*.

❖ This section refers to Section 2603.10 of the IBC for large-scale testing criteria for approval of foam plastic insulation products. That section lists such tests as FM 4880, NFPA 286, UL 1040 and UL 1715. The intent is to require testing based on the proposed end-use configuration of the foam-plastic assembly and on an exposing fire that is appropriate in size and location for the proposed application.

602.2.1.5.3 Covering. The foam plastic insulation shall be covered by corrosion-resistant steel having a base metal thickness of not less than 0.0160 inch (0.4 mm) and shall exhibit a flame spread index of 75 or less and a smoke-developed index of 450 or less when tested in accordance with ASTM E 84 or UL 723 at the thickness and density intended for use.

❖ The required metal covering is intended to act as a barrier against ignition of the foam plastic. It is not intended to serve as a thermal barrier such as is required in Section 602.2.1.5.1.

602.3 Stud cavity and joist space plenums. Stud wall cavities and the spaces between solid floor joists to be utilized as air plenums shall comply with the following conditions:

1. Such cavities or spaces shall not be utilized as a *plenum* for supply air.

2. Such cavities or spaces shall not be part of a required fire-resistance-rated assembly.

3. Stud wall cavities shall not convey air from more than one floor level.

4. Stud wall cavities and joist space plenums shall comply with the floor penetration protection requirements of the *International Building Code*.

5. Stud wall cavities and joist space plenums shall be isolated from adjacent concealed spaces by *approved* fireblocking as required in the *International Building Code*.

6. Stud wall cavities in the outside walls of building envelope assemblies shall not be utilized as air plenums.

❖ Literally speaking, stud and joist cavities are not plenums because Section 602.1 does not include them, but, for the purpose of this section, they are referred to as plenums. Common in residential construction, stud and joist space cavities (plenums) may be used in buildings of any occupancy. These spaces are limited to use for return air only because the negative pressures within the return air plenum with respect to surrounding spaces will decrease the likelihood of spreading smoke to other spaces via the plenum. Also, the temperature and moisture content of heated, cooled and conditioned supply air could cause a fire hazard or deterioration of the construction materials exposed in the spaces. Although referred to as "plenums," stud cavities are actually used as ducts, whereas joist spaces are likely to be

plenums because they typically join multiple stud cavity ducts into a common pathway back to an air handler.

The space must not be a part of a fire-resistance-rated assembly because the ASTM E 119 test does not consider the impact of air movement within the assembly on the fire-resistance rating. This restriction is a concession to the convenience and cost-savings potential of this method of moving air. The use of stud spaces inherently means the interconnection of different floor levels by the concealed space. Because of the hazard of such an interconnection, the use of this type of plenum is limited to return air from one floor level only for each independent stud cavity. All cavities not used for air movement must be isolated from the plenum by fireblocking constructed and installed to comply with the IBC.

Commentary Figure 602.3(1) shows an example of an acceptable stud and joist space installation. The bottom plate of the wall is cut away for the plenum to function, and fireblocking is installed in the joist and stud space to limit communication of the plenum with other spaces. The stud cavity shown is being used to return air from one floor level only, conducting it to the space below where the air-handling equipment is located.

Whether viewed as a shaft or as a duct, a stud cavity plenum penetrates floor assemblies and is, therefore, subject to the floor penetration protection requirements of the IBC, which may require floor penetration protection or a fire-resistance-rated shaft enclosure for stud cavity plenums, depending on the number of floors penetrated and the fire-resistance rating of the floor/ceiling assemblies.

Commentary Figure 602.3(2) shows an unacceptable stud and joist space plenum installation. Because air is returned from more than one floor level through the same stud cavity, the use is prohibited by this section.

The intent is to prohibit arrangements that create a direct connection from one floor level to another by means of a concealed cavity. Such direct connections would act as a chase or chimney, allowing fire and smoke to spread quickly upward through the building. Thus, stud cavity return air plenums are subject to the same restrictions and requirements as floor openings and penetrations regulated by the IBC. Stud and joist space plenums must be viewed as an exception to the fireblocking provisions of the IBC because one or more fireblocks (wall plates) must be removed or relocated to construct the plenums. Note that the code does not mention the type of materials allowed for "panning" the bottom of open joists to create joist space plenums. Traditionally, sheet metal has been used; however, composite materials are also used. The code official must determine what materials are acceptable for joist panning.

Item 6 disallows the use of stud wall cavities in outside walls because of outdoor air infiltration and the obvious loss of thermal insulation value in hollow cavities. See Section R403.2.3 of the *International Energy Conservation Code®* (IECC®), which prohibits the use of all framing cavities as ducts or plenums. The IECC is more stringent in this case.

[B] 602.4 Flood hazard. For structures located in flood hazard areas, plenum spaces shall be located above the elevation required by Section 1612 of the *International Building Code* for utilities and attendant equipment or shall be designed and constructed to prevent water from entering or accumulating within the plenum spaces during floods up to such elevation. If the plenum spaces are located below the elevation required by Section 1612 of the *International Building Code* for utilities and attendant equipment, they shall be capable of resisting hydrostatic and hydrodynamic loads and stresses, including the effects of buoyancy, during the occurrence of flooding up to such elevation.

❖ In areas designated by the IBC as flood hazard areas, plenums must be installed above the elevation specified by the IBC or, if allowed below that elevation, plenums must be designed to be water tight. Water in plenums could damage the plenum and related equipment. If allowed below the required elevation, water-tight plenums will be exposed to hydrostatic and hydrodynamic loads and stresses, including the effects of buoyancy, which will exert a net upward force on the structure. In flood hazard areas, under-floor spaces and crawl spaces cannot be used as plenums because the foundation walls that form those spaces must have flood openings (see ASCE 24 and *International Residential Code®* (IRC®) Section R322.2.2. For additional guidance, refer to FEMA 348.

SECTION 603
DUCT CONSTRUCTION AND INSTALLATION

603.1 General. An air distribution system shall be designed and installed to supply the required distribution of air. The installation of an air distribution system shall not affect the fire protection requirements specified in the *International Building Code*. Ducts shall be constructed, braced, reinforced and installed to provide structural strength and durability.

❖ The function of a duct is to convey air between specific points in an air distribution or exhaust system. Section 603 contains requirements for the construction of ducts used to transport environmental air for HVAC systems and for general exhaust purposes such as toilet, noncommercial kitchen and nonhazardous exhaust. This section is not intended to regulate the design of duct-system flow rates required to achieve space temperature control, comfort conditioning, contaminant control or space pressurization. The focus of Section 603 is on building and installing duct systems that have been designed for function by the responsible designer. This section does, however, require that duct systems be capable of handling design loads.

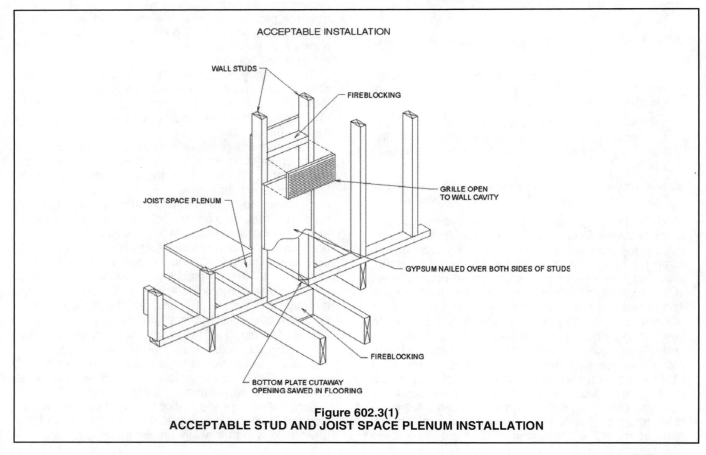

Figure 602.3(1)
ACCEPTABLE STUD AND JOIST SPACE PLENUM INSTALLATION

A duct system by its very nature penetrates floor assemblies and wall assemblies, both fire-resistance rated and nonrated assemblies. It is therefore subject to the fire protection requirements of the IBC, which may require fire dampers, smoke dampers, floor penetration protection or a fire-resistance-rated shaft enclosure. The construction of clothes dryer exhaust systems, commercial kitchen exhaust systems, hazardous exhaust systems, conveyor systems and combustion air ducts is regulated by Sections 504, 506, 510, 511 and 708, respectively. Section 603 addresses the performance aspects of ductwork relating to dimensional and structural stability, leak-

age control, pressure, temperature durability, support, vapor permeance and airflow resistance.

603.2 Duct sizing. Ducts installed within a single *dwelling unit* shall be sized in accordance with ACCA Manual D or other *approved* methods. Ducts installed within all other buildings shall be sized in accordance with the ASHRAE *Handbook of Fundamentals* or other equivalent computation procedure.

❖ Properly sized ducts are essential to the safe operation of heating appliances and to ensure the distribution of the prescribed amount of ventilation air required by Chapter 4. Improper duct sizing can cause inadequate cooling or heating, inadequate

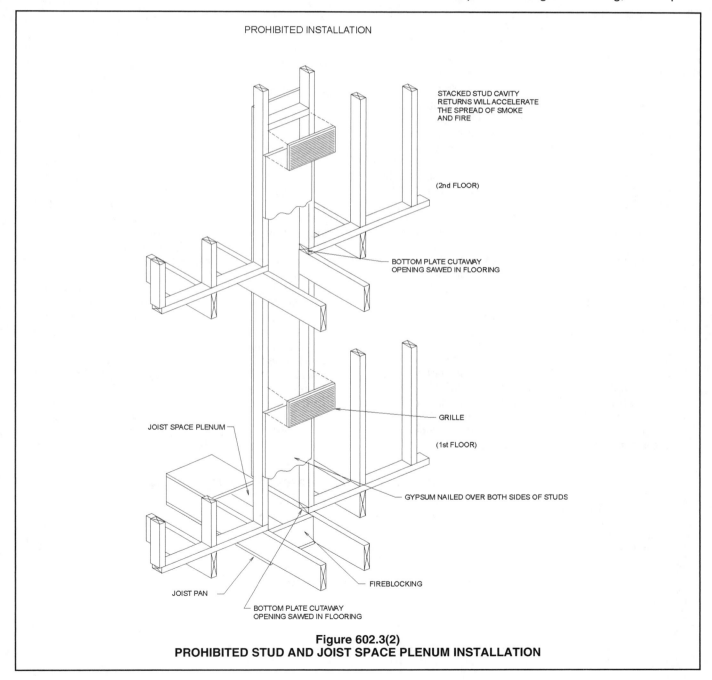

Figure 602.3(2)
PROHIBITED STUD AND JOIST SPACE PLENUM INSTALLATION

dehumidification and excess energy losses. This section does not include prescriptive sizing requirements, but directs the user to the applicable industry standard, depending on whether the system to be sized is residential or commercial. It is important to note that the design objectives and criteria are significantly different for residential and commercial duct systems, making it critical to choose the proper industry standard for duct sizing. For example, ACCA Manual D limits the velocity in a residential duct system to 900 feet per minute (4.6 m/s). The velocity in a commercial duct system can be much higher. The designer must submit documentation that a system has been sized in an approved manner.

603.3 Duct classification. Ducts shall be classified based on the maximum operating pressure of the duct at pressures of positive or negative 0.5, 1.0, 2.0, 3.0, 4.0, 6.0 or 10.0 inches (1 inch w.c. = 248.7 Pa) of water column. The pressure classification of ducts shall equal or exceed the design pressure of the air distribution in which the ducts are utilized.

❖ The pressure classification of ducts must equal or exceed the static pressure of the air distribution system in which the ducts will be installed to reduce air leakage and the possibility of developing a structural failure in the air distribution ductwork. The classification is based on the maximum positive or negative operating pressure the duct is designed to withstand, ranging from 0.5 to 10 inches of water column (0.12 to 63 kPa). Excessive positive or negative pressures can cause a duct or plenum to rupture, deform or collapse.

The static pressure is established by the designer of the air distribution system and should be stated on the plans and specifications. Ducts that are identified as complying with UL 181 will be marked to indicate their positive or negative pressure classification and velocity rating. Ducts that are fabricated in the field may not be so marked.

Duct construction must conform to the requirements of SMACNA *HVAC Duct Construction Standards—Metal and Flexible*, or SMACNA *Fibrous Glass Duct Construction Standards*, which state duct thicknesses and bracing requirements based on static pressure requirements of the system and include pressure and velocity limits.

603.4 Metallic ducts. All metallic ducts shall be constructed as specified in the SMACNA *HVAC Duct Construction Standards-Metal and Flexible*.

Exception: Ducts installed within single *dwelling units* shall have a minimum thickness as specified in Table 603.4.

❖ Metallic ducts are usually constructed using galvanized sheet steel. Duct size is based on required airflow, system pressure, flow velocity and pressure losses caused by friction. Duct material thickness is determined by duct size, static pressure of the system, distance between supports and whether the duct is reinforced. Metallic ducts must be constructed with the minimum thicknesses specified in the SMACNA *HVAC Duct Construction Standards—Metal and Flexible*, which bases the minimum required duct thickness on the geometry of the duct, the material used and the major dimension of the duct (the diameter for round ducts, and the widest side for rectangular ducts). In addition to those parameters, the SMACNA standard also includes such necessary information as the design static pressure of the air distribution system and whether reinforcement is required for duct support.

Metallic rigid air ducts could be approved by the code official in accordance with Section 105.2 if the ducts have been shown to meet the requirements of UL 181 for a Class 0 or Class 1 rigid air duct (see commentary, Section 603.5). Although UL 181 is used primarily to evaluate nonmetallic duct materials, the use of the standard for evaluating metallic ductwork is not precluded by its scope. UL 181 tests samples of the duct to determine fire performance characteristics; corrosion and erosion resistance; leakage resistance; mold growth and humidity resistance; and structural integrity. Air ducts that conform to the requirements of UL 181 are identified by the manufacturer's or vendor's name, rated velocity and rated negative and positive pressures.

Metallic ducts must be constructed to comply with the requirements contained in SMACNA *HVAC Duct Construction Standards—Metal and Flexible*. Besides the minimum thicknesses of duct materials required, this standard contains information on duct reinforcement, joints, fittings, hangers and supports and other

TABLE 603.4
DUCT CONSTRUCTION MINIMUM SHEET METAL THICKNESSES FOR SINGLE DWELLING UNITS

DUCT SIZE	GALVANIZED		ALUMINUM MINIMUM THICKNESS (in.)
	Minimum thickness (in.)	Equivalent galvanized gage no.	
Round ducts and enclosed rectangular ducts			
14 inches or less	0.0157	28	0.0175
16 and 18 inches	0.0187	26	0.018
20 inches and over	0.0236	24	0.023
Exposed rectangular ducts			
14 inches or less	0.0157	28	0.0175
Over 14 inches[a]	0.0187	26	0.018

For SI: 1 inch = 25.4 mm, 1 inch water gage = 249 Pa.

a. For duct gages and reinforcement requirements at static pressures of $^1/_2$-inch, 1-inch and 2-inch w.g., SMACNA *HVAC Duct Construction Standards*, Tables 2-1, 2-2 and 2-3, shall apply.

pertinent design information needed to achieve a stable, efficient and durable installation of ductwork.

The exception to this section allows ducts that are installed within a single dwelling unit to be constructed with the minimum thickness specified in Table 603.4. The required minimum thickness of the duct materials is less than the required thickness in SMACNA *HVAC Duct Construction Standards—Metal and Flexible* because airflow volume, system pressure, flow velocity and pressure losses caused by friction are lower in a typical dwelling unit because of the small size of the system.

The metal thicknesses allowed by Table 603.4 have been used in dwelling unit construction without evidence of failure. Therefore, the lighter gages are justified in this limited application.

603.4.1 Minimum fasteners. Round metallic ducts shall be mechanically fastened by means of at least three sheet metal screws or rivets spaced equally around the joint.

Exception: Where a duct connection is made that is partially inaccessible, three screws or rivets shall be equally spaced on the exposed portion so as to prevent a hinge effect.

❖ The code specifies a minimum number of fasteners when joining sections of round metal ducts. A minimum of three sheet metal screws is what is specified in the SMACNA Duct Construction Standards to accomplish this. Three screws located in close proximity to each other are not compliant. The screws must be spaced equidistant from one another around the duct to make a rigid connection. Although not a requirement of the code, a commonly used general rule of thumb for the number of screws to make a rigid round duct joint is the duct diameter in inches divided by two. For example, a 16-inch duct should have at least 8 screws equally spaced around the duct.

Equally spacing the screws around a duct is not always possible to achieve where ducts are located in chases or against walls or ceilings. The exception allows for some deviation when it is not possible to comply with the strict letter of the code. A "hinge effect" results from fasteners not being properly spaced along the circumference of the duct thus allowing the joint to sag, flex or move like a hinged joint.

603.5 Nonmetallic ducts. Nonmetallic ducts shall be constructed with Class 0 or Class 1 duct material and shall comply with UL 181. Fibrous duct construction shall conform to the SMACNA *Fibrous Glass Duct Construction Standards* or NAIMA *Fibrous Glass Duct Construction Standards*. The air temperature within nonmetallic ducts shall not exceed 250°F (121°C).

❖ Nonmetallic ducts and duct materials must be tested and classified in accordance with the provisions of UL 181. Only Class 0 and Class 1 ducts may be used. Class 0 indicates flame spread and smoke-developed indexes of zero; Class 1 indicates a flame spread index not greater than 25 and a smoke-developed index of not greater than 50 when tested to ASTM E 84.

UL 181 requires that a nonmetallic duct be tested to determine its fire performance characteristics, corrosion resistance, mold growth resistance, humidity resistance, leakage resistance, temperature resistance, erosion resistance and structural performance. Air ducts that conform to the requirements of UL 181 are identified by the manufacturer's or vendor's name, rated velocity, negative and positive pressure classification and duct material class.

Fibrous ducts are constructed of a composite material of rigid (high density) fiberglass board and a factory-applied facing (typically reinforced aluminum). The surface of the fibrous duct that is exposed to the airflow is sealed with a fiber-bonding adhesive that prevents erosion of the fiberglass material. The factory-applied exterior duct board facing contributes to the strength and rigidity of the composite material, acts as a heat reflector, serves as a vapor barrier and is an integral component of the joining method used to construct fibrous ducts. The material is available in board form for shop or field fabrication into rectangular sections, or in "10-sided duct" form, which approximates a circular cross section.

Fibrous ducts take advantage of the inherent insulating qualities of the glass fiber material. The air friction factors for fibrous ducts are greater than those for sheet metal because of the relatively rough surface finish.

Construction of fibrous glass ducts must conform to the requirements of SMACNA *Fibrous Glass Duct Construction Standards*, which contains details for the design and fabrication of air distribution systems using fibrous glass ducts. The SMACNA standards referenced in this section and the previous section are enforceable extensions of the code.

The maximum discharge temperature permitted by industry standards for warm air-heating systems is 250°F (121°C). This section prohibits nonmetallic ducts from being used in applications in which the air temperature would exceed 250°F (121°C) because the material has not been tested to withstand higher temperatures and high temperatures will cause accelerated aging of the duct material.

603.5.1 Gypsum ducts. The use of gypsum boards to form air shafts (ducts) shall be limited to return air systems where the air temperatures do not exceed 125°F (52°C) and the gypsum board surface temperature is maintained above the airstream dew-point temperature. Air ducts formed by gypsum boards shall not be incorporated in air-handling systems utilizing evaporative coolers.

❖ Gypsum board is a composite material commonly used for the construction of air plenums and shafts which can reduce construction costs because gypsum board is a common component of building construction assemblies. By serving a dual purpose, gypsum board eliminates the need for independent duct construction. The use of gypsum board to form ducts and plenums is specifically regulated to prevent deterioration of the gypsum board material. Air temperatures that exceed 125°F (52°C) will, over time, dry both the paper facing and the gypsum of the gypsum board, leading to deterioration of the panel.

Gypsum board can also deteriorate when exposed to moisture, which will happen if the surface temperature of the gypsum board is lower than the airstream dew point temperature, causing water to condense on the surface of the gypsum board. For these reasons, gypsum board may not be used for air distribution systems using evaporative cooling equipment. It is further restricted to return air system applications only, a maximum airstream temperature of 125°F (52°C) and an airstream dew point temperature continuously below the temperature of the gypsum board surface. Evaporative cooling equipment such as "swamp coolers" uses water as a refrigerant. The resulting addition of moisture to the airstream could cause deterioration of the gypsum board.

603.6 Flexible air ducts and flexible air connectors. Flexible air ducts, both metallic and nonmetallic, shall comply with Sections 603.6.1, 603.6.1.1, 603.6.3 and 603.6.4. Flexible air connectors, both metallic and nonmetallic, shall comply with Sections 603.6.2 through 603.6.4.

❖ Flexible air ducts and connectors are typically factory-made assemblies consisting of an inner duct with or without a thermal insulation covering and an outer vapor barrier jacket. The inner duct may be a synthetic membrane supported by a spiral wire frame or a corrugated flexible metal pipe. The duct covering is commonly composed of an insulation blanket covered by an outer plastic or metal foil vapor barrier. Nonmetallic flexible connectors terminate in sheet metal collars.

Flexible air ducts and connectors are usually used in air distribution systems for relatively short runs of duct because they are easy to install and they can be used in areas of tight construction that may require frequent changes of direction for the duct material. Ceiling supply and return grilles, diffusers and registers connect easily after the final ceiling plan has been determined [see Commentary Figure 603.6(1)]. Exact air duct termination locations are often unknown when the main ductwork is being installed.

One drawback of using flexible air ducts and connectors is the higher pressure losses caused by the friction of the irregular inside duct surface that could result in the need for larger ducts or air-handling equipment.

Flexible air ducts and flexible air connectors can look exactly alike. However, the markings on the material will identify it as either a flexible air duct or a flexible air connector. Even though flexible air ducts and connectors may appear to be identical and may be used in the same applications, there can be a difference in the material properties. Commentary Table 603.6 was taken from UL 181, which is the standard used to investigate the performance of flexible air

ducts and air connectors. Flexible air ducts require more extensive testing (the flame penetration test, puncture test and impact test) than flexible air connectors. This difference in testing is what determines the markings on the material and whether it is classified as a flexible air duct or a flexible air connector [see Commentary Figures 603.6(2) and 603.6(3) for example markings of each].

Commentary Figure 603.6(4) provides guidelines for installing flexible ducts and connectors.

603.6.1 Flexible air ducts. Flexible air ducts, both metallic and nonmetallic, shall be tested in accordance with UL 181. Such ducts shall be *listed* and *labeled* as Class 0 or Class 1 flexible air ducts and shall be installed in accordance with Section 304.1.

❖ Flexible air ducts must comply with the requirements of UL 181 for Class 0 or Class 1 flexible air ducts. Commentary Table 603.6 was taken from UL 181, which is the standard used to investigate the performance of flexible air ducts and connectors. The table indicates the testing protocol for flexible air ducts and whether or not the duct has undergone that particular test. A Class 0 air duct has flame spread and smoke-developed indexes of zero. A Class 1 air duct has a flame spread index not greater than 25 and a smoke-developed index not greater than 50 when tested to ASTM E 84. Air ducts that conform to the requirements of UL 181 are identified with the manufacturer's or vendors' name, the rated velocity, rated negative and positive pressures and the information related to fabrication and joining of materials. Commentary Figure 603.6(2) shows a sample marking of a Class 1 flexible air duct. Note that the manufacturer's name does not appear, but the number in the upper left corner is the third-party inspection agency's file number of the manufacturer. Also note that air duct labels are rectangular (see Section 607.7).

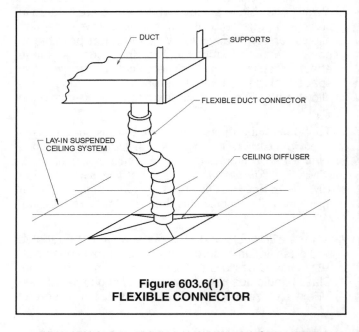

Figure 603.6(1)
FLEXIBLE CONNECTOR

ATCO RUBBER PRODUCTS, INC.

Thermal Performance

Figure 603.6(2)
SAMPLE LABELS
(Courtesy of ATCO Rubber Products, Inc.)

Figure 603.6(3)
SAMPLE LABELS
(Courtesy of ATCO Rubber Products, Inc.)

Table 603.6
TESTING OF FLEXIBLE AIR DUCTS AND CONNECTORS
(Source: UL 181)

Test	Flexible Air Ducts	Flexible Connectors
Surface-burning Characteristics	Yes	Yes
Flame Penetration	Yes	No
Burning	Yes	Yes
Corrosion	Yes	Yes
Mold Growth and Humidity	Yes	Yes
Temperature	Yes	Yes
Puncture	Yes	No
Impact	Yes	No
Erosion	Yes	Yes
Pressure	Yes	Yes
Collapse	Yes	Yes
Tension	Yes	Yes
Torsion	Yes	Yes
Bending	Yes	Yes
Leakage	Yes	Yes

4

Installation Guidelines

4.1 Code Reference

The "authority having jurisdiction" should be referenced to determine what law, ordinance or code shall apply in the use of flexible duct.

Ducts conforming to NFPA 90A or 90B shall meet the following requirements:

 a. Shall be tested in accordance with Sections 5-21 of Underwriters Laboratories Standard for Factory-Made Air Ducts and Air Connectors, UL 181.
 b. Shall be installed in accordance with the conditions of their listing.
 c. Shall be installed within the limitations of the applicable NFPA 90A or 90B Standard.

4.2 General

The routing of flexible duct, the number of bends, the number or degrees in each bend and the amount of sag allowed between support joints will have serious effects on system performance due to the increased resistance each introduces. Use the minimum length of flexible duct to make connections. It is not recommended that excess length of ducts be installed to allow for possible future relocations of air terminal devices.

Avoid installations where exposure to direct sunlight can occur, e.g. turbine vents, sky lights, canopy windows, etc. Prolonged exposure to sunlight will cause degradation of the vapor barrier. Direct exposure to UV light from a source lamp installed within the HVAC system will cause degradation of some inner core/liner materials.

Terminal devices shall be supported independently of the flexible duct.

Repair torn or damaged vapor barrier/jacket with duct tape listed and labeled to Standard UL 181B. If internal core is penetrated, replace flexible duct or treat as a connection.

4.3 Installation and Usage

Install duct fully extended, do not install in the compressed state or use excess lengths. This will noticeably increase friction losses.

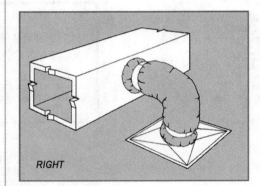

RIGHT

Figure 6

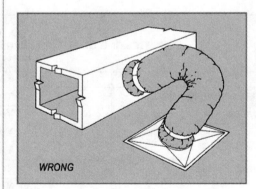

WRONG

Figure 7

12

ADC Flexible Duct Performance & Installation Standards, 4th Edition

Figure 603.6(4)
INSTALLATION INSTRUCTIONS FOR FLEXIBLE AIR DUCTS AND AIR CONNECTORS
(Courtesy of Air Diffusion Council)

Installation Guidelines . . . continued

Avoid bending ducts across sharp corners or incidental contact with metal fixtures, pipes or conduits. Radius at center line shall not be less than one duct diameter.

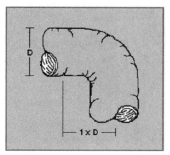

Figure 8

Do not install near hot equipment (e.g. furnaces, boilers, steam pipes, etc.) that is above the recommended flexible duct use temperature.

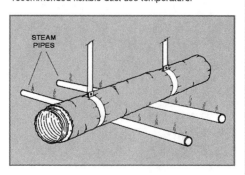

Figure 9

4.4 Connecting, Joining and Splicing Flexible Ducts

All connections, joints and splices shall be made in accordance with the manufacturer's installation instructions.

For flexible ducts with plain ends, standardized installation instructions conforming to this standard are shown in Sections 4.5 "Nonmetallic With Plain Ends" (uses tape and clamp to seal/secure the core to the fitting), 4.6 "Alternate Nonmetallic With Plain Ends" (uses mastic and clamp to seal/secure the core to the fitting), and 4.7 "Metallic With Plain Ends (optional use of tape or mastic and metal screws to seal/secure the core to the fitting).

Due to the wide variety of ducts and duct assemblies with special end treatments, e.g. factory installed fittings, taped ends, crimped metal ends, etc., no standardized installation instructions are shown. Reference manufacturer's installation instructions.

All tapes, mastics, and nonmetallic clamps used for field installation of flexible ducts shall be listed and labeled to Standard UL 181B - Closure Systems for Use With Flexible Air Ducts and Air Connectors.

Sheet metal fittings to which flexible ducts with plain ends are attached shall be beaded and have a minimum of 2 inches [50 mm] collar length. Beads are optional for fittings when attaching *metallic* flexible ducts.

Sheet metal sleeves used for joining two sections of flexible duct with plain ends shall be a minimum of 4 inches [100 mm] in length and beaded on each end. Beads are optional for sleeves when joining *metallic* flexible ducts.

Flexible ducts secured with nonmetallic clamps shall be limited to 6 inches w.g. [1500 Pa] positive pressure.

13

ADC Flexible Duct Performance & Installation Standards, 4th Edition

Figure 603.6(4)—continued
INSTALLATION INSTRUCTIONS FOR FLEXIBLE AIR DUCTS AND AIR CONNECTORS
(Courtesy of Air Diffusion Council)
(continued)

Installation Guidelines . . . continued

4.5 Installation Instructions for Air Ducts and Air Connectors - Nonmetallic with Plain Ends

Connections

1. After desired length is determined, cut completely around and through duct with knife or scissors. Cut wire with wire cutters. Fold back jacket and insulation.

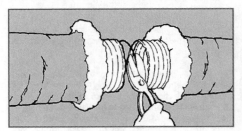

2. Slide at least 1" [25 mm] of core over fitting and past the bead. Seal core to collar with at least 2 wraps of duct tape. Secure connection with clamp placed over the core and tape and past the bead.

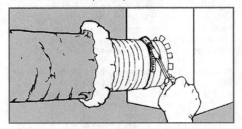

3. Pull jacket and insulation back over core. Tape jacket with at least 2 wraps of duct tape. A clamp may be used in place of or in combination with the duct tape.

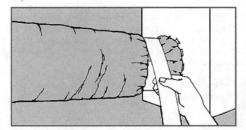

Splices

1. Fold back jacket and insulation from core. Butt two cores together on a 4" [100 mm] length metal sleeve.

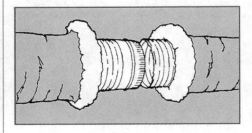

2. Tape cores together with at least 2 wraps of duct tape. Secure connection with 2 clamps placed over the taped core ends and past the beads.

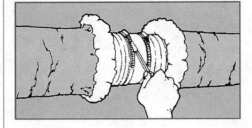

3. Pull jacket and insulation back over cores. Tape jackets together with at least 2 wraps of duct tape.

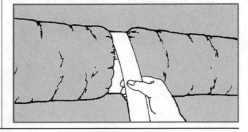

NOTES:
1. For uninsulated air ducts and air connectors, disregard references to insulation and jacket.
2. Use beaded sheet metal fittings and sleeves.
3. Use tapes listed and labeled in accordance with Standard UL 181B and marked "181B-FX".
4. Nonmetallic clamps shall be listed and labeled in accordance with Standard UL 181B and marked "181B-C". Use of nonmetallic clamps shall be limited to 6 in. w.g. [1500 Pa] positive pressure.

**Figure 603.6(4)—continued
INSTALLATION INSTRUCTIONS FOR FLEXIBLE AIR DUCTS AND AIR CONNECTORS
(Courtesy of Air Diffusion Council)**
(continued)

14

ADC Flexible Duct Performance & Installation Standards, 4th Edition

---

Installation Guidelines . . . continued

4.6 Alternate Installation Instructions for Air Ducts and Air Connectors - Nonmetallic with Plain Ends

Connections and Splices

Step 1
After desired length is determined, cut completely around and through duct with knife or scissors. Cut wire with wire cutters. Pull back jacket and insulation from core.

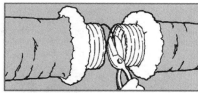

Step 2
Apply mastic approximately 2" [50 mm] wide uniformly around the collar of the metal fitting or over the ends of a 4" [100 mm] metal sleeve. Reference data on mastic container for application rate, application thickness, cure times and handling information.

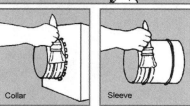

Collar   Sleeve

Step 3
Slide at least 2" [50 mm] of core over the fitting or sleeve ends and past the bead.

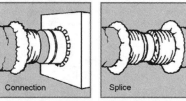

Connection   Splice

Step 4
Secure core to collar with a clamp applied past the bead. Secure cores to sleeve ends with 2 clamps applied past the beads.

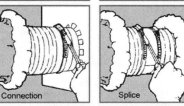

Connection   Splice

Step 5
Pull jacket and insulation back over core ends. Tape jacket(s) with at least 2 wraps of duct tape. A clamp may be used in place of or in combination with the duct tape.

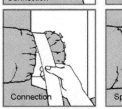

Connection   Splice

NOTES:
1. For uninsulated air ducts and air connectors, disregard references to insulation and jacket.
2. Use beaded sheet metal fittings and sleeves.
3. Use mastics listed and labeled in accordance with Standard UL 181B and marked "181B-M" on container.
4. Use tapes listed and labeled in accordance with Standard UL 181B and marked "181B-FX".
5. Nonmetallic clamps shall be listed and labeled in accordance with standard UL 181B and marked "181B-C". Use of nonmetallic clamps shall be limited to 6 in. w.g. [1500 Pa] positive pressure.

15

ADC Flexible Duct Performance & Installation Standards, 4th Edition

Figure 603.6(4)—continued
INSTALLATION INSTRUCTIONS FOR FLEXIBLE AIR DUCTS AND AIR CONNECTORS
(Courtesy of Air Diffusion Council)
(continued)

Installation Guidelines . . . continued

4.7 Installation Instruction for Air Ducts and Air Connectors - Metallic with Plain Ends

Connections and Splices

1. After cutting duct to desired length, fold back jacket and insulation exposing core. Trim core ends squarely using suitable metal shears. Determine optional sealing method (Steps 2 or 5) before proceeding.

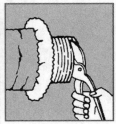

4. Secure to collar/sleeve using #8 sheet metal screws spaced equally around circumference. Use 3 screws for diameters under 12" [300 mm] and 5 screws for diameters 12" [300 mm] and over.

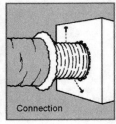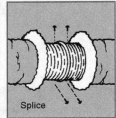

Connection — Splice

2. When mastics are required and for pressures 4" w.g. [1000 Pa] and over, seal joint with mastic applied uniformly to the outside surface of collar/sleeve. (Disregard this step when not using mastics and proceed to Step 3).

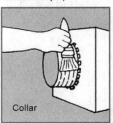

Collar — Sleeve

5. For pressures under 4" w.g. [1000 Pa] seal joint using 2 wraps of duct tape applied over screw heads and spirally lapping tape to collar/sleeve. (Disregard this step when using mastics per Step 2).

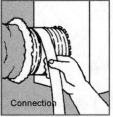

Connection — Splice

3. Slide at least 1" [25 mm] of core over metal collar for attaching duct to take off or over ends of a 4" [100 mm] metal sleeve for splicing 2 lengths of duct.

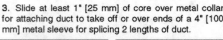
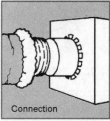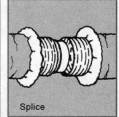

Connection — Splice

6. Pull jacket and insulation back over core. Tape jacket with 2 wraps of duct tape. A clamp may be used in place of or in combination with the duct tape.

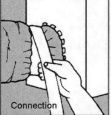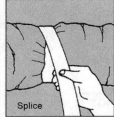

Connection — Splice

NOTES:
1. For uninsulated air ducts and air connectors, disregard references to insulation and jacket.
2. Use mastics listed and labeled to Standard UL 181B and marked"181B-M" on container.
3. Use tapes listed and labeled to Standard UL 181B and marked "181B-FX".
4. Nonmetallic clamps shall be listed and labeled in accordance with Standard UL 181B and marked "181B-C".

Figure 603.6(4)—continued
INSTALLATION INSTRUCTIONS FOR FLEXIBLE AIR DUCTS AND AIR CONNECTORS
(Courtesy of Air Diffusion Council)
(continued)

16

ADC Flexible Duct Performance & Installation Standards, 4th Edition

Installation Guidelines . . . continued

4.8 Supporting Flexible Duct

Flexible duct shall be supported at manufacturer's recommended intervals, but at no greater distance than 5' [1.5 m]. Maximum permissible sag is ½" per foot [42 mm per meter] of spacing between supports.

A connection to rigid duct or equipment shall be considered a support joint. Long horizontal duct runs with sharp bends shall have additional supports before and after the bend approximately one duct diameter from the center line of the bend.

Hanger or saddle material in contact with the flexible duct shall be of sufficient width to prevent any restriction of the internal diameter of the duct when the weight of the supported section rests on the hanger or saddle material. In no case will the material contacting the flexible duct be less than 1½" [38 mm] wide.

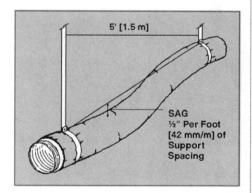

Figure 10

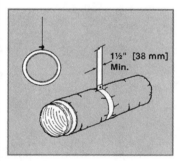

Figure 11

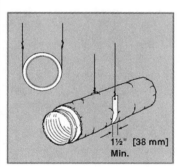

Figure 12

Figure 603.6(4)—continued
INSTALLATION INSTRUCTIONS FOR FLEXIBLE AIR DUCTS AND AIR CONNECTORS
(Courtesy of Air Diffusion Council)
(continued)

Installation Guidelines . . . continued

Factory installed suspension systems integral to the flexible duct are an acceptable alternative hanging method when manufacturer's recommended procedures are followed.

Support the duct between a metal connection and bend by allowing the duct to extend straight for a few Inches before making the bend. This will avoid possible damage of the flexible duct by the edge of the metal collar.

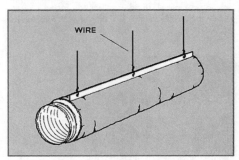

Figure 13

Flexible ducts may rest on ceiling joists or truss supports. Maximum spacing between supports shall not exceed the maximum spacing per manufacturer's installation instruction.

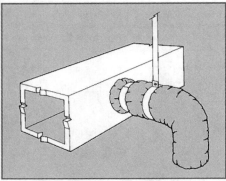

Figure 15

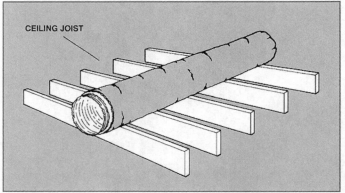

Figure 14

Note:
Factory-made air ducts may not be used for vertical risers in air duct systems serving more than two stories.

Vertically installed duct shall be stabilized by support straps at a max. of 6' [1.8 m] on center.

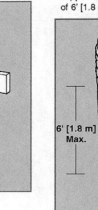

6' [1.8 m] Max.

Figure 16

19

ADC Flexible Duct Performance & Installation Standards, 4th Edition

Figure 603.6(4)—continued
INSTALLATION INSTRUCTIONS FOR FLEXIBLE AIR DUCTS AND AIR CONNECTORS
(Courtesy of Air Diffusion Council)
(continued)

603.6.1.1 Duct length. Flexible air ducts shall not be limited in length.

❖ The length of flexible air ducts is not limited, unlike flexible air connectors, which are limited to 14 feet (4267 mm) (see commentary, Section 603.6.2.1).

603.6.2 Flexible air connectors. Flexible air connectors, both metallic and nonmetallic, shall be tested in accordance with UL 181. Such connectors shall be *listed* and *labeled* as Class 0 or Class 1 flexible air connectors and shall be installed in accordance with Section 304.1.

❖ Flexible air connectors must comply with the requirements of UL 181 for Class 0 or Class 1 flexible air connectors. Table 603.6 indicates the testing protocol used for flexible air connectors. A Class 0 connector has flame spread and smoke-developed indexes of zero. A Class 1 connector has a flame spread index not greater than 25 and a smoke-developed index not greater than 50 when tested to ASTM E 84. Connectors that conform to the requirements of UL 181 are identified by the manufacturer's or vendor's name, rated velocity, negative and positive pressure classification and information related to fabrication and joining of materials. Commentary Figure 603.6(3) shows a sample marking of a Class 1 flexible air connector. See the commentary to Section 603.6 for a general discussion on flexible air ducts and connectors. Current air connector labels are round or oval, making them readily distinguishable from rectangular air duct labels. See the commentary to Section 603.6 for additional information on flexible air ducts and flexible air connectors (see Section 607.7).

603.6.2.1 Connector length. Flexible air connectors shall be limited in length to 14 feet (4267 mm).

❖ Because the testing protocol for flexible air connector material is less stringent than it is for flexible air duct material, the length of flexible air connectors is limited [see commentary, Section 603.6 and Commentary Figure 603.6(2)]. Commentary Figure 603.6(3) shows that the 14-foot (4267 mm) length limitation will be included on the material identification markings. The length is limited to 14 feet (4267 mm) to reduce the probability of the material being ruptured by impact or puncture, and to control the amount of the material that could be exposed to fire.

603.6.2.2 Connector penetration limitations. Flexible air connectors shall not pass through any wall, floor or ceiling.

❖ Flexible air connectors must not pass through any wall, floor or ceiling, whether the assembly is fire-resistance rated or not. An inadequate seal at the assembly penetration could allow smoke or flame to penetrate the assembly. Flexible air connectors can be constructed of both combustible and noncombustible components; therefore, the duct's resistance to the passage of fire could be less than the resistance of the penetrated assembly. All construction assemblies, whether fire-resistance rated or not, have some inherent resistance to the spread of fire; duct penetra-

tions can significantly affect that fire resistance (see Section 607.7).

603.6.3 Air temperature. The design temperature of air to be conveyed in flexible air ducts and flexible air connectors shall be less than 250°F (121°C).

❖ The maximum discharge temperature permitted by industry standards for warm air heating systems is 250°F (121°C). This section prohibits use of flexible air duct and connector materials in applications in which the design air temperature is 250°F (121°C) or higher because the material has not been tested or designed for higher temperatures.

603.6.4 Flexible air duct and air connector clearance. Flexible air ducts and air connectors shall be installed with a minimum *clearance* to an *appliance* as specified in the *appliance* manufacturer's installation instructions.

❖ Flexible air ducts and connectors could contain combustible materials; therefore, they must be installed with adequate clearance from heat-producing appliances in accordance with the appliance manufacturer's installation instructions. The manufacturer's instructions for heating equipment will specify minimum distances between the equipment duct connections and flexible air ducts and connectors.

603.7 Rigid duct penetrations. Duct system penetrations of walls, floors, ceilings and roofs and air transfer openings in such building components shall be protected as required by Section 607. Ducts in a private garage that penetrate a wall or ceiling that separates a dwelling from a private garage shall be continuous, shall be constructed of sheet steel having a thickness of not less than 0.0187 inch (0.4712 mm) (No.26 gage) and shall not have openings into the garage. Fire and smoke dampers are not required in such ducts passing through the wall or ceiling separating a dwelling from a private garage except where required by Chapter 7 of the *International Building Code*.

❖ All construction assemblies, whether fire-resistance rated or not, have some inherent resistance to the spread of fire, and duct penetrations can significantly affect that fire resistance. To maintain the integrity of the construction assemblies, penetrations must be properly protected. Acceptable protection methods for various penetrations of fire-resistance-rated and unrated assemblies are identified in Section 607 (see commentary, Section 607).

The term "penetration" typically describes an object such as a duct passing through an assembly, whereas the term "opening" typically describes a hole in an assembly or a termination opening such as an outlet or diffuser.

This section has an exception that allows ducts to penetrate the wall or ceiling separating a dwelling and a private garage without installing fire and smoke dampers. This exception only applies if the ducts are continuous, are a minimum of 26 gage galvanized sheet metal (0.0187 inches) (0.048 mm), and there are no openings in the ducts into the garage. It should be noted that this exception only applies to private garages. Based on Section 406.1 in the IBC, a private garage is a Group U occupancy. Therefore this exception would not apply if the garage were an S-2 occupancy (see Commentary Figure 603.7).

This section does not prohibit duct openings into the garage where the ducts serve only the garage and do not pass through any walls, floors or ceiling of the garage.

603.8 Underground ducts. Ducts shall be *approved* for underground installation. Metallic ducts not having an *approved* protective coating shall be completely encased in a minimum of 2 inches (51 mm) of concrete.

❖ Ducts installed underground must be able to resist the forces imposed on them by the materials that encase them, the forces created by floodwaters in and around them and corrosion. The ASHRAE *Handbook of HVAC Systems and Equipment* recommends that underground ducts and fittings be round for optimum structural performance. Unlike round ducts, square or rectangular ducts offer little resistance to deformation or collapse caused by the structural loads associated with burial.

Metal ducts must either have a protective coating to resist corrosion or be completely encased in concrete, a minimum of 2 inches (51 mm) thick all around. Concrete-encased ducts may eventually corrode; however, the air passageway will be maintained because of the remaining concrete enclosure. Nonmetallic ducts and metallic ducts with factory-applied protective coatings must be approved and installed to

comply with the manufacturer's installation instructions. Application of any field-applied protective coating must be approved. Great care must be taken to protect underground ducts from damage prior to placing the concrete or installing the permanent structure above them. Plastic duct and fitting systems designed for underground applications are available that allow corrosion-resistant and waterproof installations.

UL 181 applies to flexible nonmetallic underground ducts. Section 603.6 requires nonmetallic flexible air duct to conform to the material requirements of UL 181, but the standard does not provide assurance that flexible ducts are rated for underground installation. Suitability must be evaluated based on the manufacturer's installation instructions for the intended underground application and the corresponding supporting evidence.

603.8.1 Slope. Ducts shall have a minimum slope of $^1/_8$ inch per foot (10.4 mm/m) to allow drainage to a point provided with access.

❖ Underground ducts must be sloped to drain to an accessible point in the event that water enters the duct through duct openings or from the surrounding soil. Water can cause corrosion, deterioration of duct materials and duct blockages. Sloping the duct to drain to a collection point will allow removal of water. Specifying minimum slope provides the inspector with a criterion for enforcement. Note that the code does not state that ducts must be water tight.

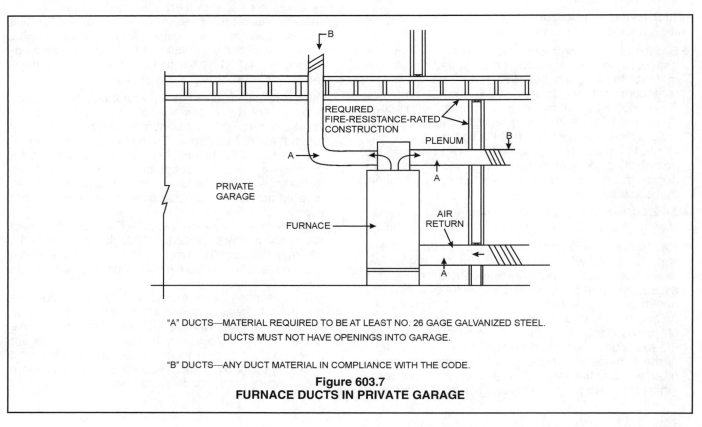

"A" DUCTS—MATERIAL REQUIRED TO BE AT LEAST NO. 26 GAGE GALVANIZED STEEL.
DUCTS MUST NOT HAVE OPENINGS INTO GARAGE.

"B" DUCTS—ANY DUCT MATERIAL IN COMPLIANCE WITH THE CODE.

Figure 603.7
FURNACE DUCTS IN PRIVATE GARAGE

603.8.2 Sealing. Ducts shall be sealed and secured prior to pouring the concrete encasement.

❖ Underground ducts need not be water tight. This section requires sealing of the duct before the concrete is poured to prevent concrete from entering the duct.

603.8.3 Plastic ducts and fittings. Plastic ducts shall be constructed of PVC having a minimum pipe stiffness of 8 psi (55 kPa) at 5-percent deflection when tested in accordance with ASTM D 2412. Plastic duct fittings shall be constructed of either PVC or high-density polyethylene. Plastic duct and fittings shall be utilized in underground installations only. The maximum design temperature for systems utilizing plastic duct and fittings shall be 150°F (66°C).

❖ Plastic ducts have the advantage of being corrosion resistant. The pipe stiffness criterion gives the pipe the ability to resist deformation from the loads associated with direct burial. Plastic ducts rapidly lose strength as their temperature approaches their maximum service temperature. At temperatures above 150°F (66°C), PVC pipe is substantially weakened, and deformation and/or collapse is possible.

603.9 Joints, seams and connections. All longitudinal and transverse joints, seams and connections in metallic and nonmetallic ducts shall be constructed as specified in SMACNA *HVAC Duct Construction Standards—Metal and Flexible* and NAIMA *Fibrous Glass Duct Construction Standards*. All joints, longitudinal and transverse seams and connections in ductwork shall be securely fastened and sealed with welds, gaskets, mastics (adhesives), mastic-plus-embedded-fabric systems, liquid sealants or tapes. Closure systems used to seal ductwork *listed* and *labeled* in accordance with UL 181A shall be marked "181A-P" for pressure-sensitive tape, "181A-M" for mastic or "181A-H" for heat-sensitive tape. Closure systems used to seal flexible air ducts and flexible air connectors shall comply with UL 181B and shall be marked "181B-FX" for pressure-sensitive tape or "181B-M" for mastic. Duct connections to flanges of air distribution system *equipment* shall be sealed and mechanically fastened. Mechanical fasteners for use with flexible nonmetallic air ducts shall comply with UL 181B and shall be marked "181B-C." Closure systems used to seal metal ductwork shall be installed in accordance with the manufacturer's installation instructions. Unlisted duct tape is not permitted as a sealant on any duct.

Exception: Continuously welded and locking-type longitudinal joints and seams in ducts operating at static pressures less than 2 inches of water column (500 Pa) pressure classification shall not require additional closure systems.

❖ Duct sealing is commonly overlooked or poorly performed. The U.S. Environmental Protection Agency (EPA) estimates that 20 percent of the energy efficiency of heating and cooling systems can be lost because of duct air leaks located outside of the conditioned space. With the increased focus on lowering energy use in all types of buildings, substantial gains in energy efficiency can be obtained by making sure that ducts are "substantially" air tight. While approved duct materials have low permeability, the joints in

these materials, as well as the joints in the connections of these materials to fittings/equipment, must be carefully sealed to achieve a "substantially" air-tight condition.

In general, joints must be sealed using tapes, mastics, liquid sealants, gasketing or other approved closure systems. A "closure system" consists of the materials and an installation method used to make the joint substantially air tight. This section specifically addresses three types of duct material and requires the sealing to be accomplished as follows:

• Rigid fibrous glass ductwork listed in accordance with UL 181A. Where pressure-sensitive tape is used, the exterior of the tape must be factory marked with the designation "181A-P." Where mastic is used, the container label must indicate "181A-M." Where heat sensitive tape is used, the exterior of the tape must be factory marked with the designation "181A-H."

• Flexible air ducts and flexible air connectors listed in accordance with UL 181B. Where pressure-sensitive tape is used, the exterior of the tape must be factory marked with the designation "181B-FX." Where mastic is used, the container label must indicate "181B-M."

• Metal ducts must be sealed in accordance with the closure system manufacturer's installation instructions. Although there are no requirements for metal duct closure systems to be listed to any standard, the use of unlisted "duct tape" for metal duct sealing is prohibited. Even though many in the HVAC industry have long considered general-purpose duct tapes to be unsuitable for the intended service conditions and required length of service, the code did not always prohibit such use. This does not imply that tapes which comply with UL 181A or UL 181B must be used for sealing metal ducts; however, because these tapes do possess high performance qualities necessary for similar sealing applications, these tapes are commonly used for metal duct sealing. Types of high-quality tape such as metal foil tape with acrylic adhesive are being used for metal duct sealing. Generally, closure systems (including tapes) that are listed will be of high quality.

In addition to sealing, joints must also be mechanically fastened. In other words, sealing by itself is not a substitute for the mechanical fastening of a duct joint. Two examples are (1) taping a flexible air duct to a fitting and (2) "gluing" a round metal duct joint with mastic or caulking. Both practices are not acceptable fastening methods. Mechanical fasteners must be used. This section specifically addresses the mechanical fastening of three types of duct material as follows:

• Rigid fibrous glass ductwork must be mechanically attached to flanges of air distribution equipment or sheet metal fittings. For example, where a rigid fibrous glass duct connects to a sheet metal flange of an air handler, sheet metal screws (with flat washers) can be used to attach the duct board to the metal flange. Where fittings for other ducts must attach to the duct, a "spin in," "twist lock" or "tabbed" collar provides for the required mechanical fastening of the joint.

- Flexible air ducts and flexible air connectors are fastened to collar stubs or duct connector sleeves using a "cable tie" that complies with UL 181B and is marked "181-C" or a worm-gear clamp. These fittings should have a convex bead formed on the circumference of the fitting to prevent the flexible duct from being pulled off of the connector. Commentary Figure 603.6(4) shows installation guidelines for flexible air ducts and flexible air connectors.

- Round metal duct with crimp joints must have an overlap of at least $1^1/_2$ inches (38.1 mm). At least three sheet metal screws or rivets must be installed, equally spaced around the duct, for mechanical fastening of the joint.

The exception allows for longitudinal welded joints and locking-type joints (including spiral type) and seams to be considered as sealed. A welded or locking-type (e.g., snap lock, Pittsburgh lock) joint is sufficiently air tight at low pressures such that joint sealing would be of limited value. S-slip and drive joints and lap joints must always be sealed (see Commentary Figure 603.9).

603.10 Supports. Ducts shall be supported at intervals not to exceed 12 feet (3658 mm) and shall be in accordance with SMACNA *HVAC Duct Construction Standards—Metal and*

Flexible. Flexible and other factory-made ducts shall be supported in accordance with the manufacturer's instructions.

❖ Proper support of the duct system requires designing for the weight of the duct as well as for dynamic loads caused by vibrations, and is necessary to maintain proper alignment of the duct system and to prevent stresses that could lead to joint failures. A sagging duct will increase the internal resistance to airflow and reduce the efficiency of the system. As with piping systems, ducts and ductwork systems that are improperly supported will deteriorate structurally and functionally.

Duct support spacing is limited to not more than 12 feet (3658 mm) between supports because of the inability of commonly used duct materials to resist deflection at larger spacing intervals. This spacing criterion is primarily applicable to rigid ducts constructed of sheet metal; some duct configurations and duct materials will require smaller spacing intervals. For example, a 12-foot (3658 mm) support spacing would be inadequate for flexible ducts, and some duct materials would require support at each joint.

SMACNA *HVAC Duct Construction Standards— Metal and Flexible* contains information for choosing support systems for ducts. Additionally, the IBC could require designing the duct support system to resist earthquake loads, depending on the location and use of the building.

If duct hangers are not used and a portion of the building structure is used to support the duct, the support system must be designed by a registered professional engineer or architect to comply with the IBC. Duct hangers are to be approved for the application and must be designed for the loads to be carried. For

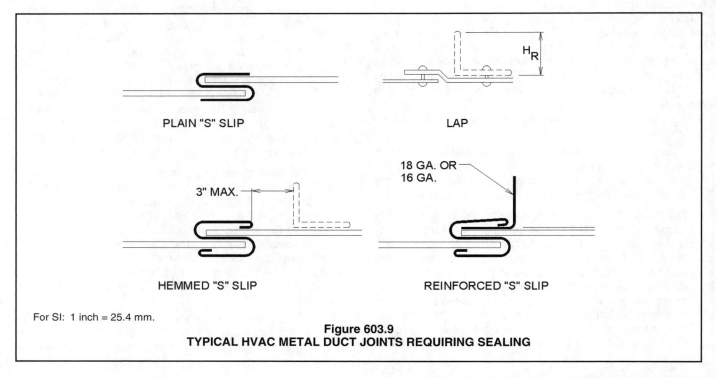

For SI: 1 inch = 25.4 mm.

Figure 603.9
TYPICAL HVAC METAL DUCT JOINTS REQUIRING SEALING

example, hangers for nonmetallic ducts and flexible ducts must be of proper width or must have saddles to prevent damage to the ducts.

Improper support of rigid and flexible ducts is a common problem and a primary cause of failed ducts and duct connections.

603.11 Furnace connections. Ducts connecting to a furnace shall have a *clearance* to combustibles in accordance with the furnace manufacturer's installation instructions.

❖ Where air ducts connect to a furnace, the minimum required clearance from the duct to combustible material must be maintained as specified in the manufacturer's installation instructions. This, of course, requires constructing the duct of a noncombustible material for the length of duct required to have a clearance to combustibles. Because of the high air temperatures and radiant heat from furnace surfaces, proper clearance between duct connections to furnaces and combustible materials must be maintained to reduce the potential for ignition of the duct material. Typically, duct clearances must be maintained from the point of connection to the furnace to a specified distance or length of duct run from the furnace. As an example, this requirement may prohibit the direct connection of a Class I flexible duct to a furnace return or supply connection and could require noncombustible ductwork to extend from the furnace for a specified distance (see Section 603.6.4).

603.12 Condensation. Provisions shall be made to prevent the formation of condensation on the exterior of any duct.

❖ Condensation can form on a duct when the temperature of the air in the duct is near the dewpoint of the air around the duct. The application of insulation with a vapor barrier covering prevents the duct from "sweating" by preventing moisture vapor from penetrating the insulation and reaching the duct surface. Duct sweating (condensation formation) is typically a problem for cooling ducts that pass through unconditioned areas where the humidity is not controlled. Condensation can cause insulation damage, corrosion or duct failure, and the accumulated water can cause damage to the building. Condensation can also occur inside of ducts, such as when ducts conveying warm, moist air from a bathroom exhaust pass through an attic space or similar unconditioned area. Condensation in duct interiors can degrade the duct, promote mold and fungus growth and cause structural damage; however, this section does not address the protection of ducts from condensation forming on the interior of the duct.

[B] 603.13 Flood hazard areas. For structures in flood hazard areas, ducts shall be located above the elevation required by Section 1612 of the *International Building Code* for utilities and attendant equipment or shall be designed and constructed to prevent water from entering or accumulating within the ducts during floods up to such elevation. If the ducts are located below the elevation required by Section 1612 of the *International Building Code* for utilities and attendant equipment, the ducts shall be capable of resisting hydrostatic and hydrodynamic loads and stresses, including the effects of buoyancy, during the occurrence of flooding up to such elevation.

❖ In areas designated by the IBC as flood hazard areas, ducts must be installed above the elevation specified by the IBC or, if allowed below that elevation, ducts must be designed to be water tight. Water in ducts could damage the ducts, cause corrosion and deterioration of the duct materials and affect related equipment. If allowed below the required elevation, water-tight ducts will be exposed to hydrostatic and hydrodynamic loads and stresses, including the effects of buoyancy, which will exert a net upward force on the structure. For additional guidance, refer to FEMA 348.

603.14 Location. Ducts shall not be installed in or within 4 inches (102 mm) of the earth, except where such ducts comply with Section 603.8.

❖ Unless a duct is approved for underground installation, it must not be located in or within 4 inches (102 mm) of the ground. The 4-inch (102 mm) clearance is considered adequate to keep the duct from contacting the ground and possible moisture, which can cause duct deterioration.

603.15 Mechanical protection. Ducts installed in locations where they are exposed to mechanical damage by vehicles or from other causes shall be protected by *approved* barriers.

❖ This section is applicable to all duct locations, outdoors and indoors, and includes ducts in crawl spaces. For example, ducts in garages, in warehouses and on building exteriors can be subject to impact by vehicles and machinery. Where ducts cannot be protected by their location, barriers are required.

603.16 Weather protection. All ducts including linings, coverings and vibration isolation connectors installed on the exterior of the building shall be protected against the elements.

❖ Although most ducts are constructed of galvanized steel or aluminum that will resist corrosion, these materials may not withstand long-term exposure to weather. Ducts and their associated materials (duct liners, coverings, insulation and vibration isolation connectors) must be protected from exposure to moisture, sunlight, wind and contaminants. A weatherproof covering installed to completely cover the duct system is typically used to protect outdoor ductwork.

603.17 Air dispersion systems. Air dispersion systems shall:

1. Be installed entirely in exposed locations.

2. Be utilized in systems under positive pressure.

3. Not pass through or penetrate fire-resistant-rated construction.

4. Be listed and labeled in compliance with UL 2518.

❖ Essentially, an air dispersion system is a tubular-shaped exposed supply air register/diffuser that is made of air-impermeable fabric material. The tube is fitted with air holes or nozzles to direct air into the area served. The requirements for using air dispersion systems are that they not be installed in concealed locations, be supplied with positive air pressure, not penetrate fire-resistance-rated construction and be listed and labeled to UL 2518. Commentary Figures 603.17(1) and 603.17(2) show a typical system. Note that air dispersion systems are not ducts.

603.18 Registers, grilles and diffusers. Duct registers, grilles and diffusers shall be installed in accordance with the manufacturer's installation instructions. Volume dampers or other means of supply air adjustment shall be provided in the branch ducts or at each individual duct register, grille or diffuser. Each volume damper or other means of supply air adjustment used in balancing shall be provided with access.

❖ Air is supplied and returned to an air distribution system through registers, grilles and diffusers. From the definitions contained in the ASHRAE *Handbook of Fundamentals*, a "grille" is a covering over any opening through which air passes; a "diffuser" is an outlet through which air discharges into a room in various directions and planes; and a "register" is a grille equipped with a damper or volume control. The code requires that these devices be installed in accordance with the manufacturer's installation instructions so that they operate properly and do not adversely affect the performance of the air distribution system.

To be able to control the conditions within a room or space, the code requires that air supply terminals (outlets) or branch supply ducts have controls to regulate the volume of air supplied to the room. The means of control can be located anywhere between the beginning of the supply branch duct and the branch outlet. In order to adjust the settings of the volume dampers, they must be provided with access to allow the occupants to easily make the adjustments. This requirement is intended to allow the air distribution system to be properly balanced. In some cases where the system has controls that are designed to balance the air distribution system, such as variable air volume units, the dampers and controls required by this section would not be necessary.

603.18.1 Floor registers. Floor registers shall resist, without structural failure, a 200-pound (90.8 kg) concentrated load on a 2-inch-diameter (51 mm) disc applied to the most critical area of the exposed face.

❖ Registers installed in a floor must be capable of supporting a 200-pound (890 N) load distributed over a 2-inch-diameter (51 mm) circular area. The test must be conducted with the register installed in accordance with the manufacturer's installation instructions with the circular disk placed at the most structurally critical location on the register. This test simulates a person stepping on the floor register or a furniture leg placed on the floor register.

603.18.2 Prohibited locations. Diffusers, registers and grilles shall be prohibited in the floor or its upward extension within toilet and bathing rooms required by the *International Building Code* to have smooth, hard, nonabsorbent surfaces.

Exception: *Dwelling units.*

❖ The IBC requires nonabsorbent floor surfaces in toilet and bathing rooms, other than in private dwellings, including extending the surface of the wall at least 6 inches (152 mm). This is required to facilitate cleaning of the floor surface. Placing an HVAC register or grille in the floor or the upturned extension will allow water to accumulate in the duct during normal cleaning. This moisture can promote the growth of mold and bacteria in the ducts, creating a health hazard for the occupants.

SECTION 604
INSULATION

604.1 General. Duct insulation shall conform to the requirements of Sections 604.2 through 604.13 and the *International Energy Conservation Code.*

❖ To conserve nonrenewable energy resources, the code requires that air ducts be thermally insulated in accordance with the *International Energy Conservation Code®* (IECC®), which determines where insulation is required and what minimum thermal resistance (*R*-value) is required. Other installation requirements for insulation are specified in Sections 604.2 through 604.13.

The insulating value (*R*) of insulation is typically expressed as thermal resistance (resistance to heat transfer), which is expressed in units of (h × ft² × °F)/ Btu [(m² × K)/W]. As the number of Btus transferred decreases (the denominator), the quotient (*R*) increases. An *R*-value (thermal resistance) is the reciprocal of thermal transmittance (*U*). Thermal transmittance is the time rate of heat flow [Btu/h × (*W*)] per unit area [square feet (m²)] and unit temperature difference [°F × (K)] between the warm-side and cold-side air films. The thermal resistance required for the insulation is a function of the design temperature difference between the air in the duct and the air in the space surrounding the duct. The IECC specifies insulation *R*-values based on the location of the duct (conditioned space, unconditioned space, outdoors). Note that in many cases, the *R*-values required by the IECC have increased from the traditionally installed insulation *R*-values.

**603.17(1)
AIR DISPERSION SYSTEM
(Photo courtesy of DuctSox, Inc.)**

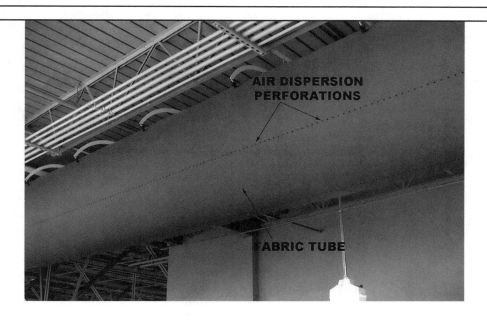

**603.17(2)
AIR DISPERSION SYSTEM
(Photo courtesy of DuctSox, Inc.)**

604.2 Surface temperature. Ducts that operate at temperatures exceeding 120°F (49°C) shall have sufficient thermal insulation to limit the exposed surface temperature to 120°F (49°C).

❖ The intent of this section is to protect occupants from burn injury and to prevent a fire hazard where combustibles could be close to high-temperature duct surfaces.

604.3 Coverings and linings. Coverings and linings, including adhesives when used, shall have a flame spread index not more than 25 and a smoke-developed index not more than 50, when tested in accordance with ASTM E 84 or UL 723, using the specimen preparation and mounting procedures of ASTM E 2231. Duct coverings and linings shall not flame, glow, smolder or smoke when tested in accordance with ASTM C 411 at the temperature to which they are exposed in service. The test temperature shall not fall below 250°F (121°C). Coverings and linings shall be listed and labeled.

❖ Air distribution systems connect most rooms and spaces within a building, thereby providing a path for fire and smoke to spread quickly throughout the structure. Duct coverings, linings, tape and vibration isolation connectors are exposed to the surrounding environment or to the airstream in the duct and could contribute to the spread of fire and development of smoke (see Commentary Figure 604.3). To reduce the possible contribution to the spread of fire and smoke, these materials are limited to a maximum flame spread index of 25 and a smoke-developed index of 50 when tested to ASTM E 84 or UL 723,

which corresponds to a Class 1 material as defined in the IBC. ASTM E 84 and UL 723 provide the test methods required for duct coverings and linings, including a requirement for testing of systems representative of the actual field installation. ASTM E 2231 is included in this code section because it provides the specimen preparation and mounting procedures necessary to ensure that the specimen tested in the laboratory is as close as possible to the actual field installation. This will result in a safer field installation where the actual performance of the material can be more accurately predicted.

Duct coverings and linings must be rated for the design temperatures of the air distribution system to avoid degradation of the materials. To verify that duct coverings and linings will not present a fire hazard, they must be tested at their rated temperatures or to 250°F (121°C), whichever is higher, in accordance with the procedures of ASTM C 411. The minimum of 250°F (121°C) for testing represents the maximum temperature that industry standards will permit in the airstream of a warm air heating appliance.

604.4 Foam plastic insulation. Foam plastic used as duct coverings and linings shall conform to the requirements of Section 604.

❖ Foam plastic is a general term given to insulating products that have been manufactured by injecting a gas into a raw plastic product. Extruded polystyrene, expanded polyurethane (also called EPS or beadboard), isocyanurate, open-cell isocyanurate, phenolic foam and polyurethane are among the many types of foam insulating products. Foam plastic insulation must meet the requirements of Section 604 (see commentary, Section 604.3).

604.5 Appliance insulation. *Listed* and *labeled* appliances that are internally insulated shall be considered as conforming to the requirements of Section 604.

❖ Factory-built appliances must be listed and labeled in accordance with Section 301.7. Insulation that is part of mechanical appliances that are factory built and listed and labeled is considered to be in compliance with the requirements of Section 604 (see commentary, Sections 301.7 and 301.8).

604.6 Penetration of assemblies. Duct coverings shall not penetrate a wall or floor required to have a fire-resistance rating or required to be fireblocked.

❖ Because duct coverings do not comply with the IBC for fireblocking materials or opening protectives, these materials must not penetrate wall, roof/ceiling or floor/ceiling assemblies that must be fire-resistance-rated or fireblocked. Note that the IBC could also require penetration protection for unrated floor/ceiling assemblies. The nature of duct coverings would make it difficult, if not impossible, to properly protect the annular space around a duct penetration; therefore, duct coverings must terminate at the penetration to allow installation of the annular space treatment.

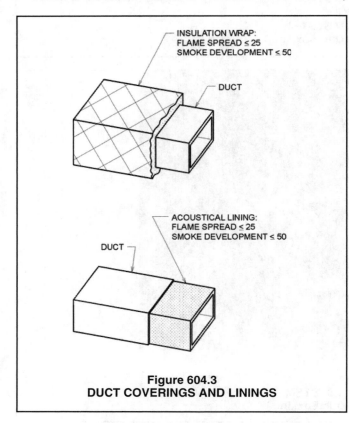

INSULATION WRAP:
FLAME SPREAD ≤ 25
SMOKE DEVELOPMENT ≤ 50

DUCT

ACOUSTICAL LINING:
FLAME SPREAD ≤ 25
SMOKE DEVELOPMENT ≤ 50

DUCT

Figure 604.3
DUCT COVERINGS AND LININGS

604.7 Identification. External duct insulation, except spray polyurethane foam, and factory-insulated flexible duct shall be legibly printed or identified at intervals not greater than 36 inches (914 mm) with the name of the manufacturer, the thermal resistance R-value at the specified installed thickness and the flame spread and smoke-developed indexes of the composite materials. All duct insulation product R-values shall be based on insulation only, excluding air films, vapor retarders or other duct components, and shall be based on tested C-values at 75°F (24°C) mean temperature at the installed thickness, in accordance with recognized industry procedures. The installed thickness of duct insulation used to determine its R-value shall be determined as follows:

1. For duct board, duct liner and factory-made rigid ducts not normally subjected to compression, the nominal insulation thickness shall be used.

2. For duct wrap, the installed thickness shall be assumed to be 75 percent (25 percent compression) of nominal thickness.

3. For factory-made flexible air ducts, the installed thickness shall be determined by dividing the difference between the actual outside diameter and nominal inside diameter by two.

4. For spray polyurethane foam, the aged R-value per inch, measured in accordance with recognized industry standards, shall be provided to the customer in writing at the time of foam application.

❖ To aid the inspection process, duct insulation must have a label affixed to it by the manufacturer with the manufacturer's name, the thermal resistance (R) at the specified installation thickness and the flame spread and smoke-developed indexes. A third-party agency must do quality control inspections at the manufacturer's facility in accordance with the requirements for labeling contained in Sections 301.7 and 301.8. Testing by an independent agency must determine the insulating R-value, the flame spread index and the smoke-developed index. The 36-inch (914 mm) label intervals are intended to increase the likelihood that every cut piece of insulation and flexible duct will have a label.

The thermal performance of duct insulation depends on its "installed" condition, including the compression that is the case for duct wrap insulations. This section is intended to provide manufacturers, installers and inspectors with specific guidance for meeting the intent of the code. For example, installed duct wrap is assumed to have a thickness of 75 percent of the nominal uninstalled thickness. The R-value on the product will account for the decreased thermal resistance caused by compression of the product.

Because labeling of the exterior of spray polyurethane insulation is not feasible, the code requires that the same information be given, in writing, to the customer at the time of application.

604.8 Lining installation. Linings shall be interrupted at the area of operation of a fire damper and at a minimum of 6 inches (152 mm) upstream of and 6 inches (152 mm) downstream of electric-resistance and fuel-burning heaters in a duct system. Metal nosings or sleeves shall be installed over exposed duct liner edges that face opposite the direction of airflow.

❖ Duct lining materials must not be located where they could interfere with the operation of fire dampers. Duct liners must terminate on both sides of a fire damper to ensure clearance to any moving components. Linings within a duct must be interrupted a minimum of 6 inches (152 mm) upstream and downstream of a heat-producing appliance. The 6-inch (152 mm) clearance is considered adequate because linings must have a Class 0 or Class 1 rating (see commentary, Section 604.3). Metal nosings or sleeves are used to cover exposed duct liner edges that face into the flow of air because airflow can erode the insulation fibers and can lift the edge, causing liner displacement and duct blockage or disruption of airflow.

604.9 Thermal continuity. Where a duct liner has been interrupted, a duct covering of equal thermal performance shall be installed.

❖ To maintain energy efficiency, a covering of equal thermal resistance must be installed where a duct liner has been interrupted. This is intended to prevent gaps or weak links in the required thermal performance of ducts.

604.10 Service openings. Service openings shall not be concealed by duct coverings unless the exact location of the opening is properly identified.

❖ Duct coverings must not impede recognition of or access to service openings for components of the air distribution system; therefore, duct coverings must be interrupted at those locations or the service opening locations must be clearly identified to allow access by removing the covering.

604.11 Vapor retarders. Where ducts used for cooling are externally insulated, the insulation shall be covered with a vapor retarder having a maximum permeance of 0.05 perm [2.87 ng/(Pa · s · m²)] or aluminum foil having a minimum thickness of 2 mils (0.051 mm). Insulations having a permeance of 0.05 perm [2.87 ng/(Pa · s · m²)] or less shall not be required to be covered. All joints and seams shall be sealed to maintain the continuity of the vapor retarder.

❖ Vapor retarders are required to prevent moisture from deteriorating the insulating material and duct. Insulation alone will not prevent condensation unless the insulation is impervious (impermeable) to moisture. Insulations having a permeance of 0.05 perm or less do not have to be covered because they serve as their own vapor barrier.

The continuity of the vapor barrier cannot be maintained if the joints and seams are not properly sealed against moisture vapor penetration.

604.12 Weatherproof barriers. Insulated exterior ducts shall be protected with an *approved* weatherproof barrier.

❖ Insulated exterior ducts must have a weatherproof covering to prevent the outdoor elements from deteriorating the materials (see commentary, Section 603.16).

604.13 Internal insulation. Materials used as internal insulation and exposed to the airstream in ducts shall be shown to be durable when tested in accordance with UL 181. Exposed internal insulation that is not impermeable to water shall not be used to line ducts or plenums from the exit of a cooling coil to the downstream end of the drain pan.

❖ Insulation that is not durable can eventually erode, causing increased energy use, condensation and associated water damage. Insulation that is not impermeable to water must not be used to line ducts or plenums at the exit of a cooling coil because there is a potential for wetting the duct insulation. This can cause insulation degradation, loss of insulating thermal properties and the growth of mold or fungus, which can cause adverse health effects. Depending on air velocity and turbulence, water droplets can be carried from the coil surfaces well downstream of the coil drain pan.

SECTION 605
AIR FILTERS

605.1 General. Heating and air-conditioning systems of the central type shall be provided with *approved* air filters. Filters shall be installed in the return air system, upstream from any heat exchanger or coil, in an *approved* convenient location. Liquid adhesive coatings used on filters shall have a flash point not lower than 325°F (163°C).

❖ The accumulation of combustible dust and debris in and around air filters can create a potential hazard. The combustibility of the filter medium and any adhesive coatings used to improve filter efficiency may also create a hazard. Approval of filters, except those installed within dwelling units, must be based on listing and testing in accordance with Section 605.2.

Air filters must be made accessible for inspection, cleaning and replacement. Typically, air filters will be located behind doors, panels, cover plates or return air grilles. Filters are supported in frames, tracks or cabinets and may be located within air-handling equipment, return air ducts or return grille assemblies. Filters should be installed in convenient locations to encourage frequent inspection and maintenance.

Chapter 2 defines "Access" as "that which enables a device, appliance or equipment to be reached by ready access or by a means that first requires the removal or movement of a panel, door or similar obstruction." As filters become loaded with dust and debris, the resistance to airflow through them increases, which will reduce system airflow, causing a loss in heating and cooling system efficiency and causing excessive temperature rise across a heat exchanger (such as in a forced air furnace). The resultant rise in discharge air temperature and elevated equipment operating temperature can cause a fire hazard. For example, a forced air furnace with poorly maintained air filters can cause the furnace to cycle its high-limit control, resulting in excessively high operating temperatures and an unsafe condition.

As an air filter becomes loaded, air pathways are blocked, which illustrates the need for more frequent filter inspections as the filter nears the time for replacement or cleaning. Poorly maintained air filters are a common problem and also create a fire hazard because of the accumulation of combustible dust and debris on the filters and within the mechanical equipment.

605.2 Approval. Media-type and electrostatic-type air filters shall be *listed* and *labeled*. Media-type air filters shall comply with UL 900. High efficiency particulate air filters shall comply with UL 586. Electrostatic-type air filters shall comply with UL 867. Air filters utilized within *dwelling units* shall be designed for the intended application and shall not be required to be *listed* and *labeled*.

❖ This section specifies specific standards for testing of various types of filters. UL 900 is the recognized standard for evaluating media-type filters, UL 586 is the recognized standard for evaluating high-efficiency particulate air filters and UL 867 is the recognized standard for evaluating electrostatic air filters.

UL 900 contains tests to establish the flammability characteristics of air filter units while the filters are clean. Class 1 units are those filters that, when clean, do not contribute fuel when attacked by flame and give off only negligible amounts of smoke. Class 2 units are those that, when clean, burn moderately when attacked by flame or create moderate amounts of smoke, or both.

UL 586 addresses high-efficiency particulate air filter units (HEPA), which are intended for the removal of very fine particulate matter (retaining not less than 99.97 percent of particles 0.3 micron in diameter and larger).

UL 867 addresses electrostatic filters rated at 600 volts or less, intended to remove dust and other nonhazardous particles from the air by electrostatic precipitation. Electricity is used to charge particles and capture the particles on oppositely charged wire grids.

For residential occupancies, filter units must be suitable for the application and designed to serve the air-handling equipment under the expected conditions of operation. Air filters for air-handling systems in dwelling units are generally small, low-efficiency, media-type filters that are located within individual dwelling units. They need not be listed and labeled.

605.3 Airflow over the filter. Ducts shall be constructed to allow an even distribution of air over the entire filter.

❖ Air ducts must be constructed to allow an even distribution of air over the entire face of a filter. This will extend the life of the filter because of the increase in its capacity to trap and retain particulates over more of its surface area. Also, maximizing the useable surface area of the filter minimizes the pressure drop across it. Equipment manufacturers' instructions and duct design handbooks provide information on proper filter placement and location in duct systems. For example, filters are ideally located where there is little or no airflow turbulence and filter enclosures are commonly enlarged or the filters are placed at angles other than perpendicular to the airflow to expose more filter surface area to the airstream.

SECTION 606
SMOKE DETECTION SYSTEMS CONTROL

606.1 Controls required. Air distribution systems shall be equipped with smoke detectors *listed* and *labeled* for installation in air distribution systems, as required by this section. Duct smoke detectors shall comply with UL 268A. Other smoke detectors shall comply with UL 268.

❖ Section 606 contains requirements for protection against the spread of smoke throughout a building through an air distribution system. Generally, smoke spreads through the air distribution system when the return air system conducts smoke from a fire in a room or space back to the air-handling equipment where the smoke is fed into the supply distribution system. The smoke may be diluted by mixing it with return air from other parts of the building and by the introduction of outdoor air; however, smoke will still be routed through the supply air distribution system to other parts of the building that may not be involved with the fire.

The intent of these provisions is to prevent ducted air distribution systems from distributing smoke from the area of origin to other areas or spaces in a building. An air distribution system can distribute smoke throughout a building much faster than the smoke would have traveled naturally and, of course, ducts can carry smoke across boundaries that otherwise would have stopped the natural migration of smoke. For these reasons, duct system smoke detectors are used to shut down a ducted air distribution system before it can threaten the building occupants by spreading smoke.

This section applies to ducted air distribution systems (see the definition of "Air distribution system") and, therefore, does not apply to installations of air handlers and HVAC equipment that do not involve ductwork and extensive plenums. For example, rooftop HVAC units, suspended unit heaters, suspended horizontal furnaces and blower/coil combination units are commonly installed without distribution ductwork. In these examples, the extent of what might be considered as ductwork is typically limited to short supply and return box plenums (drops) extending from the roof unit, or simple supply and return plenum extensions of the suspended equipment cabinets that are used only to accommodate return grilles, filter racks, flow splitters and similar directional discharge fittings and devices. Because no distribution ductwork extends to areas and spaces beyond the immediate location of the HVAC equipment, the HVAC equipment cannot contribute significantly to the spread of smoke in the building; therefore, a duct-mounted smoke detector would be of little value.

Also, it is not the intent of Section 606 to require duct smoke detectors in systems that function only as exhaust systems or only as makeup air supply systems. A makeup air supply system that discharges 100-percent outdoor air into a building does not withdraw air from the building and, therefore, cannot contribute to the movement of smoke. Likewise, an exhaust-only system discharges all air to the outdoors and, therefore, in most circumstances will not contribute to the movement of smoke into other areas of the building. In fact, the operation of an exhaust system in the area of fire origin could be considered beneficial because it provides some smoke removal capacity. Note that exhaust fans can create negative pressures within a building, thereby causing smoke migration.

It is more important to keep in mind the intended application of Section 606, which is to address the potential hazard caused by ducted air distribution systems that link together rooms and spaces within a building, thereby providing the pathway to distribute smoke to the linked rooms and spaces. Air-handling systems of any type that cannot transport smoke beyond the area of fire origin are exempt from the provisions of this section.

Smoke detectors installed in air distribution systems must be labeled for that application. Because the moving-air environment within the duct differs from a still-air environment, a duct smoke detector must be specially designed and tested to sense smoke within the air ducts at the design airflow rates. UL 268A is used to evaluate duct smoke detectors. The UL *Fire Protection Equipment Directory* contains listings of smoke detectors that have been evaluated for use in air duct systems. Some smoke detectors are mounted within the duct and sense smoke in the airstream; other designs incorporate air-sampling tubes that supply an air sample to smoke detectors mounted in a chamber attached to the exterior of the duct. It is important to observe the airstream velocity limitations in the detector's installation instructions to maintain acceptable detector sensitivity. For detectors that use air-sampling tubes, a maximum pressure differential between tubes will also be given.

606.2 Where required. Smoke detectors shall be installed where indicated in Sections 606.2.1 through 606.2.3.

Exception: Smoke detectors shall not be required where air distribution systems are incapable of spreading smoke beyond the enclosing walls, floors and ceilings of the room or space in which the smoke is generated.

❖ Sections 606.2.1 through 606.2.3 state where and under what conditions duct smoke detectors are required in air distribution systems.

The design capacity of an air-handling system consisting of two or more air handlers paralleled to function as a unit would be the sum of the capacities of the individual air handlers. On the other hand, a building may have multiple air handlers, with each air handler serving entirely independent duct systems. In this case, the capacities of the individual independent air handlers would not be summed. Each system would be evaluated separately. If multiple air-handling units share any ductwork or plenum space, which occurs in a multiple-zone variable air volume system having individual zone air handler units that draw air from a common return air plenum, the capacities of all of the zone air handlers would be summed because the common return air plenum creates a single air distribution system. In such a system, the main air handler serving the return plenum would be evaluated as another distinct air distribution system (see commentary, Section 606.2.2).

The exception clarifies the intent of this section to prevent air distribution systems from distributing smoke to other rooms and spaces where the occupants might not be aware of the fire. Air distribution systems that serve only a single room or space would not require smoke detectors. The typical warehouse-type retail store served by roof top units that serve a single open space would be an example. In such spaces, the occupants would be aware of a fire anywhere in the space and could determine the safest

egress path. Commentary Figure 606.2(1) shows an air distribution system that is greater than 2,000 cfm (0.9 m³/s) but is not capable of moving smoke beyond the walls, floors or ceilings of the space. In this installation a smoke detector would not be required in the return air duct or plenum. Commentary Figure 606.2(2) shows the same air distribution system that is greater than 2,000 cfm (0.9 m³/s) and is serving two separate spaces. In this installation smoke that might be generated in one space could be circulated through the air distribution system to another space. This system would require a smoke detector in the return air plenum.

606.2.1 Return air systems. Smoke detectors shall be installed in return air systems with a design capacity greater than 2,000 cfm (0.9 m³/s), in the return air duct or *plenum* upstream of any filters, *exhaust air* connections, outdoor air connections, or decontamination *equipment* and appliances.

Exception: Smoke detectors are not required in the return air system where all portions of the building served by the air distribution system are protected by area smoke detectors connected to a fire alarm system in accordance with the *International Fire Code*. The area smoke detection system shall comply with Section 606.4.

❖ Smoke detectors must be installed in return air systems having design capacities exceeding 2,000 cubic feet per minute (0.9 m³/s) (see Commentary Figure 606.2.1). Return air systems with design capacities equal to or less than 2,000 cfm (0.9 m³/s) are exempt from this requirement because their small size limits their capacity for spreading smoke to parts of the building not already involved with fire. The area that could be served by a 2,000 cfm (0.9 m³/s) system (approximately 5 tons of cooling capacity) is compar-

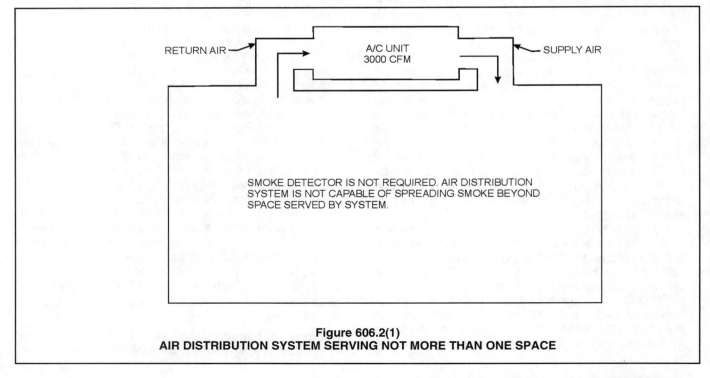

RETURN AIR   A/C UNIT   SUPPLY AIR
                    3000 CFM

SMOKE DETECTOR IS NOT REQUIRED. AIR DISTRIBUTION SYSTEM IS NOT CAPABLE OF SPREADING SMOKE BEYOND SPACE SERVED BY SYSTEM.

Figure 606.2(1)
AIR DISTRIBUTION SYSTEM SERVING NOT MORE THAN ONE SPACE

atively small. Therefore, the distribution of smoke in a system of that size would be minimal.

To maximize the effectiveness of smoke detectors in the return air duct, the detectors must be installed in the area of the highest concentration of smoke in the system. Therefore, the detectors must be installed in the path of airflow upstream of any filters, exhaust air connections, outdoor air connections or decontamination equipment in the system.

Filters and decontamination (air cleaning) equipment can remove some of the smoke from the air stream, and exhaust air and outdoor air connections can bleed off or dilute smoke in the air stream, all of which can delay the response time of the detector. Where a single detector would be unable to sample the total airflow at all times, detectors would be required for both the return and the exhaust ducts or plenums. If a return air system has a takeoff for exhausting a portion of the return air, and the takeoff is located upstream of the duct smoke detector, smoke detectors should be installed in both the return air and the exhaust air ducts or plenums. In this configuration, the detector is sampling only a fraction of the total airflow, because some of the total return air is exhausted before passing by the smoke detector (see commentary, Section 606.3). It is not uncommon for the exhaust (relief) duct to be independent of the return air system. In this case, airflow through the exhaust duct would not be sampled by the detector. This limited sampling by a smoke detector would delay the detector's response because of the reduced concentration of smoke.

An exception to the requirements of this section occurs where a system of area smoke detectors is

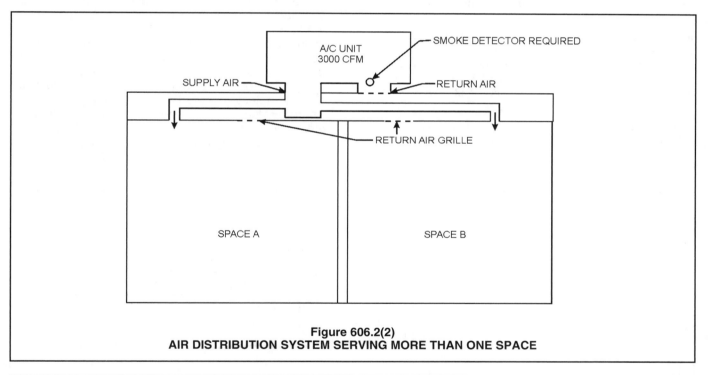

Figure 606.2(2)
AIR DISTRIBUTION SYSTEM SERVING MORE THAN ONE SPACE

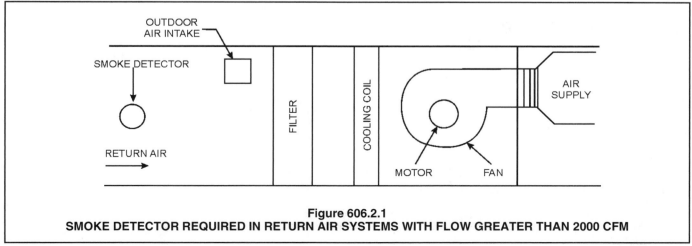

Figure 606.2.1
SMOKE DETECTOR REQUIRED IN RETURN AIR SYSTEMS WITH FLOW GREATER THAN 2000 CFM

used to protect all spaces served by the air distribution system. The smoke detectors must be installed as required by the *International Fire Code®* (IFC®). The logic of the exception is that a fire in the space served by the air distribution system will be detected by the area smoke detectors and the alarm will be processed as required by the IFC, thereby eliminating the need for smoke detectors in the return air system.

It is also important to note that installing duct smoke detectors does not waive any other requirements for smoke detectors within a room or space because duct smoke detectors are effective only if the air distribution system is operating. Additionally, duct detectors cannot respond until the concentration of smoke in a duct is detectable, whereas an area detector can respond before smoke even reaches the air distribution system. This exception still requires the air handler to shut down when smoke is detected (see commentary, Section 606.4).

606.2.2 Common supply and return air systems. Where multiple air-handling systems share common supply or return air ducts or plenums with a combined design capacity greater than 2,000 cfm (0.9 m³/s), the return air system shall be provided with smoke detectors in accordance with Section 606.2.1.

Exception: Individual smoke detectors shall not be required for each fan-powered terminal unit, provided that such units do not have an individual design capacity

greater than 2,000 cfm (0.9 m³/s) and will be shut down by activation of one of the following:

1. Smoke detectors required by Sections 606.2.1 and 606.2.3.

2. An *approved* area smoke detector system located in the return air *plenum* serving such units.

3. An area smoke detector system as prescribed in the exception to Section 606.2.1.

In all cases, the smoke detectors shall comply with Sections 606.4 and 606.4.1.

❖ It is not uncommon to have multiple air-handling systems of less than 2,000 cfm (0.9 m³/s) each share a common return or supply air duct or plenum. When the combined capacity of all air-handling systems sharing a common duct or plenum is greater than 2,000 cfm (0.9 m³/s), a smoke detector must be installed as required by Section 606.2.1 (see commentary, Section 606.2.1). Because multiple air-handling units are involved, a smoke detector installed in accordance with Section 606.2.1 at each air-handling unit may be required. For instance, if multiple air-handling units are located in different parts of a building, a single detector located at some point in a common return air duct might not be able to sample all of the airflow. A detector at each unit could be required. In this case, upon activation the smoke detector would shut down only the air-handling unit it serves [see Commentary Figure 606.2.2(1)].

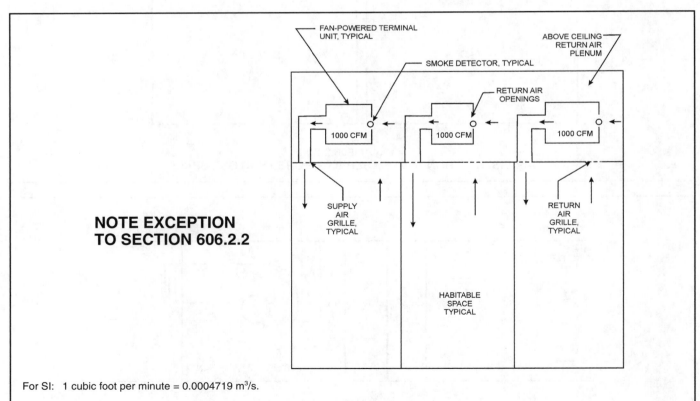

NOTE EXCEPTION TO SECTION 606.2.2

For SI: 1 cubic foot per minute = 0.0004719 m³/s.

Figure 606.2.2(1)
LOCATION OF SMOKE DETECTORS IN COMMON RETURN AIR PLENUM

If multiple air-handling units are located in a common area and are served by a common return air duct that allows a single detector to sample all the airflow, multiple smoke detectors at each unit would not be required. In this case, upon activation, the smoke detector must shut down all the air-handling units served by the detector (see commentary, Section 606.2). Where multiple air-handling systems serve a common space but do not share common ducts or plenums, the systems are treated as independent (stand-alone) systems, even though the combined capacity might exceed 2,000 cfm (0.9 m³/s).

The exception states that individual smoke detectors are not required for fan-powered terminal units that are part of a larger air distribution system that has a method of smoke shutdown installed. The individual capacity of these units cannot exceed 2,000 cfm (0.9 m³/s) and they must be shut down by one of the three means listed in the exception. In all cases, the fan-powered terminal units must be shut down, but not necessarily by their own dedicated smoke detectors. If the terminal unit design capacity exceeds 2,000 cfm (0.9 m³/s), the unit is treated as an independent system and an individual smoke detector would be required. The air distribution system shown in Commentary Figure 606.2.2(2) has a supply air fan and return air fan along with two fan-powered terminal units. The return air fan is 3,000 cfm (1.42 m³/s) and has a smoke detector ahead of the fan and filter. The two fan-powered terminal units are 500 cfm (0.24 m³/s) each. The smoke detector in this system would have to be connected to the two

fan-powered terminal units or the fan-powered terminal units would have to be shut down by a smoke detector that complies with one of the three items in the exception to Section 606.2.2. Commentary Figure 606.2.2(3) shows a typical fan-powered terminal unit.

606.2.3 Return air risers. Where return air risers serve two or more stories and serve any portion of a return air system having a design capacity greater than 15,000 cfm (7.1 m³/s), smoke detectors shall be installed at each story. Such smoke detectors shall be located upstream of the connection between the return air riser and any air ducts or plenums.

❖ Where a return air system with a design capacity greater than 15,000 cfm (7.1 m³/s) serves more than one story, the return air from each story must be monitored before intermixing the return in the common riser. This results in early smoke detection and is a means for determining on which story a fire has occurred.

Determining the story of fire origin provides valuable information to fire-fighting personnel and to smoke control systems. For example, if a smoke control system is installed in a building, detection of smoke in a particular story may cause the HVAC system to switch to a smoke control mode that could supply 100-percent exhaust in that story while the adjacent stories are supplied with 100-percent outdoor air and no return or exhaust, thereby creating a positive pressure on those stories.

To monitor each story accurately and to prevent detection delays, the duct smoke detectors must be installed within the return air duct or plenum of the

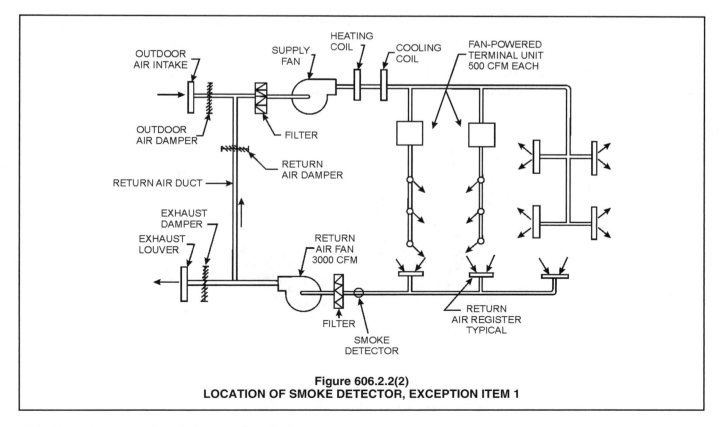

Figure 606.2.2(2)
LOCATION OF SMOKE DETECTOR, EXCEPTION ITEM 1

story served upstream of its connection to the return air riser. This location ensures that the detectors will be sampling return air from the story they serve before the air is mixed with return air from other stories, thus resulting in early detection (see Commentary Figure 606.2.3).

If smoke detection occurred only in the main return (common) duct riser in a multiple-story building, it is obvious that a fire could progress for quite some time before the diluted smoke in the common return duct reached a detectable concentration. This section applies to return air risers that convey any portion or all of the design capacity of the system.

[F] 606.3 Installation. Smoke detectors required by this section shall be installed in accordance with NFPA 72. The required smoke detectors shall be installed to monitor the entire airflow conveyed by the system including return air and exhaust or relief air. Access shall be provided to smoke detectors for inspection and maintenance.

❖ Smoke detectors required by Section 606.2 must be installed in accordance with the requirements of NFPA 72. This standard specifies performance requirements and characteristics for smoke detectors for duct installation as well as location and installation requirements for those detectors. Smoke detectors must be installed to monitor the entire airflow within the system (see commentary, Section 606.2.1). Typical air distribution systems may have a number of branches for return air, exhaust or relief air. Where a single detector would be unable to sample the total airflow at all times, multiple detectors would be required.

Chapter 2 defines "Access" as "that which enables a device, appliance or equipment to be reached by ready access or by a means that first requires the removal or movement of a panel, door or similar obstruction." Duct smoke detectors must be made

accessible for maintenance and inspection. Many failures and false alarms are caused by a lack of maintenance and cleaning of the smoke detectors. Smoke detectors must be maintained in accordance with the manufacturer's installation instructions and inspected in accordance with the IFC.

[F] 606.4 Controls operation. Upon activation, the smoke detectors shall shut down all operational capabilities of the air distribution system in accordance with the listing and label-

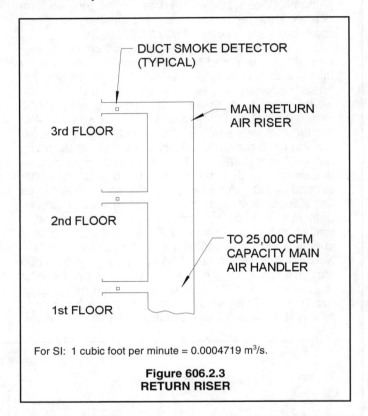

For SI: 1 cubic foot per minute = 0.0004719 m³/s.

Figure 606.2.3
RETURN RISER

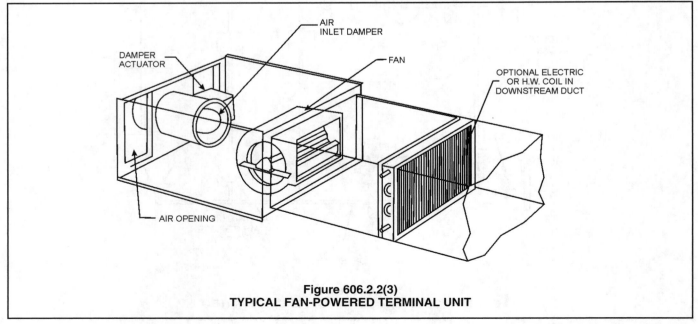

Figure 606.2.2(3)
TYPICAL FAN-POWERED TERMINAL UNIT

ing of appliances used in the system. Air distribution systems that are part of a smoke control system shall switch to the smoke control mode upon activation of a detector.

❖ Smoke detectors are required within the air distribution system as required by Section 606 to prevent the air distribution system from contributing to the spread of smoke within a building by stopping (shutting down) the air handlers (blowers and fans) upon detection of smoke. Smoke is spread through an air distribution system when smoke enters the duct system and is transported to other areas of the building through the ducts. Therefore, upon activation of a smoke detector in the duct, the air distribution system must be shut down or, if the air distribution system is part of a smoke control system, it must switch to the smoke control mode of operation. The IBC contains requirements for the installation of smoke control systems in buildings. The smoke detectors required by Section 606 are relied upon to automatically initiate air distribution system shutdown or smoke control system operation.

This section has been open to interpretation as to how to shut down the air distribution system. Many inspectors, contractors and system designers require rewiring an appliance to only shut down the blower. This practice can nullify the listing of the appliance as well as promote unsafe operation in cases where the appliance is being fired at the time the blower is shut down. Another practice has been to break the 24-volt wire to the thermostat, which may not in some cases shut down the appliance. An example of this would be if the high-limit switch were to trip, the fan would still run. The only way of ensuring that the air distribution system is shut down is to interrupt the power supply to the appliance. There is still the potential of shutting down the appliance in the middle of a firing cycle, creating the potential for overheating in the appliance.

[F] 606.4.1 Supervision. The duct smoke detectors shall be connected to a fire alarm system where a fire alarm system is required by Section 907.2 of the *International Fire Code*. The actuation of a duct smoke detector shall activate a visible and audible supervisory signal at a constantly attended location.

Exceptions:

1. The supervisory signal at a constantly attended location is not required where the duct smoke detector activates the building's alarm-indicating appliances.

2. In occupancies not required to be equipped with a fire alarm system, actuation of a smoke detector shall activate a visible and audible signal in an *approved* location. Duct smoke detector trouble conditions shall activate a visible or audible signal in an *approved* location and shall be identified as air duct detector trouble.

❖ This section does not require sending a signal to the fire department or activating the alarm notification devices within a building. Instead, this section requires that a supervisory signal be sent to a constantly attended location. Where a building is required to have a fire alarm system, duct smoke detectors must be connected to the fire alarm system. Fire alarm systems required by the IBC must be installed in accordance with the provisions of NFPA 72. A supervisory signal is defined by NFPA 72, as "a signal indicating the need of action in connection with the supervision of guard tours, fire suppression systems or equipment, or maintenance features of related systems." Connection to the fire alarm system will activate a visible and audible supervisory signal at a constantly attended location to alert building supervisory personnel that a smoke alarm has activated, and also to provide electronic supervision of the duct detectors, thereby indicating any problems that may develop in the detector system circuitry or power supply. The signal must be sent to a location that is constantly attended by supervisory personnel while the building is occupied.

Exception 1 allows activation of the building's alarm-indicating appliances in place of a supervisory signal. Causing the building's fire alarm system to sound and indicate an alarm would alert the occupants of the building that an alarm condition existed within the air distribution system, thereby performing the same function as a supervisory signal sent to a constantly attended location.

Exception 2 recognizes that a fire alarm is not required in all buildings. A visible and an audible signal must be activated at an approved location that will alert building supervisory personnel to take appropriate action. Additionally, the duct smoke detectors must be electronically supervised to indicate trouble (system fault) in the detector system circuitry or power supply. A trouble condition must activate a distinct visible or audible signal at a location that will alert the responsible personnel.

[B] SECTION 607
DUCT AND TRANSFER OPENINGS

607.1 General. The provisions of this section shall govern the protection of duct penetrations and air transfer openings in assemblies required to be protected.

❖ The term "assemblies" in this section refers to floor, wall, ceiling and shaft assemblies that are designed to have a fire-resistance rating or to limit the passage of smoke from one area to another. An assembly can also be a floor that is not required to have a fire-resistance rating (see Section 607.6.3). The design of most buildings necessitates that these assemblies be penetrated by ducts and air transfer openings that serve the ventilating systems for the building spaces. These penetrations are potential points where fire and smoke can pass through the assembly and quickly move to other parts of the building that are intended to otherwise be protected by the assembly.

Fire dampers, smoke dampers, combination fire/ smoke dampers and ceiling radiation dampers are devices that are installed in air transfer openings and ducts to provide protection against fire or smoke passing across the assembly at the penetration points in the assembly. This section concerns the location, design, testing and installation of these devices. Commentary Figures 607.1(1), 607.1(2), 607.1(3) and 607.1(4) show typical fire, smoke, combination fire/smoke and ceiling radiation dampers. Commentary Figure 607.1(5) shows a fire damper installed in a frame that is intended to be attached to the face of a fire-resistance-rated wall assembly. Note the required access panel on the bottom of the frame for inspection and servicing of the damper (see Section 607.4).

607.1.1 Ducts that penetrate fire-resistance-rated assemblies without dampers. Ducts that penetrate fire-resistance-rated assemblies and are not required by this section to have dampers shall comply with the requirements of Sections 714.2 through 714.3.3 of the *International Building Code*. Ducts that penetrate horizontal assemblies not required to be contained within a shaft and not required by this section to have dampers shall comply with the requirements of Sections 714.4 through 714.4.2.2 of the *International Building Code*.

❖ In some situations, horizontal ducts that penetrate fire-resistance-rated wall assemblies are not required to have or are prohibited from having fire dampers at the point of penetration (see exception conditions for Sections 607.5.2, 607.5.3, 607.5.4 and 607.5.5). The annular space between the outside of the duct and the inside of the opening in the assembly could be a path where fire and smoke could pass from one side of the assembly to the other. Blocking off this annular space is required in accordance with Sections 714.2 through 714.3.3 of the IBC, in order to provide continuity of the fire-resistance-rated assembly at these duct penetration points.

In some situations, vertical ducts that are not contained in fire-resistance-rated shaft enclosures and that penetrate fire-resistance-rated horizontal assemblies are not required to have or are prohibited from having fire dampers at the point of penetration of the assembly (see exception conditions for Sections 607.6.1 and Section 607.2.1). The annular space between the outside of the duct and the inside of the opening in the assembly could be a path where fire and smoke could pass from one side of the assembly to the other. Blocking off this annular space is required in accordance with Sections 714.4 through 714.4.2.2 of the IBC, in order to provide continuity of the fire-resistance-rated assembly at these duct penetration points.

607.1.1.1 Ducts that penetrate nonfire-resistance-rated assemblies. The space around a duct penetrating a nonfire-resistance-rated floor assembly shall comply with Section 717.6.3 of the *International Building Code*.

❖ There are situations where a floor assembly is not required to have a fire-resistance rating. Even though a floor assembly design is not intended to withstand fire exposure for a specified period of time, a nonfire-resistance-rated floor will provide some resistance to the passage of heat, smoke and flame from one floor to the next. Therefore, where a duct penetrates a nonfire-resistance-rated floor assembly, the annular space between the duct and the opening in the assembly needs to be blocked off to provide some resistance to the passage of heat, smoke and flame. Section 717.6.3 only requires the annular space to be "plugged" with a noncombustible material. There are no requirements for the material to be in accordance with any standard or for the material to be listed or labeled. However, the material to be used must be approved by the code official. Because Section 607.6.3 is controlled by the IBC, Section 717.6.3 of the IBC is identical to Section 607.6.3. Take special note of Item 3 in the referenced section that requires a fire damper at each floor line. This item essentially reads as follows: "Where the duct connects more than two stories but not more than three stories (i.e., two floors are penetrated), a fire damper is required at each floor line" (see commentary, Section 607.6.3).

If more than two nonfire-resistance-rated floors are penetrated by a duct, a shaft is required for the duct even though the floor assemblies being penetrated by the shaft are not required to have a fire-resistance rating.

607.2 Installation. Fire dampers, smoke dampers, combination fire/smoke dampers and ceiling radiation dampers located within air distribution and smoke control systems shall be installed in accordance with the requirements of this section, and the manufacturer's installation instructions and listing.

❖ Dampers must be installed in accordance with the manufacturer's installation instructions and listing (see commentary, Section 304.1). Those instructions will result in an installation that not only protects the penetration when the damper is actuated but will also result in a protected opening should the duct fail.

The fire damper's installation instructions include requirements, such as application, expansion clearances, mounting, fastening means, duct attachment, damper sleeve and frame construction and damper placement/orientation. Ductwork connected to fire dampers must be designed to prevent damper displacement if duct failure occurs. This is intended to keep the fire damper in place so that the opening in the fire-resistance-rated assembly is not left unprotected.

Figure 607.1(1)
FIRE DAMPER
(Courtesy of Ruskin Air and Sound Control)

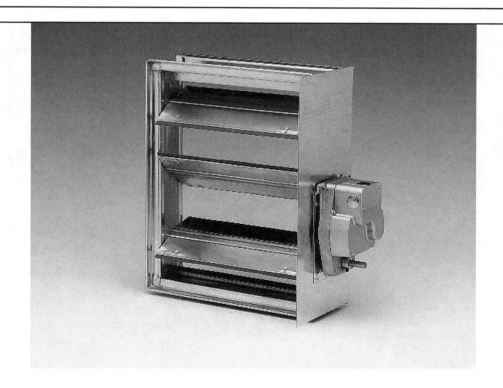

Figure 607.1(2)
SMOKE DAMPER
(Courtesy of Ruskin Air and Sound Control)

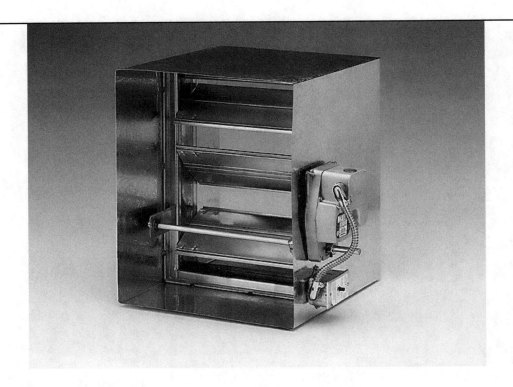

Figure 607.1(3)
COMBINATION FIRE/SMOKE DAMPER
(Courtesy of Ruskin Air and Sound Control)

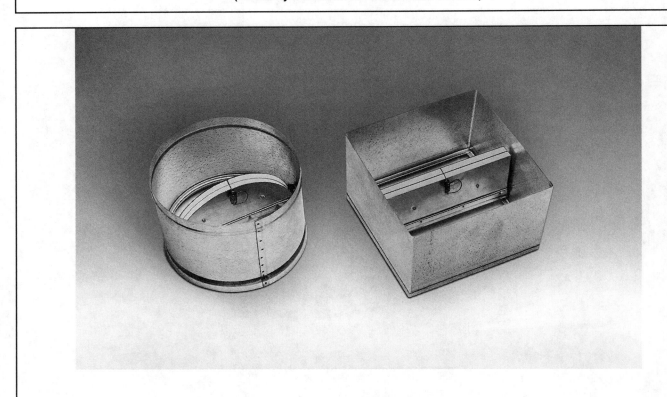

Figure 607.1(4)
CEILING DAMPERS
(Courtesy of Ruskin Air and Sound Control)

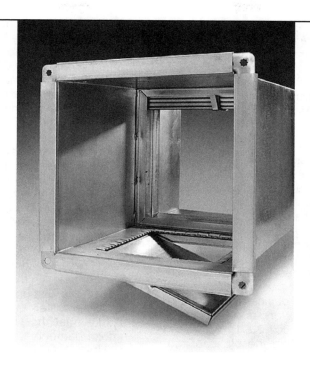

Figure 607.1(5)
FIRE DAMPER IN WALL FRAME
(Courtesy of Ruskin Air and Sound Control)

607.2.1 Smoke control system. Where the installation of a fire damper will interfere with the operation of a required smoke control system in accordance with Section 909 of the *International Building Code, approved* alternative protection shall be used. Where mechanical systems including ducts and dampers used for normal building ventilation serve as part of the smoke control system, the expected performance of these systems in smoke control mode shall be addressed in the rational analysis required by Section 909.4 of the *International Building Code.*

❖ Fire dampers are prohibited where they will interfere with, obstruct or inhibit the proper operation of a required smoke control system. Fire dampers interfering with smoke control could do more harm than good. However, the absence of a fire damper at a location where it is normally required does not relieve the designer of providing protection against the travel of heat and fire through the duct or transfer opening. An alternative means of protection must be approved by the code official. Typically, the alternative means of protection could involve vertical subducts, shaft or horizontal enclosure assemblies, listed through-penetration protection assemblies or special duct materials.

A required smoke control system is an important life safety system. The intent is to make it as reliable as possible by eliminating devices or system designs that could lead to system failure. If a smoke control system not required by the code is installed by designer choice, the system is still subject to the same fire damper requirements as if the system was required by the code. However, the unrequired system could use fire dampers instead of the mandatory alternative protection prescribed in this section for required smoke control systems.

Of course, it is logical and desirable to design and install all life safety protection systems as though they were required because their presence gives building occupants a sense of security. Regardless of the reasons for installing the system, it is expected to function as intended.

Note that this section does more than simply allow omission of fire dampers in required smoke control systems. This section actually mandates the omission of fire dampers where the operation of a required smoke control system could be jeopardized by the presence of fire dampers in the system.

607.2.2 Hazardous exhaust ducts. Fire dampers for hazardous exhaust duct systems shall comply with Section 510.

❖ See the commentary to Sections 510.6 through 510.6.3.

607.3 Damper testing, ratings and actuation. Damper testing, ratings and actuation shall be in accordance with Sections 607.3.1 through 607.3.3.

❖ See the commentary to Sections 607.3.1 through 607.3.3.

607.3.1 Damper testing. Dampers shall be listed and labeled in accordance with the standards in this section. Fire dampers shall comply with the requirements of UL 555. Only fire dampers *labeled* for use in dynamic systems shall be installed in heating, ventilation and *air-conditioning* systems designed to operate with fans on during a fire. Smoke dampers shall comply with the requirements of UL 555S. Combination fire/smoke dampers shall comply with the requirements of both UL 555 and UL 555S. Ceiling radiation dampers shall comply with the requirements of UL 555C or shall be tested as part of a fire-resistance-rated floor/ceiling or roof/ceiling assembly in accordance with ASTM E119 or UL 263.

❖ A fire damper is a device designed to automatically close upon the detection of a specific level of heat within the air-handling system at the damper location. While some fire dampers are designed with a heat-sensing element integral to the damper (e.g., a fusible link), others are triggered to close by remote heat-sensing devices or systems. The purpose of a fire damper is to close to restrict the flow of heat and prevent the passage of flame past the damper location. Once closed, a fire damper can only be opened by manually moving the shutter device and installing a new fusible link. Fire dampers are rated for either $1^1/_2$ hours or 3 hours (see Table 607.3.2.1).

Fire dampers are located within the plane of the fire-resistance-rated floor, ceiling or wall assembly that is penetrated by the duct or transfer opening. Although the code prescribes the required locations for fire dampers (see Section 607.5), the exact positioning of the damper assembly in relation to the faces of specific fire-resistance-rated assemblies is detailed in the manufacturer's installation instructions for each damper. This is critical as the installed position of the damper in relation to the faces of the fire-resistance-rated assembly must duplicate the conditions under which the damper was tested in order for the damper to be listed and labeled for the application.

Fire dampers must be designed and tested in accordance with UL 555. The UL 555 acceptance criteria for fire dampers are that they (1) remain in the duct or transfer opening for the fire exposure period for period of time for which they are rated; (2) close and latch automatically within 60 seconds of being triggered to close; and (3) remain in position without warping beyond prescribed limitations.

UL 555 also requires that fire dampers be labeled with specific information regarding the damper's capacity to operate under static or dynamic conditions, installation orientation, fire-resistance rating, airflow rating and closure pressure rating. Fire dampers designed for static service are only intended to be able to close when the flow in the system has been stopped by some other means, such as fan shut down, a balance of pressure in the system or the closure of another damper. Fire dampers designed for dynamic conditions must be able to close while maximum airflow is occurring in the system. Closure under airflow conditions requires significantly more actuating effort from the mechanism that closes the damper. Therefore, it is very important to never install a fire damper designed for closure under static condi-

tions in a location that requires a fire damper designed to operate under dynamic conditions as the damper might fail to operate as intended (see Commentary Figure 607.3). However, fire dampers labeled for use under dynamic conditions can be installed where a fire damper is required only to operate under static conditions.

As some fire dampers close using only gravity as the means of closure power, proper installation orientation of the damper is critical. Where a fire damper label indicates that the damper orientation is for only vertical (V) or only horizontal (H), the damper must be installed in the vertical or horizontal orientation, respectively. Fire dampers having both "H" and "V" on the label can be installed in either orientation. Dampers are also marked with an indication as to the top of the damper as the proper operation of the damper often depends on a specific face or side of the damper being in the "up" position.

A smoke damper is a device installed in a duct or an air transfer opening to resist the passage of smoke past the location of the damper. Generally, smoke dampers are located immediately adjacent to or within the plane of the floor, ceiling or wall assembly that acts as a smoke barrier. However, smoke dampers might also be located in other parts of an air distribution system as part of a system used to purposely direct smoke flow during fire-fighting efforts. Smoke dampers are designed and tested in accordance with the requirements of UL 555S to provide for a low level of air leakage past the closed damper. Smoke damper operation is controlled by a signal from an adjacent smoke detector in the duct, a smoke detector in the area served by the duct or a remote (fire fighter) command station that allows for manual override of automatic control of the smoke dampers in the building. Electric or pneumatic actuators are used to move the damper blade(s). Smoke dampers are always designed to operate under conditions of maximum duct airflow condition when the damper is open and maximum differential pressure developed when the damper is closed. Note that a fire damper tested to UL 555 cannot serve as a smoke control damper as it will not have the required low level leakage characteristics required of a smoke damper. Labeling requirements for smoke dampers are the same as for fire dampers.

A combination fire/smoke damper is a device installed in a duct or an air transfer opening to resist the passage of smoke, restrict the flow of heat and prevent the passage of flame past the damper location. This type of damper must located within the plane of the fire-resistance-rated floor, ceiling or wall assembly that is penetrated by the duct or transfer opening. A combination fire/smoke damper must comply with both UL 555 and UL 555S. Fire/smoke damper operation is controlled by a signal from an adjacent heat and smoke detector in the duct, a heat and smoke detector in the area served by the duct or a remote (fire fighter) command station that allows for manual override of automatic control of the fire/ smoke dampers in the building. Electric or pneumatic actuators are used to move the damper blade(s). Combination fire/smoke dampers are always designed to operate under conditions of maximum duct airflow condition when the damper is open and maximum differential pressure developed when the damper is closed. Labeling requirements for combination fire/smoke dampers are the same as for fire dampers.

A ceiling radiation damper (or ceiling damper) is a device installed in an air inlet/outlet opening in a fire-resistance-rated floor/ceiling or roof/ceiling assembly. The purpose of a ceiling radiation damper is to close to restrict the flow of heat and prevent the passage of flame past the damper location and maintain continuity of the fire-resistance rating of the floor/ceiling or roof/ceiling assembly at the opening. A heat sensing element integral to the damper (e.g., a fusible link) causes the damper to close using gravity force. Ceiling radiation dampers are designed and tested in accordance with the requirements of UL 555C. Although the purpose of a ceiling radiation damper is the same as a fire damper, the ceiling radiation damper is specifically designed and tested to also resist heat radiating from a fire below the damper location. Because a UL 555 fire damper is not

designed and tested to resist this type of radiant heat exposure, a fire damper must not be used in place of a ceiling radiation damper. However, it is possible that fire damper could be designed and tested in accordance with UL 555C in order to be used in a ceiling radiation damper application. Labeling requirements for ceiling radiation dampers are the same as for fire dampers.

607.3.2 Damper rating. Damper ratings shall be in accordance with Sections 607.3.2.1 through 607.3.2.3.

❖ See the commentary to Sections 607.3.2.1 through 607.3.2.3.

607.3.2.1 Fire damper ratings. Fire dampers shall have the minimum fire protection rating specified in Table 607.3.2.1 for the type of penetration.

❖ See the commentary to Table 607.3.2.1.

TABLE 607.3.2.1
FIRE DAMPER RATING

TYPE OF PENETRATION	MINIMUM DAMPER RATING (hour)
Less than 3-hour fire-resistance-rated assemblies	$1^1/_2$
3-hour or greater fire-resistance-rated assemblies	3

❖ This table summarizes the required hourly ratings for fire dampers based on the fire-resistance-rated

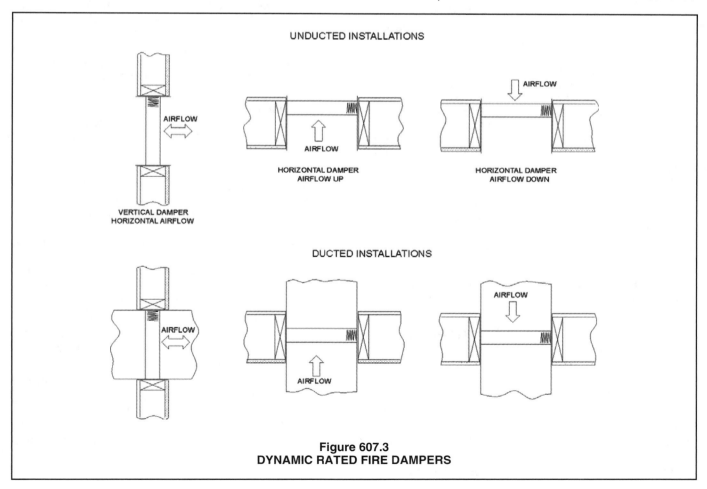

UNDUCTED INSTALLATIONS

VERTICAL DAMPER
HORIZONTAL AIRFLOW

HORIZONTAL DAMPER
AIRFLOW UP

HORIZONTAL DAMPER
AIRFLOW DOWN

DUCTED INSTALLATIONS

Figure 607.3
DYNAMIC RATED FIRE DAMPERS

assembly that is being penetrated by the air distribution system. The left-hand column lists the associated hourly ratings of either vertical or horizontal assemblies and the right-hand column lists the applicable damper rating. These fire damper ratings, obtained from the UL 555 test standard, represent the ratings necessary to maintain the integrity of the rated wall, floor/ceiling or roof/ceiling assembly.

607.3.2.2 Smoke damper ratings. Smoke damper leakage ratings shall be Class I or II. Elevated temperature ratings shall not be less than 250°F (121°C).

❖ The standard for smoke dampers, UL 555S (see Section 607.3.1), requires smoke dampers to be marked with a temperature rating and a leakage rating. The range of temperature ratings for smoke dampers starts at 250°F (121°C) and rises in increments of 100°F (38°C). Leakage ratings range from Classes I (lowest leakage rate) through IV (highest leakage rate) at an indicated test pressure that ranges from 1.0 to 12 inches of water column (0.25 to 3.0 kPa).

607.3.2.3 Combination fire/smoke damper ratings. Combination fire/smoke dampers shall have the minimum fire protection rating specified for fire dampers in Table 717.3.2.1 of the *International Building Code* for the type of penetration and shall also have a minimum smoke damper rating as specified in Section 717.3.2.2 of the *International Building Code*.

❖ As the name implies, combination fire/smoke dampers provide protection against flame spread, as well as the passage of smoke past a point in the duct or air transfer opening. The fire protection rating of a combination fire/smoke damper must be in accordance with Section 717.3.2.1 of the IBC, which requires a 1.5 hour rating for fire-resistance-rated assemblies of less than 3 hours and a 3-hour rating for fire-resistance-rated assemblies of 3 hours or greater. The leakage rating of a combination fire/smoke damper must be in accordance with Section 717.3.2.2 of the IBC, which requires a leakage rating of either Class I or II.

607.3.3 Damper actuation. Damper actuation shall be in accordance with Sections 607.3.3.1 through 607.3.3.4 as applicable.

❖ See the commentary to Sections 607.3.3.1 through 607.3.3.4.

607.3.3.1 Fire damper actuation device. The fire damper actuation device shall meet one of the following requirements:

1. The operating temperature shall be approximately 50°F (28°C) above the normal temperature within the duct system, but not less than 160°F (71°C).

2. The operating temperature shall be not more than 350°F (177°C) where located in a smoke control system complying with Section 909 of the *International Building Code*.

❖ Item 1 of this section establishes the minimum temperature at which the heat sensor for a fire damper can cause the fire damper to close. The minimum

"trigger" temperature must not be less than 160°F (71°C). If the normal air temperature in the duct is greater than 110°F, then the minimum trigger temperature must not be less than 50°F (27.8°C) greater than the normal air temperature in the duct. In accordance with UL 555, where the fire damper is designed to close only under static conditions, the maximum trigger temperature is limited to 215°F (102°C). Establishing a minimum trigger temperature prevents fire dampers from inadvertently closing when fire conditions do not exist. The maximum trigger temperature for a fire damper designed to operate under static conditions ensures that the fire damper will reliably close long before excessively high heat conditions occur at the damper location.

Item 2 of this section concerns fire dampers located within a required smoke control system and establishes the maximum temperature at which a heat sensor can cause closure of fire dampers in this application. A fire damper located in a required smoke control system must be designed to close under dynamic conditions. In accordance with UL 555, fire dampers designed for operation under dynamic conditions must close at a temperature not greater than 350°F (177°C). The operation of a smoke control system will cause hot smoke and air to be conveyed through the ductwork. Fire dampers in these systems must have a higher limit for the trigger temperature so that the smoke control system can remain operational until such time that the temperature of the hot smoke and air becomes so high that the conveyed heat could begin to adversely affect other parts of the system or building beyond the area where the fire is located.

607.3.3.2 Smoke damper actuation. The smoke damper shall close upon actuation of a *listed* smoke detector or detectors installed in accordance with Section 907.3 of the *International Building Code* and one of the following methods, as applicable:

1. Where a smoke damper is installed within a duct, a smoke detector shall be installed in the duct within 5 feet (1524 mm) of the damper with no air outlets or inlets between the detector and the damper. The detector shall be *listed* for the air velocity, temperature and humidity anticipated at the point where it is installed. Other than in mechanical smoke control systems, dampers shall be closed upon fan shutdown where local smoke detectors require a minimum velocity to operate.

2. Where a smoke damper is installed above smoke barrier doors in a smoke barrier, a spot-type detector *listed* for releasing service shall be installed on either side of the smoke barrier door opening.

3. Where a smoke damper is installed within an unducted opening in a wall, a spot-type detector *listed* for releasing service shall be installed within 5 feet (1524 mm) horizontally of the damper.

4. Where a smoke damper is installed in a corridor wall or ceiling, the damper shall be permitted to be controlled by a smoke detection system installed in the corridor.

5. Where a total-coverage smoke detector system is provided within areas served by a heating, ventilation and air-conditioning (HVAC) system, smoke dampers shall be permitted to be controlled by the smoke detection system.

❖ This section presents specific information for proper placement of a smoke detector to ensure that it will most efficiently detect smoke (see Item 1). Where there are multiple smoke dampers in a duct system, Item 1 would require locating a smoke detector within 5 feet (1524 mm) of each damper, as opposed to one detector operating multiple dampers. Additionally, for passive systems (not part of a smoke control system), detector location is critical to their performance. Smoke dampers are installed in transfer openings as well as duct penetrations [see Commentary Figure 607.3.3.2(1)].

Item 2 requires a spot detector on one side of the smoke barrier door opening, not one on each side of the door opening.

Item 3 specifies a spot-type detector within 5 feet of a smoke damper, installed in a wall, without any ducts attached to the damper. This type of installation usually occurs in return air transfer openings in walls that require a smoke damper. The spot-type detector that operates the smoke damper has to comply with Sections 907.3 and 907.3.1 in the IBC. These two sections in the IBC require the smoke detector to be connected to the fire alarm control panel where a fire alarm system is required by Section 907.2 in the IBC. The location of the smoke detector shall be in accordance with NFPA 72 [see Commentary Figure 607.3.3.2(2)].

Items 4 and 5 allow for a smoke damper to be activated by a smoke detection system installed in a corridor or where a total coverage system is installed in a building. Both of these systems would have to comply with Section 907 in the IBC or IFC.

Smoke dampers may also be operated remotely, especially when used as fire dampers and as part of a smoke removal system. Combination fire and smoke dampers can be activated by either the heat sensor or the smoke sensor. Once the primary heat-sensing device has been activated, a remote control system can be used to open the damper to permit its use as a smoke damper in the smoke removal system. Then, if the temperature rises to the damper's maximum degradation test temperature, a secondary heat sensor closes the damper again.

607.3.3.3 Combination fire/smoke damper actuation. Combination fire/smoke damper actuation shall be in accordance with Sections 607.3.3.1 and 607.3.3.2. Combination fire/smoke dampers installed in smoke control system shaft penetrations shall not be activated by local area smoke detection unless it is secondary to the smoke management system controls.

❖ This section primarily serves the purpose of requiring the actuation of combination fire/smoke dampers to comply with the requirements for both fire and smoke dampers, including the requirements in Sections 607.3.3.1 and 607.3.3.2. Further, when these dampers are used as part of a smoke control system, this section prohibits their activation by local area smoke detectors because this could render the smoke control system inoperable.

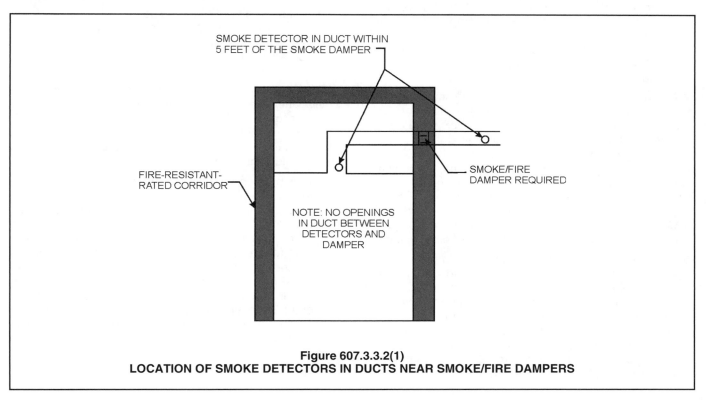

SMOKE DETECTOR IN DUCT WITHIN 5 FEET OF THE SMOKE DAMPER

FIRE-RESISTANT-RATED CORRIDOR

SMOKE/FIRE DAMPER REQUIRED

NOTE: NO OPENINGS IN DUCT BETWEEN DETECTORS AND DAMPER

Figure 607.3.3.2(1)
LOCATION OF SMOKE DETECTORS IN DUCTS NEAR SMOKE/FIRE DAMPERS

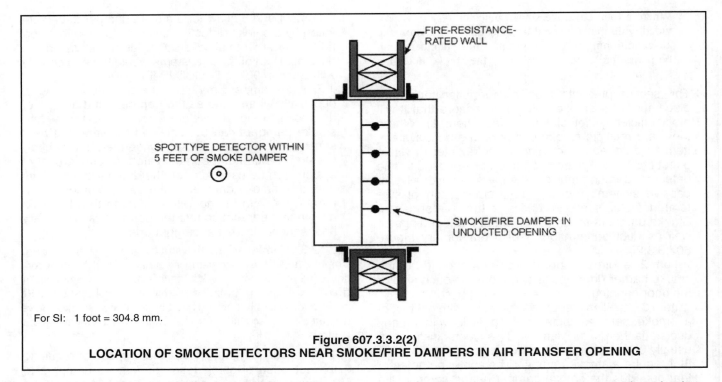

For SI: 1 foot = 304.8 mm.

Figure 607.3.3.2(2)
LOCATION OF SMOKE DETECTORS NEAR SMOKE/FIRE DAMPERS IN AIR TRANSFER OPENING

607.3.3.4 Ceiling radiation damper actuation. The operating temperature of a ceiling radiation damper actuation device shall be 50°F (28°C) above the normal temperature within the duct system, but not less than 160°F (71°C).

❖ This section establishes the minimum temperature at which the heat sensor for a ceiling radiation damper can cause the ceiling radiation damper to close. The minimum "trigger" temperature must not be less than 160°F (71°C). If the normal air temperature in the duct is greater than 110°F (43.3°C), then the minimum trigger temperature must not be less than 50°F (27.8°C) greater than the normal air temperature in the duct.

607.4 Access and identification. Fire and smoke dampers shall be provided with an *approved* means of access, large enough to permit inspection and maintenance of the damper and its operating parts. The access shall not affect the integrity of fire-resistance-rated assemblies. The access openings shall not reduce the fire-resistance rating of the assembly. Access points shall be permanently identified on the exterior by a label having letters not less than 0.5 inch (12.7 mm) in height reading: FIRE/SMOKE DAMPER, SMOKE DAMPER or FIRE DAMPER. Access doors in ducts shall be tight fitting and suitable for the required duct construction.

❖ Fire and smoke dampers must be maintained to ensure that they will operate as intended. The need to maintain dampers, as well as reset dampers and replace fusible links after operation requires that dampers be accessible. To maintain a fire-resistance rating, access doors may need to be fire doors in accordance with Section 714 of the IBC, depending on the location of the door. Access doors in the duct itself need not have a fire-resistance rating.

Where installed at duct openings, such as intakes, supply outlets and transfer openings, fire dampers can be accessed by the removal of grilles, louvers, diffusers or registers. Access doors in ducts must be made reasonably air tight to allow the air distribution system to function as intended and to maintain energy efficiency by controlling the infiltration of unconditioned air into the ducts and the exfiltration of conditioned air out of the ducts.

To make access points readily apparent, a label meeting the requirements of this section must be mounted at the point of access (see commentary, Section 604.10).

607.5 Where required. Fire dampers, smoke dampers and combination fire/smoke dampers shall be provided at the locations prescribed in Sections 607.5.1 through 607.5.7. Where an assembly is required to have both fire dampers and smoke dampers, combination fire/smoke dampers or a fire damper and smoke damper shall be required.

❖ To maintain the fire-resistance rating of fire walls, fire barriers, fire partitions and floor/ceiling and roof/ceiling assemblies, fire dampers and smoke dampers must be installed for all duct penetrations as dictated by Sections 607.5.1 through 607.5.7. The requirements for fire dampers and ceiling radiation dampers in floor/ceiling assemblies and the ceiling membrane of fire-resistance-rated roof/ceiling assemblies are found in Sections 607.6 through 607.6.3. The manufacturer's installation instructions must be followed so that the operation and functioning of the installed device will be consistent with that required by the code.

Commentary Figure 607.5 shows the difference between where a fire damper is required and where one is not. In this example, the designer has chosen to separate the mixed occupancies in accordance with the option of Section 508.4.4 of the IBC. The occupancy separation wall (fire barrier) that must be fire-resistance rated must have the duct penetration protected by a fire damper, while the unrated wall requires no protection.

Duct penetrations and transfer openings in certain assemblies can require protection with both fire and smoke dampers (see Sections 607.5.4 and 607.5.5). Section 607.2.1 requires alternative protection where fire dampers will interfere with the operation of a smoke control system.

607.5.1 Fire walls. Ducts and air transfer openings permitted in fire walls in accordance with Section 706.11 of the *International Building Code* shall be protected with *listed* fire dampers installed in accordance with their listing.

❖ Fire walls create separate buildings that might be on a single parcel of land or, in the case of "zero lot line" construction, the fire wall would be located on the property (lot) line. In the case where the fire wall separates different buildings on different lots, the code does not permit openings (see Sections 706.1.1 and 706.11 of the IBC). In such instances, a duct penetration is not permitted, whether protected or not. Where the fire wall is not located on a lot line, the wall may be penetrated by a duct or transfer opening if the opening is protected with a fire damper listed for that application (see IBC Section 706.8).

607.5.1.1 Horizontal exits. A *listed smoke damper* designed to resist the passage of smoke shall be provided at each point that a duct or air transfer opening penetrates a *fire wall* that serves as a horizontal *exit*.

❖ The purpose of this section is to recognize that horizontal exits are required to resist the passage of smoke, as well as fire. This section, therefore, requires that smoke dampers be installed at ducts or transfer openings that penetrate a fire wall.

607.5.2 Fire barriers. Ducts and air transfer openings that penetrate fire barriers shall be protected with *listed* fire dampers installed in accordance with their listing. Ducts and air transfer openings shall not penetrate exit enclosures and exit passageways except as permitted by Sections 1022.5 and 1023.6, respectively, of the *International Building Code*.

Exception: Fire dampers are not required at penetrations of fire barriers where any of the following apply:

1. Penetrations are tested in accordance with ASTM E 119 or UL 263 as part of the fire-resistance-rated assembly.

2. Ducts are used as part of an *approved* smoke control system in accordance with Section 513 and where the fire damper would interfere with the operation of the smoke control system.

3. Such walls are penetrated by ducted HVAC systems, have a required fire-resistance rating of 1 hour or less, are in areas of other than Group H and are in buildings equipped throughout with an automatic sprinkler system in accordance with Section 903.3.1.1 or 903.3.1.2 of the *International Building Code*. For the purposes of this exception, a ducted HVAC system shall be a duct system for the structure's HVAC system. Such a duct system shall be constructed of sheet steel not less than 26 gage

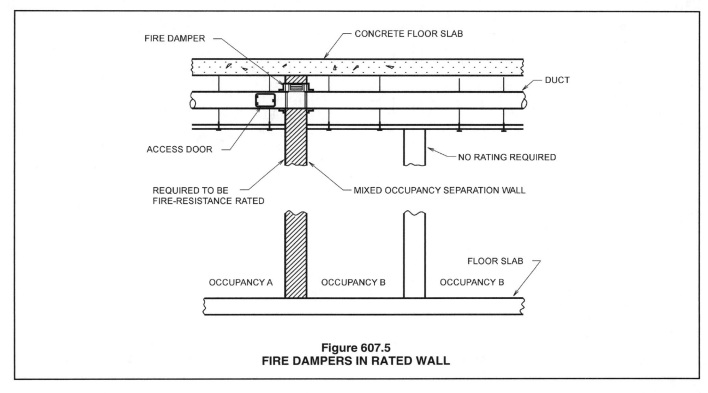

Figure 607.5
FIRE DAMPERS IN RATED WALL

[0.0217 inch (0.55 mm)] thickness and shall be continuous from the air-handling *appliance* or *equipment* to the air outlet and inlet terminals.

❖ It is important that the fire dampers installed where a duct or air transfer opening penetrates a fire barrier are listed for the fire rating of the barrier wall. Penetrations of exit enclosures and exit passageways by ducts and air transfer openings are prohibited to maintain the integrity of the exit path. Such penetrations have the potential to allow smoke or flames to be introduced into the enclosure from other parts of the building, effectively blocking the exit path. This prohibition mirrors the requirements of Sections 707.7.1 and 713.7.1 of the IBC. Sections 1022.5 and 1023.6 of the IBC provide a list of acceptable penetrations into the exit enclosure.

This section also includes three exceptions to the general requirement.

Exception 1 addresses penetrations that were tested as part of the entire tested assembly. Any penetrations will be specified in the design detail for the tested assembly.

Exception 2 reinforces the provisions of Section 607.2.1 (see commentary for Section 607.2.1). A fire damper could interfere with the operation of a smoke control system; however, some form of alternative protection must be installed.

Exception 3 states that, in occupancies other than Group H, fire dampers may be omitted in duct penetrations of walls having a fire-resistance rating of 1 hour or less when penetration is part of a ducted HVAC system in a building protected throughout by an automatic sprinkler system meeting the requirements of NFPA 13 or 13R. Fire dampers would still be required for penetrations of walls requiring a rating greater than 1 hour and for fire-resistance-rated floor/ceiling assemblies. Fire dampers would also be required for transfer openings (openings without ducts) through fire barriers. This section defines "ducted HVAC system" to emphasize that the duct system must be constructed of sheet steel not less than 26 gage thickness and must be continuous from the air-handling appliance or equipment to the air outlet and inlet terminals. This is to ensure that the integrity of the fire separation will not be compromised by discontinuous ducts or the failure of thin-walled metal ducts or nonmetallic ducts [see Commentary Figure 607.5.2].

607.5.2.1 Horizontal exits. A *listed smoke damper* designed to resist the passage of smoke shall be provided at each point that a duct or air transfer opening penetrates a *fire barrier* that serves as a horizontal *exit*.

❖ The purpose of this section is to recognize that horizontal exits are required to resist the passage of

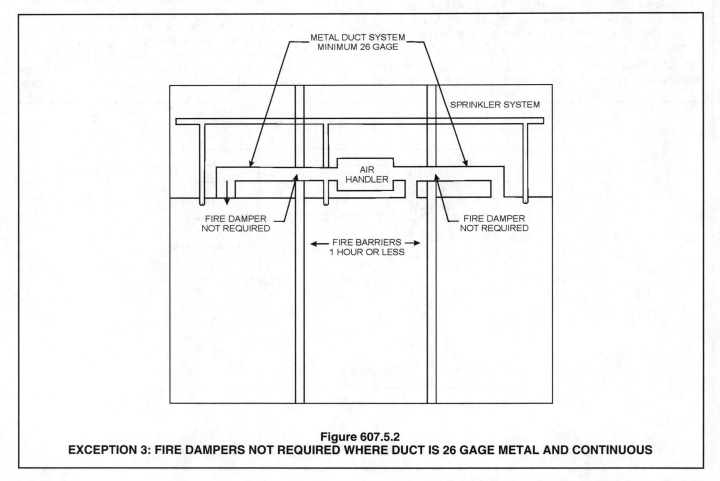

Figure 607.5.2
EXCEPTION 3: FIRE DAMPERS NOT REQUIRED WHERE DUCT IS 26 GAGE METAL AND CONTINUOUS

smoke, as well as fire. This section, therefore, requires that smoke dampers be installed at duct penetrations and transfer openings in a fire barrier.

607.5.3 Fire partitions. Ducts and air transfer openings that penetrate fire partitions shall be protected with *listed* fire dampers installed in accordance with their listing.

Exception: In occupancies other than Group H, fire dampers are not required where any of the following apply:

1. Corridor walls in buildings equipped throughout with an automatic sprinkler system in accordance with Section 903.3.1.1 or 903.3.1.2 of the *International Building Code* and the duct is protected as a through penetration in accordance with Section 714 of the *International Building Code*.

2. The partitions are tenant partitions in covered and open mall buildings where the walls are not required by provisions elsewhere in the *International Building Code* to extend to the underside of the floor or roof sheathing, slab or deck above.

3. The duct system is constructed of *approved* materials in accordance with Section 603 and the duct penetrating the wall complies with all of the following requirements:

   3.1. The duct shall not exceed 100 square inches (0.06 m²).

   3.2. The duct shall be constructed of steel a minimum of 0.0217 inch (0.55 mm) in thickness.

   3.3. The duct shall not have openings that communicate the corridor with adjacent spaces or rooms.

   3.4. The duct shall be installed above a ceiling.

   3.5. The duct shall not terminate at a wall register in the fire-resistance-rated wall.

   3.6. A minimum 12-inch-long (305 mm) by 0.060-inch-thick (1.52 mm) steel sleeve shall be centered in each duct opening. The sleeve shall be secured to both sides of the wall and all four sides of the sleeve with minimum 1¹/₂-inch by 1¹/₂-inch by 0.060-inch (38 mm by 38 mm by 1.52 mm) steel retaining angles. The retaining angles shall be secured to the sleeve and the wall with No. 10 (M5) screws. The annular space between the steel sleeve and the wall opening shall be filled with rock (mineral) wool batting on all sides.

4. Such walls are penetrated by ducted HVAC systems, have a required fire-resistance rating of 1 hour or less, and are in areas of other than Group H and are in buildings equipped throughout with an automatic sprinkler system in accordance with Section 903.3.1.1 or 903.3.1.2 of the *International Building Code*. For the purposes of this exception, a ducted HVAC system shall be a duct system for conveying supply, return or exhaust air as part of the structure's HVAC system. Such a duct system shall be constructed of sheet steel not less than 26 gage in thickness and shall be continuous from the air-handling appliance or equipment to the air outlet and inlet terminals.

❖ Fire dampers installed where a duct or air transfer opening penetrates a fire partition or corridor wall must be listed for the fire rating of the partition and installed in accordance with the listing. It should be noted that based on Section 708.1 in the IBC only the corridor walls are fire partitions. The corridor ceiling or floor is not a fire partition. This section includes three exceptions to the general requirement for occupancies other than Group H.

Because an automatic fire sprinkler system reduces the potential for duct failure by controlling the fire, Exception 1 states that dampers are not required to protect duct penetrations of corridor walls in buildings protected throughout with an automatic sprinkler system [see Commentary Figure 607.5.3(1)]. The reference to Sections 903.3.1.1 and 903.3.1.2 of the IBC establishes that the exception applies only to buildings equipped throughout with an automatic sprinkler system designed and installed in accordance with either NFPA 13 or 13R. This exception further requires protecting the duct as a through penetration in accordance with Section 714 of the IBC.

Exception 2 recognizes that walls separating tenants in a covered mall are required to be rated but are allowed to stop at unrated ceilings and unrated storefronts, providing no true separation. Therefore, dampers in penetrating ducts or air transfer openings are not necessary. If the tenant separation wall is required to be continuous to a roof sheathing, slab, deck or rated floor assembly for another purpose, the fire dampers would obviously be required.

Exception 3 states that fire dampers are not required where a steel duct of 26 gage minimum thickness penetrates a wall, the duct does not have openings that would allow smoke or fire to enter the corridor, the duct is secured in the opening by a steel sleeve and the annular space is filled with mineral wool batts to maintain the thermal barrier. Duct openings would be allowed in the corridor or in adjacent rooms but not in both locations [see Commentary Figure 607.5.3(2)]. These ducts are limited in size, must be protected by a ceiling and must not terminate at the wall.

See the commentary to Exception 3 of Section 607.5.2.

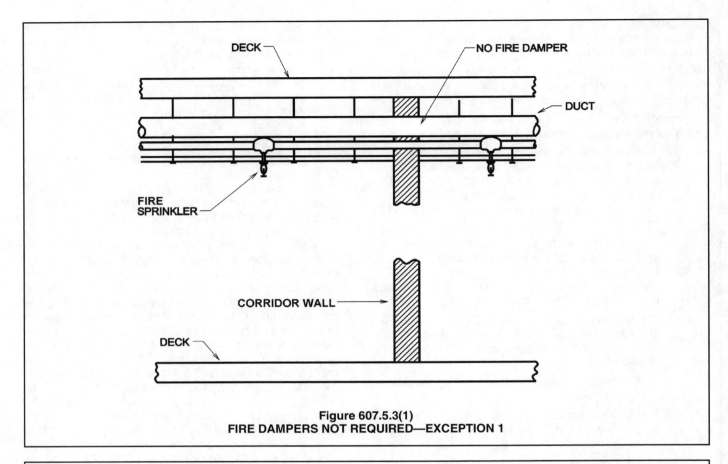

Figure 607.5.3(1)
FIRE DAMPERS NOT REQUIRED—EXCEPTION 1

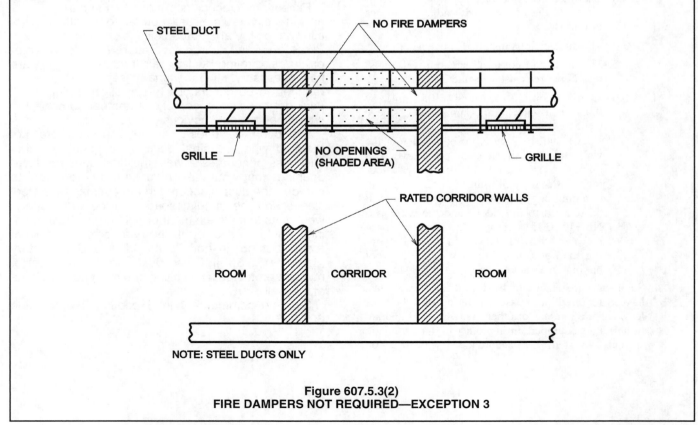

Figure 607.5.3(2)
FIRE DAMPERS NOT REQUIRED—EXCEPTION 3

607.5.4 Corridors/smoke barriers. A *listed* smoke damper designed to resist the passage of smoke shall be provided at each point a duct or air transfer opening penetrates a smoke barrier wall or a corridor enclosure required to have smoke and draft control doors in accordance with the *International Building Code*. Smoke dampers and smoke damper actuation methods shall comply with Section 607.5.4.1.

Exceptions:

1. Smoke dampers are not required in corridor penetrations where the building is equipped throughout with an *approved* smoke control system in accordance with Section 513 and smoke dampers are not necessary for the operation and control of the system.

2. Smoke dampers are not required in smoke barrier penetrations where the openings in ducts are limited to a single smoke compartment and the ducts are constructed of steel.

3. Smoke dampers are not required in corridor penetrations where the duct is constructed of steel not less than 0.019 inch (0.48 mm) in thickness and there are no openings serving the corridor.

❖ To prevent smoke migration across corridor walls and smoke barriers through duct penetrations and transfer openings, a damper designed to resist the passage of smoke must be installed. The damper must close upon detection of smoke by an approved smoke detector (see Section 607.5.4.1). Where smoke barriers and corridor wall fire partitions must be fire-resistance rated, a fire damper is also required, except as allowed by applicable exceptions. Where a fire damper is required, a combination fire and smoke damper or separate fire and smoke dampers must be installed.

Section 716.5.3 of the IBC requires that fire doors in corridor walls and smoke barriers also be tested for smoke and draft control. Similarly, this section requires that duct and air transfer openings be protected with smoke dampers. Unless the exceptions of Section 607.5.3 are applicable, this type of opening would also require a fire damper.

Exception 1 recognizes the protection afforded by a smoke control system and allows omission of smoke dampers only in corridor wall penetrations, if the smoke control system does not, in any way, depend on dampers.

In accordance with Exception 2, smoke dampers are not required in a fully ducted (steel ducts) system that has no openings that would allow the spread of smoke across the smoke barrier.

Exception 3 allows the omission of smoke dampers in corridor wall penetrations where the ducts do not communicate with the corridor through openings in the ducts and the ducts are constructed of 26 gage or heavier sheet steel. It should be noted that Section 607.5.4 requires a smoke damper at each point a duct or air transfer opening penetrates a corridor enclosure. Unlike Section 607.5.3 that requires a fire

damper at duct and air transfer openings in the fire partitions, Section 607.5.4 includes the ceiling and floor of the corridor.

607.5.4.1 Smoke damper. The smoke damper shall close upon actuation of a *listed* smoke detector or detectors installed in accordance with the *International Building Code* and one of the following methods, as applicable:

1. Where a damper is installed within a duct, a smoke detector shall be installed in the duct within 5 feet (1524 mm) of the damper with no air outlets or inlets between the detector and the damper. The detector shall be *listed* for the air velocity, temperature and humidity anticipated at the point where it is installed.

2. Where a damper is installed above smoke barrier doors in a smoke barrier, a spot-type detector *listed* for releasing service shall be installed on either side of the smoke barrier door opening.

3. Where a damper is installed within an unducted opening in a wall, a spot-type detector *listed* for releasing service shall be installed within 5 feet (1524 mm) horizontally of the damper.

4. Where a damper is installed in a corridor wall, the damper shall be permitted to be controlled by a smoke detection system installed in the corridor.

5. Where a total-coverage smoke detector system is provided within all areas served by an HVAC system, dampers shall be permitted to be controlled by the smoke detection system.

❖ This section states specific requirements for the control of smoke dampers. Any one of the five methods, as applicable, can be used. The location of smoke detectors is critical to their performance. Smoke dampers are installed not only in duct penetrations, but also in transfer openings. Item 3 describes a method for controlling dampers in transfer openings. Where smoke detection is installed throughout a corridor or the entire area is served by the duct system (Items 4 and 5, respectively), the detection system is allowed to operate the dampers.

607.5.5 Shaft enclosures. Shaft enclosures that are permitted to be penetrated by ducts and air transfer openings shall be protected with *approved* fire and smoke dampers installed in accordance with their listing.

Exceptions:

1. Fire dampers are not required at penetrations of shafts where:

    1.1. Steel exhaust subducts extend at least 22 inches (559 mm) vertically in exhaust shafts provided that there is a continuous airflow upward to the outdoors; or

    1.2. Penetrations are tested in accordance with ASTM E 119 or UL 263 as part of the fire-resistance-rated assembly; or

    1.3. Ducts are used as part of an *approved* smoke control system in accordance with Section

909 of the *International Building Code*, and where the fire damper will interfere with the operation of the smoke control system; or

  1.4. The penetrations are in parking garage exhaust or supply shafts that are separated from other building shafts by not less than 2-hour fire-resistance-rated construction.

2. In Group B and R occupancies equipped throughout with an automatic sprinkler system in accordance with Section 903.3.1.1 of the *International Building Code*, smoke dampers are not required at penetrations of shafts where kitchen, clothes dryer, bathroom and toilet room exhaust openings with steel exhaust subducts, having a minimum thickness of 0.0187 inch (0.4712 mm) (No. 26 gage), extend at least 22 inches (559 mm) vertically and the exhaust fan at the upper terminus is powered continuously in accordance with the provisions of Section 909.11 of the *International Building Code*, and maintains airflow upward to the outdoors.

3. Smoke dampers are not required at penetrations of exhaust or supply shafts in parking garages that are separated from other building shafts by not less than 2-hour fire-resistance-rated construction.

4. Smoke dampers are not required at penetrations of shafts where ducts are used as part of an *approved* mechanical smoke control system designed in accordance with Section 909 of the *International Building Code* and where the smoke damper will interfere with the operation of the smoke control system.

5. Fire dampers and combination fire/smoke dampers are not required in kitchen and clothes dryer exhaust systems installed in accordance with this code.

❖ This section requires both fire and smoke dampers at duct and air transfer openings in the shaft wall. The fire damper is required because of the penetration of a fire-resistance-rated wall. The smoke damper is required to limit the migration of smoke to other parts of the building through the shaft and the chimney effect (see commentary, IBC Section 713.14.1). Exception 1 allows omission of the fire damper only; the smoke damper is still required. Exceptions 2, 3 and 4 allow omission of smoke dampers only; the fire dampers are still required.

Exception 1, Part 1.1 recognizes that the presence of a vertical subduct in a shaft will also offer some degree of fire resistance should the duct outside of the shaft collapse. Steel exhaust ducts that consist of a 22-inch (559 mm) vertical upturn in the shaft need not be protected with a fire damper if there is a continuous upward airflow to the outside. The continuous airflow will create a negative pressure in the shaft compared to adjacent spaces, thereby minimizing the spread of hot gases from the shaft.

Exception 1, Part 1.2 states fire dampers are not required for duct penetrations of fire barriers when the assembly has been tested in accordance with

ASTM E 119 or UL 263 without fire dampers and the required fire-resistance rating is obtained. In this case, the penetration assembly has been shown to preserve the fire-resistance rating of the wall assembly.

Exception 1, Part 1.3 reinforces the provisions of Section 717.2.1 of the IBC. A fire damper might interfere with the operation of the smoke control system; however, some form of alternative protection must be installed because the duct penetrates the fire barrier.

Exception 1, Part 1.4 states that fire dampers may also be omitted in exhaust and supply shafts that serve a garage and are separated from all other shafts in the building by a 2-hour fire-separation assembly. Requiring fire dampers in garage exhaust and supply shafts would not significantly reduce the spread of smoke and fire within the garage because the vehicle ramp from floor to floor is a much greater conduit for smoke and fire in a garage.

Exception 2 recognizes the advantages of automatic sprinkler protection in Group B and R buildings and that requiring smoke dampers in steel exhaust subducts extended at least 22 inches (559 mm) vertically in the exhaust shafts would not significantly increase safety to the building occupants. There must be a continuously powered upward airflow to the outside, similar to Exception 1, Part 1.1, above. In business and residential occupancies only, this exception combined with Exception 1, Part 1.1, above allows the omission of both the fire and smoke dampers because both exceptions rely on the subduct methodology [see Commentary Figure 607.5.5(1)]. Commentary Figure 607.5.5(2) illustrates an unacceptable method of subducting. The closure of a fire/smoke damper would create the flow from the lower subducts in the shaft to the higher subducts in the shaft resulting in heat and smoke being transferred to upper floors through the subducts.

Exception 3 recognizes that requiring smoke dampers in parking garage exhaust and supply shafts would not significantly prevent the spread of smoke within the garage because the unenclosed vehicle ramp between floors is a much greater conduit for smoke between floors (see Exception 1, Part 1.4).

Exception 4 recognizes that smoke dampers are not required where the ducts are used as part of a smoke control system and where smoke dampers are not required for or would impede the operation of the smoke control system. Requiring smoke dampers in all ducts that are used as part of a smoke control system would not significantly increase safety to the building occupants.

Exception 5 is referring to commercial kitchen exhaust systems and all dryer exhaust systems. The intention of this exception is to not require fire dampers or combination smoke/fire dampers in commercial kitchen exhaust ducts when they comply with Section 506.3.10. It is also intended that dryer exhaust systems that comply with Section 504.2 will not require fire dampers or combination smoke/fire dampers.

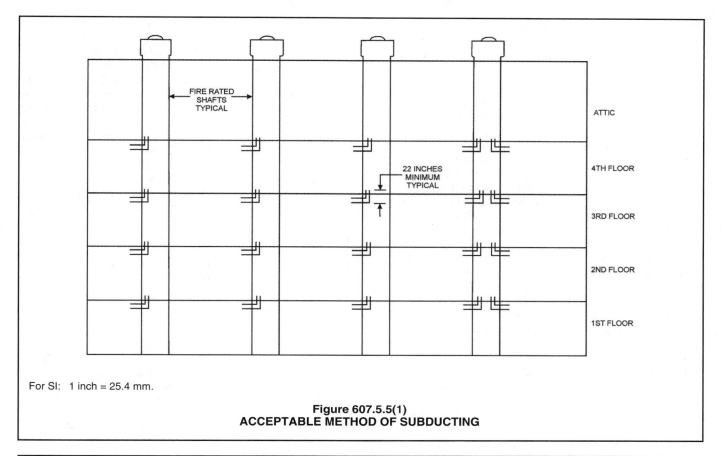

For SI: 1 inch = 25.4 mm.

Figure 607.5.5(1)
ACCEPTABLE METHOD OF SUBDUCTING

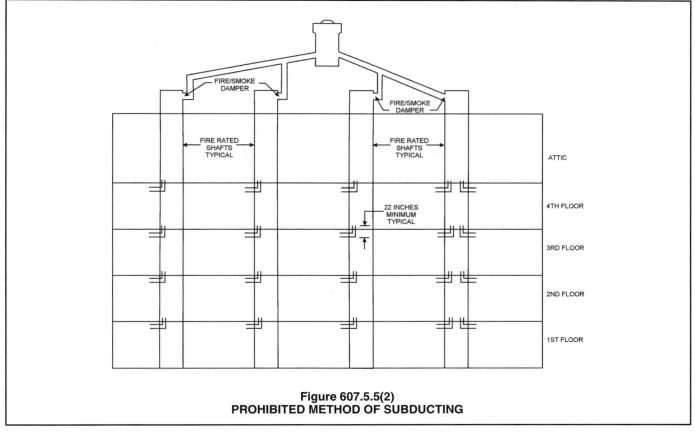

Figure 607.5.5(2)
PROHIBITED METHOD OF SUBDUCTING

607.5.5.1 Enclosure at the bottom. Shaft enclosures that do not extend to the bottom of the building or structure shall be protected in accordance with Section 713.11 of the *International Building Code*.

❖ Proper shaft enclosures must include all sides, as well as the top and bottom, unless the bottom of the shaft is at the bottom of the structure. Where the bottom of the shaft is not at the bottom of the structure, the fire-resistance rating for the bottom of the shaft must be the same as that rating required for the shaft enclosure [see Commentary Figure 607.5.5.1(1)]. The bottom of the shaft must be constructed using a horizontal assembly (IBC Section 713) with the proper fire-resistance rating. It is not permissible to simply take a fire barrier, such as the assembly used for the shaft wall, and turn it horizontally. If a duct extends through the bottom of the shaft, a fire damper is required at the penetration (see Item 3, IBC Section 713.11). Although the fire damper at the floor line will not limit the temperature transmission to the level above, the fire-resistance-rated walls of the shaft will provide the protection from any increased temperature. This section also recognizes that the purpose of some shafts cannot be accomplished if the bottom must be enclosed; therefore, a room is permitted at the bottom of the shaft (see Item 2, IBC Section 713.11), but it must relate to the shaft's purpose and have the same enclosure as the shaft [see Commentary Figure 607.5.5.1(2)]. Two common examples of this situation are either mechanical rooms, which need vents from the equipment to the exterior, or a rubbish or laundry chute, which terminates into a room at the lower level where it is collected. Two of the three exceptions enlarge the "open-shaft-bottom" concept to include two other traditional shaft systems.

Exception 1 provides for an open shaft bottom where the shaft is to be utilized for a purpose on the bottom floor, such as an exhaust system. The shaft walls are not to be interrupted with protected or unprotected openings or penetrations. Bottom closure by draftstopping is required around items entering the base of the shaft [see Commentary Figure 607.5.5.1(3)], or the room that contains the shaft bottom is to be equipped throughout with an automatic fire suppression system complying with the requirements of Chapter 9 of the IBC [see Commentary Figure 607.5.5.1(4)]. Either protection method will reduce the migration of products of combustion into

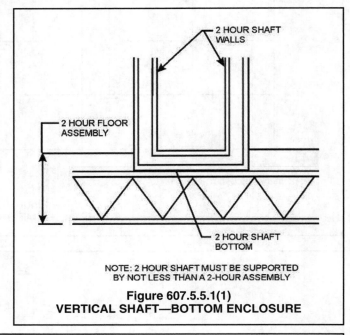

NOTE: 2 HOUR SHAFT MUST BE SUPPORTED BY NOT LESS THAN A 2-HOUR ASSEMBLY

Figure 607.5.5.1(1)
VERTICAL SHAFT—BOTTOM ENCLOSURE

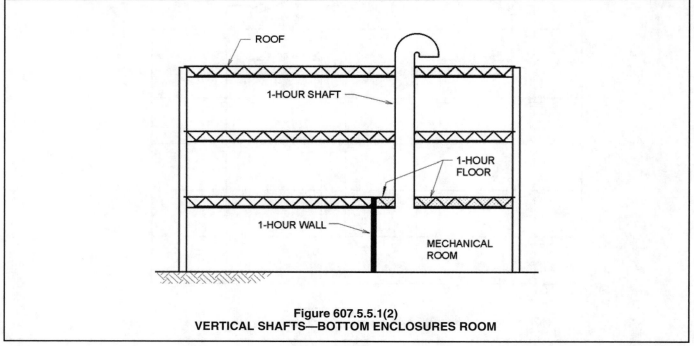

Figure 607.5.5.1(2)
VERTICAL SHAFTS—BOTTOM ENCLOSURES ROOM

the shaft. The exception differs from the main text of the section in that the room use is not required to be related to the purpose of the shaft. For example, a toilet room at the bottom of an air shaft would be permitted by this exception, even though the use of the toilet room does not relate to the purpose of the shaft to the same extent as an HVAC room. The exception also differs from the requirement found in Item 2, Section 713.11 of the IBC because the room is not required to have a fire-resistance rating equal to that of the shaft.

Exception 2. Section 713.11 of the IBC is unique to refuse or laundry chute shafts and prohibits the options afforded in this section. This exception mandates compliance with Section 713.13.4 of the IBC. The purpose of this exception is to limit the use of the shaft to the specific purpose of serving a refuse or laundry chute. Therefore, the shaft enclosure around these chutes cannot be used to permit duct work or items such as electrical cables to be run within the shaft.

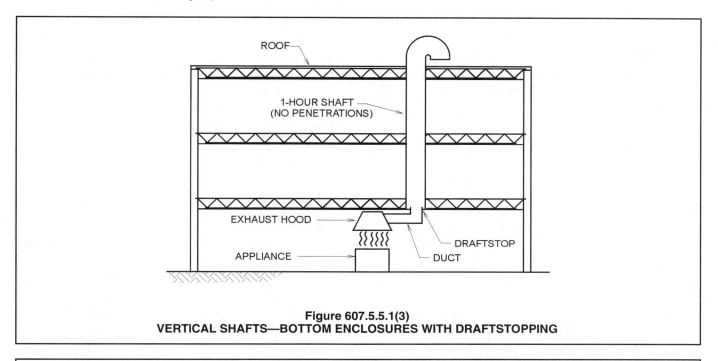

Figure 607.5.5.1(3)
VERTICAL SHAFTS—BOTTOM ENCLOSURES WITH DRAFTSTOPPING

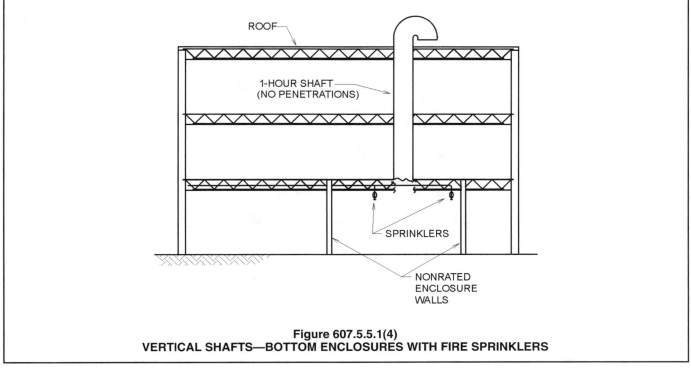

Figure 607.5.5.1(4)
VERTICAL SHAFTS—BOTTOM ENCLOSURES WITH FIRE SPRINKLERS

Exception 3. Section 713.11 of the IBC is intended to address a shaft, such as a light well, where the shaft volume can be considered as part of the story in which the bottom of the shaft terminates. The conditions for the exception are based on the fact that the shaft contents will not contribute to the fuel load and that the shaft serves and connects to only one story— the story in which it terminates [see Commentary Figure 607.5.5.1(5)].

607.5.6 Exterior walls. Ducts and air transfer openings in fire-resistance-rated exterior walls required to have protected openings in accordance with Section 705.10 of the *International Building Code* shall be protected with *listed* fire dampers installed in accordance with their listing.

❖ Fire dampers are required in ducts and air transfer openings in fire-resistance-rated exterior walls. The requirement for this can be found in Section 705.10 in the IBC. The fire resistance rating of the exterior wall can be found in Section 602.1 in the IBC.

607.5.7 Smoke partitions. A *listed* smoke damper designed to resist the passage of smoke shall be provided at each point where an air transfer opening penetrates a smoke partition. Smoke dampers and smoke damper actuation methods shall comply with Section 607.3.3.2.

> **Exception:** Where the installation of a smoke damper will interfere with the operation of a required smoke control system in accordance with Section 513, *approved* alternate protection shall be used.

❖ Smoke dampers are required in air transfer openings that penetrate smoke partitions. It is important to know that there is not a definition of an air transfer opening in the code. The intention of this code section is to provide a smoke damper in an unducted opening that is designed to move or transfer air from one area of a building to another [see Commentary Figure 607.5.7(1)]. If the opening has ducts attached to one or both sides of the smoke partition, a smoke damper is not required [see Commentary Figures 607.5.7(2) and 607.5.7(3)]. Where an opening in a smoke partition has ducts attached to it, the ducted opening will have to be protected as required in Section 710.8 of the IBC. Note that smoke partitions are not always required to have a fire-resistance rating and therefore do not always require fire dampers in a duct or air transfer penetration.

The exception to this section would allow for the smoke damper to be eliminated when it would interfere with the operation of a required smoke control system that complies with Section 513. If the smoke damper is eliminated, an approved alternate protection would have to be utilized.

607.6 Horizontal assemblies. Penetrations by air ducts of a floor, floor/ceiling assembly or the ceiling membrane of a roof/ceiling assembly shall be protected by a shaft enclosure that complies with Section 713 and Sections 717.6.1 through 717.6.3 of the *International Building Code* or shall comply with Sections 607.6.1 through 607.6.3.

❖ In general, floor openings and floor penetrations that connect two or more stories must be enclosed in shafts constructed in accordance with the IBC. Sections 607.6.1 through 607.6.3 identify those installations where a shaft is not required if the prescribed alternative protection is installed. Horizontal assemblies include both fire-resistance-rated and unrated assemblies. This section applies to penetrations of floor membranes, ceiling membranes and floor/ceiling assemblies. Penetration of a single floor/ceiling assembly connects two stories; penetration of two floor/ceiling assemblies connects three stories, etc.

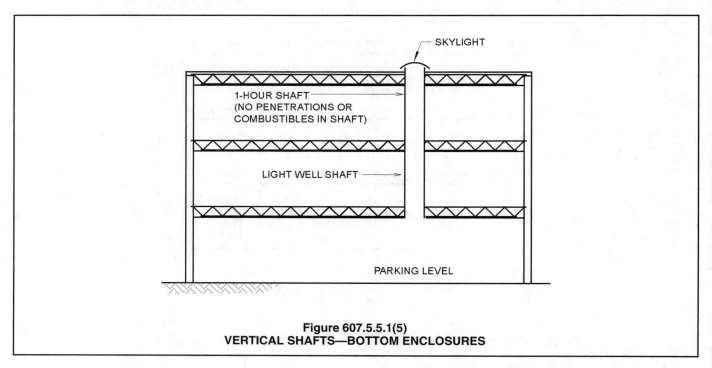

Figure 607.5.5.1(5)
VERTICAL SHAFTS—BOTTOM ENCLOSURES

607.6.1 Through penetrations. In occupancies other than Groups I-2 and I-3, a duct constructed of *approved* materials in accordance with Section 603 that penetrates a fire-resistance-rated floor/ceiling assembly that connects not more than two stories is permitted without shaft enclosure protection provided that a *listed* fire damper is installed at the floor line or the duct is protected in accordance with Section 714.4 of the *International Building Code*. For air transfer openings, see Item 7, Section 712.1.8 of the *International Building Code*.

Exception: A duct is permitted to penetrate three floors or less without a fire damper at each floor provided it meets all of the following requirements.

    1. The duct shall be contained and located within the cavity of a wall and shall be constructed of steel having a minimum thickness of 0.0187 inch (0.4712 mm) (No. 26 gage).

    2. The duct shall open into only one *dwelling unit* or *sleeping unit* and the duct system shall be continuous from the unit to the exterior of the building.

    3. The duct shall not exceed 4-inch (102 mm) nominal diameter and the total area of such ducts shall not exceed 100 square inches for any 100 square feet (64 516 mm^2 per 9.3 m^2) of the floor area.

    4. The annular space around the duct is protected with materials that prevent the passage of flame and hot gases sufficient to ignite cotton waste when subjected to ASTM E 119 or UL 263 time-temperature conditions under a minimum positive pressure differential of 0.01 inch (2.49 Pa) of water at the location of the penetration for the time period equivalent

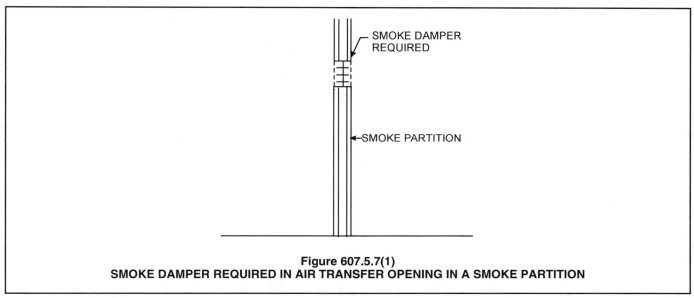

Figure 607.5.7(1)
SMOKE DAMPER REQUIRED IN AIR TRANSFER OPENING IN A SMOKE PARTITION

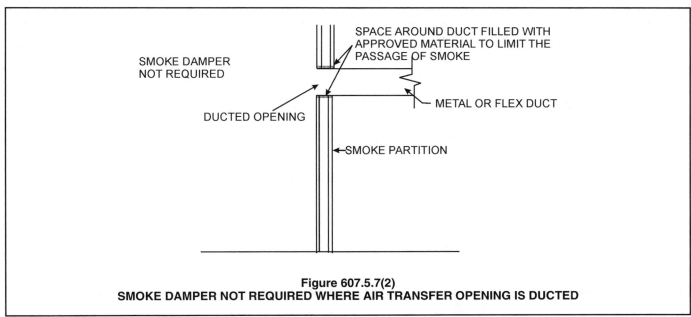

Figure 607.5.7(2)
SMOKE DAMPER NOT REQUIRED WHERE AIR TRANSFER OPENING IS DUCTED

to the fire-resistance rating of the construction penetrated.

5. Grille openings located in a ceiling of a fire-resistance-rated floor/ceiling or roof/ceiling assembly shall be protected with a *listed* ceiling radiation damper installed in accordance with Section 607.6.2.1.

❖ A fire-resistance-rated floor/ceiling assembly may be penetrated by air ducts where not more than two stories are connected (that is, the duct or opening penetrates a single floor) and a fire damper is installed at the floor line [see Commentary Figure 607.6.1(1)]. Because only two adjacent stories are involved, a fire damper is considered sufficient protection.

An air transfer opening that connects two floors is no different than any other opening in the floor as described in Section 712.1.8 of the IBC. A fire damper is not required to be installed if all seven of the criteria of that section of the IBC are satisfied.

The exception allows duct penetration of not more than three floors (four stories connected) without fire dampers at each floor penetration if all five requirements are satisfied. Item 1 specifies steel ducts and limits their location to wall cavities where they are protected by wall membranes and thus are less likely to be damaged and develop leaks. Item 2 limits the duct to serving only one dwelling or sleeping unit to lessen the potential for spreading fire and smoke from one unit to another. The duct must extend to the outdoors. Item 3 limits the size of the duct to 4 inches (102 mm) and a maximum area that is more restrictive than the requirements for pipes, vents and tubes in Section 714.4.1 of the IBC. Items 4 and 5 describe

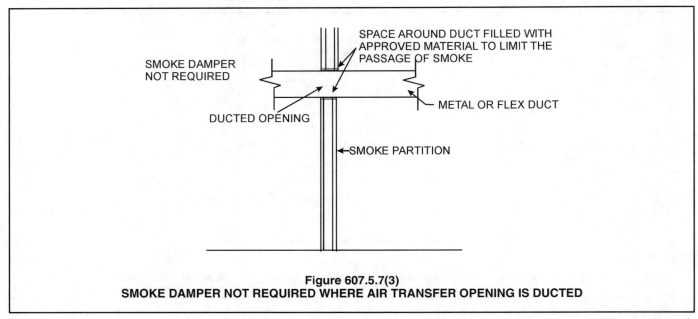

Figure 607.5.7(3)
SMOKE DAMPER NOT REQUIRED WHERE AIR TRANSFER OPENING IS DUCTED

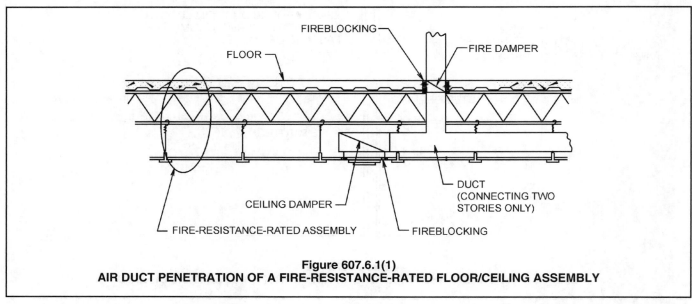

Figure 607.6.1(1)
AIR DUCT PENETRATION OF A FIRE-RESISTANCE-RATED FLOOR/CEILING ASSEMBLY

a method of maintaining the fire separation at the ceiling membrane by sealing the annular space around the duct with approved sealant material and protecting grille/register openings with a ceiling radiation damper. Note that a dryer exhaust duct is not exiting the space through the ceiling membrane and, therefore, does not require a ceiling damper [see Commentary Figure 607.6.1(2)].

607.6.2 Membrane penetrations. Ducts and air transfer openings constructed of *approved* materials, in accordance with Section 603, that penetrate the ceiling membrane of a fire-resistance-rated floor/ceiling or roof/ceiling assembly shall be protected with one of the following:

1. A shaft enclosure in accordance with Section 713 of the *International Building Code*.

2. A *listed* ceiling radiation damper installed at the ceiling line where a duct penetrates the ceiling of a fire-resistance-rated floor/ceiling or roof/ceiling assembly.

3. A *listed* ceiling radiation damper installed at the ceiling line where a diffuser with no duct attached penetrates the ceiling of a fire-resistance-rated floor/ceiling or roof/ceiling assembly.

❖ Unless the duct system is protected with a shaft enclosure in accordance with Sections 713 and 714.4, a ceiling radiation damper, tested in accordance with UL 555C, must be installed at the point where the ceiling membrane is penetrated by an air duct or air transfer opening if the ceiling is an integral component of a fire-resistance-rated assembly. This requirement applies to membranes of both floor/ceiling and roof/ceiling assemblies. The requirement for a ceiling radiation damper applies to ductwork that penetrates a ceiling membrane from above, or where a diffuser with no ductwork attached penetrates a ceiling membrane. Section 607.6.1 contains the requirements for penetrations through the entire floor/ceiling assembly [see Commentary Figure 607.6.1(1)].

Ceiling radiation dampers are designed to limit radiant heat transfer through an air inlet or outlet opening in the lower membrane of a fire-resistance-rated floor/ceiling or roof/ceiling assembly. The

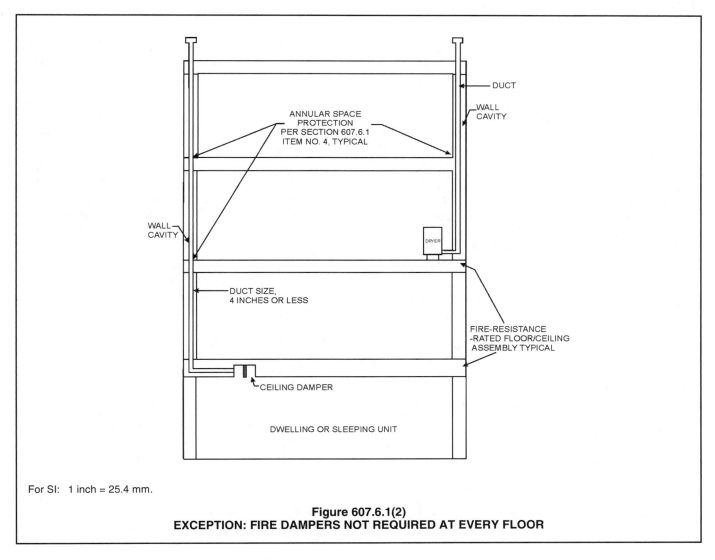

For SI: 1 inch = 25.4 mm.

Figure 607.6.1(2)
EXCEPTION: FIRE DAMPERS NOT REQUIRED AT EVERY FLOOR

description of the tested assembly will include a description of the ceiling radiation damper. On the other hand, a fire damper is a device designed to close automatically on detection of heat, to interrupt airflow and to restrict the passage of flame through duct penetrations of rated assemblies.

607.6.2.1 Ceiling radiation dampers. *Ceiling radiation dampers shall be tested in accordance with Section 607.3.1. Ceiling radiation dampers shall be installed in accordance with the details listed in the fire-resistance-rated assembly and the manufacturer's installation instructions and the listing. Ceiling radiation dampers are not required where either of the following applies:*

1. Tests in accordance with ASTM E 119 or UL 263 have shown that ceiling radiation dampers are not necessary to maintain the fire-resistance rating of the assembly.

2. Where exhaust duct penetrations are protected in accordance with Section 714.4.1.2 of the *International Building Code*, are located within the cavity of a wall and do not pass through another *dwelling unit* or tenant space.

❖ Ceiling radiation dampers, formerly known as "ceiling dampers," are evaluated either in accordance with UL 555C (see commentary, Section 607.3) or as part of the floor/ceiling or roof/ceiling assembly tested in accordance with ASTM E 119 or UL 263. Ceiling radiation dampers evaluated in accordance with UL 555C are investigated for use in lieu of hinged-door-type dampers commonly specified in the listing of fire-resistance-rated floor/ceiling or roof/ceiling assemblies. UL 555C also provides criteria for the construction of the ceiling radiation dampers. If the fire-resistance-rated floor/ceiling or roof/ceiling assembly does not incorporate a hinged-door-type damper, a ceiling radiation damper may not be utilized in the assembly. Ceiling radiation dampers investigated as part of the fire-resistance-rated floor/ceiling or roof/ceiling assembly are specified in the listing of the floor/ceiling or roof/ceiling assembly. The description of the tested assembly will include a description of the ceiling radiation damper and its installation. Both types of ceiling radiation dampers must be installed in accordance with the details listed in the fire-resistance-rated assembly and the manufacturer's installation instructions and listing.

The "Design Information Section" of the UL Fire Resistance Directory also includes two duct outlet protection systems. Duct Outlet Protection System A may be used where specified in the fire-resistance-rated floor/ceiling or roof/ceiling assembly. Duct Outlet Protection System B may be used in lieu of hinged-door-type dampers commonly specified in the listing of fire-resistance-rated floor/ceiling or roof/ceiling assemblies.

Ceiling radiation dampers need not be installed if a fire-resistance-rated assembly is tested in accordance with ASTM E 119 and meets the performance required for the assembly without ceiling dampers.

Ceiling radiation dampers are not required to be installed if the exhaust ducts comply with Section 714.4.1.2 of the IBC, are limited to a single dwelling unit or tenant spaces and are contained within a wall. Where this criteria is met, Section 714.4.1.2 of the IBC would allow either assemblies tested to ASTM E 119 or the use of a through-penetration firestop system tested to ASTM E 814 or UL 1479.

There are through-penetration firestop systems that could be approved in accordance with Section 714.4.1.2 of the IBC. Such systems are tested as alternatives to ceiling dampers.

607.6.3 Nonfire-resistance-rated floor assemblies. Duct systems constructed of *approved* materials in accordance with Section 603 that penetrate nonfire-resistance-rated floor assemblies shall be protected by any of the following methods:

1. A shaft enclosure in accordance with Section 713 of the *International Building Code*.

2. The duct connects not more than two stories, and the annular space around the penetrating duct is protected with an *approved* noncombustible material that resists the free passage of flame and the products of *combustion*.

3. The duct connects not more than three stories, and the annular space around the penetrating duct is protected with an *approved* noncombustible material that resists the free passage of flame and the products of *combustion*, and a *listed* fire damper is installed at each floor line.

Exception: Fire dampers are not required in ducts within individual residential *dwelling units*.

❖ This section applies to duct systems that penetrate floor assemblies which are not required to have a fire-resistance rating. Even though a floor assembly is not designed and tested to withstand fire exposure for a specified period of time, the floor will provide for some resistance to the passage of heat, smoke and flame from one floor to the next. If a shaft enclosure is built for the duct to pass through the floors, then the shaft construction and floor opening construction details will eliminate any annular space outside of the shaft exterior to the floor opening.

Where only one or two floors are penetrated by a duct, this section allows elimination of the shaft enclosure.

Where only one floor is penetrated by the duct, the annular space between the outside of the duct and the opening in the floor, the space is required to be "plugged" with a noncombustible material. There are no requirements for the material to be in accordance with any standard or be listed or labeled. However, the material to be used must be approved by the code official [see Commentary Figure 607.6.3(1)].

Where only two floors are penetrated by the duct, the annular space between the outside of the duct and the opening in the floor, is required to be "plugged" with an approved noncombustible material and a fire damper is required at each floor line [see Commentary Figure 607.6.3(2)]. It should be noted that the only fire damper listed to be installed in a floor is a horizontal fire damper. Horizontal fire dampers have only been tested and listed for masonry and concrete floor assemblies.

This section does not apply to transfer openings or transfer openings with a sleeve because plugging of the annular space between an assembly and the perimeter of the opening in the floor assembly would be pointless.

607.7 Flexible ducts and air connectors. Flexible ducts and air connectors shall not pass through any fire-resistance-rated assembly.

❖ Flexible air ducts are prohibited from penetrating fire-resistance-rated assemblies. Flexible air connectors are not permitted to pass through any wall, floor or ceiling whether the assembly is fire-resistance rated or not (see Section 603.6.2.2). An inadequate seal at the assembly penetration could allow smoke or flame to penetrate the assembly. Flexible air ducts and connectors can be constructed of both combustible and noncombustible components; therefore, the duct's resistance to the passage of fire could be less than the resistance of the penetrated assembly. All construction assemblies, whether fire-resistance rated or not, have some inherent resistance to the spread of fire; duct penetrations can significantly affect that fire resistance. Also, the surface contour of the duct or the presence of insulation and vapor barrier materials on the duct exterior make it difficult to effectively seal (fireblock) the annular spaces around flexible air duct and connector penetrations. Sections 714.3.3 and 714.4.13 of the IBC are interpreted to prevent combustible flexible ducts from converting to a noncombustible duct for the sole purpose of penetrating a rated assembly and circumventing the prohibition of Section 607.7.

Bibliography

The following resource materials were used in the preparation of the commentary for this chapter of the code:

IBC-12, *International Building Code.* Washington, DC: International Code Council, 2011.

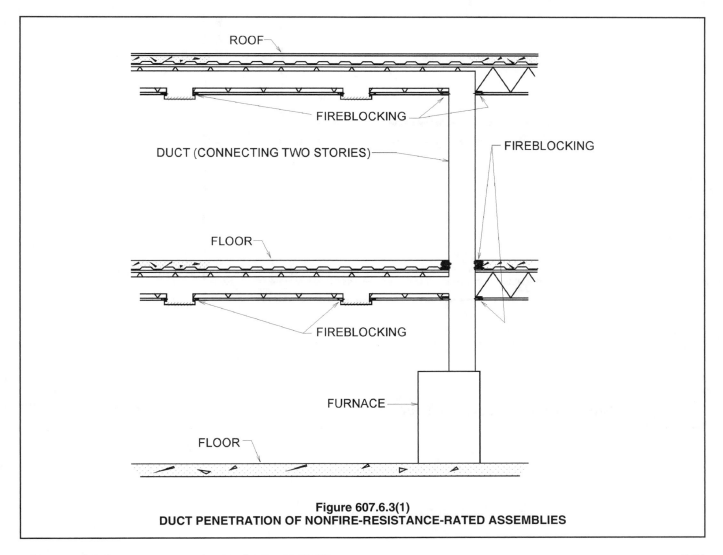

Figure 607.6.3(1)
DUCT PENETRATION OF NONFIRE-RESISTANCE-RATED ASSEMBLIES

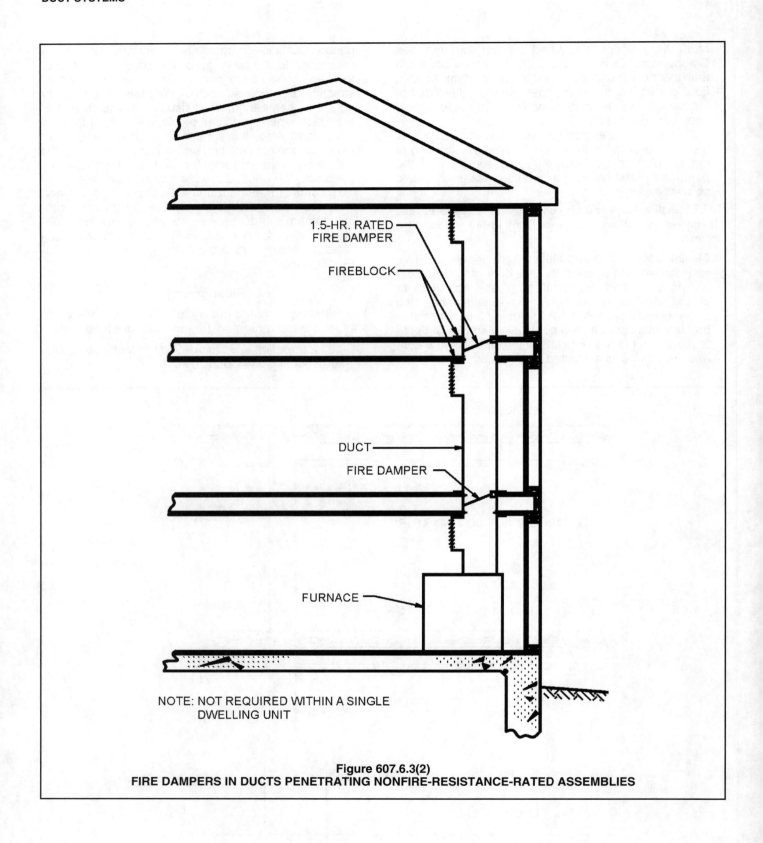

Figure 607.6.3(2)
FIRE DAMPERS IN DUCTS PENETRATING NONFIRE-RESISTANCE-RATED ASSEMBLIES

Chapter 7:
Combustion Air

General Comments

Complete combustion of fuel is essential for the proper operation of appliances, for control of harmful emissions and for achieving maximum fuel efficiency. Combustion air supplies, among other things, the oxygen necessary for the complete and efficient burning of fuel. If insufficient quantities of oxygen are supplied, the combustion process will be incomplete, creating dangerous byproducts and wasting energy in the form of unburned fuel (hydrocarbons). The byproducts of incomplete combustion are poisonous, corrosive and combustible and can cause serious appliance or equipment malfunctions that pose fire or explosion hazards.

Combustion air also serves other purposes in addition to supplying oxygen. It ventilates and cools appliances and the rooms or spaces that enclose them, and it plays an important role in producing and controlling draft in vents and chimneys. An insufficient combustion air supply could cause an appliance to overheat and discharge combustion byproducts into the building.

A combustion air supply is also necessary to prevent oxygen depletion, which threatens the safety of the building occupants. Both building-occupant respiration and the combustion process of fuel-burning appliances consume the available oxygen in a room or space. The depletion of available oxygen promotes incomplete fuel combustion.

Adequate combustion air supply is extremely important, but it is one aspect of mechanical installation that is often overlooked, ignored or compromised. Depending on the appliance type, location and building construction, supplying combustion air can be easy, or it can involve complex designs and extraordinary methods. In any case, the importance of a proper combustion air supply cannot be overemphasized.

The combustion air requirements in this chapter were completely revised in the 2009 edition of the code. It was recognized that the requirements in this chapter were based upon dated combustion air requirements for fuel-gas-burning appliances. Liquid and solid fuel-burning appliances do not have the same requirements for combustion air. Therefore, all of the requirements in this chapter were deleted and replaced with instructions on where to find fuel specific combustion air requirements for solid, liquid and fuel gas-fired appliances.

Purpose

This chapter directs the reader to appropriate combustion air requirements provided in other publications or chapters of the code.

SECTION 701
GENERAL

701.1 Scope. Solid fuel-burning *appliances* shall be provided with *combustion air* in accordance with the appliance manufacturer's installation instructions. Oil-fired *appliances* shall be provided with *combustion air* in accordance with NFPA 31. The methods of providing *combustion air* in this chapter do not apply to fireplaces, fireplace stoves and direct-vent *appliances*. The requirements for combustion and dilution air for gas-fired *appliances* shall be in accordance with the *International Fuel Gas Code*.

❖ Combustion is necessary for complete fuel combustion. Air is also required for draft hood dilution (if the appliance is equipped with a draft hood) and in some situations, air necessary for ventilation of the appliance enclosure. A lack of combustion air results in incomplete fuel combustion, which causes soot production, increased carbon monoxide production, serious appliance malfunction and the risk of fire or explosion. Lack of draft-hood dilution air can result in improper draft and appliance venting. Incomplete fuel combustion and improper draft and venting compound each other and greatly increase the risk of the release of carbon monoxide, thereby endangering the occupants of the space or building. Lack of the air necessary for ventilation of the appliance enclosure can result in excessive temperatures in the enclosure, thus introducing the risks of overheating the appliance and fire.

Potential sources of combustion air are (1) the indoor atmosphere in which the appliance is located, (2) outdoor openings and ducts, (3) a combination of the two and (4) any other approved arrangement for the dependable introduction of outdoor air into the building or directly to the fuel-burning appliance.

Combustion air requirements for a fuel-burning appliance are dependent on the type of fuel burned by the appliance. As such, this section directs the reader to other publications for the requirements specific to the type of fuel that the appliance burns.

As stated in the general comments at the beginning

of this chapter, previous editions of the code did not properly address the combustion air requirements for solid and liquid fuel-burning appliances as this chapter was originally developed based upon the combustion air requirements for fuel gas-fired appliances. Up until the 2000 edition of the code, this chapter was applicable to liquid, solid and fuel gas-fired appliances. In the 2000 edition of the code, language that was specific to fuel gas-fired appliances was removed as a new fuel gas code had been developed [the *International Fuel Gas Code*®, (IFGC®)] and introduced to cover requirements specific to fuel gas-fired appliances, including combustion air requirements. With the introduction of the IFGC, it was intended that this chapter be applicable only to liquid and solid fuel-fired appliances. Even though the first section of Chapter 7 (from the 2000 edition of the code forward) attempted to redefine the scope of the chapter to be for liquid and solid fuel-fired appliances only, many users of this code continued to apply this chapter to fuel gas-fired appliance installations. Replacing all of the combustion air requirements in Chapter 7 with instructions on where to find combustion air requirements for the specific type of fuel-burning appliance eliminates the confusion about fuel gas-fired appliance combustion air requirements and provides the location for appropriate combustion air requirements for solid and liquid fuel-fired appliances.

Because of the unique configurations and variety of solid fuel-burning appliances (e.g., wood, pellet, biomass-fired), no one set of combustion air requirements could adequately cover all solid fuel-fired appliance installations. Therefore, the best information must be obtained from the manufacturer's installation instructions for the specific appliance. NFPA 31 has long been recognized for many aspects of oil-fired appliance installations that were not covered by this code. Because NFPA 31 has combustion air requirements, the standard is the appropriate source for this information. For the benefit of code users who are curious to know the major differences between the combustion air requirements that were in this chapter and those required by NFPA 31, the following information is provided.

NFPA 31 has different requirements for installations in commercial/industrial buildings versus installations in residential buildings. Oil-burning appliances in commercial/industrial buildings must receive combustion air directly from the outdoors and not from building infiltration. Oil-burning appliances in residential and "similar" installations can receive combustion air from building infiltration or if the building is of "tight construction," receive air directly from the outdoors. The definition of "Confined space" is the same as before (50 cubic feet per 1,000 Btu/h input rating).

Commercial and Industrial Buildings

- Where the installation of oil-burning appliances is in a furnace or boiler room adjacent to an exterior wall, the appliance must receive ventilation and combustion air directly from the outdoors. Although the free area of the opening in the exterior wall is the same as before (1 square inch per 4,000 Btu/h input rating of all of the oil-burning appliances in the room), the opening is required to have at least 35 square inches (3.25 m²) of free area.

- Where the installation of oil-burning appliances in commercial and industrial buildings is in a furnace or boiler room that is not adjacent to an exterior wall, the method of supplying combustion air (from the outdoors) to the appliances must be approved by the code official.

Residential and "Similar" Buildings

- Where oil-burning appliances are installed in an unconfined space but in a building of "tight" construction, the free area of the opening (or the total of all openings) direct to the outdoors (or to spaces that communicate freely with the outdoors) must be at least 1 square inch per 5000 Btu/hour input rating of the all of the oil-burning appliances in the unconfined space. (Chapter 7 previously required 1 square inch per 4,000 Btu/h input rating.)

- Where oil-burning appliances are installed in a confined space and air is being taken from inside of a building having "adequate" air infiltration, there is not a minimum size required for the two openings to the adjacent space (Chapter 7 previously required a minimum free area of 100 square inches for each opening).

- Where oil-burning appliances are installed in a confined space and in a building of "tight" construction, and where air is to be supplied only through one opening (direct or ducted) to the outdoors (or to a space that freely communicates with the outdoors), the opening must have a free area at least 1 square inch per 5,000 Btu/h input rating of all of the oil-burning appliances in the confined space. The confined space must also have two openings (one near the top of the space, one near the bottom of the space) to provide ventilation air from an adjacent inside space. The openings must each have a free area of 1 square inch per 1000 Btu per hour input rating of the all of the oil-burning appliances in the confined space.

Chapter 8:
Chimneys and Vents

General Comments

This chapter is intended to regulate the design, construction, installation, maintenance, repair and approval of chimneys, vents and their connections to fuel-burning appliances. A properly designed chimney or vent system is needed to conduct the flue gases produced by a fuel-burning appliance to the outdoors. In the case of natural draft appliances, the chimney or vent system serving the appliance is expected to produce a draft at the appliance connection.

Draft refers to a phenomenon created by the buoyancy of lighter (less dense) gases in the presence of heavier (more dense) gases. Draft is measured as pressure in inches of water column (kPa) and is a negative pressure—meaning that it is less than the atmospheric pressure at the point of measurement. Draft is produced by the temperature difference between the combustion gases (flue gases) and the ambient atmosphere. Because hotter gases are less dense, they are buoyant and will rise in the chimney, causing negative pressures to develop that are directly proportional to the height of the chimney or vent and directly proportional to the temperature difference between the flue gases and the ambient air. Draft produced by a chimney or vent is the motivating force that conveys combustion products from natural draft appliances to the outside atmosphere through the chimney or vent passageway.

The type of vent or chimney used in a given installation is generally based on the type of fuel burned by the appliance, the temperature of the flue gases produced by the combustion process and the appliance category.

Because the code classifies appliances into three temperature categories (low, medium and high heat), the fuel type and characteristics of the flue gas must be analyzed for the design of a chimney or vent system. The development of higher efficiency appliances continues to produce lower flue gas temperatures, and the diversity of mechanical venting systems continues to expand. See the definition of "Appliance type" in Chapter 2.

With the advent of new technology and higher efficiency appliances/equipment, venting means are becoming increasingly diversified and complex. Many of today's appliances are designed to be vented by more than one type of venting system.

Purpose

The requirements of this chapter are intended to achieve the complete removal of the products of combustion from fuel-burning appliances and equipment. This chapter includes regulations for the proper selection, design, construction and installation of a chimney or vent, along with appropriate measures to minimize the related potential fire hazards. A chimney or vent must be designed for the type of appliance or equipment it serves. Chimneys and vents are designed for specific applications depending on the flue gas temperatures and the type of fuel being burned in the appliance.

The primary hazards associated with chimneys and vents are high temperatures and the toxic and corrosive nature of combustion byproducts.

SECTION 801
GENERAL

801.1 Scope. This chapter shall govern the installation, maintenance, repair and approval of factory-built chimneys, *chimney* liners, vents and connectors. This chapter shall also govern the utilization of masonry chimneys. Gas-fired *appliances* shall be vented in accordance with the *International Fuel Gas Code*.

❖ Chapter 8 contains provisions for the installation, maintenance, repair and approval of residential and commercial chimney and venting systems that convey the products of combustion from fuel-burning appliances regulated by the code to the outside atmosphere. This chapter also addresses metal chimneys, also known as smoke stacks, used in factory/industrial-type facilities. Even though factory-built chimneys (see Section 805) are constructed of metal, they are not considered to be metal chimneys (smoke stacks) (see Section 806).

Venting requirements for fuel gas-fired appliances are regulated by the *International Fuel Gas Code®* (IFGC®).

801.2 General. Every fuel-burning *appliance* shall discharge the products of *combustion* to a vent, factory-built *chimney* or masonry *chimney*, except for *appliances* vented in accordance with Section 804. The *chimney* or vent shall be designed for the type of *appliance* being vented.

> **Exception:** Commercial cooking *appliances* vented by a Type I hood installed in accordance with Section 507.

❖ Every fuel-burning appliance must be vented to a chimney or vent unless it is direct-vented, integrally vented, mechanically vented in accordance with Section 804 or vented by means of a commercial kitchen exhaust hood in accordance with Section 507. Although chimneys and vents have the same basic purpose, they are different creatures with differing code requirements. Some code sections pertain to chimneys only, some to vents only and some to both. Commonly, the code is misapplied because people refer to vents as chimneys and vice versa.

The combustion byproducts produced from the burning of fuels consist primarily of carbon dioxide, nitrogen, water vapor and small amounts of carbon monoxide, nitrous oxide and other compounds that vary with the type and purity of the fuel and the purity of the combustion air. Some of the compounds, especially carbon monoxide, are poisonous and some are corrosive. The amount of carbon monoxide in combustion gases varies with the type of fuel, the type of burner and the condition and adjustment of the burner. See the commentary to Chapter 7 for a further discussion of fuel combustion.

Many fatalities have resulted from carbon monoxide poisoning caused by malfunctioning fuel-burning appliances or malfunctioning or the lack of a chimney or vent. Because of the potential danger of asphyxiation, the products of combustion must be conducted to the outdoors so that operation of the appliance does not pose a threat to the building occupants. Fuel-burning appliances have widely varying operating characteristics, which means that the method of venting the combustion products must be designed for the particular type of appliance served. The flue gases (combustion byproducts) produced by appliances differ primarily in temperature and chemical composition. The methods of venting differ in the temperature and pressure ranges over which they operate and whether or not they can accommodate condensation.

A mismatch between an appliance and a vent or chimney can result in a hazardous operating condition. For example, an appliance that produces high-temperature flue gas could cause a vent material to degrade or melt or cause excessively high temperatures on the outer surface of the vent. An appliance that produces low-temperature flue gas can cause condensation to occur in the venting system that could corrode and deteriorate the vent material.

801.2.1 Oil-fired appliances. Oil-fired *appliances* shall be vented in accordance with this code and NFPA 31.

❖ Vents for oil-fired appliances must comply with NFPA 31 and the requirements of this chapter. NFPA 31 references NFPA 211 and contains, among other provisions, requirements for draft regulators, Type L vents,

vent and chimney terminations, chimneys and connectors.

801.3 Masonry chimneys. Masonry *chimneys* shall be constructed in accordance with the *International Building Code*.

❖ Masonry construction is more appropriately addressed in the *International Building Code*® (IBC®) (see commentary, Section 902.1). The IBC contains requirements for the construction of masonry chimneys, whereas this code contains the requirements for the sizing and use of masonry chimneys.

A masonry chimney is a field-constructed assembly that can consist of solid masonry units, reinforced concrete, rubble stone, fire-clay liners and mortars. A masonry chimney can serve low-, medium- and high-heat appliances. The IBC outlines the general code requirements regarding the construction details for all masonry chimneys, including those serving masonry fireplaces (see IBC Chapter 21).

801.4 Positive flow. Venting systems shall be designed and constructed so as to develop a positive flow adequate to convey all *combustion* products to the outside atmosphere.

❖ This section requires that chimneys and vents be designed to develop the required draft or allow the required positive pressure flow to prevent combustion byproducts from spilling into the building's interior.

801.5 Design. Venting systems shall be designed in accordance with this chapter or shall be *approved* engineered systems.

❖ An engineered chimney or vent system refers to a system designed in accordance with the formulas, data and tables found in design manuals and standards such as ASHRAE handbooks and chimney and vent manufacturers' design guides. Engineered systems should be designed in accordance with standard engineering practices and must be approved by the code official.

801.6 Minimum size of chimney or vent. Except as otherwise provided for in this chapter, the size of the *chimney* or vent, serving a single *appliance*, except engineered systems, shall have a minimum area equal to the area of the *appliance* connection.

❖ This section requires a chimney or vent serving a single appliance to have a minimum cross-sectional area equal to the cross-sectional area of the appliance outlet connection. Although not regulated by this section, the maximum size of a chimney or vent is also an important design consideration. Oversized chimneys and vents can fail to produce sufficient draft and can cause condensation of the flue gases. Existing chimneys and vents might become oversized as a result of the removal of an appliance that was previously connected (see commentary, Section 801.18). In excessively large chimneys and vents, the flue gases are cooled by expansion and the increased heat loss through the larger flue passage surface area. The cooling effect reduces the temperature differences necessary to produce draft and can cause the flue gases to reach the dew point temperature. The resulting condensation is highly corrosive and can cause deterioration of the chimney, vent, connected appliances and the building.

This section does not apply to the sizing of engineered systems and multiple appliance connections. The sizing requirements for vents serving appliances/equipment that require special vent systems are specified in the manufacturer's instructions. See the commentary to Sections 801.5 and 802.3 for a further discussion of special vent systems. Any design conflicting with the sizing criteria of this section must be considered as an engineered system.

801.7 Solid fuel appliance flues. The cross-sectional area of a flue serving a solid-fuel-burning *appliance* shall be not greater than three times the cross-sectional area of the *appliance* flue collar or flue outlet.

❖ Oversized chimneys can fail to produce sufficient draft and can encourage the production of water vapor condensation and creosote. The 3:1 ratio limitation is intended to prevent chimneys and chimney flues from being too large for the appliance served.

801.8 Abandoned inlet openings. Abandoned inlet openings in chimneys and vents shall be closed by an *approved* method.

❖ Abandoned inlet openings result from appliances being disconnected or connected at different elevations. Unused openings in chimneys and vents can allow combustion gases to enter the building; can cause loss of draft at other appliances connected to the chimney or vent; can allow conditioned air to escape, resulting in energy losses; and can allow the entry of birds and rodents. For example, an appliance that was connected to a chimney has been replaced with an appliance that vents through an exterior wall. The opening in the chimney left by the disconnection of the appliance must be properly closed to prevent a hazardous condition (see commentary, Section 801.18).

The method used to close an unused opening must not create a protrusion in the flue passageway that could restrict flue gas flow or provide a surface for the accumulation of creosote or soot.

801.9 Positive pressure. Where an *appliance* equipped with a forced or induced draft system creates a positive pressure in the venting system, the venting system shall be designed and *listed* for positive pressure applications.

❖ Commonly used chimney and vent systems are not designed for positive pressure (above atmospheric). These chimneys and vents are intended to produce a draft and thus operate with negative internal pressures. Specialized vents and chimneys are available that are designed for positive pressure applications. It is important that such vents and chimneys are listed for positive-pressure applications to prevent the misapplication of materials, which can result in leakage of combustion gases and can cause chimney or vent

deterioration from condensate corrosion (see Sections 801.14, 804.3.1 and 804.3.6).

801.10 Connection to fireplace. Connection of *appliances* to *chimney* flues serving fireplaces shall be in accordance with Sections 801.10.1 through 801.10.3.

❖ This section regulates installations where fireplace chimneys are used to vent appliances, such as room heaters and fireplace-insert heaters.

801.10.1 Closure and access. A noncombustible seal shall be provided below the point of connection to prevent entry of room air into the flue. Means shall be provided for *access* to the flue for inspection and cleaning.

❖ Without an air-tight connection between the chimney flue and the appliance chimney connector, chimney draft will draw room air into the flue, thus weakening the draft drawn through the appliance. The chimney connection must be designed for access to allow inspection and cleaning of the chimney because the chimney flue is no longer open to the fireplace firebox.

801.10.2 Connection to factory-built fireplace flue. An *appliance* shall not be connected to a flue serving a factory-built fireplace unless the *appliance* is specifically *listed* for such installation. The connection shall be made in accordance with the *appliance* manufacturer's installation instructions.

❖ Factory-built fireplaces may or may not be tested for any use other than as a traditional fireplace. Connecting an appliance or fireplace-insert heater could result in abnormally high fireplace or chimney temperatures, component deterioration and hazardous operation. The fireplace manufacturer will provide specific details stating what appliance connections are allowed. Typically, the manufacturer's instructions will prohibit any modifications and any use of the fireplace other than what it was designed for.

801.10.3 Connection to masonry fireplace flue. A connector shall extend from the *appliance* to the flue serving a masonry fireplace such that the flue gases are exhausted directly into the flue. The connector shall be provided with access or shall be removable for inspection and cleaning of both the connector and the flue. *Listed* direct connection devices shall be installed in accordance with their listing.

❖ The appliance chimney connector must extend up through the fireplace damper and smoke chamber and terminate in the chimney flue. This helps produce an adequate draft and reduces the possibility of forming combustible creosote deposits within the fireplace throat and smoke chamber.

801.11 Multiple solid fuel prohibited. A solid fuel-burning *appliance* or fireplace shall not connect to a *chimney* passageway venting another *appliance*.

❖ Each solid fuel-burning appliance or fireplace must have a dedicated independent chimney, or a dedicated independent flue in multiple-flue chimney constructions. Solid fuel appliances and fireplaces cannot share a common chimney or flueway with any other appliance or fireplace. For example, a solid fuel appliance cannot be connected to an existing masonry chimney that serves other appliances or a fireplace.

Solid fuel-burning appliances produce creosote deposits on the interior walls of chimney liners. The creosote formation is highly combustible and creates a potential fire hazard. Because of the potential for a chimney fire, other connections to the chimney could allow fire to break out of the chimney into the building. Also, chimney passageways could become blocked by the creosote formations. If other appliances were vented to those chimneys, combustion products would be discharged into the building. This section is also intended to prevent the possibility of creosote leaking out of the chimney through appliance connectors.

Combination (dual fuel) gas- or oil- and solid fuel-burning appliances are designed to be connected to a single chimney passageway. Dual-fuel appliances must be listed, labeled and installed in compliance with the manufacturer's instructions.

801.12 Chimney entrance. Connectors shall connect to a *chimney* flue at a point not less than 12 inches (305 mm) above the lowest portion of the interior of the *chimney* flue.

❖ This provision is intended to prevent blockage of the connector opening by debris that has collected at the bottom of the flue passage. The 12-inch (305 mm) deep space below the connector functions as a trap (dirt leg) for debris, such as dead animals, leaves, flaking tile, mortar and soot. This section is not intended to apply to factory-built chimneys because such chimneys are protected from the entry of debris by the required cap.

801.13 Cleanouts. Masonry *chimney* flues shall be provided with a cleanout opening having a minimum height of 6 inches (152 mm). The upper edge of the opening shall be located not less than 6 inches (152 mm) below the lowest *chimney* inlet opening. The cleanout shall be provided with a tight-fitting, noncombustible cover.

> **Exception:** Cleanouts shall not be required for *chimney* flues serving masonry fireplaces, if such flues are provided with access through the fireplace opening.

❖ This section requires installation of a cleanout in a chimney to facilitate cleaning and inspection. A fireplace inherently provides access to its chimney through the firebox, throat and smoke chamber. The cleanout cover and opening frame must be of an approved material, such as cast iron, precast cement or other noncombustible material, and must be arranged to remain tightly closed. A loose-fitting or unsecured cleanout can allow air to flow into the chimney, affecting draft at the appliance or fireplace, and can result in an energy loss in the building. Under certain conditions, combustion products could escape into the building through an ill-fitting cleanout door or cover. The requirement for placing the cleanout at least 6 inches (152 mm) below the lowest

connection to the chimney is intended to minimize the possibility of combustion products exiting the chimney through the cleanout.

801.14 Connections to exhauster. All *appliance* connections to a *chimney* or vent equipped with a power exhauster shall be made on the inlet side of the exhauster. All joints and piping on the positive pressure side of the exhauster shall be *listed* for positive pressure applications as specified by the manufacturer's installation instructions for the exhauster.

❖ It is important that any piping on the positive-pressure (discharge) side of the exhauster be designed and listed for such an application to prevent the misapplication of materials, which can result in leakage of combustion gases and can cause chimney or vent deterioration from condensate corrosion (see Sections 801.9, 804.3.1 and 804.3.6) (see commentary, Section 804.3.6).

801.15 Fuel-fired appliances. Masonry chimneys utilized to vent fuel-fired *appliances* shall be located, constructed and sized as specified in the manufacture's installation instructions for the *appliances* being vented.

❖ Manufacturers' instructions for today's fuel-fired appliances specify the conditions under which the appliance is allowed to vent to a masonry chimney. The conditions include chimney size, condition, location and construction. Because of being oversized, unlined, in poor structural condition or partially exposed to the outdoors, masonry chimneys might not be suitable for some applications. Once common for appliances operating at 75-percent efficiency or less, masonry chimneys used to vent modern fuel-fired appliances are becoming more rare, and in many cases a masonry chimney will have to be relined with a retrofit metal liner or a refractory lining (see commentary, Section 801.18).

801.16 Flue lining. Masonry chimneys shall be lined. The lining material shall be compatible with the type of *appliance* connected, in accordance with the *appliance* listing and manufacturer's installation instructions. *Listed* materials used as flue linings shall be installed in accordance with their listings and the manufacturer's installation instructions.

❖ The liner forms the flue passageway and is the actual conductor of combustion products. The chimney liner must be able to withstand exposure to high temperatures and corrosive chemicals. The chimney lining protects the masonry construction of the chimney walls and allows the chimney to be constructed gas tight. The IBC regulates chimney construction. This section regulates liners and relining systems. Liners are often used to reline an existing chimney, to salvage a masonry chimney or allow connection of higher-efficiency appliances (see Section 801.18.4).

801.16.1 Residential and low-heat appliances (general). Flue lining systems for use with residential-type and low-heat appliances shall be limited to the following:

1. Clay flue lining complying with the requirements of ASTM C 315 or equivalent. Clay flue lining shall be

installed in accordance with the *International Building Code*.

2. *Listed* and *labeled* chimney lining systems complying with UL 1777.

3. Other *approved* materials that will resist, without cracking, softening or corrosion, flue gases and condensate at temperatures up to 1,800°F (982°C).

❖ Flue lining systems for masonry chimneys include clay tile, poured-in-place refractory materials and stainless steel pipe.

801.17 Space around lining. The space surrounding a flue lining system or other vent installed within a masonry *chimney* shall not be used to vent any other *appliance*. This shall not prevent the installation of a separate flue lining in accordance with the manufacturer's installation instructions and this code.

❖ If a chimney lining system, such as flexible metallic pipe is installed within a chimney flueway, an annular space will exist between the pipe and the interior wall surfaces of the chimney. This annular space has an irregular shape that is not conducive to the flow of combustion gases; therefore, the space cannot be used to vent an additional appliance. Also, venting into the annular space could have a detrimental effect on the lining of the flueway.

801.18 Existing chimneys and vents. Where an *appliance* is permanently disconnected from an existing *chimney* or vent, or where an *appliance* is connected to an existing *chimney* or vent during the process of a new installation, the *chimney* or vent shall comply with Sections 801.18.1 through 801.18.4.

❖ Existing chimneys and vents should be reevaluated for continued suitability whenever the conditions of use change. Size, which is covered in Section 801.18.1, is of primary importance. Other considerations are the presence of liner obstructions, combustible deposits in the liner, the structural condition of the liner, the provision of a cleanout and clearances to combustibles, which are addressed in Sections 801.18.2 through 801.18.4. Whenever a new appliance is connected or an existing appliance is disconnected from a chimney or vent, the chimney or vent is subject to the requirements in Sections 801.18.1 through 801.18.4, which include requirements for inspection, cleaning, possible repair, the installation of a cleanout if one is not already there and the establishment of clearances to combustibles required for new chimneys or vent installations.

Bear in mind that chimneys, vents and the appliances served are all designed to function together as a system. Any change to an existing chimney or vent system will have an impact on the performance of that system. Something as simple as disconnecting an appliance from a chimney can upset the system's balance and cause the venting system to fail to produce a draft for the remaining appliances and could cause the venting system to produce harmful condensation.

801.18.1 Size. The *chimney* or vent shall be resized as necessary to control flue gas condensation in the interior of the *chimney* or vent and to provide the *appliance* or *appliances* served with the required draft. For the venting of oil-fired *appliances* to masonry chimneys, the resizing shall be in accordance with NFPA 31.

❖ The combined input from multiple appliances, especially older lower-efficiency appliances, can maintain chimney or vent temperatures that are high enough to provide the necessary draft and avoid condensation. Changing an existing configuration by disconnecting and eliminating an appliance or by substituting a higher-efficiency appliance can cause a decrease in flue gas temperature, resulting in poor draft or condensation. Also, the elimination of one or more draft-hood-equipped appliances will reduce the amount of dilution air in a venting system, increasing the likelihood of condensation. Often, a chimney or vent must be replaced with a smaller-size chimney or vent or be resized by installation of a liner system (see Commentary Figure 801.18.1). The requirements of NFPA 31 must be followed when resizing a masonry chimney that vents an oil-fired appliance.

A common scenario involves removing chimney-vented furnaces or boilers and leaving a water heater as the only appliance vented to the chimney. In this case, the chimney would typically be grossly oversized for the water heater, could fail to produce adequate draft and could be subject to continuous condensation. This scenario has received much attention and has created the phrase "orphaned water heaters."

801.18.2 Flue passageways. The flue gas passageway shall be free of obstructions and combustible deposits and shall be cleaned if previously used for venting a solid or liquid fuel-burning *appliance* or fireplace. The flue liner, *chimney* inner wall or vent inner wall shall be continuous and shall be free of cracks, gaps, perforations or other damage or deterioration which would allow the escape of *combustion* products, including gases, moisture and creosote. Where an oil-fired *appliance* is connected to an existing masonry *chimney*, such *chimney* flue shall be repaired or relined in accordance with NFPA 31.

❖ Cleaning of the chimney or flue is necessary to allow examination of the flue liner surface, to eliminate fuel for a possible chimney fire, to reduce frictional resistance to flow and to prevent any residual deposits from falling into lower parts of the venting system. Installation or removal of an appliance without an inspection of the existing flue is prohibited. If the flue served a solid or liquid fuel-burning appliance or fireplace, cleaning is required before the required inspection to remove any combustible deposits. In that case, the addition or removal of an appliance without both cleaning and inspection of the existing flue would be a code violation. If inspection shows that the flueway of the vent or chimney is incapable of preventing the leakage of combustion gases and liquids, the chimney or vent would have to be repaired or relined. Repair would have to adhere to the requirements of NFPA 31 where an oil-fired appliance is vented through the chimney flue.

801.18.3 Cleanout. Masonry chimneys shall be provided with a cleanout opening complying with Section 801.13.

❖ A cleanout opening must be installed, if one is not already there, for existing masonry chimney flues whenever another appliance is installed or removed (see commentary, Section 801.13).

801.18.4 Clearances. Chimneys and vents shall have airspace *clearance* to combustibles in accordance with the *International Building Code* and the *chimney* or vent manufacturer's installation instructions.

> **Exception:** Masonry chimneys without the required airspace *clearances* shall be permitted to be used if lined or relined with a *chimney* lining system *listed* for use in chimneys with reduced *clearances* in accordance with UL 1777. The *chimney clearance* shall be not less than permitted by the terms of the *chimney* liner listing and the manufacturer's instructions.

❖ Because it is usually not cost effective to increase clearances to combustibles around an existing chimney, the section provides an exception to allow existing chimneys to be lined or relined with a system complying with UL 1777 that enables the chimney to be used with reduced clearances to combustibles. The listing of the liner system, as well as the manufacturer's installation instructions specify the allowable clearances to combustibles for the masonry chimney lined or relined with a UL 1777 system.

801.18.4.1 Fireblocking. Noncombustible fireblocking shall be provided in accordance with the *International Building Code*.

❖ Prior to the 2009 edition of the code, this language was included as part of the exception for what is currently Section 801.18.4. Thus, unless the exception of Section 801.18.4 was invoked, the requirement for fireblocking was not obvious. The requirement was moved to this new section to clarify that the fireblocking requirements are applicable, irrespective of whether the masonry chimney is lined or relined according to the exception or not. Fireblocking of any annular spaces, where the chimney passes through ceilings or floors, is essential for controlling the passage of hot gases generated from a fire on a level below the annular space. Fireblocking must be installed to comply with Chapters 7 and 21 of the IBC.

801.19 Multistory prohibited. Common venting systems for appliances located on more than one floor level shall be prohibited, except where all of the appliances served by the common vent are located in rooms or spaces that are accessed only from the outdoors. The *appliance* enclosures shall not communicate with the occupiable areas of the building.

❖ This section addresses chimneys or vents that serve multiple appliances located on different floor levels. For example, a common vent may be designed to serve one or more appliances per floor of a multistory building, or a vent that rises up through the furnace rooms of single-story apartments in a multistory apartment building would be a common vent for the "stacked" apartments. At each level there may be a single appliance or several appliances entering into

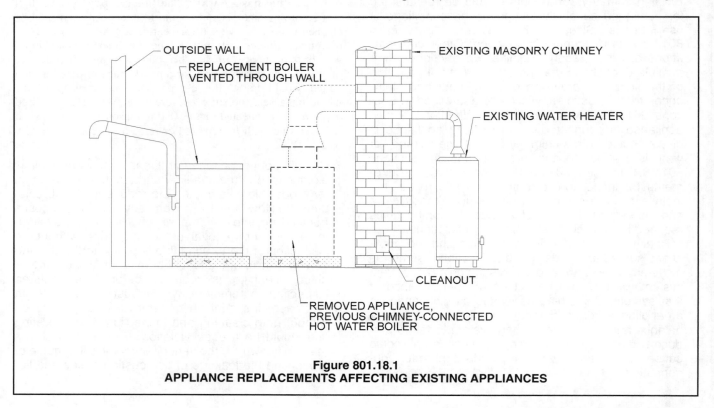

Figure 801.18.1
APPLIANCE REPLACEMENTS AFFECTING EXISTING APPLIANCES

the passageway. Multistory venting involves special design criteria and poses a unique hazard. When a vent is overloaded, blocked or otherwise producing insufficient draft, the flue gases from lower-story appliances could discharge into upper stories through the upper-story appliance connections. The lower-story vent could appear to be functioning normally while the upper-story occupants are exposed to a life-threatening hazard.

To avoid this potential hazard, multistory vent systems are designed to isolate the vent system and the appliances from the occupied portions of the building. This isolation is commonly accomplished by locating the common vent and all of the appliances served in mechanical rooms that are accessed from the outdoors only. These rooms are usually on outside walls and accessible only from a balcony. Multistory common chimney and vent systems must be designed in accordance with the vent or chimney manufacturer's installation and design instructions.

801.20 Plastic vent joints. Plastic pipe and fittings used to vent appliances shall be installed in accordance with the *appliance* manufacturer's installation instructions.

❖ Manufacturers of fuel-fired appliances are required to include installation instructions for the venting systems that the appliance is listed to use for venting. Where the vent system is required to be made of plastic pipe and fittings, the joints are required to be sealed because of the carbon monoxide, condensate and other products of combustion that could leak into the occupied space. The joints must also be joined in a manner that will prevent separation. Plastic pipe, such as PVC, CPVC and ABS are typically used to vent high efficiency appliances and the manufacturer's instruction will require the joints to be solvent welded (see Section 503.4.1.1 of the IFGC). The appliance manufacturer's installation instructions are part of the appliance listing and must be followed for the appliance installation.

SECTION 802
VENTS

802.1 General. All vent systems shall be *listed* and *labeled*. Type L vents and pellet vents shall be tested in accordance with UL 641.

❖ Vents and chimneys are distinct systems. The provisions of this section apply only to vents. The vents addressed in this section are natural-draft venting systems that produce a draft using the same principles that produce draft in chimneys (see the "General Comments" to this chapter).

Vents regulated by this section are factory fabricated and must be listed and labeled. The labeling requirement applies to all components of the system, including the sections of pipe, fittings, terminal caps, supports and spacers.

A vent system can be used only with oil-burning and pellet-burning appliances that are listed and

labeled for use with a vent. Solid fuel-burning appliances must not be connected to a vent system because those appliances produce flue-gas temperatures that are much higher than the temperatures for which vents are designed. The burning of solid fuels also produces creosote. Creosote formation leaves a combustible deposit on the surface of the flue passageway that, if ignited, can produce a fire with temperatures exceeding 2,000°F (1093°C), far exceeding the maximum safe temperature of a vent system.

Type L vents and pellet vents must be tested in accordance with UL 641 and are designed to vent oil-burning and pellet fuel-burning appliances, respectively. Oil-burning appliances produce a higher flue gas temperature than gas-fired appliances. Type L vents are typically double-wall, air-insulated vent piping systems constructed of galvanized and stainless steel.

Type L vents are designed for natural-draft applications only and must not be used to convey combustion gases under a positive pressure. For example, a Type L vent must not be used with Category III or IV gas-fired appliances and must not be used on the discharge side of an exhauster or power-vented appliance. Positive pressures will cause a Type L vent to leak combustion gases into the building and the leakage can cause condensation to form within the vent system pipe, fittings and joints, resulting in corrosion damage to the vent and eventual vent failure.

802.2 Vent application. The application of vents shall be in accordance with Table 802.2.

❖ Table 802.2 matches vent types to appliance types. The table lists types of vents and the corresponding types of appliances that can be served by the vents. The vent system must be tested and specifically approved for use with the approved appliance. If the vent system is not a tested and labeled component of the appliance, the material must be approved for use with the appliance and installed in accordance with the manufacturer's installation instructions.

Some high-efficiency oil-fired appliances might require special vent systems that are specific to the type of appliance. Special vent systems are associated with high-efficiency appliances, and could include vent materials such as PVC, CPVC and special alloys of stainless steel. There are high-efficiency (90 percent plus) condensing oil-fired appliances on the market that vent with plastic pipe.

Special vent systems must be designed and installed in compliance with the manufacturer's instructions. The manufacturer's instructions for appliances requiring special vent systems will specify installation requirements that are specific to that type of vent and appliance.

Only gas-burning, oil-burning and pellet-burning appliances can be used with a vent system if the appliance is labeled and approved for use with vent systems. The type and size of the vent must be as dictated by the manufacturer's installation instructions for the appliance. The design and installation instructions provided by the vent manufacturer must also be consulted when designing a vent system for any particular application.

TABLE 802.2
VENT APPLICATION

VENT TYPES	APPLIANCE TYPES
Type L oil vents	Oil-burning appliances listed and labeled for venting with Type L vents; gas appliances listed and labeled for venting with Type B vents.
Pellet vents	Pellet fuel-burning appliances listed and labeled for venting with pellet vents.

802.3 Installation. Vent systems shall be sized, installed and terminated in accordance with the vent and *appliance* manufacturer's installation instructions.

❖ The standards for vents and appliances require that venting instructions be supplied by the manufacturer of the appliance and the manufacturer of the vent system. These instructions are part of the labeling requirements. Any deviation from them is a violation of the code.

The clearance to combustibles for a vent system is determined by the testing agency and stated on the component labels and in the manufacturer's instructions. Not all vents have the same required clearances to combustibles. The clearances are determined by vent performance during testing in accordance with the applicable standard. Different vent materials and designs impact the vent's ability to control the amount of heat transmitted to surrounding combustibles. Installation of the vent system to comply with the clearances listed on the vent's label and in the manufacturer's installation instructions is critical.

The termination of a natural-draft vent must comply with the requirements of the manufacturer's installation instructions and Sections 802.4, 802.5 and 802.6. A vent used in conjunction with a mechanical exhauster must meet the termination requirements established in Section 804.3.

Physical protection of the vent system is required to prevent damage to the vent and to prevent combustibles from coming into contact with or being placed too close to the vents. This protection is typically achieved by enclosing the vent in chases, shafts or cavities in the building construction. Physical protection is not required in the room or space where the vent originates (at the appliance connection) and would not be required in such locations as attics that are not occupied or used for storage. For example, assume that a vent is installed in an existing building and that it extends from the basement, through a first-floor closet, through the attic and through the roof. The portion of the vent passing through the closet must be protected from damage. The means of physical damage protection should also be designed to maintain separation between the vent and any combustible storage. To prevent the passage of fire and smoke through the annular space around a vent pen-

etration through a floor or ceiling, the vent must be fireblocked with a noncombustible material in accordance with the IBC. Vent manufacturers provide installation instructions and factory-built components for fireblocking penetrations.

802.4 Vent termination caps required. Type L vents shall terminate with a *listed* and *labeled* cap in accordance with the vent manufacturer's installation instructions.

❖ This section is redundant with typical vent manufacturers' instructions and emphasizes that a vent is a system of components that is necessary for proper functioning. The cap must be as specified by the vent manufacturer and is a listed component. A vent cap not only keeps out moisture, debris and animals, it also serves to prevent wind interference that could negatively affect the vent's ability to produce the required draft. Vent caps are tested for flow resistance and performance in wind.

802.5 Type L vent terminations. Type L vents shall terminate not less than 2 feet (610 mm) above the highest point of the roof penetration and not less than 2 feet (610 mm) higher than any portion of a building within 10 feet (3048 mm).

❖ The termination heights are consistent with the requirements of NFPA 31.

Consideration should be given to vent location in a building in the design/installation phase so that the vent termination will not be difficult or require long extensions of vent pipe exposed to the outdoors. Vent piping exposed to the outdoors encourages condensation in cold weather, could require guy wires or braces, depending on height, and is considered aesthetically unattractive. Another good reason to plan for vent location during the design phase of a building is to allow straight vertical runs of vent to avoid problematic offsets. Good planning can also prevent vents from passing through the roof on the street side of the building. Quite often, problematic vent offsets in attics are installed for the sole purpose of penetrating the roof on the back side of a building.

802.6 Minimum vent heights. Vents shall terminate not less than 5 feet (1524 mm) in vertical height above the highest connected *appliance* flue collar.

Exceptions:

1. Venting systems of direct vent *appliances* shall be installed in accordance with the *appliance* and the vent manufacturer's instructions.

2. Appliances *listed* for outdoor installations incorporating integral venting means shall be installed in accordance with their listings and the manufacturer's installation instructions.

3. Pellet vents shall be installed in accordance with the *appliance* and the vent manufacturer's installation instructions.

❖ The amount of draft produced by a vent is directly proportional to the height of the vent. A minimum height must be established to produce the minimum draft necessary for the appliance served. Generally speaking, vent capacity increases as the vent height increases because of the increase in draft and vent flow velocity.

Exception 1 recognizes that direct-vent appliances use a vent system that is integral with the appliance or supplied by the appliance manufacturer for field assembly.

Exception 2 recognizes that roof-top units and outdoor appliances such as swimming pool heaters use integral venting means for which minimum heights are irrelevant.

Exception 3 recognizes that pellet vents are unique systems subject to the requirements of the appliance and vent listings.

802.7 Support of vents. All portions of vents shall be adequately supported for the design and weight of the materials employed.

❖ Vent manufacturers supply support parts and manufacturers' instructions contain detailed requirements for support of vent systems. Improper support can cause strain on vent components, appliance connectors and fittings, resulting in vent, appliance or connector damage that could cause loss of required clearance to combustibles and proper pitch. Frequently, venting installations suffer from lack of or improperly installed supports, brackets and hangers.

802.8 Insulation shield. Where vents pass through insulated assemblies, an insulation shield constructed of not less than No. 26 gage sheet metal shall be installed to provide *clearance* between the vent and the insulation material. The *clearance* shall be not less than the *clearance* to combustibles specified by the vent manufacturer's installation instructions. Where vents pass through attic space, the shield shall terminate not less than 2 inches (51 mm) above the insulation materials and shall be secured in place to prevent displacement. Insulation shields provided as part of a *listed* vent system shall be installed in accordance with the manufacturer's installation instructions.

❖ Loose insulation in attic floor assemblies or roof assemblies can fall against vents and create a fire hazard and can also cause abnormally high vent temperatures. Even though clearances to combustible construction such as wood framing are maintained, the continued heating of combustible insulation materials that have fallen against a vent could be a source of ignition. This section applies regardless of the combustibility of the insulation. Shields constructed in the field must be sufficient to serve their purpose, and must be securely attached to building construction because shields that are merely wedged into place could be accidentally dislodged or removed during maintenance activities.

SECTION 803
CONNECTORS

803.1 Connectors required. Connectors shall be used to connect *appliances* to the vertical *chimney* or vent, except

where the *chimney* or vent is attached directly to the *appliance*.

❖ This section establishes the minimum requirements for the design, construction and installation of chimney and vent connectors. Unless the chimney or vent is connected directly to the appliance, a connector is necessary to connect the appliance to its chimney or vent. A "Connector" is defined in Chapter 2 as a pipe used to connect an approved fuel-burning appliance to a chimney or vent. This includes the necessary fittings to make a connection or change in direction. The connector is usually a single-wall metal pipe, but it is also common practice to use listed and labeled chimney and vent pipe or listed factory-built single-wall connectors.

Many factors can affect the design and configuration of a connector. The most important of these is the appliance location with respect to the chimney or vent system. This impacts connector size, length and rise. Another important factor is the number of appliances being vented.

The appliance manufacturer's installation instructions may prohibit the use of single-wall connectors. For example, a single-wall vent connector may not be appropriate because of the possibility of condensation, corrosion and leakage. They are also not permitted in attic or crawl space installations subject to cold temperatures, which increase the possibility of condensation that could lead to accelerated material failure.

Connectors designed, constructed and installed in accordance with Section 803 are required except where a chimney or vent serves a single appliance and connects directly to that appliance. In such cases, a connector does not exist.

803.2 Location. Connectors shall be located entirely within the room in which the connecting *appliance* is located, except as provided for in Section 803.10.4. Where passing through an unheated space, a connector shall not be constructed of single-wall pipe.

❖ A single-wall connector has more than double the heat loss of a listed metal vent or chimney system, producing high temperatures on the outside surface and cooling of the flue gases on the inside of the connector. The cooling of flue gases produces condensation that can cause connector deterioration; the excessive surface temperatures pose a potential fire hazard. For these reasons, a connector can be located only within the room or space in which the appliance is located. A connector is prohibited from passing through any wall, ceiling or roof.

This section does not intend to prohibit the penetration of a combustible enclosure of a masonry chimney as permitted in Section 803.10.4 (see commentary, Section 803.10.4). Connectors that pass through walls, floors or ceilings might not be readily observable, and a potential fire hazard or connector deterioration could go undetected. Also, such pass-throughs increase the likelihood that the connector will be excessively long, exposed to low ambient temperatures or subject to contact with combustibles. Unlike double-wall pipe, single-wall pipe has a high heat loss and when installed in an unconditioned space, condensation can occur, deteriorating the pipe. This section effectively prohibits the use of single-wall connectors in garages, attics and crawl spaces where the spaces are unheated and occur in all but extremely moderate climates.

803.3 Size. The connector shall not be smaller than the size of the flue collar supplied by the manufacturer of the *appliance*. Where the *appliance* has more than one flue outlet, and in the absence of the manufacturer's specific instructions, the connector area shall be not less than the combined area of the flue outlets for which it acts as a common connector.

❖ Some appliances are designed with more than one venting (flue) outlet. For example, some large furnaces have dual heat exchangers and burner sections and thus have two flue-collar outlets. The manufacturer's instructions must be followed. However, in the event that manufacturer's instructions are not provided, this section specifies required sizing criteria.

803.4 Branch connections. All branch connections to the vent connector shall be made in accordance with the vent manufacturer's instructions.

❖ Joints, connections and manifold-type connections in venting systems are addressed in detail in the chimney or vent manufacturer's instructions, which include fitting usage, joint fastening, sizing and connector rise requirements.

803.5 Manual dampers. Manual dampers shall not be installed in connectors except in *chimney* connectors serving solid fuel-burning *appliances*.

❖ Manual dampers are intended for installation in a chimney system downstream of a solid fuel-burning appliance. The damper is intended to control the rate of combustion by regulating the amount of draft and allows closing the flue-gas passageways when the appliance is not operating to reduce the loss of conditioned air through the chimney. The installation of a manual flue damper in conjunction with an oil-burning appliance is strictly prohibited (see commentary, Section 803.6). Manual dampers can be used only with solid fuel-burning appliances.

Fireplaces require a method of closing off the chimney, usually a manual damper that will be opened during the process of manually starting a fire in a solid fuel-burning appliance. If the damper is not opened, the user will be aware of that fact because of the smoking and improper combustion. In the case of liquid fuel-fired appliances, however, the user may not be aware of a partially or completely closed damper and a hazardous condition could develop.

This section does not prohibit the use of fixed baffles that are components of listed appliances.

803.6 Automatic dampers. Automatic dampers shall be *listed* and *labeled* in accordance with UL 17 for oil-fired heating appliances. The dampers shall be installed in accordance with the manufacturer's installation instructions. An automatic vent damper device shall not be installed on an existing *appliance* unless the *appliance* is *listed* and *labeled* and the device is installed in accordance with the terms of its listing. The name of the installer and date of installation shall be marked on a label affixed to the damper device.

❖ Automatic dampers are intended for use only with liquid fuel-fired natural-draft appliances. Automatic dampers must be installed in strict accordance with the manufacturer's installation instructions. Because automatic damper failure could result in a hazardous condition, automatic dampers must be listed and labeled and must not serve more than one appliance.

The manufacturer's installation instructions require that the damper be installed by a qualified installer in accordance with the terms of the listing and the manufacturer's instructions. Because the purpose of a flue damper is to close or restrict the flue passageway of an appliance, the device must be properly installed to minimize the possibility of failure. A malfunctioning or improperly installed flue damper could cause the appliance to malfunction and discharge combustion products directly into the building interior.

Automatic flue dampers are energy-saving devices designed to close off or restrict an appliance flue passageway when the appliance is not operating or is in its "off" cycle. These devices save energy by trapping residual heat in a heat exchanger after the burners shut off and by preventing the escape of conditioned room air up the chimney or vent. Thus, appliance efficiency can be boosted and building air infiltration can be reduced.

Automatic fire dampers can be field-installed additions to existing equipment or can be a factory-supplied component of an appliance. The most common type of flue damper uses an electric motor to rotate a damper blade. The device uses switches that prove the damper position and allow the appliance to start only after the damper is opened and verified. The sequence of operation is the same as that described in the commentary to Section 804.3.2.

803.7 Connectors serving two or more appliances. Where two or more connectors enter a common vent or *chimney*, the smaller connector shall enter at the highest level consistent with available headroom or *clearance* to combustible material.

❖ A common vent is somewhat oversized when only a single appliance is operating; therefore, every effort is made to achieve proper vent/chimney operation during all possible operating circumstances. Placing the smallest appliance connection at the highest elevation in the common vent/chimney will allow the greatest connector rise for the smaller appliance connector and will take advantage of any draft priming effect caused by the lower connector. For example, it has been common practice to connect a domestic water heater connector above the furnace or boiler connector in a chimney or vent system.

803.8 Vent connector construction. Vent connectors shall be constructed of metal. The minimum thickness of the connector shall be 0.0136 inch (0.345 mm) (No. 28 gage) for galvanized steel, 0.022 inch (0.6 mm) (No. 26 B & S gage) for copper, and 0.020 inch (0.5 mm) (No. 24 B & S gage) for aluminum.

❖ The vent connector must be constructed of noncombustible, corrosion-resistant material capable of withstanding the flue-gas temperatures produced by the appliance it is serving. Specifically, vent connectors must be constructed of galvanized steel, copper or aluminum with minimum thicknesses as specified. The connector material must be as specified by the appliance manufacturer. The thickness for galvanized sheet steel was revised to reflect the low end of the thickness tolerance range for No. 28 gage. This allows for all possible thicknesses of No. 28 gage to be acceptable as such.

This section does not apply to chimney or vent connectors that are components of a listed and labeled vent or chimney. Those chimney and vent pipe and fittings are regulated by the manufacturer's installation instructions. The requirements found in this section are intended to regulate only untested and unlabeled materials connecting the appliance to an appropriate chimney or vent. Factory-constructed listed and labeled single-wall and double-wall connectors are available. These connectors are not subject to the construction requirements of Sections 803.8 and 803.9.

803.9 Chimney connector construction. *Chimney* connectors for low-heat *appliances* shall be of sheet steel pipe having resistance to corrosion and heat not less than that of galvanized steel specified in Table 803.9(1). Connectors for medium-heat *appliances* and high-heat appliances shall be of sheet steel not less than the thickness specified in Table 803.9(2).

❖ The requirements for chimney connectors are more stringent than those for vent connectors because they are subjected to higher flue-gas temperatures and corrosive flue gases from solid fuels. Chimney connectors must be constructed of sheet steel conforming to the thickness requirements of Table 803.9(1) for connectors serving low-heat appliances and Table 803.9(2) for connectors serving medium- and high-heat appliances.

Connector wall thickness increases as the connector's cross-sectional area increases to maintain structural integrity and rigidity. Sheet metal thicknesses are expressed in fractions of an inch (mm) and in industry standard gage designations. Medium- and high-heat appliance connectors are often rectangular; therefore, both areas and round pipe diameters are given in the table.

TABLE 803.9(1)
MINIMUM CHIMNEY CONNECTOR
THICKNESS FOR LOW-HEAT APPLIANCES

DIAMETER OF CONNECTOR (inches)	MINIMUM NOMINAL THICKNESS (galvanized) (inches)
5 and smaller	0.022 (No. 26 gage)
Larger than 5 and up to 10	0.028 (No. 24 gage)
Larger than 10 and up to 16	0.034 (No. 22 gage)
Larger than 16	0.064 (No. 16 gage)

For SI: 1 inch = 25.4 mm.

TABLE 803.9(2)
MINIMUM CHIMNEY CONNECTOR
THICKNESS FOR MEDIUM- AND HIGH-HEAT APPLIANCE

AREA (square inches)	EQUIVALENT ROUND DIAMETER (inches)	MINIMUM THICKNESS (inches)
0-154	0-14	0.0575 (No. 16 gage)
155-201	15-16	0.075 (No. 14 gage)
202-254	17-18	0.0994 (No. 12 gage)
Greater than 254	Greater than 18	0.1292 (No. 10 gage)

For SI: 1 inch = 25.4 mm, 1 square inch = 645.16 mm^2.

❖ The thicknesses for Nos. 16, 14 and 10 gage were revised to reflect the low end of the thickness tolerance range for those gages. This allows for all possible thicknesses of those gages to be acceptable.

803.10 Installation. Connectors shall be installed in accordance with Sections 803.10.1 through 803.10.6.

❖ Sections 803.10.1 through 803.10.6 contain the installation requirements for vent and chimney connectors. If the requirements of this section conflict with those of the manufacturer's installation instructions, the more stringent requirements apply.

803.10.1 Supports and joints. Connectors shall be supported in an *approved* manner, and joints shall be fastened with sheet metal screws, rivets or other *approved* means.

❖ A connector must be supported for the design and weight of the material used. Proper support is necessary to maintain the clearances required by Section 803.10.6, to maintain the required pitch and to prevent physical damage and separation of joints. The joints between connectors and appliance draft hood outlets and flue collars must be secured with screws, rivets or other approved means. The joints between pipe sections must also be secured with screws, rivets or other approved means. Some connectors are available with a proprietary fastening method. The special proprietary joints do not require screws, rivets or other fasteners if they provide equivalent resistance to disengagement or displacement. This section addresses connectors, not chimneys and vents. Most chimney and vent manufacturers do not require or might even prohibit the use of screws or rivets to secure joints for chimney and vent pipe sections and fittings.

803.10.2 Length. The maximum horizontal length of a single-wall connector shall be 75 percent of the height of the *chimney* or vent.

❖ The appliance and chimney or vent must be located to keep the connector length as short as practicable. This section establishes the maximum allowable length for uninsulated chimney and vent connectors. These requirements are based on the heat loss of the connector and the ability of the vent or chimney system to produce a draft. An insulated connector (double wall) reduces the amount of heat transfer through the connector pipe, thus keeping the flue gas at a higher temperature. The amount of draft is directly proportional to the chimney or vent height and the difference in temperature between the flue gases and the ambient air. An uninsulated connector is a run of single-wall pipe.

The length limitations of this section are based on the total vertical rise of the chimney or vent and on the developed length of horizontal connectors within the chimney or vent system. The connector length limitation of 75 percent of the height of the vent or chimney is based on field experience and laboratory testing.

Connector lengths are limited because of the flow resistance of the connector pipe and because heat loss through the connector is directly proportional to its length. For a venting system to work, the draft produced by the vertical vent or chimney must be able to overcome the resistance to flow created by the connector. The longer the connector, the longer it takes to prime a cold venting system and develop draft. Connectors should always be as short as the installation conditions will permit.

803.10.3 Connection. The connector shall extend to the inner face of the *chimney* or vent liner, but not beyond. A connector entering a masonry *chimney* shall be cemented to masonry in an *approved* manner. Where thimbles are installed to facilitate removal of the connector from the masonry *chimney*, the thimble shall be permanently cemented in place with high-temperature cement.

❖ All of the precautions taken to ensure adequate chimney or vent design may be ineffective if the appliance and the chimney or vent are not properly connected. Improper connections can lead to appliance or connector failure and the leakage of harmful flue gases into the building. This section contains requirements for the connection between a chimney or vent and the appliance connector. The connection to a vent is properly accomplished by vent fittings and likewise for factory-built chimneys.

Chimney connectors must pass through a masonry chimney wall to the inner face of the liner but not beyond. A connector that extends into a chimney passageway can restrict the flow of flue gases and provide a ledge for creosote, soot and debris to accumulate. The joint between the connector and the chimney must meet the requirements of Section 803.10.1. If the connector enters a masonry chimney, it must be cemented in place with an approved material such as refractory mortar or another heat-resistant cement [see Commentary Figure 803.10.3(1)].

This section also permits the use of thimbles at a masonry chimney opening to provide for easy removal of the connector to facilitate cleaning. When a thimble is installed, it is to be permanently cemented in place with an approved high-temperature cement [see Commentary Figure 803.10.3(2)]. A connector must be attached to a thimble in an approved manner to prevent displacement.

803.10.4 Connector pass-through. *Chimney* connectors shall not pass through any floor or ceiling, nor through a fire-resistance-rated wall assembly. *Chimney* connectors for domestic-type *appliances* shall not pass through walls or par-

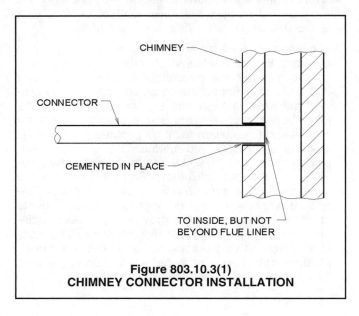

Figure 803.10.3(1)
CHIMNEY CONNECTOR INSTALLATION

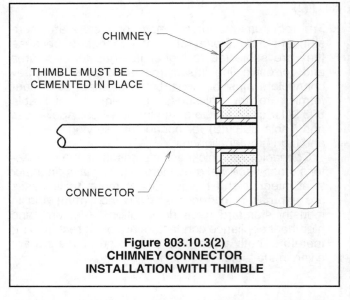

Figure 803.10.3(2)
CHIMNEY CONNECTOR
INSTALLATION WITH THIMBLE

titions constructed of combustible material to reach a masonry *chimney* unless:

1. The connector is *labeled* for wall pass-through and is installed in accordance with the manufacturer's instructions;

2. The connector is put through a device *labeled* for wall pass-through; or

3. The connector has a diameter not larger than 10 inches (254 mm) and is installed in accordance with one of the methods in Table 803.10.4. Concealed metal parts of the pass-through system in contact with flue gases shall be of stainless steel or equivalent material that resists corrosion, softening or cracking up to 1,800°F (980°C).

❖ The requirements contained in this section address the fire hazards associated with chimney connectors that serve domestic-type appliances and that penetrate combustible walls or partitions. A domestic-type appliance, as defined by NFPA 211, is a fuel-burning heating appliance for heating building spaces having a volume of not more than 25,000 cubic feet (708 m³) or other heat-producing appliances of the type mainly used in residences, but which may also be used in other buildings.

Evidence has shown that many house fires have been caused by chimney connectors that serve solid fuel-burning appliances and that penetrate a combustible wall or partition. Commonly, chimneys are enclosed by walls of combustible construction and a wall must be penetrated by a chimney connector in order to connect to the chimney.

The following penetration methods have been tested and have been shown to adequately control temperature rise on adjacent combustible surfaces. These methods are prescriptive and must not be altered or modified in any way.

The method found in Item 1 permits installation of a connector labeled for combustible wall pass-through to be installed in accordance with the manufacturer's installation instructions.

Item 2 permits a connector to penetrate a combustible wall or partition by passing through a device that is labeled for pass-through.

Item 3 permits a connector to penetrate a combustible wall or partition if installed in accordance with System A, B, C or D in Table 803.10.4.

It is important to note that these provisions do not conflict with the requirements found in Section 803.2. Chimney and vent connectors are not permitted to leave the room in which the appliance served is located. The provisions of this section are applicable only to combustible walls or partitions that enclose a masonry chimney (see commentary, Section 803.2).

TABLE 803.10.4. See next column.

❖ Table 803.10.4 contains specifications for four different methods of constructing a chimney connector wall penetration assembly (pass-through). The methods are designed to protect the combustible wall assembly from the high temperatures associated with

single-wall chimney connectors serving domestic-type solid fuel appliances. These connectors have a required airspace clearance to combustibles of 18 inches (457 mm). Commentary Figure 803.10.4 helps explain the complex specifications of Table 803.10.4.

TABLE 803.10.4
CHIMNEY CONNECTOR SYSTEMS AND CLEARANCES TO COMBUSTIBLE WALL MATERIALS FOR DOMESTIC HEATING APPLIANCES[a, b, c, d]

System A (12-inch clearance)	A 3.5-inch-thick brick wall shall be framed into the combustible wall. An 0.625-inch-thick fire-clay liner (ASTM C 315 or equivalent)[e] shall be firmly cemented in the center of the brick wall maintaining a 12-inch clearance to combustibles. The clay liner shall run from the outer surface of the bricks to the inner surface of the chimney liner.
System B (9-inch clearance)	A labeled solid-insulated factory-built chimney section (1-inch insulation) the same inside diameter as the connector shall be utilized. Sheet steel supports cut to maintain a 9-inch clearance to combustibles shall be fastened to the wall surface and to the chimney section. Fasteners shall not penetrate the chimney flue liner. The chimney length shall be flush with the masonry chimney liner and sealed to the masonry with water-insoluble refractory cement. Chimney manufacturers' parts shall be utilized to securely fasten the chimney connector to the chimney section.
System C (6-inch clearance)	A steel ventilated thimble having a minimum thickness of 0.0236 inch (No. 24 gage) having two 1-inch air channels shall be installed with a steel chimney connector. Steel supports shall be cut to maintain a 6-inch clearance between the thimble and combustibles. The chimney connector and steel supports shall have a minimum thickness of 0.0236 inch (No. 24 gage). One side of the support shall be fastened to the wall on all sides. Glass-fiber insulation shall fill the 6-inch space between the thimble and the supports.
System D (2-inch clearance)	A labeled solid-insulated factory-built chimney section (1-inch insulation) with a diameter 2 inches larger than the chimney connector shall be installed with a steel chimney connector having a minimum thickness of 0.0236 inch (24 gage). Sheet steel supports shall be positioned to maintain a 2-inch clearance to combustibles and to hold the chimney connector to ensure that a 1-inch airspace surrounds the chimney connector through the chimney section. The steel support shall be fastened to the wall on all sides and the chimney section shall be fastened to the supports. Fasteners shall not penetrate the liner of the chimney section.

For SI: 1 inch = 25.4 mm, 1.0 Btu x in/ft² · h · °F = 0.144 W/m² · K.

a. Insulation material that is part of the wall pass-through system shall be noncombustible and shall have a thermal conductivity of 1.0 Btu x in/ft² · h · °F or less.

b. All clearances and thicknesses are minimums.

c. Materials utilized to seal penetrations for the connector shall be noncombustible.

d. Connectors for all systems except System B shall extend through the wall pass-through system to the inner face of the flue liner.

e. ASTM C 315.

The thickness for galvanized sheet steel was revised to reflect the low end of the thickness tolerance range for No. 24 gage. This allows for all possible thicknesses of No. 24 gage to be acceptable as such.

803.10.5 Pitch. Connectors shall rise vertically to the *chimney* or vent with a minimum pitch equal to one-fourth unit vertical in 12 units horizontal (2-percent slope).

❖ The ideal chimney or vent configuration is a totally vertical system, even though it is not always practical. This section requires all portions of a chimney or vent connector to rise vertically a minimum of $^{1}/_{4}$ inch per foot (21 mm/m) of its horizontal length. The connector slope is intended to induce the flow of flue gases by using the natural buoyancy of the hot gases. Connector slope can promote the priming of a cold venting system and can partially compensate for short connector vertical rise.

803.10.6 Clearances. Connectors shall have a minimum *clearance* to combustibles in accordance with Table 803.10.6. The clearances specified in Table 803.10.6 apply, except where the listing and labeling of an *appliance* specifies a different *clearance*, in which case the *labeled clearance* shall apply. The *clearance* to combustibles for connectors shall be reduced only in accordance with Section 308.

❖ Flue-gas passageways must have minimum clearances to ignitable materials. The clearance requirements in Table 803.10.6 apply to unlabeled single-wall connectors serving a listed appliance. Connectors that are listed and labeled must maintain a clearance to combustibles as required by the connector manufacturer's installation instructions. If the appliance or connector manufacturer's installation instructions specify larger clearances than those prescribed by this section, the manufacturer's installation instructions govern.

Section 308 contains methods to allow the reduction of the required clearances. It is important to remember that all required clearances are airspace clearances measured from the chimney or vent connector to the combustible surface.

TABLE 803.10.6
CONNECTOR CLEARANCES TO COMBUSTIBLES

TYPE OF APPLIANCE	MINIMUM CLEARANCE (inches)
Domestic-type appliances	
Chimney and vent connectors	
Electric and oil incinerators	18
Oil and solid-fuel appliances	18
Oil appliances labeled for venting with Type L vents	9
Commercial, industrial-type appliances	
Low-heat appliances	
Chimney connectors	
Oil and solid-fuel boilers, furnace and water heaters	18
Oil unit heaters	18
Other low-heat industrial appliances	18
Medium-heat appliances	
Chimney connectors	
All oil and solid-fuel appliances	36
High-heat appliances Masonry or metal connectors All oil and solid-fuel appliances	(As determined by the code official)

For SI: 1 inch = 25.4 mm.

❖ Table 803.10.6 prescribes connector clearances to combustibles based on the type of appliance.

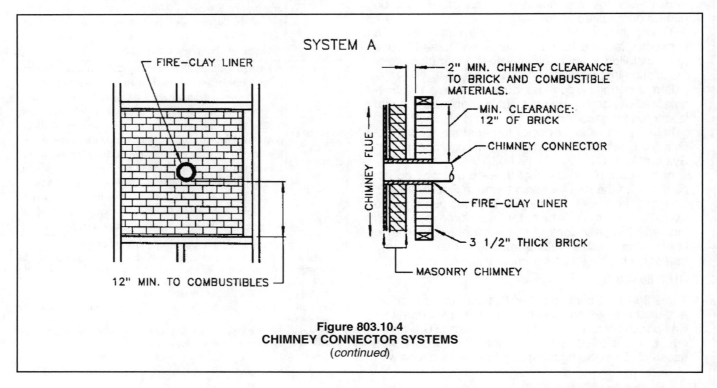

SYSTEM A

FIRE-CLAY LINER

12" MIN. TO COMBUSTIBLES

2" MIN. CHIMNEY CLEARANCE TO BRICK AND COMBUSTIBLE MATERIALS.

MIN. CLEARANCE: 12" OF BRICK

CHIMNEY CONNECTOR

FIRE-CLAY LINER

3 1/2" THICK BRICK

CHIMNEY FLUE

MASONRY CHIMNEY

Figure 803.10.4
CHIMNEY CONNECTOR SYSTEMS
(continued)

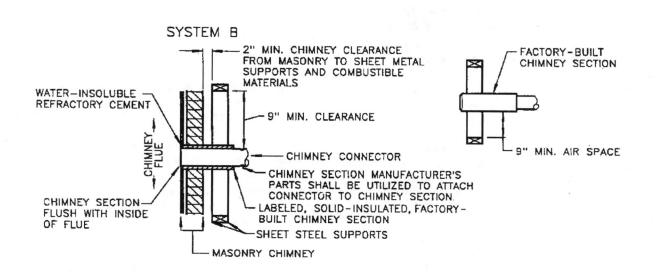

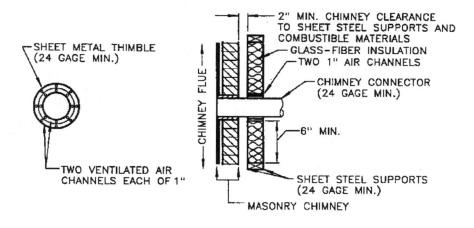

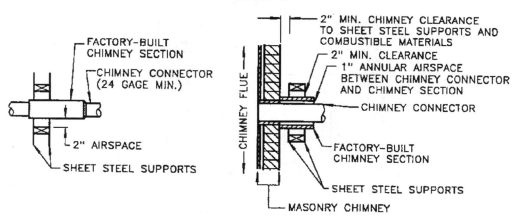

Figure 803.10.4—continued
CHIMNEY CONNECTOR SYSTEMS

SECTION 804
DIRECT-VENT, INTEGRAL VENT AND
MECHANICAL DRAFT SYSTEMS

804.1 Direct-vent terminations. Vent terminals for *direct-vent appliances* shall be installed in accordance with the manufacturer's installation instructions

❖ Direct-vent appliances include those appliances having a closed combustion chamber and a conduit for bringing all combustion air directly from the outdoors (see the definition of "Direct-vent appliance") (see Commentary Figure 804.1). Direct-vent appliances are often referred to as "sealed combustion" or "separated combustion" appliances.

The code states prescriptive termination requirements for these appliances and refers to the manufacturer's installation instructions.

804.2 Appliances with integral vents. *Appliances* incorporating integral venting means shall be installed in accordance with their listings and the manufacturer's installation instructions.

❖ This section refers to appliances such as rooftop heating ventilating and air-conditioning (HVAC) units, outdoor furnaces, through-the-wall space conditioning units and outdoor swimming pool heaters, all of which have built-in gravity or power venting means and must be installed to comply with their listings and the manufacturer's installation instructions.

804.2.1 Terminal clearances. *Appliances* designed for natural draft venting and incorporating integral venting means shall be located so that a minimum *clearance* of 9 inches (229 mm) is maintained between vent terminals and from any openings through which *combustion* products enter the building. *Appliances* using forced draft venting shall be located so that a minimum clearance of 12 inches (305 mm) is maintained between vent terminals and from any openings through which *combustion* products enter the building.

❖ Discharge terminal dispersion determines the required distance to openings such as openable windows, doors, intake louvers and ducts. This section is not intended to apply to the distance between an integral vent terminal on a rooftop HVAC unit and the outdoor air intake on the same unit. This distance has been evaluated as part of the rooftop unit's design and listing.

804.3 Mechanical draft systems. Mechanical draft systems of either forced or induced draft design shall be listed and labeled in accordance with UL 378 and shall comply with Sections 804.3.1 through 804.3.7.

❖ The appliance installations addressed in this section use listed and labeled auxiliary or integral fans and blowers to force the flow of combustion products to the outdoors. This section applies to externally installed power exhausters, integrally power-exhausted appliances and venting systems equipped with draft inducers. This section does not address direct-vent appliances.

Power exhausters are field-installed pieces of equipment that are independent of, but used in conjunction with, appliances [see Commentary Figures 804.3(1) through 804.3(4)].

804.3.1 Forced draft systems. Forced draft systems and all portions of induced draft systems under positive pressure dur-

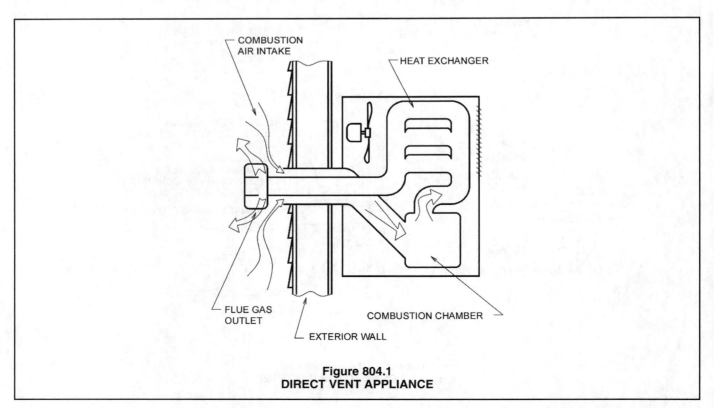

Figure 804.1
DIRECT VENT APPLIANCE

ing operation shall be designed and installed so as to be gas tight to prevent leakage of *combustion* products into a building.

❖ Power exhausters are typically used where other means of venting are impractical, impossible or uneconomical. Power exhausters are typically designed for use with gas-fired and oil-fired natural-draft appliances (see the IFGC for gas-fired appliances). Power exhausters produce negative pressures at their inlet connection and produce positive pressures at their outlet (discharge connection). The most common installation locates the exhauster at the point of termination of the vent or chimney. In such installations, the vent or chimney between the appliance and the exhauster operates under negative pressure.

Induced draft systems use separate field-installed units designed to boost draft in a natural (gravity) draft chimney or vent. Draft inducers may or may not produce positive pressures in the vent or chimney system [see Commentary Figure 804.3(4)].

Power-vented (self-venting) appliances are equipped with factory-installed integral blowers that force the combustion products through special venting systems. Power-vented appliances include some integral-vent rooftop units and some direct-vent appliances. Appliances for outdoor installation, addressed in Section 804.2, are typically power vented and discharge directly to the atmosphere. Solid fuel-burning appliances are not vented by power exhausters because of the high temperatures and corrosive nature of the combustion products and because solid

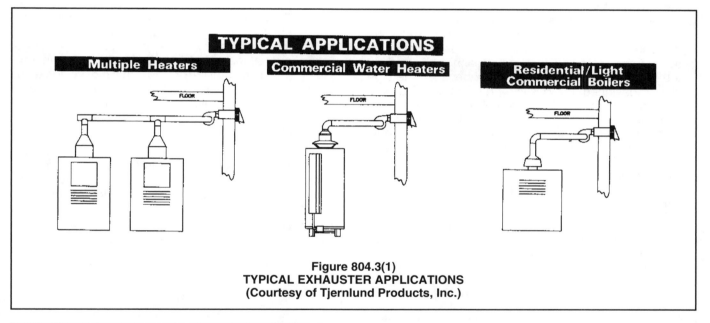

Figure 804.3(1)
TYPICAL EXHAUSTER APPLICATIONS
(Courtesy of Tjernlund Products, Inc.)

Figure 804.3(2)
TYPICAL POWER EXHAUSTER
(Courtesy of Tjernlund Products, Inc.)

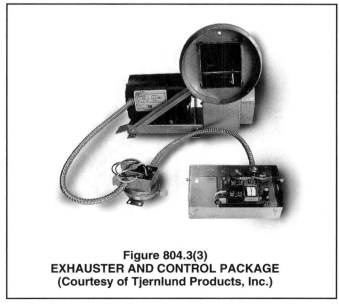

Figure 804.3(3)
EXHAUSTER AND CONTROL PACKAGE
(Courtesy of Tjernlund Products, Inc.)

fuel appliances cannot be immediately shut off in the event of exhauster failure; however, specialized solid fuel chimney exhausters are available for mounting on the tops of existing chimneys that produce insufficient draft. These solid fuel chimney exhausters are intended for remedial application on existing chimneys; they are not intended to substitute for proper chimney design. Although not specifically referred to as an appliance, a power exhauster is considered a mechanical appliance attachment and, therefore, should bear the label of an approved agency in accordance with Section 301.7 (see the definition of "Appliance"). A power exhauster is an essential component of the appliance installation it serves and must be installed in accordance with the manufacturer's installation instructions. Mechanical-draft devices that are an integral part of an appliance are covered by the appliance listing.

Vent or chimney systems installed downstream (discharge side) of a power exhauster must be designed and approved for positive-pressure applications. For example, Type L vent cannot be used on the discharge side of an exhauster because the pipe is not designed for positive-pressure applications. The application and installation of exhausters (power venters) must comply with the manufacturer's installation instructions for the exhauster and the appliance(s) served by the exhauster (see commentary, Section 801.9).

Three distinct variations of power exhaust blowers at the combustion chamber inlet working in conjunction with the fuel burner are in use:

1. *Blowers that supply turbulent combustion air to aid fuel-air mixing in a combustion chamber that is under negative pressure.* Residential pressure-atomizing oil burners are examples. Obtaining proper negative overfire draft (which optimizes combustion) also requires steady

negative (below atmospheric) pressure at the flue outlet. This negative pressure will be produced by a natural-draft chimney and may be controlled by a barometric draft regulator. This is tabulated on line "A" in Commentary Figure 804.3.1.

2. *Blowers that supply sufficient combustion air and pressure to produce flow through the combustion chamber, but the combustion process does not need additional vent or chimney draft.* This permits use of gravity or neutral-draft venting products, such as chimneys, for equipment burning oil, coal, or a draft regulator may be used for such equipment to prevent excess draft from affecting combustion efficiency. This is tabulated on line "B" in Commentary Figure 804.3.1.

3. *Blowers with enough power to overcome internal flue passage pressure losses (that is, in fire-tube boilers) that also produce positive pressure at the outlet.* This outlet pressure must be added to gravity draft as the motive force for flow in a chimney. If positive outlet pressure exists, a sealed pressure-tight chimney is required. This is tabulated on line "C" in Commentary Figure 804.3.1.

The above types of equipment usually have integral blower/burner systems and all could be considered forced combustion systems. Only those described in Item 3 truly produce forced chimney draft.

804.3.2 Automatic shutoff. Power exhausters serving automatically fired *appliances* shall be electrically connected to

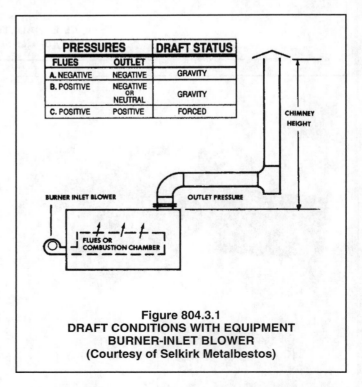

PRESSURES		DRAFT STATUS
FLUES	**OUTLET**	
A. NEGATIVE	NEGATIVE	GRAVITY
B. POSITIVE	NEGATIVE OR NEUTRAL	GRAVITY
C. POSITIVE	POSITIVE	FORCED

Figure 804.3.1
DRAFT CONDITIONS WITH EQUIPMENT
BURNER-INLET BLOWER
(Courtesy of Selkirk Metalbestos)

Figure 804.3(4)
DRAFT INDUCER
(Courtesy of Tjernlund Products, Inc.)

each *appliance* to prevent operation of the *appliance* when the power exhauster is not in operation.

❖ When power-venting equipment is installed, it becomes an essential part of the appliance it serves. The appliance relies on the power-venting equipment (exhauster) to provide sufficient draft for proper appliance operation and for venting of the combustion products. Power exhausters are electrically interlocked with the appliance or appliances they serve to ensure that the appliances will not operate if there is insufficient draft to vent the products of combustion. If an exhauster fails, the appliances served by the exhauster could discharge the products of combustion into the building and/or operate dangerously. An improper or lacking interlock can cause the appliance(s) to malfunction and spill harmful flue gases into the building space.

The electrical interlock (interconnection) is typically accomplished with controls supplied by the exhauster manufacturer [see Commentary Figure 804.3(3)]. The usual sequence of operation is as follows: the call for heat from the appliance operating control starts the exhauster, and pressure controls start the appliance only after adequate draft has been proven to exist. In some cases, temperature sensors are used in addition to pressure controls (see Commentary Figure 804.3.2).

804.3.3 Termination. The termination of *chimneys* or vents equipped with power exhausters shall be located a minimum of 10 feet (3048 mm) from the lot line or from adjacent buildings. The exhaust shall be directed away from the building.

❖ The terminations of chimney and vent systems equipped with power exhausters must be located a minimum of 10 feet (3048 mm) from a lot line and from any adjacent buildings to allow for the dilution of noxious and toxic flue gases that might otherwise affect occupants of other buildings. The exhaust discharge can also be harmful to building surfaces, vehicles and vegetation.

804.3.4 Horizontal terminations. Horizontal terminations shall comply with the following requirements:

1. Where located adjacent to walkways, the termination of mechanical draft systems shall be not less than 7 feet (2134 mm) above the level of the walkway.

2. Vents shall terminate at least 3 feet (914 mm) above any forced air inlet located within 10 feet (3048 mm).

3. The vent system shall terminate at least 4 feet (1219 mm) below, 4 feet (1219 mm) horizontally from or 1 foot (305 mm) above any door, window or gravity air inlet into the building.

4. The vent termination point shall not be located closer than 3 feet (914 mm) to an interior corner formed by two walls perpendicular to each other.

5. The vent termination shall not be mounted directly above or within 3 feet (914 mm) horizontally from an oil tank vent or gas meter.

6. The bottom of the vent termination shall be located at least 12 inches (305 mm) above finished grade.

❖ This section establishes the requirements for exhaust pipes and power exhausters that terminate horizontally or through a side wall; thus the name "horizontal" vent. Horizontal vents are typically used for appliance installations in existing buildings or in new construction where a long vertical vent runs or terminating above the roof is impractical.

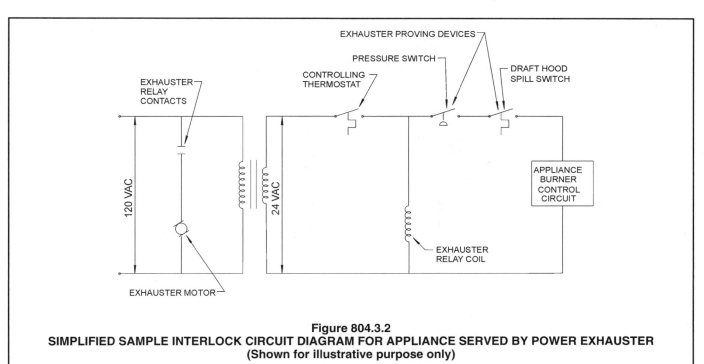

Figure 804.3.2
SIMPLIFIED SAMPLE INTERLOCK CIRCUIT DIAGRAM FOR APPLIANCE SERVED BY POWER EXHAUSTER
(Shown for illustrative purpose only)

Commentary Figures 804.3.4(1) and 804.3.4(2) illustrate some of the termination requirements for a horizontal vent. To prevent combustion products from entering the building, the vent terminal must be located a minimum distance from any gravity air inlet such as a louver, door or window. A more restrictive clearance requirement is required for forced-air inlets because of the increased chance that flue gases will reenter the building through a powered intake. These requirements apply to the specific applications of this section and supersede the general requirements of Section 401.4. The general provisions of Section 401.4 are applicable to chimneys and vents that are not equipped with power exhausters and that are not components of power-vented appliances as addressed in this section.

The vent terminal must be located at least 1 foot (305 mm) above grade to prevent the vent terminal from being blocked by snow and vegetation, and to eliminate a potential fire hazard from igniting plants and debris. When a walkway is present, the vent has to be located at least 7 feet (2134 mm) above the walking surface to protect any passersby from harmful flue gases. A walkway may be present even though a paved or other prepared walking surface does not exist. Any area that is normally expected to serve as a path of travel is considered a walkway for the purposes of protecting individuals from exposure to the vent's discharge. The code does not specify a minimum horizontal separation distance between a walkway and the termination of a horizontal or vertical vent equipped with a power exhauster before the 7-foot (2134 mm) vertical separation takes effect. The code official must evaluate the manufacturer's installation instructions, as well as each installation to determine an appropriate distance to minimize the hazards.

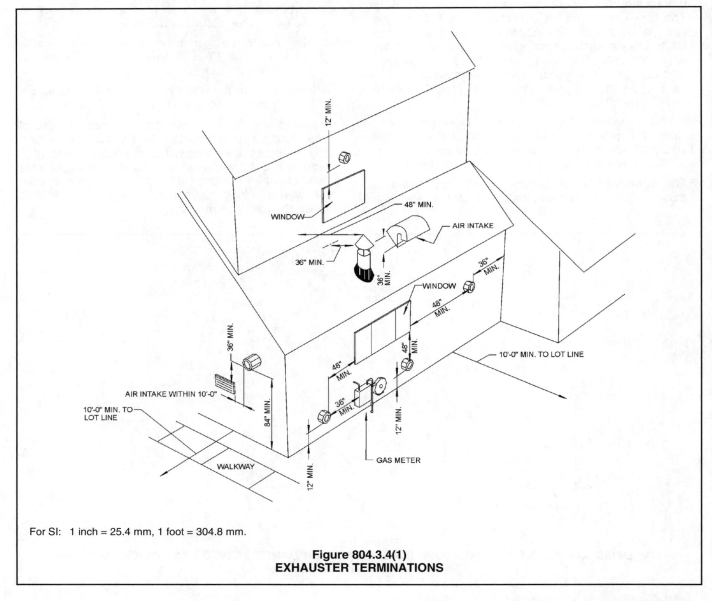

For SI: 1 inch = 25.4 mm, 1 foot = 304.8 mm.

Figure 804.3.4(1)
EXHAUSTER TERMINATIONS

The high-temperature discharge of the horizontal vent is a potential source of ignition for any flammable gases that may leak out of the gas meter, service pressure regulator vent or fuel oil tank vent. Dripping condensate could cause corrosion or ice damage to such components. Commentary Figure 804.3.4(2) illustrates the area around the gas meter and service pressure regulator where the horizontal vent terminal is prohibited.

Lastly, this section requires locating the vent terminal a minimum of 3 feet (914 mm) away from an interior corner of the building created by two intersecting walls. This prevents the vent terminal from being located in an area of stagnation where the flue gases cannot be properly diluted, protects adjacent walls from the heat and corrosive effects of flue gases and helps protect the vent terminal from wind-induced pressure zones.

804.3.5 Vertical terminations. Vertical terminations shall comply with the following requirements:

1. Where located adjacent to walkways, the termination of mechanical draft systems shall be not less than 7 feet (2134 mm) above the level of the walkway.

2. Vents shall terminate at least 3 feet (914 mm) above any forced air inlet located within 10 feet (3048 mm) horizontally.

3. Where the vent termination is located below an adjacent roof structure, the termination point shall be located at least 3 feet (914 mm) from such structure.

4. The vent shall terminate at least 4 feet (1219 mm) below, 4 feet (1219 mm) horizontally from or 1 foot (305 mm) above any door, window or gravity air inlet for the building.

5. A vent cap shall be installed to prevent rain from entering the vent system.

6. The vent termination shall be located at least 3 feet (914 mm) horizontally from any portion of the roof structure.

❖ This section establishes the requirements for exhaust pipes and power exhausters that terminate vertically; thus the name "vertical" vent. Although most vents and chimneys terminate vertically, this section is applicable only to mechanical (power) venting systems that do not rely on natural draft to vent flue gases. The requirements of this section are similar to those of Section 804.3.4. See the commentary to Section 804.3.4 for a discussion of the vent termination requirements. It should be noted that the 10-foot (3048 mm) distance between the exhauster outlet and any forced-air inlet must be measured horizontally rather than parallel to the sloped roof or around a

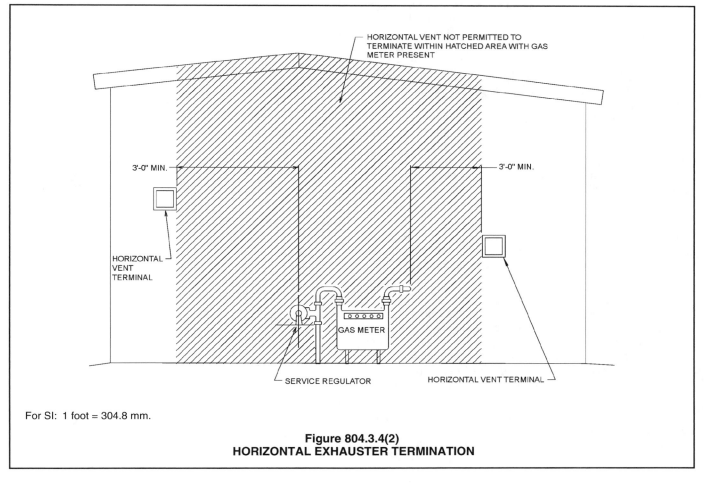

For SI: 1 foot = 304.8 mm.

Figure 804.3.4(2)
HORIZONTAL EXHAUSTER TERMINATION

corner. This provides a consistent measurement for the designer and the code official to avoid misunderstandings.

The vertical vent has additional location requirements with respect to adjacent roof structures. Commentary Figure 804.3.4(1) depicts a vertical vent termination. When the vent terminal is located below an adjacent roof structure, the vent must terminate a minimum of 3 feet (914 mm) from the roof structure to protect the structure from the hot flue gases and to protect the vent from wind-induced pressure zones. When the roof is subjected to human traffic, as in the case of a sun deck, the vent must terminate a minimum of 7 feet (2134 mm) above the deck surface. The vent terminal must have a cap to protect it from the entry of precipitation and animals (see commentary, Section 802.4).

804.3.6 Exhauster connections. An *appliance* vented by natural draft shall not be connected into a vent, *chimney* or vent connector on the discharge side of a mechanical flue exhauster.

❖ When a mechanical draft device or power exhauster is installed, the resulting "forced draft" creates a positive pressure inside the chimney or vent on the discharge or outlet side of the fan. A hazardous condition will result if an appliance is connected on the outlet or discharge side of a power exhauster. Commentary Figure 804.3.6 shows an improper appliance connection to an exhauster vented system. The appliance connector provides an alternative path for the flue gases to travel back into the building. Also, the appliance connected on the discharge side (outlet) of the exhauster will not have draft; its combustion products will spill into the building interior (see Section 801.14).

A hazardous condition can also occur when a vent connector serving an appliance vented by natural draft is connected to the vent of an appliance equipped with integral power-venting means.

Any vent, chimney or connector piping into which an exhauster discharges must be rated and approved for positive pressure and properly installed to prevent leakage of combustion gases.

804.3.7 Exhauster sizing. Mechanical flue exhausters and the vent system served shall be sized and installed in accordance with the manufacturer's installation instructions.

❖ Exhauster manufacturer's instructions include design and sizing criteria that must be used for those systems. The general vent sizing information in this chapter does not apply to power-vented systems. Some exhausters are listed for use with specific types and models of appliances. In some cases, the exhauster installation instructions will require the installation of a draft regulator control in the appliance vent connector.

804.3.8 Mechanical draft systems for manually fired appliances and fireplaces. A mechanical draft system shall be permitted to be used with manually fired appliances and

fireplaces where such system complies with all of the following requirements:

1. The mechanical draft device shall be listed and labeled in accordance with UL 378, and shall be installed in accordance with the manufacturer's instructions.

2. A device shall be installed that produces visible and audible warning upon failure of the mechanical draft device or loss of electrical power, at any time that the mechanical draft device is turned on. This device shall be equipped with a battery backup if it receives power from the building wiring.

3. A smoke detector shall be installed in the room with the *appliance* or fireplace. This device shall be equipped with a battery backup if it receives power from the building wiring.

❖ This section contains a remedy for existing chimneys that do not produce sufficient draft. Some chimneys fail to produce sufficient draft intermittently because of wind speed and direction or outdoor temperatures.

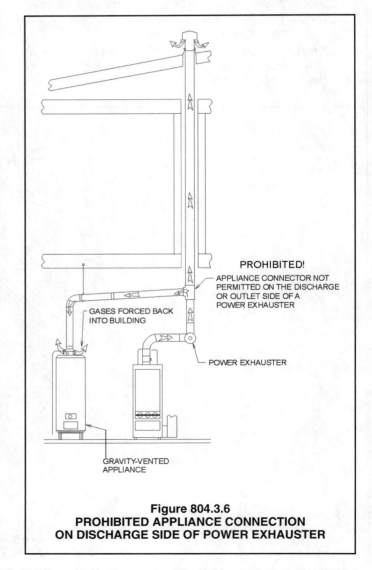

PROHIBITED!
APPLIANCE CONNECTOR NOT PERMITTED ON THE DISCHARGE OR OUTLET SIDE OF A POWER EXHAUSTER

GASES FORCED BACK INTO BUILDING

POWER EXHAUSTER

GRAVITY-VENTED APPLIANCE

Figure 804.3.6
PROHIBITED APPLIANCE CONNECTION
ON DISCHARGE SIDE OF POWER EXHAUSTER

Gas-fired and oil-fired appliances can be interlocked to the exhauster to immediately shut off the flow of fuel if there is a power failure or malfunction of the exhauster. Obviously, the same interlock cannot be used in a wood-burning fireplace or stove. This section allows the use of mechanical draft systems with solid fuel appliances and fireplaces if certain requirements are met.

The first requirement is that the draft device must be listed and labeled in accordance with UL 378 for this application and installed in compliance with the manufacturer's installation instructions. This will ensure that the draft system is installed in the same way it was tested in the laboratory of the listing agency.

The second requirement is that a visible and audible alarm be installed to warn occupants upon failure of the mechanical-draft device or loss of electrical power. If the exhauster fails to operate, the solid fuel will continue to burn and produce smoke and other products of combustion. The occupants of the building must be warned that the potentially deadly products of combustion might be spilling into the living space. This is especially important in residential occupancies where the occupants may have gone to sleep with the fireplace or appliance still burning.

The third requirement is that a smoke detector must be installed in the room with the appliance to provide further warning should the exhauster fail to operate properly. This section is not intended to allow mechanical-draft systems (chimney exhausters) as a substitute for proper chimney design and construc-

tion. It is intended to apply to existing chimneys that fail to produce the required draft.

SECTION 805
FACTORY-BUILT CHIMNEYS

805.1 Listing. Factory-built *chimneys* shall be *listed* and *labeled* and shall be installed and terminated in accordance with the manufacturer's installation instructions.

❖ Chimneys and vents are distinct systems. This section applies only to chimneys. Prefabricated chimney systems must bear the label of an approved agency. Commentary Figure 805.1(1) shows sample labels for factory-built chimney systems. A label is required on all components of the chimney system, such as the pipe sections, shields, fireblocks, fittings, termination caps and supports. The label states information such as the type of appliance the chimney was tested for use with, a reference to the manufacturer's installation instructions and the minimum required clearances to combustibles.

Most factory-built chimneys are either of the double-wall fiber-insulated design or the triple-wall air-cooled design. Factory-built chimneys are constructed of stainless steel inner liners with stainless or galvanized steel outer walls. Factory-built chimneys, like vent systems, are composed of components that must be installed as a complete system. Components from different manufacturers are not designed to be mixed and installed together [see Commentary Figure 805.1(2)].

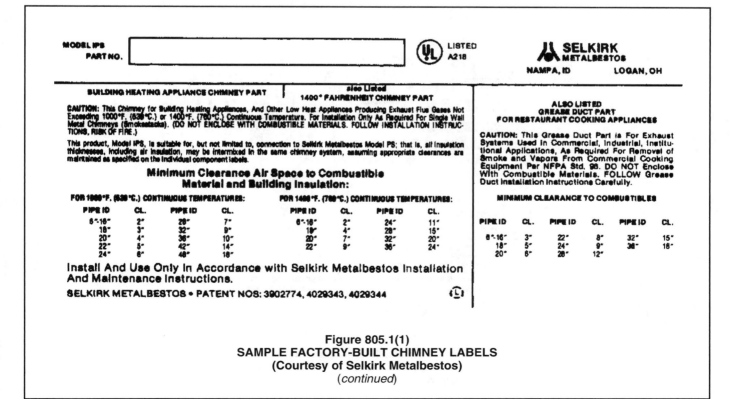

Figure 805.1(1)
SAMPLE FACTORY-BUILT CHIMNEY LABELS
(Courtesy of Selkirk Metalbestos)
(continued)

The manufacturer's instructions contain sizing criteria and the requirements for every aspect of a factory-built chimney installation. The requirements include component assembly, clearances to combustibles, support, terminations, connections, protection from damage and fireblocking.

Where a chimney for a residential or building heating appliance penetrates a floor and ceiling assembly, the IBC requires fireblocking around the opening in compliance with UL 103. The fireblock assembly is required to establish and maintain the required minimum clearances between the chimney and combustible construction.

Additionally, an insulation shield is required in attics to prevent batt or loose-fill insulation from contacting the chimney. The attic shield maintains the airspace clearance between the chimney and the insulation material. Some insulation materials are combustible and all insulation materials placed in contact with a chimney can cause excessive chimney temperatures by retarding radiant heat transmission from the chimney surfaces. The shield must extend a

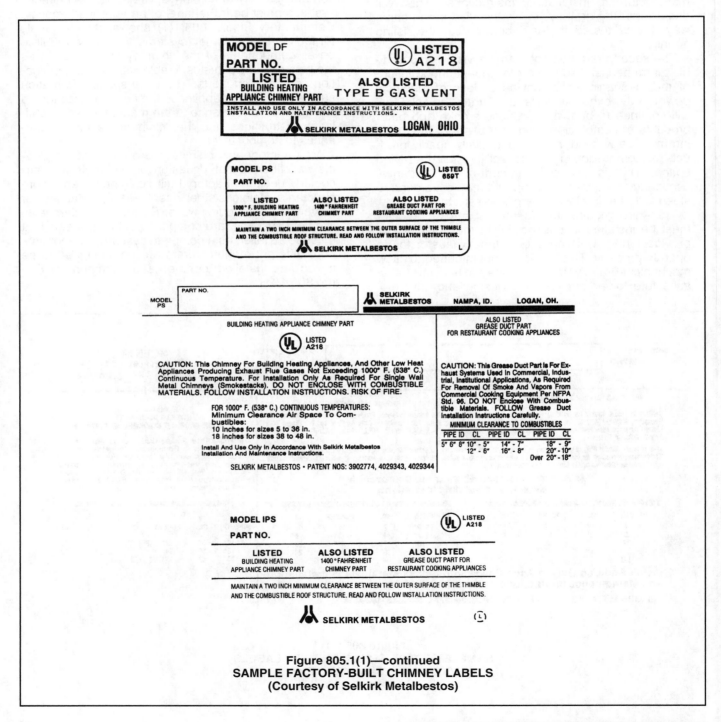

Figure 805.1(1)—continued
SAMPLE FACTORY-BUILT CHIMNEY LABELS
(Courtesy of Selkirk Metalbestos)

minimum distance above the attic floor to accommodate the maximum anticipated insulation depth.

The IBC requires fireblocking of chimneys serving factory-built fireplaces in accordance with UL 127. The assembly must provide fireblocking and establish and maintain clearances to combustibles determined as required by the manufacturer's installation instructions.

Factory-built chimneys must extend at least 3 feet (914 mm) above the roof measured from the highest

point of the roof penetration on pitched roofs. They must also be at least 2 feet (610 mm) higher than any portion of the roof within a 10-foot (3048 mm) horizontal distance [see Commentary Figure 805.1(3)].

The 3-foot (914 mm) termination height has long been recognized as the minimum height necessary to allow small burning embers and sparks emitted by the chimney to extinguish before landing on a combustible roof surface. The 2-foot (610 mm) termination requirement is intended to prevent wind and

Figure 805.1(2)
FACTORY-BUILT CHIMNEY COMPONENTS
(Courtesy of Selkirk Metalbestos)

bustible roof surface. The 2-foot (610 mm) termination requirement is intended to prevent wind and pressure zones from reducing the amount of draft produced by the chimney.

Wind and wind-induced eddy currents can react with building structural surfaces, thereby creating air pressure zones that can diminish chimney draft and cause reverse flow (backdraft) in places served by the chimney to discharge combustion products into the building interior. Locating the chimney outlet well into the undisturbed wind stream and away from the cavity and wake (eddy) zones around the building can counteract the adverse effects and also prevent the reentry of flue gases into the building through openings and fresh air intakes. Commentary Figure 805.1(4) shows a chimney improperly located in the eddy current area above the roof surface. Terminating a chimney in the eddy current area recirculates

the combustion products and allows them to enter the building via infiltration, wall openings and air intakes. A chimney terminal properly located above the eddy current area allows the wind to carry the combustion products away from the building.

Factory-built chimneys are equipped with termination caps that are part of the labeled chimney system and that must be installed. Like vent systems, the chimney cap is designed to keep out precipitation and animals and minimize the negative effects of wind.

805.2 Solid fuel appliances. Factory-built *chimneys* installed in *dwelling units* with solid fuel-burning appliances shall comply with the Type HT requirements of UL 103 and shall be marked "Type HT" and "Residential Type and Building Heating *Appliance Chimney.*"

Exception: *Chimneys* for use with open *combustion* chamber fireplaces shall comply with the requirements of

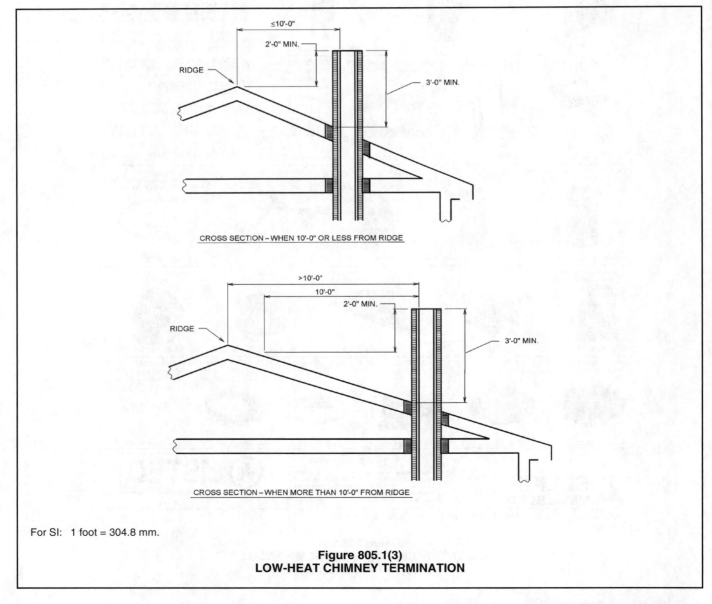

CROSS SECTION – WHEN 10'-0" OR LESS FROM RIDGE

CROSS SECTION – WHEN MORE THAN 10'-0" FROM RIDGE

For SI: 1 foot = 304.8 mm.

Figure 805.1(3)
LOW-HEAT CHIMNEY TERMINATION

UL 103 and shall be marked "Residential Type and Building Heating *Appliance Chimney.*"

Chimneys for use with open *combustion* chamber appliances installed in buildings other than *dwelling units* shall comply with the requirements of UL 103 and shall be marked "Building Heating *Appliance Chimney*" or "Residential Type and Building Heating *Appliance Chimney.*"

❖ Factory-built chimneys designed to vent solid fuel-burning residential or building heating appliances having continuous flue-gas outlet temperatures not exceeding 1,000°F (538°C) must be tested in accordance with UL 103.

In some cases, chimneys may serve appliances that are designed for use with vents but the reverse is not true. Vents can never be used to vent appliances that require connection to a chimney.

When a wood- or solid fuel-burning appliance is in operation, it produces tar and organic compounds that combine with moisture to form creosote. The creosote vapors can condense and form a deposit in the relatively cool chimney of a slow-burning fire. When a very hot fire is burning in the appliance, accumulated creosote can ignite, causing a chimney fire. A creosote fire burns very hot [in excess of 1,700°F (927°C)], which can cause accelerated deterioration of the chimney. The chimney should be inspected periodically during the heating season to determine whether creosote has accumulated. If there is a buildup of creosote, the chimney should be cleaned.

The requirements found in this section address specific solid fuel-burning appliances served by factory-built chimneys. Closed-combustion chamber wood-burning appliances for residential or building heating applications, such as wood stoves and room heaters, are capable of producing very hot fires by precisely controlling and directing the flow of combustion air into the combustion chamber. These appliances are equipped with tight-fitting doors that are closed when the appliance is operating. Adjustable shutters admit combustion air.

Because closed-combustion appliances are designed for precise control of combustion air, they can be adjusted for long, slow burning to make the fuel last longer. The disadvantage of a slow-burning, less-intense fire is that creosote formation will be greatly accelerated as a result of incomplete combustion of the fuel and lower flue-gas temperatures. Creosote fires can result in chimney buckling, distortion and seam splitting, which would make the chimney unsafe for continued use. The increased likelihood of creosote fires is another reason that high-temperature-rated chimneys are required for closed-combustion appliances.

The test described in UL 103 added an optional temperature test consisting of three 10-minute cycles at 2,030°F (1128°C) above room temperature that allows the manufacturer to designate the chimney as Type HT. Factory-built chimneys used with closed-combustion wood-burning appliances must comply with the Type HT requirements of UL 103 and be marked "Type HT" and "Residential Type and Building Heating Appliance Chimney."

The Type HT requirement does not pertain to a chimney used with a factory-built fireplace because the fireplace has an open combustion chamber. A chimney serving a factory-built fireplace has to be tested in accordance with UL 127. This standard con-

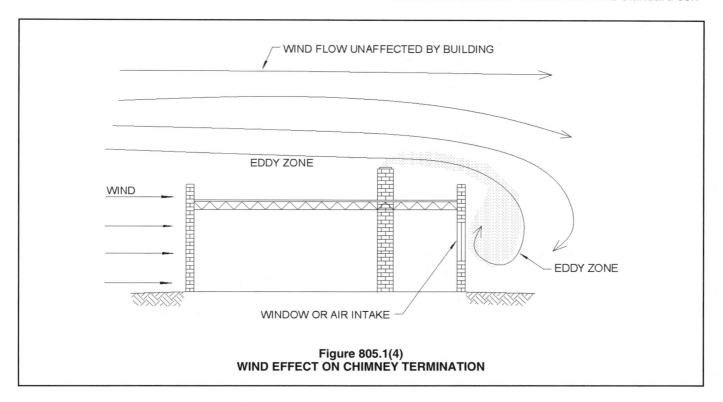

Figure 805.1(4)
WIND EFFECT ON CHIMNEY TERMINATION

tains the requirements for all factory-built fireplaces and components, including the chimney. Factory-built fireplaces must be served by the chimney supplied or required by the fireplace manufacturer.

Additionally, the Type HT requirement does not apply to a free-standing fireplace stove with an open combustion chamber and conforming to UL 737.

The two exceptions provide marking requirements for open-combustion chamber fireplace stoves and fireplaces that operate without tight-fitting doors. This allows much greater quantities of combustion air and dilution air to be drawn into the combustion chamber and into the chimney. Because of the excess air and the fact that the combustion air is not directed into the fuel bed, open chamber appliances and fireplaces do not produce fires as hot as those produced by the closed appliance and, therefore, do not require the "Type HT" marking. This is one reason why a Type HT (high-temperature rated) chimney is required for closed, solid-fuel appliances.

805.3 Factory-built chimney offsets. Where a factory-built chimney assembly incorporates offsets, no part of the chimney shall be at an angle of more than 30 degrees (0.52 rad) from vertical at any point in the assembly and the chimney assembly shall not include more than four elbows.

❖ These requirements are part of the requirements of UL 103 as referenced in Section 805.2. Limiting the angle of elbows and the number of elbows helps maintain draft in the chimney and allows for ease of cleaning (see Commentary Figure 805.3). Also, angles greater than 30 degrees are more difficult to support as the center of gravity shifts.

805.4 Support. Where factory-built *chimneys* are supported by structural members, such as joists and rafters, such members shall be designed to support the additional load.

❖ Factory-built chimneys are supported by the building structure and can impose considerable weight on structural components. Joists, rafters and other structural members must be designed to support the additional loading. Structural evaluation may be necessary, especially where framing for a chimney requires cutting and heading of joists and rafters.

805.5 Medium-heat appliances. Factory-built *chimneys* for medium-heat appliances producing flue gases having a temperature above 1,000°F (538°C), measured at the entrance to the *chimney*, shall comply with UL 959.

❖ Factory-built chimneys serving medium-heat appliances in which the maximum continuous flue-gas temperature does not exceed 2,000°F (1093°C) must be tested in accordance with UL 959. Such chimneys are associated with factory and industrial equipment and incinerators.

Chimneys serving medium-heat appliances may require greater termination heights and clearances to combustibles.

805.6 Decorative shrouds. Decorative shrouds shall not be installed at the termination of factory-built *chimneys* except where such shrouds are *listed* and *labeled* for use with the specific factory-built *chimney* system and are installed in accordance with Section 304.1.

❖ Decorative shrouds (see Commentary Figure 805.6) have become a popular architectural feature that, in some cases, have proven to be a fire hazard. They are designed to be aesthetically pleasing and to hide chimney and vent terminations. These shrouds have caused fires resulting from overheated combustible construction and the accumulation of debris and animal nesting. Shrouds can also interfere with the functioning of a chimney or vent system. Shrouds are allowed only where they are listed and labeled for use with the specific factory-built chimney system. Unlisted, field-constructed shrouds are prohibited.

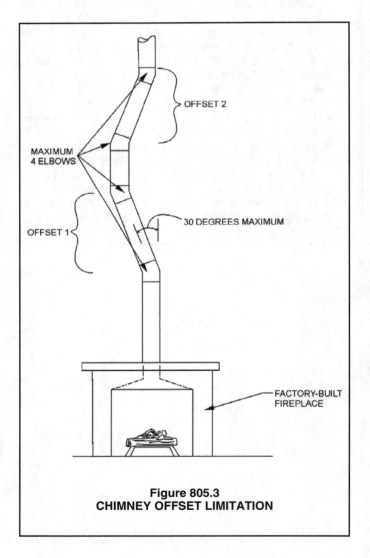

Figure 805.3
CHIMNEY OFFSET LIMITATION

SECTION 806
METAL CHIMNEYS

806.1 General. Metal *chimneys* shall be constructed and installed in accordance with NFPA 211.

❖ Metal chimneys, or smokestacks, should not be confused with vents and factory-built chimneys, even though they are constructed of metal. Metal chimneys are field constructed and are not listed products. They are found almost exclusively in industrial buildings.

Installation and construction requirements for metal chimneys are covered in NFPA 211.

Bibliography

The following resource materials were used in the preparation of the commentary for this chapter of the code:

IBC-12, *International Building Code.* Washington, DC: International Code Council, 2011.

Figure 805.6
DECORATIVE SHROUDS

Chapter 9:
Specific Appliances, Fireplaces and Solid Fuel-Burning Equipment

General Comments

Chapter 9 regulates the design and construction of fireplaces, solid fuel-burning appliances, barbecue appliances and the host of specifically named appliances. It reflects the code's intent to specifically address all of the types of appliances that the code intends to regulate (see the definition of "Appliance" and Section 101.2). Incinerators and crematories are considered to be solid fuel-burning appliances and are also addressed. Other regulations affecting the installation of solid fuel-burning fireplaces, appliances and accessory appliances are found in Chapters 3, 6, 7, 8, 10, 11, 12, 13 and 14.

Solid fuel-burning fireplaces and appliances are the oldest form of space heating still in use today. In earlier times, the fireplace was the central focus of residential living as a means for cooking meals and as a main or sole source of heat and light.

Today, in this country, the fireplace is used principally as a decorative feature of residential construction, although energy-efficient fireplaces and solid fuel-burning stoves and heaters are often used to supplement the primary heating system in dwelling units. It is also not uncommon in some parts of the world to rely on solid fuel as the sole heating fuel for dwelling units. Before the widespread use of oil, gas and electricity for home heating, coal-fired furnaces, boilers, stoves and room heaters were common.

For application of this chapter, the term "fireplace" refers to solid fuel-burning equipment and does not refer to gas-fired appliances that are marketed as gas fireplaces. Solid fuels include wood, coal and manufactured (formed) wood composite products, such as wood pellets and artificial logs.

Solid fuel-burning appliances include wood and coal stoves, room heaters, fireplace inserts, wood and coal furnaces, wood and coal boilers and dual-fuel furnaces and boilers that are equipped to burn a solid fuel and a gaseous or liquid fuel. Dual-fuel appliances are not designed to burn more than one type of fuel at a time and must be changed over to operate with the chosen fuel type.

The chapters of the code, with the exception of Chapter 3 and this chapter, are dedicated to single subjects. This chapter, however, is a collection of requirements for various appliances, equipment, solid fuel-burning equipment and fireplaces. The only commonality in this chapter is that all subjects use energy to perform some task or function. Because none of the subjects in this chapter have the volume of text to warrant an individual chapter, they have been combined into a single chapter.

Purpose

This chapter sets minimum construction and performance criteria for fireplaces, appliances and equipment and provides for the safe installation of these items.

SECTION 901
GENERAL

901.1 Scope. This chapter shall govern the approval, design, installation, construction, maintenance, *alteration* and repair of the appliances and *equipment* specifically identified herein and factory-built fireplaces. The approval, design, installation, construction, maintenance, *alteration* and repair of gas-fired appliances shall be regulated by the *International Fuel Gas Code*.

❖ This chapter regulates all aspects of appliances, equipment and fireplaces to the extent found in each of Sections 902 through 925. It also directs the user to the *International Fuel Gas Code*® (IFGC®) for the regulation of gas-fired appliances and equipment. This chapter covers solid fuel-burning, liquid fuel-burning and electrically heated appliances.

901.2 General. The requirements of this chapter shall apply to the mechanical *equipment* and appliances regulated by this chapter, in addition to the other requirements of this code.

❖ This section states that other chapters of the code contain requirements applicable to the subjects of this chapter. The other chapters include Chapters 3, 5, 6, 7, 8, 10, 11, 12, 13 and 14.

901.3 Hazardous locations. Fireplaces and solid fuel-burning appliances shall not be installed in hazardous locations.

❖ This section prohibits the installation of fireplaces and solid fuel-burning appliances in hazardous locations. A hazardous location is defined as any location considered to be a fire hazard because of the presence of flammable vapors, dust, combustible fibers or other highly combustible substances. Repair garages, painting shops and similar occupancies that house special processes and operations that involve flammable or combustible liquids, vapors or dusts are considered hazardous locations and are subject to the provisions of this section.

Solid fuel-burning appliances and fireplaces must not be installed in hazardous locations because the combustion process in this kind of equipment is hard to control and contain. Both gas- and oil-burning appliances are equipped with control devices that can shut off the supply of fuel to the appliance and stop combustion almost immediately. With solid fuel-burning equipment, however, the combustion process can be stopped only by eliminating the combustion air supply, and combustion would stop slowly. Should flammable or combustible vapors be detected in a room or space, there would be no quick means to eliminate the potential of the equipment/appliance being a source of ignition.

Additionally, the openings of the combustion chamber and the nature of solid fuels make it difficult to contain the burning fuel and hot ashes or cinders. Even if the combustion chamber is properly elevated above the floor, sparks, embers or burning fuel can fall from an open combustion chamber (firebox) or from an enclosed combustion chamber during lighting, refueling and ash removal. Also, the strong draft created by solid fuel appliances and fireplaces can draw flammable vapors into the combustion chamber.

For these reasons, it is obvious that fireplaces and solid fuel-burning appliances can create an unsafe condition when installed in hazardous locations.

901.4 Fireplace accessories. Listed and labeled fireplace accessories shall be installed in accordance with the conditions of the listing and the manufacturer's instructions. Fireplace accessories shall comply with UL 907.

❖ This section addresses fireplace doors, screens, heat exchangers, blower units and similar items not classified as appliances. Such accessories must be listed and labeled in accordance with UL 907. Installation must be performed in accordance with the listing and the manufacturer's instructions as improper installation could result in abnormal operation of the fireplace and a resultant fire hazard.

SECTION 902
MASONRY FIREPLACES

902.1 General. Masonry fireplaces shall be constructed in accordance with the *International Building Code*.

❖ Masonry construction is regulated by the *International Building Code*® (IBC®) and typically inspected by the building official; therefore, the code refers to the IBC for the construction of masonry fireplaces. The IBC contains construction and chimney sizing requirements for masonry fireplaces. Factory-built fireplaces are regulated in Section 903.

SECTION 903
FACTORY-BUILT FIREPLACES

903.1 General. Factory-built fireplaces shall be *listed* and *labeled* and shall be installed in accordance with the conditions of the listing. Factory-built fireplaces shall be tested in accordance with UL 127.

❖ This section addresses factory-constructed fireplace units that are served by factory-constructed chimneys. Factory-built fireplaces are solid fuel-burning units having a fire chamber that is intended to be either open to the room or, if equipped with doors, operated with the doors either open or closed. Commentary Figures 903.1(1) and 903.1(2) show two common styles of factory-built fireplaces: the recessed wall fireplace and the free-standing fireplace stove. Note that fireplaces are not referred to as appliances. The term "fireplace" describes a complete assembly that includes the hearth, a fire chamber and a chimney. A factory-built fireplace is composed of factory-built components representative of the prototypes tested, and is installed in accordance with the manufacturer's instructions to form the completed fireplace.

Factory-built fireplaces have an integral chimney as part of the total system package. Interchanging or substituting different types of chimneys is not permitted and could create an unsafe installation. Therefore, factory-built fireplaces must be installed with compatible components as identified and supplied by the fireplace manufacturer in strict accordance with its installation instructions.

Factory-built chimneys, designed for use with factory-built fireplaces, are tested and evaluated under

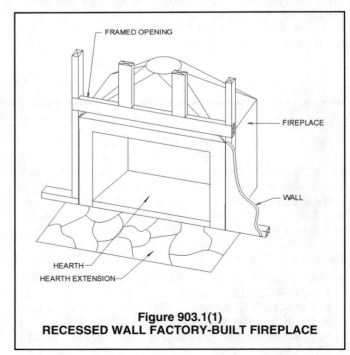

Figure 903.1(1)
RECESSED WALL FACTORY-BUILT FIREPLACE

the same standard as fireplaces, UL 127, and are not addressed in Chapter 8. A chimney intended for a specific factory-built fireplace is integral with the fireplace and is not considered as a separate or distinct system. Chimneys for factory-built fireplaces are either air insulated, air cooled or mineral-fiber insulated, and have either two or three walls. The inner flue wall is stainless steel; the outer wall(s) may be stainless or galvanized steel. The manufacturer's installation instructions contain detailed requirements for the chimney installation, including maximum offset angles, firestopping, support, clearances, section assembly and termination (see Section 805.6).

Factory-built fireplaces are constructed of corrosion-resistant sheet metal and refractory materials designed to withstand the high temperatures and corrosive effects of combustion byproducts. Some fireplaces may be equipped with air-circulating components that pass room air through a heat exchanger that partially surrounds the firebox. Fireplace stoves are free-standing assemblies having fire chambers intended to be open to the room or, if equipped with doors, to be operated with the doors either open or closed. Fireplace stoves differ from room heaters in that fireplace stoves need not be operated with the fire chamber (doors) closed.

Both UL 127 and UL 737 address wood-burning and coal-burning units. Both standards address construction, durability, containment of fuel, corrosion resistance, mechanical strength, heat conductance/transfer, thermal radiation, clearances to combustibles, protection of electrical components (blowers, fans and controls), electrical grounding/bonding, current leakage, voltage-withstand (dielectric), control function, operator protection and other safety aspects. The standards also contain requirements for the installation instructions and for marking and identifying the units. All units are tested under abnormal firing conditions and temperature rise is monitored at several locations on and around the unit and test structure.

For fireplaces tested in accordance with UL 127, all of the components necessary for the installation of the fireplace and chimney must be supplied by the manufacturer.

In addition to testing requirements, fireplaces and fireplace stoves must be labeled by an approved agency. Sections 301.7 and 301.8 state the requirements for approved agencies and provide details for the information that must appear on a label. Factory-built fireplaces must be installed in accordance with the manufacturer's installation instructions. This is important because the manufacturer's installation instructions are used by the testing agency when the test installation is constructed. Therefore, a unit with its compatible components, that is designed, constructed and tested in accordance with the applicable standard and installed in accordance with the manufacturer's instructions, is expected to perform the same as the representative sample unit tested.

A factory-built fireplace is a complete package that includes its chimney and all of the related parts necessary for a complete installation. Fireplace stoves are not supplied with a chimney, factory-built chimney or a masonry chimney as specified by the manufacturer. Section 805.2 requires use of fireplace stoves with low-heat chimneys that are designed, constructed and tested in accordance with UL 103 (see commentary, Section 805).

The term "zero clearance," when describing factory-built fireplace units, refers to the fact that zero clearance is required between the bottom surface of the unit and a combustible floor. Zero clearance also refers to the clearance to combustibles from shields, standoff brackets, spacers and other factory-supplied components that have been designed and tested for contact with combustibles. In other words, a so-called "zero clearance" factory-built fireplace unit actually has a required clearance distance to combustibles from any of its surfaces and components except the bottom, and the term "zero clearance" simply refers to the allowed clearance from factory-applied spacers and brackets that maintain a fixed distance between heat-producing surfaces and any combustible building material.

Many surfaces of factory-built fireplace units are marked or labeled with minimum clearance to combustible dimensions that must be observed. Note that gypsum wallboard is considered to be a combustible material in the code. Gypsum wallboard, often referred to as "drywall" or "sheetrock," will degrade and increase in combustibility when exposed to continuous heat from appliances and fireplaces.

All clearances to combustibles, as specified in the unit and its label, refer to airspace clearances. An air-

Figure 903.1(2)
FREE-STANDING FACTORY-BUILT FIREPLACE

space clearance cannot be filled with insulation materials or any other material regardless of its combustibility or noncombustibility. An airspace clearance relies on the high insulating value of free air coupled with convection currents that serve to control the temperature rise on adjacent material surfaces.

Factory-built fireplaces come with detailed installation instructions that must be followed to achieve a code-complying installation. Some commonly overlooked requirements in the manufacturer's installation instructions include firestop installation; shield and joint treatment between the hearth and hearth extension; hearth elevation requirements and limitations; combustion air supplies; ductwork installations when so equipped; hearth extension materials; chimney supports and construction and framing member clearances.

Factory-built fireplaces cannot be fitted or equipped with fireplace inserts or otherwise converted or altered, except where specifically permitted by the fireplace manufacturer (based on test results) and where installed in accordance with the manufacturer's instructions (see Sections 903.3 and 805.6).

903.2 Hearth extensions. Hearth extensions of approved factory-built fireplaces shall be installed in accordance with the listing of the fireplace. The hearth extension shall be readily distinguishable from the surrounding floor area. Listed and labeled hearth extensions shall comply with UL 1618.

❖ As discussed in Section 903.1, factory-built fireplaces must be installed in strict accordance with the manufacturer's installation instructions. Those instructions will include specifications for the materials, dimensions and installation of hearth extensions. As shown in Figure 903.1(1), the fireplace hearth consists of two parts. The hearth, commonly called the inner hearth, is the floor of the combustion chamber (firebox) and is constructed of masonry or concrete and masonry. The outer hearth, referred to as the hearth extension, projects beyond the face of the fireplace opening into the room and is constructed of masonry, concrete or factory-specified elements.

Hearth extensions are commonly decorated with cut stone, ceramic tile and other similar noncombustible, aesthetically pleasing materials.

The hearth extension protects the floor structure in front of the fireplace from radiant heat from the fire and from sparks, embers, coals or burning solid fuel that may escape from the firebox. The nature of solid fuel can cause burning material to be expelled from the combustion chamber and large pieces of the fuel can fall from a grate and roll out of the firebox. In addition to the required hearth extensions, screens should be used as the initial and primary protection against sparks and embers from the fire.

The hearth extension is typically at the same elevation as the hearth and is simply constructed as a horizontal extension of the firebox floor. The code does not prohibit a hearth extension from being at a lower elevation than the hearth; however, this configuration

could cause burning fuel, embers or coals to bounce or travel beyond the hearth extension because of the energy gained from the vertical drop. The required minimum dimensions of the hearth extension are intended to protect the floor structure where burning fuel, embers or coals would most likely fall and where the radiant heat energy could be great enough to risk ignition of the floor structure.

Factory-built fireplaces must have hearth extensions as prescribed in the manufacturer's installation instructions. If the hearth extension is distinguishable from the surrounding floor surface, the hearth extension will likely be recognized for its importance, will be readily observable by the occupants and will probably not be covered by rugs, carpeting or other floor-covering materials. Note that the typical fireplace instructions will require some insulating value or minimum thermal mass for hearth extension assemblies. Although commonly used, the fireplace instructions generally do not allow ceramic tile or stone slabs to be placed directly on wood supports as a form of hearth extension construction.

Factory-built hearth extensions are available and have to be listed and labeled and installed in accordance with the manufacturer's instructions.

903.3 Unvented gas log heaters. An unvented gas log heater shall not be installed in a factory-built fireplace unless the fireplace system has been specifically tested, *listed* and *labeled* for such use in accordance with UL 127.

❖ Unvented gas-log heaters are not gas logs in the traditional sense. They are considered to be room heaters. The key difference between a solid fuel fire and use of an unvented gas log unit is that the fireplace chimney damper is typically closed during operation of the unvented gas-log heater. Because there is no draft, dilution air movement or combustion gas venting, the firebox can reach higher temperatures than with a solid-fuel fire.

Not all factory-built fireplaces can be converted for use with unvented gas logs. Wood-burning factory-built fireplaces must be tested and listed in accordance with the requirements of UL 127, which requires the fireplace system to be fired in an over-fired condition during testing until the temperatures of the combustible framing surrounding the firebox have reached a stable maximum. Unvented gas-log heaters are tested and listed in accordance with the requirements of ANSI Z21.11.2, which allows a less severe test in which the heater is cycled on and off on a set schedule that results in lower temperatures in the surrounding combustible framing. However, tests have shown that unvented gas-log heaters that are not cycled on and off and are operated with the fireplace chimney damper closed can cause higher temperatures in the surrounding combustible materials than the limits allowed by UL 127.

The latest edition of UL 127 contains provisions for testing factory-built fireplaces for use with unvented gas-log heaters that take into account the higher operating temperatures in the firebox. Unvented gas-log heaters must not be installed in either existing or new factory-built fireplaces without proof that the fireplace system has been tested as required by the latest edition of UL 127. The fireplace manufacturer must state in the installation instructions whether or not the unit is listed for use with an unvented gas-log heater. Traditionally, fireplace manufacturers have referred only to vented gas logs tested to ANSI Z21.60. If the fireplace installation instructions do not state that the fireplace has been tested for use with unvented gas logs, the unvented gas logs must not be installed in that fireplace.

SECTION 904
PELLET FUEL-BURNING APPLIANCES

904.1 General. Pellet fuel-burning appliances shall be *listed* and *labeled* in accordance with ASTM E 1509 and shall be installed in accordance with the terms of the listing.

❖ Pellet fuel-burning appliances, as defined in Chapter 2, use relatively new technology that allows the use of waste products that have been converted to a usable fuel. Although considered solid fuel-burning appliances, pellet appliances connect to vents instead of chimneys (see commentary, Section 802.2).

SECTION 905
FIREPLACE STOVES AND ROOM HEATERS

905.1 General. Fireplace stoves and solid-fuel-type room heaters shall be *listed* and *labeled* and shall be installed in accordance with the conditions of the listing. Fireplace stoves shall be tested in accordance with UL 737. Solid-fuel-type room heaters shall be tested in accordance with UL 1482. Fireplace inserts intended for installation in fireplaces shall be *listed* and *labeled* in accordance with the requirements of UL 1482 and shall be installed in accordance with the manufacturer's installation instructions.

❖ This section requires testing of solid-fuel-burning room heaters in accordance with UL 1482, which addresses solid fuel-burning, free-standing fire chamber assemblies that heat space by direct radiation, circulated heated air or a combination of both. Solid-fuel room heaters are chimney connected, and are designed for operation with the fire chamber (firebox) closed.

Commentary Figure 905.1(1) shows a typical room heater. The standard addresses the construction and performance of room heaters from the standpoint of fire and life safety, and includes the performance of electrical components for units equipped with fans, blowers and controls. The standard also has requirements for marking and identifying the units, and requirements for the content of installation and operating instructions. The testing procedures for wood-burning units differ from those for coal-burning units. Therefore, a label (plate) must be attached to the unit that indicates the permitted type of fuel.

All units are installed in a test structure that simulates an actual room installation in accordance with the manufacturer's installation instructions. The units undergo several fire tests intended to produce extreme operating conditions. Temperature-rise measurements are taken at multiple locations throughout the test installation to determine whether the clearances to combustibles are sufficient to prevent the temperature rise of combustibles from exceeding the maximum specified in the standard. Commentary Figure 905.1(2) shows the typical specifications included on appliance labels. In addition to meeting testing requirements, room heaters must be labeled by an approved agency.

Mechanical appliances and equipment must be installed in accordance with the manufacturer's installation instructions. This is important because the manufacturer's installation instructions are used by the testing agency when the test installation is constructed. Therefore, a unit tested in compliance with UL 1482 and installed in accordance with the manufacturer's installation instructions is expected to perform the same as the representative sample unit that was tested.

In addition to being the standard for testing room heaters, UL 1482 sets guidelines for the testing of fireplace inserts. Fireplace inserts are basically wood stoves that fit into a fireplace opening and use the fireplace chimney, converting the fireplace into a room heater (see commentary, Section 801.10). Fireplace inserts must not be installed in a factory-built fireplace unless the fireplace listing specifically allows

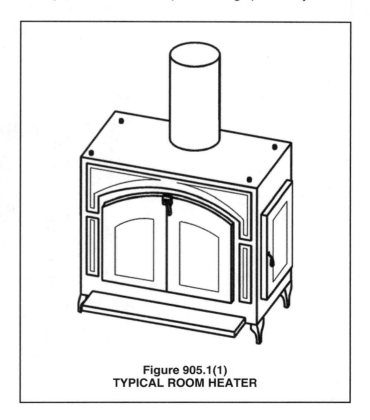

Figure 905.1(1)
TYPICAL ROOM HEATER

Figure 905.1(2)
SAMPLE APPLIANCE LABEL
(Courtesy of Vermont Casting, Inc.)

such an installation (see commentary, Section 903). Fireplace inserts must also be listed and labeled in accordance with the requirements of UL 1482 and installed in accordance with the manufacturer's installation instructions.

UL 127 includes an optional test procedure for fireplace units that can accommodate the installation of unvented room heaters (unvented gas logs). Fireplace units may or may not be tested for use with the chimney damper closed, such as is the case where an unvented room heater is installed in the firebox. Section 903.3 contains text that requires a fireplace in which an unvented gas log is installed to be listed and labeled for that use (see commentary, Section 903.3).

905.2 Connection to fireplace. The connection of solid fuel appliances to *chimney* flues serving fireplaces shall comply with Sections 801.7 and 801.10.

❖ Sections 801.7 and 801.10 contain chimney sizing limitations and installation requirements for appliances connected to fireplace chimneys (see commentary, Sections 801.7 and 801.10).

905.3 Hearth extensions. Hearth extensions for fireplace stoves shall be installed in accordance with the listing of the fireplace stove. The hearth extension shall be readily distinguishable from the surrounding floor area. Listed and labeled hearth extensions shall comply with UL 1618.

❖ The listing of fireplace stoves requires a hearth extension to be installed. The hearth extension protects the floor structure in front of the fireplace stove from radiant heat from the fire and from sparks, embers, coals or burning solid fuel that may escape from the firebox. The nature of solid fuel can cause burning material to be expelled from the combustion chamber and large pieces of the fuel can fall from a grate and roll out of the firebox.

Factory-built hearth extensions are available and have to be listed and labeled and installed in accordance with the manufacturer's instructions.

SECTION 906
FACTORY-BUILT BARBECUE APPLIANCES

906.1 General. Factory-built barbecue appliances shall be of an *approved* type and shall be installed in accordance with the manufacturer's installation instructions, this chapter and Chapters 3, 5, 7, 8 and the *International Fuel Gas Code*.

❖ This section addresses factory-constructed appliances designed to burn wood or charcoal with or without an auxiliary fuel. Some appliances use solid fuel only and some use a supplemental fuel such as gas to create heat for cooking and/or to start the solid-fuel fire. It is becoming more common to find factory-built solid fuel-fired appliances and equipment

in restaurants. Such equipment would require a Type I hood and fire suppression system in accordance with Chapter 5 and could also involve a chimney or an auxiliary fuel supply.

SECTION 907
INCINERATORS AND CREMATORIES

907.1 General. Incinerators and crematories shall be *listed* and *labeled* in accordance with UL 791 and shall be installed in accordance with the manufacturer's installation instructions.

❖ This section addresses solid fuel-burning appliances and equipment designed to consume waste and organic tissue. Incinerators used today are installed in hospitals, universities and other relatively large institutions that must dispose of medical wastes, toxic and contaminated substances, pathological waste and human and animal tissues. To achieve a more complete incineration and thereby reduce the amount of pollutants released into the atmosphere, multiple-chamber incinerators are used rather than single-chamber incinerators. There are two basic designs of multiple-chamber incinerators: the in-line type and the retort type. The in-line type is characterized by the flow of combustion gases always traveling straight through the incinerator, with 90-degree (1.57 rad) turns only in the vertical direction. The retort-type incinerator is characterized by a design that causes the combustion gases to flow through 90-degree (1.57 rad) turns in both lateral and vertical directions. Modern incinerator installations take advantage of the heat produced by incineration by passing the flue gases through boilers or other heat exchangers for space heating, cooling or processing operations. The reclamation of heat produced by incinerators can substantially reduce the energy costs of the facility served by the incinerator.

A crematory is a furnace that is used exclusively to reduce human and animal cadavers to ashes. This section requires factory-built incinerators and crematories to conform to the requirements of Chapter 3 for equipment approval and installation. The incinerator or crematory must be tested to the standard that governs its use, and evidence must be submitted indicating that the manufacturer has made provisions for a third-party inspection of the manufacturer's facility to determine that the units being produced are the same as those that were tested.

When tested in the laboratory, the equipment is installed in accordance with the manufacturer's installation instructions. For the equipment to perform as designed and tested, the field installation of the equipment must follow the same instructions. Therefore, the manufacturer's installation instructions become an integral part of the approval process.

SECTION 908
COOLING TOWERS, EVAPORATIVE CONDENSERS AND FLUID COOLERS

908.1 General. A cooling tower used in conjunction with an air-conditioning *appliance* shall be installed in accordance with the manufacturer's installation instructions. Factory-built cooling towers shall be listed in accordance with UL 1995.

❖ This section addresses outdoor heat exchangers that expose a heat transfer fluid (usually water) directly to the ambient air. Cooling towers are the heat rejection portion of a condenser circuit. They operate by creating a large surface area over which the condenser cooling water flows. Air is passed through the cooling tower, by gravity or fan power, and by heat transfer and some evaporative cooling the condenser water is cooled and recirculated back to the condenser.

Factory-built cooling towers must be listed in accordance with UL 1995.

908.2 Access. Cooling towers, evaporative condensers and fluid coolers shall be provided with ready access.

❖ Typically installed on roof tops and outdoors on grade-level pads, these heat exchangers require routine cleaning and maintenance and therefore must be easily accessed by personnel (see the definition of "Ready access").

908.3 Location. Cooling towers, evaporative condensers and fluid coolers shall be located to prevent the discharge vapor plumes from entering occupied spaces. Plume discharges shall be not less than 5 feet (1524 mm) above or 20 feet (6096 mm) away from any ventilation inlet to a building. Location on the property shall be as required for buildings in accordance with the *International Building Code*.

❖ Liquid-to-air heat exchangers often produce vapors that must be prevented from entering the building through windows, doors and air intake openings. The moisture concentration and possible presence of water treatment chemicals or disease-causing bacteria could create a health hazard for building occupants. The IBC regulates the location of these structures on a lot and the distance of the structures from lot lines and other structures.

908.4 Support and anchorage. Supports for cooling towers, evaporative condensers and fluid coolers shall be designed in accordance with the *International Building Code*. Seismic restraints shall be as required by the *International Building Code*.

❖ The IBC regulates the construction of supporting elements for cooling equipment. These heat exchangers, which can be quite large, are subject to wind and seismic forces and can impose significant dead and dynamic loads on roof structures where so installed.

908.5 Water supply. Water supplies and protection shall be as required by the *International Plumbing Code*.

❖ Liquid-to-air heat exchangers use makeup water to replace evaporation and blow-down losses. The

makeup water connection could be a source of contamination to the potable water supply and therefore must be installed in accordance with the *International Plumbing Code*® (IPC®). Water-conditioning chemicals are typically used to control the growth of algae and other organisms that would lessen efficiency and lead to system cleaning and flushing. Also, cooling towers and similar equipment produce an ideal environment for the growth of disease-causing organisms such as Legionella pneumophila (Legionnaires' disease).

908.6 Drainage. Drains, overflows and blowdown provisions shall be indirectly connected to an *approved* disposal location. Discharge of chemical waste shall be *approved* by the appropriate regulatory authority.

❖ Liquid-to-air heat exchangers have drains for excess, waste and flushing water that must discharge to an approved location, such as the sanitary sewer system. Because chemicals can be involved, drainage should not be allowed to discharge to grade or to a storm sewer.

908.7 Refrigerants and hazardous fluids. Heat exchange *equipment* that contains a refrigerant and that is part of a closed refrigeration system shall comply with Chapter 11. Heat exchange *equipment* containing heat transfer fluids which are flammable, combustible or hazardous shall comply with the *International Fire Code*.

❖ This section states that any refrigerant contained in a closed refrigerant system governed by Section 908 must meet the same requirements as refrigerants governed by Chapter 11 of this code. If the heat transfer fluids are flammable, combustible or hazardous, they must meet the requirements of various sections of the *International Fire Code*® (IFC®) for storage, handling and fire prevention.

SECTION 909
VENTED WALL FURNACES

909.1 General. Vented wall furnaces shall be installed in accordance with their listing and the manufacturer's installation instructions. Oil-fired furnaces shall be tested in accordance with UL 730.

❖ Vented wall furnaces are a type of room heater typically designed to be installed within a 2 x 4 stud cavity in frame construction. Wall furnaces are typically used in cottages, room additions and homes in mild climates. Some units are designed to serve a single room and others are designed as through-the-wall units to serve adjacent rooms. Wall furnaces are ductless; however, some units are listed for use with a surface-mounted supply outlet extension. Wall furnaces can be either gravity or forced-air type.

909.2 Location. Vented wall furnaces shall be located so as not to cause a fire hazard to walls, floors, combustible furnishings or doors. Vented wall furnaces installed between bathrooms and adjoining rooms shall not circulate air from bathrooms to other parts of the building.

❖ Wall furnaces, like all room heaters, can present a fire hazard if improperly located. The heat discharged or directly radiated from such units can ignite nearby wall or floor surfaces, furniture, trim items, window treatments and doors (see commentary, Section 909.3). A through-the-wall unit serving both a bathroom and an adjacent room must not recirculate air between the two spaces (see commentary, Section 303.3).

909.3 Door swing. Vented wall furnaces shall be located so that a door cannot swing within 12 inches (305 mm) of an air inlet or air outlet of such furnace measured at right angles to the opening. Doorstops or door closers shall not be installed to obtain this *clearance.*

❖ A combustible door that swings close to a wall furnace could be a fire hazard if at some position the door would be within 12 inches (305 mm) of an air inlet or outlet. A door could also interfere with airflow through the furnace, thereby causing appliance overheating. Because door closures and door stops are easily defeated, they must not be depended upon to secure the required clearance. If the door swing cannot comply with this section, the door would have to be removed or rehung to change its swing direction.

909.4 Ducts prohibited. Ducts shall not be attached to wall furnaces. Casing extension boots shall not be installed unless *listed* as part of the *appliance.*

❖ Most wall furnaces are not designed to force air through ducts, especially those gravity types that do not use fans. Attachment of ducts would add resistance to the flow of air through the furnace, thereby causing abnormally high temperature rise across the heat exchanger and abnormally high temperatures on furnace surfaces. Some wall furnaces are listed and designed for use with supply duct extensions intended for wall mounting and intended to improve heat distribution. These duct extensions are factory-built and supplied only by the furnace manufacturer.

909.5 Manual shutoff valve. A manual shutoff valve shall be installed ahead of all controls.

❖ The fuel line connection to every appliance must be equipped with an individual shutoff valve to permit maintenance, repair or replacement of the appliance or its components. The shutoff valve must be located in accordance with Section 409 of the IFGC.

909.6 Access. Vented wall furnaces shall be provided with access for cleaning of heating surfaces, removal of burners, replacement of sections, motors, controls, filters and other working parts, and for adjustments and lubrication of parts requiring such attention. Panels, grilles and access doors that must be removed for normal servicing operations shall not be attached to the building construction.

❖ This section contains the same provisions found in Section 306.1 (see commentary, Section 306.1).

Access panels and doors must be removable as intended by the manufacturer of the furnace.

SECTION 910
FLOOR FURNACES

910.1 General. Floor furnaces shall be installed in accordance with their listing and the manufacturer's installation instructions. Oil-fired furnaces shall be tested in accordance with UL 729.

❖ Floor furnaces are vented appliances that are installed in an opening in the floor. These units supply heat to the room by gravity convection and direct radiation and typically serve as the sole source of space heating. Such furnaces are common in cottages, small homes, seasonally occupied structures and rural homes. Because the floor grille can become very hot, extreme care must be exercised to prevent occupants, especially children, from contacting the grille by walking or falling on it. Also, care must be taken to avoid a fire hazard caused by placement of materials or furnishings on or near the furnace floor grille.

910.2 Placement. Floor furnaces shall not be installed in the floor of any aisle or passageway of any auditorium, public hall, place of assembly, or in any egress element from any such room or space.

With the exception of wall register models, a floor furnace shall not be placed closer than 6 inches (152 mm) to the nearest wall, and wall register models shall not be placed closer than 6 inches (152 mm) to a corner.

The furnace shall be placed such that a drapery or similar combustible object will not be nearer than 12 inches (305 mm) to any portion of the register of the furnace. Floor furnaces shall not be installed in concrete floor construction built on grade. The controlling thermostat for a floor furnace shall be located within the same room or space as the floor furnace or shall be located in an adjacent room or space that is permanently open to the room or space containing the floor furnace.

❖ Floor furnaces must not interfere with or impede egress. In an emergency egress situation, a floor furnace could be a tripping hazard and could collapse under the live load of many occupants. As stated in Section 909.2, the furnace location must not create a fire hazard by being too close to wall surfaces, trim items, furnishings and window treatments.

A floor furnace installed in a slab on grade would have to be in a pit, would be subject to flooding and corrosion and would be inaccessible for service and inspection. If the controlling thermostat does not sense the air temperature in the room in which the furnace is installed, dangerous overheating could result. A thermostat isolated from the source of heat it controls would not respond to the condition in the space served by the furnace.

910.3 Bracing. The floor around the furnace shall be braced and headed with a support framework design in accordance with the *International Building Code.*

❖ The framing around the floor opening must be capable of supporting the floor system, the anticipated floor loads and the weight of the furnace. The structural requirements of the IBC must be complied with.

910.4 Clearance. The lowest portion of the floor furnace shall have not less than a 6-inch (152 mm) clearance from the grade level; except where the lower 6-inch (152 mm) portion of the floor furnace is sealed by the manufacturer to prevent entrance of water, the minimum clearance shall be reduced to not less than 2 inches (51 mm). Where these clearances are not present, the ground below and to the sides shall be excavated to form a pit under the furnace so that the required clearance is provided beneath the lowest portion of the furnace. A 12-inch (305 mm) minimum clearance shall be provided on all sides except the control side, which shall have an 18-inch (457 mm) minimum clearance.

❖ This section specifies clearances between the ground and the furnace that apply to crawl space installations. The clearances allow access for service and inspection and help prevent corrosion of the furnace assembly.

SECTION 911
DUCT FURNACES

911.1 General. Duct furnaces shall be installed in accordance with the manufacturer's installation instructions. Electric duct furnaces shall comply with UL 1996.

❖ Duct furnaces are designed for "in-line" installation in an air duct served by an external blower. Such units are basically a heat exchanger and burner assembly in a cabinet that rely on an independent air handler.

SECTION 912
INFRARED RADIANT HEATERS

912.1 General. Electric infrared radiant heaters shall comply with UL 499.

❖ Infrared heaters function by creating a very hot surface area from which heat energy is directly radiated. Infrared radiant heaters are usually suspended from ceilings or roofs. Such heaters are typically used for "spot" heating in spaces that are otherwise unconditioned or that are not conditioned to human comfort levels. Radiant heaters have the advantage of being able to heat objects and personnel without having to heat the surrounding air. Supports must prevent the heaters from falling, losing the required clearance to combustibles, putting tension or pressure on electrical connections and dislocating so as to redirect the radiant output to where a fire hazard would result.

912.2 Support. Infrared radiant heaters shall be fixed in a position independent of fuel and electric supply lines. Hangers and brackets shall be noncombustible material.

❖ This section addresses radiant heaters including ceramic-element and steel tube-type designs.

Infrared heaters are produced in both vented and unvented types and function by creating a very hot surface area from which heat energy is directly radiated. Infrared radiant heaters are usually suspended from ceilings or roofs. Such heaters are typically used for "spot" heating in spaces that are otherwise unconditioned or that are not conditioned to human comfort levels. Radiant heaters have the advantage of being able to heat objects and personnel without having to heat the surrounding air. Supports must prevent the heaters from falling, losing the required clearance to combustibles, putting tension or pressure on electrical, fuel and vent connections and dislocating so as to redirect the radiant output to where a fire hazard would result.

912.3 Clearances. Heaters shall be installed with clearances from combustible material in accordance with the manufacturer's installation instructions.

❖ This section is redundant with Section 304.9 and serves to remind the installer and code official of the paramount importance of maintaining the clearance to combustibles for radiant heaters. Direct radiation is a very effective method of transferring heat energy. Improperly located combustible materials can be readily ignited.

SECTION 913
CLOTHES DRYERS

913.1 General. Clothes dryers shall be installed in accordance with the manufacturer's installation instructions. Electric residential clothes dryers shall be tested in accordance with UL 2158. Electric coin-operated clothes dryers shall be tested in accordance with UL 2158. Electric commercial clothes dryers shall be tested in accordance with UL 1240.

❖ This section addresses clothes dryer appliances, both Types 1 and 2 (see the definition of "Clothes dryer"). Clothes dryers must be exhausted in accordance with the manufacturer's installation instructions. Dryers are tested to the safety standard appropriate for the appliance. The manufacturer's installation instructions include the information needed to duplicate the installation configuration that was tested and found to meet the requirements of the safety standard. The manufacturer's installation instructions are evaluated by the agency responsible for testing, listing and labeling the appliance. Following these instructions will result in safe and satisfactory installation. Residential dryers are required to be tested in accordance with UL 2158.

913.2 Exhaust required. Clothes dryers shall be exhausted in accordance with Section 504.

❖ Clothes dryer exhaust systems must convey moisture directly to the exterior of the building (outdoors). Note that the code does not use the term "dryer vent." Clothes dryers are not "vented" (see IFGC, Section 501.8) as are other appliances, but are exhausted.

Clothes dryer exhaust systems cannot terminate in or discharge to an enclosed space, such as an attic or crawl space, regardless of whether or not the space is ventilated through openings to the outdoors. The high levels of moisture in the exhaust air can cause condensation to form on exposed surfaces or in insulation materials. Water vapor condensation can cause structural damage, deterioration of building materials and mold and fungus growth. Clothes dryer exhausts that discharge to enclosed spaces will also cause an accumulation of highly combustible lint and debris, creating a significant fire hazard. An improperly installed clothes dryer exhaust system cannot only reduce dryer efficiency and increase running time, but it can also cause a significant increase in exhaust temperature, causing the dryer to cycle on its high limit control, which is an unsafe operating condition (see commentary, Section 504).

913.3 Clearances. Clothes dryers shall be installed with *clearance* to combustibles in accordance with the manufacturer's instructions.

❖ This section contains the same provisions found in Sections 304.1 and 304.9 (see commentary, Sections 304.1 and 304.9).

SECTION 914
SAUNA HEATERS

914.1 Location and protection. Sauna heaters shall be located so as to minimize the possibility of accidental contact by a person in the room.

❖ Sauna heaters produce very high heat. The choice of locations must consider the fact that a heat-producing appliance will be exposed within a small room with unclothed occupants.

914.1.1 Guards. Sauna heaters shall be protected from accidental contact by an *approved* guard or barrier of material having a low coefficient of thermal conductivity. The guard shall not substantially affect the transfer of heat from the heater to the room.

❖ Guards are required to protect the occupants from being burned. The guards must be constructed of a material that is a poor conductor of heat, such as wood, so that the guard itself will not present a burn hazard.

914.2 Installation. Sauna heaters shall be *listed* and *labeled* in accordance with UL 875 and shall be installed in accor-

dance with their listing and the manufacturer's installation instructions.

❖ This section references UL 875 as the standard for listing and labeling of sauna heaters. The scope of UL 875 covers electric dry-bath heating equipment and other equipment rated at 600 volts or less, intended to be installed in accordance with NFPA 70. The relative humidity in the heated environment is in the range of 10-25 percent, and the purpose of the heated environment is to promote perspiration in a short time by means of a relatively warm and dry atmosphere. Sauna heaters must have an automatic temperature-regulating control that is either integral with the heater or wall-mounted along with an integral manual-reset limit control and a timer. Because the listing and labeling requires that the sauna heater be installed in the same conditions under which the heater was tested, the manufacturer's installation instructions must be followed when the heater is installed at the job site.

914.3 Access. Panels, grilles and access doors that are required to be removed for normal servicing operations shall not be attached to the building.

❖ Access panels, covers and doors must not be made unusable by trim, woodwork or room enclosures that prevent or interfere with access (see commentary, Sections 306.1 and 909.6).

914.4 Heat and time controls. Sauna heaters shall be equipped with a thermostat that will limit room temperature to 194°F (90°C). If the thermostat is not an integral part of the sauna heater, the heat-sensing element shall be located within 6 inches (152 mm) of the ceiling. If the heat-sensing element is a capillary tube and bulb, the assembly shall be attached to the wall or other support, and shall be protected against physical damage.

❖ A thermostat is required to limit the temperature in the sauna for fire safety. The control must sense the warmest air near the ceiling and must be protected from physical damage.

914.4.1 Timers. A timer, if provided to control main burner operation, shall have a maximum operating time of 1 hour. The control for the timer shall be located outside the sauna room.

❖ To protect both the occupants and the building, timers must limit the heater operating time to 1 hour. Resetting the timer would require the occupant to exit the sauna, thus lessening the chances of overexposure.

914.5 Sauna room. A ventilation opening into the sauna room shall be provided. The opening shall be not less than 4 inches by 8 inches (102 mm by 203 mm) located near the top of the door into the sauna room.

❖ A ventilation opening is required to allow the escape of heat and to provide ventilation for the occupants.

914.5.1 Warning notice. The following permanent notice, constructed of *approved* material, shall be mechanically attached to the sauna room on the outside:

WARNING: DO NOT EXCEED 30 MINUTES IN SAUNA. EXCESSIVE EXPOSURE CAN BE HARMFUL TO HEALTH. ANY PERSON WITH POOR HEALTH SHOULD CONSULT A PHYSICIAN BEFORE USING SAUNA.

The words shall contrast with the background and the wording shall be in letters not less than $^1/_4$-inch (6.4 mm) high.

> **Exception:** This section shall not apply to one- and two-family dwellings.

❖ A warning sign is required to alert users of the potential hazard associated with prolonged exposure to high temperature levels.

SECTION 915
ENGINE AND GAS TURBINE-POWERED EQUIPMENT AND APPLIANCES

915.1 General. The installation of liquid-fueled stationary internal *combustion* engines and gas turbines, including exhaust, fuel storage and piping, shall meet the requirements of NFPA 37. Stationary engine generator assemblies shall meet the requirements of UL 2200.

❖ This section addresses liquid-fueled internal combustion engines and turbines. Engine-driven electrical generators for private use are becoming more popular as are engine-driven cooling appliances and heat pumps. Such equipment is used to power fire pumps, generators, water pumps, refrigeration machines and other stationary equipment. NFPA 37 addresses the fire safety for this kind of equipment including requirements for enclosures, controls, fuel supplies, exhaust systems, cooling systems and combustion air. Engine generator units must be listed and labeled to UL 2200.

915.2 Powered equipment and appliances. Permanently installed *equipment* and appliances powered by internal *combustion* engines and turbines shall be installed in accordance with the manufacturer's installation instructions and NFPA 37.

❖ This section contains the same provisions found in Section 915.1 (see commentary, Section 915.1).

SECTION 916
POOL AND SPA HEATERS

916.1 General. Pool and spa heaters shall be installed in accordance with the manufacturer's installation instructions. Oil-fired pool and spa heaters shall be tested in accordance with UL 726. Electric pool and spa heaters shall be tested in accordance with UL 1261.

❖ This section addresses specialized types of water heaters used with swimming pools, recreational or therapeutic spas and hot tubs. Pool and spa heaters

are specialized water heaters similar in design to hot water boilers. Typically, these heaters are of the water-tube type and are designed for either indoor or outdoor installation. This section contains test standards to be used as acceptance criteria for oil-fired and electric pool and spa heaters. The IFGC references ANSI Z21.56 as the test standard for gas-fired heaters.

SECTION 917
COOKING APPLIANCES

917.1 Cooking appliances. Cooking appliances that are designed for permanent installation, including ranges, ovens, stoves, broilers, grills, fryers, griddles and barbecues, shall be *listed*, *labeled* and installed in accordance with the manufacturer's installation instructions. Commercial electric cooking appliances shall be *listed* and *labeled* in accordance with UL 197. Household electric ranges shall be *listed* and *labeled* in accordance with UL 858. Microwave cooking appliances shall be *listed* and *labeled* in accordance with UL 923. Oil-burning stoves shall be *listed* and *labeled* in accordance with UL 896. Solid-fuel-fired ovens shall be *listed* and *labeled* in accordance with UL 2162.

❖ The code intends to regulate the design, construction and installation of cooking appliances that are designed for permanent installation in all occupancies. The code does not regulate portable appliances such as electrically heated countertop appliances. Appliances that are not readily movable to another location because of an electric-power or liquid-fuel supply connection would be considered as permanently installed even if on casters. Line equipment under a Type I hood, for example, is typically on casters and connected with cord and plug supply lines to allow movement for routine cleaning. Such equipment would be considered permanently installed.

917.2 Prohibited location. Cooking appliances designed, tested, *listed* and *labeled* for use in commercial occupancies shall not be installed within *dwelling units* or within any area where domestic cooking operations occur.

❖ Commercial cooking appliances are tested and labeled to different standards than those listed for domestic use. Commercial cooking appliances generally are not insulated to the same level as domestic cooking appliances, have higher surface operating temperatures than domestic appliances and require a much greater clearance to combustible material. The safety measures inherent to household cooking appliances, such as child-safe push-to-turn knobs and insulated oven doors, are not usually installed in commercial cooking appliances.

Commercial cooking appliances also have a greater ventilation air requirement for safe operation than household-type cooking appliances. For this reason, installation of commercial-type cooking appliances in dwellings is prohibited (see commentary, Section 917.3). Note that the code would not prohibit an appliance that is dual listed for both commercial and domestic use, as this would meet the intent of the code.

917.3 Domestic appliances. Cooking appliances installed within *dwelling units* and within areas where domestic cooking operations occur shall be *listed* and *labeled* as household-type appliances for domestic use.

❖ Cooking appliances used in dwelling units or in areas where domestic cooking operations occur require a greater degree of user protection and must be listed and labeled as household-type appliances for domestic use (see commentary, Section 917.2). To satisfy consumer demand for commercial appliances in the home, some manufacturers are producing listed household-type appliances that have the appearance of commercial cooking appliances.

SECTION 918
FORCED-AIR WARM-AIR FURNACES

918.1 Forced-air furnaces. Oil-fired furnaces shall be tested in accordance with UL 727. Electric furnaces shall be tested in accordance with UL 1995. Solid fuel furnaces shall be tested in accordance with UL 391. Forced-air furnaces shall be installed in accordance with the listings and the manufacturer's installation instructions.

❖ Forced-air warm-air furnaces are considered to be central heating units. They consist of burners or heating elements, heat exchangers, blowers and associated controls. Forced-air furnaces are made in many different configurations, including upflow, counterflow (down flow), horizontal flow and indoor and outdoor units. This section defines test standards to be used as acceptance criteria for oil-fired, solid-fuel and electric furnaces. The IFGC references ANSI Z21.47 and UL 795 as the test standards for gas-fired forced-air warm-air furnaces.

918.2 Minimum duct sizes. The minimum unobstructed total area of the outdoor and return air ducts or openings to a forced-air warm-air furnace shall be not less than 2 square inches per 1,000 Btu/h (4402 mm²/kW) output rating capacity of the furnace and not less than that specified in the furnace manufacturer's installation instructions. The minimum unobstructed total area of supply ducts from a forced-air warm-air furnace shall not be less than 2 square inches for each 1,000 Btu/h (4402 mm²/kW) output rating capacity of the furnace and not less than that specified in the furnace manufacturer's installation instructions.

Exception: The total area of the supply air ducts and outdoor and return air ducts shall not be required to be larger than the minimum size required by the furnace manufacturer's installation instructions.

❖ The aggregate area of all ducts or openings that convey supply air from the furnace or return air back to the furnace must be adequate to allow the required airflow through the furnace. A furnace that is "starved" for return air or is restricted by an inadequate supply-air duct size will produce an abnormal

temperature rise across the heat exchanger, which is a fire hazard and may damage the furnace. The furnace output rating is not usually indicated on the label. It would be determined as approximately the input rating in British thermal units per hour [Btu/h] (W) times the efficiency rating of the furnace. As an example, 100,000 Btu/h (29 310 W) input times 0.80 (80-percent efficiency) is 80,000 Btu/h (23 448 W) output. Return or supply air openings required by this section must not be less than those specified by the furnace manufacturer's installation instructions.

The exception is a reminder that if the furnace installation instructions specify a lesser return or supply area than this section, that lesser area is permitted. The code-specified minimum area applies when the furnace manufacturer does not specify a minimum area.

918.3 Heat pumps. The minimum unobstructed total area of the outdoor and return air ducts or openings to a heat pump shall be not less than 6 square inches per 1,000 Btu/h (13 208 mm²/kW) output rating or as indicated by the conditions of listing of the heat pump. Electric heat pumps shall be tested in accordance with UL 1995.

❖ Heat pumps are central heating units that use the condenser of a refrigeration circuit as the heat source, often "backed up" by auxiliary electric resistance heating elements. Heat pumps require greater airflow through the unit because of the significant difference in intensity of the heat source compared to fuel-fired and electrical-resistance heat sources (see commentary, Section 918.1).

918.4 Dampers. Volume dampers shall not be placed in the air inlet to a furnace in a manner that will reduce the required air to the furnace.

❖ Dampers are usually avoided in return air ducts and openings because of the risk of starving the furnace for return air. If dampers are installed, the total unrestricted return air duct or opening area must be as required by the code with all of the dampers in the fully closed position. Dampered openings cannot count toward the required duct area.

918.5 Circulating air ducts for forced-air warm-air furnaces. Circulating air for fuel-burning, forced-air-type, warm-air furnaces shall be conducted into the blower housing from outside the furnace enclosure by continuous air-tight ducts.

❖ This section parallels the manufacturer's installation instructions, which require return air ducts to extend from the furnace cabinet to the exterior of any closet, alcove or furnace room that encloses a furnace. If return air was drawn from within a furnace enclosure, negative pressures could be produced in that room. Negative pressure could cause combustion byproducts to be drawn from the combustion chamber or venting system into the circulating airflow.

Negative pressure differentials between the interior and exterior of a furnace enclosure could also cause hazardous burner and venting system malfunctions.

918.6 Prohibited sources. Outdoor or return air for forced-air heating and cooling systems shall not be taken from the following locations:

1. Less than 10 feet (3048 mm) from an *appliance* vent outlet, a vent opening from a plumbing drainage system or the discharge outlet of an exhaust fan, unless the outlet is 3 feet (914 mm) above the outdoor air inlet.

2. Where there is the presence of objectionable odors, fumes or flammable vapors; or where located less than 10 feet (3048 mm) above the surface of any abutting public way or driveway; or where located at grade level by a sidewalk, street, alley or driveway.

3. A hazardous or insanitary location or a refrigeration *machinery room* as defined in this code.

4. A room or space, the volume of which is less than 25 percent of the entire volume served by such system. Where connected by a permanent opening having an area sized in accordance with Sections 918.2 and 918.3, adjoining rooms or spaces shall be considered as a single room or space for the purpose of determining the volume of such rooms or spaces.

   Exception: The minimum volume requirement shall not apply where the amount of return air taken from a room or space is less than or equal to the amount of supply air delivered to such room or space.

5. A closet, bathroom, toilet room, kitchen, garage, boiler room, furnace room or unconditioned attic.

   Exceptions:

   5.1. Where return air intakes are located not less than 10 feet (3048 mm) from cooking appliances, and serve the kitchen area only, taking return air from a kitchen shall not be prohibited.

   5.2. Dedicated forced-air systems serving only a garage shall not be prohibited from obtaining return air from the garage.

6. An unconditioned crawl space by means of direct connection to the return side of a forced air system. Transfer openings in the crawl space enclosure shall not be prohibited.

7. A room or space containing a fuel-burning *appliance* where such room or space serves as the sole source of return air.

   Exceptions:

   7.1. This shall not apply where the fuel-burning *appliance* is a direct-vent *appliance*.

   7.2. This shall not apply where the room or space complies with the following requirements:

   7.2.1. The return air shall be taken from a room or space having a volume exceeding 1 cubic foot for each 10 Btu/h (9.6 L/W) of combined input rating of all fuel-burning appliances therein.

   7.2.2. The volume of supply air discharged back into the same space shall be approximately equal to the volume of return air taken from the space.

   7.2.3. Return-air inlets shall not be located within 10 feet (3048 mm) of any *appliance* firebox or draft hood in the same room or space.

   7.3. This shall not apply to rooms or spaces containing solid-fuel-burning appliances, provided that return-air inlets are located not less than 10 feet (3048 mm) from the firebox of the appliances.

❖ This section prohibits outdoor air and return air from being taken from locations that would be potential sources of contamination, odor, flammable vapors or toxic substances and also from locations that would negatively affect the operation of the furnace itself or other fuel-burning appliances.

The intent of Item 4 is to prevent the system from being starved for return air by placing the main or only return air intake in an area not meeting the volume requirements of this item. This item is airflow-balance related as opposed to Items 1, 2, 3, 5 and 6, which are contaminant related. This item does not prohibit the common practice of installing return air intakes in bedrooms and similarly sized rooms that typically have a volume that is far less than 25 percent of the total volume of the space served by the furnace. The return air system must be able to move the required air volume to the furnace regardless of the position of any doors to any rooms in the building served by the furnace.

The prohibition in Item 4 is intended to avoid arrangements that cause an air pressure imbalance. Air pressure imbalances can cause fuel-fired appliances to spill combustion products into the occupied space. Pressure imbalances can be avoided by making sure that the amount of supply air discharge to a room or space is approximately equal to the amount of return air taken from the room or space. Exception 5.2 clarifies that forced air systems that serve only a garage necessarily must take return air from the garage.

918.7 Outside opening protection. Outdoor air intake openings shall be protected in accordance with Section 401.5.

❖ Outside air intake openings must be protected as prescribed in Section 401.6 (see commentary, Section 401.6).

918.8 Return-air limitation. Return air from one *dwelling unit* shall not be discharged into another *dwelling unit*.

❖ This section prohibits a forced-air heating/cooling system from serving more than one dwelling unit. Any arrangement in which dwelling units share all or part of an air distribution system would allow a communication of atmospheres in the units. This communication would spread odors, smoke, contaminants and disease-causing organisms from one dwelling unit to another and therefore must be avoided (see also Section 403.2.1).

SECTION 919
CONVERSION BURNERS

919.1 Conversion burners. The installation of conversion burners shall conform to ANSI Z21.8.

❖ The referenced standard is an installation standard that, in addition to the manufacturer's instructions, governs the installation of conversion burners. Conversion burners are an assembly of components including burners, gas controls, blowers, safety devices and supporting structures. The units are designed to convert an existing appliance from another fuel, commonly from fuel oil or coal, to gas (see commentary, Section 301.12). Conversion of an existing appliance to a different fuel can involve much more than installation of a conversion burner, such as the addition of safety controls and limits; combustion air and secondary air provisions; fuel gas piping installation; chimney or vent alterations; and other modifications to the existing appliance and its control system.

SECTION 920
UNIT HEATERS

920.1 General. Unit heaters shall be installed in accordance with the listing and the manufacturer's installation instructions. Oil-fired unit heaters shall be tested in accordance with UL 731.

❖ This section addresses self-contained space-heating appliances, including forced-air types. Unit heaters are ductless warm-air space heaters that are self-contained and usually suspended from a ceiling or roof structure. Garages, workshops, warehouses, factories, gymnasiums, mercantile spaces and similar large open buildings are the most common locations for unit heaters.

920.2 Support. Suspended-type unit heaters shall be supported by elements that are designed and constructed to accommodate the weight and dynamic loads. Hangers and brackets shall be of noncombustible material. Suspended-type oil-fired unit heaters shall be installed in accordance with NFPA 31.

❖ As with all suspended fuel-fired appliances, a support failure could result in a fire, explosion or injury to building occupants. The supports themselves must be properly designed. Equally important are the structural members to which the supports are attached, such as rafters, beams, joists and purlins. Brackets, pipes, rods, angle irons, structural members and fasteners must be designed for the dead and dynamic loads of the suspended appliance.

920.3 Ductwork. A unit heater shall not be attached to a warm-air duct system unless *listed* for such installation.

❖ Unit heaters are usually not designed to move air through ductwork. Unless specifically listed for the application, the fans or blowers on unit heaters are designed only for moving air across the heat exchanger without the added friction of ductwork. The addition of ductwork could create a hazard by restricting airflow through the heater.

SECTION 921
VENTED ROOM HEATERS

921.1 General. Vented room heaters shall be *listed* and *labeled* and shall be installed in accordance with the conditions of the listing and the manufacturer's instructions.

❖ This section addresses vented space/room heaters, including direct-vent and vent- or chimney-connected appliances. As required by the appliance standard, such appliances are equipped with safety controls that will prevent fuel flow to the burners in the event of ignition system failure.

SECTION 922
KEROSENE AND OIL-FIRED STOVES

922.1 General. Kerosene and oil-fired stoves shall be listed and labeled and shall be installed in accordance with the conditions of the listing and the manufacturer's installation instructions. Kerosene and oil-fired stoves shall comply with NFPA 31 and UL 896.

❖ This section addresses permanently installed kerosene- and oil-fired appliances such as furnaces and room heaters. NFPA 31 contains installation requirements for such appliances. These appliances must be tested in accordance with UL 896.

SECTION 923
SMALL CERAMIC KILNS

923.1 General. The provisions of this section shall apply to kilns that are used for ceramics, have a maximum interior volume of 20 cubic feet (0.566 m³) and are used for hobby and noncommercial purposes. Electric kilns shall comply with UL 499.

❖ Small ceramic kilns used for hobby and noncommercial purposes operate at high temperatures and, therefore, should be regulated by the code as are other heat-producing appliances. Such kilns must comply with UL 499.

§23.1.1 Installation. Kilns shall be installed in accordance with the manufacturer's installation instructions and the provisions of this code.

❖ The minimum level of safety required for installation of small kilns is compliance with the kiln manufacturer's installation instructions.

SECTION 924
STATIONARY FUEL CELL POWER SYSTEMS

924.1 General. Stationary fuel cell power systems having a power output not exceeding 10 MW shall be tested in accordance with ANSI/CSA America FC 1 and shall be installed in accordance with the manufacturer's installation instructions, NFPA 853, the *International Building Code* and the *International Fire Code.*

❖ The private generation of electricity for commercial and residential use is an emerging technology that is expected to gain widespread acceptance as future technological improvements result in lower costs. Fuel cells have been included in this code to provide guidance to the building community for those future installations.

The code requires fuel cell power plants with an output of 10 MW or less to be listed in accordance with ANSI/CSA America FC1 and requires installation of the units in accordance with the manufacturer's instructions and NFPA 853. NFPA 853 contains guidance on location of the power plants, fuel system supplies, ventilation and exhaust for the space containing the unit and fire protection requirements.

Although fuel cell technology has been available since the 1800s, its first practical use was by NASA to generate electricity and water for the Gemini and Apollo spacecraft and today's space shuttle. Because the installation of fuel cells in privately owned buildings is relatively new, the following discussion of the fuel cell process is included to provide a basic understanding of the technology.

A fuel cell is an electrochemical device that converts the chemical energy of a hydrocarbon fuel (hydrogen, natural gas, coal-bed methane, methanol, gasoline, etc.) and an oxidant (air or oxygen) into useable electricity. The fuel cell consists of a fuel electrode (anode) and an oxidant electrode (cathode) separated by an ion-conducting membrane (electrolyte). When a hydrocarbon fuel is introduced into the fuel cell, the catalyst surface of the membrane converts the hydrogen gas molecules into hydrogen ions and electrons. The hydrogen ions pass through the membrane to react with oxygen and electrons to form water as a reaction byproduct. The electrons, which cannot pass through the membrane, must travel around it through an external circuit, thereby creating direct current (dc) electricity [see Commentary Figure 924.1(1)].

When fueled by pure hydrogen, the only byproducts of the fuel cell process are heat and water. Low levels of CO_2 are produced when other fuels are used. Unlike traditional fossil-fuel electric generating stations that burn the fuels, the electrochemical fuel cell process produces no particulate matter, nitrogen oxides or sulfur oxides.

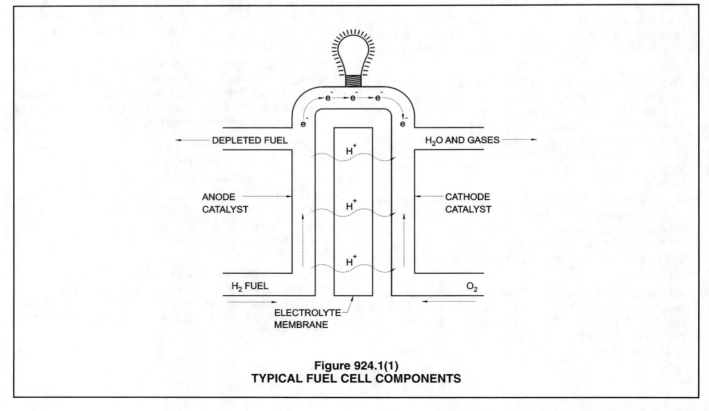

Figure 924.1(1)
TYPICAL FUEL CELL COMPONENTS

When the waste heat is captured for use in heating water or a cogeneration process, fuel cell efficiency can reach 80 percent.

A typical fuel cell consists of three sections: a fuel processor, a fuel stack (power generator) and a power conditioner [see Commentary Figure 924.1(2)].

The fuel processor section has two components: the fuel reformer, which processes a hydrocarbon fuel, such as methane, into a hydrogen-rich formate gas and a carbon monoxide cleanup system to reduce CO concentrations to acceptable levels.

The fuel cell stack consists of many individual fuel cells that operate as previously described to generate electricity and usable heat. The power output of the unit increases based on the number and arrangement of individual cells stacked in series and/or in parallel in the unit. There are several types of fuel cell technologies, including phosphoric acid (PA), molten carbonate (MC), solid oxide (SO) and proton exchange membrane (PEM), which are named based on the type of electrolyte employed in the membrane section of the cell.

The power conditioner (inverter) section converts the direct current (dc) produced in the fuel cell stack into alternating current (ac). Batteries are used to smooth out power surge demands, such as heating, ventilating and air-conditioning (HVAC) start-ups, and to provide additional power when demand exceeds the peak output of the cell stack.

Although currently available fuel cell units are not cost competitive with established electrical suppliers, the initial and operating costs are expected to decrease as more manufacturers enter the market. Where expansion, in remote locations or in areas where reduced emission of pollutants is critical, fuel cells could be seriously considered as an energy source. Fuel cells can also be attractive where the quality and dependability of commercial power is a concern.

SECTION 925
MASONRY HEATERS

925.1 General. Masonry heaters shall be constructed in accordance with the *International Building Code.*

❖ Masonry heaters are appliances designed to absorb and store heat from a relatively small fire and to radiate that heat into the building interior. They are thermally more efficient than traditional fireplaces because of their design. Interior passageways through the heater allow hot exhaust gases from the fire to transfer heat into the masonry, which then radiates into the building. Section 2112 of the IBC requires listing and installation of masonry heaters in accordance with ASTM E 1602. Guidance for seismic reinforcing and clearance to combustibles is also provided.

SECTION 926
GASEOUS HYDROGEN SYSTEMS

926.1 Installation. The installation of gaseous hydrogen systems shall be in accordance with the applicable requirements of this code, the *International Fire Code,* the *International Fuel Gas Code* and the *International Building Code.*

❖ Systems that use or produce hydrogen gas can create potentially explosive conditions where installation and ventilation instructions are not adhered to strictly. Chapter 7 of the IFGC is the primary source of regulations related to gaseous hydrogen systems, but the IBC, the IFC and Section 304.4 of this code also provide code requirements. Chapters 4 and 5 of the IBC provide requirements for hydrogen cutoff rooms and Chapter 23 of the IFC provides requirements for hydrogen fueling stations.

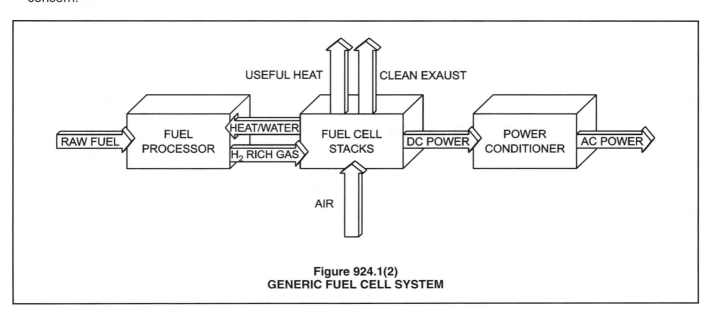

Figure 924.1(2)
GENERIC FUEL CELL SYSTEM

SECTION 927
RADIANT HEATING SYSTEMS

927.1 General. Electric radiant heating systems shall be installed in accordance with the manufacturer's instructions and shall be listed for the application.

❖ Radiant heating systems contain electrical elements that create heat when an electric current passes through them. Radiant heating systems consist of heating cables and radiant heating panels that have been evaluated for contact with specific surfaces. If the system is not installed properly, electricity and the heat emitted could create hazards for the occupants of the dwelling. The manufacturer's installation instructions will have details on the different type of surfaces and the temperature limits for each surface. These requirements are in addition to the requirements found in Sections 927.2 through 927.5.

927.2 Clearances. Clearances for radiant heating panels or elements to any wiring, outlet boxes and junction boxes used for installing electrical devices or mounting luminaires shall be in accordance with the *International Building Code* and NFPA 70.

❖ This section references certain minimum clearances so that the heating elements are not damaged or the electrical circuit continuity is not interrupted, thus rendering portions or all of the radiant heating systems inoperable. This section is also intended to protect the electrical systems, equipment, and components from damage from the heating elements.

927.3 Installation on wood or steel framing. Radiant panels installed on wood or steel framing shall conform to the following requirements:

1. Heating panels shall be installed parallel to framing members and secured to the surface of framing members or shall be mounted between framing members.

2. Mechanical fasteners shall penetrate only the unheated portions provided for this purpose. Panels shall not be fastened at any point closer than $^1/_4$ inch (7 mm) to an element. Other methods of attachment of the panels shall be in accordance with the panel installation instructions.

3. Unless listed and labeled for field cutting, heating panels shall be installed as complete units.

❖ Radiant heat panels mounted on wood framing may act as an ignition source if improperly installed. Because these systems generate heat, they must be mounted to wood framing by fastening through the unheated portions of each panel. Metallic fasteners, such as nails or screws, are excellent conductors of heat as well as electricity; therefore, they must not be driven through any portion of a panel closer than $^1/_4$ inch (6.4 mm) to a heating element. The manufacturer's installation instructions may contain additional methods that can be used to attach the panels.

927.4 Installation in concrete or masonry. Radiant heating systems installed in concrete or masonry shall conform to the following requirements:

1. Radiant heating systems shall be identified as being suitable for the installation, and shall be secured in place as specified in the manufacturer's instructions.

2. Radiant heating panels and radiant heating panel sets shall not be installed where they bridge expansion joints unless they are protected from expansion and contraction.

❖ Fire hazards created by electric radiant heating systems installed in masonry construction are minimal, but these installations require special mounting hardware and must be installed in accordance with the manufacturer's installation instructions. Radiant panels must be protected from expansion and contraction when mounted over masonry expansion joints to avoid structural damage.

927.5 Finish surfaces. Finish materials installed over radiant heating panels and systems shall be installed in accordance with the manufacturer's instructions. Surfaces shall be secured so that fasteners do not pierce the radiant heating elements.

❖ Finish materials must be carefully installed over electric radiant heating systems so that nails or screws do not penetrate the heating elements. Fasteners that are accidentally driven through these elements could cause short circuits, ground faults, destroy the elements or cause the surfaces to overheat and start a fire. The finish surfaces must be installed in accordance with the instructions supplied by the manufacturer of the radiant heating system.

SECTION 928
EVAPORATIVE COOLING EQUIPMENT

928.1 General. Evaporative cooling equipment shall:

1. Be installed in accordance with the manufacturer's instructions.

2. Be installed on level platforms in accordance with Section 304.10.

3. Have openings in exterior walls or roofs flashed in accordance with the *International Building Code*.

4. Be provided with potable water backflow protection in accordance with Section 608 of the *International Plumbing Code*.

5. Have air intake opening locations in accordance with Section 401.4.

❖ Evaporative coolers are commonly known as "swamp coolers." This type of air cooling equipment is used in climates where the relative humidity is low. Outside air is drawn in through water-wetted pads and discharged inside the building. The evaporation of water into the dry air reduces the temperature of the air.

Where the outside air is very dry (very low relative humidity), the relative humidity of the air discharged to the building is well within a pleasant range. However, where the outside air has moderate relative humidity, the indoor air relative humidity can become high enough such that the air feels muggy or "swamp like," hence the term "swamp cooler."

Evaporative coolers must be installed in accordance with the manufacturer's instructions and must be installed on a level platform. The discharge duct going into the building must be properly flashed so that rainwater does not enter the building. Evaporative coolers require a potable water connection and as such, the connection must be protected by a backflow protection device. The location of air intake openings of evaporative coolers must consider the restrictions listed in Section 401.4.

Chapter 10:
Boilers, Water Heaters and Pressure Vessels

General Comments

Chapter 10 applies to all types of boilers and pressure vessels, regardless of size, heat input, operating pressure or operating temperature [note that gas-fired boilers and pressure vessels are regulated by the *International Fuel Gas Code*® (IFGC®) as stated in Section 101.2]. When a water heater is considered a boiler and vice versa, it has been confusing. Traditionally, the definitions of "Boiler" and "Water heater" have been linked to specific operating parameters, including maximum temperature, pressure, heat input and storage volume. The thresholds for determining boiler categories or for deciding whether the equipment is a boiler or a water heater have been determined by the appliance/equipment standards. The application of a water boiler and the end use of the hot water produced create additional categories. For example, a hot water boiler recirculates hot water used for space heating or processing (closed system), and a hot water supply boiler heats water to be used without recirculation, thus requiring complete makeup (open system).

In the context of this code, the definitions are simple, and make the distinction between boilers and water heaters clear. A water heater is any appliance that heats potable water and then supplies it to the plumbing hot water distribution system. Note, however, that some water heaters may also be used for space heating, depending on the appliance listing and label. A boiler either heats water or generates steam for space heating, processing or developing power, and is generally a closed system. A hot water supply boiler, however, serves an open system in which the hot water is not returned to the boiler. In all cases, the appliance/equipment label always describes the unit accurately, and references the applicable standard, in which the title also indicates the appliance/equipment category.

Because pressure vessels are closed containers designed to contain liquids, gases or both under pressure, the addition of heat to, or improper installation of, these pieces of mechanical equipment can cause expansion beyond the pressure limitation of the vessel. Certain safety features are therefore required to reduce the potential for explosion hazards.

Within the context of this code, pressure vessels are containers having internal pressures greater than atmospheric pressure and include tanks, accumulators and vessels containing liquids, gases or both. Although boilers and water heaters are considered pressure vessels, this section is not intended to be redundant. It is intended to place in the scope of this code pressure equipment such as: air compressor tanks, expansion tanks, hydro-pneumatic tanks, accumulators, steam generators and containment vessels associated with mechanical systems. Again, the code intends to avert the loss of human life and damage to property associated with the potential explosion of such vessels.

Purpose

This chapter presents regulations for the proper installation of boilers, water heaters and pressure vessels to protect life and property from the hazards associated with those appliances and vessels.

SECTION 1001
GENERAL

1001.1 Scope. This chapter shall govern the installation, *alteration* and repair of boilers, water heaters and pressure vessels.

Exceptions:

1. Pressure vessels used for unheated water supply.

2. Portable unfired pressure vessels and Interstate Commerce Commission containers.

3. Containers for bulk oxygen and medical gas.

4. Unfired pressure vessels having a volume of 5 cubic feet (0.14 m^3) or less operating at pressures not exceeding 250 pounds per square inch (psi) (1724 kPa) and located within occupancies of Groups B, F, H, M, R, S and U.

5. Pressure vessels used in refrigeration systems that are regulated by Chapter 11 of this code.

6. Pressure tanks used in conjunction with coaxial cables, telephone cables, power cables and other similar humidity control systems.

7. Any boiler or pressure vessel subject to inspection by federal or state inspectors.

❖ Chapter 10 contains code text that is specific to boilers and water heaters. All other related sections, such as those found in Chapters 3, 7, 8, 12 and 13 must also be applied. The provisions of this chapter are applicable to new installations, replacements and all repair and alteration work involving any of the components regulated by the code. Throughout the code, the common intent is for all repairs and alterations to be performed in the same manner as the code prescribes for new installations (see commentary, Section 102.4). For example, if repair work involves the replacement of a safety device, such as a pressure relief valve, the replacement device must comply with the requirements of this chapter.

The exceptions recognize a variety of situations in which containers, pressure tanks, boilers and pressure vessels are regulated by a federal, state or utility-run agency. Furthermore, the exceptions address circumstances in which the equipment or appliance is either unfired, unheated or otherwise not subject to explosion. Therefore, regulation of these items in the code would be considered unnecessary for equipment or appliances that pose little to no risk to human life or property.

SECTION 1002
WATER HEATERS

1002.1 General. Potable water heaters and hot water storage tanks shall be listed and labeled and installed in accordance with the manufacturer's installation instructions, the *International Plumbing Code* and this code. All water heaters shall be capable of being removed without first removing a permanent portion of the building structure. The potable water connections and relief valves for all water heaters shall conform to the requirements of the *International Plumbing Code*. Domestic electric water heaters shall comply with UL 174 or UL 1453. Commercial electric water heaters shall comply with UL 1453. Oil-fired water heaters shall comply with UL 732. Solid-fuel-fired water heaters shall comply with UL 2523. Thermal solar water heaters shall comply with Chapter 14 and UL 174 or UL 1453.

❖ This is the only section in Chapter 10 specifically addressing water heaters and hot water storage tanks. In the context of this section, hot water storage tanks are used in conjunction with circulating-type water heaters or other exterior heat sources.

Because all water heaters have a far shorter working life than the building in which they are located, they will have to be replaced one or more times during the lifetime of the building. This future need has often been overlooked, and water heaters have been installed in locations that would require the dismantling or destruction of a permanent portion of the structure for replacement. For this reason, any room, space or alcove in which a water heater is to be installed must have an entry or doorway through which the water heater will pass. Additionally, other equipment or appliances installed in the same space as the water heater must be arranged to allow removal of the water heater without removal of other equipment or appliances.

This section recognizes that water heaters and hot water storage tanks must be considered as both mechanical appliances (equipment) and plumbing appliances and, therefore, must comply with both this code and the *International Plumbing Code*® (IPC®).

When a water heater is tested to obtain a listing and label, the approved agency installs the water heater in accordance with the manufacturer's installation instructions. The water heater is then tested under these conditions; thus, the installation instructions become an integral part of the listing and labeling process. Manufacturer's installation instructions are thoroughly evaluated by the listing and labeling agency to establish that a safe installation is prescribed. The listing and labeling agency can require the manufacturer to alter, delete or add information in the installation instructions as necessary to achieve compliance with the applicable standards and code requirements.

The manufacturer's installation instructions must be available to the code official because they are an enforceable extension of the code and are necessary for determining that the water heater has been properly installed. Simply put, the listing and labeling process indicates that the water heater and its installation instructions are in compliance with applicable standards. Therefore, an installation in accordance with the manufacturer's installation instructions is required, except where the code requirements are more stringent. The inspector must carefully and completely read and comprehend the manufacturer's instructions to properly perform an installation inspection.

In some cases, the code will specifically address an installation requirement that is also addressed in the manufacturer's installation instructions. The code requirement may be the same or may exceed the requirement in the manufacturer's installation instructions. The manufacturer's installation instructions could contain requirements that exceed those in the code. In all such cases, the more restrictive requirements must apply (see commentary, Section 304.2).

Even if an installation appears to be in compliance with the manufacturer's instructions, the installation cannot be completed or approved until all associated components, connections and systems that serve the water heater are also in compliance with the applicable code provisions.

A water heater installation is complex in that it has a fuel or power supply; a chimney or vent connection, if fuel fired; a combustion air supply, if fuel fired; connections to the plumbing potable water distribution system; and controls and devices to prevent a multitude of potential hazards from conditions such as excessively high temperatures, pressures and ignition failure. Applicable sections are found in this code, the IPC and the IFGC.

It is not uncommon for jurisdictions to issue both plumbing and mechanical permits for water heater installations, or to require that the installer be licensed in both the plumbing and mechanical trades when performing installations.

1002.2 Water heaters utilized for space heating. Water heaters utilized both to supply potable hot water and provide hot water for space-heating applications shall be *listed* and *labeled* for such applications by the manufacturer and shall be installed in accordance with the manufacturer's installation instructions and the *International Plumbing Code.*

❖ Water heaters serving the dual purpose of supplying potable hot water and serving as a heat source for a hot water space-heating system must be listed and labeled for that dual application. This section does not address water heaters used solely for space-heating applications, but rather addresses water heaters that serve a secondary purpose of space heating. Water heater labels indicate whether or not the appliance is suitable for space heating.

Chapter 5 of the IPC contains additional requirements for the proper installation of this kind of water heater.

1002.2.1 Sizing. Water heaters utilized for both potable water heating and space-heating applications shall be sized to prevent the space-heating load from diminishing the required potable water-heating capacity.

❖ In the United States, the average person uses 20 gallons (75.7 L) of hot water per day. Thus, the ability of a hot water supply system to meet this demand without necessitating dangerously high water temperatures is essential to our lifestyle and the interior environments in which we live and work. In designing a potable water heating system used also for space-heating purposes, the registered design professional must recognize that space comfort conditions should not be achieved at the expense of potable hot water capacity.

A typical installation of a water heater used for space heating might be an underfloor radiant-heating system, a fan/coil unit or a baseboard convector system. The water heater must be sized or the system functions prioritized to prevent the space-heating load from compromising the potable water-heating capacity of the heater. The first priority is the potable water supply. Also, where the design water temperature is inadequate to supply sufficient hot water for domestic purposes and heating, separate heating systems or tempering equipment must be installed.

However, the code does not regulate the calculations or methods necessary for sizing such equipment or appliances. Although methods for sizing water heaters vary, the specific requirements generally differ in total volume, demand flow rate, duration of peak load period and temperature. Water heaters and systems should be selected based on these requirements and the anticipated space-heating loads. Whichever method is chosen to size these types of systems, the system must supply adequate potable hot water.

1002.2.2 Temperature limitation. Where a combination potable water-heating and space-heating system requires water for space heating at temperatures higher than 140°F (60°C), a temperature actuated mixing valve that conforms to ASSE 1017 shall be provided to temper the water supplied to the potable hot water distribution system to a temperature of 140°F (60°C) or less.

❖ Chapter 5 of the IPC regulates the outlet water temperature from a water heater that supplies both potable hot water and hot water for space heating. Scalding accidents can easily occur when the potable hot water exceeds a temperature of 140°F (60°C). A temperature actuated mixing valve is required to limit the temperature of hot water to be used for bathing and other domestic purposes to 140°F (60°C) or less when the water heater is used for both potable hot water and hot water for space heating.

Regardless of the water supply demand downstream from the valve or supply pressure fluctuations upstream from the valve, the user will be provided some protection from scalding injury because the temperature of the water supplied will not exceed 140°F (60°C). Section 424.3 of the IPC requires further reduction of water temperature for bathtubs and showers to 120°F (49°C).

1002.3 Supplemental water-heating devices. Potable water-heating devices that utilize refrigerant-to-water heat exchangers shall be *approved* and installed in accordance with the *International Plumbing Code* and the manufacturer's installation instructions.

❖ Waste heat from refrigeration systems can be used to heat potable water and can result in substantial energy savings over time. This section allows the use of supplemental systems if the equipment is approved and installed in accordance with the manufacturer's installation instructions and the requirements of the IPC. A serious concern for these systems is the possibility of contamination of the potable water.

SECTION 1003
PRESSURE VESSELS

1003.1 General. All pressure vessels shall be in accordance with the ASME *Boiler and Pressure Vessel Code,* shall bear the label of an *approved* agency and shall be installed in accordance with the manufacturer's installation instructions.

❖ Regulatory requirements for the approval and installation of pressure vessels are the same as for the approval of all other mechanical equipment and appliances, and are therefore subject to the provisions of Chapter 3 and this section. All pressure vessels must be designed and constructed in accordance with the ASME *Boiler and Pressure Vessel Code* (see commentary, Sections 301.7 and 304.1).

1003.2 Piping. All piping materials, fittings, joints, connections and devices associated with systems utilized in conjunction with pressure vessels shall be designed for the specific application and shall be *approved*.

❖ This section addresses the requirements for the types of piping materials, joints, fittings, connections and devices associated with pressure vessels and their ancillary systems.

Compatibility between the pressure vessel system equipment and the medium (gas, liquid or both) contained within the system is essential because the medium under containment or the processes involved can be corrosive. Additionally, the piping, joints, fittings, connections and devices must be capable of withstanding the full range of operating pressures, temperatures and vibrations commonly associated with the pressure vessel and its system(s), bearing in mind that operating conditions range from the normal (intended) to the abnormal (unintended). Improperly rated system components could cause catastrophic failure of the system.

Pipe joints and connections represent a potential weak link in the piping of a system under high internal pressure. For this reason, the piping materials, joints, fittings, connections and devices associated with pressure vessels must be of an approved type suitable for the system's characteristics.

1003.3 Welding. Welding on pressure vessels shall be performed by *approved* welders in compliance with nationally recognized standards.

❖ The applicable sections of ASME B31 and the ASME *Boiler and Pressure Vessel Code* specify proper welding methods. ASME B31 requires that all welders and welding procedure specifications be qualified. Separate welding procedure specifications are needed for different welding methods and materials. Qualifying tests and the variables requiring separate procedure specifications are set forth in Section IX of the ASME *Boiler and Pressure Vessel Code*. The manufacturer, fabricator or contractor is responsible for the welding procedure and the welders. ASME B31 requires visual examination of welds and outlines limitations of acceptability.

The American Society of Mechanical Engineers (ASME) has established a series of symbols for the marking of boilers, pressure vessels and certain appurtenances that are constructed and inspected in accordance with the ASME *Boiler and Pressure Vessel Code*. Welding procedures, welders and all types of manual and machine arc and gas welding operations must comply with the rules of Section IX of the ASME *Boiler and Pressure Vessel Code* if they are to be used in the fabrication of boilers and pressure vessels in compliance with that code. For more information on welding, refer to AWS D1.1.

SECTION 1004
BOILERS

1004.1 Standards. Oil-fired boilers and their control systems shall be listed and labeled in accordance with UL 726. Elec-

tric boilers and their control systems shall be listed and labeled in accordance with UL 834. Solid-fuel-fired boilers shall be listed and labeled in accordance with UL 2523. Boilers shall be designed and constructed in accordance with the requirements of ASME CSD-1 and as applicable, the ASME *Boiler and Pressure Vessel Code*, Section I or IV; NFPA 8501; NFPA 8502 or NFPA 8504.

❖ The history of boilers shows that they are potentially dangerous if not properly designed and constructed.

Along with the code, several industry standards are referenced for the design and construction of boilers. Equipment manufacturers produce boilers in accordance with the referenced standards (see Commentary Figure 1004.1). Only the design and construction of boilers are regulated by these references.

1004.2 Installation. In addition to the requirements of this code, the installation of boilers shall conform to the manufacturer's instructions. Operating instructions of a permanent type shall be attached to the boiler. Boilers shall have all controls set, adjusted and tested by the installer. The manufacturer's rating data and the nameplate shall be attached to the boiler.

❖ Section 1004.2 governs the installation and commissioning of boilers and their control systems. The mechanical equipment requirements for approval, labeling, installation, maintenance, repair and alteration are regulated by Chapters 1 and 3. Applicable installation requirements are also located in Chapters 7, 8, 12 and 13.

Complete operating instructions must be permanently affixed to the boiler upon completion of installation by the contractor. Boiler systems can be complex and generally require coordination among several pieces of equipment. The proper operating procedures, set points, etc., must be specified and included in these instructions. This is usually done in the control system design documentation, including such things as diagrams, system schematics and control sequence descriptions.

Typically an operating and maintenance (O&M) manual is given to the building owner/operator upon completion of the project. Along with operating and control procedures, the O&M manual will clearly identify routine maintenance and calibration information. It is also typical for operation sequences and instructions to be placed under glass in a conspicuous place in the boiler room for use by operating and service personnel. Furthermore, the installing contractor must commission the boiler upon installation by calibrating, setting, adjusting and testing all operating and control functions (see commentary, Section 1004.7). The intent is to give the owner/operator all of the information necessary to properly operate the boiler and its associated controls and equipment.

1004.3 Working clearance. Clearances shall be maintained around boilers, generators, heaters, tanks and related *equipment* and appliances so as to permit inspection, servicing, repair, replacement and visibility of all gauges. When boilers are installed or replaced, clearance shall be provided to allow access for inspection, maintenance and repair. Passageways around all sides of boilers shall have an unobstructed width of not less than 18 inches (457 mm), unless otherwise *approved*.

❖ Because boilers, related appliances and mechanical equipment require inspection, observation, routine maintenance, repairs and possible replacement, working clearance is required around the equipment and appliances. Access recommendations or requirements are usually stated in the manufacturer's installation instructions. As a result, the provisions stated here are intended to supplement the manufacturer's installation instructions and are in addition to the requirements of the manufacturer. The manufacturer's installation instructions could contain requirements that exceed those in the code. Where differences or conflicts occur, the more restrictive requirements apply.

Addressed specifically is access to specific equipment and appliances, such as boilers, generators,

heaters and tanks and specific components such as gauges. The intent is to provide access to all components that require observation, inspection, adjustment, servicing, repair and replacement. Access is also necessary to conduct operating procedures such as startup or shutdown.

The code defines "Access" as being the ability to reach an item with or without removal of a panel, door or similar obstruction. An appliance or piece of equipment is not accessible if any portion of the structure's permanent finish materials such as drywall, plaster, paneling, built-in furniture or cabinets or any other similar permanently affixed building component must be removed.

1004.3.1 Top clearance. Clearances from the tops of boilers to the ceiling or other overhead obstruction shall be in accordance with Table 1004.3.1.

❖ Boilers are manufactured with manholes, handholes, washouts or other openings for examination, clean-

WEIL-McLAIN
MODEL LGB SERIES 2 BOILER

TO DETERMINE BOILER SIZE, COUNT THE NUMBER OF SECTIONS OR MEASURE THE JACKET LENGTH. CHECK BOX NEXT TO BOILER SIZE INSTALLED.

MODEL NUMBER	NUMBER OF SECTIONS	JACKET LENGTH INCHES	MIN. RELIEF VALVE CAP. LBS/HR OR MBH	INPUT BTU/HR	MINIMUM INPUT BTU/HR	CSA GROSS OUTPUT BTU/HR	NET I=B=R OUTPUT Steam Sq. Ft.	Steam MBH	Water MBH
☐ LGB-4	4	21	325	400,000	---	324,000	1013	243	282
☐ LGB-5	5	26	422	520,000	---	421,200	1317	316	366
☐ LGB-6	6	31	527	650,000	325,000	526,500	1646	395	458
☐ LGB-7	7	36	632	780,000	390,000	631,800	1975	474	549
☐ LGB-8	8	41	738	910,000	455,000	737,100	2304	553	641
☐ LGB-9	9	46	843	1,040,000	520,000	842,400	2633	632	733
☐ LGB-10	10	51	948	1,170,000	585,000	947,700	2965	711	824
☐ LGB-11	11	56	1053	1,300,000	650,000	1,053,000	3292	790	916
☐ LGB-12	12	61	1159	1,430,000	715,000	1,158,300	3621	869	1007
☐ LGB-13	13	66	1264	1,560,000	780,000	1,263,600	3954	949	1099
☐ LGB-14	14	71	1369	1,690,000	845,000	1,368,900	4313	1035	1190
☐ LGB-15	15	76	1475	1,820,000	910,000	1,474,200	4679	1123	1282
☐ LGB-16	16	81	1580	1,950,000	975,000	1,579,500	5046	1211	1373
☐ LGB-17	17	86	1685	2,080,000	1,040,000	1,684,800	5408	1298	1465
☐ LGB-18	18	91	1791	2,210,000	1,105,000	1,790,100	5775	1386	1557
☐ LGB-19	19	96	1896	2,340,000	1,170,000	1,895,400	6125	1470	1648
☐ LGB-20	20	101	2001	2,470,000	1,235,000	2,000,700	6471	1553	1740
☐ LGB-21	21	106	2106	2,600,000	1,300,000	2,106,000	6813	1635	1831
☐ LGB-22	22	111	2212	2,730,000	1,365,000	2,211,300	7155	1717	1923
☐ LGB-23	23	116	2317	2,860,000	1,430,000	2,316,600	7496	1799	2014

CERTIFIED BY
WEIL-McLAIN
523 S New Street
Eden, North Carolina
27288-3623

MAWP, WATER 50 PSI
MAWP, STEAM 15 PSI
MAX. WATER TEMP. 250°F

Electrical Input Less Than 12 Amperes, 120 Volts, & 60 Hertz.
Canadian Registration No. H7268.51234679T
MEA Number: 333-85-E
Design Certified Under ANSI Z21.13a 2005 • CSA 4.9a 2005
Low Pressure Boiler

IMPORTANT
Installation not complete unless gas information label attached here.
Labels are attached to gas train in gas control carton.

• For installation on noncombustible flooring.
• Provide service clearances and minimum 24" between jacket and any combustible wall(s) and ceiling. Install in space large in comparison to size of boiler.
• Vent Category I.

550-223-880 (0606)

Figure 1004.1
STEAM AND HOT WATER BOILER LABEL
(Courtesy of Weil-McLain)

ing, maintenance and repair. Top clearances are required because the shell top is typically where access is provided if the head or shell is more than 40 inches (1016 mm) in diameter. Where such openings are not needed or used, such as when the boiler is too small to make entrance practicable, manholes are, of course, omitted. For extremely large boilers, ancillary equipment in the form of a block and tackle, hoist or crane may be integral to facilitate maintenance, adjustment, servicing, repair or replacement processes further justifying the code's provisions for top clearance. Table 1004.3.1 summarizes the top clearance requirements.

The location of manholes, handholes and washout plug openings shall comply with the ASME *Boiler and Pressure Vessel Code*.

TABLE 1004.3.1
BOILER TOP CLEARANCES

BOILER TYPE	MINIMUM CLEARANCES FROM TOP OF BOILER TO CEILING OR OTHER OVERHEAD OBSTRUCTION (feet)
All boilers with manholes on top of the boiler except where a greater clearance is required in this table.	3
All boilers without manholes on top of the boiler except high-pressure steam boilers and where a greater clearance is required in this table.	2
High-pressure steam boilers with steam generating capacity not exceeding 5,000 pounds per hour.	3
High-pressure steam boilers with steam generating capacity exceeding 5,000 pounds per hour.	7
High-pressure steam boilers having heating surface not exceeding 1,000 square feet (93 m²).	3
High-pressure steam boilers having heating surface in excess of 1,000 square feet (93 m²).	7
High-pressure steam boilers with input not exceeding 5,000,000 Btu/h (1465 kW).	3
High-pressure steam boilers with input in excess of 5,000,000 Btu/h (1465 kW).	7
Steam-heating boilers and hot water-heating boilers with input exceeding 5,000,000 Btu/h (1465 kW).	3
Steam-heating boilers exceeding 5,000 pounds of steam per hour (2268 kg/h).	3
Steam-heating boilers and hot water-heating boilers having heating surface exceeding 1,000 square feet (93 m²).	3

For SI: 1 foot = 304.8 mm, 1 square foot = 0.0929 m², 1 pound per hour = 0.4536 Kg/h.

1004.4 Mounting. *Equipment* shall be set or mounted on a level base capable of supporting and distributing the weight contained thereon. Boilers, tanks and *equipment* shall be secured in accordance with the manufacturer's installation instructions.

❖ The provisions of Section 1004.4 are intended to control the design and construction of the structural ele-

ments that support and distribute the weight and vibratory energy of boilers, their related appliances and mechanical equipment. Because boilers and other pressure vessels are usually quite heavy, the support structure must be strong enough to carry all superimposed loads. These structural elements must be designed to comply with the *International Building Code®* (IBC®).

Typically, boilers will not be anchored to the structure because of their weight. The manufacturer's installation instructions will dictate whether or not the appliance or equipment must be secured to the structure. Equipment, such as boilers, chillers, pumps, blowers, cooling towers, ducts, piping and rigid electrical conduits that are rigidly bolted to a structure, transmits 100 percent of its vibratory energy. The introduction of properly selected vibration isolators will reduce this transmitted energy to a level that is no longer annoying to the occupants, or structurally destructive or detrimental to connected systems. As with the anchoring requirements, the manufacturer will mandate whether or not vibration isolation is required. Because loads caused by vibration are not easily or exactly determined, the services of a registered design professional may be required to design job-specific vibration isolation.

1004.5 Floors. Boilers shall be mounted on floors of noncombustible construction, unless *listed* for mounting on combustible flooring.

❖ As part of the labeling requirement, the manufacturer's installation instructions will specify the type of flooring on which the boiler may be installed. Some types of boilers must be installed on special subbases designed to protect combustible floors and provide clearance from heat-radiating components. Depending on the design, some types of boilers must be installed on a noncombustible floor to eliminate a fire hazard caused by the heat radiated from the unit. The manufacturer's installation instructions will specify the minimum distance that a noncombustible floor must extend beyond the boiler housing, combustion chamber access, access openings or combustion air inlets. The risk of fire is greater at these locations because of heat radiation and the potential for flame roll-out, sparks, embers or burning debris. Unless expressly listed otherwise in the manufacturer's installation instructions, all boilers must be mounted on floors of noncombustible construction.

1004.6 Boiler rooms and enclosures. Boiler rooms and enclosures and access thereto shall comply with the *International Building Code* and Chapter 3 of this code. Boiler rooms shall be equipped with a floor drain or other *approved* means for disposing of liquid waste.

❖ This section addresses the enclosures that house or contain boilers and their related mechanical equipment.

A "Boiler room" is defined as a room primarily used for the installation of a boiler. Although the code defines this term, it is not the intent to require installation of boilers in rooms used primarily for that purpose. A room housing a boiler does not necessarily constitute a boiler room. See the commentary on boiler rooms in Chapter 2.

When a boiler is installed in a room that is used primarily for housing the boiler, the resulting boiler room might have to be separated or protected in accordance with the IBC. See the IBC for specific details on boiler room enclosures and fire suppression systems.

Either a fire-resistance-rated enclosure or fire suppression could be required because a boiler installed in an enclosed space is considered to be a greater hazard than if the appliance were out in the open and not enclosed. This is based on experience that such rooms are often used for unapproved storage and a fire hazard is created where combustible materials are stored near the fuel-fired appliances. It is not uncommon to find highly combustible materials such as paper, cardboard, cleaning chemicals and similar materials placed directly against appliances, vent or chimney connectors, draft hoods and other components having a required minimum clearance to combustibles.

Furthermore, the boiler room must have a floor drain or other means to dispose of waste water. The floor drain requirement is intended for floor systems, such as concrete, that can be made to slope to the drain. Floor drains are typically installed as an emergency fixture, preventing flooding of a room or space. For boiler rooms in particular, a floor drain conveniently permits the disposal of standing water or the liquid waste that can be expected from the operation and maintenance of boilers (blow-downs, skimmers and relief valves). Uncontained and idle liquids on the boiler room floor may represent a hazard or cause damage to the building's finish or structure. Any other means for disposing of liquid wastes from boiler rooms would have to be approved as an alternative in accordance with Section 105.2. Material standards and trap and strainer requirements, along with sizing provisions for floor drains, are in the IPC.

1004.7 Operating adjustments and instructions. Hot water and steam boilers shall have all operating and safety controls set and operationally tested by the installing contractor. A complete control diagram and boiler operating instructions shall be furnished by the installer for each installation.

❖ As part of the final approval process, this section requires careful calibration and testing of the operating and safety control systems for hot water and steam boilers by the installing contractor. Startup can be the most critical period in the life of the boiler. The manufacturer's instructions must be consulted and the contents thoroughly understood by all parties involved during initial startup and eventual operation of the boiler (see commentary, Section 1004.2). It is critical that the operating and safety controls be tested operationally to verify the proper functioning of the control system. Owners and operating and maintenance personnel should be present during these procedures. Boiler operation instructions and control system design documentation, including such things as diagrams, system schematics and control sequence descriptions, must be furnished by the installing contractor for use by operating and maintenance personnel. Typically, an O&M manual is given to the building owner/operator upon completion of the project. Along with operating and control procedures, the O&M manual will clearly identify routine maintenance and calibration procedures. It is also typical for operation sequences and instructions to be placed under glass in a conspicuous place in the boiler room for use by operating and service personnel. The intent is to give the owner/operator all of the information necessary to properly operate the boiler and its associated controls and equipment.

SECTION 1005
BOILER CONNECTIONS

1005.1 Valves. Every boiler or modular boiler shall have a shutoff valve in the supply and return piping. For multiple boiler or multiple modular boiler installations, each boiler or modular boiler shall have individual shutoff valves in the supply and return piping.

> **Exception:** Shutoff valves are not required in a system having a single low-pressure steam boiler.

❖ Shutoff valves are required to facilitate maintenance and repairs by isolating a boiler from the system that it serves. Without isolation valves, the entire system of piping and components would have to be taken out of service, drained and relieved of pressure before a boiler or water-side/steam-side component could be removed for servicing, repair or replacement. Where systems have multiple boilers, isolation valves permit removal of one or more boilers from service without affecting the remaining units, thus preserving partial system operation [see Commentary Figure 1005.1(1)]. A modular boiler unit refers to a field-assembled boiler composed of multiple interdependent modules. Modular units are designed to permit close matching of boiler capacity to building loads and to allow for future expansion of capacity by the addition of modules to a boiler "package." Each independent boiler or modular boiler, whether single or in groups, must be valved individually [see Commentary Figure 1005.1(2)].

The exception states that shutoff valves are not required for systems with a single low-pressure steam boiler. Shutoff valves would not be of value for a single low-pressure steam boiler installation because of a negligible amount of heat transfer medium that needs to be relieved from the system. Also, maintaining a partial capacity system operation is impossible with only a single unit.

1005.2 Potable water supply. The water supply to all boilers shall be connected in accordance with the *International Plumbing Code.*

❖ The fill and makeup water supply to both steam and hot water boilers is generally taken directly from the potable water supply of the building. Water supply connections to boilers are referred to as "makeup water supplies," and are installed to replace, either manually or automatically, boiler water that is lost from boiler blowdown, relief/safety valve discharge, evaporation, control flushing, air purging or leakage. This arrangement causes the potable water supply to be highly susceptible to contamination by backflow. Where any direct connection is made between two pressurized systems, the direction of flow through that connection or interface is dependent on the pressure differential between the systems. If the boiler pressure is higher than the potable water supply pressure, which it often is, boiler water will flow into

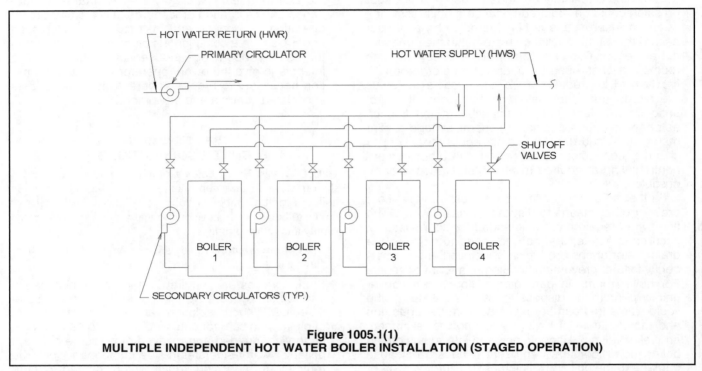

Figure 1005.1(1)
MULTIPLE INDEPENDENT HOT WATER BOILER INSTALLATION (STAGED OPERATION)

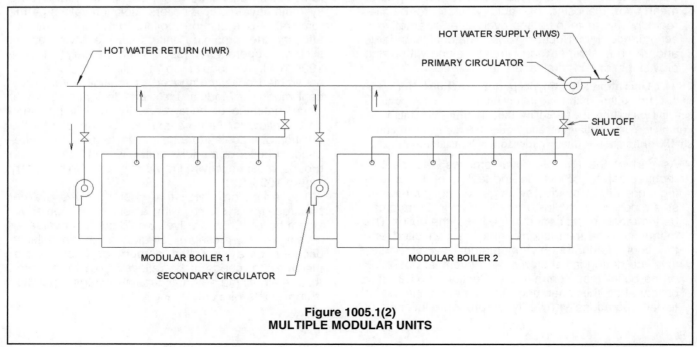

Figure 1005.1(2)
MULTIPLE MODULAR UNITS

the potable water distribution system. Boiler water is not potable and it often contains conditioning and cleaning chemicals; therefore, any backflow into the potable water supply system could produce a health hazard. As a result, the potable water supply must be fitted with backflow prevention devices as prescribed in the IPC. Three examples of makeup water connections are shown in Commentary Figure 1005.2.

SECTION 1006
SAFETY AND PRESSURE RELIEF VALVES AND CONTROLS

1006.1 Safety valves for steam boilers. All steam boilers shall be protected with a safety valve.

❖ A steam boiler is a pressure vessel designed to generate steam by heating water above its boiling point. The developed steam pressure is directly proportional to both the amount of heat input and the temperature of the boiling water. The desired steam pressure is achieved and maintained by controlling the rate of energy input. If the boiler controls malfunction, it is possible for the steam pressure to exceed the maximum pressure for which the boiler and piping distribution system was designed. The result could be a violent rupture or an explosion accompanied by the

instantaneous release of the tremendous thermal energy stored in the superheated water. To prevent such a result, steam boilers must be protected by one or more safety valves designed to relieve pressure at a selected setpoint, thus preventing the buildup of dangerously high pressures. A safety valve is designed to open fully whenever its opening pressure (setpoint) is reached and to close after a predetermined reduction in pressure. The differential between the opening and closing pressure of a safety valve is known as its "blowdown" (see Commentary Figure 1006.1).

1006.2 Safety relief valves for hot water boilers. Hot water boilers shall be protected with a safety relief valve.

❖ Although hot water boilers are not designed to generate steam, hydrostatic pressures proportional to the water temperature will develop. For the same reasons as stated in the commentary to Section 1006.1, hot water boilers must be protected from overpressure by one or more relief valves. Relief valves perform the same function as safety valves, but they do not operate identically. A relief valve is designed to open in direct proportion to the water pressure force acting on its closure disk. The higher the pressure, the greater the force, and the more the valve opens (see Commentary Figure 1006.2).

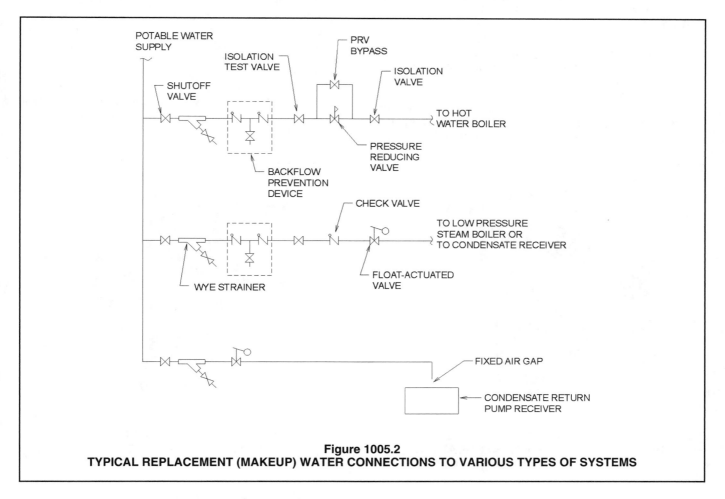

Figure 1005.2
TYPICAL REPLACEMENT (MAKEUP) WATER CONNECTIONS TO VARIOUS TYPES OF SYSTEMS

1006.3 Pressure relief for pressure vessels. All pressure vessels shall be protected with a pressure relief valve or pressure-limiting device as required by the manufacturer's installation instructions for the pressure vessel.

❖ Installed in accordance with the manufacturer's installation instructions, a pressure relief valve or a pressure limiting device protects the pressure vessel against possible explosion. Pressure vessels installed without one of these protection devices can produce devastating explosions and have been responsible for loss of life and damage to property. Relief valves also play an important role in protecting aging equipment that may have deteriorated or weakened as a result of exposure and use. Furthermore, in the event of control system malfunction or improper adjustments, these devices are a safe and effective means of preventing a violent rupture of the system.

1006.4 Approval of safety and safety relief valves. Safety and safety relief valves shall be *listed* and *labeled*, and shall have a minimum rated capacity for the *equipment* or appliances served. Safety and safety relief valves shall be set at a maximum of the nameplate pressure rating of the boiler or pressure vessel.

❖ Because of their importance, safety and relief valves must be tested and certified by an approved agency to determine that each device complies with the applicable standard, is dependable and functions within its specifications. Safety and relief valves for steam and hot water boilers have a rated relief capacity and a fixed or adjustable opening pressure. All boiler labels will specify the boiler's maximum working pressure. The safety and relief valve must be set to open at or below this pressure.

The labels on boilers will also specify the minimum required safety and relief valve capacity. A safety and relief valve must be able to dissipate energy at a rate equal to or greater than the rate of energy input to the boiler. An undersized (insufficient capacity) valve would be unable to prevent the boiler pressure from exceeding the maximum capacity, resulting in a dangerous vessel failure. Safety and relief valve capacity is expressed in terms of pounds of steam per hour or British thermal units per hour [Btu/h] (W) of boiler input rating.

1006.5 Installation. Safety or relief valves shall be installed directly into the safety or relief valve opening on the boiler or pressure vessel. Valves shall not be located on either side of a safety or relief valve connection. The relief valve shall discharge by gravity.

❖ Boilers are manufactured with either factory-installed safety or relief valves or with the provisions for the field installation of such valves. In all cases, the safety or relief valves must be installed directly into the opening provided for that purpose by the boiler manufacturer. The intent is to eliminate any possibility of blockage or restriction between the source of the pressure (boiler) and the safety or relief valve. Where safety and relief valves are located any distance from the boiler, the possibility exists for blockages or restrictions to develop in stagnant lines, increasing the risk of creating a piping arrangement in which the safety or relief valve could be isolated (valved off) from the boiler by control valves.

Additionally, any length of piping and fittings would cause a pressure drop between the boiler and the safety or relief valve, allowing the boiler pressure to rise above the opening pressure of the safety and relief valve. A shutoff or control valve of any type located upstream (inlet side) or downstream (discharge side) of a safety or relief valve could prevent or impede the operation of the device, creating an extremely hazardous condition.

Figure 1006.1
SAFETY VALVE FOR STEAM APPLICATIONS
(Courtesy of Watts Regulator Co.)

Figure 1006.2
PRESSURE RELIEF VALVE FOR
HOT WATER APPLICATIONS
(Courtesy of Watts Regulator Co.)

The requirement to drain by gravity is intended to prevent water from standing in the discharge pipe and causing corrosion damage to the device and/or creating a blockage or restriction. Water or condensate trapped in a discharge pipe would also act as an obstruction by creating a backpressure that would affect the capacity and operation of the relief valve. In the case of a steam safety valve, any standing water in the discharge pipe could cause severe water hammer and damage because of the velocity of the steam and the rapid expansion of the standing water.

1006.6 Safety and relief valve discharge. Safety and relief valve discharge pipes shall be of rigid pipe that is *approved* for the temperature of the system. The discharge pipe shall be the same diameter as the safety or relief valve outlet. Safety and relief valves shall not discharge so as to be a hazard, a potential cause of damage or otherwise a nuisance. High-pressure-steam safety valves shall be vented to the outside of the structure. Where a low-pressure safety valve or a relief valve discharges to the drainage system, the installation shall conform to the *International Plumbing Code*.

❖ Safety or relief valve discharge pipe is designed to direct the discharge to a location where it cannot cause injury or property damage. A discharge pipe could be conducting high-temperature or high-pressure water or steam; therefore, the material from which the discharge pipe is constructed must be able to withstand such pressures and temperatures, as well as be able to resist the forces developed during discharge that would tend to dislocate the discharge pipe.

If a discharge pipe were smaller in its internal cross-sectional area than the safety or relief valve outlet, the resulting restriction would reduce the relieving capacity, adversely affecting the operation of the device.

Because the discharge from any safety or relief valve is a threat to both the building and its occupants, each installation must be individually evaluated to prevent the potential discharge from being hazardous.

Because a high-pressure steam safety valve discharge into a building interior would be especially dangerous, the discharge must be directed to the outdoors. Discharge pipes must be as short as possible to prevent friction losses affecting the device's operation and capacity. Discharge piping must be sized and arranged so the relieving capacity of the device is not affected.

Safety and relief valve discharge pipes must never be located where they are subject to freezing because ice could cause a complete or partial blockage. Where discharge pipes must extend to the outdoors, as in the case of high-pressure steam safety valves, the discharge pipe must be connected indirectly through a special air-gap-type fitting or other arrangement that will not cause the safety valve outlet to be blocked in the event of a discharge pipe blockage.

This section does not require any safety or relief valve to discharge to a building's drainage system;

however, an installation that does must conform to the IPC.

Because safety and relief discharge is a symptom of a boiler or system abnormality or malfunction, the discharge pipe should be installed so any discharge will be noticeable, serving as a warning to begin remedial action.

1006.7 Boiler safety devices. Boilers shall be equipped with controls and limit devices as required by the manufacturer's installation instructions and the conditions of the listing.

❖ The boiler manufacturer's installation instructions, the standard to which the boiler is listed and the conditions of listing will determine what controls and safety devices are required for any boiler. These include high pressure limits, high temperature limits, low-water cutoffs, fuel pressure limits, flame safeguard controls, purge and ignition sequence controllers and combustion airflow sensors. Devices used to protect and control heating appliances are mandated by national standards such as those of ASME, the International Approval Services (IAS), Underwriters Laboratories (UL) or the National Fire Protection Association (NFPA). Standards may also be mandated by insurance carriers and state and local authorities (see commentary, Section 1004.1).

Probably the most important safety device on a boiler is the flame-safeguard (flame-supervision) control. Its function is to monitor the pilot burner or main burner flames and open or close the fuel valves, depending on the presence or absence of flame, respectively (see commentary, IFGC Section 602.2).

A flame-safeguard control establishes safe conditions for initiating and sustaining combustion. Often, the flame-safeguard control function is provided by a package called the "primary control." The primary control starts the burner events in the proper sequence, proves that combustion air is available, purges the combustion chambers, proves that an ignition source is present, proves that the burner flame is established and supervises the flame during burner operation. It triggers safety shutdown after a set time interval, identified as the equipment's safety control timing, when the pilot or main burner fails to ignite or upon loss of flame. In addition, the primary control checks its own circuitry for proper function. Typically, a check for simulated flame failure and a continuity check of safety-switch circuitry are made. The most sophisticated self-checking feature is a dynamic self-check system that checks internal circuitry roughly one to four times each second during operation.

On larger-size boilers, additional timing functions are required; the package is called a "flame-safeguard programming control." Added functions include purging with a minimum combustion airflow rate before light-off, sequencing of ignition and pilot light-off, proof of pilot before main valve opening and, sometimes, interrupting the pilot to prove the main flame before commencing normal operation.

The function of proving the existence of flame is common to all safeguard controllers. Methods of detecting the presence of a pilot flame have included thermocouples, liquid-filled capillary tubes, bimetal mechanisms, power pile generators, infrared and ultraviolet radiation detectors and flame rectification circuits. Three general approaches are found in most flame-sensing devices: thermal sensing, flame-rod sensing and optical detection. The most commonly used type of flame-safeguard device uses a thermocouple and an electromechanical device that function together to supervise a standing pilot flame. Thermal sensing, as it is called, is typical for small-input burners. If the pilot flame is extinguished, the drop in thermocouple output voltage will cause the control valve to "lock out" in the closed position, thereby preventing the flow of fuel to the main burner and the pilot burner.

Modern energy-efficient domestic equipment and appliances seldom use standing pilots any more, and instead use electric-spark-ignited intermittent pilots or direct ignition devices such as hot surface igniters and glow coils. Flame-rod sensing (flame rectification) relies on the ability of a flame to conduct current through the ionized combustion gases in the flame. The flame rod and its associated opposite polarity electrode function together to rectify a weak alternating current to create a detectable direct current, measured in microamperes. The presence of the dc current proves the existence of a flame at the electrodes. Flame-rod authentication that passes current in only one direction is used because it provides protection against the high-resistance leakage to ground giving a false indication that flame is present. This method of sensing is suitable for gas, but not typically used for oil-fired boilers because carbon and ash deposits tend to form on the rod. Optical sensing is implemented in three methods that sense visible, infrared (IR) or ultraviolet (UV) light. The visible-light optical method uses a rectifying photocell, typically cadmium sulfide, and is suitable only for oil-fired boilers because gas flames generally do not provide sufficient visible light. Infrared optical detection is accomplished by use of a lead sulfide cell and is suitable for gas- or oil-fired boilers. The sensor package reacts to flickering detection as opposed to a steady signal from a hot refractory surface. Ultraviolet detectors must be positioned so that the detector is not subject to oil vapor that absorbs UV light and does not see ignition spark or hot refractory over 2,500°F (1371°C). See Table 1006.7 for characteristics of flame rod and optical- or IR/UV-sensing methods.

The application of a flame-safeguard control depends on boiler and burner design. The system therefore is normally supplied as a complete package by the boiler-burner manufacturer. Only main burner (as opposed to pilot burner) monitoring is necessary where the ignition is pilotless (direct electric or hot surface), where the pilot is dropped out (interrupted) after main burner ignition or where combustion is not involved (electric boilers).

Safety shutoff fuel valves are typically motorized and are usually installed in series with another automatic fuel valve. Oil-fired boilers must use more than one fuel valve, installed in series, to control the flow of oil to the burner where pressures greater than 10 pounds per square inch (psi) (69 kPa) are possible. Motorized fuel valves may also be used to achieve certain light-off characteristics or to allow the installation of valve-closure-proving devices. Generally, hand-operated valves (such as lubricated plug cocks) are also installed so that the fuel supply to the burner can be shut off manually.

Safety shutoff valves must have temperature and pressure ratings compatible with the type of fuel being burned and must be interlocked with the required limit devices and controllers. Typically, this interlocking control arrangement is done through the primary or programming control provided as a package by the burner manufacturer.

Gas-fired boilers have safety shutoff valves that respond to the action of the required limit controls and combustion safeguards. It is important to reinforce the point that fuel-burning appliances must never be operated outside of the safe limits for those appliances. Operation cannot be tolerated at fuel pressures that are excessively high or low, ultimately resulting in burner malfunction or overfiring. If fuel pressure were allowed to go above the normal operating range, control damage could result or the boiler could be overfired. If the fuel pressure is too low, combustion could be affected and efficiency and flame properties would be adversely affected, possibly causing flame roll-out or flashback within burners. Pressure-sensitive switches monitor fuel input pressure and close the fuel valve(s) if an abnormally high- or low-fuel pressure is detected. Once the high or low fuel pressure limit control activates, it locks out and the limit control must be manually reset (nonrecycling) before the burner will fire.

Oil-fired boilers have safety shutoff valves that respond to the action of the required limit controls and combustion safeguards. As with the gas-fired boilers above, it is important to reinforce the point that fuel-burning appliances must not be operated outside of their safe limits. Oil-fired boilers incorporate metering valves, orifices and/or air-pressurized systems to atomize the fuel into an acceptable air-fuel mixture for combustion.

Oil is sometimes preheated to lower its viscosity and improve atomization. In such burner systems, proper combustion is dependent on oil preheating; therefore, lower-than-required oil temperature must trigger a burner shutdown.

When the pressure of the steam or air supply used for fuel atomization is inadequate, a fuel supply shutdown and lockout must occur. Controls for this pressure monitoring must be of the manual reset type that require service personnel to rectify the problem before the burner is restarted.

Forced- or induced-draft systems and forced-combustion-air systems must use one or more control devices to shut down the burner in the event of fan failure, inadequate draft or inadequate combustion air supply. For other than natural draft, atmospheric-burner-type boilers, draft monitoring is required to verify the function of blowers or fans used to produce draft in the boiler and/or in the chimney or vent. Power burners are equipped with blowers that force combustion air into the burner where it is mixed with the fuel. Generally, an air-proving switch is wired to the burner flame-safeguard interlock circuit. The device monitors airflow and, in the event of inadequate flow, closes the fuel valve(s), resulting in burner shutdown. A time-delay relay device wired into

the automatic shutdown circuit is often used to prevent the nuisance of frequent burner shutdowns caused by improper draft conditions. The installation of an air-proving switch or similar control device is not necessary for combustion airflow monitoring for an oil burner when a single dual-shaft motor directly drives both the fan and the oil pump. In this case, failure of the fan will also cause loss of oil supply and result in a shutdown.

A low-fire start prevents burner ignition at full input capacity of the boiler. There are three major firing rate control methods: on/off, off/low/high and modulating. In the on/off method, the burner is either off or firing at its full rated capacity. This type of control is used primarily on smaller boilers when control of output temperature or pressure is not critical. The off/low/high method allows the boiler to more closely

Table 1006.7
COMPARISON OF FLAME DETECTION METHODS
(Source – Handbook of HVAC Design)

	FLAME ROD	RECTIFYING PHOTOCELL	INFRARED (lead sulfide) DETECTOR	ULTRAVIOLET DETECTOR
Will supervise oil	no	yes	yes	yes
Will supervise gas	yes	no	yes	yes
Will detect ignition spark	no	no	no	yes
Type of signal	dc	dc	ac	dc
Light responsive	none	visible	infrared	ultraviolet
Max. ambient temperature at the sensor or cell	500°F (260°C)[a]	165°F (74°C)	125°F (52°C)	125-250°F (52-121°C)[b]
False flame signals by:				
Inductive pickup	no	no	no[c]	no[d]
Capacitive pickup	no	no	no[e]	no[d]
Refractory glow	no	yes	no[f]	no[g]
Checkout tests required:				
Hot refractory saturation	no	no	yes	no
Hot refractory hold-in	no	yes	yes	no
Ignition interference test	yes	no	no	yes[h]
Pilot turndown test	yes	yes	yes	yes

For SI: 1 foot = 304.8 mm, °C = (°F -32)/1.8.

a. Maximum insulator temperature is 500°F. Maximum rod temperatures are 2,000°F for Jellit Alloy "K," 2,462°F for Kanthal A-1 and 2,600°F for Globar.

b. C7012E and F - 135°F; C7012A and C - 135°F; C7044A - 212°F; C7027A - 215°F; C7035A - 250°F.

c. Leads must be run alone in grounded conduit all the way to the wiring subbase.

d. C7027 and C7035 should use number 14TW leadwire when running long distances of up to 1,100 feet. C7012E should use coaxial cable; RG62U will provide satisfactory performance with lengths of the order of 500 feet.

e. Leads must be BX cable, shielded cable or twisted pair. Conduit must be rigid or fastened securely to minimize vibration.

f. Will not sense the hot refractory alone, but turbulent hot air, steam, smoke or oil spray may cause the radiation to fluctuate, stimulating flame.

g. An ultraviolet detector will respond to hot refractory at different temperatures: C7076 maximum hot refractory at 2,200°F; C7012 maximum hot refractory at 2,300°F; and C7027, C7035 and C7044 maximum hot refractory at 2,800°F.

h. An ultraviolet detector will sense ignition spark; ignition spark response test is required.

match the load. Given a demand for heat, the boiler begins firing at approximately 30- to 50-percent capacity (low fire). Upon a further increase in demand, the burner fires at maximum capacity (high fire).

The modulating firing rate control method provides the greatest flexibility in matching the boiler output to load. On call for heat, the boiler is brought to low fire, and as a greater load is imposed on the boiler, an increasing amount of fuel and air is introduced to the burner. The output of the boiler varies continuously between low and high fire, matching any load between 30- and 100-percent of boiler capacity.

Some boilers are required to incorporate a purging cycle prior to burner ignition. The purging cycle removes any unburned fuel/air mixture. During the purging cycle, a forced- or induced-draft fan, a combustion-air fan or natural draft sweeps the unburned fuel/air mixture from the boiler combustion chamber and heat exchanger passes. Even though this type of cycling takes energy out of the system, it permits smooth light-off and prevents internal explosions.

If burners have closable secondary air openings, the closing mechanism must be designed to open and hold open the secondary air openings for a period of not less than 4 minutes prior to burner ignition.

Dual high-temperature limit controls provide redundant protection in the event of control device failure. A manual reset on the high limit control is intended to alert the maintenance or service personnel to an extreme condition in which the lower setting device has failed to shut down the heat input to the boiler. Low-water cutoffs are not typically installed with isolation valves between them and the boiler because closed isolation valves would prevent the device from acting. Forced-circulation (water coil and water tube) boilers use flow switches to prove circulation. These devices would detect a low-water condition, because that condition would result in circulation failure.

Steam boilers also use dual low-water limit controls and dual high-pressure limit controls. Dual low-water limit (low-water cutoff) controls are installed to achieve backup (redundant) protection (see commentary, Section 1007). Dual high-pressure limit controls are installed to achieve redundant (backup) protection. These control devices are set at some value below that of the steam safety valve so that the burner shuts down before the safety valve opens. The high limit control with the higher setting of the two controllers must be equipped with a manual reset that requires remedial action before the boiler can be restarted. Once this high limit control locks out, it must be manually reset (nonrecycling) before the burner will fire again.

1006.8 Electrical requirements. The power supply to the electrical control system shall be from a two-wire branch cir-

cuit that has a grounded conductor, or from an isolation transformer with a two-wire secondary. Where an isolation transformer is provided, one conductor of the secondary winding shall be grounded. Control voltage shall not exceed 150 volts nominal, line to line. Control and limit devices shall interrupt the ungrounded side of the circuit. A means of manually disconnecting the control circuit shall be provided and controls shall be arranged so that when deenergized, the burner shall be inoperative. Such disconnecting means shall be capable of being locked in the off position and shall be provided with ready access.

❖ Field-installed control wiring for appliances and equipment must be installed in accordance with NFPA 70 and this code. NFPA 70 is referenced in Section 301.10 (see commentary, Section 301.10) and governs all electrical power and control wiring for mechanical equipment and appliances regulated in the code. Although field-installed power wiring is not included within the scope of this section, wiring, disconnects, overcurrent protection devices, starters and related hardware used to supply electrical power to the appliance or equipment must comply with Section 301.10.

Control wiring includes the wiring, devices and related hardware that connect the main unit to external controls and accessories, such as temperature and pressure sensors, water level controls and safety shutoff valves. The internal factory wiring of appliances and equipment is not covered by this section unless it is specifically addressed in NFPA 70, however, this wiring is covered by the testing and review performed by an approved agency as part of the labeling process.

The code official responsible for the inspection of appliances and equipment must be familiar with the applicable sections of NFPA 70.

The intent of this section is to require wiring and a grounding arrangement that will not allow control devices to be bypassed as a result of a conductor that has faulted to ground. In circuits up to 150 volts, one conductor will be ungrounded (hot) and the other will be grounded. By grounding all devices and raceways and by switching (breaking) only ungrounded (hot) conductors, unintended fault current paths can be avoided. A fault to ground would open an overcurrent protective device, indicating a fault and preventing boiler operation. Such unintended paths can effectively bypass a safety control, resulting in a dangerous condition.

The disconnect switch is a safety device that allows the service technician to shut off and lock out the power to the unit and its controls during service, maintenance and repair operations to protect the technician and to comply with the requirements of the Occupational Safety and Health Association (OSHA) for isolation of equipment. The disconnect switch must be readily accessible.

SECTION 1007
BOILER LOW-WATER CUTOFF

1007.1 General. All steam and hot water boilers shall be protected with a low-water cutoff control.

❖ If a steam or hot water boiler is operated (fired) without water or with water below the minimum level, overheating and severe boiler damage can occur. In addition to damage or destruction of the boiler, a severe fire or explosion hazard can result from the overheating or failure of the combustion chamber enclosure. Burner malfunction can also occur as a result of leakage and destruction of the boiler heat exchange surfaces. The heat exchange surfaces of fuel-fired boilers absorb radiated heat from the flames and heat from the combustion gases produced by the combustion of fuel. This heat is conducted to the water at a rate that prevents the temperature of the heat exchanger material from exceeding the maximum for which it was designed. As the water level drops in a boiler, a greater area of heat exchange surface becomes dry and thus subjected to overheating. All boiler manufacturers specify and mark on the unit the minimum water level to be maintained in the boiler, below which damage would be possible. All boilers are subject to a loss of water. This is especially true for steam boilers because they are dependent on the timely return of condensate water and the proper operation and sequencing of controls for the return and makeup water feed systems.

A low-water cutoff is an essential control device designed to prevent boiler operation when the water level is too low. Because steam boilers, as opposed to hot water boilers, which are completely filled, maintain a water level, a low-water condition is much more likely to occur in a steam system. However, the hazards resulting from low-water levels are the same for both boilers. Also, a flash steam explosion can occur if makeup water is introduced into an over-heated steam or hot water boiler. For these reasons, hot water boilers must be protected by a low-water cutoff control. This section requires low-water cutoffs without regard to heat energy input. Some low-water cutoff controls are intended to be field installed as part of the boiler "trim" (appurtenances) and others are factory installed directly into tapped openings on the boiler.

Traditional low-water cutoff controls use float-actuated mechanisms. Electronic probe-type controls, which are commonly used for hot water boiler protection, detect the water level with an electrode that relies on the slight conductivity of the boiler water to complete a relay circuit. In the case of coil-type (water-tube) boilers requiring forced circulation to prevent coil or tube overheating, a flow-sensing device, which detects flow and verifies that the boiler and system are full of water, is required. Low-water cutoff controls do not sense flow and, therefore, cannot protect a forced circulation coil-type/watertube boiler from overheating as a result of loss of circulation.

Electric boilers are generally protected with factory-installed water-level sensing devices to prevent heating element failure, which could occur if the elements were not submerged [see Commentary Figures 1007.1(1) through (3)].

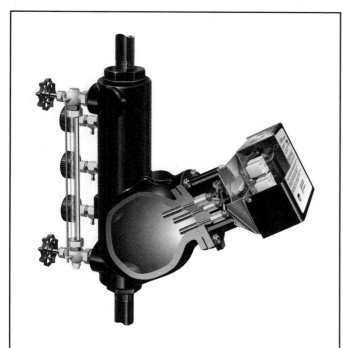

Figure 1007.1(1)
ELECTRONIC PROBE-TYPE LOW-WATER
CUTOFF CONTROL WITH GAUGE GLASS AND
TRI-COCKS FOR STEAM BOILER APPLICATION
(Photo courtesy of McDonnell & Miller)

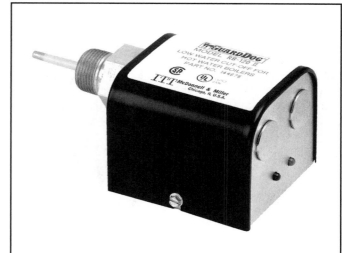

Figure 1007.1(2)
ELECTRONIC PROBE-TYPE LOW-WATER CUTOFF
CONTROL FOR HOT WATER BOILER APPLICATION
(Photo courtesy of McDonnell & Miller)

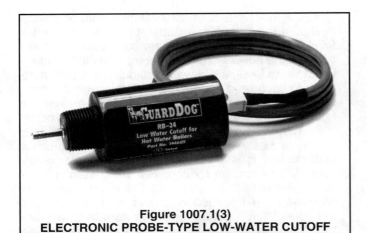

Figure 1007.1(3)
ELECTRONIC PROBE-TYPE LOW-WATER CUTOFF
CONTROL FOR HOT WATER BOILER APPLICATION
(Photo courtesy of McDonnell & Miller)

1007.2 Operation. The low-water cutoff shall automatically stop the *combustion* operation of the *appliance* when the water level drops below the lowest safe water level as established by the manufacturer.

❖ The sole purpose of a low-water cutoff control is to stop (cut off) the heat input to the boiler whenever the water level is dangerously low. These devices automatically interrupt the power supply to the burner controls or heating elements to cause boiler shutdown. Commentary Figure 1007.2 shows a sample boiler safety control schematic.

SECTION 1008
STEAM BLOWOFF VALVE

1008.1 General. Every steam boiler shall be equipped with a quick-opening blowoff valve. The valve shall be installed in the opening provided on the boiler. The minimum size of the valve shall be the size specified by the boiler manufacturer or the size of the boiler blowoff-valve opening.

❖ Steam boilers collect sediment from the water, system piping and equipment. If allowed to build up, this sediment causes boiler overheating, control fouling, loss of efficiency and premature boiler failure.

The sediment consists of minerals from makeup water, piping scale, chemical cleaner and conditioner precipitate, rust and similar materials. Blowoff valves are required to allow the periodic elimination or purging of harmful sediment. The quick-opening feature permits an almost instantaneous full opening of the valve to induce a sudden sweeping flow in the lowest part of the boiler where sediment collects. The fast opening and closing feature also prevents the waste of steam and heated water, which would occur with a slow-acting shutoff or gate valve. The operation must be fast enough to prevent significant reduction of the water level and to allow sediment to be flushed out before steam is relieved through the blowoff valve. The blowoff valve is sized to comply with the manufacturer's recommendations or, in the absence of a recommendation, must not be less than the size of the blowoff valve opening on the boiler. Undersized valves do not allow the necessary flow volume and velocity to entrain and discharge the sediment.

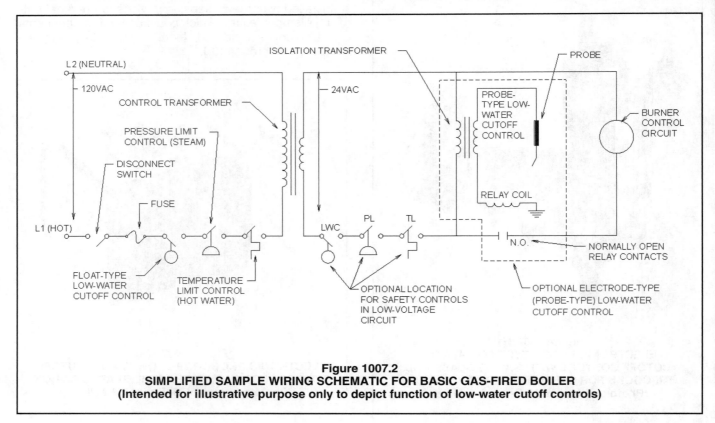

Figure 1007.2
SIMPLIFIED SAMPLE WIRING SCHEMATIC FOR BASIC GAS-FIRED BOILER
(Intended for illustrative purpose only to depict function of low-water cutoff controls)

1008.2 Discharge. Blowoff valves shall discharge to a safe place of disposal. Where discharging to the drainage system, the installation shall conform to the *International Plumbing Code*.

❖ The discharge from a blowoff valve can consist of high-temperature water and steam. Therefore, the discharge must be directed to a place of disposal in a manner that does not threaten people or property. Blowoff discharge is generally drained to the sanitary sewer system of the building or premises. Those installations must comply with the IPC.

SECTION 1009
HOT WATER BOILER EXPANSION TANK

1009.1 Where required. An expansion tank shall be installed in every hot water system. For multiple boiler installations, a minimum of one expansion tank is required. Expansion tanks shall be of the closed or open type. Tanks shall be rated for the pressure of the hot water system.

❖ Hot water heating systems are closed systems that are completely filled with water and purged of entrapped and dissolved gases to the greatest extent possible. When the water in such a system is heated, it will expand and, because it is a closed system, quickly cause hydrostatic pressure that can be relieved only by system failure or the opening of the safety relief valve. Obviously, system failure must be avoided. Because relief valves are intended to open only in the event of an emergency, the continuous opening of a relief valve to accommodate expansion is not acceptable.

Expansion tanks are used to absorb harmlessly the additional system water volume caused by expansion, thus avoiding relief valve opening and preventing wide variations in system pressure. Expansion tanks are either sealed vessels or open tank reservoirs. Open tank reservoirs are, of course, not pressurized except for the static elevation head they impose on the boiler. Closed expansion tanks, however, are pressurized vessels. Because they are subjected to the same pressures as the boiler system, closed expansion tanks must have a pressure rating greater than or equal to the maximum boiler system operating pressure.

Closed expansion tanks contain a cushion of air that compresses as water expands into the tank. The pressure in a hot water system will increase several pounds per square inch as the temperature increases. However, the compression of the air in a properly sized expansion tank will prevent the pressure from exceeding the maximum system operating pressure. An undersized expansion tank would not prevent the system pressure from reaching the relief valve opening pressure.

Rarely used today, open expansion tanks are nothing more than elevated reservoirs holding the expanded hot water at atmospheric pressure.

Multiple boiler installations serving a single common piping system can use either a single expansion tank of the proper size or multiple tanks with the proper aggregate capacity. Any arrangement of one or more tanks capable of controlling system pressure by compensating for expansion is acceptable.

1009.2 Closed-type expansion tanks. Closed-type expansion tanks shall be installed in accordance with the manufacturer's instructions. The size of the tank shall be based on the capacity of the hot-water-heating system. The minimum size of the tank shall be determined in accordance with the following equation:

$$V_t = \frac{(0.00041\,T - 0.0466)\,V_s}{\left(\dfrac{P_a}{P_f}\right) - \left(\dfrac{P_a}{P_o}\right)} \qquad \text{(Equation 10-1)}$$

For SI:

$$V_t = \frac{(0.000738\,T - 0.03348)\,V_s}{\left(\dfrac{P_a}{P_f}\right) - \left(\dfrac{P_a}{P_o}\right)}$$

where:

V_t = Minimum volume of tanks (gallons) (L).

V_s = Volume of system, not including expansion tanks (gallons) (L).

T = Average operating temperature (°F) (°C).

P_a = Atmospheric pressure (psi) (kPa).

P_f = Fill pressure (psi) (kPa).

P_o = Maximum operating pressure (psi) (kPa).

❖ Closed-type expansion tanks can be subdivided into two categories: simple air/water interface compression tanks and air-charged diaphragm (bladder) tanks [see Commentary Figures 1009.2(1) and 1009.2(2)]. Diaphragm-type tanks have become the choice for most residential hydronic systems because they have

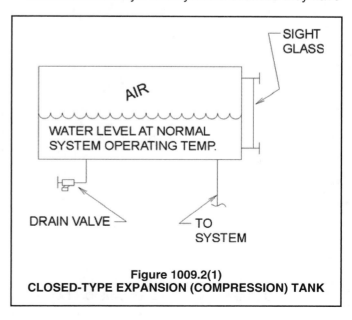

Figure 1009.2(1)
CLOSED-TYPE EXPANSION (COMPRESSION) TANK

the advantage of being maintenance free. Unlike the simple compression tank, a diaphragm-type tank does not have air and water in contact and, therefore, the air cushion cannot be absorbed by the water and cause a loss of capacity.

Traditional compression tanks allowing the air and water to interface directly will slowly become "water logged" as the air is absorbed into the water. For this reason, such tanks require maintenance and special valve and drain fittings to permit the restoration of the air. Compression tanks are often equipped with a sight glass used to observe the water level in the tank, and to determine the need for maintenance. A diaphragm-type tank will not contain water when the system is at its normal fill pressure [typically 12 to 15 psi (83 to 103 kPa) for residential systems] because the air charge on the air side of the diaphragm is intended to be equal to the system pressure. Nondiaphragm-type compression tanks will always have some water below the trapped air cushion.

The amount of expansion in any hot water system is directly proportional to the temperature rise and volume of the water in the system. For example, an old gravity circulation system with cast-iron radiators will contain a large volume of water, requiring a much larger expansion capacity than a modern forced-circulation fin-tube system with the same design operating temperature. Likewise, hydronic systems with a design water temperature of 140°F (60°C) will require less expansion capacity than a system with the same water volume and a design water temperature of 200°F (93°C).

Compression tanks must have a minimum volume equal to or greater than the volume determined using

Equation 10-1. This equation does not apply to diaphragm-type expansion tanks. The required volume is a function of the system's total water volume, average design operating temperature, cold fill pressure, maximum operating pressure at the tank, and the atmospheric pressure at the system's location (altitude).

Example:

Determine the minimum required volume of a compression tank with:

V_s = 42 gallons

T = 200°F

P_a = 14.7 psi

P_f = 12 psig + 14.7 psi = 26.7 psia P_o = 25 psig + 14.7 psi = 39.7 psia

where:

V_s = Total volume of boiler(s), system piping and heating elements.

T = Average design operating temperature.

P_a = Atmospheric pressure at system altitude.

P_f = Required system (cold) fill pressure (absolute).

P_o = Maximum system operating pressure (absolute) at the tank location.

V_t = Minimum volume of expansion tank in gallons.

[For SI: 1 gallon = 3.785 L, 1 pound per square inch = 6.895 kPa, °C = (°F-32)/1.8, 1 pound per square inch gauge = 6.895 kPa.]

$$V_t = \frac{(0.00041\,T - 0.0466)\,V_s}{(P_a/P_f) - (P_a/P_o)}$$

$$= \frac{[(0.00041 \times 200) - (0.0466)] \times 42}{(14.7/26.7) - (14.7/39.7)} = 8.26\,\text{gallons}$$

Note that this formula is not to be used for diaphragm-type expansion tanks. Diaphragm-type expansion tanks are pre-engineered units with a factory-charged compressed air cushion on the air side of the diaphragm. These tanks are sized by the tank manufacturer or are specified by the boiler manufacturer.

1009.3 Open-type expansion tanks. Open-type expansion tanks shall be located a minimum of 4 feet (1219 mm) above the highest heating element. The tank shall be adequately sized for the hot water system. An overflow with a minimum diameter of 1 inch (25 mm) shall be installed at the top of the tank. The overflow shall discharge to the drainage system in accordance with the *International Plumbing Code*.

❖ Open-type expansion tanks were once common for old gravity circulation systems but are rarely encountered in modern system designs (see Commentary Figure 1009.3). An adequately sized open-type tank would have the capacity to hold, without overflowing

**Figure 1009.2(2)
CLOSED-DIAPHRAGM-TYPE EXPANSION TANK**

and admitting air into the system, the maximum volume of water that would expand from the system at the maximum operating temperature. Because this type of expansion tank is open to the atmosphere, it must be located at an elevation above the highest radiator or other water-filled system component. The 4-foot (1219 mm) minimum height will produce a slight head pressure in the top-most radiators, which assists in the purging of air from them. Tank overflow could cause property damage or personal injury. An overflow pipe is necessary, therefore, to prevent tank overflow. The tank's overflow pipe is intended to conduct water only during system fill and purging operations, and any other overflow should be recognized as an indicator of a system problem in need of immediate correction. Under normal operating conditions, the water level in an open expansion tank should not rise to the overflow pipe or drop below the bottom of the tank. Loss of water from any hot water heating system is unacceptable because it will result in the introduction of oxygen and mineral-bearing makeup water. The result should be avoided to prolong the life of the system and maintain heat transfer efficiency. The overflow pipe must not be subjected to freezing and is required to discharge to the building drainage system in accordance with the IPC.

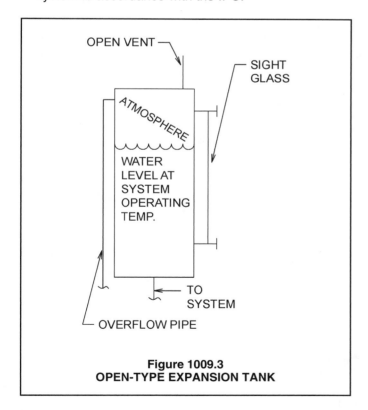

Figure 1009.3
OPEN-TYPE EXPANSION TANK

SECTION 1010
GAUGES

1010.1 Hot water boiler gauges. Every hot water boiler shall have a pressure gauge and a temperature gauge, or a combination pressure and temperature gauge. The gauges shall indicate the temperature and pressure within the normal range of the system's operation.

❖ All hot water boilers operate within certain pressure and temperature limits. Problems and hazards can develop when these limits are exceeded. Pressure and temperature gauges are necessary to allow boiler owners, operators and service technicians to monitor system operating conditions. Gauges not only indicate hazardous conditions and causes of problems or malfunctions, but also help predict future developing hazards or problems by indicating the need for system repairs and service. Pressure and temperature gauges are essential boiler components that demonstrate both normal and abnormal operating conditions. They are typically supplied by the boiler manufacturer as part of the factory- or field-installed "trim" (appurtenances).

Gauges must have a range and scale of values that are proportional to the boiler's normal operating range to allow reasonably accurate readings. For example, a pressure gauge with a range of 0 to 250 psi (0 to 1724 kPa) would barely deflect the needle when installed on a 12-psi (83 kPa) system and pressure readings would be grossly inaccurate. Typical gauges are inherently less accurate at the extreme low and high end of their range. The intent of the code is to require temperature and pressure gauges to be suitably matched to the operating ranges of the boiler for legible and accurate readings.

A good rule-of-thumb would be that the highest reading on the pressure gauge scale should not be more than three times nor less than one and one-half times the opening pressure setting of the safety relief valve.

Hot water boiler gauges are often combination pressure and temperature gauges and may also include an inactive "altitude" needle, recording the pressure necessary to maintain desired minimum pressure at the highest elevation heating element.

An altitude scale, on pressure gauges so equipped, indicates pressure head of water in feet (mm).

1010.2 Steam boiler gauges. Every steam boiler shall have a water-gauge glass and a pressure gauge. The pressure gauge shall indicate the pressure within the normal range of the system's operation.

❖ All steam boilers operate within certain pressure and water level limits that must be observed. However, unlike hot water boilers, water temperature is not a concern for steam systems. Whenever steam is present, the water temperature in a steam boiler will be a function of steam pressure and is easily determined when the steam pressure is known. A steam pressure gauge indicates both normal and abnormal operating conditions and the presence of a hazardous condition, and is necessary to observe and monitor boiler operation. The water level in a boiler fluctuates within a narrow margin between too low and too high. Because proper boiler operation is dependent on maintaining the water level within the high and low parameters, a water-gauge glass is necessary to indicate visually the actual water level in the boiler.

A water-gauge glass, commonly referred to as a "sight glass," indicates the following: boiler flooding, low-water conditions, condensate return operation, unstable water levels resulting from foaming and surging, boiler water condition and the amount of automatic or manual makeup water feed (see Commentary Figure 1010.2). If either or both of the gauge-glass cocks are closed, the water-level reading will be erroneous and the boiler could be empty, even though the gauge glass reads "normal."

Ideally, a steam boiler pressure gauge should indicate the normal system operating pressure at approximately the first quarter to midpoint of the scale. For example, a 0 to 100 psi (0 to 690 kPa) pressure gauge would be a very poor indicator for a low-pressure [15 psi (103 kPa) or less] steam heating boiler. It would also be a substandard option for a boiler whose normal operating pressure is at or near 100 psi (690 kPa) because there would be no headroom on the scale to indicate abnormally high pressures

(see commentary, Section 1010.1). Low-pressure steam gauges are often designed to read both pressure and vacuum because some steam systems will develop a partial vacuum as steam condenses in the distribution system.

1010.2.1 Water-gauge glass. The gauge glass shall be installed so that the midpoint is at the normal boiler water level.

❖ A water-gauge glass is always installed on a boiler with the bottom of the glass connected below the normal water level and the top of the glass connected above the normal water level. When connected in this manner, the boiler water level will be accurately represented in the column of water in the gauge glass. It has been an industry tradition to equate the midpoint of a gauge glass with the "normal" water level in a boiler. The coincidence of the gauge-glass midpoint and normal water level also allows the greatest amount of indicator range above and below normal, thus allowing clear indication of abnormal water levels.

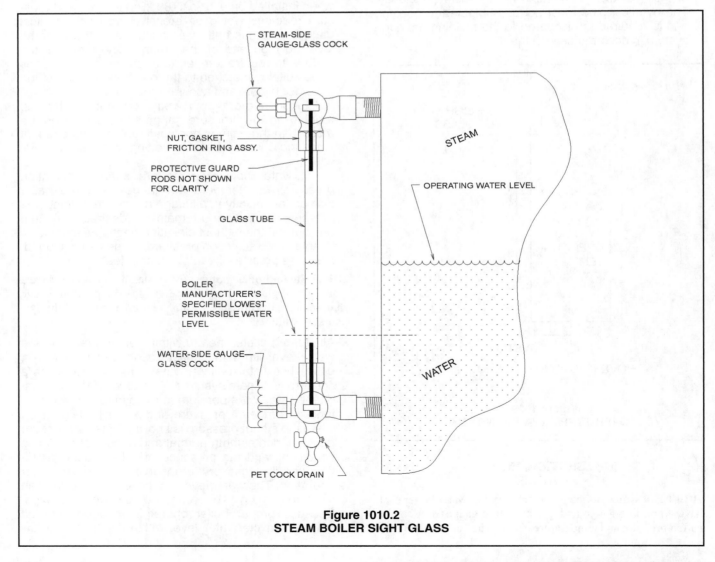

Figure 1010.2
STEAM BOILER SIGHT GLASS

SECTION 1011
TESTS

1011.1 Tests. Upon completion of the assembly and installation of boilers and pressure vessels, acceptance tests shall be conducted in accordance with the requirements of the ASME *Boiler and Pressure Vessel Code*. Where field assembly of pressure vessels or boilers is required, a copy of the completed U-1 Manufacturer's Data Report required by the ASME *Boiler and Pressure Vessel Code* shall be submitted to the code official.

❖ The history of boilers and pressure vessels shows that such equipment is potentially dangerous if not properly designed, installed, protected, tested, maintained and monitored. Such equipment must be equipped with various safety controls and safeguards. However, safety devices cannot prevent failures that result from improper installation. Testing verifies the structural integrity of the vessel and the proper operation of all controls, safety devices, burners, venting systems and appurtenances. Some boilers, such as large sectional cast-iron boilers, are shipped in sections and assembled on site. Therefore, as part of the building commissioning process, Section 1011 requires that boilers and pressure vessels be properly inspected, tested and commissioned upon installation.

The ASME *Boiler and Pressure Vessel Code* is referenced for the testing of boilers and pressure vessels. Testing procedures for specific types of boilers and pressure vessels are found in the respective sections of the referenced standard. These testing procedures may require the use of test formulas or other calculation methods. Calculations for the procedures identified in the referenced standard should be performed by qualified people. Error in certifying test procedures could result in the hazardous operation of equipment.

1011.2 Test gauges. An indicating test gauge shall be connected directly to the boiler or pressure vessel where it is visible to the operator throughout the duration of the test. The pressure gauge scale shall be graduated over a range of not less than one and one-half times and not greater than four times the maximum test pressure. All gauges utilized for testing shall be calibrated and certified by the test operator.

❖ Pressure gauges used for testing must be accurate and therefore must be calibrated to a known reference pressure (see commentary, Section 1010.2).

Bibliography

The following resource materials were used in the preparation of the commentary for this chapter of the code:

IBC-12, *International Building Code*. Washington, DC: International Code Council, 2011.

IFGC-12, *International Fuel Gas Code*. Washington, DC: International Code Council, 2011.

IPC-12, *International Plumbing Code*. Washington, DC: International Code Council, 2011.

NFPA 70-11, *National Electrical Code*. Quincy, MA: National Fire Protection Association, 2008.

Chapter 11: Refrigeration

General Comments

The purpose of this chapter is to regulate the use of refrigerants and protect refrigeration systems, property and life from the hazards associated with the refrigerants and their related equipment. The hazards include, but are not limited to, the physical impacts, toxic effects and the combustible, flammable and decomposing effects of refrigerants.

Refrigerants create a hazard because they are liquified gas under pressure in a mechanical system and many refrigerant vapors cannot be seen, tasted or smelled, so there is no natural warning of a hazard occurring. Refrigerants and refrigeration systems that are not maintained or are improperly used or installed could result in a release of pressurized refrigerant resulting in physical injuries, damage to the building and an impact on the environment.

Physical injuries include, but are not limited to, exposure that can cause frostbite and asphyxiation. Building damage includes, but is not limited to, fires, explosions and loss of property. Environmental impacts include, but are not limited to, the potential depletion of the ozone layer, global warming and increase in exposure to ultraviolet radiation levels [see Commentary Figure 11(1)].

Typically, refrigerants used with mechanical systems, that are regulated via the code, undergo toxicity testing before being released for use with the data usually indicated on Material Safety Data Sheets (MSDS). The testing involves a range of exposure levels that are indicated as valves to establish how much of a specific refrigerant a person can regularly be exposed to without adverse effects.

Some refrigerants burn with oxygen, but only at higher pressures or temperatures and never in air at atmospheric conditions (combustible). Some refrigerants, when combined with air at atmospheric pressure, ignite causing a flame and possibly an explosion (flammable).

In addition to the immediate safety concern related to refrigeration equipment and the physical and chemical characteristics of refrigerants, there has been a serious concern over the effect of chlorofluorocarbon (CFC) refrigerants and their contribution to the depletion of the ozone layer in the earth's atmosphere. This concern has resulted in rapid changes in the refrigeration industry as alternatives to CFC refrigerants are developed.

The Montreal Protocol, the U.S. Clean Air Act, Department of Energy (DOE) and environmental issues have been the driving forces behind the regulation and evolution in the chemicals used as refrigerants and the design and use of refrigeration systems. For example, new alternative refrigerants continue to appear in the marketplace, and there has been renewed interest in the rules by which all refrigerants are classified and safely applied; therefore, refrigeration systems continue to be improved for the use of refrigerants. Thus, the user of this code will find many new elements, and should expect that the list of refrigerants will continue to grow with time.

It is worth noting that the Montreal Protocol, the U.S. Clean Air Act, Department of Energy (DOE) and environmental issues have been focused on hydrofluorocarbons (HFCs) and hydrochloro-fluorocarbons (HCFCs). In these compounds, hydrogen atoms are substituted for halogen atoms (that is flourine, chlorine, bromine, astatine and iodine). HFCs are considered "ozone friendly" in that they are dechlorinated (there is no chlorine present to be released into the stratosphere). These compounds appear to be the best alternative refrigerants that have been developed to date. R-134a is an HFC refrigerant that is already in use and is permitted by the code. HCFCs are compounds that are still ozone depleting, but much less so than CFCs. These are transitional refrigerants and have been undergoing a gradual phaseout. Alternative refrigerants are being developed continuously. HCFCs will bridge the gap for the industry until permanent replacements are developed. R-22 and R-123 are HCFC refrigerants that are currently permitted by the code.

The U.S. Clean Air Act also makes venting of CFCs and HCFCs into the atmosphere unlawful. This means that the common practice of discharging halogenated refrigerants into the air is not allowed in the United States and that all used CFC refrigerants must be reclaimed, recycled or destroyed. The termination of CFC production and the prohibition of venting to the atmosphere have been based on the examination of data concluding the effect of CFCs on the ozone layer.

CFC molecules deplete the ozone layer when they reach the stratosphere, the second of four layers that make up the earth's atmosphere. The ozone layer is located at the top of the stratosphere and protects the earth by blocking most of the sun's harmful ultraviolet radiation. This ozone layer consists of molecules made up of three oxygen atoms, or O_3. CFC molecules remain stable in the first layer, but once they reach the stratosphere, solar energy in the form of ultraviolet radiation causes the chlorine atoms to break free and attach themselves to the ozone molecules. The resulting compounds break down into chlorine monoxide (ClO) and O_2, with destruction of the ozone molecule. The ultraviolet radiation further breaks down O_2 which then reacts with the chlorine monoxide to produce O_2 + Cl. The reliberated chlorine molecules are now free to attach themselves to another ozone molecule. It has been estimated that each chlorine atom can destroy up to 100,000 molecules of ozone. Therefore, even if the total ban of CFC refrigerants were to happen today, their ozone-destroying effects are presumed to continue for years [see Commentary Figure 11(1)].

Some of the consequences of the depletion of the ozone layer include an increase in skin cancer; autoimmune deficiencies; increase in cataracts; and an impact on world food production based on the potential for ultraviolet radiation to impede plant-life photosynthesis and protein formation.

As new refrigerants are developed and ozone-depleting ones are phased out, changes to this chapter to keep pace with the industry are inevitable. For example, the formerly used safety group classifications are not adequate to classify the new refrigerants being developed. ASHRAE 34 has been increasing safety group classifications intended to adequately cover current refrigerants and those yet to be developed.

As defined in ASHRAE 15, refrigeration systems are a combination of interconnected components and piping

assembled to form a closed circuit in which a refrigerant is circulated. The system's function is to extract heat from a location or medium, and to reject that heat to a different location or medium. The term "heat pump" is an appropriate description of a refrigeration system. Heat is a form of energy that can be measured; "cold" is not. The terms "cool" and "cold" are used to describe the relative quantity of heat in a substance or space [see Commentary Figures 11(2) and 11(3)].

Purpose

Chapter 11 contains regulations pertaining to the life safety of building occupants. These regulations establish minimum requirements to achieve the proper design, construction, installation and operation of refrigerating systems. This chapter establishes reasonable safeguards for the occupants by defining and mandating practices that are consistent with the practices and experience of the industry.

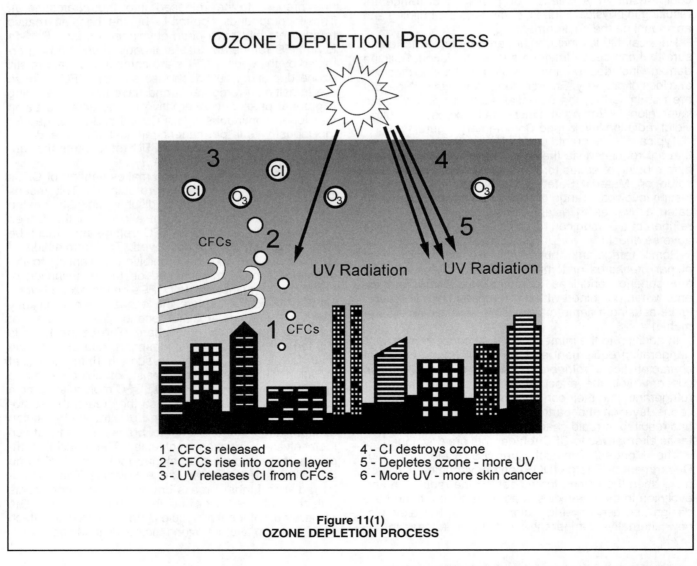

Figure 11(1)
OZONE DEPLETION PROCESS

1 - CFCs released
2 - CFCs rise into ozone layer
3 - UV releases Cl from CFCs
4 - Cl destroys ozone
5 - Depletes ozone - more UV
6 - More UV - more skin cancer

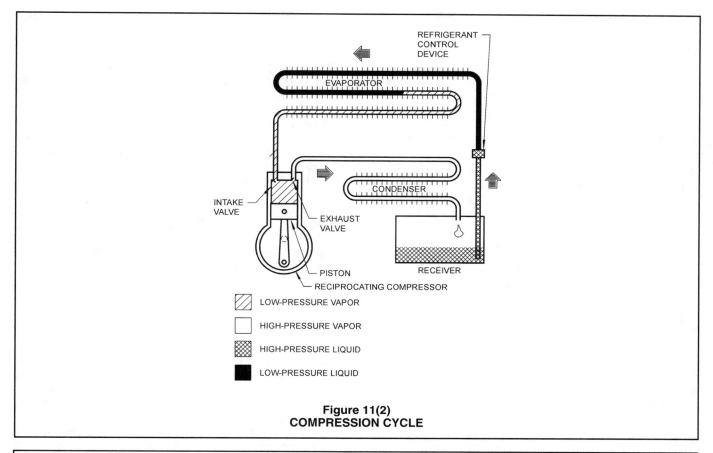

Figure 11(2)
COMPRESSION CYCLE

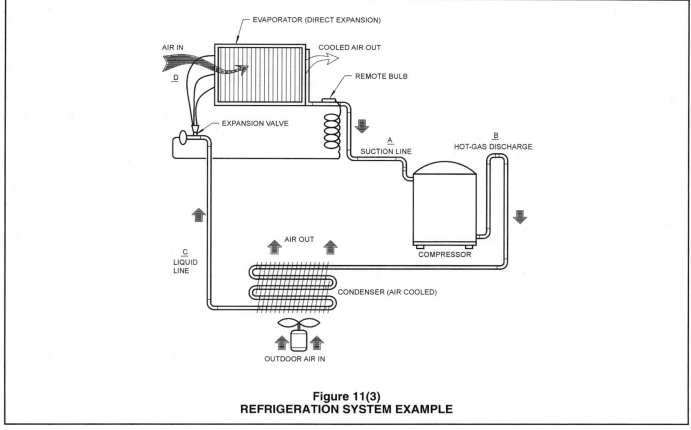

Figure 11(3)
REFRIGERATION SYSTEM EXAMPLE

SECTION 1101
GENERAL

1101.1 Scope. This chapter shall govern the design, installation, construction and repair of refrigeration systems that vaporize and liquefy a fluid during the refrigerating cycle. Refrigerant piping design and installation, including pressure vessels and pressure relief devices, shall conform to this code. Permanently installed refrigerant storage systems and other components shall be considered as part of the refrigeration system to which they are attached.

❖ While federal regulations for a safer environment have influenced the development of many provisions in this chapter, the hazards associated with the use of refrigerants still include, but are not limited to, frostbite, fire, inhalation hazards, chemical burns, long-term health problems and death. Chapter 11 is intended to minimize the hazards to the building occupants and the environment by requiring that all refrigeration systems be designed, constructed and installed in compliance with the provisions of this chapter.

1101.2 Factory-built equipment and appliances. *Listed* and *labeled* self-contained, factory-built *equipment* and appliances shall be tested in accordance with UL 207, 412, 471 or 1995. Such *equipment* and appliances are deemed to meet the design, manufacture and factory test requirements of this code if installed in accordance with their listing and the manufacturer's installation instructions.

❖ Self-contained, factory-built mechanical refrigeration equipment and appliances can be defined as factory-made and tested equipment that is fabricated and shipped in one or more sections, and in which the refrigerant-containing parts are not connected in the field other than by companion or block valves.

Section 301.4 requires mechanical equipment to bear the label of an approved agency unless otherwise approved in accordance with Section 105. Therefore, the label that appears on self-contained, factory-built equipment or appliances demonstrates that the equipment or appliances have been tested and evaluated by an approved agency in accordance with specific standards and test methods. While there are other standards for testing self-contained, factory-built equipment and appliances, the code specifically requires testing in accordance with UL 207, 412, 471 or 1995. It is worth noting that UL 207 is for nonelectric refrigerant-containing components and accessories, UL 412 is for refrigeration unit coolers, UL 471 is for commercial refrigerators and coolers and UL 1995 is for heating and cooling equipment. The information on the label and the standards to which the equipment was evaluated and tested enables the code official to determine conformance to Sections 1104, 1107 and 1108.

1101.3 Protection. Any portion of a refrigeration system that is subject to physical damage shall be protected in an *approved* manner.

❖ Dents, scratches, rust, exposure to excessive heat or other physical abuses can reduce the strength of refrigeration system materials. Damage or weakened materials may fail at typical refrigerant pressures lower than originally constructed, designed and specified to handle. Some refrigeration system materials are designed with encasements that provide protection from physical abuses. Other system materials such as refrigerant piping and cylinders must be protected from damage. A common-sense approach will indicate when and where physical protection is needed to prevent damage from vehicular impact, occupant activities and similar causes. The same common-sense approach must be relied upon to determine the type of protection that is warranted.

1101.4 Water connection. Water supply and discharge connections associated with refrigeration systems shall be made in accordance with this code and the *International Plumbing Code*.

❖ Some types of refrigeration equipment use potable water as a cooling medium and therefore connect to the potable water distribution system and discharge the waste water to the drainage system. Water chillers, cooling towers and thermal storage systems may have potable makeup water connections, while ice-making machines may have potable water supply connections and drainage system connections. All such connections must be protected against contamination as directed by the *International Plumbing Code*® (IPC®).

1101.5 Fuel gas connection. Fuel gas devices, *equipment* and appliances used with refrigeration systems shall be installed in accordance with the *International Fuel Gas Code*.

❖ Many refrigeration machines are driven by internal combustion engines or are powered by the heat of combustion. Examples include absorption equipment and natural gas engine-driven condensing units. As with all fuel-gas-fired equipment and appliances, the provisions of the *International Fuel Gas Code*® (IFGC®) apply as expressed within the scope of the code.

1101.6 General. Refrigeration systems shall comply with the requirements of this code and, except as modified by this code, ASHRAE 15. Ammonia-refrigerating systems shall comply with this code and, except as modified by this code, ASHRAE 15 and IIAR 2.

❖ This section requires full compliance with ASHRAE 15 and IIAR 2, except for any instances where the code has intentionally modified or exceeded the requirements of the standards. Chapter 11 is based on ASHRAE 15; however, there are requirements that are unique to both documents and the code official and designer must be familiar with both documents. Therefore, while the exception to Section 1106.5.1 conflicts with ASHRAE 15, the provisions of this code shall apply pursuant to Section 102.8. Furthermore, many provisions in ASHRAE 15 are not contained in the code, including general system and component location restrictions, general installation requirements, material usage restrictions, system design pressure specifications, pressure vessel requirements, pressure relief protection, pressure vessel protection, relief valve discharge provisions, pressure limiting controls, system service provisions, factory testing requirements and general requirements for system signage and identification, refrigerant handling/storage and emergency procedures. IIAR 2 is similar in format and content to ASHRAE 15, but it is ammonia (R-717) specific.

1101.7 Maintenance. Mechanical refrigeration systems shall be maintained in proper operating condition, free from accumulations of oil, dirt, waste, excessive corrosion, other debris and leaks.

❖ Periodic maintenance is essential for the proper operation of mechanical refrigeration equipment. Typically, routine maintenance inspections occur when periodic testing is performed in accordance with Section 1109.1. However, maintenance should be performed any time damaged or weakened refrigeration systems are observed. Damaged refrigeration systems cannot be relied upon for protection since improperly maintained mechanical equipment will operate below peak efficiency, waste energy and eventually fail prematurely, adding unnecessary costs to the owner. The build up of oil, dirt and corrosion can adversely affect the heat exchange function of heat exchangers, and can contribute to motor failure and component overheating. The refrigeration systems must be kept clean in accordance with the manufacturer's recommendations, and any damaged or weakened component or device must be repaired or replaced and discarded. In essence, if the refrigerant stays contained in the refrigeration system, the hazards to occupants and the environment are greatly reduced; the hazards increase when the refrigerant becomes exposed outside of the system, often quickly and unexpectedly.

1101.8 Change in refrigerant type. The type of refrigerant in refrigeration systems having a refrigerant circuit containing more than 220 pounds (99.8 kg) of Group A1 or 30 pounds (13.6 kg) of any other group refrigerant shall not be changed without prior notification to the code official and compliance with the applicable code provisions for the new refrigerant type.

❖ The intent of this section is to keep the local fire code official or code official up to date on the status of large refrigeration systems and systems containing toxic and/or flammable refrigerants. Section 606 of the *International Fire Code®* (IFC®) contains several related requirements, including maintaining fire department access, posting emergency signs and labels, storage requirements and periodic system testing and maintenance of the test records (see commentary, Sections 1101.9 and 1106.6).

[F] 1101.9 Refrigerant discharge. Notification of refrigerant discharge shall be provided in accordance with the *International Fire Code*.

❖ Section 606 of the IFC requires notification of the fire department of any time there is a refrigerant discharge, either accidental or planned, from a refrigerant circuit containing more than 220 pounds (100 kg) of Group A1 or 30 pounds (14 kg) of refrigerant from any other group. Exceptions are stated for minor discharges associated with automatic pressure relief valves, service operations after system pump-down and systems operating below atmospheric pressure, which include an automatic purge system.

1101.10 Locking access port caps. Refrigerant circuit access ports located outdoors shall be fitted with locking-type tamper-resistant caps or shall be otherwise secured to prevent unauthorized access.

❖ This provision intends to address a relatively new type and method of substance abuse where people intentionally inhale refrigerant gases for the intoxicating effect. In some cases, this inhalant abuse has resulted in the death of the individual. The typical condensing unit or heat pump unit is located outdoors and is equipped with access ports on the vapor and liquid refrigerant lines. These access ports are necessary for several purposes, such as to allow the connection of diagnostic gauges and to allow refrigerant to be added to or taken from the unit during servicing of the unit. Some of these access ports require back-seated valves to be opened with a wrench to allow refrigerant to escape and many of these access ports are equipped with simple "Schraeder" valves that are similar to the valve cores used on car and truck tires. All access ports are provided with threaded caps to keep out debris and moisture and to also guard against valve leakage. Individuals intent on inhaling refrigerants have found that they can withdraw refrigerant from these units by using simple hand tools or their bare hands to remove the caps and open the valves.

The purpose of this section is to prevent this dangerous form of substance abuse by making it difficult if not impossible to gain access to the access ports. Special locking-type caps are available such that wrenches, pliers and fingers cannot be used to remove the caps without a special tool/key [see Commentary Figures 1101.10(1) through 1101.10(4)]. Methods other than special locking caps can be used such as installing the equipment in a chain link fence "cage" that is locked to prevent access to the equipment. Equipment installed on roofs that cannot be accessed except through a locked access hatch is yet another way to prevent access. All newly installed equipment, including replacement equipment, must be provided with a method of protection as part of such installations.

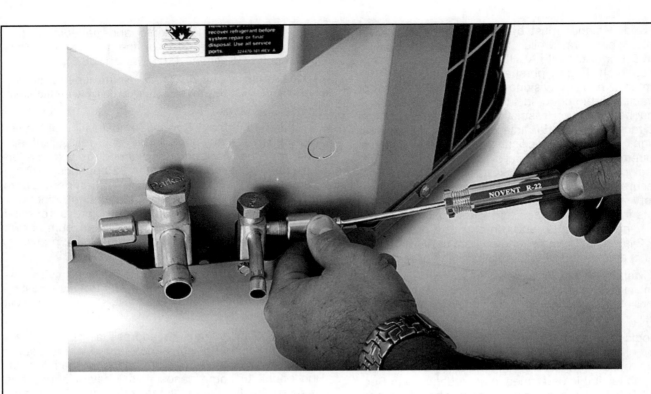

Figure 1101.10(1)
INSTALLATION OF LOCKING CAPS FOR REFRIGERANT ACCESS PORT
(Photo Courtesy of Novent LLC. & Airtec Products Corporation, Inc.)

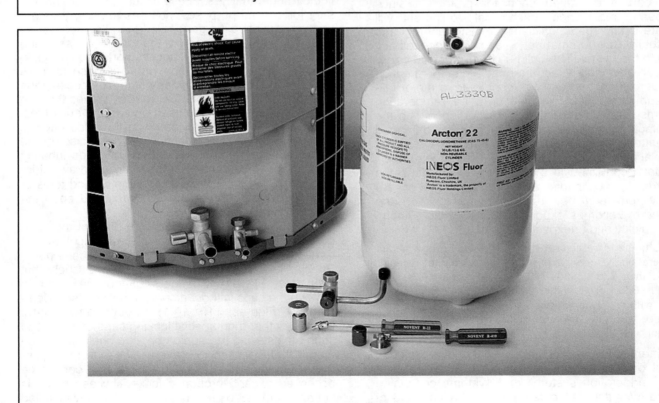

Figure 1101.10(2)
REFRIGERANT LOCKING CAPS AND TOOLS
(Photo Courtesy of Novent LLC. & Airtec Products Corporation, Inc.)

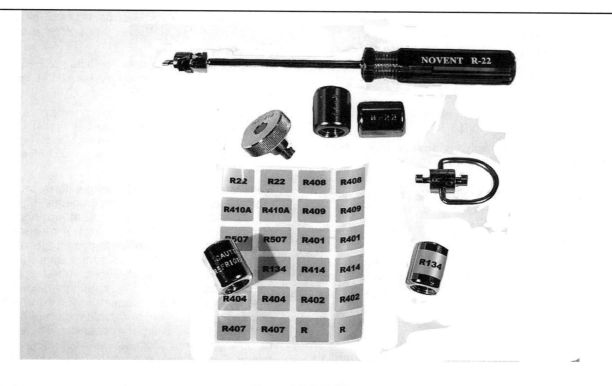

Figure 1101.10(3)
REFRIGERANT LABELS FOR LOCKING CAPS
(Photo Courtesy of Novent LLC. & Airtec Products Corporation, Inc.)

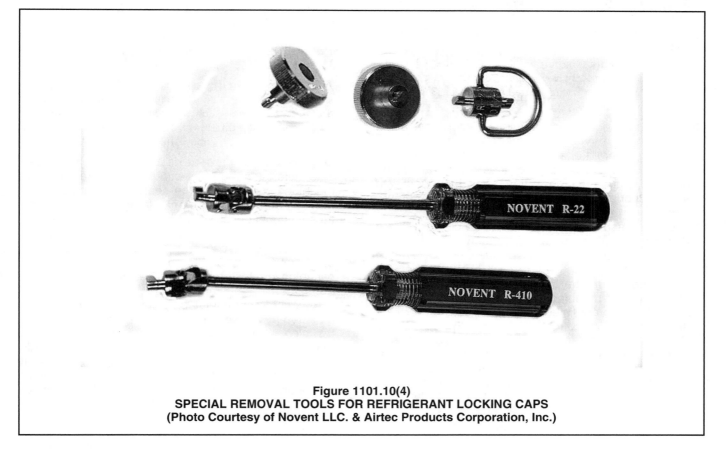

Figure 1101.10(4)
SPECIAL REMOVAL TOOLS FOR REFRIGERANT LOCKING CAPS
(Photo Courtesy of Novent LLC. & Airtec Products Corporation, Inc.)

SECTION 1102
SYSTEM REQUIREMENTS

1102.1 General. The system classification, allowable refrigerants, maximum quantity, enclosure requirements, location limitations, and field pressure test requirements shall be determined as follows:

1. Determine the refrigeration system's classification, in accordance with Section 1103.3.

2. Determine the refrigerant classification in accordance with Section 1103.1.

3. Determine the maximum allowable quantity of refrigerant in accordance with Section 1104, based on type of refrigerant, system classification and *occupancy*.

4. Determine the system enclosure requirements in accordance with Section 1104.

5. Refrigeration *equipment* and *appliance* location and installation shall be subject to the limitations of Chapter 3.

6. Nonfactory-tested, field-erected *equipment* and appliances shall be pressure tested in accordance with Section 1108.

❖ This section contains general requirements regarding the purity and compatibility of refrigerants, serves to coordinate the provisions of Chapter 11 and includes basic user guidance for the application of those provisions. Items 1 through 4 provide logically ordered instruction on the application of the interdependent Sections 1103, 1104, 1105 and 1106 (see commentary, Sections 301, 1103, 1104 and 1108). These requirements apply to all systems, regardless of classification.

1102.2 Refrigerants. The refrigerant shall be that which the *equipment* or *appliance* was designed to utilize or converted to utilize. Refrigerants not identified in Table 1103.1 shall be *approved* before use.

❖ Refrigeration equipment is designed to operate with a specific type or types of refrigerant. Using the wrong type of refrigerant or a contaminated refrigerant could cause equipment damage or loss of operating efficiency. Refrigerant types differ in how they react with system lubricants, seals, gaskets and other components.

Existing equipment may have to be converted to a different type of refrigerant or charged with refrigerant recovered from it or another system; however, it would be an unnecessary risk to charge new equipment with any refrigerant other than that specified by the equipment manufacturer.

Refrigerants not listed in Table 1103.1 are prohibited unless approved in accordance with Section 105. Because of constantly evolving technology and a never-ending search for better and safer refrigerants, new types of refrigerants will likely be entering the market place regularly. Once these new products are classified in accordance with ASHRAE 34, the code

official can evaluate them based on their properties and comparisons to those refrigerants listed in Table 1103.1.

1102.2.1 Mixing. Refrigerants, including refrigerant blends, with different designations in ASHRAE 34 shall not be mixed in a system.

Exception: Addition of a second refrigerant is allowed where permitted by the *equipment* or *appliance* manufacturer to improve oil return at low temperatures. The refrigerant and amount added shall be in accordance with the manufacturer's instructions.

❖ Refrigerants of different designations must not be mixed together in a system because the chemical combinations would have different and unpredictable or possibly hazardous properties. A mixture of refrigerants could cause equipment damage and loss of system performance. Also combinations would trigger the requirements of Section 1104 applicable to each separate type of refrigerant in a mixture.

Another significant problem with mixing refrigerant types is that recovery and reclamation of the refrigerants would be difficult or impossible. There might not be any practical process for separating refrigerant mixtures into their component parts. In some cases, destruction of the refrigerant mixture is the only practical solution for handling cross-contaminated (mixed) refrigerants.

1102.2.2 Purity. Refrigerants used in refrigeration systems shall be new, recovered or *reclaimed refrigerants* in accordance with Section 1102.2.2.1, 1102.2.2.2 or 1102.2.2.3. Where required by the *equipment* or *appliance* owner or the code official, the installer shall furnish a signed declaration that the refrigerant used meets the requirements of Section 1102.2.2.1, 1102.2.2.2 or 1102.2.2.3.

Exception: The refrigerant used shall meet the purity specifications set by the manufacturer of the *equipment* or *appliance* in which such refrigerant is used where such specifications are different from that specified in Sections 1102.2.2.1, 1102.2.2.2 and 1102.2.2.3.

❖ Oils, waxes, moisture, acids and residues in refrigerants taken from burned-out or retired refrigeration machines can damage the seals and motor parts of new refrigeration equipment. Refrigerants containing these impurities can cause equipment damage or poor performance, and may lead to equipment leaks. Equipment leaks can endanger the health of occupants and contribute to the amount of refrigerant escaping into the atmosphere, causing environmental problems.

Because contaminated refrigerant is not easily detected by the equipment owner or the code official, the signed declaration serves as some assurance that clean refrigerant, whether recovered (temporarily removed from a system with the intent of putting it back in) or reclaimed (reprocessed for use in compatible equipment), is being used. Unethical installers or service personnel might take shortcuts by not filtering recovered refrigerant or processing reclaimed refrigerant in accordance with ARI 700 (referenced in Section 1102.2.2.3). Because contaminated refrigerant could go undetected for some time, the written declaration is a safeguard that can serve to make the installer liable if the equipment eventually fails because of contaminated refrigerant.

1102.2.2.1 New refrigerants. Refrigerants shall be of a purity level specified by the *equipment* or *appliance* manufacturer.

❖ It is unlikely that virgin refrigerants would be produced that do not meet most manufacturers' purity specifications. However, it is quite possible that recovered and reclaimed refrigerants would not meet purity standards (see commentary, Sections 1102.2.2.2 and 1102.2.2.3). Equally important is verifying that the refrigerant specified by the manufacturer of the equipment is used. In other words, there is a greater possibility that the wrong refrigerant could be used than that new refrigerant would somehow not meet purity standards.

1102.2.2.2 Recovered refrigerants. Refrigerants that are recovered from refrigeration and air-conditioning systems shall not be reused in other than the system from which they were recovered and in other systems of the same owner. *Recovered refrigerants* shall be filtered and dried before reuse. *Recovered refrigerants* that show clear signs of contamination shall not be reused unless reclaimed in accordance with Section 1102.2.2.3.

❖ Recovered refrigerants are withdrawn from existing systems that are being retired or repaired. Recovering the refrigerant prevents its escape into the atmosphere and allows it to be recycled. This is economically and environmentally sound and required by federal law.

Transferring refrigerant to other equipment could cause system contamination in much the same way as a blood transfusion could introduce disease into another creature. This section prohibits the reuse of recovered refrigerant in systems of different ownership than the equipment from which it was withdrawn, unless reprocessed in accordance with the requirements for reclaimed refrigerants (see Section 1102.2.2.3). In all cases, recovered refrigerants must not be reused until filtered and treated for moisture removal. Any recovered refrigerant suspected of being or shown to be contaminated (such as in the case of a compressor motor burnout) must be processed as necessary to meet the specified purity standard. Kits are often used in the field to test for contaminants, such as acids and moisture.

1102.2.2.3 Reclaimed refrigerants. Used refrigerants shall not be reused in a different owner's *equipment* or appliances unless tested and found to meet the purity requirements of ARI 700. Contaminated refrigerants shall not be used unless reclaimed and found to meet the purity requirements of ARI 700.

❖ This section relates to Section 1102.2.2.2 and requires testing of used refrigerants for purity before they are introduced into someone else's equipment. While many refrigerants meet the purity requirements of ARI 700, contaminated refrigerant is sometimes reused, marketed and sold as new. The reuse of the contaminated refrigerant has resulted in equipment failure and subsequent venting of toxic, combustible, decomposing, pressurized and ozone-depleting refrigerants. If the test indicates that it does not meet the purity standards of ARI 700, it must be reprocessed in accordance with that standard before it is reused. Note that purification required to meet ARI 700 could involve much more processing than the simple filtering and drying required by Section 1102.2.2.2. Contaminated refrigerants must not be introduced into any equipment, regardless of ownership.

SECTION 1103
REFRIGERATION SYSTEM CLASSIFICATION

1103.1 Refrigerant classification. Refrigerants shall be classified in accordance with ASHRAE 34 as listed in Table 1103.1.

❖ Because the classification of refrigeration systems is a necessary step in the application of Section 1104, the code addresses the hazards of refrigeration systems to building occupants by considering three things: the type of refrigerant, the type of system (Section 1103.3) and the type of building occupancy (Section 1103.2). Certain systems are more hazardous in terms of possible exposure to escaping refrigerants (see commentary, Section 1103.3). Certain occupancies are more hazardous in terms of the number of people who could be exposed or who are, for various reasons, particularly susceptible to injury because of disability, detention or incapacity (see commentary, Section 1103.2).

The classification of refrigerants is based on ASHRAE 34, which numbers and classifies refrigerants in accordance with their potential hazards (see Commentary Figure 1103.1). In the code, refrigerants are identified by their number, which is preceded by the letter "R," for example, R-22. Trademark names of manufacturers, such as "Freon," are not used in the code. The refrigerant number, name and chemical formula can be found in Table 1103.1.

Each refrigerant is classified into a safety group that is based on two factors: flammability and toxicity. The safety group in which an individual refrigerant is classified relates to the potential hazard to the building occupants and is one of the factors used to determine the maximum quantities of refrigerants allowed by the code.

The new safety classification designations of ASHRAE 34 are necessary to provide for regulations applicable to the newly developed and implemented "ozone-friendly" refrigerants intended to replace the ozone-depleting refrigerants such as R-11 and R-12 now used in many existing refrigeration systems. Note e to Table 1103.1 identifies refrigerants that have been designated as "Class I ozone depleting substances" and are banned by federal law for use in new installations.

TABLE 1103.1. See page 11-11.

❖ Table 1103.1 indicates the TLV-TWA (threshold limit value-time weighted average), which is the amount of refrigerant a person can be exposed to for 8 hours a day, 40 hours a week, without adverse effects unless noted "C" for "ceiling." Ceiling is an exposure level (as in PEL-C, REL-C or TLV-C) that should not be exceeded during any part of the day and assessed as a 15-minute TWA unless otherwise specified. Note that Table 1103.1 establishes the maximum value for any refrigerant chemical as 1,000 parts per million (ppm). Furthermore, there is also the short-term exposure limit (STEL), which is a term used in the refrigeration industry. STEL is based on a 15-minute exposure time in any given day in addition to the value Immediately Dangerous to Life or Health (IDLH). The IDLH is referenced in Note c of Table 1103.1 and is used to provide guidance for the machinery room requirements of Sections 1104, 1105 and 1106. IDLH is set by the U.S. National Insti-

tute of Occupational Safety and Health (NIOSH) to establish the maximum concentration of air-borne contaminants, normally expressed as ppm, from which one could escape within 30 minutes without a respirator and without experiencing any escape impairing (e.g., severe eye irritation) or irreversible health effects.

As a point of information, the American Conference of Government and Industrial Hygienists (ACGIH) sets the TLV-TWA for chemicals, while the permissible exposure limit (PEL), which is referenced in Note e of Table 1103.1, is set by the Occupational Safety and Health Association (OSHA). Refrigerants not classified in ASHRAE 34 should be reviewed with suppliers to make sure enough is known about their toxicity properties. Some blends may not be classified, but contain classified components. Therefore, it is worth noting that other refrigerant chemical producers and organizations have similar exposure level indexes based on similar criteria as PEL and TWA. For example, workplace environmental exposure limit (WEEL), acceptable exposure limit (AEL), industrial exposure limit (IEL), occupational exposure limit (OEL) and Programme for Alternative Fluorocarbon Toxicity Testing (PAFT) are generally used for substances for which a PEL has not been established. Note that a separate MSDS produced by the manufacturer of the chemical refrigerant will often clearly indicate the toxicity of refrigerants used in conjunction with the mechanical systems.

Note that Section 1103.1 indicates that ASHRAE 34 must be used to classify the combustible, flammable and decomposing effects of refrigerants.

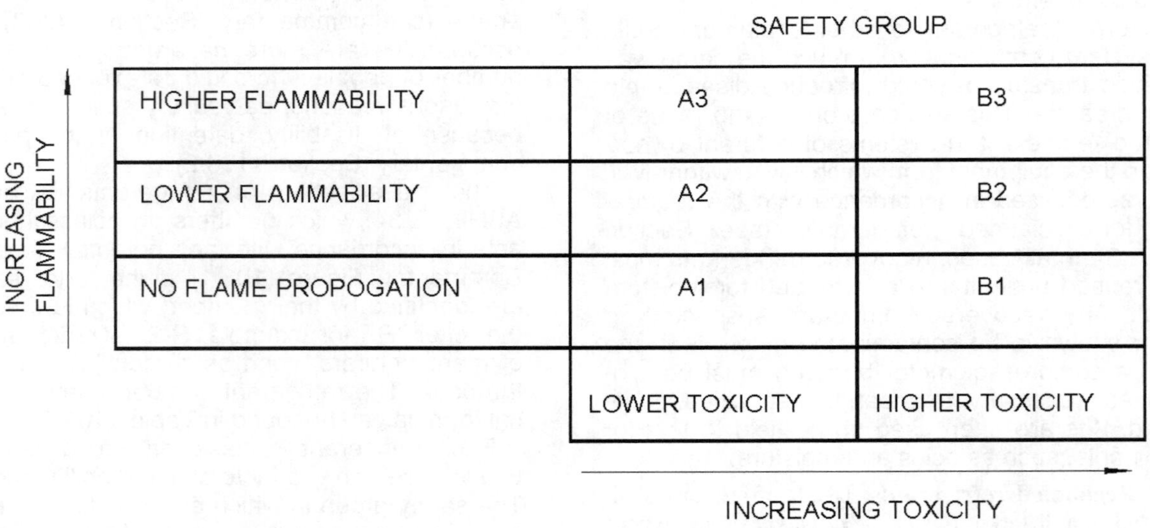

SAFETY GROUP		
HIGHER FLAMMABILITY	A3	B3
LOWER FLAMMABILITY	A2	B2
NO FLAME PROPOGATION	A1	B1
	LOWER TOXICITY	HIGHER TOXICITY

INCREASING FLAMMABILITY ↑

INCREASING TOXICITY →

Figure 1103.1
CLASSIFICATION OF REFRIGERANTS
(Source: ASHRAE 15-94)

[F] TABLE 1103.1
REFRIGERANT CLASSIFICATION, AMOUNT AND OEL

CHEMICAL REFRIGERANT	FORMULA	CHEMICAL NAME OF BLEND	REFRIGERANT CLASSIFICATION	DEGREES OF HAZARD[a]	[M] AMOUNT OF REFRIGERANT PER OCCUPIED SPACE				
					Pounds per 1,000 cubic feet	ppm	g/m³	OEL[e]	
R-11[d]	CCl_3F	trichlorofluoromethane	A1	2-0-0[b]	0.39	1,100	6.2	C1,000
R-12[d]	CCl_2F_2	dichlorodifluoromethane	A1	2-0-0[b]	5.6	18,000	90	1,000
R-13[d]	$CClF_3$	chlorotrifluoromethane	A1	2-0-0[b]	—	—	—	1,000
R-13B1[d]	$CBrF_3$	bromotrifluoromethane	A1	2-0-0[b]	—	—	—	1,000
R-14	CF_4	tetrafluoromethane (carbon tetrafluoride)	A1	2-0-0[b]	25	110,000	400	1,000
R-22	$CHClF_2$	chlorodifluoromethane	A1	2-0-0[b]	13	59,000	210	1,000
R-23	CHF_3	trifluoromethane (fluoroform)	A1	2-0-0[b]	7.3	41,000	120	1,000
R-32	CH_2F_2	difluoromethane (methylene fluoride)	A2	—	4.8	36,000	77	1,000
R-113[d]	CCl_2FCClF_2	1,1,2-trichloro-1,2,2-trifluoroethane	A1	2-0-0[b]	1.2	2,600	20	1,000
R-114[d]	$CClF_2CClF_2$	1,2-dichloro-1,2,2-tetrafluoroethane	A1	2-0-0[b]	8.7	20,000	140	1,000
R-115	$CClF_2CF_3$	chloropentafluoroethane	A1	—	47	120,000	760	1,000
R-116	CF_3CF_3	hexafluoroethane	A1	1-0-0	34	97,000	550	1,000
R-123	$CHCl_2CF_3$	2,2-dichloro-1,1,1-trifluoroethane	B1	2-0-0[b]	3.5	9,100	57	50
R-124	$CHClFCF_3$	2-chloro-1,1,1,2-tetrafluoroethane	A1	2-0-0[b]	3.5	10,000	56	1,000
R-125	CHF_2CF_3	pentafluoroethane	A1	2-0-0[b]	23	75,000	370	1,000
R-134a	CH_2FCF_3	1,1,1,2-tetrafluoroethane	A1	2-0-0[b]	13	50,000	210	1,000
R-141b	CH_3CCl_2F	1,1-dichloro-1-fluoroethane	—	—	0.78	2,600	12	500
R-142b	CH_3CClF_2	1-chloro-1,1-difluoroethane	A2	—	5.1	20,000	83	1,000
R-143a	CH_3CF_3	1,1,1-trifluoroethane	A2	2-0-0[b]	4.5	21,000	70	1,000
R-152a	CH_3CHF_2	1,1-difluoroethane	A2	1-4-0	2	12,000	32	1,000
R-170	CH_3CH_3	ethane	A3	2-4-0	0.54	7,000	8.7	1,000
R-E170	CH_3OCH_3	Methoxymethane (dimethyl ether)	A3	—	1	8,500	16	1,000
R-218	$CF_3CF_2CF_3$	octafluoropropane	A1	2-0-0[b]	43	90,000	690	1,000
R-227ea	CF_3CHFCF_3	1,1,1,2,3,3,3-heptafluoropropane	A1	—	36	84,000	580	1,000
R-236fa	$CF_3CH_2CF_3$	1,1,1,3,3,3-hexafluoropropane	A1	2-0-0[b]	21	55,000	340	1,000
R-245fa	$CHF_2CH_2CF_3$	1,1,1,3,3-pentafluoropropane	B1	2-0-0[b]	12	34,000	190	300
R-290	$CH_3CH_2CH_3$	propane	A3	2-4-0	0.56	5,300	9.5	1,000
R-C318	$-(CF_2)_4-$	octafluorocyclobutane	A1	—	41	80,000	660	1,000

(continued)

[F] TABLE 1103.1—continued
REFRIGERANT CLASSIFICATION, AMOUNT AND OEL

CHEMICAL REFRIGERANT	FORMULA	CHEMICAL NAME OF BLEND	REFRIGERANT CLASSIFICATION	DEGREES OF HAZARD[a]	[M] AMOUNT OF REFRIGERANT PER OCCUPIED SPACE			OEL[e]
					Pounds per 1,000 cubic feet	ppm	g/m^3	
R400[d]	zeotrope	R-12/114 (50/50)	A1	2-0-0[b]	10	28,000	160	1,000
R-400[d]	zeotrope	R-12/114 (60/40)	A1		11	30,000	170	1,000
R-401A	zeotrope	R-22/152a/124 (53/13/34)	A1	2-0-0[b]	6.6	27,000	110	1,000
R-401B	zeotrope	R-22/152a/124 (61/11/28)	A1	2-0-0[b]	7.2	30,000	120	1,000
R-401C	zeotrope	R-22/152a/124 (33/15/52)	A1	2-0-0[b]	5.2	20,000	84	1,000
R-402A	zeotrope	R-125/290/22 (60/2/38)	A1	2-0-0[b]	8.5	33,000	140	1,000
R-402B	zeotrope	R-125/290/22 (38/2/60)	A1	2-0-0[b]	15	63,000	240	1,000
R-403A	zeotrope	R-290/22/218 (5.0/75.0/20.0)	A2	2-0-0[b]	7.6	33,000	120	1,000
R-403B	zeotrope	R-290/22/218 (5/56/39)	A1	2-0-0[b]	18	70,000	290	1,000
R-404A	zeotrope	R-125/143a/134a (44/52/4)	A1	2-0-0[b]	31	130,000	500	1,000
R-405A	zeotrope	R-22/152a/142b/C318 (45.0/7.0/5.5/2.5)	—	—	16	57,000	260	1,000
R-406A	zeotrope	R-22/600a/142b (55/4/41)	A2	—	4.7	21,000	25	1,000
R-407A	zeotrope	R-32/125/134a (20/40/40)	A1	2-0-0[b]	18	78,000	290	1,000
R-407B	zeotrope	R-32/125/134a (10/70/20)	A1	2-0-0[b]	20	77,000	320	1,000
R-407C	zeotrope	R-32/125/134a (23/25/52)	A1	2-0-0[b]	17	76,000	270	1,000
R-407D	zeotrope	R-32/125/134a (15/15/70)	A1	2-0-0[b]	15	65,000	240	1,000
R-407E	zeotrope	R-32/125/134a (25/15/60)	A1	2-0-0[b]	16	75,000	260	1,000
R-408A	zeotrope	R-125/143a/22 (7/46/47)	A1	2-0-0[b]	21	95,000	340	1,000
R-409A	zeotrope	R-22/124/142b (60/25/15)	A1	2-0-0[b]	7.1	29,000	110	1,000
R-409B	zeotrope	R-22/124/142b (65/25/10)	A1	2-0-0[b]	7.3	30,000	120	1,000
R-410A	zeotrope	R-32/125 (50/50)	A1	2-0-0[b]	25	130,000	390	1,000
R-410B	zeotrope	R-32/125 (45/55)	A1	2-0-0[b]	24	130,000	390	1,000
R-411A	zeotrope	R-127/22/152a (1.5/87.5/11.0)	A2	—	2.9	14,000	46	990
R-411B	zeotrope	R-1270/22/152a (3/94/3)	A2	—	2.8	13,000	45	980
R-412A	zeotrope	R-22/318/142b (70/5/25)	A2	—	5.1	22,000	82	1,000
R-413A	zeotrope	R-218/134a/600a (9/88/3)	A2	—	5.8	22,000	94	1,000
R-414A	zeotrope	R-22/124/600a/142b (51/28.5/4/16.5)	A1	—	6.4	26,000	100	1,000
R-414B	zeotrope	R-22/124/600a/142b (50/39/1.5/9.5)	A1	—	6	23,000	95	1,000

(continued)

[F] TABLE 1103.1—continued
REFRIGERANT CLASSIFICATION, AMOUNT AND OEL

CHEMICAL REFRIGERANT	FORMULA	CHEMICAL NAME OF BLEND	REFRIGERANT CLASSIFICATION	DEGREES OF HAZARD[a]	[M] AMOUNT OF REFRIGERANT PER OCCUPIED SPACE			
					Pounds per 1,000 cubic feet	ppm	g/m³	OEL[e]
R-415A	zeotrope	R-22/152a (82.0/18.0)	A2	—	12	57,000	190	1,000
R-415B	zeotrope	R-22/152a (25.0/75.0)	A2	—	9.3	52,000	120	1,000
R-416A	zeotrope	R-134a/124/600 (59/39.5/1.5)	A1	2-0-0[b]	3.9	14,000	62	1,000
R-417A	zeotrope	R-125/134a/600	A1	2-0-0[b]	3.5	13,000	56	1,000
R-418A	zeotrope	R-290/22/152a (1.5/96.0/2.5)	A2	—	13	59,000	200	1,000
R-419A	zeotrope	R-125/134a/E170 (77.0/19.0/4.0)	A2	—	19	70,000	310	1,000
R-420A	zeotrope	R-134a/142b (88.0/0)	A1	2-0-0[b]	12	45,000	190	1,000
R-421A	zeotrope	R-125/134a (58.0/42.0)	A1	2-0-0[b]	17	61,000	280	1,000
R-421B	zeotrope	R-125/134a (85.0/15.0)	A1	2-0-0[b]	21	69,000	330	1,000
R-422A	zeotrope	R-125/134a/600a (85.1/11.5/3.4)	A1	2-0-0[b]	18	63,000	290	1,000
R-422B	zeotrope	R-125/134a/600a (55.0/42.0/3.0)	A1	2-0-0[b]	16	26,000	250	1,000
R-422C	zeotrope	R-125/134a/600a (82.0/15.0/3.0)	A1	2-0-0[b]	18	62,000	290	1,000
R-422D	zeotrope	R-125/134a/600a (65.1/31.5/3.4)	A1	2-0-0[b]	16	58,000	260	1,000
R-423A	zeotrope	R-134a/227ea (52.5/47.5)	A1	2-0-0[c]	19	59,000	310	1,000
R-424A	zeotrope	R-125/134a/600a/600/601a (50.5/47.0/1.0/0.6)	A1	2-0-0[b]	6.2	23,000	100	970
R-425A	zoetrope	R-32/134a/227ea (18.5/69.5/0)	A1	2-0-0[b]	16	67,000	250	1,000
R-426A	zeotrope	R-125/134a/600a/601a (5.1/93.0/1.3/0.6)	A1	—	5.2	20,000	83	990
R-427A	zeotrope	R-32/125/143a/134a (15.0/25.0/10.0/50.0)	A1	—	18	76,000	280	1,000
R-428A	zeotrope	R-125/143a/290/600a (77.5/20.0/0.6/1.9)	A1	—	23	83,000	370	1,000
R-429A	zeotrope	R-E170/152a/600a (60.0/10.0/30.0)	A3	—	0.81	6,300	13	1,000
R-430A	zeotrope	R-152a/600a (76.0/24.0)	A3	—	1.3	8,000	21	1,000
R-431A	zeotrope	R-290/152a (71.0/29.0)	A3	—	0.69	5,500	11	1,000
R-432A	zeotrope	R-1270/E170 (80.0/20.0)	A3	—	0.13	1,200	2.1	710
R-433A	zeotrope	R-1270/290 (30.0/70.0)	A3	—	0.34	3,100	5.5	880
R-433B	zeotrope	R-1270/290 (5.0/95.0)	A3	—	0.51	4,500	8.1	950
R-433C	zeotrope	R-1270/290 (25.0-75.0)	A3	—	0.41	3,600	6.6	790
R-434A	zeotrope	R-125/143a/600a (63.2/18.0/16.0/2.8)	A1	—	20	73,000	320	1,000
R-435A	zeotrope	R-E170/152a (80.0/20.0)	A3	—	1.1	8,500	17	1,000
R-436A	zeotrope	R-290/600a (56.0/44.0)	A3	—	0.5	4,000	8	1,000
R-436B	zeotrope	R-290/600a (52.0/48.0)	A3	—	0.5	4,000	8	1,000
R-437A	zeotrope	R-125/134a/600/601 (19.5/78.5/1.4/0.6)	A1	—	5	19,000	81	990
R-438A	zeotrope	R-32/125/134a/600/601a (8.5/45.0/44.2/1.7/0.6)	A1	—	4.9	19,000	79	990

(continued)

[F] TABLE 1103.1—continued
REFRIGERANT CLASSIFICATION, AMOUNT AND OEL

CHEMICAL REFRIGERANT	FORMULA	CHEMICAL NAME OF BLEND	REFRIGERANT CLASSIFICATION	DEGREES OF HAZARD[a]	[M] AMOUNT OF REFRIGERANT PER OCCUPIED SPACE			
					Pounds per 1,000 cubic feet	ppm	g/m³	OEL[e]
R-500[e]	azeotrope	R-12/152a (73.8/26.2)	A1	2-0-0[b]	7.6	30,000	120	1,000
R-501[d]	azeotrope	R-22/12 (75.0/25.0)	A1	—	13	54,000	210	1,000
R-502[e]	azeotrope	R-22/115 (48.8/51.2)	A1	2-0-0[b]	21	73,000	330	1,000
R-503[e]	azeotrope	R-23/13 (40.1/59.9)	—	2-0-0[b]	—	—	—	1,000
R-504[d]	azeotrope	R-32/115 (48.2/51.8)	—	—	29	140,000	460	1,000
R-507A	azeotrope	R-125/143a (50/50)	A1	2-0-0[b]	32	130,000	520	1,000
R-508A	azeotrope	R-23/116 (39/61)	A1	2-0-0[b]	14	55,000	220	1,000
R-508B	azeotrope	R-23/116 (46/54)	A1	2-0-0[b]	13	52,000	200	1,000
R-509A	azeotrope	R-22/218 (44/56)	A1	2-0-0[b]	24	75,000	390	1,000
R-510A	azeotrope	R-E170/600a (88.0/0)	A3	—	0.87	7,300	14	1,000
R-600	$CH_3CH_2CH_2CH_3$	butane	A3	1-4-0	0.1	1,000	2.4	1,000
R-600a	$CH(CH_3)_2CH_3$	2-methylpropane (isobutane)	A3	2-4-0	0.6	4,000	9.6	1,000
R-601	$CH_3CH_2CH_2\,CH_2\,CH_3$	pentane	A3	—	0.2	1,000	2.9	600
R-601a	$(CH_3)_2CHCH_2CH_3$	2-methylbutane (isopentane)	A3	—	0.2	1,000	2.9	600
R-717	NH_3	ammonia	B2	3-3-0[c]	0.014	320	0.22	25
R-718	H_2O	water	A1	0-0-0	—	—	—	—
R-744	CO_2	carbon dioxide	A1	2-0-0[b]	4.5	40,000	72	5,000
R-1150	$CH_2{=}CH_2$	ethene (ethylene)	A3	1-4-2	—	—	—	200
R-1234yf	$CF_3CF{=}CH_2$	2,3,3,3-tetrafluoro-1 propene	A2	—	4.7	16,000	75	500
R-1270	$CH_3CH{=}CH_2$	Propene (propylene)	A3	1-4-1	0.1	1,000	1.7	500

For SI: 1 pound = 0.454 kg, 1 cubic foot = 0.0283 m³.

a. Degrees of hazard are for health, fire, and reactivity, respectively, in accordance with NFPA 704.

b. Reduction to 1-0-0 is allowed if analysis satisfactory to the code official shows that the maximum concentration for a rupture or full loss of refrigerant charge would not exceed the IDLH, considering both the refrigerant quantity and room volume.

c. For installations that are entirely outdoors, use 3-1-0.

d. Class I ozone depleting substance; prohibited for new installations.

e. Occupational Exposure Limit based on the OSHA PEL, ACGIH TLV-TWA, the AIHA WEEL or consistent value on a time-weighted average (TWA) basis (unless noted C for ceiling) for an 8 hr/d and 40 hr/wk.

1103.2 Occupancy classification. Locations of refrigerating systems are described by *occupancy* classifications that consider the ability of people to respond to potential exposure to refrigerants. Where *equipment* or appliances, other than piping, are located outside a building and within 20 feet (6096 mm) of any building opening, such *equipment* or appliances shall be governed by the *occupancy* classification of the building. *Occupancy* classifications shall be defined as follows:

1. Institutional *occupancy* is that portion of premises from which, because they are disabled, debilitated or confined, occupants cannot readily leave without the assistance of others. Institutional occupancies include, among others, hospitals, nursing homes, asylums and spaces containing locked cells.

2. Public assembly *occupancy* is that portion of premises where large numbers of people congregate and from which occupants cannot quickly vacate the space. Public assembly occupancies include, among others, auditoriums, ballrooms, classrooms, passenger depots, restaurants and theaters.

3. Residential *occupancy* is that portion of premises that provides the occupants with complete independent living facilities, including permanent provisions for living, sleeping, eating, cooking and sanitation. Residential occupancies include, among others, dormitories, hotels, multiunit apartments and private residences.

4. Commercial *occupancy* is that portion of premises where people transact business, receive personal service or purchase food and other goods. Commercial occupancies include, among others, office and professional buildings, markets (but not large mercantile occupancies) and work or storage areas that do not qualify as industrial occupancies.

5. Large mercantile *occupancy* is that portion of premises where more than 100 persons congregate on levels above or below street level to purchase personal merchandise.

6. Industrial *occupancy* is that portion of premises that is not open to the public, where access by authorized persons is controlled, and that is used to manufacture, process or store goods such as chemicals, food, ice, meat or petroleum.

7. Mixed *occupancy* occurs when two or more occupancies are located within the same building. When each *occupancy* is isolated from the rest of the building by tight walls, floors and ceilings and by self-closing doors, the requirements for each *occupancy* shall apply to its portion of the building. When the various occupancies are not so isolated, the *occupancy* having the most stringent requirements shall be the governing *occupancy*.

❖ These occupancy definitions are unique to the code and may or may not coincide with the occupancy definitions in Chapter 3 of the *International Building Code®* (IBC®). Building occupancy classification is one of the three main factors that determine the quantity and application limits for refrigerants. If outdoor refrigeration equipment is placed within 20 feet (6096 mm) of a door, window, air intake or any other opening through which refrigerant could enter the building, that equipment is treated as if it were installed inside the building. This section recognizes the inherent safety of equipment located outdoors and offers exemption from limitations for such equipment; however, location within 20 feet (6096 mm) of a building opening voids those exemptions. Note that ASHRAE 15 exempts building openings connecting to a machine room.

1103.3 System classification. Refrigeration systems shall be classified according to the degree of probability that refrigerant leaked from a failed connection, seal or component could enter an occupied area. The distinction is based on the basic design or location of the components.

❖ The method of cooling or heating determines the probability of the material leaking into the occupied space. Direct systems have coils containing primary refrigerant over which the room air passes. A leak in the heat exchanger could place refrigerant directly in the occupied space. Such systems are high-probability systems. Other systems cool an intermediate fluid that is piped to various parts of the building. A leak in the refrigerant heat exchanger might result in refrigerant leaking into the cooling fluid, but not into the occupied space. These systems are most often low-probability systems.

There are two basic types of refrigeration systems: direct and indirect. A direct system is one in which the air supplied to the occupied space passes over the evaporator or condenser that contains the refrigerant. An indirect system is one in which a secondary heat transfer medium, such as brine or water, is heated or cooled by the refrigeration system and the cooled or heated brine or water is circulated to another heat exchanger to cool or heat the air within the space. Therefore, an indirect system is one in which the refrigeration system does not extract heat directly from or add heat directly to the air within the occupied space, whereas a direct system extracts heat directly from or adds heat directly to the air within the occupied space. One characteristic of a direct system that represents a higher potential hazard than an indirect system is that a leak will cause the refrigerant to be dispersed throughout the space served by the system. Refrigeration systems, although thought of primarily as cooling or freezing systems, can also be designed for the purpose of heating an occupied space or a commodity. In many systems, heat is extracted from a location for the purpose of cooling, and the extracted heat is delivered to another location.

Grocery stores commonly use such systems to cool food products and simultaneously heat the building. Heat pump loop systems pick up heat from zones cooling and reject that heat to zones calling for heating.

1103.3.1 Low-probability systems. Double-indirect open-spray systems, indirect closed systems and indirect-vented closed systems shall be classified as low-probability systems, provided that all refrigerant-containing piping and fittings are isolated when the quantities in Table 1103.1 are exceeded.

❖ The nature of low-probability systems makes refrigerant leakage into occupied areas unlikely. In an indirect system, the refrigerant-containing components, evaporators, piping, heat exchangers, etc., are isolated from the occupied areas of a building and only a heat-transfer fluid, often called "brine," is circulated within the occupied area (see Commentary Figure 1103.3.1). The term "brine" does not imply that the heat transfer fluid is necessarily a salt solution. If a system's refrigerant quantity exceeds the limits of Table 1103.1, the refrigerant-containing portions of the systems will be located in a machinery room of some type. Where the quantity of refrigerant in a system is within the limits of Table 1103.1, the system could be located entirely within the occupied building areas and the distinction between low-probability systems and high-probability systems would be meaningless. In other words, the low-probability system exists only if the refrigerant-containing portions of the overall system are isolated from all areas of a building except machinery rooms.

1103.3.2 High-probability systems. Direct systems and indirect open-spray systems shall be classified as high-probability systems.

Exception: An indirect open-spray system shall not be required to be classified as a high-probability system if the pressure of the secondary coolant is at all times (operating and standby) greater than the pressure of the refrigerant.

❖ The probability classifications are simply a means of assessing the potential hazard posed by refrigeration systems. In a high-probability system, chances are good that system leakage would expose building occupants to a refrigerant (see the definitions of "Direct refrigeration system," "Indirect refrigeration system" and "High-probability systems").

The exception recognizes a type of system that substantially reduces the likelihood of refrigerant leakage into the occupied space. The typical split system heat pump; DX coil in an air handler, furnace or split system air conditioner; package terminal units and window air-conditioning units are all high-probability systems. In these units, the room air passes directly over coils containing refrigerant.

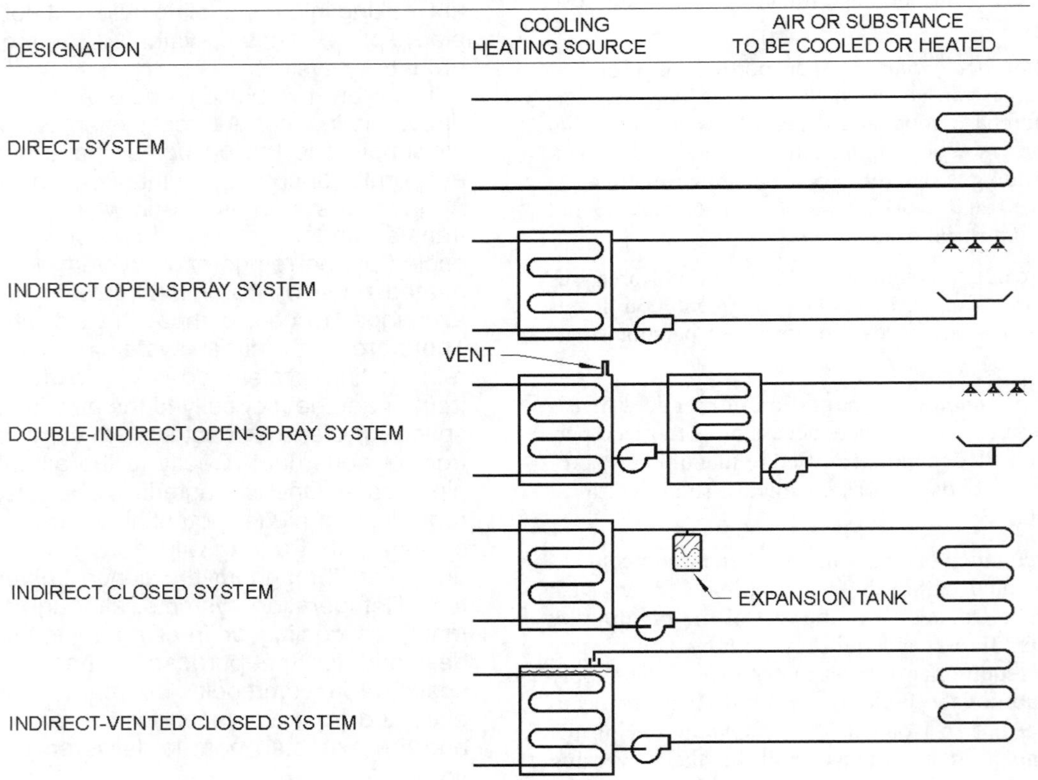

Figure 1103.3.1
REFRIGERATION SYSTEM CLASSIFICATION
(Source: ASHRAE 15-89)

SECTION 1104
SYSTEM APPLICATION REQUIREMENTS

1104.1 General. The refrigerant, *occupancy* and system classification cited in this section shall be determined in accordance with Sections 1103.1, 1103.2 and 1103.3, respectively. For refrigerant blends assigned dual classifications, as formulated and for the worst case of fractionation, the classifications for the worst case of fractionation shall be used.

❖ Whereas Section 1103 deals with classification of systems, Section 1104 deals with specific requirements based on these classifications. These include when a machinery room is required to house refrigeration equipment (Section 1104.2) and when equipment can be located outside of machinery rooms (Section 1104.2.2). The requirements in Section 1104 parallel ASHRAE 15, referenced in Section 1101.6, and have become a part of the code for ease of reference.

Before the specific application requirements of Section 1104 can be applied, the system must be classified in accordance with Section 1103. "Fractionation" is the chemical separation of blended refrigerants into various ratios of the components that compose the blend. Fractionation results in compounds of different flammability or toxicity than the original blend. Fractionation of, for example, zeotropic blends (R-400 series refrigerants) can occur if a leak develops in the system. For refrigerant blends, a classification is given to the composition based on its formulated ratio of components. The blend is also given a classification based on the ratio of components at the worst case of fractionation. The worst-case classification must be used because the refrigerant blend can change properties as its composition changes and the most hazardous composition could exist under certain conditions, such as system leakage. A refrigerant blend with the classification A1/A2, for example, would be subject to the code requirements for A2 refrigerants. For more information on fractionation, see the appendix to ASHRAE 15.

1104.2 Machinery room. Except as provided in Sections 1104.2.1 and 1104.2.2, all components containing the refrigerant shall be located either outdoors or in a *machinery room* where the quantity of refrigerant in an independent circuit of a system exceeds the amounts shown in Table 1103.1. For refrigerant blends not listed in Table 1103.1, the same requirement shall apply when the amount for any blend component exceeds that indicated in Table 1103.1 for that component. This requirement shall also apply when the combined amount of the blend components exceeds a limit of 69,100 parts per million (ppm) by volume. Machinery rooms required by this section shall be constructed and maintained in accordance with Section 1105 for Group A1 and B1 refrigerants and in accordance with Sections 1105 and 1106 for Group A2, B2, A3 and B3 refrigerants.

Exceptions:

1. Machinery rooms are not required for *listed equipment* and appliances containing not more than 6.6 pounds (3 kg) of refrigerant, regardless of the refrigerant's safety classification, where installed in accordance with the equipment's or appliance's listing and the *equipment* or *appliance* manufacturer's installation instructions.

2. Piping in conformance with Section 1107 is allowed in other locations to connect components installed in a *machinery room* with those installed outdoors.

❖ This section establishes when a machinery room is required to house refrigeration systems if the amount of refrigerant in the system exceeds the amount indicated in Table 1103.1. Sections 1105 and 1106 contain the construction requirements for the machinery rooms.

A machinery room provides a buffer between building occupants and the system, so that if something goes wrong, the hazard will be isolated from normally occupied areas. Section 1105 requires protected openings, refrigerant detectors and special ventilation for the room.

Exception 1 provides relief from the requirement for a machinery room when a system contains only 6.6 pounds (3 kg) of refrigerant because the quantity is insignificant and the machine is tested, listed and labeled. Exception 2 allows the piping connecting components in a machinery room to components on the exterior of the building to pass through spaces that do not meet the requirements for a machinery room.

Where Table 1103.1 shows a dash instead of a refrigerant amount, that information had not been determined at the date of printing.

1104.2.1 Institutional occupancies. The amounts shown in Table 1103.1 shall be reduced by 50 percent for all areas of institutional occupancies except kitchens, laboratories and mortuaries. The total of all Group A2, B2, A3 and B3 refrigerants shall not exceed 550 pounds (250 kg) in occupied areas or machinery rooms.

❖ Because of the danger to people who may be incapable of self-preservation, the table limits are cut in half for institutional occupancies. The cumulative amount of the more hazardous refrigerants, even for systems located in machinery rooms, is also limited by this section. Systems that require greater amounts of refrigerants than would be allowed by this section would have to be located outdoors.

1104.2.2 Industrial occupancies and refrigerated rooms. This section applies only to industrial occupancies and refrigerated rooms for manufacturing, food and beverage preparation, meat cutting, other processes and storage. Machinery rooms are not required where all of the following conditions are met:

1. The space containing the machinery is separated from other occupancies by tight construction with tight-fitting doors.

2. Access is restricted to authorized personnel.

3. The floor area per occupant is not less than 100 square feet (9.3 m²) where machinery is located on floor levels

with exits more than 6.6 feet (2012 mm) above the ground. Where provided with egress directly to the outdoors or into *approved* building exits, the minimum floor area shall not apply.

4. Refrigerant detectors are installed as required for machinery rooms in accordance with Section 1105.3.

5. Surfaces having temperatures exceeding 800°F (427°C) and open flames are not present where any Group A2, B2, A3 or B3 refrigerant is used (see Section 1104.3.4).

6. All electrical *equipment* and appliances conform to Class 1, Division 2, *hazardous location* classification requirements of NFPA 70 where the quantity of any Group A2, B2, A3 or B3 refrigerant, other than ammonia, in a single independent circuit would exceed 25 percent of the lower flammability limit (LFL) upon release to the space.

7. All refrigerant-containing parts in systems exceeding 100 horsepower (hp) (74.6 kW) drive power, except evaporators used for refrigeration or dehumidification; condensers used for heating; control and pressure relief valves for either; and connecting piping, shall be located either outdoors or in a *machinery room*.

❖ This section contains provisions for industrialized food processing plants, manufacturing plants and refrigerated storage. Refrigeration systems in these spaces need not be contained in a machinery room as long as all of the seven items of this section are met. "Industrial occupancy" is defined in Section 1103.2 and does not necessarily coincide with the description in Section 306 of the IBC for a Group F occupancy. Because these occupancies are restricted to employees who are familiar with their surroundings and contain low occupant loads, the hazard is perceived to be reduced.

Item 3 specifically limits the occupant density when occupants would have to travel down open interior stairs to leave the building. This restriction does not apply if the required exit doors from the space lead directly to exterior stairs or to stairs enclosed with fire-resistance-rated construction and protected openings in accordance with the IBC for interior exit stairs.

The "lower flammability limit" mentioned in Item 6 is defined in Chapter 2 of the code as "the minimum concentration of refrigerant that is capable of propagating a flame through a homogeneous mixture of refrigerant and air" [see definition of "Lower flammability limit (LFL)"]. If a refrigerant were released into the air, thus mixing with the air (to create a homogeneous mixture), a certain concentration would eventually be reached that would make the air/refrigerant mixture flammable/explosive. The lowest concentration at which this could occur is the LFL. Item 6 limits the amount in a single circuit to 25 percent of the amount needed to attain the LFL, unless all electrical equipment and appliances in the space conform to the Class 1, Division 2 hazardous location requirements found in NFPA 70.

The flammability classification of the refrigerants used can be obtained from the manufacturer or the MSDS for the refrigerant. See also the definition of "Flammability classification" in Chapter 2.

1104.3 Refrigerant restrictions. Refrigerant applications, maximum quantities and use shall be restricted in accordance with Sections 1104.3.1 through 1104.3.4.

❖ Along with the type of refrigeration system, the physical and chemical characteristics of the refrigerant used in the system play an important role in determining the maximum permissible refrigerant quantities. Quantities of refrigerants are limited or special design requirements must be followed because of the risks refrigerants present to the building occupants, including the risk from explosion, fire or the rupture of high-pressure components or piping. Of equal concern are the health risks that result from the accidental release of refrigerant into the occupied space. These risks include the following:

• Oxygen-displacing refrigerants (that is, refrigerants that are heavier than air will displace the air near the floor level) pose the potential threat of suffocation in unventilated or poorly ventilated spaces;

• Corrosive refrigerants attack skin, eyes and other tissue;

• Body contact with liquid refrigerants can cause freezing of skin and other body tissue;

• Narcotic and cardiac sensitization effects; and

• Toxic effects from refrigerant gas or from the decomposition products resulting from refrigerant contact with flames or hot surfaces.

The maximum permissible quantities of refrigerants are based on the safety group classifications in ASHRAE 34 and Table 1103.1, the characteristics and risks associated with each individual refrigerant, the occupancy classification and the type of refrigeration system. The permissible quantities allowed by the code are based on ASHRAE 15.

1104.3.1 Air-conditioning for human comfort. In other than industrial occupancies where the quantity in a single independent circuit does not exceed the amount in Table 1103.1, Group B1, B2 and B3 refrigerants shall not be used in high-probability systems for air-conditioning for human comfort.

❖ The reasons for these restrictions are the higher toxicity levels of Group B1, B2 and B3 refrigerants and the higher exposure potential of high-probability systems. High-probability systems are described in Section 1103.3.2 (see commentary, Section 1103.3.2). Industrial occupancies are described in Section 1103.2 and the description does not necessarily correspond to the Group F classification in Chapter 3 of the IBC. Industrial occupancies typically have lower occupant loads, and are occupied only by employees who are familiar with their surroundings. Note that the restrictions for Group B1, B2 and B3 refrigerants for human comfort space conditioning would also apply to industrial occupancies if the quantity of refrigerant in any circuit exceeds the limits of Table 1103.1.

1104.3.2 Nonindustrial occupancies. Group A2 and B2 refrigerants shall not be used in high-probability systems where the quantity of refrigerant in any independent refrigerant circuit exceeds the amount shown in Table 1104.3.2. Group A3 and B3 refrigerants shall not be used except where *approved*.

> **Exception:** This section does not apply to laboratories where the floor area per occupant is not less than 100 square feet (9.3 m²).

❖ This section applies to all occupancies other than industrial occupancies. Table 1104.3.2 limits the amounts of Group A2 and B2 refrigerants in high-probability systems because of the properties of these refrigerants and the potential for release of refrigerant in high-probability systems (see commentary, Section 1103.3.2.). Group A3 and B3 refrigerants are the most flammable and therefore can be used only in industrial occupancies and where specifically approved by the code official. The exception exempts laboratories because of the specialized tasks and safety procedures typically found in laboratories.

TABLE 1104.3.2. See below.

❖ These table limits apply (1) in all occupancies other than industrial occupancies as described in Section 1103.2, and (2) only to Group A2 and B2 refrigerants in high-probability systems (see Section 1103.3.2).

The term "exit access" as used in the table means any area that occupants would have to pass through on their way to an exit (see the IBC definitions of "Exit access" and "Exit"). If the location of the room containing the equipment would allow occupants to move away from the equipment and toward the exits instead of having to pass by the equipment, the equipment would not be considered as being in an exit access.

1104.3.3 All occupancies. The total of all Group A2, B2, A3 and B3 refrigerants other than R-717, ammonia, shall not exceed 1,100 pounds (499 kg) except where *approved*.

❖ Group A2, B2, A3 and B3 refrigerants in amounts exceeding 1,100 pounds (500 kg) represent a significant hazard to occupants and therefore require spe-

cial approval by the code official. This section acts as a cap for all occupancies and all types of systems. The criteria for approval would come from safety specialists in the refrigeration industry.

Ammonia systems are exempt from this limitation because ammonia is difficult to ignite, is very stable, is self-alarming (a strong odor is easily detected) and is lighter than air so normal ventilation will, in most cases, prevent its accumulation. Ammonia systems must conform to standard IIAR 2 referenced in Section 1101.6.

1104.3.4 Protection from refrigerant decomposition. Where any device having an open flame or surface temperature greater than 800°F (427°C) is used in a room containing more than 6.6 pounds (3 kg) of refrigerant in a single independent circuit, a hood and exhaust system shall be provided in accordance with Section 510. Such exhaust system shall exhaust *combustion* products to the outdoors.

> **Exception:** A hood and exhaust system shall not be required:
>
> 1. Where the refrigerant is R-717, R-718 or R-744;
> 2. Where the *combustion* air is ducted from the outdoors in a manner that prevents leaked refrigerant from being combusted; or
> 3. Where a refrigerant detector is used to stop the *combustion* in the event of a refrigerant leak (see Sections 1105.3 and 1105.5).

❖ When most refrigerants are exposed to high temperatures, they break down chemically and can form toxic and/or corrosive byproducts. Some of these chemical byproducts are extremely toxic to humans and can cause serious problems in fuel-fired appliances. The required hood and exhaust system is intended to capture and exhaust any hazardous chemicals. The exceptions recognize that the hazard would be reduced where the refrigerant is not halogen based or where provisions are made to prevent refrigerants from coming into contact with flames or hot surfaces.

1104.4 Volume calculations. Volume calculations shall be in accordance with Sections 1104.4.1 through 1104.4.3.

❖ To apply the refrigerant-per-volume requirements of Table 1103.1, the method of determining occupied space volume must be known. The amount of refrigerant per occupied space is intended for high-probability systems. In this case, human occupants could

TABLE 1104.3.2
MAXIMUM PERMISSIBLE QUANTITIES OF REFRIGERANTS

TYPE OF REFRIGERATION SYSTEM	MAXIMUM POUNDS FOR VARIOUS OCCUPANCIES			
	Institutional	Assembly	Residential	All other occupancies
Sealed absorption system In exit access In adjacent outdoor locations In other than exit access	0 0 0	0 0 6.6	3.3 22 6.6	3.3 22 6.6
Unit systems In other than exit access	0	0	6.6	6.6

For SI: 1 pound = 0.454 kg.

be directly exposed to refrigerant if the refrigerant circuit leaks. The amounts given [pounds per 1,000 cubic feet (kg/28 m), volume percent or grams/cubic meter] are the maximum allowed in the largest single circuit in a given space, assuming that only one circuit is likely to leak refrigerant at a given time.

1104.4.1 Noncommunicating spaces. Where the refrigerant-containing parts of a system are located in one or more spaces that do not communicate through permanent openings or HVAC ducts, the volume of the smallest, enclosed occupied space shall be used to determine the permissible quantity of refrigerant in the system.

❖ A noncommunicating space is a space that is not connected to other spaces through permanent openings or HVAC ducts. Leaked refrigerant would be retained in the space in which the leakage occurred because permanent openings and ducts are not present to allow dispersion of refrigerant. This section considers a machinery room to be a space because, in the case of self-contained machines such as chillers, the only room volume to consider is the volume of the space that encloses the machines. Thus, the refrigerant quantity limits and machinery room requirements would be based on the volume of the room containing the chillers.

This section applies to direct systems and any other system having components located within occupied space.

This section intends to account for the worst-case condition that would exist if a refrigerant leak were to occur in the smallest enclosed space that contains refrigerant-containing parts of a system (see commentary, Section 1105.5).

1104.4.2 Communicating spaces. Where an evaporator or condenser is located in an air duct system, the volume of the smallest, enclosed occupied space served by the duct system shall be used to determine the maximum allowable quantity of refrigerant in the system.

Exception: If airflow to any enclosed space cannot be reduced below one-quarter of its maximum, the entire space served by the air duct system shall be used to determine the maximum allowable quantity of refrigerant in the system.

❖ This section addresses direct (high-probability) systems that are part of a ducted air distribution system that serves one or more spaces (see commentary, Section 1104.4.1). As used in this section, the term "communicating" means that the spaces connect (communicate) through a system of HVAC ducts. The duct system would be capable of spreading leaked refrigerant to all spaces served by the system.

A DX coil in an air handler connected to an air distribution duct system is the typical scenario addressed by this section. The exception allows the volumes of enclosed spaces served by the air distribution system to be summed if the spaces cannot be isolated from the air distribution system. The logic of this exception is that refrigerant leakage would be distributed throughout all spaces served by the duct system rather than concentrating in any one enclosed space. In air duct systems with multiple dampers, the volume is dependent on the isolation ability of the dampers. Commentary Figure 1104.4.2 shows such a system. The intent is to determine the worst possible case of distribution of refrigerant in the airstream in case of leakage.

Example 1: See Commentary Figure 1104.4.2:

Question 1: Which volume is used if both dampers have mechanical stops to prevent airflow from being reduced below one-quarter of its maximum?

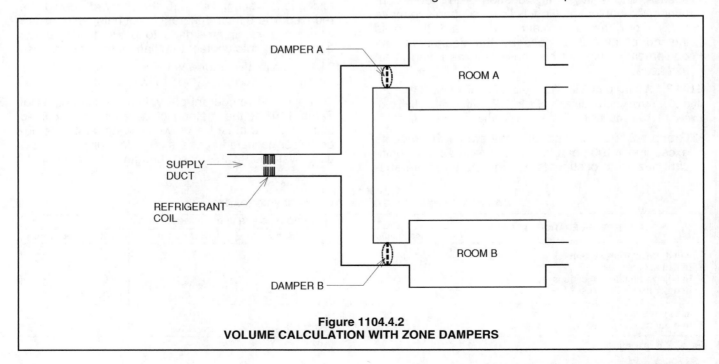

Figure 1104.4.2
VOLUME CALCULATION WITH ZONE DAMPERS

Answer 1: The entire space—Room A, plus Room B—can be used to determine the maximum quantity of refrigerant.

Question 2: Which volume is used if one damper (either A or B) can be completely closed?

Answer 2: The volume must be based only on the smaller of the two rooms because in the event of a leak, one of the dampers could be closed and concentrate all the refrigerant to the other room.

The pertinent dampers are the operating dampers in the system, not the dampers that are intended to function only in an emergency that would require evacuation of the building. Therefore, fire and smoke dampers are excluded from this analysis.

Example 2:

Question: What volume is used with a typical ducted split system serving a single-family dwelling?

Answer: The volume of the conditioned space in the whole house is used, assuming that the registers at the end of duct runs cannot be adjusted to reduce the flow below 25 percent of the maximum.

When the space being considered is served by an air distribution system, the evaporator of the direct system is usually located in the supply air system for the space. If the supply airflow to the smallest enclosed space cannot be reduced to less than one-fourth of the maximum design airflow for that space, and the same supply air system also serves another space or spaces where the supply airflow also cannot be reduced to less than one quarter of the maximum design flow for the space or spaces, the total volume of all spaces served by the supply air system is used to determine the maximum permissible quantity of refrigerant allowed in the system. Because the enclosed spaces are interconnected by an air distribution system and the airflow cannot be completely shut down, the spaces served by the supply air system can be considered as one communicating space for the purpose of applying this section. If there is a leak, the refrigerant will be distributed to all spaces served by the air supply system and, therefore, the combined volume of all the spaces will allow the refrigerant to be sufficiently diluted to minimize the hazard to the occupants. For example, the entire volume of a business occupancy with both an open-plan floor area and enclosed offices can be used when the supply airflow to the enclosed offices and the open floor area cannot be reduced to less than one-quarter of the maximum design flow and all spaces are served by a common supply air system. Additionally, the volume of a continuous air plenum used for return air located above a suspended ceiling can be included in the volume of the enclosed space.

When the space is not served by an air distribution system, the volume of the smallest enclosed occupied space in which the refrigerant-containing parts are located must be used to determine the permissible quantity of refrigerant.

1104.4.3 Plenums. Where the space above a suspended ceiling is continuous and part of the supply or return air *plenum* system, this space shall be included in calculating the volume of the enclosed space.

❖ This section addresses plenums as described in Section 602 and allows inclusion of the plenums in the volume determined by Section 1104.4.2.

SECTION 1105
MACHINERY ROOM, GENERAL REQUIREMENTS

[B] 1105.1 Design and construction. Machinery rooms shall be designed and constructed in accordance with the *International Building Code* and this section.

❖ Section 1104.2 dictates when a machinery room is required. The requirements of this section apply to all machinery rooms, whereas the requirements of Section 1106 apply only to special machinery rooms. Machinery rooms are required where refrigerant quantities exceed the limits of Table 1103.1. Machinery rooms enclose all refrigerant-containing system components except piping and components located outdoors and serve to protect building occupants from exposure to refrigerants. Section 509.4 of the IBC addresses machinery room enclosure and fire suppression, and Section 1014.4 addresses machinery room egress.

1105.2 Openings. Ducts and air handlers in the *machinery room* that operate at a lower pressure than the room shall be sealed to prevent any refrigerant leakage from entering the airstream.

❖ Return air ducts and the intake of air handlers typically operate at less than the ambient room pressures. Unless they are adequately sealed, refrigerant could enter these ducts and be spread throughout the building. Section 603.9 requires sealing of ducts to conform to the requirements of the *International Energy Conservation Code*® (IECC®).

[F] 1105.3 Refrigerant detector. Refrigerant detectors in machinery rooms shall be provided as required by Section 606.8 of the *International Fire Code*.

❖ A refrigerant-specific detector is required for leak detection, early warning and actuation of emergency exhaust systems. Detector requirements are found in Section 606 of the IFC. Depending on the density of the refrigerants, leakage may collect near the floor, near the ceiling or disperse equally throughout the space.

Refrigerant detector locations must be carefully considered. Most refrigerants are heavier than air, making floor depressions and pits natural areas for accumulation. The code does not specify the location of sensors because of the endless variety of equipment room designs. The key to properly locating a detector in the machinery room is to remember that occupant safety is the primary objective, and the danger is in breathing refrigerant. Placing the sensor below the common breathing height of 5 feet (1525 mm) results in an additional safety margin because all commonly used halocarbon refrigerants are three to five times heavier than air. When undistributed by airflow, escaping refrigerant will flow to the floor, seeking the lowest levels and filling the room from the bottom up. Because pits, stairwells or trenches are likely to fill with refrigerant first, detectors should also be placed in any of these areas that may be occupied. The alarm actuation threshold is dictated by the last column of Table 1103.1.

Manufacturers' instructions for detectors will provide installation guidance for the location of detectors and the required number of detectors for any given room size.

Because most general machinery rooms are unoccupied for long periods, a refrigeration leak may go undetected, allowing a buildup of refrigerant that can pose a threat to the building occupants and the maintenance personnel who will be required to enter the machinery room. Also, the refrigerants may or may not be detectable by the senses of smell, sight and taste, depending on the chemical nature and concentration of the refrigerant in air. This can be especially critical when a toxic refrigerant is used in the refrigeration system.

The ACGIH defines three levels of refrigerant exposure: Level 1 is the AEL, which is the level at which a person can be exposed for 8 hours per day for 40 hours per week without having an adverse effect on health. Level 2 is the STEL, which is defined as three times the AEL. At this level a person should not be exposed for more than 30 minutes at a time. Persons working in a machinery room having this concentration of refrigerant should be equipped with respiratory protection. Level 3 is the emergency exposure limit (EEL). At this level, persons should not be in the room at all without a self-contained breathing apparatus.

Early detection of leaking refrigerant depends on the location of the refrigerant detectors. If they are improperly located, a refrigerant leak could go undetected for an undesirable length of time, thus allowing a significant amount of refrigerant to escape.

Items to be considered when choosing locations for detectors are the airflow patterns of the room, the particular refrigerant density and the fact that the primary hazard to the occupants is inhalation. The detectors should be located to prevent the normal ventilation system from interfering with detection. Placing detectors between the refrigeration system and exhaust fan inlets should help to ensure that the presence of refrigerant will be detected.

Depending on the size of the machinery room and the number and type of refrigeration systems, more than one detector may be necessary. Manufacturers' installation instructions for the refrigeration detection system should be followed when choosing the location and the number of sensors for a particular machinery room application. For example, many refrigerant detection systems are capable of activating a warning system with a warning light, alarm or similar device if refrigerant exceeds a preset concentration or level due to a leak. Typically, the warning systems have the capability to alert people inside and outside the machinery room so that the leak can be quickly located and repaired.

1105.4 Tests. Periodic tests of the mechanical ventilating system shall be performed in accordance with manufacturer's specifications and as required by the code official.

❖ Detector manufacturers specify testing frequency to monitor the detector's condition, sensitivity and performance. The audible and visual alarms must also be tested periodically. The detector also functions to start the emergency ventilation system, which must be tested for proper operation.

1105.5 Fuel-burning appliances. Fuel-burning appliances and *equipment* having open flames and that use *combustion* air from the *machinery room* shall not be installed in a *machinery room*.

Exceptions:

1. Where the refrigerant is carbon dioxide or water.
2. Fuel-burning appliances shall not be prohibited in the same *machinery room* with refrigerant-containing *equipment* or appliances where *combustion* air is ducted from outside the *machinery room* and sealed in such a manner as to prevent any refrigerant leakage from entering the *combustion* chamber, or where a refrigerant vapor detector is employed to automatically shut off the *combustion* process in the event of refrigerant leakage.

❖ Refrigerant likely to be present in a machinery room could contaminate the combustion air for fuel-fired appliances. When refrigerants, especially halogenated, are subjected to the high temperatures of fuel combustion, they break down into compounds that are highly corrosive or toxic. The corrosive chemicals can cause severe damage to appliances, vents and chimneys, resulting in a hazardous condition. Also, toxic byproducts that could be produced present a health hazard to building occupants and fire-fighting personnel.

Exception 1 recognizes that carbon dioxide (CO_2) and water vapor are normally present in all combustion air.

Exception 2 allows two options, one of which prevents refrigerant from entering the combustion chamber of the appliance. The other option would prevent combustion in the event of refrigerant leakage.

1105.6 Ventilation. Machinery rooms shall be mechanically ventilated to the outdoors.

Exception: Where a refrigerating system is located outdoors more than 20 feet (6096 mm) from any building opening and is enclosed by a penthouse, lean-to or other open structure, natural or mechanical ventilation shall be provided. Location of the openings shall be based on the relative density of the refrigerant to air. The free-aperture cross section for the ventilation of the *machinery room* shall be not less than:

$$F = \sqrt{G}$$ **(Equation 11-1)**

For SI: $F = 0.138\sqrt{G}$

where:

F = The free opening area in square feet (m²).

G = The mass of refrigerant in pounds (kg) in the largest system, any part of which is located in the *machinery room.*

❖ Mechanical ventilation of the machinery room to the outdoors is required, and is specified at two levels. The lowest level (normal ventilation) is required any time the equipment room is occupied (see commentary, Section 1105.6.3). The highest level (emergency) is required when the refrigerant concentration reaches the alarm level (TLV-TWA) (see commentary, Section 1105.6.4).

The distribution of mechanical ventilation through the equipment room deserves special consideration to avoid creating areas in which refrigerant could accumulate. Supply and discharge air should be positioned with consideration that most refrigerants are heavier than air.

To qualify for the natural ventilation system exception, the structure must not connect to the occupied building by doorways, pipe tunnels, transfer grilles, electrical conduit raceways, ducts or other such openings.

The required ventilation system must be capable of functioning as required by Sections 1105.6.3 and 1105.6.4. This section does not limit the design choices for the ventilation system as long as the detector required by Section 1105.3 controls the exhaust rate and the exhaust system design creates a uniform air movement through the room to "sweep" the entire area and to avoid stagnant spots.

The exception allows natural (gravity) ventilation of the machinery room if the room is located outdoors and not closer than 20 feet (6096 mm) in any direction to any building opening through which refrigerant could enter. These machinery rooms could be separate structures or could be attached to or placed on the roof of the building served.

The formula is used to calculate the minimum net free area of required openings but does not specify how the opening area is to be distributed. As with all naturally ventilated spaces, the location and distribution of openings is critical to creating air movement through the ventilated space.

The ventilation system must operate whenever the refrigeration system is operational, including periods when the equipment is in its "off" cycle. This means that the ventilation system must operate continuously except when the refrigeration system is shut down for servicing or during periods when it is not in use.

The air within the room must be exhausted to the outdoors, with an approximately equal amount of makeup air being introduced into the room. Ventilation systems are typically designed to maintain a slight negative pressure in the room to help prevent the migration of air within the room into adjacent areas through cracks and openings in the room enclosure.

The ventilation system must be independent of all other building ventilation systems to prevent contamination of adjacent areas of the building.

Typically, the ventilation system will consist of an exhaust fan or fans that draw air through the machinery room to create a sweeping effect across the refrigeration system. Makeup air openings must be placed to induce a cross-sectional flow that will purge the entire machinery room.

1105.6.1 Discharge location. The discharge of the air shall be to the outdoors in accordance with Chapter 5. Exhaust from mechanical ventilation systems shall be discharged not less than 20 feet (6096 mm) from a property line or openings into buildings.

❖ To prevent contaminated air from reentering a building, air must be exhausted in accordance with the provisions of Chapter 5. For example, Sections 501.2, 501.4, 503 and 510, among others, address discharge location, duct construction, motors and fans and hazardous exhaust.

1105.6.2 Makeup air. Provisions shall be made for *makeup air* to replace that being exhausted. Openings for *makeup air* shall be located to avoid intake of *exhaust air*. Supply and exhaust ducts to the *machinery room* shall serve no other area, shall be constructed in accordance with Chapter 5 and shall be covered with corrosion-resistant screen of not less than ¹/₄-inch (6.4 mm) mesh.

❖ No exhaust system can function without makeup air (see commentary, Section 501.4). To reach the required exhaust rate, an approximately equal amount of makeup air must be supplied by openings to the outdoors, makeup air supply units, calculated infiltration or a combination of these. Because machinery rooms must be sealed off from all other spaces, transfer grilles cannot be used to supply makeup air from other areas. To prevent the possibility of refrigerant escape to other areas, machinery room exhaust and makeup air supply systems must be completely independent and dedicated to serving only the machinery room.

1105.6.3 Ventilation rate. For other than ammonia systems, the mechanical ventilation systems shall be capable of exhausting the minimum quantity of air both at normal operating and emergency conditions, as required by Sections

1105.6.3.1 and 1105.6.3.2. The minimum required ventilation rate for ammonia shall be 30 air changes per hour in accordance with IIAR2. Multiple fans or multispeed fans shall be allowed to produce the emergency ventilation rate and to obtain a reduced airflow for normal ventilation.

❖ The minimum ventilation rate for ammonia systems must be in accordance with IIAR-2 and at least 30 air changes per hour. For all other refrigerants, the ventilation system must be capable of exhausting the minimum quantity of air under normal conditions and under emergency conditions in accordance with Sections 1105.6.3.1 and 1105.6.3.2, respectively.

1105.6.3.1 Quantity—normal ventilation. During occupied conditions, the mechanical ventilation system shall exhaust the larger of the following:

1. Not less than 0.5 cfm per square foot (0.0025 m³/s · m²) of *machinery room* area or 20 cfm (0.009 m³/s) per person; or

2. A volume required to limit the room temperature rise to 18°F (10°C) taking into account the ambient heating effect of all machinery in the room.

❖ Normal ventilation is required only when the room is occupied. If the equipment room is not normally occupied, normal ventilation should be interlocked with the room lighting. If the equipment room is constantly occupied, the normal ventilation must be continuous. The flow rate for both Items 1 and 2 must be calculated to determine which is the larger rate. Note that Item 1 does not specify an estimated occupant load for machinery rooms. Basing the exhaust rate on the number of occupants could result in grossly inadequate ventilation. For example, a 2,000-square-foot (186 m²) machinery room might be assumed to have only one occupant. Based on occupancy, 20 cubic feet per minute (cfm) would be required but if based on square footage, 1,000 cfm would be required. Clearly, much consideration must be given to the use of occupant loads in determining the required exhaust rate. Item 2 would be very difficult to determine prior to construction of the machinery room and may have to be determined after the machinery room is in operation. Continuous ventilation (occupied or unoccupied) would provide the greatest level of personnel protection, and interlocking the exhaust system with the room lighting should be the minimum level of personnel protection.

1105.6.3.2 Quantity—emergency conditions. Upon actuation of the refrigerant detector required in Section 1105.3, the mechanical ventilation system shall *exhaust air* from the *machinery room* in the following quantity:

$$Q = 100 \times \sqrt{G} \qquad \text{(Equation 11-2)}$$

For SI: $Q = 0.07 \times \sqrt{G}$

where:

Q = The airflow in cubic feet per minute (m³/s).

G = The design mass of refrigerant in pounds (kg) in the largest system, any part of which is located in the *machinery room*.

❖ This section specifies the emergency exhaust rate to be initiated by the refrigerant detector upon detection of refrigerant leakage. The emergency exhaust rate is intended to maintain an atmosphere in the machinery room that would allow service personnel to escape and take emergency action to control leakage. The exhaust rate is based on the largest refrigerant charge in any one system. The emergency system can be a separate system or it can be an operating mode of the normal system.

1105.7 Termination of relief devices. Pressure relief devices, fusible plugs and purge systems located within the *machinery room* shall terminate outside of the structure at a location not less than 15 feet (4572 mm) above the adjoining grade level and not less than 20 feet (6096 mm) from any window, ventilation opening or exit.

❖ All devices and system outlets that are capable of discharging refrigerant to the atmosphere must be shut down to avoid affecting building occupants and people outdoors.

Discharge piping must be located to pose no undue hazard to people outside of the building or in an adjacent building. To protect people outside the building, the discharge end of the pipe must be located at least 15 feet (4572 mm) above the adjoining ground level. To protect the occupants of a building, the discharge must not be less than 20 feet (6096 mm) in any direction from any window, ventilation opening or exit in any building in the vicinity of the discharge pipe outlet. The 20-foot (6096 mm) minimum requirement is considered adequate to allow for the dissipation and dilution of the refrigerant into the atmosphere and pertains to all building openings, including those in buildings that are located across lot lines on adjacent lots.

The material used for the discharge piping must be compatible with the refrigerant in the system.

Discharge piping that is connected to a fusible plug or rupture-member relief device must be designed to prevent the piping from becoming obstructed and thereby impeding its relief capacity. A fusible plug is a temperature-sensitive device in which the orifice closure member is designed to melt at a specified temperature. An upturn member is a device in which the orifice closure member is a diaphragm-type seal that will burst at a specified pressure. In the process of operating and releasing the pressure, the closure member in the device is forced outward and into the discharge piping by the pressure of the refrigerant. The discharge piping must be designed to prevent debris from lodging in the discharge piping and thus interfering with the release of the refrigerant.

The size of discharge piping must be not less than the size of the pressure relief device outlet. Although not addressed in this section, all relief device discharge piping is subject to maximum lengths because of friction losses in the piping. When there are multiple pressure relief devices in a system and these devices discharge to a common header or manifold, the common discharge piping must be designed for the simultaneous discharge of all relief devices and must be sized to handle the total discharge capacity of all connected relief devices at the lowest pressure setting of any one device. Sizing of the discharge piping must take into account the backpressure that could develop when one or more relief devices discharge into common piping. These provisions are intended to prevent backpressure and friction loss from reducing the required capacity of the relief devices. The discharge piping must be sized so that all connected relief devices are capable of discharging refrigerant at their design capacity.

Designs must also account for the fact that the discharge of any relief device into common discharge piping will create a backpressure that will affect the opening pressure of all other relief devices connected to the common piping.

In some installations, it may be reasonable to expect two or more systems to discharge simultaneously. For example, two or more machines located in close proximity could discharge simultaneously if exposed to a fire in that room or space. In this example, the designer should consider sizing the common discharge piping to accommodate the simultaneous discharge of all systems located in close proximity.

To prevent possibly hazardous chemical reactions between different refrigerants, only refrigeration systems using refrigerants from the same group classification may be interconnected.

ASHRAE 15 contains information on determining the required capacity of pressure relief devices and for determining the maximum length of discharge piping. Section 606.12 of the IFC also regulates refrigerant discharge.

1105.8 Ammonia discharge. Pressure relief valves for ammonia systems shall discharge in accordance with ASHRAE 15.

❖ The ASHRAE 15 standard permits pressure relief valves for ammonia systems to discharge into the atmosphere. However, it also contains an alternative for discharge into tanks containing water and other treatment systems approved by the authority having jurisdiction. Section 606.12.3 of the IFC also addresses ammonia discharge.

[F] 1105.9 Emergency pressure control system. Refrigeration systems containing more than 6.6 pounds (3 kg) of flammable, toxic or highly toxic refrigerant or ammonia shall be provided with an emergency pressure control system in accordance with Section 606.10 of the *International Fire Code.*

❖ To aid emergency responders and untrained refrigeration personnel and mitigate an overpressure condition prior to operation of emergency pressure relief valves, this section establishes a requirement for a fully redundant, automatic safety control system in

lieu of a manual system. This automatic safety control system ensures that lower pressure zones will be capable of handling additional pressure added by a crossover condition without overpressurizing or opening the emergency relief vents on the lower zone. Section 606.10 of the IFC establishes provisions for automatic crossover valves, overpressure limit set point, manual operation, system design pressure, automatic emergency stop, operation of an automatic crossover valve and overpressure in low pressure zones. The 6.6-pound (3.3 kg) threshold parallels existing provisions in Section 606.12 of the IFC.

More information on the history of the evolution of this code section, as well as all other revised sections, can be found in the *Code Changes Resource Collection—2012 International Mechanical Code.*

SECTION 1106
MACHINERY ROOM, SPECIAL REQUIREMENTS

1106.1 General. Where required by Section 1104.2, the *machinery room* shall meet the requirements of this section in addition to the requirements of Section 1105.

❖ A special machinery room must comply with both Section 1105 and this section when using the higher risk refrigerant Groups A2, A3, B2 and B3.

1106.2 Elevated temperature. There shall not be an open flame-producing device or continuously operating hot surface over 800°F (427°C) permanently installed in the room.

❖ The presence of any open flame or a hot surface could create a fire or explosion hazard where refrigerants are flammable. Also, as discussed in Section 1105.5, exposing refrigerant to high temperatures can produce dangerously toxic and corrosive chemicals.

1106.3 Ammonia room ventilation. Ventilation systems in ammonia machinery rooms shall be operated continuously at the emergency ventilation rate determined in accordance with Section 1105.6.3.2.

Exceptions:

1. Machinery rooms equipped with a vapor detector that will automatically start the ventilation system at the emergency rate determined in accordance with Section 1105.6.3.2, and that will actuate an alarm at a detection level not to exceed 1,000 ppm; or

2. Machinery rooms conforming to the Class 1, Division 2, *hazardous location* classification requirements of NFPA 70.

❖ Because ammonia (R-717) is toxic and, under certain conditions, flammable, it is classified as Group B2, which requires continuous ventilation of the machinery room to reduce the hazard of an accumulation of ammonia vapor. The code requires that continuous ventilation be operated at the emergency ventilation rate in accordance with Section 1105.6.4. Section 1105.6.4 specifies the actual method for determining the ventilation rate. ASHRAE 15 requires that an alarm be indicated in the event of ventilation system failure. In designing the ventilation system, keep in mind that ammonia is lighter than air with a relative density of 0.5963.

Exception 1 allows an alternative to continuous ventilation, and establishes an exhaust rate when the ammonia level reaches the emergency level of 1,000 ppm. The intent of Exception 1 is that the exhaust system actuated at 1,000 ppm must operate at the emergency rate required by Section 1105.6.4, considering that the alarm threshold of 1,000 ppm is 40 times the TLV for ammonia.

Exception 2 would allow the exhaust system to operate in accordance with Section 1105.6.3 instead of requiring it to operate continuously. Section 1105.6.3 requires the exhaust system to operate when the room is occupied. As defined in Article 500 of NFPA 70, a Class I location is one in which flammable gases or vapors are or could be present in the air in quantities sufficient to produce explosive or ignitable mixtures. A Class I, Division 2 location includes those in which flammable gases are handled and used, but in which the gases will normally be confined within closed containers or closed systems from which they can escape only in case of system or container failure or abnormal operation.

All electrical system components, devices, fixtures, wiring methods, motors and controls must be of a type suitable for use in Class I locations as dictated in NFPA 70. The cost of compliance with the Class I, Division 2 requirements should be weighed against the cost of the ventilation system of Exception 1. Section 500-2 of NFPA 70 lists the allowable protection techniques, one of which is an explosionproof apparatus, and Article 501 of NFPA 70 states the specific requirements for all portions of the electrical systems.

1106.4 Flammable refrigerants. Where refrigerants of Groups A2, A3, B2 and B3 are used, the *machinery room* shall conform to the Class 1, Division 2, *hazardous location* classification requirements of NFPA 70.

Exception: Ammonia machinery rooms that are provided with ventilation in accordance with Section 1106.3.

❖ The intent of this section is to reduce the number of potential ignition sources in machinery rooms involving flammable refrigerants (see commentary, Section 1106.3). The exception exempts ammonia machinery rooms; however, this exception would be negated if the room was not ventilated in accordance with Section 1106.3 or its first exception.

[F] 1106.5 Remote controls. Remote control of the mechanical equipment and appliances located in the machinery room shall comply with Sections 1106.5.1 and 1106.5.2.

❖ The remote controls required by Sections 1106.5.1 and 1106.5.2 are intended to allow personnel to initiate emergency procedures without having to enter a room in which the atmosphere could be hazardous. Section 606 of the IFC also addresses remote controls.

[F] 1106.5.1 Refrigeration system emergency shutoff. A clearly identified switch of the break-glass type or with an approved tamper-resistant cover shall provide off-only control of refrigerant compressors, refrigerant pumps, and normally closed, automatic refrigerant valves located in the machinery room. Additionally, this equipment shall be automatically shut off whenever the refrigerant vapor concentration in the machinery room exceeds the vapor detector's upper detection limit or 25 percent of the LEL, whichever is lower.

❖ Shutting down compressors and related refrigeration equipment could be necessary to prevent a hazardous condition from worsening and to allow the room to be occupied. The emergency "kill" switch must be a tamper-resistant type (similar to fire alarm pull stations) that requires more than one action to actuate the switch. To prevent an accidental startup, the switch must be capable of stopping only the controlled machinery. The switch must not affect the operation of life safety systems such as detectors and exhaust equipment and should not affect room and egress lighting. Emergency shutdown controls located outside the machinery room enclosure will allow the compressors and related equipment to be shut down without requiring someone to enter the room and risk being exposed to refrigerant or fire. This arrangement would also permit equipment shutdown by fire-fighting personnel without the risk of fire spreading into or out of the fire-resistance-rated enclosure. The controls must be located near the entrance to the machinery room so that their location is conspicuous. The controls should be labeled and color coded so that their purpose is obvious. Such controls are customarily painted red to make them readily identifiable as emergency devices.

In addition to the manual emergency shutoff switch, the machinery room must have an automatic shutoff device for the machinery. The automatic shutoff device shall activate when the vapor concentration in the machinery room reaches the lower of 25 percent of the lower explosive limit (LEL) or the maximum detection level of the vapor detector. The automatic shutoff device is necessary to prevent a hazardous condition from going unnoticed and building up to an explosion condition.

Where refrigerants are nonflammable, electrically energized equipment will not cause safety concerns since the lower flammable limit is not an issue. The intent of this section is to provide a safe environment for emergency response personnel when responding to an incident in a refrigeration room. Because flammability is not an issue with nonflammable refrigerants, shutdown of area equipment and electrical devices does not factor into safety for emergency response personnel. Shutdown does, however, induce a very significant risk of unnecessary damage to business operations and infrastructure with potential losses of data and production.

While Chapter 11 is based on ASHRAE 15, there are requirements that are unique to both documents and the code official and designer must be familiar with them.

[F] 1106.5.2 Ventilation system. A clearly identified switch of the break-glass type shall provide on-only control of the *machinery room* ventilation fans.

❖ For the same reasoning as Section 1106.5.1, a remote switch is required and must not be capable of stopping the exhaust system. Although not specifically stated, the logical intent is that the remote control activates the emergency mode of operation (see commentary, Section 1105.6.4). To maximize the dependability of the exhaust systems, ASHRAE 15 requires powering such systems from independent, dedicated electrical branch circuits.

[F] 1106.6 Emergency signs and labels. Refrigeration units and systems shall be provided with *approved* emergency signs, charts, and labels in accordance with the *International Fire Code*.

❖ Section 606.7 of the IFC requires emergency signs and labels in accordance with NFPA 704 on refrigeration units or systems containing more than 220 pounds (100 kg) of Group A1 or more than 30 pounds (13.6 kg) of any other refrigerant group. These signs must be prominently displayed at each entrance to the machinery room and should show, at a minimum, the hazard classification of the refrigerant in accordance with Section 1103.

SECTION 1107
REFRIGERANT PIPING

1107.1 General. All refrigerant piping shall be installed, tested and placed in operation in accordance with this chapter.

❖ Sections 1107.1 through 1107.8.3 and 1108.1 through 1108.4 cover piping materials, pipe joining methods, piping installation requirements and field testing of piping.

1107.2 Piping location. Refrigerant piping that crosses an open space that affords passageway in any building shall be not less than 7 feet 3 inches (2210 mm) above the floor unless the piping is located against the ceiling of such space. Refrigerant piping shall not be placed in any elevator, dumbwaiter or other shaft containing a moving object or in any shaft that has openings to living quarters or to means of egress. Refrigerant piping shall not be installed in an enclosed public stairway, stair landing or means of egress.

❖ The requirements of Sections 1107.2, 1107.2.1 and 1107.2.2 are consistent with ASHRAE 15 standard. As Section 1101.6 has long required that refrigeration systems comply with ASHRAE 15, these are not new requirements. The information is repeated here to help prevent plan reviewers and installers from overlooking a few of the more important details contained within the standard.

1107.2.1 Piping in concrete floors. Refrigerant piping installed in concrete floors shall be encased in pipe ducts. The piping shall be isolated and supported to prevent damaging vibration, stress and corrosion.

❖ See commentary for Section 1107.2.

1107.2.2 Refrigerant penetrations. Refrigerant piping shall not penetrate floors, ceilings or roofs.

Exceptions:

1. Penetrations connecting the basement and the first floor.

2. Penetrations connecting the top floor and a machinery penthouse or roof installation.

3. Penetrations connecting adjacent floors served by the refrigeration system.

4. Penetrations by piping in a direct system where the refrigerant quantity does not exceed Table 1103.1 for the smallest occupied space through which the piping passes.

5. In other than industrial occupancies and where the refrigerant quantity exceeds Table 1103.1 for the smallest space, penetrations for piping that connects separate pieces of *equipment* that are either:

    5.1. Enclosed by an *approved* gas-tight, fire-resistive duct or shaft with openings to those floors served by the refrigeration system or

    5.2. Located on the exterior of the building where vented to the outdoors or to the space served by the system and not used as an air shaft, closed court or similar space.

❖ See commentary for Section 1107.2.

1107.3 Pipe enclosures. Rigid or flexible metal enclosures or pipe ducts shall be provided for soft, annealed copper tubing and used for refrigerant piping erected on the premises and containing other than Group A1 or B1 refrigerants. Enclosures shall not be required for connections between condensing units and the nearest riser box(es), provided such connections do not exceed 6 feet (1829 mm) in length.

❖ The intent of this section is to protect refrigerant piping from physical damage that could result in leakage and a hazardous condition. The protection is required for soft (coiled) bending-tempered copper tubing conveying refrigerants in Groups A2, A3, B2 and B3. Because this type of tubing is more susceptible to damage and these refrigerant groups are flammable and/or toxic, the extra protection is warranted.

Tubing runs of 6 feet (1829 mm) or less that connect to condensing units are exempt because the tubing and the units are typically located outdoors or in a machinery room.

1107.4 Condensation. All refrigerating piping and fittings, brine piping and fittings that, during normal operation, will reach a surface temperature below the dew point of the surrounding air, and are located in spaces or areas where condensation will cause a safety hazard to the building occupants, structure, electrical *equipment* or any other *equipment* or appliances, shall be protected in an *approved* manner to prevent such damage.

❖ Refrigerant suction lines and chilled water piping, for example, will produce water vapor condensation in most environments. Condensation will deteriorate some piping and insulation materials and can also cause serious moisture damage to other parts of the building. A combination of insulation and vapor barriers is used to retard condensation on piping. Note that IIAR 2 prohibits ammonia piping capable of producing condensation from passing over electrical equipment.

Condensing unit suction lines are also insulated to reduce the pickup of additional superheat after the evaporator to improve system efficiency and provide compressor cooling.

1107.5 Materials for refrigerant pipe and tubing. Piping materials shall be as set forth in Sections 1107.5.1 through 1107.5.5.

❖ Sections 1107.4.1 through 1107.4.5 specify the allowable piping materials for conveying refrigerants. All piping materials must be compatible with the type of refrigerant. For example, copper and brass pipe, tube and fittings cannot be used with ammonia refrigerant (see commentary, Sections 1107.4.2 and 1107.4.3).

1107.5.1 Steel pipe. Carbon steel pipe with a wall thickness not less than Schedule 80 shall be used for Group A2, A3, B2 or B3 refrigerant liquid lines for sizes 1.5 inches (38 mm) and smaller. Carbon steel pipe with a wall thickness not less than Schedule 40 shall be used for Group A1 or B1 refrigerant liquid lines 6 inches (152 mm) and smaller, Group A2, A3, B2 or B3 refrigerant liquid lines sizes 2 inches (51 mm) through 6 inches (152 mm) and all refrigerant suction and discharge lines 6 inches (152 mm) and smaller. Type F steel pipe shall not be used for refrigerant lines having an operating temperature less than -20°F (-29°C).

❖ This section requires a heavier wall pipe for liquid lines conveying the more hazardous refrigerant groups. The smaller a pipe is, the more susceptible it is to physical damage. Type F steel pipe is defined in ASTM A 53 as furnace butt-welded pipe. Type F steel pipe is not intended for flanging and could be susceptible to stress failures at low temperatures.

1107.5.2 Copper and brass pipe. Standard iron-pipe size, copper and red brass (not less than 80-percent copper) pipe shall conform to ASTM B 42 and ASTM B 43.

❖ Copper and brass pipe (not tubing) is expensive, rarely used and incompatible with ammonia.

1107.5.3 Copper tube. Copper tube used for refrigerant piping erected on the premises shall be seamless copper tube of Type ACR (hard or annealed) complying with ASTM B 280. Where *approved*, copper tube for refrigerant piping erected on the premises shall be seamless copper tube of Type K, L or M (drawn or annealed) in accordance with ASTM B 88.

Annealed temper copper tube shall not be used in sizes larger than a 2-inch (51 mm) nominal size. Mechanical joints shall not be used on annealed temper copper tube in sizes larger than $^7/_8$-inch (22.2 mm) OD size.

❖ Type ACR tube is designed for air-conditioning and refrigeration applications and is factory cleaned, dehydrated and shipped with end caps or plugs to prevent internal contamination. It is available in rigid or bending temper. If approved by the code official, Type K, L or M copper water tube can be used; however, this is rarely done because water tube is not protected from contamination in shipping and storage as is Type ACR. Also, Type M tube has less wall thickness than the almost exclusively used Type L tube. Note that ASHRAE 15 prohibits the use of copper and its alloys in systems with ammonia refrigerant.

The larger a soft (annealed temper) tube is in diameter, the more difficult it is to work with, form and join. Also, larger sizes tend to be "out of round," making it difficult to achieve a good seal with mechanical joints.

1107.5.4 Copper tubing joints. Copper tubing joints used in refrigerating systems containing Group A2, A3, B2 or B3 refrigerants shall be brazed. Soldered joints shall not be used in such refrigerating systems.

❖ Because of the higher risk associated with Group A2, A3, B2 and B3 refrigerants, the only type of copper joint allowed is a brazed joint (see the definition of "Brazed joint"). Because commonly used solders melt at temperatures less than 500°F (260°C), soldered joints will fail early in a fire and release refrigerant into the building, endangering occupants and fire fighters.

1107.5.5 Aluminum tube. Type 3003-0 aluminum tubing with high-pressure fittings shall not be used with methyl chloride and other refrigerants known to attack aluminum.

❖ Type 3003-0 alloy of aluminum is susceptible to chemical corrosion when used with refrigerant R-40. Piping, tubing, fittings and refrigerant-containing components must be compatible with and unaffected by the refrigerant used. ASHRAE 15 prohibits the use of aluminum, zinc, magnesium and any alloys of these metals in systems with methyl chloride refrigerant. ASHRAE 15 also states that magnesium alloy materials must not be used with any of the halogenated refrigerants.

1107.6 Joints and refrigerant-containing parts in air ducts. Joints and all refrigerant-containing parts of a refrigerating system located in an air duct of an air-conditioning system carrying conditioned air to and from human-occupied space shall be constructed to withstand, without leakage, a pressure of 150 percent of the higher of the design pressure or pressure relief device setting.

❖ Leakage of refrigerant into a duct would cause refrigerants to be spread throughout the areas served by the duct system; therefore, the integrity of pipes, tubes, coils and devices within ducts must be exceptional. The design pressure for the components must be not less than one- and one-half times the system design pressure or one- and one-half times the system relief device setting, whichever is higher.

1107.7 Exposure of refrigerant pipe joints. Refrigerant pipe joints erected on the premises shall be exposed for visual inspection prior to being covered or enclosed.

❖ This section is redundant with Section 107.1. Like all piping regulated by the code, refrigerant piping must be inspected prior to being concealed.

1107.8 Stop valves. All systems containing more than 6.6 pounds (3 kg) of a refrigerant in systems using positive-displacement compressors shall have stop valves installed as follows:

1. At the inlet of each compressor, compressor unit or condensing unit.

2. At the discharge outlet of each compressor, compressor unit or condensing unit and of each liquid receiver.

   Exceptions:

   1. Systems that have a refrigerant pumpout function capable of storing the entire refrigerant charge in a receiver or heat exchanger.

   2. Systems that are equipped with provisions for pumpout of the refrigerant using either portable or permanently installed recovery *equipment*.

   3. Self-contained systems.

❖ A compressor is the heart of a mechanical refrigeration system. It is used in the vapor refrigeration cycle to raise the pressure and enthalpy of the refrigerant into the superheated vapor state, at which point the refrigerant vapor enters the condenser and transfers heat energy to a cooler medium. There are two basic types of compressors: positive displacement and dynamic. Positive-displacement compressors increase the pressure of the refrigerant vapor by reducing the volume of the compressor chamber such as in a reciprocating, screw or rotary compressor. Dynamic compressors, such as centrifugal compressors, increase the pressure of the refrigerant vapor by a continuous transfer of angular momentum from the rotating member to the vapor.

Stop valves are gas-tight shutoff valves used to isolate portions of a refrigeration system circuit. Stop valves in the inlet (suction side) and discharge side of compressors and condensing units allow components and portions of piping to be isolated for repair and component replacement. Isolation reduces the potential for accidental discharge of refrigerant to the atmosphere, reduces the amount of refrigerant that would have to be recovered and reduces the volume of the system to be evacuated. This section does not apply to small [6 pounds (3 kg)] systems because the limited quantity of refrigerant can be recovered easily and does not apply to systems in which pressures will equalize across the compressor in the off cycle.

Exceptions 1 and 2 would allow omission of stop valves for systems in which the refrigerant charge can be isolated or extracted and stored, thus allowing the system to be opened for service or alteration.

Exception 3 applies to completely self-contained systems such as factory-built unitary equipment having no piping to external refrigerant-containing components.

1107.8.1 Liquid receivers. All systems containing 100 pounds (45 kg) or more of a refrigerant, other than systems utilizing nonpositive displacement compressors, shall have stop valves, in addition to those required by Section 1107.8, on each inlet of each liquid receiver. Stop valves shall not be required on the inlet of a receiver in a condensing unit, nor on the inlet of a receiver which is an integral part of the condenser.

❖ Systems containing 100 pounds (45 kg) or more of a refrigerant, other than systems using compressors that are not positive displacement, must have stop valves, in addition to those required by Section 1107.7, on each inlet of each liquid receiver. Stop valves are neither required on the inlet of a receiver in a condensing unit, nor on the inlet of a receiver that is an integral part of the condenser.

Receivers are pressure vessels designed to receive and store liquid refrigerant downstream of the condenser. For large systems, complete isolation of the receiver (valved inlet and outlet) would allow the majority of the refrigerant charge to be "pumped down" and held in the receiver for maintenance operations.

1107.8.2 Copper tubing. Stop valves used with soft annealed copper tubing or hard-drawn copper tubing $^7/_8$-inch (22.2 mm) OD standard size or smaller shall be securely mounted, independent of tubing fastenings or supports.

❖ The forces involved in rotating a stop valve handle will be transferred to the tubing that is connected to the valve, creating the possibility in small-diameter tubing for movement and eventual failure. On larger tubing, the strength of the material can be adequate to resist the forces applied to the valves. Valves that are independently supported will not cause stress in the tubing.

1107.8.3 Identification. Stop valves shall be identified where their intended purpose is not obvious. Numbers shall not be used to label the valves, unless a key to the numbers is located near the valves.

❖ Unless located where their purpose is obvious, valves must be identified with some means that describe their function. Because valves are used in emergency situations and during service work, their purpose must be known to the user.

SECTION 1108
FIELD TEST

1108.1 General. Every refrigerant-containing part of every system that is erected on the premises, except compressors, condensers, vessels, evaporators, safety devices, pressure gauges and control mechanisms that are *listed* and factory tested, shall be tested and proved tight after complete installation, and before operation. Tests shall include both the high-

and low-pressure sides of each system at not less than the lower of the design pressures or the setting of the pressure relief device(s). The design pressures for testing shall be those listed on the condensing unit, compressor or compressor unit nameplate, as required by ASHRAE 15.

Exceptions:

1. Gas bulk storage tanks that are not permanently connected to a refrigeration system.

2. Systems erected on the premises with copper tubing not exceeding $^5/_8$-inch (15.8 mm) OD, with wall thickness as required by ASHRAE 15, shall be tested in accordance with Section 1108.1, or by means of refrigerant charged into the system at the saturated vapor pressure of the refrigerant at 70°F (21°C) or higher.

3. Limited-charge systems equipped with a pressure relief device, erected on the premises, shall be tested at a pressure not less than one and one-half times the pressure setting of the relief device. If the *equipment* or *appliance* has been tested by the manufacturer at one and one-half times the design pressure, the test after erection on the premises shall be conducted at the design pressure.

❖ Mechanical refrigeration systems that are field constructed must be pressure tested. Factory-assembled components that have been factory tested can be excluded from this testing requirement. Self-contained, factory-built equipment that bears the label of an approved agency does not require field testing because the code official can determine from information provided on the label that the equipment has been tested at the factory.

After a refrigeration system is assembled, a pressure test using air or an inert gas is required to determine whether the system leaks. Refrigerant cannot be used for testing because the purpose of testing is to find and repair any leaks before the system is charged with refrigerant, thereby preventing exposure of anyone to the refrigerant. Also, testing with refrigerant introduces the additional and unnecessary risk of environmental damage from release of refrigerant to the atmosphere.

Dry nitrogen and carbon dioxide are suitable for testing of refrigeration equipment. Gases such as nitrogen and carbon dioxide are readily available, relatively inexpensive, noncorrosive, nontoxic and can be easily removed from the system. Even though air is listed as acceptable for testing refrigeration systems, it is generally not used because moisture in the air can contaminate the system. Although inert gases are acceptable for testing, most are very expensive, making them impractical to use.

Section 9 of ASHRAE 15 determines the required design pressure for all refrigeration systems. The design pressure for equipment such as condensing

units will be marked on the equipment. The design pressure is intended to represent the worst-case condition for systems when operating, idle or in shipment prior to installation.

This section does not require field testing of factory-tested components. Factory-tested components typically include pressure vessels, coils, condensing units and control devices. Field-connected piping can be tested only on the job site and is more likely to have defects than any factory-assembled system component. In systems having positive displacement compressors, the low side and high side of the system are typically designed for different design pressures.

Exception 1 exempts refrigerant storage containers because they are not part of the refrigeration system.

Exception 2 describes an optional test for small field-assembled systems that are constructed with small diameter copper tubing. The vapor pressure of a refrigerant varies with the temperature and the type of chemical. The test pressure must not be less than the vapor pressure at 70°F (21°C) of the refrigerant for which the system is designed. The saturated vapor pressure is the pressure exerted by a refrigerant when liquid and gaseous refrigerant are in equilibrium in a container.

Exception 3 is actually a distinct requirement in itself and applies to limited charge systems that are field assembled and equipped with a pressure relief device. In contrast, Section 1108.1 would allow testing at design pressures or relief device setting, whichever is less. The refrigerant charge in this type of system is small enough for all of the liquid refrigerant in the system to be vaporized (depending on the ambient temperature) and the system pressure must not exceed the design pressure.

1108.1.1 Booster compressor. Where a compressor is used as a booster to obtain an intermediate pressure and discharges into the suction side of another compressor, the booster compressor shall be considered a part of the low side, provided that it is protected by a pressure relief device.

❖ For field testing in accordance with Section 1108.1, the booster compressor is considered as part of the low-pressure side and the main compressor and piping on the discharge side of the main compressor are considered as part of the high-pressure side.

1108.1.2 Centrifugal/nonpositive displacement compressors. In field-testing systems using centrifugal or other nonpositive displacement compressors, the entire system shall be considered as the low-side pressure for field test purposes.

❖ The entire system is field tested as a low-pressure side where any nonpositive displacement compressor serves as the only compressor. The high-pressure side test is not necessary.

1108.2 Test gases. Tests shall be performed with an inert dried gas including, but not limited to, nitrogen and carbon dioxide. Oxygen, air, combustible gases and mixtures containing such gases shall not be used.

Exception: The use of air is allowed to test R-717, ammonia, systems provided that they are subsequently evacuated before charging with refrigerant.

❖ Dry nitrogen is the most commonly used test medium. Refrigerant should not be used. If there is a system defect, refrigerant would escape and the refrigerant would have to be recovered. The system would have to be swept with an inert gas to allow repair of the defects. Test gases must be dried to prevent the introduction of harmful moisture into a system. Oxygen, air, toxic gases and flammable gases can create an explosion, fire or health hazard. Combining flammable refrigerant with oxygen or air can create an explosive/flammable mixture.

1108.3 Test apparatus. The means used to build up the test pressure shall have either a pressure-limiting device or a pressure-reducing device and a gauge on the outlet side.

❖ The method of introducing the test pressure into a system must be controlled to prevent overpressure that could damage the system. The pressure gauge allows monitoring and a pressure relief or reducing device acts as a backup to manual control. Containers (cylinders) of test gases such as nitrogen are normally pressurized far in excess of the system test pressures; therefore, there is the potential for damage to the system under test.

1108.4 Declaration. A certificate of test shall be provided for all systems containing 55 pounds (25 kg) or more of refrigerant. The certificate shall give the name of the refrigerant and the field test pressure applied to the high side and the low side of the system. The certification of test shall be signed by the installer and shall be made part of the public record.

❖ The test certificate can be in whatever form is acceptable to the code official. This documentation must be filed with the permit and inspection records.

[F] SECTION 1109
PERIODIC TESTING

1109.1 Testing required. The following emergency devices and systems shall be periodically tested in accordance with the manufacturer's instructions and as required by the code official:

1. Treatment and flaring systems.
2. Valves and appurtenances necessary to the operation of emergency refrigeration control boxes.
3. Fans and associated *equipment* intended to operate emergency pure ventilation systems.
4. Detection and alarm systems.

❖ The devices and systems listed in this section are critical life safety and fire protection elements; therefore, it is imperative that they be tested periodically to assess their condition and dependability. The same requirements are in Section 606 of the IFC, indicating the importance to the fire department of testing these devices and systems. Failure of the safety systems could lead to deadly consequences for the building occupants and the fire personnel entering the building in an emergency situation.

Chapter 12:
Hydronic Piping

General Comments

Hydronic piping includes piping, fittings and valves used in building space conditioning systems. Applications include hot water, chilled water, steam, steam condensate, brines and water/antifreeze mixtures. Chapter 12 contains the provisions that govern the construction, installation, alteration and repair of all hydronic piping systems.

Purpose

Chapter 12 regulates hydronic systems by stating requirements that affect reliability, serviceability, energy efficiency and safety.

SECTION 1201
GENERAL

1201.1 Scope. The provisions of this chapter shall govern the construction, installation, *alteration* and repair of hydronic piping systems. This chapter shall apply to hydronic piping systems that are part of heating, ventilation and air-conditioning systems. Such piping systems shall include steam, hot water, chilled water, steam condensate and ground source heat pump loop systems. Potable cold and hot water distribution systems shall be installed in accordance with the *International Plumbing Code*.

❖ This chapter regulates piping, fittings, valves, insulation and the heat transfer fluids used in building heating, cooling and air-conditioning systems. The regulations include material quality and properties and installation requirements. This chapter does not regulate hot and cold potable water distribution piping, which is regulated by the *International Plumbing Code®* (IPC®).

Hydronic systems that are regulated by this chapter include, but are not limited to, the following:

- Low-, medium- and high-temperature hot water heating systems;
- Chilled water cooling systems;
- Dual-temperature water systems;
- Condenser and cooling tower water systems;
- Steam and steam condensate piping systems; and
- Solar heating systems.

1201.2 Sizing. Piping and piping system components for hydronic systems shall be sized for the demand of the system.

❖ This section requires that the piping network for a hydronic system be sized for the flow volume it must conduct. System design is normally based on maximum flow velocity or pressure drop and the control of noise and erosion. When piping is undersized, velocity and pressure drop are excessive, and the system

can be subjected to unacceptable amounts of stress and erosion. Noise in the piping system is dependent on velocity and is a result of turbulence, cavitation, release of entrained air and water hammer. Erosion is the gradual wearing away of the piping inner walls as a result of flow friction, turbulence and entrained air bubbles or debris in the heat transfer medium. Erosion increases as the temperature of the fluid increases. Improperly sized piping can also cause inefficient system operation, loss of system capacity and poor or unsafe operation of boilers, heat exchangers, chillers, circulators and other system components.

1201.3 Standards. As an alternative to the provisions of Sections 1202 and 1203, piping shall be designed, installed, inspected and tested in accordance with ASME B31.9.

❖ ASME B31.9 was developed in 1982 as a simplified standard for the piping systems normally encountered in commercial and multiple-family residential buildings. While the long established ASME B31.1 and ASME B31.3 standards cover everything in ASME B31.9 and more, those standards were deemed too complex for the relatively low pressures/temperatures and comparatively benign fluids conveyed by piping in buildings of commercial and residential nature. ASME B31.9 is intended to be applied to piping for water and antifreeze solutions for heating and cooling, steam and steam condensate, air, liquids and other nontoxic, nonflammable fluids contained in piping that does not exceed the following limitations:

- Dimensional limits

   Carbon steel: 30 inches (762 mm) and 0.500 inch (12.7 mm) wall
   Stainless steel: 12 inches (305 mm) and 0.500 inch (12.7 mm) wall
   Aluminum: 12 inches (305 mm)
   Brass and copper: 12 inches (305 mm), 12.125 inches (308 mm) for copper tube.

Thermoplastics: 24 inches (610 mm)

Ductile iron: 18 inches (457 mm)

Reinforced thermosetting Resin: 24 inches (610 mm)

- Pressure and temperature limits

   Compressed air, steam and steam condensate to 150 psi (1035 kPa) gauge

   Steam and steam condensate from ambient to 366°F (186°C)

   Other gases from ambient to 0 to 200°F (-18 to 93°C)

   Liquids to 350 psi (2415 kPa) gauge and from 0 to 250°F (-18 to 121°C)

   Vacuum to 14.7 psi (1 bar).

SECTION 1202
MATERIAL

1202.1 Piping. Piping material shall conform to the standards cited in this section.

Exception: Embedded piping regulated by Section 1209.

❖ The piping materials that may be used in hydronic systems are listed in Section 1202.4 along with the applicable referenced standards. This section does not prohibit the use of other materials that provide equivalent performance and that have been approved in accordance with Section 105. Conformance to the applicable standards results in a degree of assurance that the materials will adequately perform their intended function. Section 1209 specifies pipe materials for radiant panel installations.

1202.2 Used materials. Reused pipe, fittings, valves or other materials shall be clean and free of foreign materials and shall be *approved* by the code official for reuse.

❖ Used materials must meet the same requirements as new materials. Because materials that have been previously used for other than hydronic applications could contain residual contaminants that are harmful to the hydronic system, before pipe, fittings, valves or other materials can be reinstalled, they must be, at the very minimum, visually examined to determine that the material is at least equivalent to that required for new materials for the intended application. The inspection is intended to determine that the materials are free from physical defects, such as pitting, kinks, damaged threads, pipe wall thinning, corrosion and deposits of solids on the inner wall of the pipe.

1202.3 Material rating. Materials shall be rated for the operating temperature and pressure of the hydronic system. Materials shall be suitable for the type of fluid in the hydronic system.

❖ Heat transfer fluids and system operating and standby pressures and temperatures all vary widely depending on the application. All portions of a hydronic system must be compatible with the working fluid and rated for the operating temperatures and

pressures of the system. The lowest rating of any of the components establishes the operating limitations of the system. Note also that pressure and temperature can vary significantly within a system, and different materials may be needed at different locations in the system. For example, in a high-rise building, the static pressure of hydronic piping is greater in the lower floors than in the upper floors, which means that the piping and components at the lower elevations must be rated for the higher pressures.

1202.4 Piping materials standards. Hydronic pipe shall conform to the standards listed in Table 1202.4. The exterior of the pipe shall be protected from corrosion and degradation.

❖ Table 1202.4 identifies the standards for piping materials used in hydronic piping systems. Pipe manufacturers stencil the number of the standard to which the pipe conforms at specified intervals on each length of pipe, allowing for ease of field identification.

This section also requires that piping be protected from corrosion damage or other forms of degradation. When piping is buried or embedded in concrete, the surrounding conditions may cause the pipe to degrade or corrode. Where this can occur, protective measures, such as wrapping or coating the piping, or installing a material that is inherently resistant to degradation are required to shield the piping. For example, steel chilled-water piping is susceptible to rusting caused by continued exposure to condensation that can form on the pipe exterior.

TABLE 1202.4
HYDRONIC PIPE

MATERIAL	STANDARD (see Chapter 15)
Acrylonitrile butadiene styrene (ABS) plastic pipe	ASTM D 1527; ASTM D 2282
Brass pipe	ASTM B 43
Brass tubing	ASTM B 135
Copper or copper-alloy pipe	ASTM B 42; ASTM B 302
Copper or copper-alloy tube (Type K, L or M)	ASTM B 75; ASTM B 88; ASTM B 251
Chlorinated polyvinyl chloride (CPVC) plastic pipe	ASTM D 2846; ASTM F 441; ASTM F 442
Cross-linked polyethylene/ aluminum/cross-linked polyethylene (PEX-AL-PEX) pressure pipe	ASTM F 1281; CSA CAN/CSA-B-137.10
Cross-linked polyethylene (PEX) tubing	ASTM F 876; ASTM F 877
Ductile iron pipe	AWWA C151/A21.51; AWWA C115/A21.15
Lead pipe	FS WW-P-325B
Polybutylene (PB) plastic pipe and tubing	ASTM D 3309
Polyethylene/aluminum/ polyethylene (PE-AL-PE) pressure pipe	ASTM F 1282; CSA B137.9

(continued)

**TABLE 1202.4—continued
HYDRONIC PIPE**

MATERIAL	STANDARD (see Chapter 15)
Polyethylene (PE) pipe, tubing and fittings (for ground source heat pump loop systems)	ASTM D 2513; ASTM D 3035; ASTM D 2447; ASTM D 2683; ASTM F 1055; ASTM D 2837; ASTM D 3350; ASTM D 1693
Polypropylene (PP) plastic pipe	ASTM F 2389
Polyvinyl chloride (PVC) plastic pipe	ASTM D 1785; ASTM D 2241
Raised temperature polyethylene (PE-RT)	ASTM F 2623; ASTM F 2769
Steel pipe	ASTM A 53; ASTM A 106
Steel tubing	ASTM A 254

❖ Ductile iron, polyethylene/aluminum/polyethylene, polypropylene and raised temperature polyethylene piping materials were added to the table. The following outlines the basic descriptions of the pipe materials identified in Table 1202.4.

Acrylonitrile butadiene styrene (ABS) plastic pipe: ASTM D 1527 covers Schedules 40 and 80 ABS pipe. The schedule number indicates the wall thickness; the higher the schedule number, the thicker the pipe wall. ASTM D 2282 is for SDR-PR pipe. "SDR" is an abbreviation for "standard dimension ratio," the ratio of average outside pipe diameter to minimum wall thickness. The "PR" designation indicates that the pipe is pressure rated.

Brass pipe: The piping material is red brass, which is composed of approximately 85-percent copper and 15-percent zinc. The pipe is available in standard weight, extra-strong weight and double-extra-strong wall thicknesses. Extra-strong weight has the same outside diameter as standard weight pipe but has a greater pipe wall thickness.

The standard-weight pipe is commonly referred to as "Schedule 40" and extra-strong is called "Schedule 80" in the vernacular of the trade. The brass pipe dimensions are similar to the steel pipe dimensions.

Brass tubing: The brass industry no longer distinguishes between pipe and tubing; the two designations now mean the same thing.

Copper or copper-alloy pipe: There are two types of copper pipe: threaded and threadless. Both pipes are seamless and have a chemical composition of 99.9-percent copper.

Threadless copper has the same outside diameter as threaded copper pipe, steel pipe and brass pipe. The wall thickness of threadless pipe is less than that of threaded pipe, making the inside diameter greater. Threadless copper pipe is normally joined by brazing, and is continuously marked in gray, including the designation "TP" for "threadless pipe."

Threaded copper pipe is available in both standard and extra-strong weights. The pipe dimensions are similar to those of brass pipe and steel pipe.

Copper or copper-alloy tubing: Tubing is manufactured in two different tempers. The tempers are identified as drawn (also called "hard copper" having the designation "H") and annealed (called "soft copper" having the designation "O"). Hard (drawn) copper tubing comes in straight lengths. Soft (annealed) copper tubing comes in both straight lengths and coils.

Annealed copper and tempered drawn copper may be formed by bending. Tempered copper that can be bent (bending-tempered) is identified on the pipe with the designation "BT." Of the hard (drawn) copper, only bending-tempered copper tubing may be joined by flared connections.

Copper water tubing is available in Type K, L or M. The tubing type indicates the wall thickness. Type K has the greatest wall thickness followed by Types L and M, respectively. The outside diameter of the tubing is the same for all three types; only the inside diameter varies with wall thickness.

Copper tubing is joined by soldering or brazing with wrought- or cast-copper capillary socket-end fittings. Small-diameter copper tubing is also joined by flared or compression-type fittings.

The types of copper tubing are identified with a continuous, colored marking for ease of identification.

The colored markings are green for Type K, blue for Type L and red for Type M.

Chlorinated polyvinyl chloride (CPVC) plastic pipe: CPVC raw material used to produce pipe is a polyvinyl chloride that has been chlorinated to improve the material characteristics. The resulting pipe is more resistant to temperature extremes than PVC pipe. CPVC plastic pipe is typically white or milky white (cream) in color. The pipe is marked at close intervals with the manufacturer's name, the ASTM standard and "CPVC 4120" or "CPVC 41" followed by two additional numbers.

The designation "CPVC 4120" identifies the quality of the material used to produce the pipe. The pipe is made with Grade 23447 plastic material. The number 23447 is a code for the quality of the pipe. The first digit indicates the base resin is CPVC. The remaining four digits determine the impact strength, tensile strength, modulus of elasticity and deflection temperature.

Class 23447 was previously designated Type IV Grade 1, which was shortened to CPVC 41. The last two digits in CPVC 4120 indicate the hydrostatic design stress in hundreds of pounds per square inch (psi). CPVC 4120 has a hydrostatic design stress of 2,000 psi (13 790 kPa).

The hydrostatic design stress, which is the pressure exerted on the walls of the pipe by the fluid inside the pipe under no-flow conditions, is not an indication of the pressure rating of the pipe. However, the pressure rating of the pipe at 73°F (23°C) may be computed using the hydrostatic design stress by the following:

$$P = \frac{2S}{(OD/t) - 1} = \frac{2S}{R - 1}$$

where:

P = Pressure rating.

S = Hydrostatic design stress (psi).

OD = Average outside diameter (inches).

t = Minimum wall thickness (inches).

R = Standard dimension ratio (SDR) (based on outside diameter).

ASTM D 2846 is for SDR 11 CPVC plastic pipe and CPVC socket fittings.

ASTM F 441 is for Schedules 40 and 80 plastic pipe. Dimensions of the pipe are the same as Schedules 40 and 80 steel pipe.

ASTM F 442 is for SDR 13.5, 17, 21, 26 and 32.5 CPVC pipe. The pipe is designated PR.

Cross-linked PEX-AL-PEX pipe: PEX-AL-PEX is a composite pipe made of an aluminum tube laminated to interior and exterior layers of cross-linked polyethylene. The layers are bonded together with an adhesive. The cross-linked molecular structuring gives the pipe additional resistance to rupture than that of polyethylene (see the commentary for cross-linked polyethylene tubing in this section). Therefore, the pipe is suitable for hot and cold water distribution and is pressure rated for 125 psi at 180°F (862 kPa at 82°C).

Although it is partially plastic, the PEX-AL-PEX pipe resembles metal tubing in that it can be bent by hand or with a suitable bending device while maintaining its shape without fittings or supports, In other words, the pipe is referred to as "form stable." The minimum bending radius specified by manufacturers is five times the outside diameter.

Mechanical joints are the only methods currently available to join PEX-AL-PEX pipe. A number of proprietary mechanical-compression-type connectors have been developed for use with the composite pipe to perform transition to other pipes and fittings. The installation of such fittings must be done in accordance with the manufacturers' instructions (see commentary, Section 1203.3.2).

Cross-linked polyethylene (PEX) tubing: Cross-linked polyethylene (designated as PEX by ASTM) has been used extensively in Europe as an undersurface (wall, floor and ceiling) radiant heating system. PEX pipe is rated for use at 160 psi at 73.4°F (1103 kPa at 23°C), 100 psi at 180°F (689 kPa at 82°C) and 80 psi at 200°F (552 kPa at 93°C).

A specially controlled chemical reaction takes place during the manufacturing of polyethylene pipe to form cross-linked polyethylene. The cross-linked molecular structure gives the pipe additional resistance to rup-

ture over a wider range of temperatures and pressures than that of other polyolefin plastics (PE, PB and PP). Because of PEX pipe's unique molecular structure and resistance to heat, it cannot be joined by heat fusion. Because PEX is a member of the polyolefin plastic family, it is resistant to solvents and cannot be joined by solvent cementing.

Because all plastic is gas permeable to some extent, it can allow oxygen to diffuse through the pipe wall and enter the working fluid of the hydronic system, causing corrosion of metallic components such as boilers, circulators and pressure vessels. To minimize the amount of dissolved oxygen in the heating system, some manufacturers of PEX add an oxygen-impermeable polymer coating to the exterior of the pipe. PEX pipe is somewhat flexible, allowing it to be bent.

Mechanical connectors and fittings for PEX pipe are proprietary and should be used only with the pipe for which they have been designed. A number of mechanical fastening techniques have been developed for joining PEX pipe. The manufacturer's installation instructions must be consulted to identify authorized fittings for use with PEX piping (see Section 1203.11).

Ductile iron pipe: Ductile iron pipe has the characteristic that when a specimen of ductile iron is loaded in tension, the material will exhibit elongation before pulling apart (rupture). This is opposed to cast iron which under the same tension loading, does not appreciably elongate before pulling apart (rupture). Because of this property, cast iron is said to be "brittle" (not ductile). AWWA C151/A21.51 pipe is centrifugally cast, ductile iron pipe intended to be connected with push-on joints or mechanical joints. This standard covers 3 inch (76 mm) to 64 inch (1626 mm) sizes. AWWA C115/A21.15 covers the same sizes, but the pipe is made with threaded ends to accept threaded flanges for flange-type gasketed and bolted joints.

Lead pipe: Lead pipe is supplied in straight lengths, coils or reels; in sizes $^3/_8$-inch (9.5 mm) through $1^3/_4$-inch (44 mm) internal diameter (ID) and in numerous wall thicknesses. Lead tubing sizes are $^1/_{16}$-inch (1.6 mm) through $^7/_{16}$-inch (11.1 mm) ID.

Polybutylene (PB) plastic pipe and tubing: Polybutylene is classified as an inert polyolefin material, meaning that it is resistant to chemicals. This is the reason that polybutylene pipe cannot be solvent cemented like some other plastic pipes. It is joined by flared, heat-fusion or mechanical joints.

The designation PB 2110 indicates the quality of the material. The material must be a Type 2, Category 1, polybutylene, which is identified as PB 21. Type 2 indicates the density and Category 1 indicates the extrusion flow rate. The last two digits in PB 2110 indicate the hydrostatic design stress in hundreds of pounds per square inch. PB 2110 has a hydrostatic design stress of 1,000 psi (6895 kPa).

The hydrostatic design stress, which is the pressure exerted on the walls of the pipe by the fluid inside the pipe under no-flow conditions, is not an indication of the pressure rating of the pipe. However, the pressure rating of the pipe at 73°F (23°C) may be computed by using the hydrostatic design stress:

$$P = \frac{2S}{(OD/t) - 1} = \frac{2S}{R - 1}$$

where:

P = Pressure rating.

S = Hydrostatic design stress (psi).

OD = Average outside diameter (inches).

t = Minimum wall thickness (inches).

R = Standard dimension ratio (SDR) (based on outside diameter).

or:

$$P = \frac{2S}{R_i - 1}$$

where:

R_i = Standard dimension inside ratio [SDR (inside) or SDIR].

ASTM D 3309 covers polybutylene tubing of SDR 11 used for hot and cold potable water and building heating and cooling applications. The piping is rated for pressures of 100 psi and 180°F (689 kPa at 82°C).

Polyethylene (PE) pipe, tubing and fittings: Like polybutylene, polyethylene is an inert polyolefin material. It is resistant to chemical action and cannot be solvent welded. PE pipe must be heat-fusion or mechanically joined.

Polyethylene pipe is well suited and commonly used for ground source heat pump loop systems (see commentary, Sections 1203.15 and 1208.1.1). Polyethylene piping is corrosion resistant, compatible with heat exchange fluids used in such systems and is suitable for the pressures and temperatures associated with the application.

PE piping is available in very long coiled lengths and is relatively inexpensive, making it an ideal choice for underground loop systems that may be several hundred feet in length, often consisting of multiple loops.

Polyethylene of raised temperature (PE-RT) plastic tubing: Polyethylene of raised temperature is an inert polyolefin material. It is resistant to chemical action and cannot be solvent welded. PE-RT pipe must be heat-fusion or mechanically joined. Cold bending radius must not be less than six times the outside diameter of the tubing. PE-RT must not be stored in locations or used in applications where there will be exposure to sunlight (UV light).

Polyethylene/aluminum/polyethylene pipe (PE/AL/PE): PE-AL-PE is a composite pipe made of an aluminum tube laminated to interior and exterior layers of polyethylene. The layers are bonded together with an adhesive. The pipe is suitable for hot and cold water distribution and is pressure rated for 100 psi at 180°F (689 kPa at 82° C).

Although it is partially plastic, the PE-AL-PE pipe resembles metal tubing in that it can be bent by hand or with a suitable bending device while maintaining its shape without fittings or supports. In other words, it is "form stable." The minimum bending radius specified by manufacturers is five times the outside diameter.

Mechanical joints are the only method currently available to join PE-AL-PE pipe. A number of proprietary mechanical-compression-type connectors have been developed for use with the composite pipe to permit transition to other pipes and fittings. Such fittings must be installed in accordance with the manufacturer's instructions.

Raised temperature polyethylene pipe (PE-RT): ASTM F 2623 covers polyethylene, SDR 9 tubing that is outside diameter controlled, and pressure rated for water at 100 psi at 180°F (689 kPa at 82.2°C).

Polypropylene pipe (PP): Polypropylene pipe is excellent for corrosive media. It has good resistance to strong acids except highly active oxidizers, such as nitric acid. It also has excellent resistance to weak and strong alkalies and to most organic solvents. Polypropylene can be used at temperatures up to 150°F (66°C), in continuous pressure service.

Polyvinyl chloride (PVC) plastic pipe: PVC used for hydronic purposes is the same material rated for use as water service pipe in the IPC. The pipe is white and it must be continuously marked with the manufacturer's name, the ASTM standard and the grade of the PVC material.

A number of grades of PVC material are used to produce pipe. The compounds are identified as PVC 12454-B, 12454-C and 14333-D. The first digit indicates the base resin. The following four digits identify the impact strength, tensile strength, modulus of elasticity and deflection temperature. The letter suffix indicates the material's resistance to chemicals.

The compounds were previously identified by type and grade; 12454-B is Type 1, Grade 1; 12454-C is Type 1, Grade 2 and 14333-D is Type 2, Grade 1. The marking on the pipe lists the material grade by the term "PVC" followed by four digits. The first two digits use the previous type and grade numbers to identify the compound. The last two digits indicate the hydrostatic design stress in hundreds of pounds per square inch.

The hydrostatic design stress, which is the pressure exerted on the walls of the pipe by the fluid in the pipe under no-flow conditions, is not an indication of the pressure rating of the pipe. However, the pressure rating of the pipe at 73°F (23°C) may be computed by using the hydrostatic design stress:

$$P = \frac{2S}{(OD/t) - 1} = \frac{2S}{R - 1}$$

where:

P = Pressure rating.

S = Hydrostatic design stress (psi).

OD = Average outside diameter (inches).

t = Minimum wall thickness (inches).

R = Standard dimension ratio (SDR) (based on outside diameter).

ASTM D 1785 is for Schedules 40, 80 and 160 PVC plastic pipe. The dimensions of the pipe are the same as Schedules 40, 80 and 160 steel pipe.

ASTM D 2241 is for SDR-PR PVC plastic pipe. The SDR means standard dimension ratio, the ratio of average outside pipe diameter to minimum wall thickness. The PR means the pipe is pressure rated. The SDRs of the pipe are 13.5, 17, 21, 26, 32.5, 41 and 64.

Steel pipe: Steel pipe is manufactured by several processes. Seamless pipe, made by piercing or extruding, has no longitudinal seam. Other manufacturing methods roll sheet steel into a cylinder and weld a longitudinal seam. A continuous-weld (CW) furnace-buttwelding process forces and joins the edges together at high temperatures. Continuous weld is used for pipe sizes $^1/_4$ inch (6.4 mm) through 4 inches (102 mm). An electric current welds the seam of electric-resistance-welded (ERW) pipe, and is used for pipe sizes 6 inches (152 mm) and larger. ASTM A 106 and A 53 cover steel pipe. Both designate A and B grades. The A grade has a lower tensile strength and is not widely used.

Steel pipe is manufactured with wall thicknesses identified by schedule and weight. Although schedule numbers and weight designations are related, they are not constant for all pipe sizes. Standard weight (STD) and Schedule 40 pipe have the same wall thickness through 10-inch (254 mm) nominal pipe size (NPS). For 12-inch (305 mm) and larger standard-weight pipe, wall thicknesses increase with each size. A similar equality exists between extra-strong (XS) and Schedule 80 pipe up through 8 inches (203 mm); after which XS pipe has a 0.500-inch (12.7 mm) wall, while Schedule 80 increases in wall thickness.

Because the pipe wall thickness of threaded standard weight pipe is so small after deducting the allowance (A), the mechanical strength of the pipe is low. It is a good practice to limit standard-weight threaded pipe pressures to 90 pounds per square inch gauge (psig) (621 kPa) for steam and 125 psig (862 kPa) for water.

Steel tubing: ASTM A 254 covers double-walled, copper-brazed steel tubing. This tubing is manufactured by rolling steel strip into the form of a double thickness tube and copper brazing along the longitudinal seam(s). The tubing is also available with a copper coating on the inside and outside surfaces or a hot-dipped, lead-in alloy on the outside surface.

1202.5 Pipe fittings. Hydronic pipe fittings shall be *approved* for installation with the piping materials to be installed, and shall conform to the respective pipe standards or to the standards listed in Table 1202.5.

❖ To avoid chemical or corrosive action in a hydronic system that could cause premature failure, fittings must be of the same material as the pipe being used or of a compatible material.

Many of the pipe standards also apply to the pipe fittings. However, there are a number of other standards that establish criteria primarily for pipe fittings as identified in Table 1202.5. Fittings include ells [90 degrees and 45 degrees (1.57 rad and 0.79 rad)], tees, wyes, flow inducers, caps, plugs, couplings and increasers/reducers (concentric and eccentric). Fittings must be rated for the pressures and temperatures of the intended application.

TABLE 1202.5
HYDRONIC PIPE FITTINGS

MATERIAL	STANDARD (see Chapter 15)
Brass	ASTM F 1974
Bronze	ASME B16.24
Copper and copper alloys	ASME B16.15; ASME B16.18; ASME B16.22; ASME B16.23; ASME B16.26; ASME B16.29
Ductile iron and gray iron	ANSI/AWWA C110/A21.10
Ductile iron	ANSI/AWWA C153/A21.53
Gray iron	ASTM A 126
Malleable iron	ASME B16.3
PEX fittings	ASTM F 877; ASTM F 1807; ASTM F 2159
Plastic	ASTM D 2466; ASTM D 2467; ASTM D 2468; ASTM F 438; ASTM F 439; ASTM F 877; ASTM F 2389; ASTM F 2735
Steel	ASME B16.5; ASME B16.9; ASME B16.11; ASME B16.28; ASTM A 420

❖ Standards for PEX tubing fittings were added. Standard ASTM F 2735 was added to the plastic fittings row. This table contains pipe fitting standards with the corresponding pipe fitting material.

1202.6 Valves. Valves shall be constructed of materials that are compatible with the type of piping material and fluids in the system. Valves shall be rated for the temperatures and pressures of the systems in which the valves are installed.

❖ Valves are constructed to withstand a specific range of temperature, pressure and mechanical stress. The pressure ratings of valves vary with the kind of fluid or gas being conducted (steam, water, oil or gas). Valve types include gates, stops, lubricated plug cocks, ball valves, butterfly valves and variations of each of these types. The code official must evaluate the various valves for their acceptability in hydronic systems based on the many standards that regulate the rating of valves and the application limitations placed on the valve by the manufacturer. The valves should operate easily, be as liquid or gas tight as the application requires, have a reasonable service life, be accessible, be approved and be installed in the application for which they were designed and approved.

Valves must be of the same material as the pipe or of a compatible material to avoid chemical or corrosive action. Additionally, valves must be rated for the pressures and temperatures to which they are subjected.

1202.7 Flexible connectors, expansion and vibration compensators. Flexible connectors, expansion and vibration control devices and fittings shall be of an *approved* type.

❖ This section contains the requirements for devices designed to protect the components of the hydronic system from damage caused by expansion, contraction, transverse movement, angular deflection and vibration. The devices are designed to absorb pipe movement and stress, reduce system noise, isolate mechanical vibration and compensate for slight misalignment of pipes. Piping movements are produced by temperature and pressure variations in the surrounding atmosphere and working fluid causing dimensional changes. Vibrations are produced by mechanical equipment such as pumps.

Flexible connectors and expansion joints are used where the potential movements are too large to be accommodated by pipe bends or loops, or where insufficient room exists to construct a loop of adequate size.

Examples of expansion compensators are:

• Packed slip expansion joint [see Commentary Figure 1202.7(1)];

• Flexible ball joint [see Commentary Figure 1202.7(2)];

• Metal bellows expansion joint [see Commentary Figure 1202.7(3)];

• Rubber expansion joint [see Commentary Figure 1202.7(4)]; and

• Flexible hose [see Commentary Figure 1202.7(5)].

Because these devices are available in a wide variety of sizes, materials and temperature/pressure ratings, it is imperative that they be rated for the system operating temperature and pressure and be compatible with the fluid in the system.

SECTION 1203
JOINTS AND CONNECTIONS

1203.1 Approval. Joints and connections shall be of an *approved* type. Joints and connections shall be tight for the pressure of the hydronic system.

❖ Joints and connections must be fabricated using a method described in this section or must be specifically approved by the code official. This section does not prohibit the use of joints and connections that are not regulated here. Many joints and connections are proprietary and are not regulated by standards; however, a type of joint or connection other than those addressed in this section may be used only after the code official has determined the proposed joint or connection is satisfactory and complies with the intent of the code. Approval of joints and connections must consider the compatibility of the joint or connection with the working fluid of the system and the pipe materials being joined. Joints and connections must be able to withstand the maximum operating conditions of the system.

1203.1.1 Joints between different piping materials. Joints between different piping materials shall be made with *approved* adapter fittings.

❖ Numerous adapter fittings are available that are designed to join different pipe materials. When adapter fittings are used, they must be evaluated and shown to be compatible with the pipe material and working fluid (see commentary, Section 1203.1). The fittings must also be rated for the maximum operating conditions of the hydronic system.

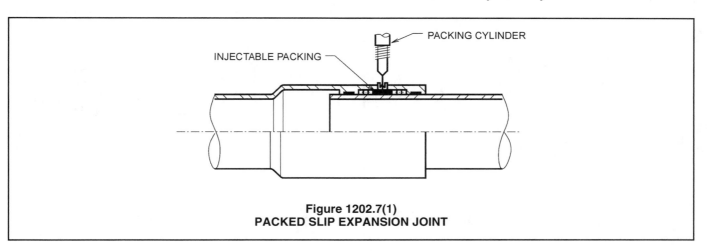

Figure 1202.7(1)
PACKED SLIP EXPANSION JOINT

When joining dissimilar metals, a method of protecting against galvanic corrosion is required. Galvanic corrosion occurs when two different metals are in contact in the presence of an electrolyte, such as water. Galvanic corrosion accelerates the natural corrosion process that occurs in all metals and, because certain metals corrode faster than others, they have been placed in a hierarchy in order of rate of corrosion.

The more reactive metal at a juncture is called the "anode" and the less reactive metal is called the "cathode." When metals of different reactivity are coupled, the more reactive metal (anode) will develop a surface layer of corrosion (such as rust on steel) or tend to dissolve in an electrolyte (such as water), thereby generating an electric current flow between the dissimilar metals.

The rate of galvanic corrosion is strongly influenced by the difference in activity between dissimilar metals. The greater the difference in metals in the hierarchy of resistance to corrosion, the greater the reactivity between them and, therefore, the faster the anodic metal will corrode.

This section prescribes two methods of protection against galvanic corrosion: dielectric fittings and brass converter fittings. These fittings are designed to provide a barrier or buffer between dissimilar metals in the waterway system or are designed to electrically isolate the metals from each other. Dielectric fittings include insulated couplings, insulated nipples and insulated unions. Cast brass male and female threaded adapters are examples of converter fittings.

1203.2 Preparation of pipe ends. Pipe shall be cut square, reamed and chamfered, and shall be free of burrs and obstructions. Pipe ends shall have full-bore openings and shall not be undercut.

❖ If pipe ends are not cut at right angles to the pipe barrel, the result can be the misalignment of the piping and insufficient insertion depth into the fitting. This can lead to joint failure.

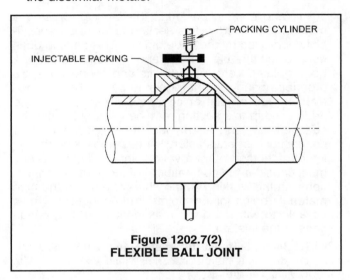

Figure 1202.7(2)
FLEXIBLE BALL JOINT

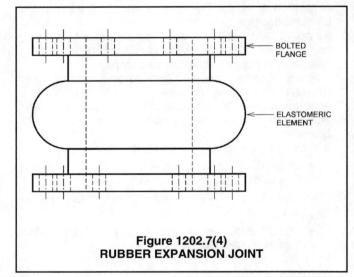

Figure 1202.7(4)
RUBBER EXPANSION JOINT

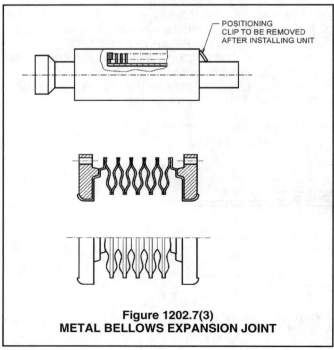

Figure 1202.7(3)
METAL BELLOWS EXPANSION JOINT

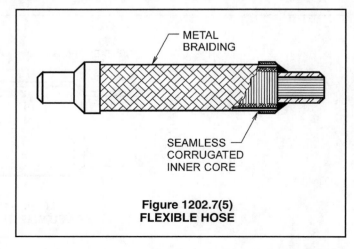

Figure 1202.7(5)
FLEXIBLE HOSE

Shoulders and burrs produced by cutting must be removed to eliminate obstruction of fluid flow. The pipe must be reamed internally and prepared externally to remove burrs, shoulders and protruding edges. Undercutting or reducing the pipe wall thickness during this process must be avoided. Undercutting of the pipe structurally weakens the pipe wall and can also weaken threads or other joints (see Commentary Figure 1203.2).

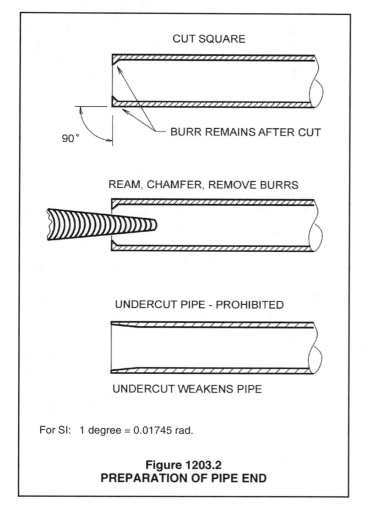

CUT SQUARE

90°

BURR REMAINS AFTER CUT

REAM, CHAMFER, REMOVE BURRS

UNDERCUT PIPE - PROHIBITED

UNDERCUT WEAKENS PIPE

For SI: 1 degree = 0.01745 rad.

**Figure 1203.2
PREPARATION OF PIPE END**

1203.3 Joint preparation and installation. When required by Sections 1203.4 through 1203.14, the preparation and installation of brazed, mechanical, soldered, solvent-cemented, threaded and welded joints shall comply with Sections 1203.3.1 through 1203.3.7.

❖ Sections 1203.3.1 through 1203.3.8 regulate the allowed joints and connections for the pipe materials listed in Sections 1203.4 through 1203.15. Joining methods that are unique to a specific pipe material are listed in the sections that apply to that pipe material. For example, Section 1203.15.2 is a subsection of Section 1203.15 because electrofusion joints apply only to polyethylene pipe and tubing.

1203.3.1 Brazed joints. Joint surfaces shall be cleaned. An *approved* flux shall be applied where required. The joint shall be brazed with a filler metal conforming to AWS A5.8.

❖ Brass tubing and pipe, and copper or copper-alloy tubing and pipe, can be joined by brazing. Brazed joints are similar to soldered joints; however, they are joined at temperatures in excess of 1,000°F (538°C). A brazed joint is much stronger than a soldered joint because of the higher temperatures.

Joint surfaces to be brazed are typically cleaned with an emery cloth and specially designed brushes. Surface oxides and other substances can interfere with filler metal flow and adhesion.

Some types of filler metals do not require flux for certain types of joints. Manufacturer's instructions must be consulted to determine whether and what type of flux is needed.

Common filler metals used for brazing contain copper alloys and phosphorus or contain 30 to 60 percent silver with zinc and copper alloys. Some of the filler metal alloys also contain cadmium, which is highly toxic.

1203.3.2 Mechanical joints. Mechanical joints shall be installed in accordance with the manufacturer's instructions.

❖ Many mechanical joints use an elastomeric material to form a seal. Because mechanical joints are typically proprietary, they must be made in accordance with the manufacturer's installation instructions.

Mechanical joints for metallic tubing include a large array of fittings, the most common of which is the compression fitting. This type of fitting uses a metallic compression ring to form the seal. The ring compresses around the pipe outside wall as the fitting is tightened. Pipe-joint compound is not required but is often used to act as a lubricant for the threads and mating metal surfaces.

Mechanical joints for plastic pipe include insert-type fittings, metallic lock ring fittings, compression fittings and crimp-type fittings.

Although an insert-type fitting can reduce the inside diameter of the pipe, the reduction may be accounted for as part of the design considerations of the hydronic system.

Mechanical joints for steel pipe include compression gasket, grooved and mechanically deformed (pressed) joints. Grooved joint systems require cutting or forming a shallow groove in the pipe wall. A segmented clamp engages the grooves, and the seal is a special gasket designed so that internal pressure expands and tightens the seal. Some clamps are designed with clearance between the tongue and groove to accommodate misalignment and thermal movements, while others are designed to limit movement and provide a nearly rigid joint. Manufacturer's data specify temperature and pressure limitations for the proper application of mechanical joints.

Another form of a mechanical joint consists of a coupling slightly larger than the outside diameter of the pipe. The pipe ends are inserted into the sleeve, and gaskets are packed into the annular space between pipe and coupling and held in place by retainer rings. This type of joint can accommodate some axial misalignment, but must be anchored or otherwise restrained to prevent axial pullout or lateral movement. Manufacturers provide pressure/temperature data for the proper application of mechanical joints.

A relatively new type of joint requires a special power tool that compresses a copper socket fitting onto the copper pipe and seals with an O-ring.

1203.3.3 Soldered joints. Joint surfaces shall be cleaned. A flux conforming to ASTM B 813 shall be applied. The joint shall be soldered with a solder conforming to ASTM B 32.

❖ Soldered joints are commonly used with brass tubing, copper or copper-alloy pipe and copper or copper-alloy tubing. ASTM B 32 covers many grades of solder. Solders are metal alloys of tin and lead, tin and antimony, tin and silver and other more complex alloys of tin, silver, copper and other metals.

The joint surfaces for a soldered joint must be cleaned using emery cloth and specially designed brushes to remove oxides and impurities. Flux is required to prevent oxidation from forming and to promote proper solder flow into the joint.

1203.3.4 Solvent-cemented joints. Joint surfaces shall be clean and free of moisture. An *approved* primer shall be applied to CPVC and PVC pipe-joint surfaces. Joints shall be made while the cement is wet. Solvent cement conforming to the following standards shall be applied to all joint surfaces:

1. ASTM D 2235 for ABS joints.
2. ASTM F 493 for CPVC joints.
3. ASTM D 2564 for PVC joints.

CPVC joints shall be made in accordance with ASTM D 2846.

Exception: For CPVC pipe joint connections, a primer is not required where all of the following conditions apply:

1. The solvent cement used is third-party certified as conforming to ASTM F 493.
2. The solvent cement is yellow in color.
3. The solvent cement is used only for joining $^1/_2$ inch (12.7 mm) through 2-inch (51 mm) diameter CPVC pipe and fittings.
4. The CPVC pipe and fittings are manufactured in accordance with ASTM D 2846.

❖ Solvent cementing is a common method of joining plastic pipe and fittings and is often referred to as "solvent welding." The referenced ASTM standards contain recommended procedures for handling solvent cement and making joints.

The pipe and fitting should be at approximately the same temperature when solvent cementing is done. The pipe end and socket fitting must be clean, dry and free from grease, oil and foreign substances. CPVC and PVC joints must first be primed. A primer

is applied to all joint surfaces before the solvent cement is applied. The primer is a solvent for the pipe and fitting material that conditions the joint surfaces for the subsequent application of solvent cement.

Failure to apply a primer to a solvent-cemented joint could result in inferior joint strength. If a clear primer is used, close examination is required to determine that primer has been applied. Primers remove the shiny finish on the surface of pipe and fittings, leaving a dull surface. Solvent cement must be applied while the primer is still active (wet).

Solvent-cemented joints must cure (set up) for the time period specified by the manufacturer of the solvent cement. Curing time is affected by relative humidity, ambient temperature and the formulation of the cement.

All-purpose solvent cement or universal solvent cement cannot be used unless it conforms to the standard for solvent cement applicable to the pipe and fitting materials being joined. Each type of solvent cement is specifically designed for a given piping material. All-purpose solvent cement may or may not conform to the applicable ASTM standards for solvent cements for each of the plastic materials. Solvent cement is identified on the container by the appropriate ASTM standard number. ABS solvent cement must conform to ASTM D 2235. The solvent cement for CPVC must conform to ASTM F 493, and the solvent cement for PVC must conform to ASTM D 2564.

The exception allows for CPVC pipe to be joined without the use of a primer. The size of the pipe must be in the range of $^1/_2$ inch to 2 inches, the solvent cement must be yellow in color and third-party certified, and the pipe and fittings must be manufactured in accordance with ASTM D 2846. Using primers, especially colored primers, can be messy and can leave indelible stains on finished surfaces. The priming step also takes time. Eliminating primer by using the "one step" cement makes the job go faster and look more professional.

1203.3.5 Threaded joints. Threads shall conform to ASME B1.20.1. Schedule 80 or heavier plastic pipe shall be threaded with dies specifically designed for plastic pipe. Thread lubricant, pipe-joint compound or tape shall be applied on the male threads only and shall be *approved* for application on the piping material.

❖ ASME B1.20.1 identifies pipe thread dimensions as "NPT." NPT is not an abbreviation for terms such as "national pipe thread," "nominal pipe thread" or "national pipe tapered thread"; rather, NPT is a coded designation. "N" stands for USA standard, "P" indicates pipe, and "T" means the threads are tapered.

Plastic pipe with a wall thickness less than that of Schedule 80 pipe may not be threaded because of the reduced wall thickness and strength at the thread location. Plastic pipe must be threaded with threading dies that are specially designed to cut plastic pipe threads to avoid damaging the pipe and the threads.

Pipe-joint compounds (dope) and tape are used to lubricate the threads and seal minute imperfections on the thread surfaces. The primary function of pipe-joint compound is to lubricate the surfaces to allow the joint to be tightened enough to obtain a tight metal-to-metal or plastic-to-plastic seal. Pipe-joint compounds must be compatible with the piping material and the fluid or gas passing through the piping. Some plastic piping materials can be damaged by various pipe-joint compound formulations.

If pipe-joint compound or tape is applied to female threads, the compound or tape will be pushed into the piping system where it could contaminate the system.

1203.3.6 Welded joints. Joint surfaces shall be cleaned by an *approved* procedure. Joints shall be welded with an *approved* filler metal.

❖ A welded joint is similar to a brazed joint. The differences between the two types of joints are the temperature at which the joint is made and the type of filler metals (see commentary, Section 1203.3.1). Welded joint uses filler metals of the same material as the pipe or fitting being welded and the welding temperatures reach the melting point of the workpiece. Welding results in a homogeneous fusion of the materials being joined.

The applicable sections of ASME B31 and the ASME *Boiler and Pressure Vessel Code* specify proper welding methods. ASME B31 requires that welders and welding procedure specifications be qualified. Separate welding procedure specifications are needed for different welding methods and materials. Qualifying tests and the variables requiring separate procedure specifications are set forth in Section IX of the ASME *Boiler and Pressure Vessel Code.* The manufacturer, fabricator or contractor is responsible for the welding procedure and welders. ASME B31 requires visual examination of welds and outlines limitations of acceptability.

Shielded metal arc welding (stick welding) is most often used for hydronic piping because of its availability and ease of operation. The molten weld metal is shielded by vaporizing the electrode flux coating while the electrode melts to add material to the weld zone.

1203.3.7 Grooved and shouldered mechanical joints. Grooved and shouldered mechanical joints shall conform to the requirements of ASTM F 1476 and shall be installed in accordance with the manufacturer's installation instructions.

❖ This method of joining pipes uses an elastomeric gasket and a bolted clamping ring assembly that interlocks with grooves or shoulders formed on the pipe ends. ASTM F 1476 covers this joining method. This joining method is commonly used in fire sprinkler systems and the fittings are often referred to as "Victaulic" fittings because this is the name of a well-known manufacturer of this type of fitting/coupling.

1203.3.8 Mechanically formed tee fittings. Mechanically extracted outlets shall have a height not less than three times the thickness of the branch tube wall.

❖ Mechanically formed tee fittings are formed in a continuous operation consisting of drilling a pilot hole and drawing out the tube surface to form a collar having a height not less than three times the thickness of the tube wall. Specialized tools are required for this operation. Tees formed in accordance with this section lack the quality control that is imposed in a manufacturer's plant to meet applicable standards. When tees are mechanically formed in the field, tolerances do not always conform to standards. Because of the shape and size of the collars produced by this method, soft soldering is deemed inadequate to ensure the required joint strength.

1203.3.8.1 Full flow assurance. Branch tubes shall not restrict the flow in the run tube. A dimple/depth stop shall be formed in the branch tube to ensure that penetration into the outlet is of the correct depth. For inspection purposes, a second dimple shall be placed $\frac{1}{4}$ inch (6.4 mm) above the first dimple. Dimples shall be aligned with the tube run.

❖ Notching and dimpling of the branch tube is required to ensure that the branch tube is not inserted too far into the formed collar, allowing it to obstruct flow through the fitting or increase friction loss beyond what can normally be expected for a tee fitting. The notching and dimpling are done in one operation to ensure that the dimples are located properly with respect to the end of the notch. The branch tube is inserted into the collar until the first dimple touches the collar. The first dimple will be covered by the brazing filler. The second dimple, located $\frac{1}{4}$-inch (6.4 mm) above the first dimple, serves as a visual inspection point above the area to be brazed [see Commentary Figures 1203.3.8.1(1) and (2)].

Figure 1203.3.8.1(1)
MECHANICALLY FORMED JOINT
(Courtesy of T-DRILL Industries, Inc.)

Figure 1203.3.8.1(2)
MECHANICALLY FORMED JOINT—DIMPLE LOCATION
(Courtesy of T-DRILL Industries, Inc.)

1203.3.8.2 Brazed joints. Mechanically formed tee fittings shall be brazed in accordance with Section 1203.3.1.

❖ See the commentary to Section 1203.3.1.

1203.4 ABS plastic pipe. Joints between ABS plastic pipe or fittings shall be solvent-cemented or threaded joints conforming to Section 1203.3.

❖ Sections 1203.4 and 1203.3.5 describe the joining methods allowed for ABS plastic pipe.

1203.5 Brass pipe. Joints between brass pipe or fittings shall be brazed, mechanical, threaded or welded joints conforming to Section 1203.3.

❖ Sections 1203.3.1, 1203.3.2, 1203.3.5 and 1203.3.6 describe the joining methods allowed for brass pipe.

1203.6 Brass tubing. Joints between brass tubing or fittings shall be brazed, mechanical or soldered joints conforming to Section 1203.3.

❖ Sections 1203.3.1, 1203.3.2 and 1203.3.3 describe the joining methods allowed for brass tubing.

1203.7 Copper or copper-alloy pipe. Joints between copper or copper-alloy pipe or fittings shall be brazed, mechanical, soldered, threaded or welded joints conforming to Section 1203.3.

❖ Sections 1203.3.1, 1203.3.2, 1203.3.3, 1203.3.5 and 1203.3.6 describe the methods allowed for joining copper and copper-alloy pipe.

1203.8 Copper or copper-alloy tubing. Joints between copper or copper-alloy tubing or fittings shall be brazed, mechanical or soldered joints conforming to Section 1203.3, flared joints conforming to Section 1203.8.1, push-fit joints conforming to Section 1203.8.2 or press-type joints conforming to Section 1203.8.3.

❖ Sections 1203.3.1, 1203.3.2, 1203.3.3, 1203.8.1, 1203.8.2 and 1203.8.3 describe the joining methods allowed for copper or copper-alloy tubing.

1203.8.1 Flared joints. Flared joints shall be made by a tool designed for that operation.

❖ Because the pipe end is expanded in a flared joint, only annealed and bending-tempered (drawn) copper tubing may be flared. Commonly used flaring tools use a screw yoke and block assembly or an expander tool that is driven into the tube with a hammer. The flared tubing end is compressed between a fitting seat and a threaded nut to form a metal-to-metal seal.

1203.8.2 Push-fit joints. Push-fit joints shall be installed in accordance with the manufacturer's instructions.

❖ Push-to-connect fittings are becoming more commonplace as they do not require soldering, special tightening procedures or extensive personnel training to ensure a leak free joint. These fittings are typically intended for connecting only hard drawn copper tube and not soft temper copper tubing. The specific manufacturer's installation instructions provide information as to the piping application that the fitting can be used in.

1203.8.3 Press joints. *Press joints* shall be installed in accordance with the manufacturer's instructions.

❖ Press joint fittings are becoming commonplace as they do not require soldering or extensive personnel training to ensure a leak free joint. A special tool is needed to properly crimp the fittings onto the piping. Once installed, the fittings cannot be removed. These fittings are typically intended for connecting only hard drawn copper tube and not soft temper copper tubing. The specific manufacturer's installation instructions provide information as to the piping application that the fitting can be used in.

1203.9 CPVC plastic pipe. Joints between CPVC plastic pipe or fittings shall be solvent-cemented or threaded joints conforming to Section 1203.3.

❖ Sections 1203.3.4 and 1203.3.5 describe the joining methods allowed for CPVC plastic pipe.

1203.10 Polybutylene plastic pipe and tubing. Joints between polybutylene plastic pipe and tubing or fittings shall be mechanical joints conforming to Section 1203.3 or heat-fusion joints conforming to Section 1203.10.1.

❖ Sections 1203.3.2 and 1203.10.1 describe the methods allowed for joining polybutylene plastic pipe and tubing.

1203.10.1 Heat-fusion joints. Joints shall be of the socket-fusion or butt-fusion type. Joint surfaces shall be clean and free of moisture. Joint surfaces shall be heated to melt temperatures and joined. The joint shall be undisturbed until cool. Joints shall be made in accordance with ASTM D 3309.

❖ Heat fusion is analogous to welding of metals. Specially designed tools, typically electrically heated, are used to heat the pipe ends and fitting sockets to the melting temperature and the materials are fused together. Certain butt-fusion tools also properly align the pipe and make the connection (see Commentary Figure 1203.10.1).

1203.11 Cross-linked polyethylene (PEX) plastic tubing. Joints between cross-linked polyethylene plastic tubing and fittings shall conform to Sections 1203.11.1 and 1203.11.2. Mechanical joints shall conform to Section 1203.3.

❖ Sections 1203.3.2, 1203.11.1 and 1203.11.2 address joining methods for PEX tubing. All joining methods for PEX tubing are considered mechanical joints that must be installed in accordance with the manufacturer's installation instructions.

1203.11.1 Compression-type fittings. When compression-type fittings include inserts and ferrules or O-rings, the fittings shall be installed without omitting the inserts and ferrules or O-rings.

❖ A number of fittings have been developed that can be described as mechanical compression fittings. Compression fittings are likely to include inserts, ferrules and O-rings that form an essential part of the fitting assembly and, therefore, cannot be omitted (see Commentary Figure 1203.11.1). Inserts act as stiffeners to resist the compression forces on the tubing wall. Ferrules and O-rings form the seal around the tube.

Some fittings use an insert that is forced into the expanded end of the tube with a compression sleeve forced over the tubing to compress the tube against the insert fitting. A specific technique requires the flaring or expanding of the pipe before the fitting is inserted. In all cases, the fittings connecting PEX pipe must be assembled and installed in accordance with the manufacturer's installation instructions.

1203.11.2 Plastic-to-metal connections. Soldering on the metal portion of the system shall be performed at least 18 inches (457 mm) from a plastic-to-metal adapter in the same water line.

❖ As is the case with all plastic pipe, special care must be taken when soldering or brazing any metallic pipe that is connected to plastic pipe. To avoid subjecting the plastic pipe to high temperatures, any soldering must be done at a minimum distance from the point of connection to the plastic pipe. Temperatures exceeding 338°F (170°C) can cause irreversible damage to the PEX piping.

1203.12 PVC plastic pipe. Joints between PVC plastic pipe and fittings shall be solvent-cemented or threaded joints conforming to Section 1203.3.

❖ Sections 1203.3.4 and 1203.3.5 describe the joining methods allowed for PVC plastic pipe.

1203.13 Steel pipe. Joints between steel pipe or fittings shall be mechanical joints that are made with an *approved* elastomeric seal, or shall be threaded or welded joints conforming to Section 1203.3.

❖ Sections 1203.3.2, 1203.3.5, 1203.3.6 and 1203.3.7 describe the joining methods allowed for steel pipe.

1203.14 Steel tubing. Joints between steel tubing or fittings shall be mechanical or welded joints conforming to Section 1203.3.

❖ Sections 1203.3.2 and 1203.3.6 describe the joining methods allowed for steel tubing.

1203.15 Polyethylene plastic pipe and tubing for ground source heat pump loop systems. Joints between polyethyl-

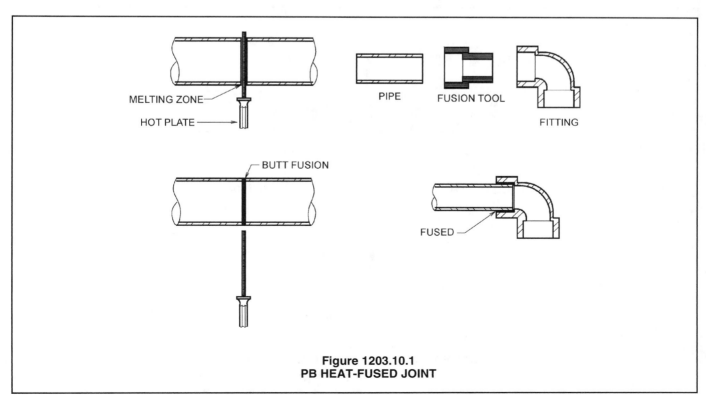

**Figure 1203.10.1
PB HEAT-FUSED JOINT**

ene plastic pipe and tubing or fittings for ground source heat pump loop systems shall be heat fusion joints conforming to Section 1203.15.1, electrofusion joints conforming to Section 1203.15.2, or stab-type insertion joints conforming to Section 1203.15.3.

❖ Polyethylene pipe and fittings are joined by heat-fusion or mechanical joints, both of which provide the dependability demanded for underground piping.

1203.15.1 Heat-fusion joints. Joints shall be of the socket-fusion, saddle-fusion or butt-fusion type, joined in accordance with ASTM D 2657. Joint surfaces shall be clean and free of moisture. Joint surfaces shall be heated to melt temperatures and joined. The joint shall be undisturbed until cool. Fittings shall be manufactured in accordance with ASTM D 2683 or ASTM D 3261.

❖ Heat-fusion joints for plastic pipe are analogous to welding of steel pipe except that a filler rod or wire is not used to add material to the joint as it is welded. The pipe-and-fitting, or butt-joined pipes are melted together (fused) using specialized heating tools. ASTM D 2657 covers the procedure for heat fusion of plastic joints. ASTM D 2683 is the standard for socket-type fusion fittings and ASTM D 3261 is the standard for butt-weld type fittings. Once cooled to ambient temperatures, these joints are permanent and irreversible.

1203.15.2 Electrofusion joints. Joints shall be of the electrofusion type. Joint surfaces shall be clean and free of moisture, and scoured to expose virgin resin. Joint surfaces shall be heated to melt temperatures for the period of time specified by the manufacturer. The joint shall be undisturbed until cool.

Fittings shall be manufactured in accordance with ASTM F 1055.

❖ Electrofusion joints are a type of heat-fusion joint using special fittings that contain an integral resistance heating element. The fitting is connected to an electrical power supply that controls the power input and duration to the fitting's integral heating element.

1203.15.3 Stab-type insert fittings. Joint surfaces shall be clean and free of moisture. Pipe ends shall be chamfered and inserted into the fittings to full depth. Fittings shall be manufactured in accordance with ASTM F 1924.

❖ Stab-type fittings are a type of mechanical joint that uses O-rings to develop a seal and gripping teeth to prevent the pipe from pulling out of the fitting. The name given to these fittings is taken from the manner in which the pipe and fittings are assembled. The pipe is "stabbed" (pushed) into the fitting until it bottoms in the socket and the joint is complete. The pipe must be chamfered; otherwise, the sharp square corner on the outside diameter of the pipe could damage the O-rings or prevent insertion. The fittings may be prelubricated or may require lubrication at time of installation. The manufacturer's installation instructions for the specific fitting must be followed as these types of fittings typically require careful preparation of the cut-end of the pipe as well as marking of the pipe to determine when full engagement of the fitting is achieved. Prior to the 2009 edition of the code, the standard referenced for this type of fitting was ASTM D 2513. However, that standard did not specifically cover stab-type mechanical fittings for PE pipe.

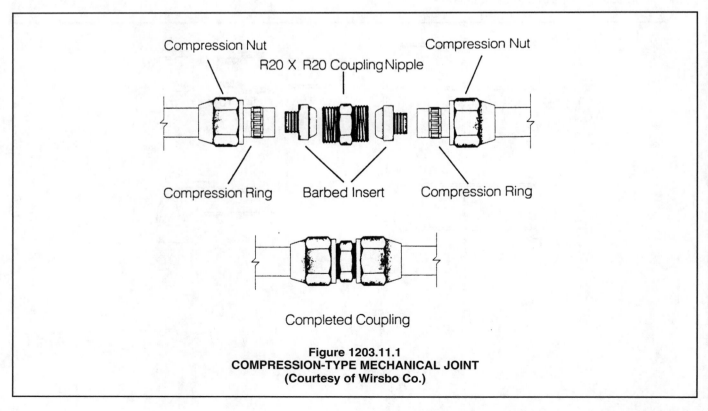

Figure 1203.11.1
COMPRESSION-TYPE MECHANICAL JOINT
(Courtesy of Wirsbo Co.)

ASTM F 1924 appropriately covers all types of mechanical fittings for PE pipe.

1203.16 Polypropylene (PP) plastic. Joints between PP plastic pipe and fittings shall comply with Sections 1203.16.1 and 1203.16.2.

❖ The fitting manufacturer's installation instructions must be followed when making connections for polypropylene (PP) pipe.

1203.16.1 Heat-fusion joints. Heat-fusion joints for polypropylene (PP) pipe and tubing joints shall be installed with socket-type heat-fused polypropylene fittings, electro-fusion polypropylene fittings or by butt fusion. Joint surfaces shall be clean and free from moisture. The joint shall be undisturbed until cool. Joints shall be made in accordance with ASTM F 2389.

❖ Heat fusion is analogous to welding of metals. Specially designed tools, typically electrically heated, are used to heat the pipe ends and fitting sockets to the melting temperature, and the materials are fused together. Certain butt-fusion tools also properly align the pipe and make the connection.

1203.16.2 Mechanical and compression sleeve joints. Mechanical and compression sleeve joints shall be installed in accordance with the manufacturer's instructions.

❖ The fitting manufacturer's installation instructions must be followed when making mechanical and compression sleeve joints for polypropylene (PP) pipe.

1203.17 Raised temperature polyethylene (PE-RT) plastic tubing. Joints between raised temperature polyethylene tubing and fittings shall conform to Sections 1203.17.1 and 1203.17.2. Mechanical joints shall conform to Section 1203.3.

❖ The fitting manufacturer's installation instructions must be followed when making connections for raised temperature polyethylene (PE-RT) tubing.

1203.17.1 Compression-type fittings. Where compression-type fittings include inserts and ferrules or O-rings, the fittings shall be installed without omitting the inserts and ferrules or O-rings.

❖ All components necessary for a particular style of compression-type connection of PE-RT tubing must be used.

1203.17.2 PE-RT-to-metal connections. Solder joints in a metal pipe shall not occur within 18 inches (457 mm) of a transition from such metal pipe to PE-RT pipe.

❖ The required heat for soldering fittings to copper pipe could melt the PE-RT pipe and cause a leak if the soldering operation was too close to an existing PE-RT pipe to copper transition. This section is not intended to prohibit a completed, cooled soldered connection from being located adjacent to where a PE-RT pipe is to be connected.

1203.18 Polyethylene/aluminum/polyethylene (PE-AL-PE) pressure pipe. Joints between polyethylene/aluminum/polyethylene pressure pipe and fittings shall conform to Sec-

tions 1203.18.1 and 1203.18.2. Mechanical joints shall comply with Section 1203.3.

❖ The fitting manufacturer's installation instructions must be followed when making connections of polyethylene/aluminum/polyethylene (PE-AL-PE) pipe.

1203.18.1 Compression-type fittings. Where compression-type fittings include inserts and ferrules or O-rings, the fittings shall be installed without omitting the inserts and ferrules or O-rings.

❖ All components necessary for a particular style of compression type connection of PE-AL-PE must be used.

1203.18.2 PE-AL-PE-to-metal connections. Solder joints in a metal pipe shall not occur within 18 inches (457 mm) of a transition from such metal pipe to PE-AL-PE pipe.

❖ The required heat for soldering fittings to copper pipe could melt the PE-AL-PE pipe and cause a leak if the soldering operation was too close to an existing PE-AL-PE to copper transition. This section is not intended to prohibit a completed, cooled soldered connection from being located adjacent to where a PE-AL-PE pipe is to be connected.

1203.19 Cross-linked polyethylene/aluminum/cross-linked polyethylene (PEX-AL-PEX) pressure pipe. Joints between cross-linked polyethylene/aluminum/cross-linked polyethylene pressure pipe and fittings shall conform to Sections 1203.19.1 and 1203.19.2. Mechanical joints shall comply with Section 1203.3.

❖ The fitting manufacturer's installation instructions must be followed when making connections for polyethylene/aluminum/polyethylene (PEX-AL-PEX) pipe.

1203.19.1 Compression-type fittings. Where compression-type fittings include inserts and ferrules or O-rings, the fittings shall be installed without omitting the inserts and ferrules or O-rings.

❖ All components necessary for a particular style of compression-type connection of PEX-AL-PEX pipe must be used.

1203.19.2 PEX-AL-PEX-to-metal connections. Solder joints in a metal pipe shall not occur within 18 inches (457 mm) of a transition from such metal pipe to PEX-AL-PEX pipe.

❖ The required heat for soldering fittings to copper pipe could melt the PEX-AL-PEX pipe and cause a leak if the soldering operation was too close to an existing PEX-AL-PEX to copper transition. This section is not intended to prohibit a completed, cooled soldered connection from being located adjacent to where a PEX-AL-PEX pipe is to be connected.

SECTION 1204
PIPE INSULATION

1204.1 Insulation characteristics. Pipe insulation installed in buildings shall conform to the requirements of the *Interna-*

tional Energy Conservation Code; shall be tested in accordance with ASTM E 84 or UL 723, using the specimen preparation and mounting procedures of ASTM E 2231; and shall have a maximum flame spread index of 25 and a smoke-developed index not exceeding 450. Insulation installed in an air *plenum* shall comply with Section 602.2.1.

Exception: The maximum flame spread index and smoke-developed index shall not apply to one- and two-family dwellings.

❖ The intent of this section is to conserve energy by controlling heat gain and loss in hydronic piping systems.

Insulation is not required on piping installed within mechanical equipment that is tested, labeled and meets the energy conservation performance requirements of the *International Energy Conservation Code*® (IECC®). The intent is to exempt piping within equipment that has been tested, and for which the piping heat losses or gains have been accounted for in the equipment ratings. Also, heat losses or gains occurring inside the equipment are assumed to add to its desired effect.

The IECC specifies piping insulation in Table 403.2.8 based on pipe size and pipe application.

The flame spread index and smoke-developed index are limited for all insulation types in all locations in all types of construction. Section 602.2.1 requires pipe insulation located in a plenum to have a flame spread index of 25 or less and a smoke-developed index of 50 or less.

ASTM E 84 and UL 723 provide the test methods required for pipe with insulation, including a requirement for testing of systems representative of the actual field installation. ASTM E 2231 provides the specimen preparation and mounting procedures necessary to ensure that the specimen tested in the laboratory is as close as possible to the actual field installation with respect to orientation and other factors of field installation. ASTM E 2231 is also required to be used in the testing of duct insulation systems in Chapter 6.

Note that the exception to this section exempts only one- and two-family dwellings from the flame spread and smoke-developed limitations for piping insulation. The requirements of the IECC for piping insulation in one- and two-family dwellings are not affected by the exception.

1204.2 Required thickness. Hydronic piping shall be insulated to the thickness required by the *International Energy Conservation Code*.

❖ See the commentary to Section 1204.1.

SECTION 1205
VALVES

1205.1 Where required. Shutoff valves shall be installed in hydronic piping systems in the locations indicated in Sections 1205.1.1 through 1205.1.6.

❖ Valves in a hydronic system have several uses. Valves are a fluid-controlling element in a hydronic system. Valves are required to isolate system components to facilitate repair, maintenance or replacement of system devices, components or piping. Valves are also used to take system components out of service temporarily.

This section addresses isolation valves only, and does not address pressure-reducing valves, relief valves (other than Section 1205.2), purge and vent valves, zone valves or flow control valves such as check valves, modulating valves, balancing valves, thermostatic control valves and similar devices. Sections 1205.1.1 through 1205.1.6 describe locations where valves are required to isolate the hydronic system components. The valves allow the isolation of system components without the need to shut down or drain the entire system.

Draining a water hydronic system causes air to enter the system, and will require that fresh water be introduced to refill the system. The time-consuming process of purging and bleeding air from the system and the corrosion problems associated with new water make it desirable to avoid system draining whenever possible.

1205.1.1 Heat exchangers. Shutoff valves shall be installed on the supply and return side of a heat exchanger.

Exception: Shutoff valves shall not be required when heat exchangers are integral with a boiler; or are a component of a manufacturer's boiler and heat exchanger packaged unit and are capable of being isolated from the hydronic system by the supply and return valves required by Section 1005.1.

❖ Shutoff valves must be installed at each connection to a heat exchanger. In a hydronic system, four valves will typically be required: one on the inlet and the outlet of the primary (supply) side of the heat exchanger, and one on the inlet and the outlet of the secondary (load) side of the heat exchanger. Heat exchangers include water-to-air, steam-to-air, water-to-water, steam-to-water and refrigerant-to-water types. Most heat exchangers will require cleaning, repair or other maintenance procedures at some time. Isolation valves facilitate these procedures (see Commentary Figure 1205.1.1).

1205.1.2 Central systems. Shutoff valves shall be installed on the building supply and return of a central utility system.

❖ Shutoff valves must be installed at each building on a multiple-building central utility system. A typical application of this requirement is a factory complex or large university campus consisting of buildings that are served by a central boiler/chiller (physical plant). Central systems typically use an underground system of distribution piping to supply steam or hot or chilled water to the buildings served. Utility-company-operated central systems were once common in the central business districts of cities (see Commentary Figure 1205.1.2).

1205.1.3 Pressure vessels. Shutoff valves shall be installed on the connection to any pressure vessel.

❖ Shutoff valves must be installed at every connection to pressure vessel, including storage tanks, expansion tanks, compression tanks, flash tanks, makeup water tanks, condensate tanks and similar tanks and receivers.

1205.1.4 Pressure-reducing valves. Shutoff valves shall be installed on both sides of a pressure-reducing valve.

❖ Shutoff valves must be used on both the inlet and the outlet side of pressure-reducing valves as shown in Commentary Figure 1205.1.4.

1205.1.5 Equipment and appliances. Shutoff valves shall be installed on connections to mechanical *equipment* and appliances. This requirement does not apply to components of a hydronic system such as pumps, air separators, metering devices and similar *equipment*.

❖ Shutoff valves must be installed at every connection to equipment in a hydronic system including boilers, chillers, condensate return systems, water conditioning equipment and cooling towers. Although pumps, circulators, air separators and metering devices are exempted from the requirement for isolation valves, these valves should be installed to allow the isolation of any device or component that will require servicing, repair or replacement at regular intervals. For example, pumps and circulators should be isolated by valves to allow repair or replacement without having to drain all or part of the entire hydronic system.

1205.1.6 Expansion tanks. Shutoff valves shall be installed at connections to nondiaphragm-type expansion tanks.

❖ A shutoff valve is required on the connection to any nondiaphragm expansion tank, as shown in Commentary Figure 1205.1.6.
An isolation valve is not required for connections to diaphragm-type expansion tanks because those tanks do not normally require service; however, in the event of tank failure, an isolation valve would simplify replacement.

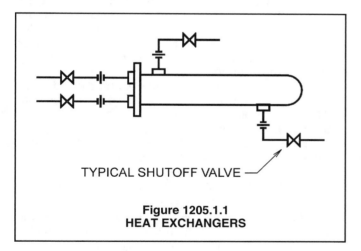

Figure 1205.1.1
HEAT EXCHANGERS

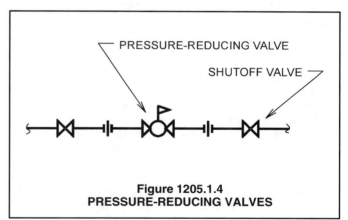

Figure 1205.1.4
PRESSURE-REDUCING VALVES

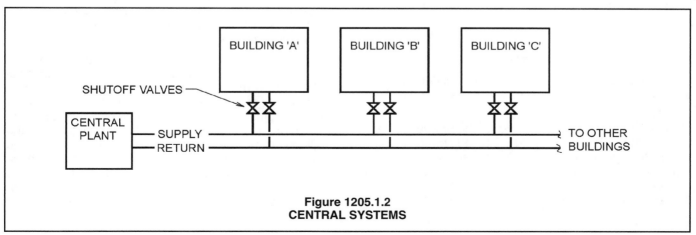

Figure 1205.1.2
CENTRAL SYSTEMS

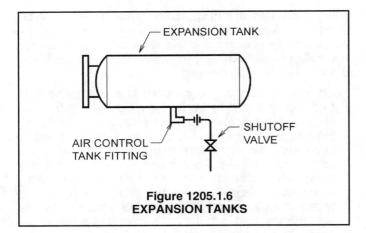

Figure 1205.1.6
EXPANSION TANKS

1205.2 Reduced pressure. A pressure relief valve shall be installed on the low-pressure side of a hydronic piping system that has been reduced in pressure. The relief valve shall be set at the maximum pressure of the system design. The valve shall be installed in accordance with Section 1006.

❖ System costs can be reduced if equipment is designed for lower working pressures. For example, in high-rise buildings, hydronic coils on lower floors would have to be constructed to withstand much higher pressures than coils on upper floors because of water column pressure (gravity). Reducing water pressure at given intervals allows use of coils constructed for lower pressure ratings, thereby reducing costs. The same would be true for the associated branch piping, fittings and valves.

Steam systems are commonly designed to supply steam at different pressures to various equipment, processes and distribution systems. A single boiler may supply high-pressure steam to process heat exchangers and may also supply reduced-pressure steam to a space-heating system.

Pressure-reducing valves can fail because of age and wear, or can be improperly adjusted and could cause the system downstream of the valve to experience higher or full supply (inlet) pressure. Excessive pressure could cause a hazardous failure of low-pressure components or piping. Installation of a pressure relief valve protects the system by relieving excess pressure before system failure could occur.

SECTION 1206
PIPING INSTALLATION

1206.1 General. Piping, valves, fittings and connections shall be installed in accordance with the conditions of approval.

❖ Conditions of approval include the manufacturer's installation instructions for labeled materials and approval granted by the code official where materials have been specifically approved for an application. Deviations from prescribed installation procedures may void the approval and be a basis for rejecting the installation.

1206.2 System drain down. Hydronic piping systems shall be designed and installed to permit the system to be drained. Where the system drains to the plumbing drainage system, the installation shall conform to the requirements of the *International Plumbing Code.*

Exception: The buried portions of systems embedded underground or under floors.

❖ To facilitate system repairs and maintenance, hydronic piping systems must be sloped and arranged to allow the transfer-medium fluids or condensate to be drained from the system. Each trapped section of the system piping must have drain cocks, unions or some other means of opening the system to drain it. Drainage discharge to the plumbing system must be by indirect connection in accordance with the IPC. The exception allows for underground or under slab portions of a hydronic system to not have a drain as this would be nearly impossible to accomplish.

1206.3 Protection of potable water. The potable water system shall be protected from backflow in accordance with the *International Plumbing Code.*

❖ Hydronic systems normally require a means of supplying fill and makeup water to replace any water lost to evaporation leakage or intentional draining. Where the system is connected directly to the potable water supply, the connections must to be isolated from the potable water source in accordance with the IPC. This provision is intended to protect the potable water system from contamination when a direct connection is made to a hydronic system.

Hydronic systems are normally pressurized, contain nonpotable water and fluids and can contain conditioning chemicals, cleaning chemicals or antifreeze solutions. Low-temperature hydronic fluids and cooling towers have also been associated with disease-causing organisms such as the Legionnaires' disease bacteria. The potable water system must be protected from potential contamination resulting from connection to hydronic systems.

1206.4 Pipe penetrations. Openings for pipe penetrations in walls, floors or ceilings shall be larger than the penetrating pipe. Openings through concrete or masonry building elements shall be sleeved. The annular space surrounding pipe penetrations shall be protected in accordance with the *International Building Code.*

❖ Openings for pipe penetrations must be larger than the penetrating pipe to prevent stress and physical damage to the pipe caused by expansion and contraction of the pipe and the building structure, and by the transmission of structural loads to the pipe. The annular space prevents abrasion and friction and allows some clearance to accommodate pipe and building movements. Fireblocking or annular space sealant must provide some degree of flexibility to accommodate expansion and contraction or other movement. Concrete and masonry penetrations must be sleeved to maintain the integrity of the material penetrated and to protect the pipe from the rough surfaces. Where required by the *International Building Code*® (IBC®), fireblocking must fill the annular space at the penetration of floors, walls and ceilings to prevent the passage of smoke and fire. When combustible piping penetrates a construction assembly, other forms of penetration protection could be required in accordance with the IBC.

1206.5 Clearance to combustibles. A pipe in a hydronic piping system in which the exterior temperature exceeds 250°F (121°C) shall have a minimum *clearance* of 1 inch (25 mm) to combustible materials.

❖ Maintaining a 1-inch (25 mm) airspace clearance allows some of the heat energy from the hydronic pipe to dissipate before reaching adjacent combustible materials. Continuous exposure to the heat produced by hydronic piping can chemically alter adjacent combustible materials, thereby lowering the ignition temperature and creating a potential fire hazard. Except for steam applications, temperatures near 250°F (121°C) are not typically found in hydronic systems. Even if the design temperature is less than 250°F (121°C), a clearance to combustibles should be maintained when higher temperatures are possible or when the maximum setpoint of system limit controls exceeds 250°F (121°C).

1206.6 Contact with building material. A hydronic piping system shall not be in direct contact with building materials that cause the piping material to degrade or corrode, or that interfere with the operation of the system.

❖ Piping must be protected to avoid external degradation resulting from corrosion, abrasion and chemical reactions. Allowing pipe to come in contact with certain construction materials can cause corrosion or degradation. For example, some plastic pipe materials are adversely affected by petroleum-based materials, and concrete can be corrosive to some metals.

Consideration must also be given to the placement of hydronic piping in contaminated or corrosive soils or environments. A protective coating is usually applied to pipe material that is sensitive to the surrounding conditions. The coating usually consists of coal tar, an epoxy or a plastic wrap or coating.

1206.7 Water hammer. The flow velocity of the hydronic piping system shall be controlled to reduce the possibility of water hammer. Where a quick-closing valve creates water hammer, an *approved* water-hammer arrestor shall be installed. The arrestor shall be located within a range as specified by the manufacturer of the quick-closing valve.

❖ Water hammer is a phenomenon that occurs when the kinetic energy of flowing fluids is dissipated in the form of shock waves.

The intensity of water hammer is directly proportional to the velocity of the flowing fluid and, therefore, the flow velocity must be controlled to reduce the intensity of water hammer.

The intent of this section is to protect the system and its components from the possible destructive forces that result from the rapid deceleration of fluid flow. The rule-of-thumb is that the force (shock wave) of water hammer is approximately equal to 60 times the change in velocity of the fluid. For example, a quick-closing valve can cause instantaneous pressures of 600 psi (4137 kPa) in a pipe with a flow velocity of 10 feet per second (3048 mm/s). The code requires that the system be designed for flow velocities that minimize the occurrence and magnitude of water hammer and, furthermore, requires installation of shock-arrestor devices where quick-closing valves are to be used.

A quick-closing valve would be any type of solenoid-actuated valve, spring-loaded self-closing valve or any other device capable of instantaneously reducing flow from full flow to no flow.

Mechanical water-hammer arresters help alleviate water hammer intensity by absorbing energy. When installed in system locations where velocity cannot be adequately controlled, mechanical arresters should be located where they have the maximum effect, which is usually achieved by locating the arrester close to the source of water hammer. Manufacturers typically supply guidelines for arrestor location.

1206.8 Steam piping pitch. Steam piping shall be installed to drain to the boiler or the steam trap. Steam systems shall not have drip pockets that reduce the capacity of the steam piping.

❖ As steam moves through piping, it loses some of its heat energy, and some of the steam will condense into liquid. The presence of liquid condensate in the steam lines interferes with proper steam flow and can cause noise, water hammer and accelerated pipe deterioration. Steam piping must drain to the boiler or to a steam trap.

A steam trap is an automatic valve that permits the passage of condensate while preventing the passage of steam through the valve. A typical steam trap design senses the presence of condensate by using a float or a thermostatic element to detect liquid accumulation or the lower temperature of the condensate. The float or thermostatic element opens and closes the valve to drain the collected condensate to the return system.

1206.9 Strains and stresses. Piping shall be installed so as to prevent detrimental strains and stresses in the pipe. Provisions shall be made to protect piping from damage resulting from expansion, contraction and structural settlement. Piping shall be installed so as to avoid structural stresses or strains within building components.

❖ Changes in temperature cause dimensional changes in all materials. The absorption of heat energy causes the atoms of a material to move apart, expanding the material in all directions with the greatest amount of expansion and contraction in piping occurring along the axial plane (its length). For systems operating at high temperatures, such as steam and hot water, the amount of expansion is high and significant movements can occur in short runs of piping.

Even though the amount of expansion per unit of length may be low for applications such as chilled and condenser water systems, which operate in the range of approximately 40°F to 100°F (4°C to 38°C), large movements can occur in long lengths of piping. Large forces can develop in restrained piping that increases or decreases in temperature. Note that the forces developed from contraction are just as large as those from expansion. See Commentary Table 1206.9 for a comparison of the amount of expansion for various piping materials.

In addition to other design requirements, piping systems must be capable of accommodating the forces resulting from thermal expansion and contraction. This section also intends to protect the structure from damage resulting from the forces of expanding and contracting piping.

Inadequate provisions to accommodate expansion and contraction can result in failure of pipe and supports, joint damage and leakage and the transmission of detrimental forces and stresses to connected equipment and building components.

Forces transmitted to rotating equipment, such as pumps or turbines, may deform the equipment case and cause bearing misalignment. Forces transmitted to the structure can cause excessive loading and damage to the members to which piping is anchored.

General practice is to never anchor or restrain a straight run of pipe at both ends. Ample movement allowance is attained by designing pipe bends and loops or supplemental devices, such as expansion joints, into the system.

Several means, such as "L," "Z" and "U" bends; packed and packless expansion joints; and flexible connectors [see Commentary Figures 1206.9(1), 1206.9(2) and 1206.9(3)] exist for accommodating expansion and contraction.

1206.9.1 Flood hazard. Piping located in a flood hazard area shall be capable of resisting hydrostatic and hydrodynamic loads and stresses, including the effects of buoyancy, during the occurrence of flooding to the *design flood elevation.*

❖ The requirements for floodproofing complement those contained in the IBC. Floodwaters can damage hydronic systems if the systems are not properly anchored and supported.

A flood hazard area is an area that is prone to flooding. Submersion in water can result in a buoyancy force being exerted on the various components of the hydronic piping system.

A flood hazard area may also be subject to wave action or high-velocity waters. In these areas, components must be protected from the dynamic forces of moving water. For additional guidance, refer to FEMA 348.

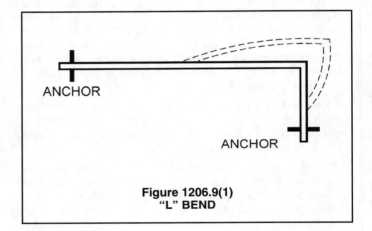

Figure 1206.9(1)
"L" BEND

Table 1206.9
PIPE EXPANSION

MATERIAL	COEFFICIENT OF LINEAR EXPANSION	EXPANSION OF 100 FEET Δ 100°F (IN.)
Cast iron	0.00000595	0.714
Steel	0.0000065	0.780
Copper	0.0000095	1.140
ABS	0.0000560	6.720
PVC Type 1	0.0000280	3.360
PVC Type II	0.0000555	6.600
Polyethylene	0.0001000	12.000
Polybutylene	0.0000710	8.520
Cross-linked polyethylene	0.0000788	9.336

For SI: 1 inch = 25.4 mm, 1 foot = 304.8 m, °C = [(°F)-32]/1.8.

$C \times L \times \Delta T$ = Expansion

where:

C = Coefficient of linear expansion.

L = Length of pipe in feet.

ΔT = Temperature difference.

1206.10 Pipe support. Pipe shall be supported in accordance with Section 305.

❖ This section contains piping support requirements. Hanger spacing intervals are given for the types of pipe presented in Section 1202.4. Cross-linked polyethylene pipe must be supported in compliance with the manufacturer's installation instructions because specialized support methods are needed to address the unique characteristics of the material. Piping supports must be installed to prevent damage to the piping caused by thermal expansion (see commentary, Section 1206.9).

As with all piping systems, support of the system is as important as any other part of the overall design. Proper supports are necessary to maintain piping alignment and slope, to support the weight of the piping and its contents, to control movement and to resist hydrodynamic loads, such as thrust.

Building design must take into consideration the structural loads created by hydronic piping systems.

Hangers or supports must not react with or be detrimental to the pipe they support. Hangers or supports for metallic pipe must be of a material that is compatible with the pipe to prevent corrosion. For example, copper, copper-clad or specially coated hangers are required if the hydronic distribution system is constructed of copper tubing (see commentary, Section 305).

1206.11 Condensation. Provisions shall be made to prevent the formation of condensation on the exterior of piping.

❖ Condensation on piping can cause corrosion to the piping as well as deterioration of pipe insulation and building materials exposed to the condensation (see commentary, Section 1206.6).

SECTION 1207
TRANSFER FLUID

1207.1 Flash point. The flash point of transfer fluid in a hydronic piping system shall be a minimum of 50°F (28°C) above the maximum system operating temperature.

❖ The "Flash point" of a fluid is defined in Chapter 2. If a heat transfer fluid with a low flash point is used in a hydronic system, a pressure relief valve discharge or a system leak could create a potential hazard. To avoid this possibility, the code requires that the flash point of the transfer fluid be at least 50°F (28°C) higher than the maximum possible temperature at which the system can operate.

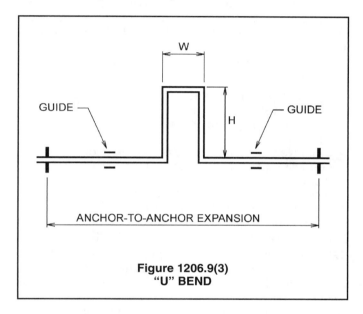

Figure 1206.9(3)
"U" BEND

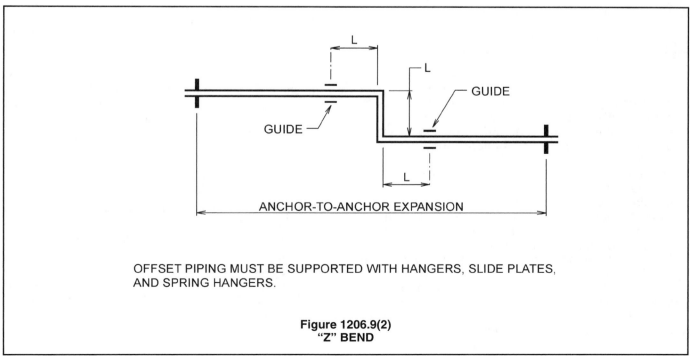

OFFSET PIPING MUST BE SUPPORTED WITH HANGERS, SLIDE PLATES, AND SPRING HANGERS.

Figure 1206.9(2)
"Z" BEND

1207.2 Makeup water. The transfer fluid shall be compatible with the makeup water supplied to the system.

❖ The transfer fluid used in a hydronic system must be compatible with the makeup water source for the system. Reactions causing excessive acidity, alkalinity or precipitation of compounds creating surface films that would impede heat transfer or which would otherwise interfere with system operation must be prevented.

SECTION 1208
TESTS

1208.1 General. Hydronic piping systems other than ground-source heat pump loop systems shall be tested hydrostatically at one and one half times the maximum system design pressure, but not less than 100 psi (689 kPa). The duration of each test shall be not less than 15 minutes. Ground-source heat pump loop systems shall be tested in accordance with Section 1208.1.1.

❖ Hydronic piping systems must be tested to determine that the system is leak free and capable of withstanding system operating pressures. This pressure will typically coincide with the setpoint pressure of the system's pressure relief devices. The code official will normally observe all required tests. Maintaining the test pressure for at least 15 minutes allows time for any leaks to be detected during the test.

Because of the way in which ground source heat pump loop systems are installed, a unique testing method is justified (see commentary, Section 1208.1.1).

1208.1.1 Ground source heat pump loop systems. Before connection (header) trenches are backfilled, the assembled loop system shall be pressure tested with water at 100 psi (689 kPa) for 30 minutes with no observed leaks. Flow and pressure loss testing shall be performed and the actual flow rates and pressure drops shall be compared to the calculated design values. If actual flow rate or pressure drop values differ from calculated design values by more than 10 percent, the problem shall be identified and corrected.

❖ Groundwater heat pump systems are either closed loop or open loop. Closed-loop systems circulate water or an antifreeze solution through plastic pipes buried in the ground. Some loop systems, although not as common, use surface water bodies or underground water as the source of heat energy. The heat transfer fluid absorbs heat from the ground (earth) and transfers it to the building in the heating season, then reverses to absorb heat from the building and transfers it to the ground in the cooling season.

As the name implies, ground source heat pumps use the mass of the soil (earth) as a source of thermal energy and as a heat sink. Below the frost line, the ground temperature remains fairly constant at average temperatures near 50°F (10°C). For a heat pump, this is a rich source of heat and an excellent sink for rejection of heat. The heat pump is similar to the air-source type except that it is coupled to the soil instead of to the ambient outdoor air.

Ground source loop systems use significant lengths of piping installed in multiple circuits (loops) to achieve the required area of heat exchange surface. Also, multiple loops have the advantage over a single-loop system because a single piping failure will not cause failure of the entire heat pump system.

Heat transfer through plastic piping is relatively slow compared to transfer through metallic piping. Larger surface areas are therefore necessary to transfer the heat into or out of the loop-system heat transfer fluid.

In the heating season, the ground source heat pump uses a refrigeration cycle machine to extract heat from the loop-system fluid, concentrates the heat to usable temperatures and transfers the heat to the air or domestic water to be heated in the building.

In the cooling season, the machine reverses the evaporator and condenser elements of the refrigeration system to extract heat from the environmental air and rejects this heat to the ground loop fluid.

Ground-source heat pump systems supply conditioned air and/or domestic water heating. Because the earth temperature is stable, these systems are not affected by changes in the ambient air temperatures and can, therefore, maintain a high coefficient of performance year round.

The significant difference between testing ground source loop systems and other hydronic systems is that ground source loops may be buried prior to testing. The connection headers (manifolds) that connect the loops (circuits) of piping must remain exposed until completion of testing. Because loops (circuits) are typically installed without joints, all of the potential sources of leaks will be located at the point of connection between the ends of the loops and the header. For this reason, the loops can be covered during testing while the headers remain exposed. Also, it is difficult to dig trenches for large loop fields without backfilling each trench before digging the adjacent trenches.

Depending on the lot, it may be necessary to backfill "as you go" to allow room for the excavator machine. Requiring that each loop be inspected prior to covering could require time-consuming multiple inspections at the same property if the excavator had to backfill each trench before digging the next.

SECTION 1209
EMBEDDED PIPING

1209.1 Materials. Piping for heating panels shall be standard-weight steel pipe, Type L copper tubing, polybutylene or other *approved* plastic pipe or tubing rated at 100 psi (689 kPa) at 180°F (82°C).

❖ This section is more limiting than Section 1202.4 because it specifies only three material choices for hydronic piping intended for embedment in concrete or plaster.

1209.2 Pressurizing during installation. Piping to be embedded in concrete shall be pressure tested prior to pouring concrete. During pouring, the pipe shall be maintained at the proposed operating pressure.

❖ Testing of piping is necessary to disclose any defects in the system. This testing is especially important where the piping is to be encased and thereby made inaccessible.

1209.3 Embedded joints. Joints of pipe or tubing that are embedded in a portion of the building, such as concrete or plaster, shall be in accordance with the requirements of Sections 1209.3.1 through 1209.3.3.

❖ Sections 1209.3.1 through 1209.3.3 address joining means for the materials specified in Section 1209.1.

1209.3.1 Steel pipe joints. Steel pipe shall be welded by electrical arc or oxygen/acetylene method.

❖ See the commentary to Section 1203.3.6.

1209.3.2 Copper tubing joints. Copper tubing shall be joined by brazing with filler metals having a melting point of not less than 1,000°F (538°C).

❖ See the commentary to Section 1203.3.1.

1209.3.3 Polybutylene joints. Polybutylene pipe and tubing shall be installed in continuous lengths or shall be joined by heat fusion in accordance with Section 1203.10.1.

❖ See the commentary to Section 1203.10.1.

1209.4 Not embedded related piping. Joints of other piping in cavities or running exposed shall be joined by *approved* methods in accordance with manufacturer's installation instructions and related sections of this code.

❖ See the commentary to Section 1203.3.

1209.5 Thermal barrier required. Radiant floor heating systems shall be provided with a thermal barrier in accordance with Sections 1209.5.1 through 1209.5.4.

Exception: Insulation shall not be required in engineered systems where it can be demonstrated that the insulation will decrease the efficiency or have a negative effect on the installation.

❖ In order for the maximum amount of heat to be transferred to the space intended to be heated, the hydronic piping system must be insulated to limit the amount of heat transferred to spaces or materials that are not intended to be heated. Sections 1209.5.1 and 1209.5.2 describe two applications where this is a concern and provides the insulation requirements.

1209.5.1 Slab-on-grade installation. Radiant piping utilized in slab-on-grade applications shall be provided with insulating materials installed beneath the piping having a minimum *R*-value of 5.

❖ Typically, rigid foam board insulation is used for insulating slabs on grade that will be heated by hydronic systems. For most manufacturers of this product, an *R*-value of 5 equates to a 1-inch (25 mm) material thickness. Other types of insulation products that are

suitable for under slab installation are also available and can be used if marked in accordance with Section 1209.5.4.

1209.5.2 Suspended floor installation. In suspended floor applications, insulation shall be installed in the joist bay cavity serving the heating space above and shall consist of materials having a minimum *R*-value of 11.

❖ Typically, fiberglass insulation is used for insulating hydronically heated floors that are suspended above an unheated space such as a crawl space or basement. Insulation is placed between the joists (or floor trusses) and should be held in place by an approved means.

1209.5.3 Thermal break required. A thermal break shall be provided consisting of asphalt expansion joint materials or similar insulating materials at a point where a heated slab meets a foundation wall or other conductive slab.

❖ The exterior edges of heated concrete slabs must be protected against heat loss. Placement of a thermal barrier (break) at the edge of the slab where it would contact the foundation wall breaks the heat conduction path to the exterior.

1209.5.4 Thermal barrier material marking. Insulating materials utilized in thermal barriers shall be installed such that the manufacturer's *R*-value mark is readily observable upon inspection.

❖ Insulation materials must be marked so that the inspector can verify that the correct *R*-value is being used for the application. Readily observable means that the inspector must not be required to remove insulation in order to see the marking. The *R*-value marking requirement might be a problem where unfaced fiberglass insulation is used or where the facing of the fiberglass insulation is required to be located such that the marking is not observable without removal of the insulation.

Bibliography

The following resource materials were used in the preparation of the commentary for this chapter of the code:

IBC-12, *International Building Code.* Washington, DC: International Code Council, 2011.

IECC-12, *International Energy Conservation Code.* Washington, DC: International Code Council, 2011.

IPC-12, *International Plumbing Code.* Washington, DC: International Code Council, 2011.

Chapter 13:
Fuel Oil Piping and Storage

General Comments

Chapter 13 regulates the design and installation of fuel oil storage and piping systems. The regulations include reference to construction standards for above-ground and underground storage tanks, material standards for piping systems (both above ground and underground) and extensive requirements for the proper assembly of system piping and components. The *International Fire Code®* (IFC®) covers items not addressed in detail here.

The importance of adequate regulations for the storage, handling and use of combustible liquid fuel cannot be overstated. A fuel-oil piping system includes the piping, valves and fittings. Also included in the fuel-oil piping system, if they are installed, are pumps, reservoirs, regulators, strainers, filters, relief valves, oil preheaters, controls and gauges. The piping materials and the method of joining sections of pipe must be approved and all other components included in the piping system should be labeled and installed in accordance with the manufacturer's instructions. Over the past 10 years, public scrutiny of flammable and combustible liquid storage installations has increased with the public's awareness of the consequences of release of these liquids into the environment. Improper installation of these storage systems has been shown by environmental studies to be a major contributing factor in system failure. Improper use of such systems also accounts for many fire losses. Although the hazards of combustible liquids are well known, accidents involving them remain one of the most common fire scenarios in the United States.

Fuel oil is considered a Class II combustible liquid, having a flash point of 100°F (38°C) or higher. Though the classification boundaries are somewhat arbitrary, flammable and combustible liquids are distinguished by their flash points. The flash point is that temperature at which the liquid produces sufficient vapor to form an ignitable vapor-air mixture above its surface. Because Class I flammable liquids all have a flash point below 100°F (38°C), it is prudent to assume that those liquids may be capable of igniting when unconfined under normal environmental conditions. On the other hand, combustible liquids, including fuel oil, are materials with flash points above 100°F (38°C) and must usually be heated above their flash points or, in the case of extremely high flashpoint liquids, above their boiling points before they will ignite.

Combustible liquids possess other characteristics besides their flash points. Significant characteristics when evaluating relative fire hazards include ignition temperature, auto-ignition temperature, flammable (explosive) range, viscosity, vapor density, vapor pressure, boiling point, evaporation rate, specific gravity and water solubility. Once the liquid is ignited, these variables have little influence over the material's heat release rate. Factors such as evaporation rate, viscosity and water solubility may profoundly affect how these fires are extinguished.

Generally, combustible liquids have low specific gravities, high vapor densities and narrow flammable ranges. These characteristics mean that liquids usually float on water, vapors will typically hug the ground and ignitable vapor-air mixtures are generally confined to a range between 6 and 15 percent in air. Thus, smothering is difficult and ignition sources near the ground are more likely to pose a hazard.

Purpose

The provisions in this chapter are intended to prevent fires, leaks and spills involving fuel oil storage and piping systems whether inside or outside of structures and above or underground.

SECTION 1301
GENERAL

1301.1 Scope. This chapter shall govern the design, installation, construction and repair of fuel-oil storage and piping systems. The storage of fuel oil and flammable and combustible liquids shall be in accordance with Chapters 6 and 57 of the *International Fire Code*.

❖ Regulations governing the design, installation and repair of combustible liquid storage and piping systems are contained in both this code and the IFC. This chapter covers only fuel-oil storage and piping systems and only within the quantity limits of the IFC.

In the context of this chapter, fuel oil is a combustible liquid used as fuel for applications, including space heating, water heating, processing and generation of electricity.

1301.2 Storage and piping systems. Fuel-oil storage systems shall comply with Section 603.3 of the *International*

Fire Code. Fuel-oil piping systems shall comply with the requirements of this code.

❖ Combustible liquids have had a variety of uses for centuries. They can also present problems in the form of fires or environmental pollution if they are not safely contained within proper storage and distribution facilities. In addition to this chapter, the provisions contained in the IFC apply to combustible liquid storage and piping systems. Also included are provisions for containers and portable tanks and storage requirements for a variety of occupancies. The IFC provides for the installation and venting of most oil-fired stationary equipment, including residential, industrial and commercial oil-fired appliances. The IFC also regulates the design and installation of above-ground tanks and piping for fuel oils, both inside and outside of structures.

1301.3 Fuel type. An *appliance* shall be designed for use with the type of fuel to which it will be connected. Such *appliance* shall not be converted from the fuel specified on the rating plate for use with a different fuel without securing reapproval from the code official.

❖ This section contains the same provisions found in Section 301.9 (see commentary, Section 301.9).

1301.4 Fuel tanks, piping and valves. The tank, piping and valves for appliances burning oil shall be installed in accordance with the requirements of this chapter. When an oil burner is served by a tank, any part of which is above the level of the burner inlet connection and where the fuel supply line is taken from the top of the tank, an *approved* antisiphon valve or other siphon-breaking device shall be installed in lieu of the shutoff valve.

❖ Section 1307.1 requires that a shutoff valve be installed in the fuel-oil supply line. This section is essentially an exception to that requirement in cases where a burner fuel pump draws from the top of the tank and when the tank (or any part of it) is above the level of the burner fuel inlet connection. In such systems, an antisiphon device must be installed to break the siphon action in the pipe in the event of a piping failure to the burner. Using the antisiphon, the amount of fuel oil subject to spillage in the event of a piping failure is limited to the contents of the piping only, not the entire tank contents.

1301.5 Tanks abandoned or removed. All exterior above-grade fill piping shall be removed when tanks are abandoned or removed. Tank abandonment and removal shall be in accordance with Section 5704.2.13 of the *International Fire Code.*

❖ Section 5704.2.13 of the IFC provides the requirements for abandoning or removing fuel oil tanks. It does not, however, require the removal of the above-grade fill piping associated with the tanks. Although the code does not specifically address vent pipes, they should also be removed because they appear to be fill pipes.

Exterior fill piping has to be removed due to the potential danger that hundreds of gallons of fuel oil could be accidentally delivered to a facility that did not request it. There have been instances of accidental filling where tanks have been removed but the fill pipe has remained. Oil delivery service personnel only see the exterior connection and may not be aware that the tank has been removed or disconnected. Piping must be removed, not just capped off; accidental filling has occurred when piping systems have been capped off. When fuel oil contamination occurs, the cost of repairs and cleanup is extremely high and, in some cases, has led to condemnation of such structures.

Commentary Figure 1301.5(1) shows a typical oil tank installation in a basement with the oil filler pipe and the vent pipe extending above ground outside the building. If the tank is removed and the fill pipe is left in place [see Commentary Figure 1301.5(2)], oil pumped through one of these pipes will fill the basement.

SECTION 1302
MATERIAL

1302.1 General. Piping materials shall conform to the standards cited in this section.

❖ This section regulates the materials that are permitted for use in combustible liquid piping systems.

1302.2 Rated for system. All materials shall be rated for the operating temperatures and pressures of the system, and shall be compatible with the type of liquid.

❖ System operating and standby pressures and temperatures all vary widely depending on the application. All portions of a combustible liquid piping and storage system must be compatible with the liquid and rated for its operating temperatures and pressures. The lowest rating of any of the components establishes the operating limitations of the system. It should also be noted that pressure and temperature can vary significantly within a system, and the specification of different materials within the system may be necessary.

1302.3 Pipe standards. Fuel oil pipe shall comply with one of the standards listed in Table 1302.3.

❖ Table 1302.3 lists the materials that are allowed for fuel oil piping. The referenced standards regulate the physical and mechanical properties of the materials used to fabricate such pipe and regulate the quality and dimensioning of the pipe.

TABLE 1302.3. See page 13-3.

❖ Table 1302.3 lists the design standards for the piping materials permitted for use with fuel oil. When selecting or approving piping materials for a given application, the material must be suitable for the temperatures and pressures generated within the system and must be compatible with the combustible liquid it will carry. In above-ground installations, additional consideration should be given to the potential fire exposure to the pipe. Brass, copper and nonmetallic pipe materials that have low melting points should be protected from failure caused by fire exposure. Installations should also take into account leak management in the event of a pipe failure caused by fire exposure (see commentary, Section 1302.4).

TABLE 1302.3
FUEL OIL PIPING

MATERIAL	STANDARD (see Chapter 15)
Brass pipe	ASTM B 43
Brass tubing	ASTM B 135
Copper or copper-alloy pipe	ASTM B 42; ASTM B 302
Copper or copper-alloy tubing (Type K, L or M)	ASTM B 75; ASTM B 88; ASTM B 280
Labeled pipe	(See Section 1302.4)
Nonmetallic pipe	ASTM D 2996
Steel pipe	ASTM A 53; ASTM A 106
Steel tubing	ASTM A 254; ASTM A 539

1302.4 Nonmetallic pipe. All nonmetallic pipe shall be *listed* and *labeled* as being acceptable for the intended application for flammable and combustible liquids. Nonmetallic pipe shall be installed only outside, underground.

❖ Not all nonmetallic piping materials are suitable for use with fuel oil. The labeling process takes compatibility issues into account and provides the code official with a sound basis for approving proposed piping. Nonmetallic pipe may be used to carry liquids or to provide secondary containment protection for metallic or nonmetallic liquid-carrying piping. Whatever its actual use or product compatibility, the agency label

will specify the pipe's suitability for use in a given application. Because of its susceptibility to softening and failure if exposed to a fire, nonmetallic piping may be used only underground. To avoid piping failure caused by loads and stresses encountered within the perimeter of a structure, the underground use of nonmetallic pipe is further limited to areas outside of structures.

1302.5 Fittings and valves. Fittings and valves shall be *approved* for the piping systems, and shall be compatible with, or shall be of the same material as, the pipe or tubing.

❖ Fittings include couplings, ells, tees, unions, caps, plugs, adapters and mechanical connectors. Fittings are used to join flammable or combustible liquid piping segments to each other or to other mechanical devices and equipment. Some fittings and couplings are proprietary and must be approved only where their use is consistent with the fitting manufacturer's instructions. Fittings must be designed for the application and the type of piping used. To avoid galvanic corrosion between dissimilar metals, fittings must be of the same material as the pipe or must be compatible with the piping material.

1302.6 Bending of pipe. Pipe shall be *approved* for bending. Pipe bends shall be made with *approved equipment*. The bend shall not exceed the structural limitations of the pipe.

❖ Pipe bending is done to accomplish changes in direction without the use of fittings. Pipe bending must be done with the appropriate bending tools and materials. Bending tools are designed to make bends without damaging the pipe. Pipe intended to be bent has a minimum bend radius that must be observed to

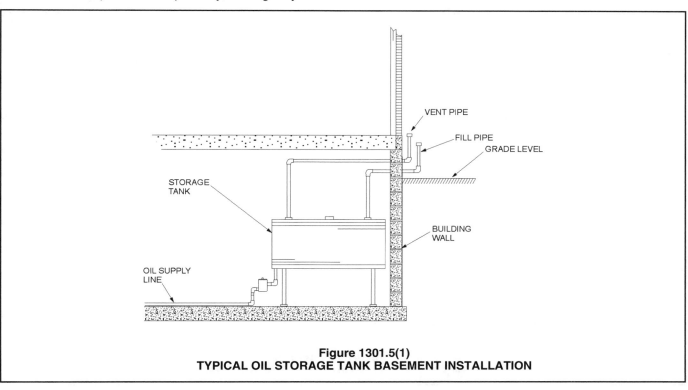

Figure 1301.5(1)
TYPICAL OIL STORAGE TANK BASEMENT INSTALLATION

avoid damage to the pipe. For example, bending welded seam pipe with the seam located outside of the neutral axis of the bend may result in a split seam because of the stresses induced by the bend. If pipe is to be bent, it must be confirmed in the pipe specifications that the pipe is suitable for bending. Rigid combustible liquid piping is not commonly bent because of the perceived risk of pipe stress failures at the bend.

1302.7 Pumps. Pumps that are not part of an *appliance* shall be of a positive-displacement type. The pump shall automatically shut off the supply when not in operation. Pumps shall be *listed* and *labeled* in accordance with UL 343.

❖ Unless they are part of a labeled appliance, pumps must be of the positive-displacement type, which will act similarly to a valve by not allowing gravity or siphon delivery of the liquid when the pump is not running. As with any mechanical appliance, pumps must be listed and labeled and must be installed in accordance with the manufacturer's instructions.

1302.8 Flexible connectors and hoses. Flexible connectors and hoses shall be *listed* and *labeled* in accordance with UL 536.

❖ Flexible connectors (either metallic or nonmetallic), flexible metal hose and hose-type pipe connectors are used in combustible liquid piping systems where flexibility is needed, such as at points where a piping system connects to a pump, tank or dispensing device. Hose-type pipe connectors for fuel-oil service and flexible connectors for motor fuel service are limited to 8 feet (2438 mm) in length and are intended for use in underground applications. Metallic flexible

connectors may be obtained with integral corrosion protection in the form of preengineered sacrificial anodes. Labeling by an approved agency provides the code official with evidence that the devices comply with UL 536 and liquid compatibility criteria.

SECTION 1303
JOINTS AND CONNECTIONS

1303.1 Approval. Joints and connections shall be *approved* and of a type *approved* for fuel-oil piping systems. All threaded joints and connections shall be made tight with suitable lubricant or pipe compound. Unions requiring gaskets or packings, right or left couplings, and sweat fittings employing solder having a melting point of less than 1,000°F (538°C) shall not be used in oil lines. Cast-iron fittings shall not be used. Joints and connections shall be tight for the pressure required by test.

❖ Joining and connecting methods and materials must be compatible with the piping used and must be approved for use in fuel-oil system applications. Acceptable joining methods for each piping material are listed under each piping material category. Thread lubricants (pipe-joint compounds) and tape are designed to lubricate the threaded joint for proper thread mating and also to fill in small imperfections on the surfaces of the threads. In choosing thread lubricants, care must be taken that the lubricant is suitable for use with the type of piping material being joined. Lubricant manufacturers' instructions are a good source for this information. Prohibition of certain types of pipe-joining methods, materials and components is intended to enhance the integrity of the pip-

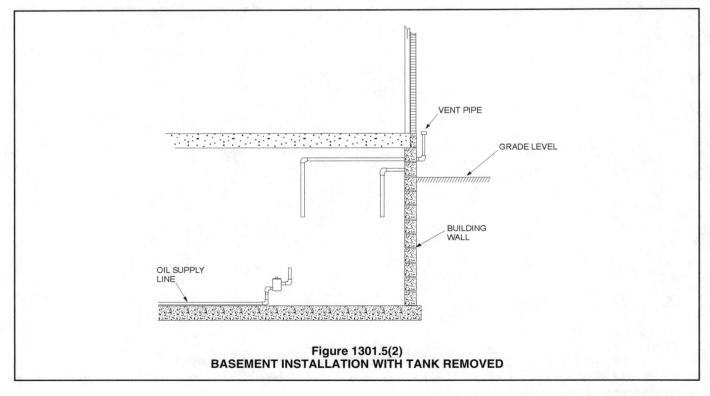

Figure 1301.5(2)
BASEMENT INSTALLATION WITH TANK REMOVED

ing system, especially where piping is installed indoors where it may be exposed to a fire, or to reduce potential maintenance problems.

Gaskets and packing materials in unions can lose their resiliency and deteriorate over time, which can lead to fuel-oil leaks. Use of right or left couplings can also lead to piping leaks as a result of their tendency to be inadvertently loosened. Solder with a melting point of less than 1,000°F (538°C) used in sweat fittings could quickly fail in a fire, allowing fuel oil to leak and contribute to the intensity of the fire. Cast-iron fittings are prohibited because of the brittleness of the material, susceptibility to impact damage and inability to withstand imposed stresses without damage. NFPA 31 contains the testing requirements for fuel-oil tank and piping systems.

1303.1.1 Joints between different piping materials. Joints between different piping materials shall be made with *approved* adapter fittings. Joints between different metallic piping materials shall be made with *approved* dielectric fittings or brass converter fittings.

❖ Galvanic corrosion results from electrical currents induced by the juncture of certain metals in the presence of an electrolyte such as water or soil. To avoid the corrosion problems with joining piping of dissimilar metals, dielectric fittings or approved adapter fittings must be used to interrupt the electrical continuity or substantially reduce galvanic currents. Dielectric unions typically use elastomeric gaskets and are therefore prohibited by Section 1303.1.

1303.2 Preparation of pipe ends. All pipe shall be cut square, reamed and chamfered and be free of all burrs and obstructions. Pipe ends shall have full-bore openings and shall not be undercut.

❖ This section prescribes what is considered to be good workmanship for the installation of piping systems. Burrs and obstructions must be removed to minimize flow resistance and to prevent interference with the method of joining the pipe (see commentary, Section 1203.2).

1303.3 Joint preparation and installation. Where required by Sections 1303.4 through 1303.10, the preparation and installation of brazed, mechanical, threaded and welded joints shall comply with Sections 1303.3.1 through 1303.3.4.

❖ The following sections regulate the installation of pipe joints.

1303.3.1 Brazed joints. All joint surfaces shall be cleaned. An *approved* flux shall be applied where required. The joints shall be brazed with a filler metal conforming to AWS A5.8.

❖ Brazing should be performed in accordance with AWS B2.2.

1303.3.2 Mechanical joints. Mechanical joints shall be installed in accordance with the manufacturer's instructions.

❖ Mechanical joints are usually proprietary joints that are developed and marketed by individual manufac-

turers. Many types of mechanical joints use a sleeve or ferrule that is compressed around the circumference of the pipe or tube. Mechanical joints must be specifically designed for and compatible with the type of pipe or tube being joined.

1303.3.3 Threaded joints. Threads shall conform to ASME B1.20.1. Pipe-joint compound or tape shall be applied on the male threads only.

❖ The referenced standard addresses tapered pipe threads that, when made up, form a metal-to-metal (interference fit) seal. Pipe-joint compound or tape is limited to application on the male threads only to decrease the possibility of tape fragments or compound entering the piping system. Such debris in piping systems can block orifices, restrict flow or interfere with the operation of controls. The primary purpose of pipe-joint compounds is to act as a lubricant to allow proper tightening and to achieve a metal-to-metal seal. They also fill in small imperfections on the threaded surfaces. Pipe-joint compounds and tapes must be compatible with both the piping material and the contents of the piping.

1303.3.4 Welded joints. All joint surfaces shall be cleaned by an *approved* procedure. The joint shall be welded with an *approved* filler metal.

❖ Welding should be done in accordance with AWS B2.1. Welding is usually performed on larger diameter, high-pressure piping systems.

1303.4 Brass pipe. Joints between brass pipe or fittings shall be brazed, mechanical, threaded or welded joints complying with Section 1303.3.

❖ Brass pipe has a high enough melting point and walls of sufficient thickness to permit threaded, welded, brazed and mechanical joints.

1303.5 Brass tubing. Joints between brass tubing or fittings shall be brazed or mechanical joints complying with Section 1303.3.

❖ Sections 1303.3.1 and 1303.3.2 describe the allowable joining methods for brass tubing.

1303.6 Copper or copper-alloy pipe. Joints between copper or copper-alloy pipe or fittings shall be brazed, mechanical, threaded or welded joints complying with Section 1303.3.

❖ Copper pipe can be joined by the same methods as brass pipe. Note that there is a distinction between copper pipe and copper tubing.

1303.7 Copper or copper-alloy tubing. Joints between copper or copper-alloy tubing or fittings shall be brazed or mechanical joints complying with Section 1303.3 or flared joints. Flared joints shall be made by a tool designed for that operation.

❖ Because tubing has a thinner wall than piping, welded joints and threaded joints are not feasible. Tubing can be flared only if it is tempered as suitable for bending. Soldered joints are not permitted.

1303.8 Nonmetallic pipe. Joints between nonmetallic pipe or fittings shall be installed in accordance with the manufacturer's instructions for the *labeled* pipe and fittings.

❖ To avoid leakage, nonmetallic piping materials and fittings must be joined in strict accordance with the manufacturer's instructions. Nonmetallic pipe and fittings are intended to be field assembled using an adhesive supplied by the pipe manufacturer. The joining methods, adhesives and instructions are all evaluated as part of the approved agency listing and labeling process.

1303.9 Steel pipe. Joints between steel pipe or fittings shall be threaded or welded joints complying with Section 1303.3 or mechanical joints complying with Section 1303.9.1.

❖ Joints for steel pipe must be threaded, welded or mechanical. The choice of a joint type is usually dictated by the application. For example, it is not practical to thread larger diameter steel pipe; therefore, a welded joint is used.

1303.9.1 Mechanical joints. Joints shall be made with an *approved* elastomeric seal. Mechanical joints shall be installed in accordance with the manufacturer's instructions. Mechanical joints shall be installed outside, underground, unless otherwise *approved*.

❖ Mechanical joints for steel pipe are usually proprietary joints using an elastomeric seal that is compressed around the circumference of the pipe. Because elastomeric seals are susceptible to failure from heat and aging, and because the joints must be restrained, mechanical joints for steel pipe must be limited to outdoor, underground applications.

1303.10 Steel tubing. Joints between steel tubing or fittings shall be mechanical or welded joints complying with Section 1303.3.

❖ Mechanical joints for steel tubing are typically made with compression fittings. A metallic ferrule is placed around the outside of the pipe and is compressed between the pipe and the fitting by the tightening of a nut.

1303.11 Piping protection. Proper allowance shall be made for expansion, contraction, jarring and vibration. Piping other than tubing, connected to underground tanks, except straight fill lines and test wells, shall be provided with flexible connectors, or otherwise arranged to permit the tanks to settle without impairing the tightness of the piping connections.

❖ Changes in temperature cause dimensional changes in all materials. The absorption of heat energy causes the atoms of a material to move apart, thus expanding the material in all directions. The greatest amount of thermal expansion and contraction in piping will occur along its length. For systems operating at high temperatures, the amount of expansion is high and significant movements can occur in short runs of piping. Even though the amount of expansion per unit of length may be low for applications that operate in the range of 40°F to 100°F (4°C to 38°C), large move-

ments can occur in long pipe runs. Large forces can develop in restrained piping that has increased or decreased in temperature. Note that the forces developed from contraction are identical to those from expansion. In addition to other design requirements, piping systems must withstand the forces resulting from thermal expansion and contraction.

This section also intends to protect the connected tank itself from damage resulting from the forces of expanding and contracting piping. Inadequate provisions for expansion and contraction can result in failure of pipe and supports, joint damage and leakage and the transmission of harmful forces and stresses to connected equipment. Forces transmitted to rotating equipment, such as pumps or turbines, may deform the equipment case and cause bearing misalignment. Forces transmitted to the structure can cause excessive loading and damage to the members to which piping is anchored. General practice is to never anchor or restrain a straight run of pipe at both ends. Ample movement allowance is attained by designing pipe bends and loops or supplemental devices, such as expansion joints, into the system. Several means exist for accommodating expansion and contraction, such as "L," "Z" and "U" bends, specially designed expansion joints and flexible connectors. Underground piping must be protected from the stress loading effects that result from settlement of tanks. Frost heave and differential movements among underground piping, structure foundations and tanks can subject piping to bending, tension, compression and shear forces.

SECTION 1304
PIPING SUPPORT

1304.1 General. Pipe supports shall be in accordance with Section 305.

❖ Piping supports are intended to reduce sag and stress in the piping system. The maximum amount of sag occurs at a point halfway between supports (see commentary, Section 305).

SECTION 1305
FUEL OIL SYSTEM INSTALLATION

1305.1 Size. The fuel oil system shall be sized for the maximum capacity of fuel oil required. The minimum size of a supply line shall be $^3/_8$-inch (9.5 mm) inside diameter nominal pipe or $^3/_8$-inch (9.5 mm) od tubing. The minimum size of a return line shall be $^1/_4$-inch (6.4 mm) inside diameter nominal pipe or $^5/_{16}$-inch (7.9 mm) outside diameter tubing. Copper tubing shall have 0.035-inch (0.9 mm) nominal and 0.032-inch (0.8 mm) minimum wall thickness.

❖ This section requires that the fuel-oil supply piping be of a size capable of delivering the maximum fuel demand of the appliance it supplies at the recommended working pressure of the appliance burner.

For purposes of flow volume and pressure control, pumps may supply more fuel oil than an appliance consumes and the excess oil must be routed back to the tank through a return line. Because the amount returning to the tank is less than what was supplied, the return line pipe diameter may be smaller than that of the supply pipe.

1305.2 Protection of pipe, equipment and appliances. All fuel oil pipe, *equipment* and appliances shall be protected from physical damage.

❖ To minimize the possibility of a hazardous fuel-oil spill, fuel system piping and the equipment and appliances it supplies must be protected from physical damage. Piping, equipment and appliances should not be located where they can be contacted by occupants or vehicles, such as cars or lawn mowers. Piping must not be located where it will be stepped on or create a tripping hazard.

1305.2.1 Flood hazard. All fuel oil pipe, equipment and appliances located in flood hazard areas shall be located above the elevation required by Section 1612 of the *International Building Code* for utilities and attendant equipment or shall be capable of resisting hydrostatic and hydrodynamic loads and stresses, including the effects of buoyancy, during the occurrence of flooding up to such elevation.

❖ See commentary for Section 301.13 and Section 1206.9.1. For additional guidance, refer to FEMA 348.

1305.3 Supply piping. Supply piping shall connect to the top of the fuel oil tank. Fuel oil shall be supplied by a transfer pump or automatic pump or by other *approved* means.

Exception: This section shall not apply to inside or above-ground fuel oil tanks.

❖ This section applies to underground tanks located outside buildings or structures. Tanks installed above ground either inside or outside a structure may supply fuel oil by gravity through approved fittings in the tank below the liquid level. Underground tanks, however, must have all piping connections made to approved fittings in the top of the tank above liquid level and must supply the appliance using a pump or other transfer device. The reasons for these differences are that above-ground tanks are limited in size by the IFC and an exposed gravity connection that might leak will be readily observable. Underground tanks, on the other hand, are not limited in size and a leak would not be readily discovered underground.

1305.4 Return piping. Return piping shall connect to the top of the fuel oil tank. Valves shall not be installed on return piping.

❖ Return piping is installed in systems fed by a supply pump taking suction from a tank. Because underground leaks are difficult to discover and repair, such connections must be made above the liquid level.

Also, a failure in a return line would not allow gravity leakage if connected to the top of a tank. A valve in a return line could inadvertently be closed causing abnormally high pressures in the piping and appliance.

1305.5 System pressure. The system shall be designed for the maximum pressure required by the fuel-oil-burning *appliance*. Air or other gases shall not be used to pressurize tanks.

❖ As with the system sizing requirements of Section 1305.1, the system must be capable of withstanding the maximum operating pressure of the equipment and appliance(s) it supplies.

1305.6 Fill piping. A fill pipe shall terminate outside of a building at a point at least 2 feet (610 mm) from any building opening at the same or lower level. A fill pipe shall terminate in a manner designed to minimize spilling when the filling hose is disconnected. Fill opening shall be equipped with a tight metal cover designed to discourage tampering.

❖ Requiring fill-pipe terminations to be outside buildings and at least 2 feet (610 mm) from building openings that are at the same level or lower than the fill pipe reduces the possibility that liquid or vapor spillage during delivery will enter the building and allow vapors to come into contact with a source of ignition. Likewise, fill pipes must be arranged in a way that will reduce the possibility of liquid spillage on the ground or pavement upon completion of the filling operation that could result in the liberation of vapors that could travel to an ignition source. One method of managing such spills is to install a spill containment device at the end of the fill pipe. These devices are typically equipped with a water-tight cover and are essentially a reservoir designed to catch any spilled fuel oil and retain it for proper disposal. Some models of these devices have a manual drain valve that allows the captured liquid to drain through the fill pipe and into the tank, while other models have a liquid level alarm to warn responsible parties on the premises that the reservoir needs to be emptied.

Although not required by the code, it is a good idea to remove abandoned (retired) fill piping to prevent fuel oil from being mistakenly put into a basement or an out-of-service (retired) storage tank.

1305.7 Vent piping. Liquid fuel vent pipes shall terminate outside of buildings at a point not less than 2 feet (610 mm) measured vertically or horizontally from any building opening. Outer ends of vent pipes shall terminate in a weatherproof vent cap or fitting or be provided with a weatherproof hood. All vent caps shall have a minimum free open area equal to the cross-sectional area of the vent pipe and shall not employ screens finer than No. 4 mesh. Vent pipes shall terminate sufficiently above the ground to avoid being obstructed with snow or ice. Vent pipes from tanks containing heaters shall be extended to a location where oil vapors discharging from the vent will be readily diffused. If the static head with a vent pipe filled with oil exceeds 10 pounds per square inch (psi) (69 kPa), the tank shall be designed for the maximum static head that will be imposed.

Liquid fuel vent pipes shall not be cross connected with fill pipes, lines from burners or overflow lines from auxiliary tanks.

❖ Vent pipe terminations must be located to allow complete and ready disposal of fuel-oil vapors that may issue from the vent during fuel delivery, by normal expansion of the fuel in the tank caused by temperature change or the installation of an oil-heating device used to keep the oil at the proper burner atomization temperature. Similar to the fill pipe clearances to building openings (see commentary, Section 1305.6), the intent of the 2-foot (610 mm) minimum clearance is to reduce the possibility of fuel-oil vapors entering the building and reaching an ignition source, and is based on an estimate of the distance that an ignitable concentration of vapors may extend from the end of the vent pipe.

Protection of the open end of the vent pipe with an approved vent cap or fitting or weatherproof hood is necessary to prevent rainwater from getting into the tank through the vent line. However, protection of the open end of the vent pipe must not be allowed to reduce the net free vent area required for the tank to less than the cross-sectional area of the vent pipe opening. Because the purpose of the vent is to allow the tank to "breathe" as liquid is drawn from it or added to it, or as the liquid volume responds to temperature fluctuations, blockage of the vent pipe or reduction of its net clear area could result in damage to the tank or cause overpressure or creation of a vacuum. Likewise, the vent pipe must terminate above grade at a level that will reduce the possibility that the opening could be obstructed by snow accumulation or an ice plug.

Although vent pipes normally convey only vapors, the code recognizes that the potential exists for the vent pipe to become filled with fuel oil during a tank filling operation. Depending on the height of the vent pipe, it is possible for such an occurrence to create sufficient static head pressure on the bottom of the full tank to damage the tank.

To reduce the potential for tank damage from excessive static head pressure, the vent pipe height must be limited to 20 feet (6096 mm) from the bottom of the tank, which would create a maximum of 10 pounds per square inch (psi) (69 kPa) static head pressure at the bottom of the tank. The hydraulic rule of thumb states that head pressure is approximately equal to 0.5 psi per foot (3.4 kPa) of elevation [10 psi ÷ 0.5 psi/ft = 20 feet (6096 mm)]. Alternatively, the tank could be specifically designed to withstand the maximum static head pressure that could be generated by vent height.

Vent lines must be free of obstructions or anything else that would interfere with their full functioning. It is critically important to the integrity of the tank they

serve that the vent lines be functional at all times. Therefore, the code prohibits vent lines from being used for any other purpose, including interconnection with any liquid-carrying piping.

SECTION 1306
OIL GAUGING

1306.1 Level indication. All tanks in which a constant oil level is not maintained by an automatic pump shall be equipped with a method of determining the oil level.

❖ In systems where an appliance supply tank or auxiliary tank is not arranged to be kept full automatically in order to maintain uniform delivery of oil to the burner, an oil-level indicating device must be installed to permit verification of tank level.

1306.2 Test wells. Test wells shall not be installed inside buildings. For outside service, test wells shall be equipped with a tight metal cover designed to discourage tampering.

❖ Test wells, also known as monitoring wells or observation wells, are installed in conjunction with underground fuel-oil storage tanks (when required by local codes) for the purpose of monitoring or sampling liquids from the tank excavation. In the event of a tank leak, a check of the test well will quickly discover the leak so that repair and cleaning can begin. Test wells installed inside buildings could permit fuel-oil vapors to collect inside the building if a tank leak develops. Test wells are best installed outside the building as close as possible to the tank and must be protected in an approved manner against vandalism, tampering or accidental introduction of petroleum product into the well.

1306.3 Inside tanks. The gauging of inside tanks by means of measuring sticks shall not be permitted. An inside tank provided with fill and vent pipes shall be provided with a device to indicate either visually or audibly at the fill point when the oil in the tank has reached a predetermined safe level.

❖ This section prohibits the manual gauging of tank liquid levels on tanks installed inside buildings. Openings for manual gauging can allow fuel-oil vapors to escape inside the building, creating a hazard the code intends to prevent. Because this chapter requires that fill and vent pipes terminate outside the building, the person delivering fuel has no way of seeing the inside tank and its liquid level gauging device, no way of knowing when the tank is approaching being full and no way of judging when to reduce the delivery flow rate to reduce the likelihood of a tank overfill and resultant fuel spill. To provide fuel delivery personnel with this vital information, this section requires that a means be installed at the fill pipe for determining when the tank is approaching full. This may be a liquid level gauge, an audible signal device that activates at a predetermined level of fuel in the tank (typically 90-percent full) or any approved equivalent device. Additional features of some overfill prevention devices include delivery flow-rate reduction when the tank reaches the predetermined level and shutting off the delivery flow at a predetermined tank level (such as 95-percent full).

1306.4 Gauging devices. Gauging devices such as liquid level indicators or signals shall be designed and installed so that oil vapor will not be discharged into a building from the liquid fuel supply system. Liquid-level indicating gauges shall comply with UL 180.

❖ This section requires that any gauging device used on a fuel-oil tank be a closed system or device that is designed to prevent the accidental escape of fuel-oil vapors inside the building.

1306.5 Gauge glass. A tank used in connection with any oil burner shall not be equipped with a glass gauge or any gauge which, when broken, will permit the escape of oil from the tank.

❖ This section prohibits the use of any gauging device, including sight glasses, that would allow the release of the tank contents in the event that the gauging device were broken, thus creating a hazardous material spill with the possibility of fuel-oil vapor ignition.

SECTION 1307
FUEL OIL VALVES

1307.1 Building shutoff. A shutoff valve shall be installed on the fuel-oil supply line at the entrance to the building. Inside or above-ground tanks are permitted to have valves installed at the tank. The valve shall be capable of stopping the flow of fuel oil to the building or to the *appliance* served where the valve is installed at a tank inside the building. Valves shall comply with UL 842.

❖ A shutoff valve allows emergency or service personnel to isolate the fuel-oil supply from equipment or appliances to reduce the hazard in an equipment/appliance fire or to enable equipment/appliance repair without a fuel spill. Valves for fuel oil service must comply with UL 842.

1307.2 Appliance shutoff. A shutoff valve shall be installed at the connection to each *appliance* where more than one fuel-oil-burning *appliance* is installed.

❖ Appliance shutoff valves enable authorized persons to isolate one (or more) appliances for service or replacement without having to use the building shutoff valves to secure all oil-fired appliances in the building.

1307.3 Pump relief valve. A relief valve shall be installed on the pump discharge line where a valve is located downstream of the pump and the pump is capable of exceeding the pressure limitations of the fuel oil system.

❖ When a shutoff valve is installed downstream of a pump, the possibility exists that the pump may be capable of overpressuring the piping system when it pumps against a closed valve. If the pump has no built-in internal bypass to prevent this from happening, a pressure relief valve must be installed in the

piping between the discharge side of the pump and the shutoff valve. The pressure setting of the relief valve must be based on the operating pressure of the system and the fuel oil discharged by the relief valve must be routed back to the tank to prevent a spill.

1307.4 Fuel-oil heater relief valve. A relief valve shall be installed on the discharge line of fuel-oil-heating appliances.

❖ Fuel-oil systems may incorporate a heater to increase the temperature of the fuel oil to enhance its flow within the piping system and to improve atomization in the burner. Oil heated by such a heater will expand and can rupture the system piping if the pressure exceeds the design operating pressure of the system and cannot be relieved. A relief valve installed in accordance with Section 1307.5 will relieve high pressures and limit the potential fire hazard of a fuel-oil spill.

1307.5 Relief valve operation. The relief valve shall discharge fuel oil when the pressure exceeds the limitations of the system. The discharge line shall connect to the fuel oil tank.

❖ The relief valve required by Section 1307.4 must be set to operate at a pressure below the design pressure of the fuel-oil piping system. The pressure setting should be as recommended by the valve manufacturer and the fuel-oil system designer. Proper disposal of the heated fuel oil discharged from the relief valve is essential because heating of the oil will not only have lowered its viscosity but will also have increased the volatility of the oil so that it has some of the fire-hazard characteristics of a flammable, rather than a combustible, liquid. To avoid the hazards of a spill, the liquid discharged from the relief valve must be routed back to the tank, either directly or through the oil return line.

SECTION 1308
TESTING

1308.1 Testing required. Fuel oil piping shall be tested in accordance with NFPA 31.

❖ A leak test is required for all installations of fuel-oil piping systems, including alterations and additions. If a leak is discovered, the leaking component must be repaired or replaced before the system is concealed or put into operation. Section 3-10 of NFPA 31 specifies test methods and test pressures for testing fuel-oil piping. Fuel-oil tanks that are not designed as pressure vessels are intended to be used at atmospheric pressure and cannot be subjected to high internal pressures when testing the piping. Hydrostatic testing and air pressure testing methods are allowed. In addition to air pressure leak testing and hydrostatic leak testing, NFPA 31 allows testing suction lines under a vacuum and specifies a minimum value and test duration. Suction lines tested in this way should be separated from the fuel-oil tanks because the tanks cannot be subjected to a vacuum.

Bibliography

The following resource materials were used in the preparation of the commentary for this chapter of the code:

IFC-12, *International Fire Code*. Washington, DC: International Code Council, 2011.

Chapter 14:
Solar Systems

General Comments

Chapter 14 contains requirements for the construction, alteration and repair of all systems and components of solar energy systems used for space heating or cooling, domestic hot water heating or processing. The provisions of this chapter are limited to those necessary to achieve safe installations that are relatively hazard free.

A solar energy system can be designed to handle 100 percent of the energy load of a building, although this is rarely accomplished. Because solar energy is a low-intensity energy source and not always available because of the weather, it is usually necessary to supplement a solar energy system with other traditional energy sources.

Solar heating and cooling systems are classified as either passive or active systems. A passive solar energy system uses solar collectors, a gas or liquid heat-transfer medium and distribution piping or ductwork. This system does not use circulators or fans but relies on natural (gravity) flow. The flow of energy (heat) is by natural convection, conduction and radiation. Commentary Figures 14(1) and 14(2) show typical passive solar heating applications.

A complete active solar energy system includes solar collectors, a gas or liquid heat transfer medium, piping, ducts, circulators, fans and controls to move the medium to the load or to an energy storage system. Commentary Figure 14(3) shows a typical active solar system used for space and domestic water heating supplemented by an auxiliary furnace and water heater.

Many design variations exist, such as hybrid systems, with components of both active and passive systems. Solar systems can either be very complex or as simple as glazing, and can act in harmony with the architectural features of the building [see Commentary Figure 14(4)].

Solar designs can be used to improve the performance of conventional space heating systems and work in conjunction with other such systems. For example, a simple passive collection system can greatly improve the performance of an air-to-air or water-to-air heat pump system.

The energy collected by a solar system can be used as it is collected or transferred to some form of thermal storage system. Typical thermal storage systems consist of large quantities of a dense mass, including stone,

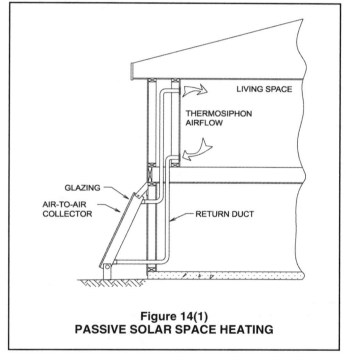

Figure 14(1)
PASSIVE SOLAR SPACE HEATING

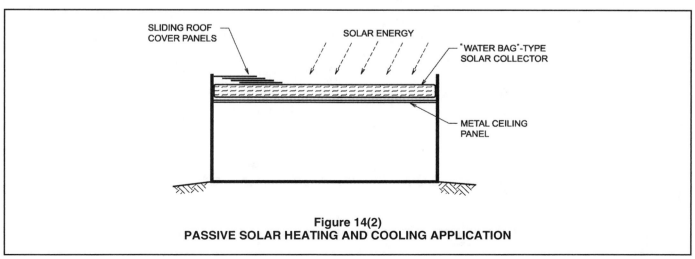

Figure 14(2)
PASSIVE SOLAR HEATING AND COOLING APPLICATION

masonry or water. Storage mass can also consist of materials in which a phase change occurs, such as chemical compounds that store the latent heat required to liquefy the compounds that are solid at room temperature.

If the energy collected by a solar system is not used or stored for later use, the collection system must either stop collecting or dump the energy to the building exterior unless the collection system is designed to withstand the higher temperatures and pressures resulting from the static (no-flow) mode of operation.

Another aspect of solar system installations that must be considered is the structural loading effect that the collectors and supports have on the building's roof sys-

tem. Collectors can add considerable weight, affect snow accumulation and increase wind loads and uplift forces. As with any roof-mounted equipment or appliances, the installation must also comply with the applicable provisions of the *International Building Code®* (IBC®).

Purpose

This chapter establishes provisions for the safe installation, operation and repair of solar energy systems. Although such systems use components similar to those of conventional mechanical equipment, many of these provisions are unique to solar energy systems.

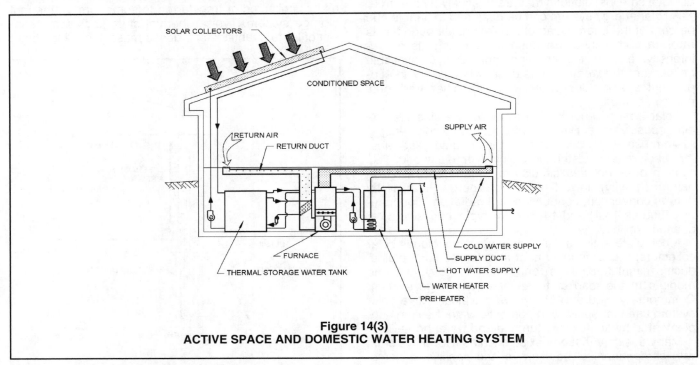

Figure 14(3)
ACTIVE SPACE AND DOMESTIC WATER HEATING SYSTEM

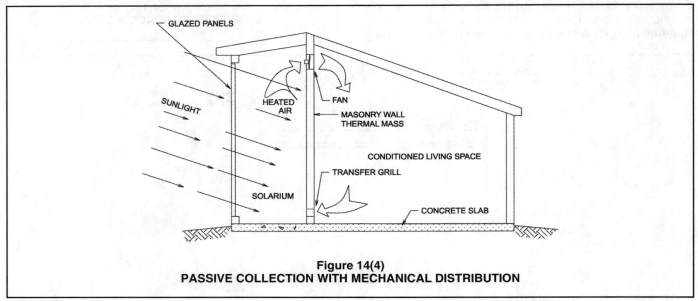

Figure 14(4)
PASSIVE COLLECTION WITH MECHANICAL DISTRIBUTION

SECTION 1401
GENERAL

1401.1 Scope. This chapter shall govern the design, construction, installation, *alteration* and repair of systems, *equipment* and appliances intended to utilize solar energy for space heating or cooling, domestic hot water heating, swimming pool heating or process heating.

❖ Solar energy can be used for a variety of purposes, including space heating, space cooling, domestic water heating, swimming pool heating and processing. Therefore, the scope encompasses all of these potential solar energy applications, and includes the design, construction, installation, alteration and repair of all systems and equipment.

1401.2 Potable water supply. Potable water supplies to solar systems shall be protected against contamination in accordance with the *International Plumbing Code*.

Exception: Where all solar system piping is a part of the potable water distribution system, in accordance with the requirements of the *International Plumbing Code*, and all components of the piping system are *listed* for potable water use, cross-connection protection measures shall not be required.

❖ Solar systems may contain conditioning chemicals, antifreeze or heat transfer fluids other than water. The fill and makeup water supply to solar hydronic systems is generally taken directly from the potable water supply of the building. Water supply connections are provided to manually or automatically replace water that is lost from relief/safety valve discharge, evaporation, draindown, control flushing, air purging or leakage. This arrangement causes the potable water supply to be highly susceptible to contamination by backflow. Where any direct connection is made between two pressurized systems, the direction of flow through the connection or interface is dependent on the pressure differential between the systems. If the solar system pressure is higher than the potable water supply pressure, boiler water will flow into the potable water distribution system. The exception allows backflow protection to be omitted where potable water flows through the solar system and all components of the solar system are listed for potable water applications.

1401.3 Heat exchangers. Heat exchangers used in domestic water-heating systems shall be *approved* for the intended use. The system shall have adequate protection to ensure that the potability of the water supply and distribution system is properly safeguarded.

❖ In accordance with the *International Plumbing Code*® (IPC®), heat exchangers using an essentially toxic transfer fluid must be double-wall construction. An air gap open to the atmosphere must be provided between the two walls. Heat exchangers using an essentially nontoxic transfer fluid may be of single-wall construction.

The extent of isolation required for a heat exchanger depends on the type of fluid used on the nonpotable side of the heat exchanger. From the definition of "Essentially nontoxic transfer fluid," the Gosselin rating of the nonpotable fluid must be evaluated. If the fluid has a Gosselin rating of 1, a single-wall heat exchanger is permitted.

The Gosselin rating is a measure of the toxicity of a substance. The name originates from one of the prime developers of the rating system, Dr. Robert E. Gosselin, Professor of Pharmacology at Dartmouth Medical School in New Hampshire.

Gosselin toxicity ratings are the values used by medical personnel to analyze poison victims. The ratings are based on the probable lethal dose for a human. The six levels of Gosselin ratings are outlined in Commentary Table 1401.3(1).

Some of the commercially available transfer fluids with a Gosselin rating of 1 are identified in Commentary Table 1401.3(2).

If the heat transfer fluid has a Gosselin rating of 2 or more, a double-wall heat exchanger is required. The double-wall heat exchanger must have an intermediate space between the walls that is open to the atmosphere. This type of construction would allow any leakage of fluid through the walls of the heat exchanger to discharge externally to the heat exchanger where it would be observable (see Commentary Figure 1401.3).

Table 1401.3(1)
GOSSELIN RATINGS

TOXICITY RATING OF CLASS	PROBABLE ORAL LETHAL DOSE (HUMAN)	
	Dose	For a 70-kg person (150 pounds)
6 Super toxic	Less than 5 mg/kg	A taste (less than 7 drops)
5 Extremely toxic	5-50 mg/kg	Between 7 drops and 1 teaspoon
4 Very toxic	50-500 mg/kg	Between 1 teaspoon and 1 ounce
3 Moderately toxic	0.5-5 gm/kg	Between 1 ounce and 1 pint (pound)
2 Slightly toxic	5-15 gm/kg	Between 1 pint and 1 quart
1 Practically nontoxic	Above 15 gm/kg	More than 1 quart (2.2 pounds)

Table 1401.3(2)
HEAT TRANSFER FLUIDS (ESSENTIALLY NONTOXIC)
GOSSELIN RATING OF 1

TRADE NAME	MANUFACTURER
Caloria Ht-43	Exxon Co.
DowFrost	Dow
Drewsol	Drew Chemical
Freeze Proof	Commonwealth
Mobiltherm 603	Mobil Oil Co.
Mobiltherm Light	Mobil Oil Co.
Nutek 835	Nuclear Technology
Nutek 876	Nuclear Technology
Process Oil 3029	Exxon Co.
Propylene Glycol U.S.P.	Union Carbide
Solargard G.	Daystar Corp.
Solar Winter Ban	Solar Alternative Inc.
Sunsol 60	Sunworks
Suntemp	Resource Technology Corp.
Syltherm 444	Dow Corning Corp.
Therminol 66	Monsanto Co.
UCAR Food Freeze 35	Union Carbide
FDA Approved Boiler	
Additives	
Freon 12	
Freon 22	
Freon 112	
Freon 114	

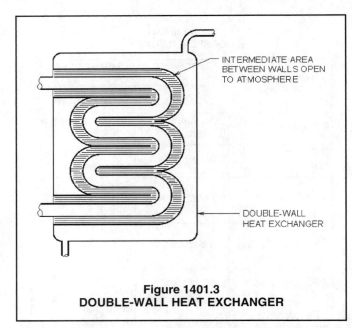

Figure 1401.3
DOUBLE-WALL HEAT EXCHANGER

1401.4 Solar energy equipment and appliances. Solar energy *equipment* and appliances shall conform to the requirements of this chapter and shall be installed in accordance with the manufacturer's installation instructions.

❖ All solar energy system components, including collectors, heat transfer fluids and thermal storage units,

are regulated by this chapter. The requirements of this chapter address only the potential hazards to life and property associated with solar installations. The system must be installed in accordance with the manufacturer's installation instructions. Components of the solar system are also required to conform to other applicable sections of the code, such as those found in Chapter 3, General Regulations; Chapter 6, Duct Systems; Chapter 10, Boilers, Water Heaters and Pressure Vessels; Chapter 12, Hydronic Piping; and Chapter 11, Refrigeration. Solar installations could also be subject to provisions in the IPC and the IBC. Included are indirect waste piping and special waste; water supply and distribution; inspection; tests and maintenance; fire-resistant and roof structures; structural loads; glass and glazing and plastic.

1401.5 Ducts. Ducts utilized in solar heating and cooling systems shall be constructed and installed in accordance with Chapter 6 of this code.

❖ Solar systems can involve hot-air-type collectors, thermal storage systems, fan-coil units and air handlers, all of which may be connected to ductwork.

SECTION 1402
INSTALLATION

1402.1 Access. Access shall be provided to solar energy *equipment* and appliances for maintenance. Solar systems and appurtenances shall not obstruct or interfere with the operation of any doors, windows or other building components requiring operation or access.

❖ In accordance with the definition of "Access" found in Chapter 2, access to solar energy equipment may be through an access panel, door or similar construction. The installation of solar equipment must also comply with the access-related installation requirements found in Section 306. The installation of solar equipment and piping must not interfere with door swings, operability of windows and access to other equipment and appliances.

1402.2 Protection of equipment. Solar *equipment* exposed to vehicular traffic shall be installed not less than 6 feet (1829 mm) above the finished floor.

Exception: This section shall not apply where the *equipment* is protected from motor vehicle impact.

❖ The same requirement exists in Section 304.6 (see commentary, Section 304.6).

1402.3 Controlling condensation. Where attics or structural spaces are part of a passive solar system, ventilation of such spaces, as required by Section 406, is not required where other *approved* means of controlling condensation are provided.

❖ As shown in Commentary Figure 1402.3, some passive solar energy systems use structural cavities, such as attics, crawl spaces and wall and floor spaces, to convey heated air from the collector to or around the conditioned space.

Attics, crawl spaces, rafter cavities and similar spaces in a building must be ventilated to control moisture and prevent structural damage that can result from high humidity and condensation. Such ventilation could interfere with or reduce the effectiveness of a solar energy system if the spaces being ventilated are also used as air plenums or ducts as part of the solar energy system. This section simply allows omitting the required ventilation of such spaces where the ventilation would affect the operation of the passive solar system and where an effective method of moisture control is substituted for the otherwise required ventilation.

Moisture in attics and structural spaces could possibly be controlled by the air currents of a passive solar system, or the space could be mechanically ventilated in accordance with Section 406 of the code. Possibly, moisture control will not be necessary where such spaces are sufficiently heated by the operation of the solar energy system. The intent is to permit any method or combination of methods that will prevent moisture accumulation and resulting damage to the structure and allow the structural space to serve as part of the solar system.

1402.4 Roof-mounted collectors. Roof-mounted solar collectors that also serve as a roof covering shall conform to the requirements for roof coverings in accordance with the *International Building Code.*

Exception: The use of plastic solar collector covers shall be limited to those *approved* plastics meeting the requirements for plastic roof panels in the *International Building Code.*

❖ Solar collectors that are an integral part of a roof covering are required to comply with the provisions for roofs and roof structures in the IBC because they function as a solar collector, as well as providing weather protection. These requirements include fire classification, weather protection, wind resistance and durability. Typical collector designs are shown in Commentary Figure 1402.4.

Solar collectors can be independent assemblies that mount directly on the roof surface or they can be built in as an integral part of the roof covering system.

The exception to this section recognizes that the IBC permits the limited use of plastic roof panels as part of a roof-covering system. Therefore, solar collectors with plastic glazing (covers) comprising part of the roof-covering system should be permitted with the same limitations as plastic roof panels.

1402.4.1 Collectors mounted above the roof. When mounted on or above the roof covering, the collector array and supporting construction shall be constructed of noncombustible materials or fire-retardant-treated wood conforming to the *International Building Code* to the extent required for the type of roof construction of the building to which the collectors are accessory.

Exception: The use of plastic solar collector covers shall be limited to those *approved* plastics meeting the require-

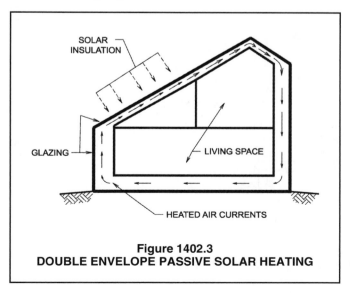

**Figure 1402.3
DOUBLE ENVELOPE PASSIVE SOLAR HEATING**

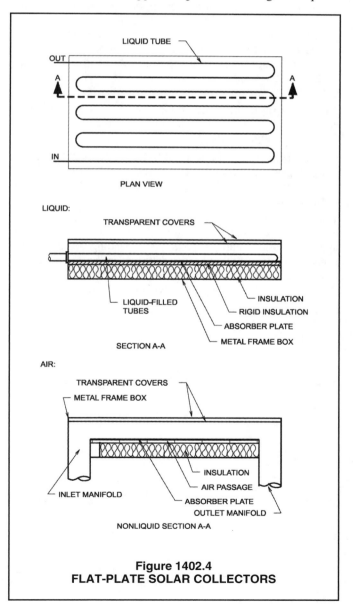

**Figure 1402.4
FLAT-PLATE SOLAR COLLECTORS**

ments for plastic roof panels in the *International Building Code*.

❖ Solar collectors and any supporting construction mounted on or above the roof covering must be constructed of materials consistent with the building's construction classification.

Rarely does the existing roof structure face the correct direction or have the desired angle of inclination for solar applications. Therefore, solar collectors or collector arrays are often mounted on support structures on or above a roof to allow the orientation of the collectors for optimum solar exposure. The collectors and any support structure must be constructed of materials permitted for the roof structure on which they are mounted.

The exception allows the use of approved plastic glazing (covers) on solar collectors, if their use falls within the parameters of the IBC. Plastic is combustible, and the intent of this exception is to allow the limited use of combustible collector covers where noncombustible or limited combustible collectors and supports would otherwise be required by the IBC. Plastic collector covers are subject to the same regulations as those for plastic-glazed skylight assemblies.

1402.5 Equipment. The solar energy system shall be equipped in accordance with the requirements of Sections 1402.5.1 through 1402.5.4.

❖ Solar energy systems must be designed and properly equipped to protect the system, the building and its occupants from damage and the potential hazards associated with such systems.

1402.5.1 Pressure and temperature. Solar energy system components containing pressurized fluids shall be protected against pressures and temperatures exceeding design limitations with a pressure and temperature relief valve. Each section of the system in which excessive pressures are capable of developing shall have a relief device located so that a section

cannot be valved off or otherwise isolated from a relief device. Relief valves shall comply with the requirements of Section 1006.4 and discharge in accordance with Section 1006.6.

❖ Any heated, closed system is capable of developing pressures that exceed its design working pressure. Closed liquid-filled systems can develop high hydrostatic pressures with even slight temperature increases. Because solar energy varies in intensity, the collector system is subject to greater temperature and pressure variations than other heating systems having precisely controlled energy inputs. Additionally, the solar energy input cannot be turned off by limit controls as can other energy sources. Therefore, a solar energy system is more likely to be subjected to extreme temperatures and pressures that could cause system failures and the associated hazards. Pressure and temperature relief valves are necessary to prevent injury and property damage that could result from the failure of pressurized vessels and piping. Typical liquid solar energy systems involve large complex piping circuits with valve arrangements that greatly increase the likelihood of portions of the piping system being isolated from the overpressure or overtemperature safety devices. Any portion of a system isolated from the relief valve or valves is unprotected from the danger of excessive pressures and temperatures. To ensure complete protection to all portions of a system, multiple relief valves at different locations in the system may be necessary.

Solar heating and cooling systems are similar in nature to hydronic heating systems; therefore, the requirements for relief valves are the same for both types of systems (see Commentary Figure 1402.5.1).

Temperature and pressure relief valves must comply with the provisions of Section 1006.4 for labeling, capacity rating and pressure setting. The discharge of relief valves must meet the requirements of Section 1006.6.

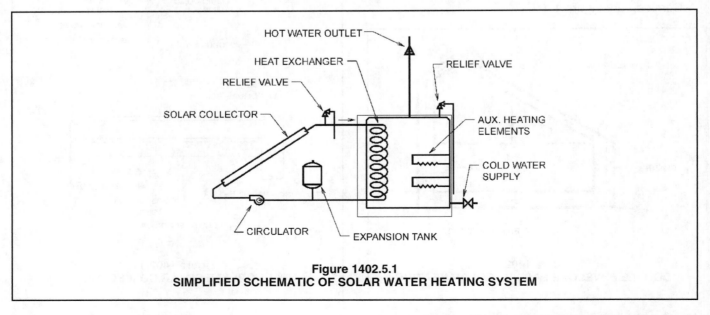

Figure 1402.5.1
SIMPLIFIED SCHEMATIC OF SOLAR WATER HEATING SYSTEM

1402.5.2 Vacuum. The solar energy system components that are subjected to a vacuum while in operation or during shutdown shall be designed to withstand such vacuum or shall be protected with vacuum relief valves.

❖ Some solar system designs using a heat transfer fluid capable of freezing employ a method of freeze protection that drains the fluid from the portions of the system exposed to freezing temperatures. Under certain conditions, this draindown function can produce significant partial vacuums in the system. Some solar system designs depend on the vaporization of a liquid heat transfer medium, possibly producing a partial vacuum that is caused by condensation of the transfer medium. These systems can also rely on a constant partial vacuum to lower the boiling point of the liquid transfer medium. Systems subject to pressures below atmospheric pressures must be able to withstand the pressure without damage; otherwise, vacuum relief devices are required where systems cannot tolerate the negative pressures that could develop in the system.

1402.5.3 Protection from freezing. System components shall be protected from damage by freezing of heat transfer liquids at the lowest ambient temperatures that will be encountered during the operation of the system.

❖ In liquid solar energy systems, freezing is a common cause of system failure. Because solar collectors can reradiate heat into the cold night sky, it is possible for systems to cool faster than the ambient air, thus freezing a water-filled system when the ambient air temperature is above 32°F (0°C). This section requires protection in areas where freezing of the heat transfer fluid can occur.

One of the most frequently used and dependable freeze protection methods is the use of an antifreeze heat transfer fluid. The most commonly used nonfreezing heat transfer fluids are water with ethylene glycol and water with propylene glycol antifreeze solutions. Other nonfreezing heat transfer fluids that can be used are silicone oils, hydrocarbon oils or

refrigerants. Some antifreeze solutions are toxic, and the use of toxic heat transfer fluids would require special precautionary means to protect the potable water system from contamination.

Other common freeze protection methods include draindown and recirculation. Draindown is a method that automatically or manually drains the heat transfer fluid from the collectors and piping where a potential freezing condition exists (see Commentary Figure 1402.5.3). In areas where freezing temperatures are rare, freeze protection might be accomplished by recirculation of the water from the heat storage system through collectors, as often or at a rate necessary to keep the exposed portions of the system above freezing. Both the draindown and recirculation means of freeze protection rely on electrical and mechanical devices and components or on human action and, therefore, are less dependable than freezeproof heat transfer fluids.

1402.5.4 Expansion tanks. Liquid single-phase solar energy systems shall be equipped with expansion tanks sized in accordance with Section 1009.

❖ In a liquid, single-phase solar energy system that uses a heat transfer fluid that does not change into a gaseous or solid phase, the heat transfer liquid expands as the temperature increases. An expansion (compression) tank is necessary to receive the expanded liquid, thus preventing excessive pressure increases that would cause system damage or opening of a pressure relief valve. The tank must be sized as required for hydronic systems.

1402.6 Penetrations. Roof and wall penetrations shall be flashed and sealed to prevent entry of water, rodents and insects.

❖ Typical solar collector arrays will involve one or more roof or wall penetrations to accommodate piping and ducts. The collector arrays could also involve roof covering penetrations caused by support framework members or fasteners. Any penetrations through the

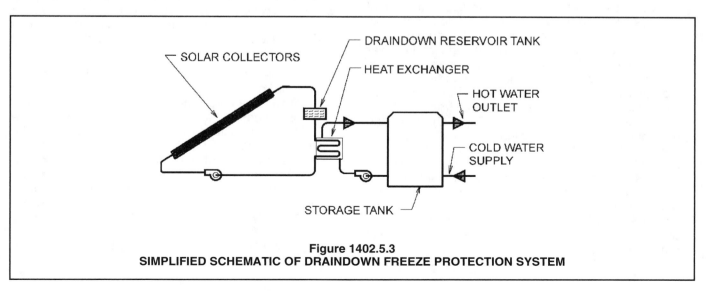

Figure 1402.5.3
SIMPLIFIED SCHEMATIC OF DRAINDOWN FREEZE PROTECTION SYSTEM

walls, roof and roof coverings must be sealed by a waterproof material or flashing that will withstand the temperatures produced by the solar energy equipment to which it is exposed and provide weather protection against water penetration.

1402.7 Filtering. Air transported to occupied spaces through rock or dust-producing materials by means other than natural convection shall be filtered at the outlet from the heat storage system.

❖ This section requires filtering to remove dust and particulates from mechanically forced air that has passed through a thermal storage unit containing pebbles, rock or other thermal mass material producing air-borne particles. In passive systems, a filter is not required because the velocity of the air is generally not sufficient to carry particulates or keep larger particles in suspension. Additionally, the resistance offered by an air filter medium would greatly impede natural convective airflow.

SECTION 1403
HEAT TRANSFER FLUIDS

1403.1 Flash point. The flash point of the actual heat transfer fluid utilized in a solar system shall be not less than 50°F (28°C) above the design maximum nonoperating (no-flow) temperature of the fluid attained in the collector.

❖ Heat transfer liquid is the "working fluid" in the solar heating or cooling system. The liquid is pumped from the thermal energy collector where heat is absorbed by the liquid, to the heat exchanger where the liquid releases the heat energy to another fluid, airstream or some form of thermal storage mass. Some of the commonly used heat transfer fluids are water, water/ethylene glycol, water/propylene glycol, silicone oils, paraffinic oils and aromatic oils. The term "flash point" applies only to flammable and combustible liquids, and is defined as the minimum temperature at which the liquid gives off enough vapor to form an ignitable mixture with air near the surface of the liquid or within the test vessel used. This section regulates the use of combustible heat transfer liquids and does not apply to flammable liquids because they are prohibited by Section 1403.2. The design parameters of the solar energy system determine the minimum allowable flash point temperature of the heat transfer fluid. See the definitions of "Combustible liquids" and "Flammable liquids" in Chapter 2. The heat transfer fluid's flash point must be at least 50°F (28°C) above the maximum design temperature that the fluid could reach in the collector during the nonoperating (no-flow) phase of the solar energy system operating cycle. This requirement establishes a safety margin by setting the flash point temperature at least 50°F (28°C) above the worst-case temperature that could be reached in the system, which is the no-flow temperature of the fluid in the solar collector at the peak radiation period.

1403.2 Flammable gases and liquids. A flammable liquid or gas shall not be utilized as a heat transfer fluid. The flash point of liquids used in occupancies classified in Group H or F shall not be lower unless *approved*.

❖ This section prohibits the use of any heat transfer fluid defined as a "flammable liquid" or a "flammable gas." Liquids that have a flash point at or above 100°F (38°C) are classified as combustible and can be used in solar energy systems within the limitations of Section 1403.1. The relatively high flash points of combustible liquids allow them to be used within the parameters of Section 1403.1. However, flammable liquids have lower flash points and could create a hazard at the temperatures associated with solar energy systems.

Because flammable liquids are not allowed, the lowest permissible flash point would be that of a combustible liquid. The flash point of heat transfer liquids used in factories or in high-hazard occupancies could be lower than the flash points of those liquids classified as combustible when specifically approved by the code official, based on the safety features incorporated into those spaces in recognition of the increased threat of fire. The code allows certain operations and systems to exist in factory, industrial and high-hazard occupancies. It is assumed that such occupancies are periodically inspected for conformance to the fire prevention code, activities are supervised by trained individuals and safety precautions are exercised or incorporated commensurate with the fire hazard. Certain materials and operations could also be necessary as part of a manufacturing process.

SECTION 1404
MATERIALS

1404.1 Collectors. Factory-built collectors shall be *listed* and *labeled*, and bear a label showing the manufacturer's name and address, model number, collector dry weight, collector maximum allowable operating and nonoperating temperatures and pressures, minimum allowable temperatures and the types of heat transfer fluids that are compatible with the collector. The label shall clarify that these specifications apply only to the collector.

❖ Solar collectors can be subjected to high temperatures and pressures, possibly leading to a hazardous condition. As a result, the quality, design and application limitations of the collector must be known to the designer, the installer, the owner and the code official. This section requires that solar collectors be labeled, indicating all maximum and minimum operating and nonoperating pressure and temperature parameters, as well as the manufacturer's name, address, model number and the collector weight, when empty. The label is not only a form of quality assurance but provides the specifications that apply only to the collector and necessary information for its use. Only fluids compatible with the collector are to be used. The misapplication of a collector could result in collector failure, injury to occupants and damage to the property or the structure.

1404.2 Thermal storage units. Pressurized thermal storage units shall be *listed* and *labeled*, and bear a label showing the manufacturer's name and address, model number, serial number, storage unit maximum and minimum allowable operating temperatures, storage unit maximum and minimum allowable operating pressures and the types of heat transfer fluids compatible with the storage unit. The label shall clarify that these specifications apply only to the thermal storage unit.

❖ Pressurized thermal storage units are containers, tanks or vessels containing some form of thermal mass used to store heat energy collected by the solar system. Thermal storage has two purposes. First, it allows the collector system to continue collecting energy when the energy cannot be transferred directly to a load. Second, thermal storage provides useful energy during periods of low or nonexistent insulation (solar radiation). Without thermal storage, any energy that is not simultaneously collected and used would have to be rejected. Thermal storage is often compared to a flywheel because of its ability to store energy temporarily. As with any pressure vessel, the storage unit must be able to withstand safely the pressures and temperatures to which it is exposed. The storage unit must also be compatible with the type of heat transfer fluid used in the system (see commentary, Section 1404.1).

Bibliography

The following resource materials were used in the preparation of the commentary for this chapter of the code:

IBC-12, *International Building Code*. Washington, DC: International Code Council, 2011.

IPC-12, *International Plumbing Code*. Washington, DC: International Code Council, 2011.

Chapter 15:
Referenced Standards

General Comments

Chapter 15 contains a comprehensive list of standards that are referenced in the code. It is organized to make locating specific document references easy.

It is important to understand that not every document related to mechanical system design, installation and construction is qualified to be a "referenced standard." The International Code Council® (ICC®) has adopted a criterion that standards referenced in the *International Codes®* (I-Codes®) and standards intended for adoption into the I-Codes must meet to qualify as a referenced standard. The policy is summarized as follows:

- Code references: The scope and application of the standard must be clearly identified in the code text.

- Standard content: The standard must be written in mandatory language and be appropriate for the subject covered. The standard cannot have the effect of requiring proprietary materials or prescribing a proprietary testing agency.

- Standard promulgation: The standard must be readily available and developed and maintained in a consensus process such as those used by ASTM or ANSI.

The ICC Code Development Procedures, of which the standards policy is a part, are updated periodically. A copy of the latest version can be obtained from the ICC offices or from the ICC web site, http://www.iccsafe.org.

Once a standard is incorporated into the code through the code development process, it becomes an enforceable part of the code. When the code is adopted by a jurisdiction, the standard also is part of that jurisdiction's adopted code. It is for this reason that the criteria were developed. Compliance with this policy means that documents incorporated into the code are developed through the use of a consensus process, are written in mandatory language and do not mandate the use of proprietary materials or agencies. The requirement that a standard be developed through a consensus process means that the standard is representative of the most current body of available knowledge on the subject as determined by a broad range of interested or affected parties without dominance by any single interest group. A true consensus process has many attributes, including but not limited to:

- An open process that has formal (published) procedures that allow for the consideration of all viewpoints;

- A definitive review period that allows for the standard to be updated and/or revised;

- A process of notification to all interested parties; and

- An appeals process.

Many available documents related to mechanical system design, installation and construction, though useful, are not "standards" and are not appropriate for reference in the code. Often, these documents are not developed or written with the intention of being used for regulatory purposes and are unsuitable for use as a standard because of extensive use of recommendations, advisory comments and nonmandatory terms. Typical examples include design guidelines and practices.

The objective of ICC's standards policy is to provide regulations that are clear, concise and enforceable; thus the requirement that standards be written in mandatory language. This requirement is not intended to mean that a standard cannot contain informational or explanatory material that will aid the user of the standard in its application. When the standard's promulgating agency wants such material to be included, however, the information must appear in a nonmandatory location, such as an annex or appendix, and be clearly identified as not being part of the standard.

Overall, standards referenced by the code must be authoritative, relevant, up to date and, most important, reasonable and enforceable. Standards that comply with ICC's standards policy fulfill these expectations.

Purpose

As a performance-based code, this code contains numerous references to documents that are used to regulate materials and methods of construction. The references to these documents within the code text consist of the promulgating agency's acronym and its publication designation (for example, AWS A5.8) and a further indication that the document being referenced is the one that is listed in Chapter 15. Chapter 15 contains all of the information that is necessary to identify the specific referenced document. Included is the following information on a document's promulgating agency (see Figure 15):

- The promulgating agency (the agency's title);

- The promulgating agency's acronym; and

- The promulgating agency's address.

For example, a reference to an ASME standard within the code indicates that the document is promulgated by the American Society of Mechanical Engineers (ASME), which is located in New York, New York. Chapter 15 lists the standards agencies alphabetically for ease of identification.

Chapter 15 also includes the following information on the referenced document itself (see Figure 15):

- The document's publication designation;
- The document's edition year;
- The document's title;
- Any addenda or revisions to the document known at the time of the code's publication; and
- Every section of the code in which the document is referenced.

For example, a reference to ASME B1.20.1 indicates that this document can be found in Chapter 15 under the heading ASME. The specific standard's designation is B1.20.1. For convenience, these designations are listed in alphanumeric order. Chapter 15 identifies that: ASME B1.20.1 is titled *Pipe Threads, General Purpose (Inch)*; the applicable edition (that is, its year of publication) is 1983; and it is referenced in two specifically identified sections of the code (see Figure 15 for an example).

Chapter 15 also indicates when a document has been discontinued or replaced by its promulgating agency. When a document is replaced by a different one, a note appears to tell the user the designation and title of the new document.

This chapter lists the standards that are referenced in various sections of this document. The standards are listed herein by the promulgating agency of the standard, the standard identification, the effective date and title and the section or sections of this document that reference the standard. The application of the referenced standards shall be as specified in Section 102.8.

The key aspect of the manner in which standards are referenced by the code is that a specific edition of a specific standard is clearly identified. The requirements for compliance can be readily determined. The basis for code compliance is, therefore, established and available to the code official, the mechanical contractor, the designer and the owner.

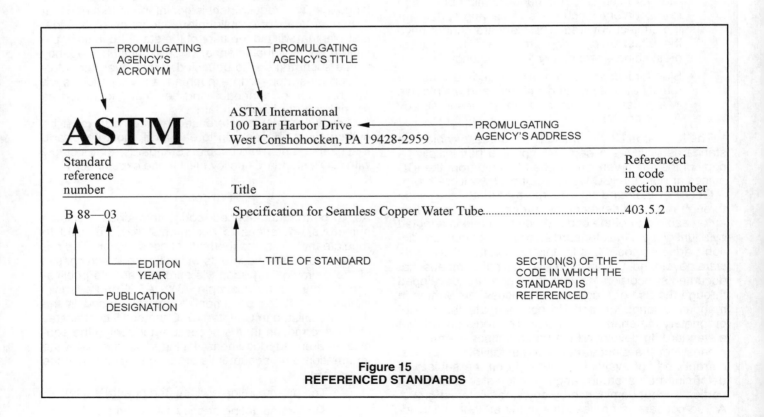

Figure 15
REFERENCED STANDARDS

This chapter lists the standards that are referenced in various sections of this document. The standards are listed herein by the promulgating agency of the standard, the standard identification, the effective date and title, and the section or sections of this document that reference the standard. The application of the referenced standards shall be as specified in Section 102.8.

ACCA
Air Conditioning Contractors of America
2800 Shirlington Road, Suite 300
Arlington, VA 22206

Standard reference number	Title	Referenced in code section number
Manual D—09	Residential Duct Systems	603.2
183—2007	Peak Cooling and Heating Load Calculations in Buildings Except Low-rise Residential Buildings	312.1

AHRI
Air-Conditioning, Heating and Refrigeration Institute
4100 North Fairfax Drive, Suite 200
Arlington, VA 22203

Standard reference number	Title	Referenced in code section number
700—2006	Purity Specifications for Fluorocarbon and Other Refrigerants	1102.2.2.3

AMCA
Air Movement and Control Association International
30 West University Drive
Arlington Heights, IL 60004

Standard reference number	Title	Referenced in code section number
550—08	Test Method for High Velocity Wind Driven Rain Resistant Louvers	401.5, 501.3.2

ANSI
American National Standards Institute
11 West 42nd Street
New York, NY 10036

Standard reference number	Title	Referenced in code section number
Z21.8—1994 (R2002)	Installation of Domestic Gas Conversion Burners	919.1

ASHRAE
American Society of Heating, Refrigerating and Air-Conditioning Engineers, Inc.
1791 Tullie Circle, NE
Atlanta, GA 30329

Standard reference number	Title	Referenced in code section number
ASHRAE—2009	ASHRAE Fundamentals Handbook	603.2
15—2010	Safety Standard for Refrigeration Systems	1101.6, 1105.8, 1108.1
34—2010	Designation and Safety Classification of Refrigerants	202, 1102.2.1, 1103.1

ASHRAE—continued

62.1—2010	Ventilation for Acceptable Indoor Air Quality	403.3.2.3.2
180—2008	Standard Practice for Inspection and Maintenance of Commercial Building HVAC Systems	102.3

ASME

American Society of Mechanical Engineers
Three Park Avenue
New York, NY 10016-5990

Standard reference number	Title	Referenced in code section number
B1.20.1—1983 (R2006)	Pipe Threads, General Purpose (Inch)	1203.3.5, 1303.3.3
B16.3—2006	Malleable Iron Threaded Fittings, Classes 150 & 300	Table 1202.5
B16.5—2003	Pipe Flanges and Flanged Fittings NPS $^1/_2$ through NPS 24	Table 1202.5
B16.9—2007	Factory Made Wrought Steel Buttwelding Fittings	Table 1202.5
B16.11—2005	Forged Fittings, Socket-welding and Threaded	Table 1202.5
B16.15—2006	Cast Bronze Threaded Fittings	Table 1202.5
B16.18—2001 (Reaffirmed 2005)	Cast Copper Alloy Solder Joint Pressure Fittings	513.13.1, Table 1202.5
B16.22—2001 (Reaffirmed 2005)	Wrought Copper and Copper Alloy Solder Joint Pressure Fittings	513.13.1, Table 1202.5
B16.23—2002 (Reaffirmed 2006)	Cast Copper Alloy Solder Joint Drainage Fittings DWV	Table 1202.5
B16.24—2006	Cast Copper Alloy Pipe Flanges and Flanged Fittings: Class 150, 300, 400, 600, 900, 1500 and 2500	Table 1202.5
B16.26—2006	Cast Copper Alloy Fittings for Flared Copper Tubes	Table 1202.5
B16.28—1994	Wrought Steel Buttwelding Short Radius Elbows and Returns	Table 1202.5
B16.29—2007	Wrought Copper and Wrought Copper Alloy Solder Joint Drainage Fittings-DWV	Table 1202.5
B31.9—08	Building Services Piping	1201.3
BPVC—2007	ASME Boiler & Pressure Vessel Code–07 Edition	1004.1, 1011.1
CSD-1—2009	Controls and Safety Devices for Automatically Fired Boilers	1004.1

ASSE

American Society of Sanitary Engineering
901 Canterbury, Suite A
Westlake, OH 44145

Standard reference number	Title	Referenced in code section number
1017—2010	Performance Requirements for Temperature Actuated Mixing Values for Hot Water Distribution Systems	1002.2.2

ASTM

ASTM International
100 Barr Harbor Drive
West Conshohocken, PA 19428

Standard reference number	Title	Referenced in code section number
A 53/A 53M—07	Specification for Pipe, Steel, Black and Hot-dipped, Zinc-coated Welded and Seamless	Table 1202.4, Table 1302.3
A 106/A106M—08	Specification for Seamless Carbon Steel Pipe for High-Temperature Service	Table 1202.4, Table 1302.3
A 126—04	Specification for Gray Iron Castings for Valves, Flanges and Pipe Fittings	Table 1202.5
A 254—97 (2007)	Specification for Copper Brazed Steel Tubing	Table 1202.4, Table 1302.3
A 420/A 420M—07	Specification for Piping Fittings of Wrought Carbon Steel and Alloy Steel for Low-Temperature Service	Table 1202.5
A 539—99	Specification for Electric-resistance-welded Coiled Steel Tubing for Gas and Fuel Oil Lines	Table 1302.3
B 32—08	Specification for Solder Metal	1203.3.3
B 42—02e01	Specification for Seamless Copper Pipe, Standard Sizes	513.13.1, 1107.5.2, Table 1202.4, Table 1302.3
B 43—98(2004)	Specification for Seamless Red Brass Pipe, Standard Sizes	513.13.1, 1107.5.2, Table 1202.4, Table 1302.3

ASTM—continued

ASTM—continued

F 1281—07	Specification for Crosslinked Polyethylene/Aluminum/Crosslinked Polyethylene (PEX-AL-PEX) Pressure Pipe	Table 1202.4
F 1282—06	Standard Specification for Polyethylene/Aluminum/Polyethylene (PE-AL-PE) Composite Pressure Pipe	Table 1202.4
F 1476—07	Specification for Performance of Gasketed Mechanical Couplings for Use in Piping Applications	1203.3.7
F 1807—08	Standard Specification for Metal Insert Fittings Utilizing a Copper Crimp Ring for SDR 9 Cross-linked Polyethylene (PEX) Tubing	Table 1202.5
F 1924—05	Standard Specification for Plastic Mechanical Fittings for Use on Outside Diameter Controlled Polyethylene Gas Distribution Pipe and Tubing	1203.15.3
F 1974—08	Standard Specification for Metal Insert Fittings for Polyethylene/Aluminum/Polyethylene and Crosslinked Polyethylene/Aluminum/Crosslinked Polyethylene Composite Pressure Pipe	Table 1202.5
F 2159—05	Standard Specification for Plastic Insert Fittings Utilizing a Copper Crimp Ring for SDR 9 Cross-linked Polyethylene (PEX) Tubing	Table 1202.5
F 2389—07e1	Specification for Pressure-rated Polypropylene Piping Systems	Table 1202.4, Table 1202.5
F 2623—08	Standard Specification for Polyethylene of Raised Temperature (PE-RT) SDR 9 Tubing1	Table 1202.4
F 2735—08a	Standard Specification for Plastic Insert Fittings for SDR 9 Cross-linked Polyethylene (PEX) and Raised Temperature (PE-RT) Tubing	Table 1202.5
F 2769—09	Polyethylene of Raised Temperature (PE-RT) Plastic Hot and Cold-water Tubing and Distribution Systems	Table 1202.4

AWS

American Welding Society
550 N.W. LeJeune Road
P.O. Box 351040
Miami, FL 33135

Standard reference number	Title	Referenced in code section number
A5.8—2004	Specifications for Filler Metals for Brazing and Braze Welding	1203.3.1, 1303.3.1

AWWA

American Water Work Association
6666 West Quincy Avenue
Denver, CO 80235

Standard reference number	Title	Referenced in code section number
C110/A21.10—03	Standard for Ductile Iron & Gray Iron Fittings, 2 inches Through 48 inches for Water	Table 1202.5
C115/A21.15—99	Standard for Flanged Ductile-iron Pipe with Ductile Iron or Grey-iron Threaded Flanges	Table 1202.4
C151/A21.51—02	Standard for Ductile-iron Pipe, Centrifugally Cast for Water	Table 1202.4
C153/A21.53—00	Standard for Ductile-iron Compact Fittings for Water Service	Table 1202.5

CSA

Canadian Standards Association
5060 Spectrum Way
Mississauga, Ontario, Canada L4W 5N6

Standard reference number	Title	Referenced in code section number
B137.9—05	Polyethylene/Aluminum/Polyethylene (PE-AL-PE) Composite Pressure-Pipe Systems	Table 1202.4
B137.10—05	Cross-linked Polyethylene/Aluminum/Cross-linked Polyethylene (PEX-AL-PEX) Composite Pressure-pipe Systems	Table 1202.4
ANSI CSA America FC1—03	Stationary Fuel Cell Power Systems	924.1

DOL

Department of Labor
Occupational Safety and Health Administration
c/o Superintendent of Documents
US Government Printing Office
Washington, DC 20402-9325

Standard reference number	Title	Referenced in code section number
29 CFR Part 1910.1000 (2009)	Air Contaminants	502.6
29 CFR Part 1910.1025 (2009)	Toxic and Hazardous Substances..	502.19

FS

Federal Specifications*
General Services Administration
7th & D Streets
Specification Section, Room 6039
Washington, DC 20407

Standard reference number	Title	Referenced in code section number
WW-P-325B (1976)	Pipe, Bends, Traps, Caps and Plugs; Lead (for Industrial Pressure and Soil and Waste Applications	Table 1202.4

*Standards are available from the Supt. of Documents, U.S. Government Printing Office, Washington, DC 20402-9325

ICC

International Code Council, Inc.
500 New Jersey Ave, NW
6th Floor
Washington, DC 20001

Standard reference number	Title	Referenced in code section number
IBC—12	International Building Code®	201.3, 202, 301.12, 301.12, 301.14, 301.15, 302.1, 302.2, 304.7, 304.10, 308.8, 308.10, 401.4, 401.6, 406.1, 502.10, 502.10.1, 504.2, 506.3.3, 506.3.10, 506.3.12.2, 506.4.1, 509.1, 510.6, 510.6.3, 510.6.2, 510.7, 511.1.5, 513.1, 513.2, 513.3, 513.4.3, 513.5, 513.5.2, 513.5.2.1, 513.6.2, 513.10.5, 513.12, 513.12.2, 513.20, 602.2.1.5.1, 602.2.1.5.2, 602.3, 603.1, 603.10, 604.5.4, 607.1.1, 607.3.2.1, 607.5.1, 607.5.2, 607.5.3, 607.5.4, 607.5.4.1, 607.5.5, 607.5.5.1, 607.6, 607.6.2, 701.4.1, 701.4.2, 801.3, 801.16.1, 801.18.4, 902.1, 908.3, 908.4, 910.3, 925.1, 1004.6, 1105.1, 1206.4, 1402.4, 1402.4.1
IEBC—12	International Existing Building Code®	101.2
IECC—12	International Energy Conservation Code®............	202, 301.2, 303.3, 312.1, 603.9, 604.1, 1204.1, 1204.2
IFC—12	International Fire Code®......................................	201.3, 310.1, 311.1, 502.4, 502.5, 502.7.2, 502.8.1, 502.9.5, 502.9.5.2, 502.9.5.3, 502.9.8.2, 502.9.8.3, 502.9.8.5, 502.9.8.6, 502.10, 502.10.3, 502.16.2, 509.1, 510.2.1, 510.2.2, 510.4, 511.1.1 513.12.3, 513.15, 513.16, 513.17, 513.18, 513.19, 513.20.2, 513.20.3, 606.2.1, 908.7, 1101.9, 1105.3, 1105.9, 1106.5, 1106.6, 1301.1, 1301.2
IFGC—12	International Fuel Gas Code®	101.2, 201.3, 301.3, 701.1, 801.1, 901.1, 906.1, 1101.5
IPC—12	International Plumbing Code®	201.3, 301.8, 512.2, 908.5, 1002.1, 1002.2, 1002.3, 1005.2, 1006.6, 1008.2, 1009.3, 1101.4, 1201.1, 1206.2, 1206.3, 1401.2
IRC—12	International Residential Code®...	101.2

IIAR

International Institute of Ammonia Refrigeration
1110 North Glebe Road
Arlington, VA 22201

Standard reference number	Title	Referenced in code section number
2—99 (with Addendum A—2005)	Addendum A to Equipment, Design, and Installation of Ammonia Mechanical Refrigerating Systems. . .	1101.6

MSS

Manufacturers Standardization Society of the Valve & Fittings Industry, Inc.
127 Park Street, N.E.
Vienna, VA 22180

Standard reference number	Title	Referenced in code section number
SP-69—2002	Pipe Hangers and Supports-Selection and Application	305.4

NAIMA

North American Insulation Manufacturers Association
44 Canal Center Plaza, Suite 310
Alexandria, VA 22314

Standard reference number	Title	Referenced in code section number
AH116—09	Fibrous Glass Duct Construction Standards	603.5, 603.9

NFPA

National Fire Protection Association
1 Batterymarch Park
Quincy, MA 02169-7471

Standard reference number	Title	Referenced in code section number
30A—12	Code for Motor Fuel-dispensing Facilities and Repair Garages	304.6
31—11	Installation of Oil-burning Equipment	801.2.1, 801.18.1, 801.18.2, 920.2, 922.1, 1308.1
37—10	Stationary Combustion Engines and Gas Turbines	915.1, 915.2
58—11	Liquefied Petroleum Gas Code	502.9.10
69—08	Explosion Prevention Systems	510.8.3
70—11	National Electrical Code	301.7, 306.3.1, 306.4.1, 511.1.1, 513.11, 513.12.1, 602.2.1.1, 1106.3, 1106.4
72—10	National Fire Alarm Code	606.3
82—09	Incinerators and Waste and Linen Handling Systems and Equipment	601.1
91—10	Exhaust Systems for Air Conveying of Vapors, Gases, and Noncombustible Particulate Solids	502.9.5.1, 502.17
92B—09	Smoke Management Systems in Malls, Atria and Large Spaces	513.8
96—08	Standard for Ventilation Control and Fire Protection Cooking Operations	507.2
211—10	Chimneys, Fireplaces, Vents and Solid Fuel-burning Appliances	806.1
262—11	Standard Method of Test for Flame Travel and Smoke of Wires and Cables for Use in Air-handling Spaces	602.2.1.1
704—12	Identification of the Hazards of Materials for Emergency Response	502.8.4, Table 1103.1, 510.1
853—10	Installation of Stationary Fuel Power Plants	924.1
850 1—97	Single Burner Boiler Operation	1004.1
8502—99	Prevention of Furnace Explosions/Implosions in Multiple Burner Boiler-furnaces	1004.1
8504—96	Atmospheric Fluidized-bed Boiler Operation	1004.1

SMACNA

Sheet Metal & Air Conditioning Contractors National Assoc., Inc.
4201 Lafayette Center Drive
Chantilly, VA 20151-1209

Standard reference number	Title	Referenced in code section number
SMACNA/ANSI—2005	HVAC Duct Construction Standards-Metal and Flexible (2005)	603.4, 603.9
SMACNA—03	Fibrous Glass Duct Construction Standards	603.5, 603.9

UL

Underwriters Laboratories, Inc.
333 Pfingsten Road
Northbrook, IL 60062-2096

Standard reference number	Title	Referenced in code section number
17—2008	Vent or Chimney Connector Dampers for Oil-fired Appliances	803.6
103—01	Factory-built Chimneys, Residential Type and Building Heating Appliance—with Revisions through March 2010	805.2
127—08	Factory-built Fireplaces—with Revisions through January 2010	805.3, 903.1, 903.3
174—04	Household Electric Storage Tank Water Heaters—with Revisions through May 2006	1002.1
180—03	Liquid-level Indicating Gauges for Oil Burner Fuels—with Revisions through March 2007	1306.4
181—05	Factory-made Air Ducts and Air Connectors—with Revisions through October 2008	512.2, 603.5, 603.6.1, 603.6.2, 604.13
181A—05	Closure Systems for Use with Rigid Air Ducts and Air Connectors—with Revisions through February 2008	603.9
181B—05	Closure Systems for Use with Flexible Air Ducts and Air Connectors—with Revisions through February 2008	603.9
207—2009	Refrigerant-containing Components and Accessories, Nonelectrical	1101.2
263—2003	Standard for Fire Test of Building Construction and Materials—with Revisions through October 2007	607.5.2, 607.5.5, 607.6.1
268—2009	Smoke Detectors for Fire Prevention Signaling	606.1
268A—2008	Smoke Detectors for Duct Application, with Revisions through September 2009	606.1
343—2008	Pumps for Oil-Burning Appliances	1302.7
378—06	Draft Equipment	804.3, 804.3.8
391—2006	Solid-fuel and Combination-fuel Central and Supplementary Furnaces—with Revisions through March 2010	918.1
412—04	Refrigeration Unit Coolers—with Revisions through January 2009	1101.2
471—06	Commercial Refrigerators and Freezers—with Revisions through October 2008	1101.2
499—05	Electric Heating Appliances—with Revisions through August 2008	912.1, 923.1
508—99	Industrial Control Equipment—with Revisions through September 2008	307.2.3
536—97	Flexible Metallic Hose—with Revisions through June 2003	1302.8
555—06	Fire Dampers—with Revisions through May 2010	607.3
555C—06	Ceiling Dampers—with Revisions through May 2010	607.3.1
555S—99	Smoke Dampers—with Revisions through May 2010	607.3.1
586—2009	High-Efficiency, Particulate, Air Filter Units	605.2
641—95	Type L Low-temperature Venting Systems—with Revisions through July 2009	802.1
710—95	Exhaust Hoods for Commercial Cooking Equipment—with Revisions through December 2009	507.1
710B—04	Recirculating Systems with Revisions through April 2006	507.1
723—2008	Standard for Test for Surface Burning Characteristics of Building Materials	510.8, 602.2.1, 602.2.1.5, 604.3, 1204.1
726—95	Oil-fired Boiler Assemblies—with Revisions through April 2010	916.1, 1004.1
727—06	Oil-fired Central Furnaces	918.1
729—03	Oil-fired Floor Furnaces—with Revisions through April 2010	910.1
730—03	Oil-fired Wall Furnaces—with Revisions through April 2010	909.1
731—95	Oil-fired Unit Heaters—with Revisions through April 2010	920.1
732—95	Oil-fired Storage Tank Water Heaters—with Revisions through April 2010	1002.1
737—2007	Fireplace Stoves—with Revisions through January 2010	905.1
762—2010	Outline of Investigation for Power Ventilators for Restaurant Exhaust Appliances	506.5.1
791—06	Residential Incinerators—with revisions through April 2010	907.1
834—04	Heating, Water Supply and Power Boilers Electric—with Revisions through December 2009	1004.1
842—07	Valves for Flammable Fluids	1307.1
858—05	Household Electric Ranges—with Revisions through May 2010	917.1
867—00	Electrostatic Air Cleaners—with Revisions through February 2010	605.2

UL—continued

Appendix A:
Chimney Connector Pass-Throughs

(This appendix is informative and is not part of the code.)

SYSTEM A

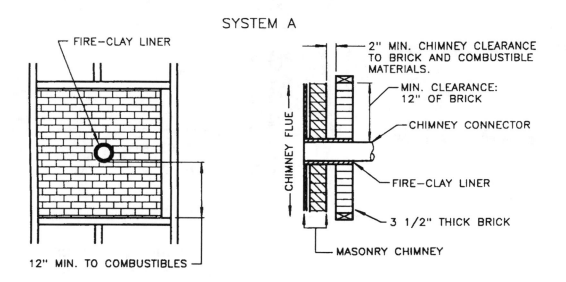

SYSTEM B

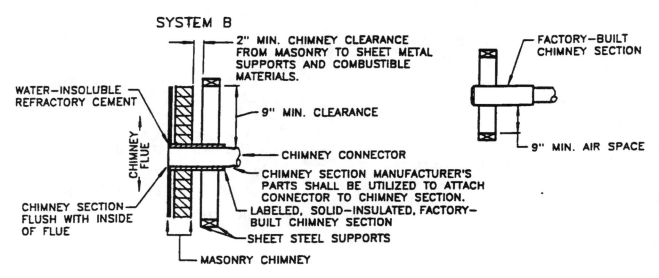

For SI: 1 inch = 25.4 mm.

FIGURE A-1
CHIMNEY CONNECTOR SYSTEMS

(continued)

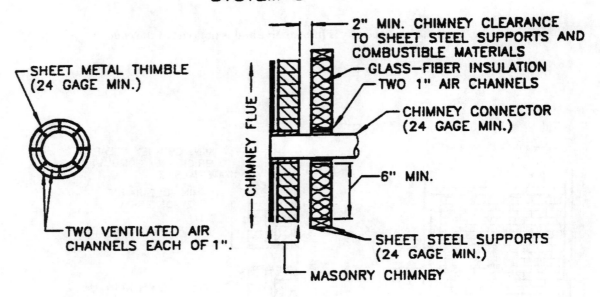

SYSTEM C

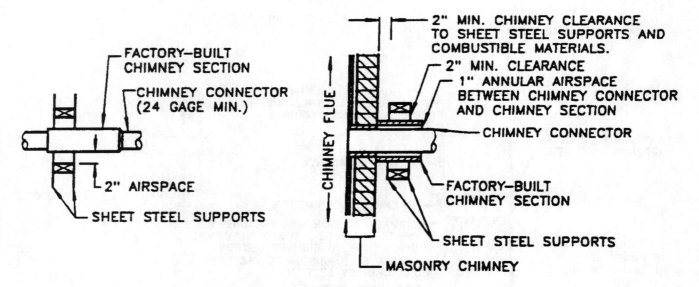

SYSTEM D

For SI: 1 inch = 25.4 mm.

FIGURE A-1—continued
CHIMNEY CONNECTOR SYSTEMS

Appendix B:
Recommended Permit Fee Schedule

(This appendix is informative and is not part of the code.)

B101
MECHANICAL WORK, OTHER THAN GAS PIPING SYSTEMS

B101.1 Initial Fee

For issuing each permit $___

B101.2 Additional Fees

B101.2.1 Fee for inspecting heating, ventilating, ductwork, air-conditioning, exhaust, venting, *combustion* air, pressure vessel, solar, fuel oil and refrigeration systems and *appliance* installations shall be $___ for the first $1,000.00, or fraction thereof, of valuation of the installation plus $___ for each additional $1,000.00 or fraction thereof.

B101.2.2 Fee for inspecting repairs, alterations and additions to an existing system shall be $___ plus $___ for each $1,000.00 or fraction thereof.

B101.2.3 Fee for inspecting boilers (based upon Btu input):

33,000 Btu (1 bhp) to 165,000 (5 bhp)	$ ___
165,001 Btu (5 bhp) to 330,000 (10 bhp)	$ ___
330,001 Btu (10 bhp) to 1,165,000 (52 bhp)	$ ___
1,165,001 Btu (52 bhp) to 3,300,000 (98 bhp)	$ ___
Over 3,300,000 Btu (98 bhp)	$ ___

For SI:1 British thermal unit = 0.2931 W, 1 bhp = 33,475 Btu/hr

B102
FEE FOR REINSPECTION

If it becomes necessary to make a reinspection of a heating, ventilation, air-conditioning or refrigeration system, or boiler installation, the installer of such *equipment* shall pay a reinspection fee of $___.

B103
TEMPORARY OPERATION INSPECTION FEE

When preliminary inspection is requested for purposes of permitting temporary operation of a heating, ventilating, refrigeration, or air-conditioning system, or portion thereof, a fee of $___ shall be paid by the contractor requesting such preliminary inspection. If the system is not *approved* for temporary operation on the first preliminary inspection, the usual reinspection fee shall be charged for each subsequent preliminary inspection for such purpose.

B104
SELF-CONTAINED UNITS LESS THAN 2 TONS

In all buildings, except one- and two-family dwellings, where self-contained air-conditioning units of less than 2 tons are to be installed, the fee charged shall be that for the total cost of all units combined (see B101.2.1 for rate).

INDEX

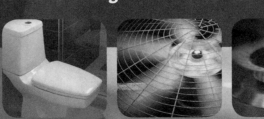